Microwave Processing of Materials III

MATERIALS RESEARCH SOCIETY SYMPOSIUM PROCEEDINGS VOLUME 269

Microwave Processing of Materials III

Symposium held April 27-May 1, 1992, San Francisco, California, U.S.A.

EDITORS:

Ronald L. Beatty

Oak Ridge National Laboratory
Oak Ridge, Tennessee, U.S.A.

Willard H. Sutton

United Technologies Research Center
East Hartford, Connecticut, U.S.A.

Magdy F. Iskander

University of Utah
Salt Lake City, Utah, U.S.A.

MRS

MATERIALS RESEARCH SOCIETY
Pittsburgh, Pennsylvania

Single article reprints from this publication are available through University Microfilms Inc., 300 North Zeeb Road, Ann Arbor, Michigan 48106

CODEN: MRSPDH

Published by:

Materials Research Society
9800 McKnight Road
Pittsburgh, Pennsylvania 15237
Telephone (412) 367-3003
Fax (412) 367-4373

Library of Congress Cataloging in Publication Data

Microwave processing of materials III : symposium L held April 27-May 1, 1992, San Francisco, California, U.S.A. / editors, Ronald L. Beatty, Willard H. Sutton, Magdy F. Iskander.
p. cm.—(Materials Research Society symposium proceedings, ISSN 0272-9172; v. 269)
Includes bibliographical references and index.
ISBN: 1-55899-164-6
1. Microwaves—Congresses. 2. Materials—Effects of ration on—Congresses. 3. Microwaves—Industrial applications—Congresses. I. Beatty, Ronald L. II. Sutton, Willard H. (Willard Holmes), 1930- III. Iskander, Magdy F. IV. Series: Materials Research Society symposium proceedings : v. 269.
TK7876.M5359 1992 92-31701
620.1'1228—dc20 CIP

Manufactured in the United States of America

Contents

*Invited Paper

PART IV: MICROWAVE/PLASMA PROCESSING

PART V: MICROWAVE/MATERIALS INTERACTIONS

PART VI: MICROWAVE PROCESSING OF CERAMICS

*Invited Paper

*Invited Paper

*Invited Paper

Preface

The third MRS Symposium on Microwave Processing of Materials was held at the 1992 MRS Spring Meeting in San Francisco, California. Ninety-one papers from ten countries were presented orally or as posters, and the symposium concluded with a panel discussion on "Key Issues for Future Developments in Microwave Processing." The symposium was organized to continue the thrust of the first two Microwave Processing of Materials symposia held in 1988 and 1990 and published as MRS Symposium Proceedings Volumes 124 and 189. This volume contains 78 peer-reviewed papers and highlights of the panel discussion from the third symposium, and will serve as an important reference source for workers in the field as well as for those seeking an update on the status of the technology. Based on symposium attendance and number of presentations, the general level of microwave/materials processing activity has maintained its vigorous pace of the past few years.

The symposium was very successful in assembling experts involved with various aspects of microwaves/materials processing. Work reported in this book is divided into ten parts. These are: Microwave Processing Overviews, Numerical Modeling Techniques, Microwave Processing System Design, Microwave/Plasma Processing, Microwave/Materials Interactions, Microwave Processing of Ceramics, Microwave Processing of Polymers, Microwave Processing of Hazardous Wastes, Microwave NDE Techniques and Dielectric Properties and Measurements. The papers contained in these ten areas show significant progress in understanding and controlling the use of microwave energy for processing and evaluating materials. Developments in equipment and modeling are particularly notable. The dielectric property data base was expanded, and progress has been very rapid in developing high temperature measurement methods.

Major applications of microwaves to high temperature materials processing, beyond the established food processing and wood drying industries, have not yet occurred. "Key Issues for Future Developments in Microwave Processing" were addressed in the panel discussion and are highlighted in Part XI.

Microwave processing symposia are planned to continue on an annual basis, alternating between the Materials Research Society and the American Ceramic Society.

Ronald L. Beatty
Magdy F. Iskander
Willard H. Sutton

September 1992

Acknowledgments

We wish to thank all of the speakers, session chairs, discussion panelists and participants who made this symposium a great success. We are especially grateful to the authors and referees who assisted in the timely publication of the proceedings. The session chairs, who did an excellent job of managing the program and who were also responsible for refereeing the papers from the respective sessions, were:

O.M. Andrade	M.F. Iskander	J.A. Pearce
J. Asmussen	M.A. Janney	J.B. Salsman
M. Barmatz	D.L. Johnson	R. Silberglitt
R.L. Beatty	V.M. Kenkre	S.S. Stuchly
J.Ch. Bolomey	J.O. Kiggans	W.H. Sutton
R.F. Decker	H.D. Kimrey	W.R. Tinga
F.D. Gac	R.J. King	C.M. Weil
W. Ho	S.J. Oda	
R.F. Hutcheon	M.C.L. Patterson	

We also thank the MRS staff for their assistance in organizing the symposium and facilities and handling the publication of the proceedings.

We extend special thanks to Fay Christie who spent many hours assisting in organizing the symposium, managing the collection and handling of manuscripts, handling of correspondence, repairing manuscripts and compiling indices and the final manuscript.

Special thanks are also extended to M.F. Iskander and H.D. Kimrey for preparing and teaching an expanded MRS Short Course, "Microwave Interactions with Dielectric Materials," in conjunction with the symposium.

We gratefully acknowledge the financial support provided for this symposium by:

U.S. DOE Assistant Secretary for Conservation and Renewable Energy, Office of Industrial Technologies, Advanced Industrial Concepts Materials Program

Alcan International Ltd.

Continental Electronics Corporation

Gerling Laboratories

Microwave Materials Technology, Inc.

This symposium was endorsed by the American Ceramic Society.

Volume 239—Thin Films: Stresses and Mechanical Properties III, W.D. Nix, J.C. Bravman, E. Arzt, L.B. Freund, 1992, ISBN: 1-55899-133-6

Volume 240—Advanced III-V Compound Semiconductor Growth, Processing and Devices, S.J. Pearton, D.K. Sadana, J.M. Zavada, 1992, ISBN: 1-55899-134-4

Volume 241—Low Temperature (LT) GaAs and Related Materials, G.L. Witt, R. Calawa, U. Mishra, E. Weber, 1992, ISBN: 1-55899-135-2

Volume 242—Wide Band Gap Semiconductors, T.D. Moustakas, J.I. Pankove, Y. Hamakawa, 1992, ISBN: 1-55899-136-0

Volume 243—Ferroelectric Thin Films II, A.I. Kingon, E.R. Myers, B. Tuttle, 1992, ISBN: 1-55899-137-9

Volume 244—Optical Waveguide Materials, M.M. Broer, G.H. Sigel, Jr., R.Th. Kersten, H. Kawazoe, 1992, ISBN: 1-55899-138-7

Volume 245—Advanced Cementitious Systems: Mechanisms and Properties, F.P. Glasser, G.J. McCarthy, J.F. Young, T.O. Mason, P.L. Pratt, 1992, ISBN: 1-55899-139-5

Volume 246—Shape-Memory Materials and Phenomena—Fundamental Aspects and Applications, C.T. Liu, H. Kunsmann, K. Otsuka, M. Wuttig, 1992, ISBN: 1-55899-140-9

Volume 247—Electrical, Optical, and Magnetic Properties of Organic Solid State Materials, L.Y. Chiang, A.F. Garito, D.J. Sandman, 1992, ISBN: 1-55899-141-7

Volume 248—Complex Fluids, E.B. Sirota, D. Weitz, T. Witten, J. Israelachvili, 1992, ISBN: 1-55899-142-5

Volume 249—Synthesis and Processing of Ceramics: Scientific Issues, W.E. Rhine, T.M. Shaw, R.J. Gottschall, Y. Chen, 1992, ISBN: 1-55899-143-3

Volume 250—Chemical Vapor Deposition of Refractory Metals and Ceramics II, T.M. Besmann, B.M. Gallois, J.W. Warren, 1992, ISBN: 1-55899-144-1

Volume 251—Pressure Effects on Materials Processing and Design, K. Ishizaki, E. Hodge, M. Concannon, 1992, ISBN: 1-55899-145-X

Volume 252—Tissue-Inducing Biomaterials, L.G. Cima, E.S. Ron, 1992, ISBN: 1-55899-146-8

Volume 253—Applications of Multiple Scattering Theory to Materials Science, W.H. Butler, P.H. Dederichs, A. Gonis, R.L. Weaver, 1992, ISBN: 1-55899-147-6

Volume 254—Specimen Preparation for Transmission Electron Microscopy of Materials-III, R. Anderson, B. Tracy, J. Bravman, 1992, ISBN: 1-55899-148-4

Volume 255—Hierarchically Structured Materials, I.A. Aksay, E. Baer, M. Sarikaya, D.A. Tirrell, 1992, ISBN: 1-55899-149-2

Volume 256—Light Emission from Silicon, S.S. Iyer, R.T. Collins, L.T. Canham, 1992, ISBN: 1-55899-150-6

Volume 257—Scientific Basis for Nuclear Waste Management XV, C.G. Sombret, 1992, ISBN: 1-55899-151-4

Volume 258—Amorphous Silicon Technology—1992, M.J. Thompson, Y. Hamakawa, P.G. LeComber, A. Madan, E. Schiff, 1992, ISBN: 1-55899-153-0

Volume 259—Chemical Surface Preparation, Passivation and Cleaning for Semiconductor Growth and Processing, R.J. Nemanich, C.R. Helms, M. Hirose, G.W. Rubloff, 1992, ISBN: 1-55899-154-9

Volume 260—Advanced Metallization and Processing for Semiconductor Devices and Circuits II, A. Katz, Y.I. Nissim, S.P. Murarka, J.M.E. Harper, 1992, ISBN: 1-55899-155-7

Volume 261—Photo-Induced Space Charge Effects in Semiconductors: Electro-optics, Photoconductivity, and the Photorefractive Effect, D.D. Nolte, N.M. Haegel, K.W. Goossen, 1992, ISBN: 1-55899-156-5

Volume 262—Defect Engineering in Semiconductor Growth, Processing and Device Technology, S. Ashok, J. Chevallier, K. Sumino, E. Weber, 1992, ISBN: 1-55899-157-3

Volume 263—Mechanisms of Heteroepitaxial Growth, M.F. Chisholm, B.J. Garrison, R. Hull, L.J. Schowalter, 1992, ISBN: 1-55899-158-1

Volume 264—Electronic Packaging Materials Science VI, P.S. Ho, K.A. Jackson, C-Y. Li, G.F. Lipscomb, 1992, ISBN: 1-55899-159-X

Volume 265—Materials Reliability in Microelectronics II, C.V. Thompson, J.R. Lloyd, 1992, ISBN: 1-55899-160-3

Volume 266—Materials Interactions Relevant to Recycling of Wood-Based Materials, R.M. Rowell, T.L. Laufenberg, J.K. Rowell, 1992, ISBN: 1-55899-161-1

Volume 267—Materials Issues in Art and Archaeology III, J.R. Druzik, P.B. Vandiver, G.S. Wheeler, I. Freestone, 1992, ISBN: 1-55899-162-X

Volume 268—Materials Modification by Energetic Atoms and Ions, K.S. Grabowski, S.A. Barnett, S.M. Rossnagel, K. Wasa, 1992, ISBN: 1-55899-163-8

Volume 269—Microwave Processing of Materials III, R.L. Beatty, W.H. Sutton, M.F. Iskander, 1992, ISBN: 1-55899-164-6

Volume 270—Novel Forms of Carbon, C.L. Renschler, J. Pouch, D. Cox, 1992, ISBN: 1-55899-165-4

Volume 271—Better Ceramics Through Chemistry V, M.J. Hampden-Smith, W.G. Klemperer, C.J. Brinker, 1992, ISBN: 1-55899-166-2

Volume 272—Chemical Processes in Inorganic Materials: Metal and Semiconductor Clusters and Colloids, P.D. Persans, J.S. Bradley, R.R. Chianelli, G. Schmid, 1992, ISBN: 1-55899-167-0

Volume 273—Intermetallic Matrix Composites II, D. Miracle, J. Graves, D. Anton, 1992, ISBN: 1-55899-168-9

Volume 274—Submicron Multiphase Materials, R. Baney, L. Gilliom, S.-I. Hirano, H. Schmidt, 1992, ISBN: 1-55899-169-7

Volume 275—Layered Superconductors: Fabrication, Properties and Applications, D.T. Shaw, C.C. Tsuei, T.R. Schneider, Y. Shiohara, 1992, ISBN: 1-55899-170-0

Volume 276—Materials for Smart Devices and Micro-Electro-Mechanical Systems, A.P. Jardine, G.C. Johnson, A. Crowson, M. Allen, 1992, ISBN: 1-55899-171-9

Volume 277—Macromolecular Host-Guest Complexes: Optical, Optoelectronic, and Photorefractive Properties and Applications, S.A. Jenekhe, 1992, ISBN: 1-55899-172-7

Volume 278—Computational Methods in Materials Science, J.E. Mark, M.E. Glicksman, S.P. Marsh, 1992, ISBN: 1-55899-173-5

Prior Materials Research Society Symposium Proceedings available by contacting Materials Research Society

PART I

Overview

MICROWAVE PROCESSING OF CERAMICS - AN OVERVIEW

Willard H. Sutton, United Technologies Research Center,
411 Silver Lane, MS 24, East Hartford, CT 06108

ABSTRACT

During the past decade, many exploratory studies and experiments have been performed on the microwave heating and processing of ceramics and composite materials. Much of this effort was stimulated by the unique and potential benefits that microwave energy can provide over conventional processing methods. While microwave processing of ceramics is still in an early developmental stage, there are many areas yet to be explored, challenges to be met, and economic and commercial payoffs to be substantiated.

Since the first MRS International Symposium on Microwave Processing in 1988, interest in this field has grown and many new developments have occurred. The purpose of this paper is to highlight some of the recent advances, to discuss the current state-of-the-art, and to suggest some future directions.

INTRODUCTION

The interaction and absorption (coupling) of many ceramic materials, such as oxides, carbides, borides, nitrides, and combinations of these materials (i.e., ceramic matrix composites and multiphase systems), have stimulated much interest in recent years, and has made possible entirely new, and often unique, routes to process these materials. With the development of radar during W.W.II, G.B.B.M. Sutherland noted almost 50 years ago that radar (microwave energy) could be used to dry large ceramic bodies [1]. However, serious efforts to process ceramics with microwave (M/W) energy did not emerge until the late 60's and early 70's, as shown in Table I.

TABLE I. Microwave Processing of Ceramics

1950's - Dielectric properties of ceramics[2] **1960's** - Patent on firing ceramics (1965)[3] Formation of IMPI (1966)[4]* M/W plasma sintering (1968)[5] **1970's** - Applications: drying, dewaxing casting molds[6,7] Melt mica, pyrex, porcelain (1970)[8] Fire 300 lbs (136 kg) Al_2O_3 castables(1975)[9] Heat Al_2O_3 > 1700°C (1976)[10] Method to sinter refractories (1979)[11] Melt UO_2[12] *IMPI = International Microwave Power Institute	**1980's** - Applications (pilot scale) • Slip casting[13,14] • Fire spark plug housings[15] • Clinkering cement[16] New areas (R&D) • Glass[17,18] • Oxides, nonoxides, composites[19] • Joining and sealing[20-24] MRS Microwave Processing Symposium (1988)[25,26] **1990's** - Microwave Processing Symposia (1990[27],1991[28],1992[29]) Rapid proliferation (R&D)[25-29] • New process applications • New materials, microstructures • Improved heating control • Modeling and measurements

While some work on heating ceramics with microwaves was reported in the 70's, a much greater effort was devoted to high temperature processing (firing, sintering, melting) during the mid to late 80's, and is continuing into the 90's, where there is now a rapid growth and proliferation in R&D efforts to explore many new processing routes and materials applications (Table I).

Although many of the potential advantages of using microwaves to process ceramics have been recognized for over 30 years, it is only during the past 5-8 years that the versatility and full potential of this process have been gaining widespread attention. The increasing awareness of the benefits that processing with microwave energy has to offer is evident from the rapid growth in the number and scope of the papers presented at technical meetings and symposia dedicated to microwave processing. The first International Symposium in the US that was devoted to the Microwave Processing of Materials was held in 1988 at the Materials Research Society's Spring Meeting, in Reno, NV. [25,26]. Since that time, there have been three additional Symposia [27,28,29] . +

Table II summarizes some of the data on the sessions, papers, and topics of the Symposia. Note the increase in the participation and number of contributions

TABLE II. MRS and ACERS Symposia on Microwave Processing of Materials

Year	Organization*	No. of papers presented (pub)	Publication (date)	No. of foreign countries (pub)	No. of sessions	Short course
1988	MRS	44 (42)	V. 184 ('88)	5 (7)	6	- -
1990	MRS	103 (64)	V. 189 ('91)	7 (22)	9	x
1991	ACerS	88 (69)	V. 21 ('91)	6 (10)	10	x
1992	***MRS***	***91 (78)***	***V. 269 ('92)***	***8 (31)***	***13***	***x***

Year	Organization	Topics: Overview	M/W - mat'l interactions	Modeling	Dielectric meas.	Equipment design	Panel discussion	Appl.
1988	MRS	2	6	- -	6	6	x	11
1990	MRS	3	- -	8	6	6	- -	16
1991	ACerS	4	7	6	8	6	- -	12
1992	***MRS***	***5***	***4***	***9***	***14***	***4***	***x***	***14***

Year	Organization	Topics: Ceramics	Polymers & compos.	Bio. med.	Reclaim. & wastes	Joining	Sinter	Other proc.	NDE
1988	MRS	14	5	1	2	3	5	9	- -
1990	MRS	10	9	5	3	- -	9	8	1
1991	ACerS	29	3	- -	6	4	15	10	- -
1992	***MRS***	***21***	***12*****	***- -***	***3***	***4***	***11***	***10***	***4***

* MRS: Materials Research Society
* ACerS: American Ceramic Society

** also - Am. Chem. Soc. Symposium on Microwave Processing of Polymers (22 papers) April 1992, San Francisco

---

+ Plans were set up to hold these Symposia annually, but to alternate the sponsorship each year between the Materials Research Society and the American Ceramic Society.

from foreign nations, and that the number of papers related to ceramic materials and processing has also increased significantly since the 1988 Symposium.

This overview focuses on the progress made in the microwave processing of ceramics during the past 5 years; discusses some of the problems and challenges encountered; and highlights some of the recent advances and future directions.

MICROWAVES FOR CERAMIC PROCESSING

Some of the primary reasons for the growing interest in using microwaves to process ceramic materials are presented in Table III(A). All of the benefits listed in Table III(A) for using microwave energy have been well documented in the literature [19,25-29]. However, as increasing numbers of researchers continue to investigate new applications, additional novel features of microwave processing are being uncovered, especially in the area of the synthesis and development of new materials.

TABLE III. Benefits and Challenges of Microwave Processing

(A) BENEFITS (Reasons for Growing Interest)	(B) PROBLEMS ENCOUNTERED (Challenges)
• Cost savings (time and energy) • Rapid, internal heating • Precise and controlled heating • Selective heating • Improved quality and properties • Synthesis of new materials • Processing capabilities not possible by conventional methods	• Inability to heat low-dielectric loss ceramics • Thermal runaway • Sample cracking • Arcing and plasmas • Nonuniform heating (low thermal conductivity, larger samples ~ 0.5-10+ kg) • Poor transfer of M/W energy to workpiece • Reaction with thermal insulation

Since microwave heating is significantly different from conventional methods, it is not surprising that several problems have been encountered, especially by new investigators who are unfamiliar with microwave processing. Some of the typical problems are listed in Table III(B). Various approaches to solving these problems will be discussed next under Processing. Both practical experience and a better understanding of how materials interact with microwaves were important factors in overcoming these problems. Several of the papers in following sections in this Volume on Numerical Modeling Techniques, Microwave Processing Equipment Design, Dielectric Properties and Measurements, and Microwave/Material Interactions provide further insight into how microwave energy can be more effectively and efficiently used to process materials.

One of the variables affecting the efficient coupling between microwaves and materials is the frequency (and wavelength) of the microwaves, which covers a range from 300 MHz to 300 GHz. The most commonly used frequency for processing ceramics is 2.45 GHz with a corresponding wavelength of 122 mm.

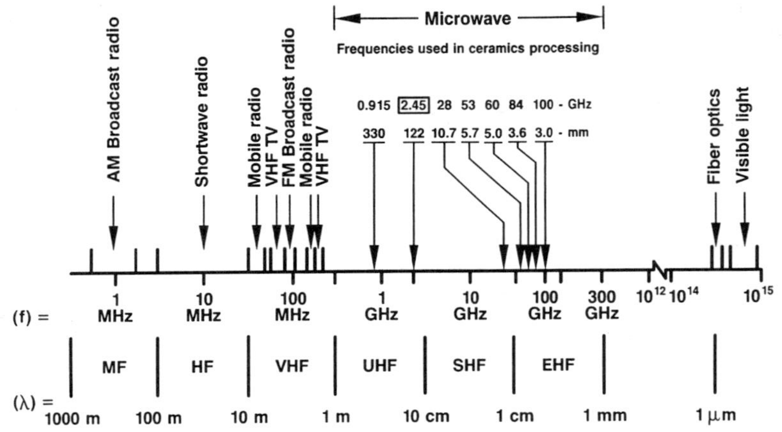

Figure 1. Electromagnetic Spectrum and Frequencies used in Microwave Processing of Ceramics.

Figure 1 shows some other microwave frequencies reported in the literature that have been used also to process ceramics. However, the 915 MHz and 2.45 GHz frequencies, which have been assigned by the FCC for medical, industrial, and scientific applications, are the two most widely used. At the highest frequencies (i.e., mm wavelengths), the microwaves can be focused by lenses or mirrors, to provide some additional flexibility to the processing of ceramics. An example will be given for a coatings application in the next section.

RECENT DEVELOPMENTS AND PROGRESS

Table IV lists some examples of recent progress and developments in the microwave processing of ceramics in the area of processing, compositional control, and applications. These examples, along with specific references, provide some measure of the range and depth of the R&D activities, especially during the past 5 years. However, it is beyond the scope of this paper to discuss each of these examples, although some will be reviewed in order to highlight a few of the significant developments.

(1.0) Processing

Six processing challenges are listed in Table IV, along with specific examples where problems have been solved, and where significant progress has been made.

The first five problems, or challenges shown in Table IV (1.1 to 1.5), are interactive and may have common causes, so that a solution to one is likely to be a solution to several or to all of these problems. For example, materials such as high purity Al_2O_3, or ZrO_2, which possess very low dielectric loss factors and low thermal conductivities at ambient temperatures, are difficult to heat initially with microwaves, and are prone to localized heating (hot spots and thermal runaway) and cracking. These problems are further compounded as the size of the sample is increased. Moreover, because the materials are poor microwave absorbers at temperatures below 400-500°C, arcing and plasma formation are more likely to occur.

TABLE IV. Challenges and Progress in Microwave (M/W) Processing Ceramics

Challenges	Progress (Examples and References)
1.0 PROCESSING	
1.1 Heating of Low M/W Absorbing Ceramics	• Efficient coupling with M/W (i.e., single mode processing with effective matching and tuning[7,30-34] • Innovations in multimode processing: M/W hybrid heating[35-44] • Frequencies: 28 GHz[35-37,45,46]; 60 GHz[47]; 84-100 GHz[48-50] • Use of M/W absorbing additives[51-52]
1.2 Nonuniform Heating and Thermal Runaway	• Improved energy deposition and thermal management[31,53] (also see examples in 1.1) • Reduce energy losses using thermal insulation[9,15,35-38] (see also 2.1) • Field modification device, multimode [43] • Very uniform grain size & structure: Al_2O_3[38,54,55], Si_3N_4[42,43]
1.3 Sample Cracking	• Reduce thermal gradients (see examples in 1.1 & 1.2) • Control phase changes [31,56]
1.4 Arcing & Plasmas	• More efficient coupling[7,30-34] • Prevent cracking [31,36,37,45,46] • Coating thermocouple with insulation[57] • Introduction of N_2 gas[57]
1.5 Sample Mass and Volume	• Heat multiple samples[13,15,58-61] • Heat large(r) samples[9,16,42,43,62]
1.6 Dual Heating	• Indirect (M/W hybrid heating)[35-44,62] • Direct (M/W plus conventional heating)[7,14,62-65]
2.0 COMPOSITION	
2.1 Contamination	• Use of proper insulation: casket approach(TiB_2, B_4C)[66-69]
2.2 Unstable Phases and Reactions	• Sintering Si_3N_4[42,43,70,71]; PZT & PLZT[41] • Annealing (oxygen) $YBa_2Cu_3O_{7-x}$ [72]
3.0 APPLICATIONS	
3.1 Binder Removal	• PZT, PLZT[41], Al_2O_3[73,74], combustion synthesis[75]
3.2 Heating Porous Materials and Preforms	• Chem. vapor infil.[76,77], infil. + decomposition[78,79] • Reaction bonded Si_3N_4[80-83]
3.3 High Pressure Processing	• M/W hot pressing[84] • M/W hot isostatic pressing[85] • Closed vessels[86,87]
3.4 Ultra-high Temperature >1900°C	• TiB_2[67,68], B_4C[68,69]
3.5 Co-firing Multiple Layers/Devices	• Optical layers SiO_2/TiO_2[88] • Ta capacitors[89], multilayer capacitors[90] • Thick film superconductor sensors[91]
3.6 Surface Treatment	• Ion exchange in glass[92,93] • Etching[94]
3.7 Waste Management	• Nuclear waste glasses[95-99]
3.8 Coating	• Al_2O_3 coating with mm-waves[48-50]
3.9 Ultra fine Structures	• Powder synthesis[99-102] • Fine grain structures: Al_2O_3[38,54,59], Si_3N_4[43], TiO_2[103]
3.10 Joining	• Oxides[23,24,107-109], carbides[104-106], nitrides[24]
3.11 Drying	• Steel plant ladles[65]
3.12 Combustible Mixes	• Self propagating high temp. synthesis, SHS[75,110]

In the following examples, multimode, microwave hybrid heating was used in solving problems 1.1 to 1.4 in Table IV. However, other approaches, such as single mode processing, when tuning and coupling were improved, also proved to be an effective solution.

Heating of Low M/W Absorbing Ceramics and Nonuniform Heating

Hybrid (indirect) heating was developed to heat materials with low dielectric losses and low thermal conductivities. Hybrid heating involves the use of a microwave absorbing material, which is in turn used to thermally heat a poorly absorbing sample until it reaches a temperature where it will absorb microwaves effectively. Thus, at temperatures below 300-500°C, samples are heated indirectly by the heat from the microwave suscepting material, and at higher temperatures they are directly heated by the microwaves. Examples of several types of hybrid heating have been reported[35-44] and will be discussed.

Figure 2 shows the use of a "picket fence" hybrid heating arrangement that was developed at the Oak Ridge National Laboratory[35-37]. In the example shown, a ZrO_2 sample is surrounded by SiC susceptor rods. At low temperatures, the SiC absorbs most of the microwave energy and heats faster than the ZrO_2 . Since the ZrO_2 is heated primarily by conduction in this critical low-temperature range, the development of hot spots and localized thermal runaway is avoided. As the chart in Fig. 2 shows, above a cross-over temperature, the dielectric loss factor of the ZrO_2 suddenly increases and exceeds that of the SiC, so that it can be directly heated by the microwaves with increasing efficiency as the temperature rises. This process essentially eliminates the development of hot spots, and with a controlled power increase, the ZrO_2 can be readily heated directly to the sintering temperature and be sintered without cracking.

The microwave power deposition pattern of the "picket fence" arrangement has been modeled by Smith and coworkers, and is reported in this Volume[44].

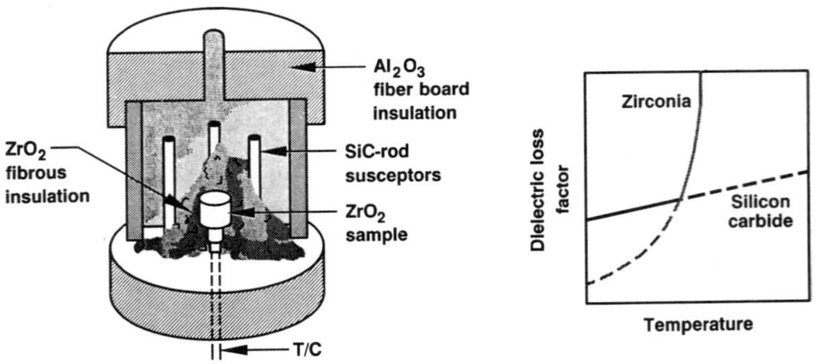

"Picket fence" susceptor/thermal insulation arrangement

Change in dielectric loss factor with temperature

Figure 2. Microwave hybrid heating of ZrO_2 (8 m/o Y_2O_3) utilizing the "picket fence" susceptor for the low temperature stage (Janney, Calhoun, Kimrey)[36,37].

The computer simulation and visualization of the effects of temperature and other variables provide a dramatic insight into how and where the energy is absorbed, and how the SiC rods influence the power deposition pattern.

Another type of hybrid, multimode furnace was developed at the University of Florida[38]. This is shown in Figure 3, where a SiC-lined M/W susceptor is placed over a poorly absorbing sample - such as high-purity Al_2O_3, which is essentially microwave (2.45 GHz) transparent at ambient temperatures. This furnace also makes use of the radiation and conduction (from the SiC) to rapidly heat the Al_2O_3 through the low temperature regime to a critical temperature, above which the microwaves couple readily with the sample. The sintered samples exhibited a more uniform microstructure and superior properties to those sintered conventionally using a similar heating schedule. It appears that microwave hybrid heating allows possibly for the highest heating rate (750°C/min.) attainable for pure, undoped Al_2O_3 in a 2.45 GHz, untuned, multimode cavity[38].

Investigators at Alcan International Ltd. have devised an elegant means to sinter multiple specimens of Si_3N_4 tool bits, which is described in this Volume[43]. Ninety tool bits were spaced in a regular pattern within an alumina crucible, which was filled with a powder mixture of SiC, BN, and Si_3N_4. The powder was used to support the specimens and also served other functions, such as being an initial susceptor; protecting the Si_3N_4 from reacting with the atmosphere so that sintering could be performed in air (vs. in N_2 by conventional or other microwave processing); and also preventing degradation (loss of desired sintering additives) at the high processing temperatures. The ninety tool bits were uniformly heated by carefully controlling the electric field pattern within the crucible by the use of conductive rings. Additional insulation was placed around the crucible and conductive rings in order to minimize heat losses and to maintain thermal uniformity within the crucible. Tool bits with superior microstructural uniformity and fine grain size were produced by balancing the M/W power input with the time required to dissipate the heat in order to achieve uniform sintering.

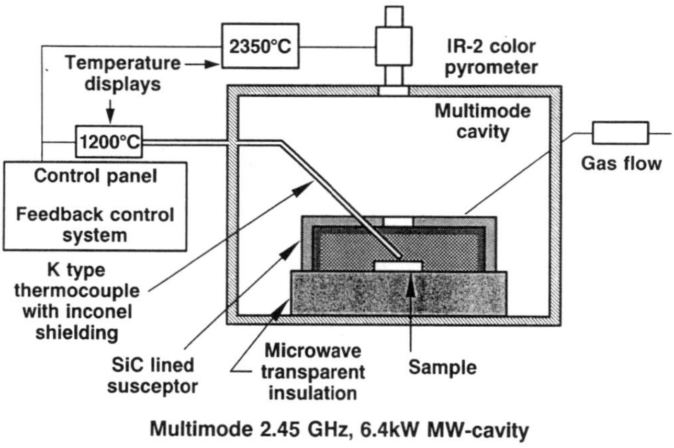

Figure 3. Use of SiC lining (susceptor) to microwave hybrid heat low loss ceramics (Dé, Ahmad, Whitney, Clark)[38].

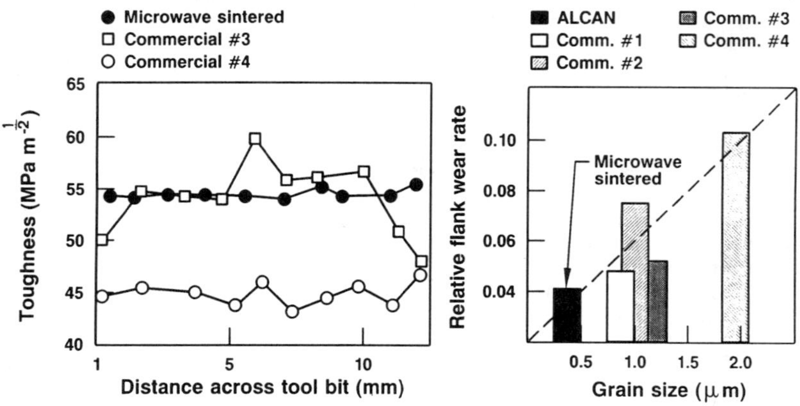

Figure 4. Comparison of microwave vs. conventionally sintered Si_3N_4 tool bit properties (Patterson, Apté, Kimber, & Roy)[42,43].

Figure 4 compares the toughness across the tool bit and also the performance (relative flank wear) in machining tests of the Alcan M/W-produced tool bits with those from several commercial sources. The superior performance of the microwave processed tool bits is clearly evident. This example also shows how both an innovative approach and familiarity with microwave/material/equipment interactions can be combined to satisfy essentially all of the processing challenges cited in Table IV.

Cracking

In addition to hybrid heating , another approach has been demonstrated by H. Kimrey and M. Janney[36,46], where the severe cracking of a ZrO_2 specimen sintered in a 2.45 GHz multimode furnace was eliminated by sintering in a 28 GHz M/W furnace. A photograph of ZrO_2 specimens processed at the two frequencies is shown in Figure 5A. The primary benefit of heating in the untuned, multimode 28 GHz cavity is that the electric field strengths are far more uniform (± 4% of the mean value) than in a 2.45 GHz cavity of similar dimensions.

Tian and coworkers[56] reported the successful sintering of 3% Y_2O_3 - ZrO_2 (TZP) in a single mode applicator. Specimens with ultrafine grains (0.2 - 0.35 μm) were sintered to densities in the range of 98-99.5% of theoretical without cracking. Critical coupling and tuning were achieved for the direct heating of the samples throughout the entire process.

Larger Sample Mass

Relatively large samples (about 1 kg) of high-purity, undoped Al_2O_3 have been sintered successfully to about 93% theoretical density in a 2.45 GHz multimode cavity by J. Katz[58]. This is a significant development because the

A) B)

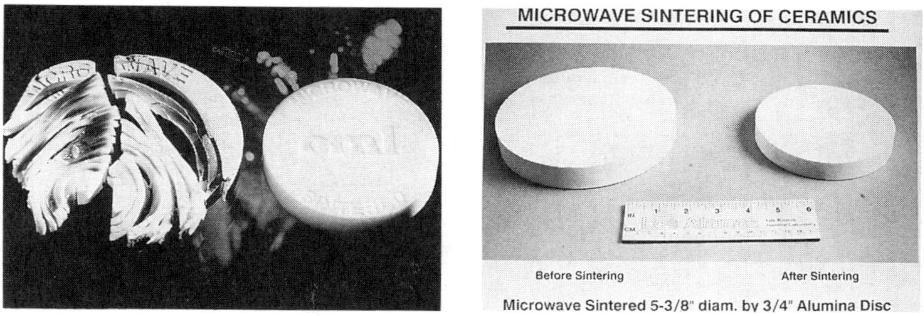

Figure 5. Results of processes developed to sinter crack-free, dense ceramics. (A) ZrO_2 specimens using @ 2.45 GHz (left) and 28 GHz (right), (courtesy H. Kimrey)[36,46], and (B) pure, nearly 1 kg Al_2O_3 specimen heated @ 2.45 GHz (courtesy J. Katz).

high purity of the sample results in a nearly transparent material to the microwaves at low temperatures and because the relatively large mass had to be heated uniformly to the sintering temperature. The specimen in Figure 5B shrank 17% during sintering, and showed no evidence of cracking. In order to maintain a uniform heating pattern in the sample, hybrid heating was used[58].

(2.0) Composition

Contamination from insulation and the surrounding environment, as well as sample decomposition, become serious problems when very high processing temperatures are required. Thus, the choice of insulation surrounding the samples or workpiece becomes of critical importance. Holcombe and coworkers[66-68] have developed a means to insulate and protect TiB_2 and B_4C samples from oxidation during sintering at temperatures in the 1900 - 2200°C range. Figure 6A shows a TiB_2 pellet, which is "casketed" (encapsulated) in a bed of Y_2O_3 grit. The hole shown in the side of the crucible is for optical temperature measurements and is not present during normal sintering, where the TiB_2 is sintered to a crack-free, ≥ 95% dense specimen. In addition to providing thermal insulation, the Y_2O_3 grit also protects the sample from oxidation, and "contaminates" the TiB_2 with about 8 wt. % Y_2O_3, which is beneficial in reduced grain size, higher hardness and density, and slightly improved fracture toughness when compared with yttria-free microwave or conventionally processed TiB_2[67].

A multilayered casket arrangement for sintering a B_4C sample is shown in Fig. 6B. The multilayered compositions shown in this figure proved to be the most effective in rapidly (hybrid) heating B_4C to the 2000-2200°C sintering temperatures, while also preventing reaction with the Y_2O_3 grit and allowing microwave transmission to the sample[68].

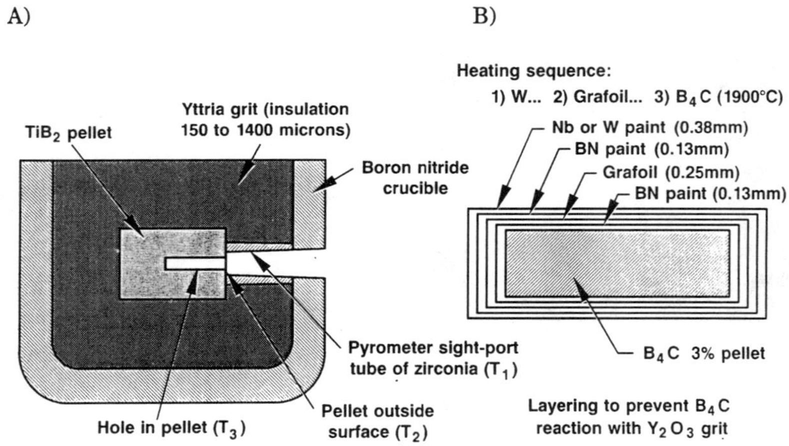

Figure 6. Use of "casketing" in microwave sintering (Holcombe and Dykes)[66-68].

(3.0) Applications

Many areas of applications and new opportunities for microwave processing of ceramics have been investigated, and several examples are presented in Table IV. These applications are especially benefited by rapid heating and by other unique features of microwave processing, such as: selective heating (binder removal, co-firing of multiple layers, surface treatment, coating, joining, and waste management); volumetric heating (CVI* and drying/decomposition of liquid infiltrants in porous preforms, reaction forming, high pressure processing, ultrafine structures, drying); and thermal runaway (ultrahigh temperature processing, and ignition of combustible mixes). Some of these examples will be described next.

Coating (mm - waves)

The radiation generated by gyrotrons produces mm waves in the form of beams which can be directed by metal mirrors, and can easily be dispersed over a considerable area or focused to the radiation wavelength[48-50]. Work at the E.O. Paton Institute (Kiev) on the mm-wave processing of powder-like materials has demonstrated the potential of this approach for synthesis, ceramic coating deposition, and processing composites[49,50]. For example, mm-wave beams, because of their shorter penetration characteristics (vs. the longer wavelengths - see Fig. 1), can be used to selectively heat and fuse pore-free coatings of highly refractory materials on lower-melting ceramic substrates.

Figure 7 illustrates the melting of Al_2O_3 (corundum) powder to form a thick coating over a lower melting refractory[49,50]. The curves show changes in the temperature profile in the coating with depth, as the mm-wave beam is passed

---

* CVI = Chemical vapor infiltration

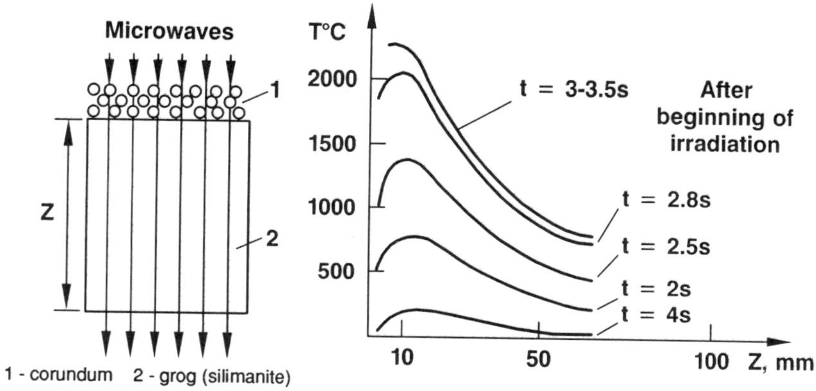

Figure 7. Temperature - depth sequence during the fusion of Al_2O_3 (corundum) powder with a passing mm-wave beam to form a dense coating over a lower melting substrate (Paton Tek, Inc., Courtesy R. Decker) [49,50].

over the surface. At a 10 mm depth from the powder (corundum) surface, the temperature of the coating rises from about 550°C in 2 sec. to over 2000°C in 2.8 sec. as the beam passes over the coating. After 4 sec. the coating has cooled to under 200°C. For greater depths from the surface (Fig. 7), the temperatures of the regions below are much lower, so that by a proper selection of beam energy, dwell time, etc. a fused, dense coating of pure Al_2O_3 can be applied to a lower melting (and lower cost) refractory without melting it.

Joining

The selective heating characteristics of microwave processing makes possible the localized heating and joining of ceramic materials, (oxides, nonoxides, and combinations of materials), where the M/W power can be concentrated primarily at the regions to be joined. This unique feature introduces a wide range of opportunities in ceramic joining that are not possible by conventional means. Moreover, the quality and strength of the joined regions is excellent, and often in self-joined materials, the original joint interface "disappears" during processing; i.e., the joint is no longer distinguishable by optical or SEM microscopy. Due to the nature of the rapid heating by microwaves, the joining process can be many times faster than by conventional methods.

Selective heating can be achieved by using single mode applicators[23,24], for joining oxides, carbides or nitrides, or in either single or multimode applicators [104-109] where interlayers or bonding materials are used that are better microwave absorbers than the materials to be joined. In either case, the joint interface becomes significantly hotter than the adjacent materials being bonded.

Figure 8 illustrates schematically the joining of two ceramic rods butted together. The rods are placed within a single mode applicator, where they are supported at their ends, and where a slight compressive load can be applied during the joining process. Either 2.45 GHz [23] or 6 GHz[24] microwaves have been used to successfully self-join oxides and nonoxides. An additional feature described by

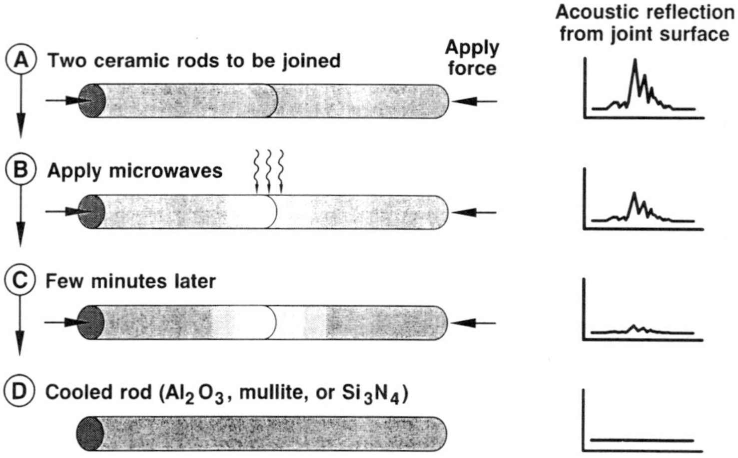

Figure 8. Sequence of events during the microwave joining of ceramic rods (drawing based on Silberglitt & Palaith)[23].

Silberglitt and Palaith[23] (Fig. 8), is the sending of an acoustic pulse down the rods during the joining process, and when joining has been accomplished, the reflected pulse from the interface disappears. Fig. 8 also indicates that the joining process is very rapid, and only requires a few minutes at the joining temperature.

Since 1988, when the early joining work was reported[23,24], many new developments have occurred, including: use of metals Al[104] and Si[105,106] and combustible powder mixes (Ti + C)[105,106] to join SiC to SiC; use of green state joining of Al_2O_3[107]; and use of sol-gel (Al00H) bonding of Al_2O_3 to Al_2O_3[108,109]. The Al00H gel has the advantage of being an adhesive at room temperature, which helps to hold the samples together without the need for a binder, and on heating converts to the parent material (Al_2O_3).

Drying [also dual heating (1.6)]

An example where dual heating can be used to great advantage over conventional (gas) heating, is shown in Figure 9, where a large steel plant ladle containing 8 tons of refractory was dried using the energy from two 915 MHz microwave generators and from gas-heated air. The time and energy savings, and quality improvement of the dual process over the conventional drying process is shown in the chart in Fig. 9. The drying time was reduced by a factor of 3; and while the total electrical power consumption was greater (some electrical power was required to drive the blowers for the hot gas), the total energy consumption was reduced by 81%. In addition, the quality of the lining dried by the dual process was greatly improved; there was no evidence of bursting or peeling.

Combustible Mixes

Volumetric (internal) heating by microwaves provides an ideal and controlled means to ignite and sustain the moving reaction (combustion) wave front in the self-propagating high temperature synthesis (SHS) of reactive mixtures.

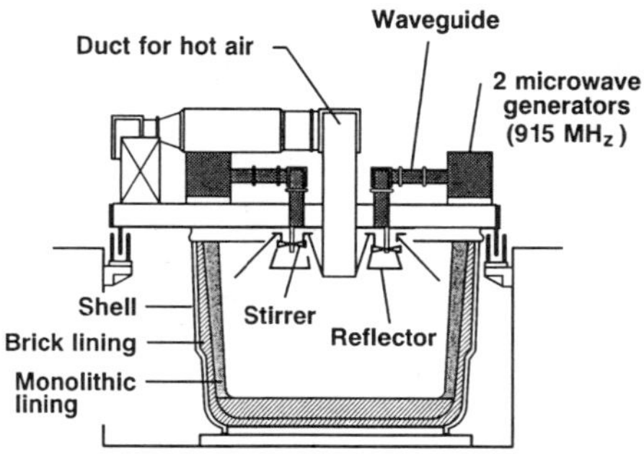

Comparison	Hybrid (Hot Air- 250°C + M/W drying)	Conventional (gas burner)
Drying Time	8	24
Electric Power (kWh/T - ref.)	67.3	15.5
Total Energy (kcal/T - ref.)	284×10^3	1468×10^3
Bursting/ Peeling During Drying	None Observed	Significant Damage

Figure 9. Advantage of dual (microwave + gas burner) vs. conventional processing for drying 8 tons of refractory in a steel plant ladle (Nippon Steel)[65].

Figure 10 shows the temperature profiles resulting from the wave front propagation of the SHS process which are ignited by different methods[75,110]. The effect of the internal ignition due to microwave penetration, and the unique radial wave front is indicated in this figure. In this process, a central hot spot and thermal runaway are both desirable features. In addition to igniting high bulk density samples, microwaves can be used to burn out binders as well as to control the propagation rate of the combustion wave front[110].

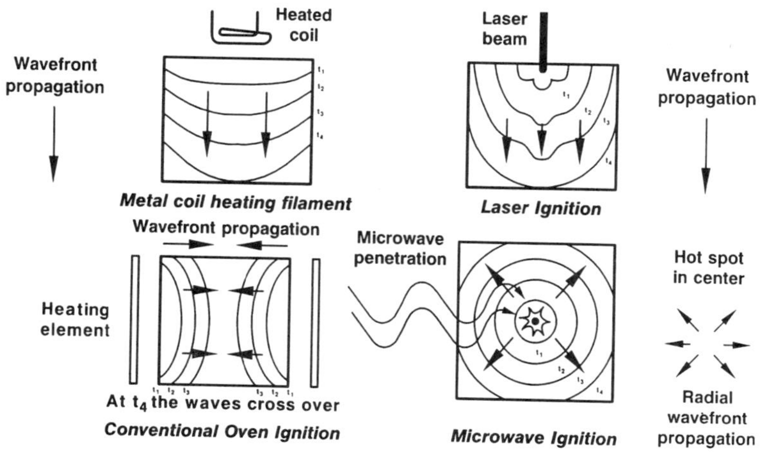

Figure 10. Various ignition methods used in self-propagating high temperature synthesis, SHS (University of Florida)[75,110].

SUMMARY AND CONCLUSIONS

A review of recent progress in the microwave processing of ceramic materials was presented. During the past 5 years, there has been considerable growth and activity in R&D efforts, both in the US and abroad, which have resulted in many significant improvements in processing (techniques, equipment, control, etc.), in exploring new areas of applications, and in developing and applying a better understanding of microwave-material interactions (including process modeling). Over 100 references were cited to illustrate areas where specific advances had been made, especially in process development and applications. Many of these developments are very recent; over 20 papers were referenced in this Volume.

In conclusion, significant advances in the state-of-the-art of microwave processing ceramic materials have been achieved recently, and more are foreseen for the future, which include:

- **During the past 5 years, major strides have been made in:**
  - *Process development and applications*
  - *Processing and synthesis of many types of ceramics*
  - *Equipment design and process control*
  - *Process modeling*
  - *More sophisticated R&D studies*
- **Next 5 - 10 years:**
  - *More effective integration of microwave engineering, materials science and processing, and equipment design*
  - *Greater (precise) process control and automation*
  - *Growth of pilot plant and larger scale studies*
  - *Emergence of industrial applications*
  - *Demonstration of M/W benefits (rapid processing, cost and energy savings, improved quality, etc.) on an industrial scale*
  - *Emergence of many new, unique, discoveries and applications*

REFERENCES

1. R. Roy, "Dedication", Microwave Processing of Materials, edited by W.H. Sutton, M.H. Brooks and I.J. Chabinsky (Matls. Res. Soc. Proc. 124, Pittsburgh, PA. 1988).
2. A. von Hippel, ed., Dielectric Materials and Applications, Technology Press of MIT and Wiley, New York, 1954, 422 pp.
3. M.L. Levinson, US. Pat. No. 3, 585, 258 (15 June 1971), orig. application 19 Oct. 1965.
4. J.M. Osepchuk, "A History of Microwave Heating Applications, " IEEE Trans, Microwave Theory Tech., MIT-32 (9), 1200-1224 (1984).
5. C.E.G. Bennett, N.A. McKinnon, and L.S. Williams, "Sintering in Gas Discharges", Nature (London), 217, 1287-1288 (1968).
6. R.E. Schroeder and W.S. Hackett, "Microwave Energy in the Foundry", Br. Foundryman, 293-98 (Aug. 1971).
7. A.C. Metaxas and R.J. Meredith, Industrial Microwave Heating, Peter Peregrinus Ltd., London, 1983, 357 pp.
8. W.R. Tinga, "Interactions of Microwave and Materials", Int'l Microwave Power Inst., Proc. on Short Course for Users with Microwave Power, Nov. 1970.
9. W.H. Sutton, "Microwave Firing of High Alumina Castables", in Ref. 1, 287-295, (1988). Work performed in 1975.

10. A.J. Berteaud and J.C. Badot, "High Temperature Microwave Heating in Refractory Materials", J. Microwave Power, 11 (4), 315-20 (1976).
11. T. Nishitani, US. Pat. No. 4,147,911 (3 April 1979)
12. P.A. Haas, "Heating of Uranium Oxides in a Microwave Oven", Am. Ceram. Soc. Bull., 58 (9), 873 (1979).
13. I.J. Chabinsky and E.E. Eves III, "The Application of Microwave Energy in Drying, Calcining, and Firing of Ceramics", InterCeram, 6 (Dec. 1986), 30-35 .
14. S.J. Oda and I.S. Balbaa, "Microwave Processing at Ontario Hydro Research Division", in Ref. 1, 303-309, (1988).
15. N.W. Schubring, "Microwave Sintering of Alumina Spark Plug Insulators", Tech. Rept., GMR-4471, General Motors Research Laboratory, 1983, pp. 1-27.
16. L. Quéменéur, J. Choisnet, B. Raveau, J.M. Thiebaut, and G. Roussy, "Microwave Clinkering with a Grooved Resonant Applicator", J. Am. Cer. Soc., 66 (12), 855-59 (1983).
17. J.F. MacDowell, "Microwave Heating of Nepheline Glass-Ceramics", Am. Cer. Soc. Bull., 63 (2), 282-286, (1984).
18. Y. Hässler and L. Johansen, "Microwave Heating of Fused Quartz to High Temperatures in the Fabrication Process of Optical Fibers", in Ref. 1, 273-278 (1988).
19. W.H. Sutton, "Microwave Processing of Ceramics", Am. Cer. Soc. Bull., 68 (2), 376-386 (1989).
20. Masao, Japanese Patent 153793/86 (30 Jun 86).
21. T.T. Meek and R.D. Blake, US. Patent 4,529,857 (16 July 1985).
22. T.T. Meek and R.D. Blake, "Ceramic-Ceramic Seals by Microwave Heating", J. Matls. Sci. Lett., 5 270-74 (1986).
23. D. Palaith, R. Silberglitt, C.C. M. Wu, R. Kleiner, and E. Libelo, "Microwave Joining of Ceramics", in Ref. 1, 255-266 (1988), Bull. Am. Cer. Soc. 68 (9), 1601 (1989).
24. H. Fukushima, T. Yamaka, and M. Matsui, "Microwave Heating of Ceramics and Its Application to Joining", in Ref. 1, 267-262 (1988), J. Matls. Res. 5 (2), 397-405 (1990).
25. L. Sheppard, "Manufacturing Ceramics with Microwaves: The Potential for Economic Production", Am. Cer. Soc. Bull., 67 (10), 1556-61 (1988).
26. W.H. Sutton, M.H. Brooks, and I.J. Chabinsky, editors, Microwave Processing of Materials, (Matls. Res. Soc. Proc. 124, Pittsburgh, PA, 1988).
27. W.B. Snyder, Jr., W.H. Sutton, M.F. Iskander, and D. Lynn Johnson, editors, Microwave Processing of Materials II, (Matls. Res. Soc. Proc. 169, Pittsburgh, PA 1990).
28. D.E. Clark, F.D. Gac, and W.H. Sutton, editors, Microwave: Theory and Application in Materials Processing, Cer. Trans. 21, Amer. Cer. Soc. (1991).
29. R.L. Beatty, W.H. Sutton, and M.F. Iskander, editors, Microwave Processing of Materials III, (Matls. Res. Soc. Proc. 269, 1992, this Volume).
30. J. Asmussen and R. Garard, "Precision Microwave Applicators and Systems for Plasma and Materials Processing", in Ref. 26, 347-352 (1988).
31. Y-L. Tian, "Practices of Ultra-Rapid Sintering of Ceramics Using Single Mode Applicators", in Ref. 28, 283-300 (1991).
32. B.Q. Tian and W.R. Tinga, "A Wide Range Tunable and Matchable High Temperature Applicator", in Ref. 28, 647-654 (1991).
33. J.F. Gerling, "Techniques to Improve the Performance of Microwave Process Systems Which Will Utilize High Q Cavities", in Ref. 28, 667-674 (1991).
34. H.S. Sa'adaldin, N.M. Black, I. Ahmad and R. Silberglitt, "Efficient Coupling with an Adjustable Compound Iris for Joining of SiC in a Single Mode Applicator", this Volume.
35. H.D. Kimrey, J.O. Kiggans, M.A. Janney, and R.L. Beatty, "Microwave Sintering of Zirconia-Toughened Alumina Composite", in Ref. 27, 243-255 (1991).
36. M.A. Janney, C.L. Calhoun, and H.D. Kimrey, "Microwave Sintering of Zirconia - 8 mol % Yttria", in Ref. 28, 311-318 (1991).

37. M.A. Janney, C.L. Calhoun and H.D. Kimrey, "Microwave Sintering of Solid Oxide Fuel Cell Materials, I: Zirconia - 8 mol % Yttria", J. Am. Cer. Soc., 75 (2), 341-346 (1992).

38. A. Dé, I. Ahmad, E.D. Whitney, and D.E. Clark, "Microwave (Hybrid) Heating of Alumina at 2.45 GHz: I. Microstructural Uniformity and Homogeneity", in Ref. 28, 319-328 (1991); II. Effect of Processing Variables, Heating Rates and Particle Size", ibid, 329-339.

39. A.D. Cozzi, D.K. Jones, Z. Fathi, and D.C. Clark, "Microstructural Evaluation of $YBa_2Cu_3O_{7-x}$ Using Microwave Energy", Ref. 28, 357-364 (1991).

40. Y. Fang, D.K. Agrawal, D. M. Roy, and R. Roy, "Rapid Sintering of Hydoxyapatite Ceramics by Microwave Processing", in Ref. 28, 349-356 (1991).

41. W.B. Harrison, M.R.B. Hanson, and B.G. Koepke, "Microwave Processing and Sintering of PZT and PLZT Ceramics", in Ref. 26, 279-295 (1988).

42. M.C.L. Patterson, P.S. Apté, R.M. Kimber, and R. Roy, "A High Performance Tool Bit Produced by Microwave Heating", presented at UIE Microwaves and Radio Frequency Conference, pp. 33-36; Nice, France, Oct. 8-10, 1991.

43. M.C.L. Patterson, P.S. Apté, R.M. Kimber, and R. Roy, "Batch Process for Microwave Sintering of Si_3N_4", this Volume; and "Mechanical and Physical Properties of Microwave Sintered Si_3N_4", this Volume.

44. R. Smith, M.F. Iskander, O.M. Andrade, and H. Kimrey, "Finite Difference Time-Domain (FDTD) Simulation of Microwave Sintering in Multimode Cavities", this Volume.

45. M.A. Janney and H.D. Kimrey, "Microwave Sintering of Alumina at 28 GHz", Cer. Trans. Vol. I, Ceramic Powder Science II-B, Am. Cer. Soc., 919-924 (1988).

46. B. Swain, "Microwave Sintering of Ceramics", Adv. Matls. and Processes, 134 (3), 76-81 (1988).

47. T.T. Meek, R.D. Blake, and J.J. Petrovic, "Microwave Sintering of Al_2O_3 and Al_2O_3 - SiC - Whisker Composites", Cer. Eng. Sci. Proc., 8 (7-8), 861-71 (1987).

48. Y. Bykov, A.F.L. Goldenberg, and V.A. Flyagin, "The Possibilities of Material Processing by Intense Millimeter-Wave Radiation", in Ref. 27, 41-42 (1991).

49. B.E. Paton and V.E. Sklyarevich, "Application of Gyrotron Radiation for Coating and Welding", in Ref. 27, 99-100 (1991).

50. V. Sklyarevich, A. Detkov, M. Shevelev, and R. Decker, "Interaction of Gyrotron Radiation with Powder Materials", this Volume.

51. Y-L. Tian and D. Lynn Johnson, "Microwave Sintering of Al_2O_3 - TiC Composites", Cer. Trans. Vol. I, Ceramic Powder Science II-B, Am. Cer. Society, 933-938 (1988).

52. Y.H. Kao and J.D. MacKenzie, "Magnetite as a Sintering Aid for Microwave Consolidation of Soda-Lime Glass", in Ref. 28, 341-348 (1991).

53. Y-L. Tian, B-S. Li, J-K. Guo, and D-S. Yan, "A Microwave Processing System for Ceramics with Improved Energy Efficiency and Processing Control", 25th Microwave Power Symposium, Int'l Microwave Power Inst., 13-14, Denver (Aug 1990).

54. Y-L. Tian, and D.L. Johnson, "Ultrafine Microstructure of Al_2O_3 Produced by Microwave Sintering", Cer. Trans. Vol. I, Ceramic Powder Science II-B, American Ceramic Soc., 925-931 (1991).

55. D.S. Patil, B.C. Mutsuddy, J. Gavulic, and M. Dahimene, "Microwave Sintering of Alumina Ceramics in a Single Mode Applicator", in Ref. 28, 301-309 (1991).

56. Y-L. Tian, B-S. Li, J-L. Shi, Y-P. Xu, J-K Guo, and D-S Yen, "Microwave Sintering of Y_2O_3 (3%) - ZrO_2 (TZP)", in Ref. 28, 577-584 (1991).

57. M.A. Janney, H.D. Kimrey, and J.O. Kiggans, "Microwave Processing of Ceramics: Guidelines Used at the Oak Ridge National Laboratory", this Volume.

58. J.D. Katz, and R.D. Blake, "Microwave Sintering of Multiple and Composite Components", Bull. Am. Cer. Soc. 70 (8), 1304-1308 (1991).

59. M.C.L. Patterson, R.M. Kimber, and P.S. Apté, " The Properties of Alumina Sintered in a 2.45 GHz Microwave Field", in Ref. 27, 257-266 (1991).

60. M.G. Hamlyn, and A.L. Bowden, "Microwave Processing of Earthenware Ceramics", Bull. Am. Cer. Soc., 69 (3), 368, 370, 372 (1990).

61. M.K. Krage, "Microwave Sintering of Ferrites", Am. Cer. Soc. Bull., 60 (11), 1232-34 (1981).

62. W.A. Hendrix, and T. Martin, "An Experimental Investigation of Hybrid Convective/ Microwave Drying of Electrical Porcelain", this Volume.

63. F. Smith, "Choosing the Right Electromagnetic Heat Process", Automation, 38 (12), 42-44 (1991).

64. J-C. Balomey, and J-P. Metail, "Review of the French Activity in Microwave Control and Processing of Materials", in Ref. 27, 15-26 (1991).

65. T. Ochiai, M. Ikdea, and T. Nishitani, "Application of Microwave Drying to Monolithic Ladle Linings", Taikabutsu Overseas, 1 (2), 92-96 (1981).

66. C.E. Holcombe, and N.L. Dykes, "Importance of 'Casketing' for Microwave Sintering of Materials", J. Matls. Sci. Letters, 9, 425-428 (1990).

67. C.E. Holcombe, and N.L. Dykes, "Microwave Sintering of Titanium Diboride", J. Matls. Sci., 26, 3730-3738 (1991).

68. C.E. Holcombe and N.L. Dykes, "Ultra High Temperature Microwave Sintering", in Ref. 28, 375-386 (1991).

69. J.D. Katz, R.D. Blake, J.J. Petrovic, and H. Scheinberg, "Microwave Sintering of Boron Carbide", in Ref. 26, 219-226 (1988).

70. T.N. Tiegs, J.O. Kiggans, Jr., and H.D. Kimrey, "Microwave Processing of Silicon Nitride", in Ref. 28, 267-272 (1991).

71. T.N. Tiegs, M.K. Ferber, J.O. Kiggans, K.L. More, C.M. Hubbard, and D.W. Coffey, "Microstructure Development During Microwave Annealing of Dense Silicon Nitride", in Ref. 28, 411-417 (1991).

72. I. Ahmad, G.T. Chandler, and D.E. Clark, "Processing of Superconducting Ceramics Using Microwave Energy", in Ref. 26, 239-246 (1988).

73. E.H. Moore, I. Ahmad, and D.E. Clark, "Microwave Thermogravimetric Thermal Analyzer (MTGA)", in Ref. 28, 675-681 (1991).

74. E.H. Moore, D.E. Clark, and R. Hutchen, "Polymethyl Methracrylate Binder Removal from an Alumina Compact: Microwave vs. Conventional Heating", this Volume.

75. R.C. Dalton, I. Ahmad, and D.E. Clark, "Combustion Synthesis Using Microwave Energy", Cer. Eng. Sci. Proc. 11 (10), 1729-1742 (1990).

76. J.W. Evans, and D. Gupta, "A Mathematical Model for Microwave-Assisted Chemical Vapor Infiltration", in Ref. 27, 101-107 (1991).

77. D.J. Devlin, R.P. Currier, and R.S. Barbero, "Microwave Assisted Chemical Vapor Infiltration", presented at the 1991 MRS Fall Meeting, Boston, MA (1991) (unpublished).

78. M.A. Willert-Porada and S. Vodegel, "Metallorganic and Microwave-Processing of Ceramic Powders and Compacts", in Ref. 28, 631-638 (1991).

79. M.A. Willert-Porada, "Metallorganic and Microwave-Processing of Monolithic and Polyphasic Ceramics", this Volume.

80. J.O. Kiggians, C.R. Hubbard, R.R. Steele, H.D. Kimrey, C.E. Holcombe, and T.N. Tiegs, "Characterization of Silicon Synthesized by Microwave Sintering", in Ref. 28, 403-410 (1991).

81. J.O. Kiggians and T.N. Tiegs, "Characterization of Sintered Reaction-Bonded Silicon Nitride Processed by Microwave Heating", this Volume.

82. J.J. Thomas, D.L. Johnson, H.M. Jennings, R.R. Jesse, and K.T. Faber, "Nitriding of Nonisothermal Si Compacts Using Microwave Heating", this Volume.

83. J.G.P. Binner, I.A. Al-Dawery, C. Aneziris, and T.E. Cross, "Use of the Inverse Temperature Profile in Microwave Processing of Advanced Ceramics", this Volume.

84. D.J. Adrian, US Patent No. 4,930,673 (3 July 1990).

85. P.S. Apté and A. Pant, US Patent No. 5,010,220 (23 April 1991).

86. H.M. Kingston and L.B. Jassie, "Introduction to Microwave Sample Preparation, Theory, and Practice", Amer. Chem. Soc. Ref. Book, Washington, DC (1988).

87. M. Pougnet, S. Michelson, and B. Downing, "Computer Controlled Microwave Waveguide for Sample Heating and Dissolution", J. Int'l Microwave Power Inst., 26 (3), 139-144 (1991).

88. H. Lydtin and A. Ritz, "Uniform Low-Temperature Deposition of Optical Layers by Plasma-Activated CVD", this Volume.

89. R.J. Lauf, C. Hanby, C.E. Holcombe, and W.F. Vierow, "Microwave Processing of Tantalum Capacitor Anodes", this Volume.

90. R.J. Lauf, C.E. Holcombe, and C. Hanby, "Microwave Sintering of Multilayer Ceramic Capacitors", this Volume.

91. P.J. Gielisse, H. Niculescu, and B. Roy, "Applications of Microwaves in High Temperature Superconductor Materials and Devices", in Ref. 28, 613-621 (1991).

92. Z. Fathi, I. Ahmad, J.H. Simmons, and D.E. Clark, "Surface Modification of Sodium Aluminosilicate Glasses Using Microwave Energy", in Ref. 28, 623-629 (1991).

93. Z. Fathi and D.E. Clark, "Surface Modification of Ceramics Using Microwave Energy", this Volume.

94. M. Jin and K.C. Kao, "Characterization of the ECR Microwave Plasma Etching System", this Volume.

95. E.F. Sturken, "The Use of 'Self-Heating' Ceramics as Crucibles for Microwave Melting of Metals and Nuclear Waste Glass", in Ref. 28, 433-440 (1991).

96. C.M. Jantzen and J.R. Cadieux, "Characterization of Radioactive Waste Melter Feed Vitrified by Microwave Energy", in Ref. 28, 441-449 (1991).

97. R.L. Schulz, Z. Fathi, and D.E. Clark, "Microwave Processing of Simulated Nuclear Waste Glass", in Ref. 28, 451-458 (1991).

98. S.J. Oda, "Microwave Processing of Hazardous Materials-Present and Future Opportunities", this Volume.

99. S.N. Kumar, A. Pant, R.R. Sood, J. Ng-Yelim, and R.T. Holt, "Production of Ultrafine Silicon Carbide by Fast Firing in Microwave and Resistance Furnaces", in Ref. 28, 395-402 (1991).

100. A.K. Singh, P. Mehta, and A.I. Kingon, "Synthesis of Nonoxide Ceramic Powders by Nonthermal Microwave Plasma", in Ref. 28, 421-429 (1991).

101. W.F. Kladding and J.E. Horn, "Submicron Oxide Powder Preparation by Microwave Processing", Ceramics International, 16 (1), 99-106 (1990).

102. D. Vollath, K.E. Sickafus, and R. Varma, "Electron Microscopy Study on Zirconia and Alumina Ceramic Powders Synthesized by Microwave Plasma Pyrolysis", this Volume.

103. J.A. Eastman, K.E. Sickafus, J.D. Katz, S.G. Boeke, R.D. Blake, C.R. Evans, R.B. Schwartz, and Y.X. Liao, "Microwave Sintering of Nanocrystalline TiO_2", in Ref. 27, 273-278 (1991).

104. T.Y. Yiin, V.V. Varadan, and V.K. Varadan, "Microwave Joining of Si-SiC/Al/Si-SiC", in Ref. 28, 507-514 (1991).

105. R. Silberglitt, D. Palaith, W.M. Black, H.S. Sa'adaldin, J.K. Katz, and R.D. Blake, "Investigation of Interlayer Materials for the Microwave Joining of SiC", in Ref. 28, 487-495 (1991).

106. I. Ahmad, R. Silberglitt, W.M. Black, H.S. Sa'adaldin, and J. Katz, "Microwave Joining of Silicon Carbide Using Several Different Approaches", this Volume.

107. X.D. Yu, V.V. Varadan, and V.K. Varadan, "Application of Microwave Processing to Simultaneous Sintering and Joining of Ceramics", in Ref. 28, 497-503 (1991).

108. S. Al-Assafi, I. Ahmad, Z. Fathi, and D.E. Clark, "Microwave Joining of Ceramics", in Ref. 28, 515-521 (1991).

109. S. Al-Assafi and D.E. Clark, "Microwave Joining of Ceramics: A Study on Joining Alumina Both Directly and With Alumina Gel", this Volume.

110. I. Ahmad, R. Dalton, and D. Clark, "Unique Application of Microwave Energy to the Processing of Ceramic Materials", J. Microwave Power and Electromagnetic Energy, 26 (3) 128-138 (1991).

MICROWAVE PROCESSING OF POLYMERS - AN OVERVIEW

DAVID A. LEWIS
IBM Corp., T.J.Watson Research Center,
P.O.Box 218,
Yorktown Heights, NY 10598.

ABSTRACT

This paper surveys some of the recent research into the microwave processing of polymers. The mechanism of coupling of the radiation with polymers is described and a historical perspective is presented on the development of this technology from the 1960's when the processing of rubber was first initiated, to current research in high performance materials and composites. The controversial area of microwave effects on processing is also discussed.

INTRODUCTION

The use of microwave radiation in the processing of polymers offers a number of distinct benefits over conventional processing alternatives. Because of the large penetration depth of microwave radiation in most polymeric materials (30 cm or more), very uniform temperature profiles can be obtained throughout the material being processed. This results in a minimization of the residual stresses which form as a result of non-uniform processing. The conventional solution is to use slower thermal ramp rates to allow thermal convection, however, since faster heating rates are able to be attained while maintaining uniform thermal distributions during microwave processing, the process cycle time is reduced. Furthermore, microwave radiation is also more efficient than convection heating, resulting in significant electrical savings. Akyel and Bilgen[1] have estimated that curing costs can be reduced by 55 - 72% (depending on throughput) and that the pay out period for the new technology would be only 9-12 months. There have been other reports showing similar cost advantages in this technology[2, 3].

However, despite the significant advantages of utilizing microwave processing and the widespread consumer acceptance for food processing, in the form of home microwave ovens, the only significant commercial application of this technology for polymer processing is the preheating of rubber weather seals for automobiles, etc. There are two major reasons for the failure of this technology to more effectively penetrate the industrial market - (i) insufficient understanding of the processes occurring and (ii) inadequate equipment - in the form of applicators and generators - available to exercise the level of control necessary to prevent thermal degradation of the materials being processed. Although there has been substantial development in both of these areas, there is still opportunity for considerable further development.

Recent research has focused on the processing of high performance materials, primarily composite materials which provide high strength and light weight, eg. carbon fiber / epoxy resin. The processing of very thick cross-section parts which required very slow thermal ramp rates due to the poor thermal conductivity when processed conventionally were initially targeted as ideal applications for microwave processing. Such structures are used as stressed members in aircraft and in applications requiring very high strength to weight ratios. An additional application was for the field repair of composite materials in a simple and quick manner - again relying on the efficient heat transfer properties of microwave radiation. Early studies[4, 5] indicated that processing composite materials with conducting (carbon) fibers would be limited to unidirectional composites with less than about 32 plys, due to the high reflectivity of the fibers and hence the poor penetration depth of the radiation into the composite. More recently, however, Wei et al.[6] demonstrated that 36 and 72 cross ply composites could be successfully processed using a single mode resonant cavity.

In addition to the processing of carbon fiber composite materials, the microwave processing of glass fiber and metal containing composites and pure resins have considerable research and potential commercial interest

Other areas in which microwave processing has shown promise include pultrusion and ultradrawing. In pultrusion, a glass fiber bundle is coated with a matrix material and "pulled" through a curing device. Conventionally this processing chamber consists of a heated die, which is quite long due to the slow heat transfer to the matrix polymer. Methven and Ghaffariyan[7] have shown that this process can be rapidly accomplished using microwave radiation in a significantly shorter process chamber, resulting in less force to pull the fiber bundle through the die. Applications for the rods produced by this technique include cost effective, lightweight reinforcement and corrosion resistant reinforcement. As a polymer rod or fiber is drawn, or pulled, the alignment of the molecules and hence the modulus of the fiber along the direction of pull increases. If this process is performed at room temperature, the fiber fails at relatively short draw ratios, however, if the drawing is performed at elevated temperatures, very high draw ratios of 20:1 can be obtained. When microwave radiation is utilized, Nakagawa and coworkers have shown that this draw ratio can be increased to 35:1 with a corresponding increase in the modulus from 35-40 GPa for conventional processing to 55-60 GPa for poly(oxymethylene)[8-10], with similar results for other polymers and copolymers[11, 12].

The fabrication of materials not possible by conventional processing is also an attractive aspect of microwave processing. A major limitation to the greater utilization of the epoxy resins is that they are very brittle. Hedrick et al.[13] showed that these materials could be toughened by the use of thermoplastics in a second phase. Subsequently Hedrick et al.[14, 15] showed that the phase structure of these materials could be altered using microwave radiation and ultimately, could be controlled to produce the morphology and properties desired. Similarly, Rodrigues[16] showed that the morphology of toughened sol-gel glasses could be varied by the use of microwave radiation and that the choice of solvent plays a role in whether the microwave processed material has a different morphology compared with conventional processing. These variations were obtained by trading off the rate of curing against the rate of phase separation and coalescence of the segregated phase which can not be obtained thermally due to either the heating rate of the bulk material being too high or an acceleration in the curing rate by the microwave radiation.

It was during the investigation into the processing of matrix resins for composites that a number of anomalies were found, so called "microwave effects", by a number of research groups. These anomalies include reports of accelerated kinetics[17-21], retarded kinetics[22] and dependencies of the heating rate[23, 24] and structure[25] of the cured polymer structures formed on the pulse repetition frequency. These effects are not observed for all systems investigated, but appear to be particularly evident for reactions which are slow.

This paper focuses on some aspects of the microwave processing of polymers. It is an attempt to provide a basic introduction into polymer processing and provide an understanding of the challenges involved by reviewing some of the research which has been reported.

Coupling Mechanism

The dielectric response of a polymer is shown schematically in Fig. 1, which shows the variation in ε' and ε'' with frequency. From this figure, it can be seen that the principle mechanism of coupling of microwave radiation to polymers is through a dipolar reorientation by the electric field. This relaxation has a maximum at about 10^7 Hz, but there is a high frequency tail which reaches the microwave region. The maximum value of this relaxation varies widely during a processing cycle as the temperature increases, the reaction proceeds (changing the type and concentration of dipole moments), a phase change occurs (changing the mobility of the dipoles), etc.

A dipole moment is formed by adjacent groups having different electron withdrawing / donating properties resulting in a net charge or partial charge localization on one atom or group and can be viewed as a small, weak bar magnet. Typical groups which form these dipoles include hydroxyl, amino, cyanate, etc. At a molecular level, when no electric field is present, the dipole moments are randomly oriented throughout the medium, but when an electric field is applied, there is a tendency to orient in the direction of the field. The extent to which this can be accomplished depends on the mobility of the dipole, the strength of the electric field and the time the electric field is "on", versus the relaxation time for the dipole. When an alternating electric field is imposed, the dipole moments will attempt to oscillate in phase with the electric field, thus storing the energy of the alternating electric field. At higher frequencies, the dipole moments can no longer maintain phase with the applied electric field, leading to a loss of the stored energy, or the absorption of energy. At even higher frequencies, the movement of the dipole is so far out of phase with the applied electric field that it essentially does not move. The electric field is doing work on the dipole moment, with the adjacent groups acting to inhibit the motion of the dipole and thus prevent the dipole from following the electric field, resulting in heating of the dipole moments.

The efficiency of this coupling is dependent on a number of factors, including the dipole strength, the mobility of the dipole and the mass of the dipole. This implies that small strong dipoles couple to the radiation the most efficiently. In the condensed state, the mobility of the medium around the dipole is also important. Small polar molecules (eg. solvent molecules or water) obtain translational energy in addition to the energy from reorientation (rotational energy), dramatically increasing the coupling of these molecules with the radiation compared with a segment of a large macromolecule. Similarly, a dipole that is part of a group pendant to the main chain will couple more strongly than a group in the main chain, which is less mobile. A liquid couples the strongest, followed by a rubber, glassy polymer and a crystalline material.

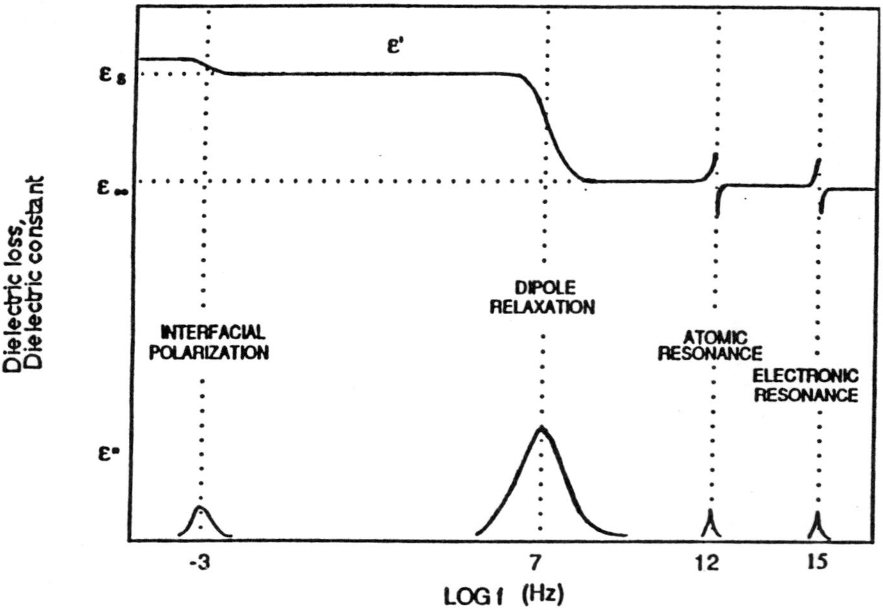

Figure 1. Typical variation in ε' and ε'' with frequency for polymers.

Since polymeric materials are typically non-magnetic, the contribution of the magnetic component to the radiation can largely be ignored. The exception occurs when magnetic particles are incorporated in the material to provide specific functions, as in radar absorbing material or for cookware to provide a "browning effect" for the food item.

Ionic conduction is not important for most polymeric systems as the concentration of free ions is typically at impurity levels in polymers. In addition, the contribution of this mechanism to the dielectric loss decreases as frequency increases, becoming a relatively small contribution by the typical processing frequency of 2.45 GHz.

Although the polymer systems that are candidates for microwave processing are typically not conductive, particles and fibers which are conductive, or partially conducting can be included to reinforce the material, or otherwise modify the mechanical properties. The presence of these fillers strongly influence the way in which the composite material interacts with the microwave radiation, with losses being substantially greater than for the matrix resin. The conductors also modify the electric field pattern in and around the composite resulting in potentially very different curing profiles compared with the neat resin. Some examples of these conductive additives include (i) carbon black (used extensively in rubbers); (ii) carbon or metal fibers and (iii) metal flakes, spheres or needles with sizes ranging from 0.1 to 100 μm. Although the final composite is not necessarily conductive, the surfaces of the conducting inclusions interact strongly with the microwave radiation by setting up 2.45 GHz alternating currents on the surface of the inclusions. The penetration of the microwave radiation into the conductive inclusions is limited to a depth referred to as the skin depth, which is controlled by the conductivity of the inclusion. The heating which results is dependent on the electrical resistivity of the inclusions and the thermal distribution is strongly linked with the distribution of the conductive inclusions. For example, carbon fibers have a relatively high resistivity and heat the surrounding matrix material very effectively, but the thermal profile has a maximum at the surface of the fibers. This has been shown to provide an enhancement of the interfacial adhesion between the fibers and the matrix resin[26], and a subsequent improvement in the fracture properties of composite materials processed in this way.

Other, non-conductive, additives include glass fibers and non-conducting metal oxides, such as titanium dioxide, which are used as pigments, couple through similar mechanisms to ceramic materials, with defects and ionic impurities having a strong role in the absorption. The microwave absorption of the glass reinforcement is dependent upon the purity of the glass, with quartz having a very small absorption and soda glass coupling quite strongly compared to the surrounding matrix resin.

Historical Perspective

The history of microwave processing of polymers has been somewhat checkered, with a great deal of interest in the late 1960's, followed by a lull until the mid 1980's. The initial work was focused on the rubber industry in a bid to cure relatively thick, poor thermally conductive cross-sections in the minimum time possible to free the expensive molds for further utilization, in addition to enhancing throughput. This work lead to the only industrial application of microwave processing of polymers - the preheating of rubber for weather seals in automobiles, etc. This technology replaced the conventional approaches which utilized salt baths which are very efficient, but messy and environmentally undesirable, and steam autoclaves used in batch processes, which are slow, inefficient and expensive to operate.

Most of the polymers used in the rubber industry do not absorb microwave radiation because of the absence of permanent dipoles. However, in the commercial application of rubbers, the materials are usually a mixture of the rubber with antioxidants, fillers, dispersants (to maintain uniform distribution of the fillers), viscosity modifiers, vulcanizing agents, etc., as shown in Table 1 for colored and black rubbers. Furthermore, each of the components listed in Table 1 may contain more than one chemical, each with specific dielectric properties, designed to provide the specific mechanical properties desired. The effect of adding each component can be seen in[27-29] which show the effect of the addition of each component on the heating rate. In cases where the final product is black, one requirement was to select the conductivity of the carbon black and produce the appropriate distribution of this filler throughout the cross-section of the rubber, thereby producing the desired heating pattern within the material.

Table 1. Typical Formulations of Commercial Rubbers.

	clear/colored	black
rubber	55%	45-50%
filler	20%	10%
carbon	-	27-35 %
plastisizer	10%	12%
dispersants	3%	1%
antioxidants	1 - 2%	1%
coloring	3%	-
sulfur	1%	1%

Materials Aspects

The two classes of polymeric materials that are suitable for microwave processing - (i) thermosetting polymers, such as epoxy resins and rubbers; and (ii) thermoplastic polymers, such as PEEK - have very different processing requirements.

Initially, thermosetting polymers are low viscosity liquids that can flow into a mold or around fibers and during processing react to increase molecular weight and viscosity, eventually becoming highly crosslinked, insoluble, infusible materials. During microwave processing, these materials show three basic regions - (i) preheating of the components; (ii) reaction with the corresponding exotherm and (iii) cooling of the cured material[30, 31] These polymers tend to be highly efficient absorbers of microwave radiation initially and due to the large reaction exotherm, can be difficult to control. After cure is complete, ε' and ε'' are significantly less due to the number of dipole moments and the mobility being significantly reduced compared with the starting materials.

Thermoplastic polymers, however, are usually difficult to heat until a critical temperature[32] is reached at which point the heating rate increases significantly. Chen and Ward[32] have shown that this critical temperature may not be the same as the glass transition temperature of the polymer. These polymers are processed approx. 50-100 °C above the glass transition temperature or melting point (if the material is polycrystalline) to reduce the melt viscosity and allow flow and adhesion. It is this rubbery domain in which the dielectric properties change the most during a typical curing cycle - adding a further complication in the control of the processing. During the processing of an epoxy resin for example, the material can pass through all of these states before becoming stable. DeMuse[33] has also shown that the crystal content of these materials is important. As expected from the forgoing discussion, amorphous polymers heat more effectively than semi-crystalline polymers, in which the crystal content approaches 35 %. The heating rate decreases significantly until about a theoretical crystal content of 45 % results in essentially no heating - this is despite the presence of 55 % of amorphous material which is expected to heat relatively efficiently. Presumably the crystal tie region is also relatively immobile and a relatively large amount of the amorphous polymer has restricted motion in these large, rigid repeat units.

One of the disadvantages of thermosetting polymers is that they are brittle materials, whereas while thermoplastics are very tough, the creep and solvent resistance is relatively poor. A compromise material is a functionally terminated thermoplastic, which provides some toughness and reduced creep and enhanced solvent resistance. These polymers (or prepolymers) undergo a combination of thermoset and thermoplastic processing, with the initial heating reducing viscosity and improving flow and ultimately reaction providing a network. Microwave processing of these polymers offers a number of advantages, particularly in reducing the processing time[15, 19, 34], but control becomes crucial if thermal degradation is to be prevented.

From the foregoing discussion, it can be seen that one of the challenges in the microwave processing of polymers is that the processing temperature for most polymers is VERY close to the thermal degradation temperature of the polymer. If the temperature is too high, the polymer undergoes undesirable crosslinking, scission and oxidation (if oxygen is present), which results in a severe change in the mechanical and optical properties of the material. This places a strict requirement on the microwave system to provide very uniform temperature distributions throughout the part being cured / processed and carefully control the temperature of the part.

Microwave Effects

There are a number of instances in which anomalies have been observed during microwave processing. These can be combined under the heading "Microwave Effects", which includes anything that is either different from thermal processing or cannot be predicted or easily explained by the different temperature profile. For example, the significantly superior mechanical properties of microwave ultradrawn poly(oxymethylene) fibers over conventionally processed fibers[8] is due to the better thermal uniformity throughout the cross-section of the fiber which is obtained during microwave processing and would NOT be categorized as a microwave effect, whereas accelerated curing kinetics in small samples would be considered a microwave effect.

This topic is extremely controversial, with some apparently conflicting results and some very strong opinion against the possibility of such effects. It remains, however, that a very large number of research groups have published results which do not correspond with the expected behavior based on conventional processing. The most prevalent reports have concerned the acceleration of reaction rates. Unfortunately, in a number of these reports, the temperature was not carefully measured - if at all. It should also be pointed out that there have been reports in which no effect of the radiation on the kinetics was observed[35-37].

A number of the reports of microwave effects are for the curing of epoxy resins and simply relate conversion with time. Unfortunately, it is difficult to analyze this data further since the reaction behaves autocatalytically. Furthermore, the epoxy resins pass through a gel point, limiting diffusion (which limits the reaction rate) and can then vitrify if the reaction temperature is not above the glass transition temperature, further complicating the analysis.

One of the difficulties in the comparison and rationalization of these effects is that the experimental conditions and the materials have differed from group to group. Even within the general class of epoxy resins, on which a large amount of work has been performed, the reactivity can vary more than an order of magnitude, depending on the actual reactants studied. Similarly, the observation of microwave effects has varied depending on the epoxy resins studied. Two generalizations which have become obvious recently are (i) slower reacting systems tend to show a greater effect under microwave radiation than faster reacting systems and (ii) the magnitude of the observed effect decreases as the temperature of the reaction is increased,.

Another cause for the variability in the observation of these effects may be in the method by which the material is processed. One important variable in kinetic analysis is temperature, but if there is any effect due to the radiation itself, the manner in which temperature is controlled may be an important variable. The simplest method of temperature control is an on/off cycle, which could be considered a long pulse, typically on the 1 - 10 second timescale. It could be expected that any "microwave effect" would be diminished as the sample thermalizes. Despite this however, there has been at least one report of such a temperature control method resulting in an acceleration in the reaction kinetics[21] and one case of a claimed retardation in the kinetics[22].

As the pulse delay decreases, the microwave system becomes a more traditional pulsed system, in which the radiation is turned on and off, but now at a relatively high rate on the timescale of a few milliseconds. However, under these conditions, Chan and Gourdenne[23] noted a dependence of the repetition frequency on the heating rate for a range of polymeric materials, making the separation of any effect on the kinetics from an effect due to pulse frequency difficult.

A more desirable method of controlling the temperature is to vary the microwave power constantly, thereby maintaining an electric field at all times. It could be expected that there will be some effect of the variable electric field on any effect observed, but this is not expected to be as severe as the on/off case. The disadvantage of this technique is that most microwave generators have some frequency dependence on the output power and as the power changes, the frequency shifts. When high Q cavity applicators are utilized, this frequency shift can lead to some detuning of the applicator, hence further complicating temperature control.

A number of these problems were avoided in a study of a solution imidization reaction which followed first order kinetics and in which the reactant and product remained soluble throughout the reaction[18, 20]. In this study, the isothermal conditions were maintained by varying the microwave power or detuning the applicator and showed a 18-35 fold enhancement in the reaction rate over the temperature range studied. This corresponded to a reduction in the activation energy for the reaction from 105kJ/mol to 55kJ/mol, assuming Arrhenius assumptions were valid.

The mechanism to explain the "microwave effect" in polymers, first proposed in 1988[18, 20], suggests a nonequilibrium, non-uniform energy distribution on the molecular level which results in certain dipole moments have a greater energy than the "average" energy of adjacent groups. For the solution imidization of a poly(amic acid), this increased energy corresponded to an increase in the "temperature" of the reacting groups of approx. 50 °C over the temperature range investigated. The energy couples directly with a reactive polar group in this system and dissipates through adjacent groups by the usual mechanisms. However, if the energy is absorbed faster than it is transferred, at least initially, there will be a steady state, non-uniformity present. This is consistent with some of the recent pulsed radiation studies in that the rate of energy transfer along the chain may be related to chain relaxations which occur on a similar timescale to the pulse repetition frequency.

Recently, there has been a calculation by Booske et al.[38] in which a small perturbation to the high energy region of the energy distribution was made which was sufficient to account for some effects that have been observed in ceramics. The results of this calculation are analogous to the mechanism proposed for polymers, although the details differ as the polymer system is not as highly ordered as the ceramic system and different coupling mechanisms apply.

Conclusions and Future Directions

The microwave processing of polymers offers a great number of benefits to industry in terms of energy savings, process time and floor space reduction, better use of assets (eg. molds) and reduced environmental impact. The understanding of the processes that occur under microwave irradiation have also been greatly enhanced recently. However, there remain a great number of challenges in this field, including (i) a better understanding of any microwave effects; (ii) improved process control; (iii) improved applicator design and (iv) modelling of the interactions of the field with complex geometries.

One of the important benefits from this form of processing is the possibility of new materials or properties which cannot be attained by other processing techniques. This area has a number of exciting possibilities, but research efforts thus far have been somewhat limited.

References

[1] C. Akyel and E. Bilgen, in: *Energy* vol. 14, (1989) 839-851.

[2] J. L. Simonian, Microwave Curing of Insulating Varnish on Armatures Stators and Transformers Proceedings of 14 th Electron Insul Conf, (IEEE, 1979) 105-109.

[3] I. J. Chabinsky, Applications of Microwave Energy - Past, Present and Future - "Brave New Worlds", eds. W. H. Sutton, M. H. Brooks, and I. J. Chabinsky, (Materials Research Society, 1988) 17-29.

[4] W. I. Lee and G. Springer, in: *J. Composite Materials* vol. 18, (1984) 257-386.

[5] W. I. Lee and G. Springer, in: *J. Composite Materials* vol. 18, (1984) 387-409.

[6] J. Wei, M. C. Hawley, J. Jow, and J. D. DeLong, in: *SAMPE J.* vol. 27, (1991) 33-39.

[7] J. M. Methven and S. R. Ghaffariyan, in: *PMSE Prepr.* vol. 66, (1992) 389 - 390.

[8] K. Nakagawa and S. Yamakawa, in: *J. Polym. Sci.,Polym. Lett. Ed.* vol. 21, (1983) 933-935.

[9] Y. Takeuchi, F. Yamamoto, K. Nakagawa, and S. Yamakawa, in: *J. Polym. Sci.,Polym. Phys. Ed.* vol. 23, (1985) 1193-1200.

[10] K. Nakagawa, T. Konaka, and S. Yamakawa, in: *Polymer* vol. 26, (1985) 84-88.

[11] K. Nakagawa and M. Amano, in: *Polymer* vol. 28, (1987) 263-268.

[12] K. Nakagawa and M. Amano, in: *Polym. Commun.* vol. 28, (1987) 119-120.

[13] J. L. Hedrick, I. Yilgor, G. L. Wilkes, and J. E. McGrath, in: *Polym. Bull.* vol. 13, (1985) 201.

[14] J. C. Hedrick, D. A. Lewis, T. C. Ward, and J. E. McGrath, in: *Polym. Prepr.* vol. 29, (1988) 363-365.

[15] J. C. Hedrick, D. A. Lewis, G. D. Lyle, T. C. Ward, and J. E. McGrath, in: *Fourth Technical Conference, Proceedings of the American Society for Composites* (1990) 167-176.

[16] D. E. Rodrigues and G. L. Wilkes, in: *PMSE Prepr.* vol. 66, (1992) 384 - 385.

[17] D. A. Lewis, J. C. Hedrick, T. C. Ward, and J. E. McGrath, in: *Polym. Prepr.* vol. 28, (1987) 330-331.

[18] D. A. Lewis, T. C. Ward, J. D. Summers, and J. E. McGrath, in: *Polym. Prepr.* vol. 29, (1988) 174-175.

[19] J. C. Hedrick, D. A. Lewis, G. D. Lyle, D. Wu, T. C. Ward, and J. E. McGrath, in: *PMSE Prepr.* vol. 60, (1989) 438-442.

[20] D. A. Lewis, T. C. Ward, J. D. Summers, and J. E. McGrath, in: *J. Polym. Sci., Polym. Chem. Ed.* vol. 30, (1992) 1647 - 1653.

[21] J. D. DeLong, J. Jow, and M. C. Hawley. presented at 2nd Topical Conference on Emerging Technologies in Materials San Francisco, 1989.

[22] J. Mijovic and J. Wijaya, in: *Macromolecules* vol. 23, (1990) 3671 - 3674.

[23] L. Chan and A. Gourdenne, in: *PMSE Prepr.* vol. 66, (1992) 382- 383.

[24] A. Gourdenne, in: *PMSE Prepr.* vol. 66, (1992) 430 - 431.

[25] F. M. Thuillier, H. Jullien, and M.-F. Grenier-Loustalot, in: *Polym. Commun.* vol. 27, (1986) 206-208.

[26] R. K. Agrawal and L. T. Drzal, in: *J. Adhesion* vol. 29, (1989) 63-79.

[27] J. Ippen, in: *Rubber Chem. Technol.* vol. 44, (1971) 294-306.

[28] H. F. Schwarz, R. G. Bosisio, M. R. Wertheimer, and D. Couderc, in: *J. Microwave Power* vol. 8, (1973) 303-322.

[29] H. F. Schwarz, R. G. Bosisio, M. R. Wertheimer, and D. Couderc, in: *Rubber Age* vol. 107, (1975) 27-38.

[30] A. Gourdenne, A. -H. Maassarani, P. Monchoux, S. Aussudre, and L. Thourel, in: *Polym. Prep.* vol. 20, (1979) 471-474.

[31] Q. L. Van and A. Gourdenne, in: *Eur. Polym. J.* vol. 23, (1987) 777-780.

[32] M. Chen, T. C. Ward, and J. E. McGrath, in: *PMSE Prepr.* vol. 60, (1989) 443-447.

[33] M. T. DeMuse, in: *PMSE Prepr.* vol. 66, (1992) 428 - 429.

[34] G. D. Lyle, J. C. Hedrick, D. A. Lewis, J. S. Senger, D. H. Chen, S. D. Wu, and J. E. McGrath, Synthesis and Characterisation of Maleimide Terminated Poly(Arylene ether ketone)s, eds. C. Feger, M.M.Khojasteh, and J.E.McGrath, (Elsevier, Amsterdam, 1989) 213-227.

[35] C. Jordan, J. Galy, J.P.Pascault, C. More, and M. Delmotte, in: *PMSE Prepr.* vol. 66, (1992) 380 - 381.

[36] H. Jullien and A. Petit, in: *PMSE Prepr.* vol. 66, (1992) 378 - 380.

[37] J. Mijovic, J. Wijaya, and A. Fishbain, in: *Poly. Prepr.* vol. 32, (1991) 366 - 367.

[38] J. H. Booske, R. F. Cooper, and I. Dobson, in: *J. Mater. Res.* vol. 7, (1992) 495 - 501.

MICROWAVE PROCESSING IN THE UK

JGP Binner† and TE Cross‡
† Department of Materials Engineering and Materials Design
‡ Department of Electrical and Electronic Engineering
University of Nottingham, University Park, Nottingham, NG7 2RD, UK

ABSTRACT

A brief review of the current status of microwave processing in the UK is provided, focussing on ceramic and polymer processing, materials synthesis, catalyst preparation and dielectric property measurement. The central role of the Materials and Microwave Processing Group is also discussed.

CERAMIC PROCESSING

Perhaps the greatest amount of research effort in the field of microwave ceramic processing in the UK is concerned with firing or sintering. A number of research programmes exist in different laboratories, for example at the University of Nottingham, EA Technology (Capenhurst), AEA Technology (Harwell/Culham) and Staffordshire Polytechnic.

At the University of Nottingham, England, a multidisciplinary approach is being taken by the Departments of Materials Engineering & Materials Design and Electrical & Electronic Engineering. Facilities include both multimode and single mode applicator furnaces operating at 2.45 GHz with power levels up to 5 kW. Controlled atmosphere capability is available and temperature measurement is by the Accufibre system of optical thermometry[1]. Research is funded by the UK government via Science and Engineering Research Council grants and the Department of Trade and Industry, and from a number of industrial sources.

The work, which is led by Dr Jon Binner, is currently focussing on manipulation of the temperature profile within advanced ceramics during microwave sintering. For example, control of the inverse temperature profile so that it is virtually uniform across the ceramic allows the use of much higher heating rates since differential expansion is not experienced. A significant degree of success has been achieved, for example, the fast heating rates attainable have allowed zirconia to be sintered to ~95% of theoretical whilst retaining a submicron grain size [1]. However it is felt that much greater benefits can be obtained from generating an inverse temperature profile in which the centre of the body becomes hotter than the surface.

In an attempt to explore these possibilities further a number of systems are currently being examined, including high-T_c superconductors and reaction bonded silicon nitride.

[1] Accufibre, Oregan, USA.

Current bulk polycrystalline ceramic superconductors suffer from the major disadvantage that only the surface layers have the correct oxygen stoichiometry and are, therefore, superconducting. This occurs because conventional sintering results in the surface densifying before the interior, cutting off the latter from the oxygen-rich atmosphere required during sintering to achieve the correct oxygen stoichiometry. This effectively creates a 'shell' of superconducting phases surrounding a central core which is non-superconducting. It is this effect which has helped drive research in recent years towards the production of thin films. Research on yttrium-barium-copper oxide (YBCO) superconductor has shown that the use of microwave energy resulted in a significantly more uniform oxygen content throughout disk shaped components by generating an inverse temperature profile during sintering [1,2].

A similar approach is being investigated with respect to reaction bonding of silicon nitride, where the necessity of obtaining nitridation throughout the compact has resulted in the need to maintain open porosity. This severely limits the mechanical strength and oxidation resistance of the material. By controlling the temperature profile within silicon compacts it has been possible to achieve nitridation from the centre of the compact [2].

Work is also underway on the use of microwave energy as an alternative energy source for the synthesis of non-oxide ceramic powders, slip casting of alumina green bodies [3] and the joining of both oxide and non-oxide ceramics. In addition, a theoretical approach to microwave heating is being developed using transmission line modelling (TLM).

A number of research programmes exist at EA Technology, based at Capenhurst, England. At present they have two microwave furnace applicators, one with a controlled atmosphere capability for inert gases and both capable of 1750-1850°C. Both furnaces operate at 2.45 GHz and a variety of power supplies are available covering power levels of 1.2 to 5 kW. Emphasis has been placed on the power and temperature control aspects with power to the furnace controlled through a programmer which responds to input from the temperature measurement system.

A major part of their work, led by Dr Jayne Samuels and Mrs Ruth Wroe, is the development of a uniform heating profile throughout samples, which they can now do successfully using a variety of techniques, some of which are being patented. In addition, they have designed and constructed a sintering dilatometer capable of operating in both conventional and microwave heated furnaces to permit monitoring of sintering continuously during firing. The results of recent work [4] have established that microwave heating enhances both densification processes in a range of ceramic systems and also the rate of conversion of alpha to beta silicon nitride.

At AEA Technology a materials processing facility is being constructed onto an existing 28 GHz microwave source at the Culham laboratory. This will be used primarily for ceramic sintering and will have controlled atmosphere capability in a programme organised by Dr Alan Hooper. At Staffordshire Polytechnic, researchers in the Department of Electrical Engineering have examined the use of microwave heating for the firing of traditional pottery materials. Whilst the work, which is led by Professor Alan Bowden, has concluded that earthenware ceramic materials are susceptible to microwave energy the mechanism of absorption of energy is not simple as it varies significantly for the individual raw materials [5].

POLYMER PROCESSING

Potential is believed to exist for the use of microwave heating within the fibre reinforced polymer industry in order to reduce cycle times through improved heat up rate of low thermal conductivity resins. To this end, investigations are underway within the Department of Mechanical Engineering at the University of Nottingham into the in-line preheating of polyester resin during resin transfer moulding. A 0-5 kW variable power microwave generator coupled to a single mode TM_{020} resonant applicator is being used to generate a heating profile, within a pipe containing the viscous resin, which is zero at the wall where the resin is stationary and a maximum along the axis where the flow is greatest. The research is led by Professor Mike Owen and funded by the Ford Motor Company Ltd.

At the University of Manchester, Division of Polymer Engineering, work is underway on microwave assisted pultrusion. Pultrusion is the only continuous process available for the manufacture of fibre reinforced thermosetting composites. In its simplest form it requires that dry reinforcement in the form of continuous rovings or fabrics is pulled through a shallow bath of catalysed resin and then through a heated metal die wherein the resin crosslinks to a rigid network. The rate of manufacture depends on two processes which occur simultaneously within the pultrusion die: the rate of heat transfer to and through the composite by conduction from the heated die wall and the rate of heat generated internally by the exothermic crosslinking reaction [6]. The advantage of using microwaves is that the temperature distribution which can be generated within the resin is significantly more uniform than that which results from relying on conduction effects. Research, led by Dr Jim Methven, has shown a resonant cavity applicator can be used to manufacture simple cylindrical profiles directly at acceptable rates and in addition, the power required is well within the range of magnetrons which are mass produced at relatively low cost for the domestic oven market.

MATERIALS SYNTHESIS

It has been established that reaction times for the synthesis of coordination and organometallic compounds may be reduced by as much as three orders of magnitude using microwave superheating techniques by researchers at the University of Oxford. Examples of organometallic and co-ordination compounds of the platinum metals, which are expensive to buy and frequently kinetically inert, have been successfully obtained in high yields and with reduced reaction times by this technique [7]. Intercalation compounds have also received attention from the group led by Dr Mike Mingos[2]. Using a teflon-based parr digestion autoclave several pyridine intercalates of the layered compound α-$VOPO_4$ have been made [7]. In a typical experiment the microwaves cause superheating to temperatures in excess of 200°C by coupling effectively with pyridine molecules which have large dipole moments and after several minutes the intercalation reactions are essentially complete. Conventional processing may take many hours.

---

[2] Dr Mike Mingos has now moved to the Department of Chemistry at Imperial College, London.

Solid state reactions have shown significant increases in their kinetics also. The research group has investigated the formation of mixed metal oxides such as yttrium-barium-copper oxide (YBCO) [7], potassium vanadate and others. Increases in the kinetics range from about 20 for the YBCO to 2 orders of magnitude for the vanadate. In addition, X-ray powder diffraction patterns of the resultant microcrystalline samples show a high degree of crystallinity.

Finely divided metal powders couple very effectively with microwaves and achieve very high temperatures within minutes. This property has been utilised in the synthesis of metal borides, sulphides, selenides and tellurides of the transition metals [7].

CATALYST PREPARATION

Microwaves have been applied to a number of catalyst preparation problems by researchers in the Department of Chemistry at Hull University. For example, the work has determined that the microwave drying of nickel catalysts produces a superior catalyst with better metal distribution [8]. This is believed to occur because of the moisture levelling effects already known to occur in the drying of 'green bodies' in the ceramics industry. An unexpected and significant additional discovery with these catalysts was that when prepared using microwave drying, the ceramic was more resistant to crushing than the comparable thermally dried material. This is believed by the researchers to be of potentially great use to industry, where catalyst attrition producing fines causes pollution problems.

With regard to catalyst characterisation, they have built a microwave heated cell for temperature-programmed reduction (TPR) and desorption (TPD) studies which can be used alongside conventional thermal TPR/TPD apparatus in order to demonstrate differences. So far it has been found that there is fine structure in the hydrogen consumption trace accompanying the reduction process in a microwave environment which is not apparent in the conventional system, a finding which is believed to be very important to the future development of the TPR/TPD technique of catalyst characterisation [9].

Microwave heating has also been used in the area of selective oxidation. Here both the selectivity and the activity of catalysts towards hydrocarbon oxidation to useful products have been altered and many curious effects have been seen which require further study.

DIELECTRIC PROPERTY MEASUREMENT

If microwave processing is to become an established fabrication route then information will be required on the dielectric properties of nonmetallic materials at elevated temperatures since these largely control the interaction of microwaves with materials. In the UK a major activity is currently underway at the University of Nottingham where two approaches are currently being taken in a programme led by Dr Tom Cross. The first makes use of a surface probe technique which has an accuracy of approximately $\pm 5\%$. Room temperature measurements have been obtainable for some time across the frequency range 40 MHz to 50 GHz and measurements at temperatures up to approximately 800°C have recently become available. This utilises a new design of high temperature surface probe which has been developed by the researchers involved [10]. Development of a second probe based on the same principles but which will have an upper temperature limit of 1500-1600°C is expected to be largely completed by the end of the summer of 1992.

The second approach concerns the use of a new design of multipurpose resonator originally developed by Tinga and co-workers in Canada [11]. Although the absolute accuracy of this technique is currently uncertain, the technique is expected to be extremely useful for obtaining values as the temperature changes very rapidly reducing the difficulties associated with thermal expansion.

Work is currently focusing on relating the effect of variables such as density and composition on dielectric properties. Ultimately it is hoped to use the information generated by this research programme to yield a simple model for changes in dielectric properties with both temperature and density - such as occurs during the sintering of a ceramic.

MATERIALS AND MICROWAVE PROCESSING GROUP

The Materials and Microwave Processing Group (MaMP) was founded in October 1990 with the principal objective of bringing together the complimentary skills required to make substantial progress in terms of understanding the fundamental nature of the interaction of microwave radiation with materials. This knowledge is then being used as a basis for achieving two goals; viz, the development of i) new materials processing routes in which microwaves form the energy source and ii) improved materials for use in components for microwave applications to meet the demands of industry.

The research programme is divided into two main parts:

a) an applications programme aimed at developing the above processing techniques and materials; and

b) a dielectric property measurement facility to provide the information necessary to underpin (a).

MaMP is led by the University of Nottingham and membership is open to companies, research institutions, higher education establishments and individuals from within the EEC and EFTA. MaMP consists of a central nucleus containing a management organisation and dielectric property measurement facility, surrounded by satellite studies into a number of applications. Each research programme is sponsored by one or more of the collaborators. The purpose of the management organisation is to ensure that the overall objectives of the Group are met. A Management Board has been formed and it consists of a Director, Chairman, Honorary Secretary and a number of Member Representatives. The Honorary Secretary, from whom further information can be obtained, is Dr Jon Binner of the Department of Materials Engineering and Materials Design at Nottingham University.

Two levels of membership exist; Associate Members receive the benefits accruing from the management organisation and dielectric property measurement facility, whilst Full Members also sponsor one or more of the satellite research studies from the Applications programme. MaMP currently has 6 Full Members and 13 Associate Members. The UK Department of Trade & Industry has contributed over £170 000 in funding over the next three years in order to significantly increase the scope and pace of the research. Total funding for the Group is nearly £500 000.

REFERENCES

1. J.G.P. Binner and T.E. Cross, to be published in proceedings of Fabrication Science and Technology, Institute of Ceramics, Manchester, England, December 1991.

2. J.G.P. Binner and T.E. Cross, to be presented at the MRS Spring Meeting, Symposium L, San Francisco, April 27 - May 01, 1992.

3. J.G.P. Binner, accepted for publication in the British Ceramic Transactions & Journal, 1992.

4. F.C.R. Wroe and J. Samuels, to be published in proceedings of Fabrication Science and Technology, Institute of Ceramics, Manchester, England, December 1991.

5. M.G. Hamlyn and A.L. Bowden, Am. Ceram. Soc. Bull. 69 [3] 368-372 (1990).

6. J.M. Methven and S.R. Ghaffariyan, in Microwave Processing of Materials, edited by J.G.P. Binner, proceedings of one day meeting, London, UK, April 1991.

7. D.M.P. Mingos, in Microwave Processing of Materials, edited by J.G.P. Binner, proceedings of one day meeting, London, UK, April 1991.

8. G. Bond, R.B. Moyes, S.D. Pollington and D.A. Whan, to be presented at the 11th International Congress on Catalysis, Budapest, July 1992.

9. R.B. Moyes (private communication)

10. M. Arai, J.G.P. Binner and T.E. Cross, to be presented at the MRS Spring Meeting, Symposium L, San Francisco, April 27 - May 01, 1992.

11. W.R. Tinga, B.Q. Tian and W.A.G. Voss, in Microwave Processing of Materials II, edited by W.B. Snyder, W.H. Sutton, M.F. Iskander and D.L. Johnson, (Mater. Res. Soc. Proc. 189, San Francisco, CA, 1991) pp 111-116.

PART II

Numerical Modeling Techniques

COMPUTER MODELING OF TWO DIMENSIONAL TEMPERATURE DISTRIBUTIONS IN MICROWAVE HEATED CERAMICS

Y.L. Tian,* J.H. Feng,** L.C. Sun,** and C.J. Tu**
*Department of Materials Science and Engineering, University of Wisconsin-Madison, Madison, WI 53706
**Department of Energy Engineering, Zhejian University, Hangzhou, China

ABSTRACT

A two dimensional model for computer simulation of microwave heating of ceramics has been developed. The effects of microwave power, properties of ceramics, and thermal insulation on dynamic temperature distributions of ceramics under various processing conditions have been simulated. The practical solutions for eliminating thermal runaway and improving temperature homogeneity under microwave rapid heating can be approached using this computation technique.

INTRODUCTION

Although microwave sintering has shown enormous potential in replacing conventional heating techniques by saving time and energy and improving performance of products, some problems still remain unsolved which hold back this emerging technique from practical applications. Nonuniform heating and thermal runaway are two of these problems frequently encountered in microwave sintering processes. It was reported from our previous experiments that nonuniform heating and thermal runaway are mainly caused by a number of factors related to both the operation parameters of microwave hardware and the properties of materials.[1,2] These factors include: (1) low thermal conductivity, (2) strong temperature dependence of dielectric loss factor, (3) large size and complex geometry of the sample, (4) high heating rate and high processing temperature, (5) nonuniform electromagnetic field. In order to improve the thermal uniformity and stability, it is important to characterize and clarify the influence of these factors on the temperature pattern of microwave heating. Unfortunately, our understanding based on the previous studies is more or less empirical and qualitative. The fundamental aspects of the nonuniform and unstable behavior of microwave heating have not been well understood.

Numerical computation is expected to be a promising approach for the solutions of these problems. A number of numerical models have been developed which have revealed some of mechanisms of microwave heating.[3-7] However, many of these computations are based on one dimensional models and an assumption of temperature independence of the properties of materials. These simplifications could significantly underestimate the heating nonuniformity. In the present study, a two dimensional numerical model has been developed to simulate the temperature distributions of ceramic samples under various microwave processing conditions. Efforts have also been made to quantitatively evaluate the influence of some factors listed above on the nonuniform heating pattern and then to find some practical solutions for improving heating uniformity and eliminating thermal runaway.

PHYSICAL AND MATHEMATICAL MODELS FOR COMPUTATION

The thermal physical model of this computation is illustrated in Fig. 1. A ceramic cylinder (diameter d, thickness h, divided into MxN grids) is placed in a microwave applicator where a uniform electric field is assumed for simplicity. The instantaneous temperature distribution inside the sample, T(r,z,t), is the solution of the equation:

$$\rho C_p(T)\frac{\partial T}{\partial t} = \frac{1}{r}\frac{\partial}{\partial r}(K(T)r\frac{\partial T}{\partial r}) + \frac{\partial}{\partial Z}(K(T)\frac{\partial T}{\partial Z}) + Q \quad (1)$$

where $C_p(T)$ and K(T) are the specific heat and the thermal conductivity of the ceramics, respectively. Both are temperature dependent. ρ is the density. The heat generated in the sample, Q, due to the absorption of microwave energy, is given by:

$$Q = 2\pi f\epsilon_0\epsilon_r \tan\delta\ E^2 \quad (2)$$

where f is the frequency and ϵ_0 the permittivity of free space. The electric field, E, is assumed as a uniform distribution. $\epsilon_r\tan\delta$ is the dielectric loss factor which is also a function of temperature. Eq(1) is subject to the following boundary condition:

$$K(T)\frac{\partial T}{\partial r}\bigg|_{r=d/2 \text{ or } z=h/2} = C(T_s - T_o)^{5/4} - \epsilon_\lambda\sigma_o(T_s^4 - T_o^4) \quad (3)$$

where the first and the second terms on the right hand side of Eq(3) are energy loss at the surfaces through convection and radiation, respectively.[8] T_s and T_o are the surface and ambient temperatures, respectively. C is the convective heat transfer coefficient. ϵ_λ is the emissivity and σ_o is the Stefan-Boltzman constant. Rewriting Eq(1) in a finite difference form by heat balance over the domain surrounding a grid point (i,j) leads the following expression:

$$T_{i,j}^{k+1} = T_{i,j}^{k} + \frac{\Delta t K(T_{i,j})}{\rho(C_p)_{i,j}(\Delta R)^2}\{(1-\frac{1}{2i})(T_{i-1,j}-T_{i,j}) + (1+\frac{1}{2i})(T_{i+1,j}-T_{i,j})$$

$$+(\frac{\Delta R}{\Delta Z})^2(T_{i,j-1}-T_{i,j}) + (\frac{\Delta R}{\Delta Z})^2(T_{i,j+1}-T_{i,j}) + (\Delta R)^2 q_{i,j}/K(T_{i,j})\} \quad (4)$$

where i and j represent the r and z coordinate indexes of a grid point, respectively. k is the index for temperature T after the last time step; and k+1 is the index for next time step. $q_{i,j}$ is the heat per unit volume generated in the domain of the grid point (i,j). This equation was numerically solved by an explicit finite difference method. Al_2O_3 was taken as the sample material because it is one of the few ceramics which have a complete database available in the literature. The temperature dependence of the thermal conductivity, K(T), heat capacity, $C_p(T)$, and dielectric loss factor, $\epsilon_r\tan\delta$ (T), were collected from the published data.[9,10]

RESULTS AND DISCUSSION

Dynamic temperature profiles

Fig.2 shows the dynamic temperature variation of a cylinder (d=16mm, h=20mm) with the coupled microwave power. At low power levels (50W to 300W) the steady state temperatures were reached. As power raised to 600W, thermal runaway

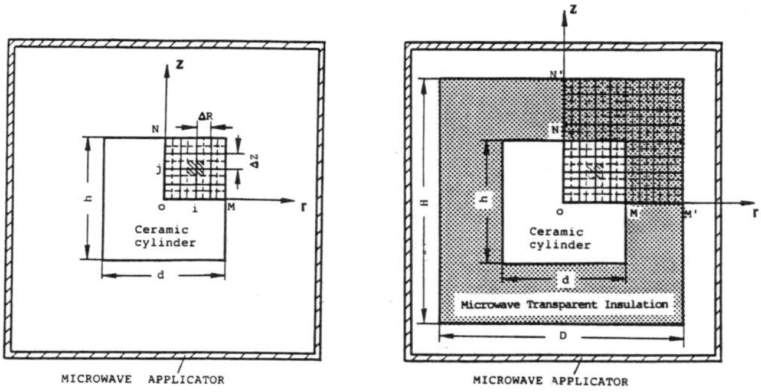

(a) (b)

Fig.1 Models for a ceramic cylinder heated in a microwave applicator. (a) no insulation, (b) enclosed in a package of thermal insulation.

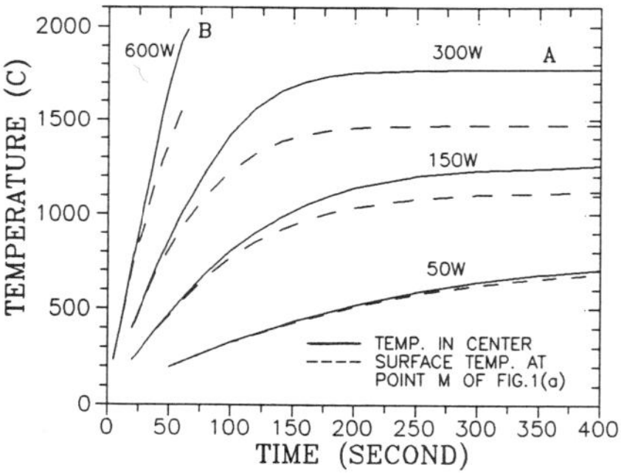

Fig.2 Variation of temperature with time for a cylinder (d = 16mm, h = 20mm) heated with different power levels.

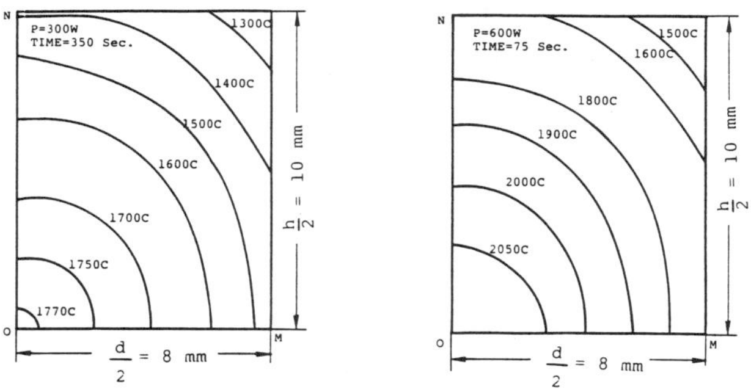

(a) (b)

Fig.3 Two dimensional distributions of temperature for a cylinder (d = 16mm, h = 20mm) heated with (a)300W for 350 seconds, a steady temperature state at point A of Fig.2,(b)600W for 75 seconds, a transient temperature at point B of Fig.2.

occurred within 75 seconds. Some examples of two dimensional temperature distributions, corresponding to the points A (a steady state) and B (a transient temperature of thermal runaway) in the curves of Fig.2, are illustrated in Fig.3. It can be seen that even in a uniform distributed electric field, nonuniform temperature distributions were observed in both radial and axial directions. The temperature gradient increased with an increase in temperature. Thermal runaway and cracking might be induced by a high temperature gradient, as shown in Fig.3(b), that the center of the sample reached the melting point while the surface was still below the normal sintering temperature.

The effects of thermal conductivity

The effects of thermal conductivity on the temperature gradient and the dynamic temperature profiles are illustrated in Fig.4. It is interesting to notice that an increase in thermal conductivity not only reduced temperature gradient but also suppressed thermal runaway. Fig.4(a) shows a proportional relationship between the decrease of temperature gradient and the increase of thermal conductivity. It can be seen from Fig.4(b) that thermal runaway (curve A) was suppressed by an increase of thermal conductivity (curves of B,C,D).

The effects of thermal insulation

Thermal insulation has been widely used in microwave sintering to prevent heat loss and to improve heating uniformity. Our computer program has been modified to simulate the heating pattern of an insulated ceramic. A model for a ceramic cylinder (d=16mm, h=20mm, divided into MxN grids) enclosed in a package of thermal insulation (D=32mm, H=40mm, divided into M'xN' grids) is illustrated in Fig.1(b). The insulation material is assumed as an ideal microwave transparent material and its density and thermal conductivity are assumed as 10% and 5% of those of alumina, respectively. Some typical computed results are illustrated in Fig.5. It is seen that the use of thermal insulation significantly reduced the energy loss of radiation and improved the temperature uniformity. For instance, the power required for heating up the sample to 1780 C dropped from 300W to 50W after the insulation was applied and the radial temperature difference between the center and the surface dropped from 200 C to 30 C.

It is known that both the high thermal conductivity and the use of thermal insulation can improve the temperature homogeneity of microwave heating. However, in many circumstances, changing the thermal conductivity may not be applicable, the use of thermal insulation would be one of the major solutions for the problem of nonuniform heating.

Optimal conditions for a rapid and uniform heating

It should be pointed out that the use of thermal insulation may significantly accelerate the heating rate at high temperature and thus increase the risk of thermal runaway. For instance, heating was stable at power levels of 150W and 300W with no insulation as shown in Fig.2, but thermal runaway occurred at the same power levels with the use of insulation as shown in Fig.5(a). A practical solution for eliminating thermal runaway during rapid heating, based on our previous experiments, is to control the heating rate by varying the input power.[11] A high input power is applied in the low temperature regime to keep high heating rate where temperature gradient and densification stress are insignificant while a low power is applied in the transition and stable sintering

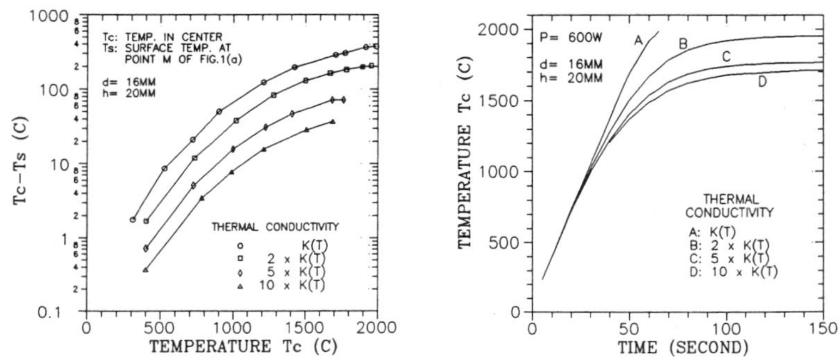

(a) (b)

Fig.4 Effects of thermal conductivity on temperature profiles, (a) variation of temperature gradient with thermal conductivity, (b) variation of temperature with time heated with 600W. (K(T) is the temperature dependence of thermal conductivity of Al_2O_3 reported from reference [10]).

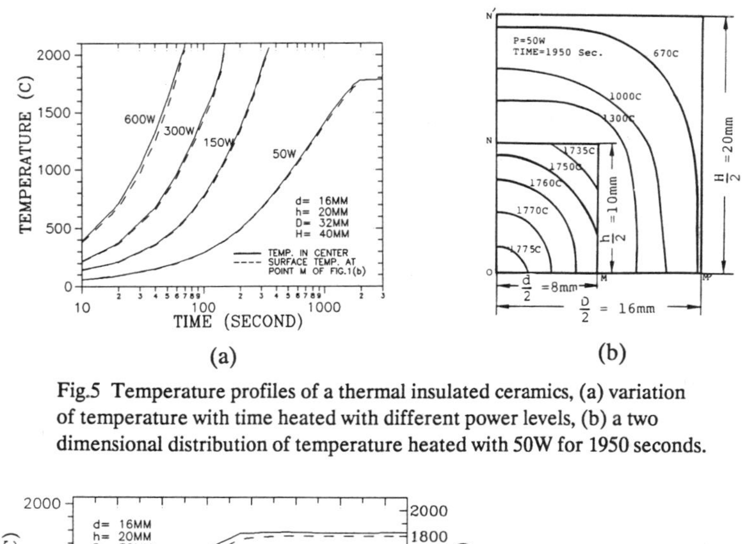

(a) (b)

Fig.5 Temperature profiles of a thermal insulated ceramics, (a) variation of temperature with time heated with different power levels, (b) a two dimensional distribution of temperature heated with 50W for 1950 seconds.

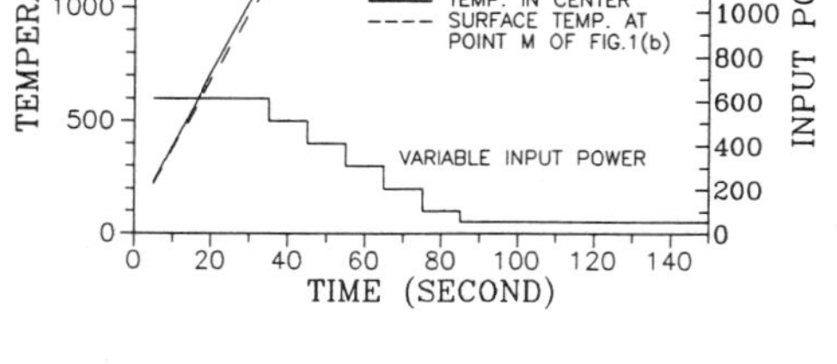

Fig.6 Dynamic temperature profile of a thermal insulated ceramics heated with a variable input power.

temperature regimes where thermal runaway and cracking may occur. Computer simulation has been carried out to find the optimal conditions for varying input power. An example of the computed results is shown in Fig.6. The sample was first heated at high heating rate to a transition temperature of 1300 C and then the input power was gradually reduced from 600W to 50W as the temperature of the sample increased from 1300 C to 1780 C where a steady sintering temperature was maintained. It can be seen from Fig.6 that the time for heating sample up to 1780 C was 80 seconds and the temperature difference between the center and the surface of the sample was less than 40 C. This example demonstrates that a rapid, uniform and stable heating is achievable in microwave sintering by the protection of insulation along with a proper control of the input power.

CONCLUSIONS

A numerical model has been developed to simulate the temperature patterns of microwave heating. The influence of different factors on thermal homogeneity and stability was quantitatively evaluated using this numerical technique. Nonuniform distributions of temperature are observed in both radial and axial directions even in a uniform electric field. High heating temperature and low thermal conductivity may exacerbate the temperature heterogeneity of the sample and then induce thermal runaway and cracking. An increase of thermal conductivity may improve the temperature uniformity and suppress thermal runaway. The use of thermal insulation may also improve the temperature uniformity, but the input power must be carefully controlled during the whole processes in order to avoid thermal runaway.

REFERENCES

1. Y.L. Tian, M.E. Brodwin and D.L. Johnson, "Sintering behavior of alumina ceramics in microwave processing", *Ceramic Transactions* Vol.7, p391-401, 1990.
2. Y.L. Tian, "Practices of ultra-rapid sintering of ceramics using single mode applicators", *Ceramic Transactions*, Vol.21, p283-300, 1991.
3. D.G. Watters, M.S. Brodwin and G.A. Kriegsmann, "Dynamic temperature profiles for a uniformly illuminated planar surface", *MRS Symp. Proc.*, Vol.124, p129-134, 1988.
4. G.A. Kriegsmann, "Microwave heating of ceramics", *Ceramic Transactions*, Vol.21, p177-183, 1991.
5. X.D. Yu, V.V. Varadan and V.K. Varadan, "Modeling microwave heating of ceramics", *Ceramic Transactions*, Vol.21, p167-176, 1991.
6. M.F. Iskander, "Computer modeling and numerical techniques for quantifying microwave interaction with materials", *MRS Symp. Proc.*, Vol.189, p149-171, 1990.
7. D.L. Johnson, "Microwave processing of ceramics at Northwestern University", *Ceramic Transactions*, Vol.21, p17-28, 1991.
8. S.M. Yang, *Heat Transfer,* Chinese Education Press, 1982.
9. H. Fukushima, T. Yamanaka and M. Matsui, " Measurements of dielectric properties of ceramics at microwave frequency", *J. Japan Soc. of Prec. Engr.*, Vol.53 [5], p1-6, 1987.
10. W.D. Kingery, H.K. Bowen and D.R. Uhlmann, *Introduction to Ceramics*, 2nd Edition, John Wiley, 1976.
11. Y.L. Tian, M.E. Brodwin and D.L. Johnson, "Ultrafine microstructure of Al_2O_3 produced by microwave sintering", *Ceramic Transactions*, Vol.1, p915-932, 1987.

FINITE-DIFFERENCE TIME-DOMAIN (FDTD) SIMULATION OF MICROWAVE SINTERING IN MULTIMODE CAVITIES

Ray L. Smith, Magdy F. Iskander, Octavio Andrade, and Hal Kimrey

The first three authors are with the Department of Electrical Engineering, University of Utah, Salt Lake City, UT 84112. Mr. Hal Kimrey is with Oak Ridge National Laboratory, Oak Ridge, TN 37831.

Abstract

Microwave sintering of ceramics in multimode cavities, particularly the use of picket-fence arrangements, has recently received considerable attention. Various types of ceramics have been successfully sintered and, in some cases, a desirable and unique "microwave effect" has been observed.

At present, various aspects of the sintering process such as sample sizes and shapes, types of insulations, and the desirability of including a process stimulus such as SiC rods are considered forms of art and highly dependent on human expertise. The simulation of realistic sintering experiments in a multimode cavity may provide a better understanding of critical parameters involved and allow for the development of guidelines towards the optimization of the sintering process.

In this paper, we utilize the FDTD technique to model various geometrical arrangements and material compatibility aspects in multimode microwave cavities and to simulate realistic sintering experiments. The FDTD procedure starts with the simulation of a field distribution in multimode microwave cavities that resembles a set of measured data using liquid crystal sheets. Also included in the simulation is the waveguide feed as well as a dielectric loading plate placed at the base of the cavity. The FDTD simulation thus provides realistic representation of a typical sintering experiment. Aspects that have been successfully simulated include types of insulation, role of SiC rods on the uniformity of the resulting fields, problems that may result from presence of thermocouples, and the possible shielding effects that may result from excessive use of SiC. These results as well as others showing the electromagnetic fields and power-deposition patterns in multiple ceramic samples are presented.

Overview of the solution procedure

For this paper the Finite-Difference Time-Domain (FDTD) technique was chosen to model and simulate realistic microwave sintering experiments using multimode cavities. The power distribution pattern in an empty cavity is initially adjusted so as to provide reasonably accurate representation of experimentally measured data using liquid crystal sheets. The liquid crystal is placed inside a microwave cavity on styrofoam supports for about 10 to 20 seconds. The resulting power distribution pattern accounts for the various modes in the cavity and accounts for the presence of the mode-stirring. The FDTD simulation then adjusts the power pattern in the cavity until the same degree of non-unformity is obtained thus modeling the mode stirring and the presence of multiple modes in the cavity. It should also be noted that this technique was not intended to produce the actual field values in the cavity, but only relative variations of power distribution in the empty cavity.These variations provide a base line to initiate simulation of microwave sintering in these cavities.

To help adjust the initial values of the power distribution on the empty cavity two FDTD modeling procedures were adopted. The first model simply consisted of choosing a set of modes that could exist independently and simultaneously within a small band of frequency centered around the cavity operating frequency of 2.45 GHz. The cavity is then forced to converge to this set of modes at the excitation planes and the solution is allowed to reach a steady-state condition elsewhere in the cavity. The resulting field pattern was then observed and the amplitudes of the modes are adjusted to help match the field patterns measured using the liquid crystal sheets. The second procedure involves the modeling of the feed waveguide in the cavity. The waveguide feed is placed horizontally along the bottom of the cavity in an arrangement that is similar to the feed arrangement used in the construction of the microwave oven used for sintering. The feed waveguide dimensions were varied and discontinuities (flares and posts) were introduced to provide some degree of mode-stirring capabilities that help reproduce the measured power distribution pattern. Reflections at the source end of the waveguide feed were minimized by introducing a first-order absorbing boundary at roughly a half guide wavelength behind the excitation plane.

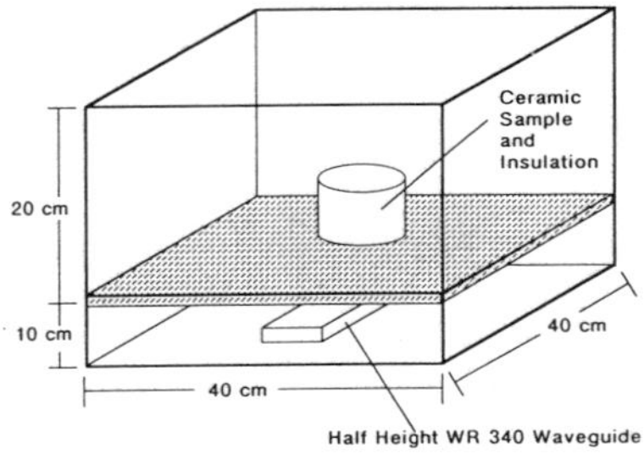

Fig. 1. Model used in the FDTD simulation of the sintering experiments. Although typical mode stirrer was not included in the simulation, the geometry and dimensions of the feed waveguide were changed to adjust the field patterns in the empty cavity so as to correlate with measured results.

Figure 1 shows the model used in the numerical simulation, while Figure 2(a) shows the obtained steady-state result for the EM field distribution in an empty cavity when a straight length of a waveguide is used as a feed at one end and terminated at the other. Figure 2(b) shows the resulting steady-state EM field pattern when the feed waveguide includes a flared transition to the empty cavity. From Figure 2 it may be seen that adjusting the feed waveguide geometry and dimensions provide variables suitable for adjusting the field patterns in the empty cavity.

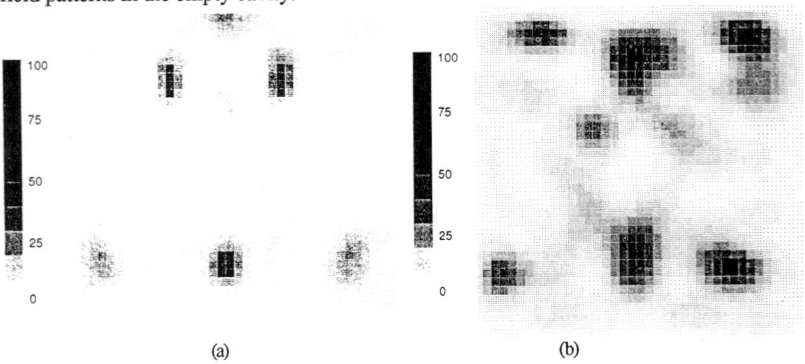

Fig. 2. Steady-state electromagnetic field patterns inside an empty cavity (a) when a straight section of the feed waveguide was used and (b) when a flared section of the feed waveguide was used. The difference in results of (a) and (b) illustrate the ability to modify field patterns in an empty cavity so as to correlate with experimentally measured results using liquid crystal sheets. The dimensions of the sintering microwave oven are $40 \times 30 \times 40$ cm^3.

Results

After obtaining an EM field pattern in the empty cavity which was representative of the measured data, simulations of a realistic microwave sintering experiment where made. Ceramic samples and surrounding insulations were placed in the cavity, and the steady-state EM power deposition pattern is calculated in both the sample and the insulation. The parameters expected to play critical roles in microwave sintering experiments are then varied to determine their effect on the steady-state EM field distribution pattern. After an extensive number of simulation runs, observations were made which can help in developing guidelines toward the optimization of the sintering process.

First, the effect of the type of insulation material and its dimensions on the resulting EM field pattern in the sample was examined. Figure 3 shows the steady-state EM field distribution in a cavity in which a green (unsintered) ceramic sample was placed surrounded by insulation of different values of ε_r. The dimensions and complex permitivities of the sample and the insulation used in this simulation run are given in Figure 3. In some cases, such as in Figure 3b, it may be seen that the fields were actually drawn into the sample while in others, such as Figure 3a, fields were concentrated just outside the sample. This change in the field pattern was clearly caused by the change in the value of the dielectric constant of the insulation from 1.537 to 10. The different locations of the field maxima in the green sample may result in nonuniform heating of the sample and possible thermal runaway problems which may result in the destruction of the sample.

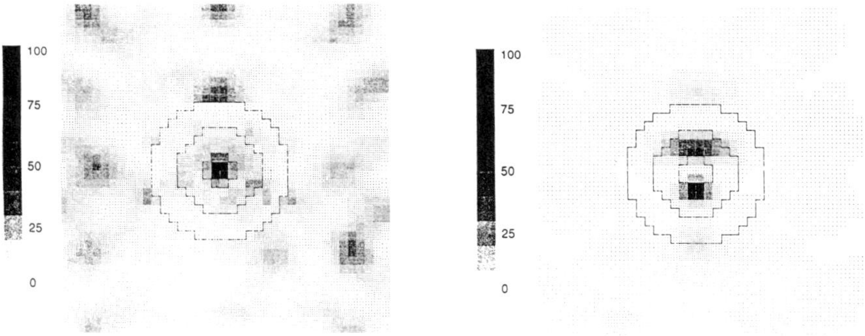

(a) (b)

Fig. 3. Steady-state electric-field (magnitude) pattern in an unsintered (green) ceramic sample surrounded by 6-cm-thick insulation. The dielectric constant and conductivity of the sample are $\varepsilon_r = 4.13$ and $\sigma = 64 \times 10^{-6}$ S/m, respectively, while those of the surrounding insulation in (a) were assumed $\varepsilon_r = 1.537$ and $\sigma = 15.374 \times 10^{-6}$ S/m, respectively. The properties of both the insulation and sample were unchanged in (b) with the exception of the dielectric constant of the insulation now having a value of 10 . The ceramic sample is of cylindrical shape of 2-cm radius and 6 cm long. The dimensions of the microwave oven are the same as those of Fig. 1.

The conductivity of the insulation can have a significant effect on the EM field distribution pattern in a typical microwave sintering experiment. Figure 4 shows the effect that different values of insulation conductivity have on the total power absorbed in the ceramic sample and the insulation. From these results it is clear that the continued increase in the conductivity of the insulation results in a decrease in the power that is delivered to the sample. For this particular simulation effective shielding of the sample started to occur when the conductivity of the insulation exceeded 0.001 S/m.

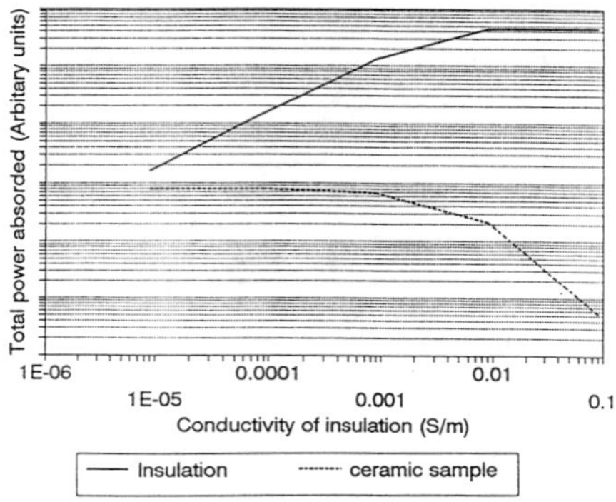

Fig. 4. Steady-state total power absorbed in sample and insulation vs. change in conductivity of the insulation. Changing the insulation conductivity to values near 0.001 S/m does not affect the total power absorbed by the sample. Further increase in the conductivity of the insulation effectively shields the sample and results in a reduction in the total power absorbed by it. The dimensions and other electrical properties of the sample and insulation are the same as those of Fig. 3a.

Some of the more successful microwave sintering experiments in multimode cavities involves the use of SiC rods in a picket-fence arrangement to help the heating process, particularly at lower temperatures. In some sintering experiments, the number and arrangement of these SiC rods is critical to the success of the experiment. The developed FDTD code was used to examine various aspects of the use of SiC rods to stimulate sintering of ceramics samples. Figure 5 shows the obtained results for the EM field distribution pattern when 4 SiC rods were used as a stimulus in the same arrangement as previously described in Figure 3a. It can be seen that the EM fields tend to concentrate around the SiC rods. Additional simulations showed that as more SiC rods are added to the model the electric-field distribution in the sample decreased as the energy is mostly absorbed by and concentrated around the SiC instead of the ceramic sample.

A drawback of the use of the stimulus arrangement is, therefore, related to the fact that an excessive use of SiC rods and the fact that they have high electrical conductivity may hinder the penetration of the EM energy to the ceramic samples. If such reduction occurs, the sintering may in fact be the result of conventional heating (heat transfer from SiC rods to sample), and this may in turn result in the reduction in the often-desirable "microwave effect." It is, therefore, suggested that a balance should be observed and the number of SiC rods should be kept to a minimum, say 4 to 6 rods, depending on the size and shape of the sample.

In some sintering experiments, multiple samples might be used. Figure 6a shows the steady-state electric-field distribution when four samples were used. From Figure 6b it may be seen that for such an arrangement, excessive electric field is localized in the insulation layer and that the inclusion of appropriately located SiC rods may result in more favorable distribution of the electric-field pattern. From Figures. 6a and 6b, we conclude that SiC rods may play a role in the sintering experiment both by absorbing EM power and subsequently conventionally heating the sample until sufficiently higher conductivity is achieved at elevated temperatures, and also by modifying the field distribution in the sintering cavity so as to provide better uniformity in the field values at the locations of the ceramic samples.

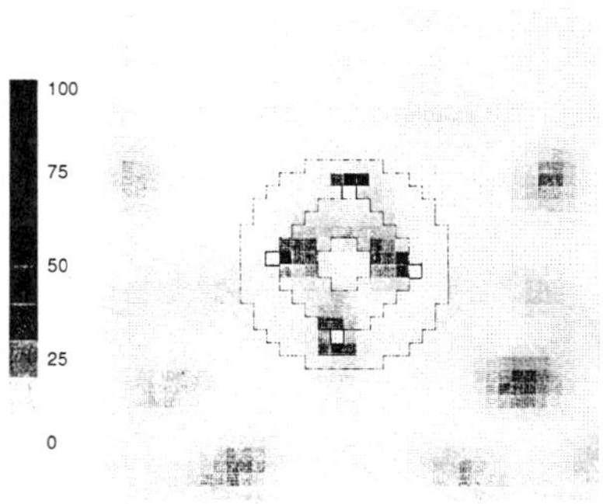

Fig. 5. Steady-state electric-field (magnitude) distribution pattern in a green ceramic sample surrounded by 6-cm-thick insulation when four SiC rods were placed in the insulation to stimulate the sintering process at lower temperatures. The dimensions of the sample and the insulation are the same as those of Fig. 6. The SiC rods are of 0.6 cm radius and 10 cm long. The dielectric constant ε_r and conductivity σ of the SiC rods are 29.363 and 0.9 S/m, respectively.

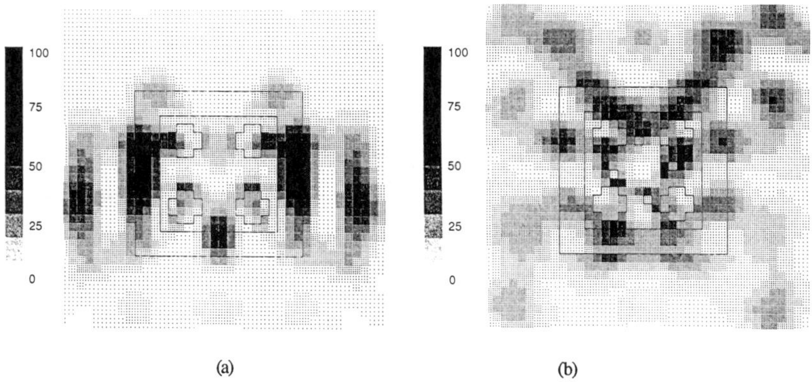

Fig. 6 Steady-state electric-field (magnitude) distribution in four ceramic samples placed in the microwave sintering cavity of Fig. 3a. Each sample is cylindrical in shape with a radius of 2 cm and 6 cm long. The insulation is 4 cm thick. No SiC rods were used in the simulation shown in (a), while four SiC rods were used in (b). The four SiC rods used in (b) showed a significant change in the field distribution pattern that was observed. The dielectric properties of the sample and insulation are the same as those in Fig. 3a.

Conclusions

A 3D-FDTD code was developed and successfully used to model some of the many factors that influence a realistic microwave sintering process in multimode cavities. The simulation procedure is based on an initial power distribution in an empty cavity that is similar to an experimentally measured pattern using liquid crystal sheets. The dimensions of the waveguide feed were adjusted and discontinuities were added to provide some control on the resulting EM power distribution pattern in the empty cavity so as to resemble the measured data.

The developed FDTD code was then used to simulate and examine several parameters of interest in a typical microwave sintering experiment. This includes the conductivity of the insulation surrounding the sample and the role of the SiC rods in modifying the EM field distribution pattern in the sample. It is shown that while an increase in the conductivity of the insulation may stimulate the sintering process by raising the temperature of the insulation and subsequently the sample, excessive increase in the conductivity of the insulation decreases the penetration of the EM fields to the sample and hence reduces the desirable "microwave effect." It is also shown that a similar effect may result from an excessive use of SiC rods as a stimulus.

Future plans include the continued use of the FDTD code to develop guidelines regarding the optimization of the sintering process in multimode cavities. Additional efforts to integrate the FDTD code with a heat-transfer computer program described in a companion paper to calculate the temperature distribution pattern in a sintering experiment are also underway.

Acknowledgements

Research sponsored by the U.S. Department of Energy Assistant Secretary for Conservation and Renewable Energy, Office of Industrial Technologies, Advanced Industrial Concepts (AIC) Materials Program, under subcontract 19X-SJ125V and under contract DE-AC05-84OR21400 with Martin Marietta Energy Systems, Inc.

References

1. M. A. Janney and H. D. Kimrey, "Microwave Sintering Of Alumina at 28 GHz," in Ceramic Power Science II, edited by G. L. Messing, E. R. Fuller, and H. Hausner, (American Ceramic Society, Westerville, Ohio, 1988) pp. 919-924.
2. Y. L. Tian, D. L. Johnson, and M. E. Brodwin, "Ultrafine Microstructure of Al_2O_3 Produced by Microwave Sintering," in Ceramic Power Science II, edited by G. L. Messing, E. R. Fuller, and H. Hausner, (American Ceramic Society, Westerville, Ohio, 1988)pp. 925-932.
3. H. D. Kimrey and M. A. Janney, "Design Principles for High Frequency Microwave Cavities," in Microwave Processing of Materials, edited by W. H. Sutton, M. H. Brooks, and I. J. Chabinsky (Matter. Res. Soc. Proc. 124, Pittsburgh, PA 1988) pp. 367-372.
4. M. F. Iskander, "Computer Modeling and Numerical Techniques for Quantifying Microwave Interactions with Materials," in Microwave Processing of Materials II, edited by W. B. Snyder, W. H. Sutton, M. F. Iskander, and D. L. Johnson, (Matter. Res. Soc. Proc. 124, Pittsburgh, PA 1990) pp. 149-171.
5. P. C. Cherry and M. F. Iskander, "FDTD Analysis of Power Deposition Patterns of an Array of Interstitial Antennas for Use in Microwave Hyperthermia," IEEE Transactions on Microwave Theory and Techniques, August, (1992).
6. K. S. Yee, "Numerical Solution of Initial Boundary Value Problems involving Maxwell's Equations in Isotropic Media," IEEE Transactions on Antennas and Propagation, 14, 302 (1966).
7. A. Taflove and K. R. Umashankar, "The Finite-Difference Time-Domain Method for Numerical Modeling of Electromagnetic Wave Interactions with Arbitrary Structures," in Finite Element and Finite-Difference Methods in Electromagnetic Scatter, edited by M. A. Morgan, (Elsevier, Amsterdam, 1990).
8. G. Mur, "Absorbing Boundary Conditions for the Finite-Difference Approximation of the Time-Domain Electromagnetic Field Equations," IEEE Transactions on Electromagnetic Compatibility, 23, 377 (1981).

SIMULATION OF SINTERING EXPERIMENTS IN SINGLE-MODE CAVITIES

B. CHAPMAN, M.F. ISKANDER, R. L. SMITH, AND O. M. ANDRADE
*Department of Electrical Engineering, University of Utah, Salt Lake City, UT, 84112

ABSTRACT

The use of single-mode cavities has been recognized as a viable technique for sintering and joining ceramics and for creating microwave plasma for materials processing. It is often reported, however, that tuning these cavities at lower and higher temperatures critically depends on the types of materials being processed, the variation of ε^* as a function of temperature, the size and shape of samples, and the overall heating system configuration with and without insulation.

In this paper we use the finite-difference time-domain technique to address the performance of the single-mode cavity heating system under a wide variety of heating conditions and system configurations. Specifically, practical issues such as limiting values of sample sizes, the dependence of optimal sample sizes on the complex permittivity of materials under test, the role of sample shape on enhancing uniformity of heating patterns, and various tradeoffs involved in utilizing various cavity modes and/or SiC rods as stimulus for heating large samples are evaluated.

OVERVIEW OF SOLUTION PROCEDURE

The Finite-Difference Time-Domain (FDTD) Method was used in these simulations. The FDTD method has several advantages such as: (1) The ability to model objects of complex geometries; (2) Provides adequate accuracy at dielectric interfaces; and (3) The simulation accounts for parameters of interest in the sintering process like the resonant frequency and Q shifts due to the coupling iris, the geometry and type of ceramic samples and insulations.

The computer code is used to first calculate the true resonant frequency taking into account the type, size and shape of the sample. Then the specific absorption rates (SAR) in the ceramic sample and the insulation are calculated at this frequency. The SAR is then used to determine the heating pattern using a heat transfer program described in a companion paper [1].

BRIEF DESCRIPTION OF FDTD SOLUTION

The FDTD algorithm, developed by Yee [2], uses a central difference (second order accuracy) method for approximating Maxwell's curl equations. In our case the FDTD computation space is naturally terminated by the metallic walls of the microwave cavity. At the open end of the feed waveguide a Mur first order boundary condition [3] is used to terminate the computation space.

To initiate the solution, the cavity is first excited at a frequency equal to the resonant frequency of the empty cavity as calculated based on its dimensions. The resonant frequency of the loaded cavity is calculated (as described below) and the cavity is excited from the waveguide via a slit iris at this "true" resonant frequency. From these excited fields the cavity Q and the SAR patterns are calculated.

The cavity used throughout these simulations has dimensions 4.6 x 7.7 x 9.2 cm^3 (see figure 1). The empty cavity without a coupling iris resonates at 2.45GHz. With the coupling slit iris the empty cavity resonates at 2.43GHz. All samples were assumed centered in this cavity unsupported.

RESONANT FREQUENCY CALCULATIONS

The resonant frequency of the modeled cavity is determined by establishing an initial TE_{101} mode within the cavity based on the dimensions of the empty cavity. The FDTD code is then used to determine the characteristics of the fields as they settle into an oscillating mode that satisfies the boundary conditions and thus takes into account the presence of the sample and the size and shape of the coupling iris. The resonant frequency is based on the period of the settled field oscillation.

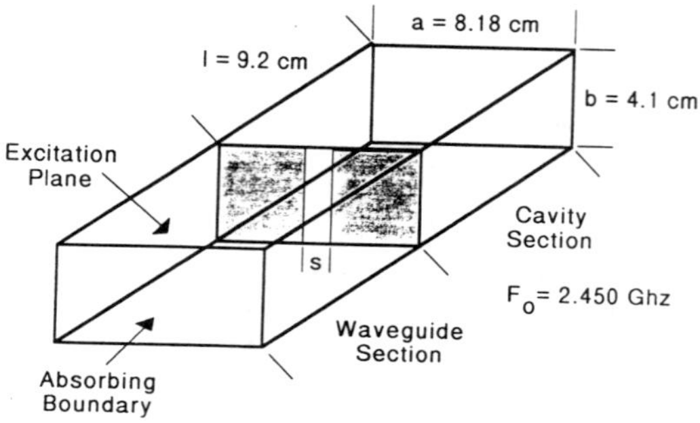

Figure 1. Schematic of microwave cavity and waveguide feed used in simulations.

To check the accuracy of the resonant frequencies calculated in this way, we excite the cavity from the waveguide side at this frequency and the adequacy of coupling is checked by observing sustained (non-decaying) oscillations. As a further check, we compare the frequencies thus calculated to those predicted by perturbation theory for electrically small samples. For a small cylindrical sample oriented with the axial electric field and running from the top to the bottom of the cavity, perturbation theory gives the resonant frequency f_s in terms of f_o of an empty cavity and ε_r' of the material by

$$\varepsilon_r' = 1 + \frac{f_o - f_s}{f_s} \cdot \frac{a \cdot L}{2 \cdot \pi \cdot r^2} \tag{1}$$

Where L and a are the dimensions of the rectangular cavity and r is the radius of the sample.

We modeled a small parallelepiped sample with dimensions .256 x .256 x 4.01 cm^3 with real permittivity ranging from 3 to 15. For simplicity the cylindrical perturbation formula is used for the comparison. The results from the two methods compare closely (see figure 2).

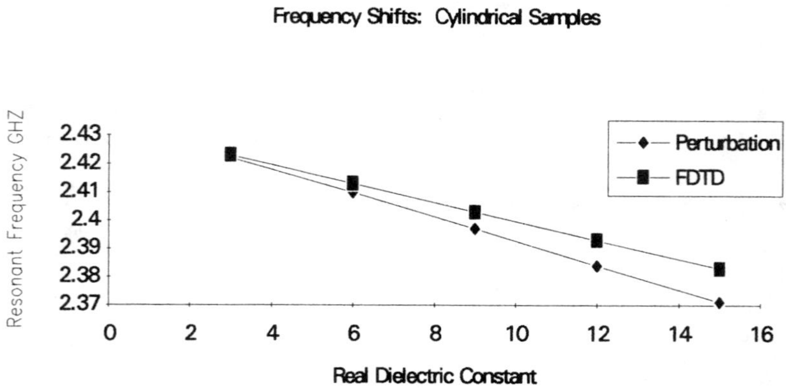

Figure 2. Comparison of resonant frequency shifts calculated using FDTD and perturbation methods

CALCULATIONS OF LOADED CAVITY Q

Maintaining reasonable values of the cavity Q is necessary for efficient coupling of energy to the sample. The cavity Q can be determined by observing the E field envelope's rate of growth when the cavity is excited by an external source. The change in the envelope's magnitude is related to the Q by

$$\frac{|E_1|}{|E_2|} = \frac{\left[1 - e^{-t_1 \cdot \omega / Q}\right]}{\left[1 - e^{-t_2 \cdot \omega / Q}\right]} \tag{2}$$

Where ω is the angular frequency, E_1, and E_2 are the electric field values at times t_1, and t_2. This equation can be solved iteratively for Q. By averaging values obtained at several different time intervals, a fairly accurate value of the Q is obtained.

We compare the values obtained for the Q using this method to those obtained using perturbation techniques. For electrically small samples the loaded Q (Q_s) can be obtained from Q_s of the unloaded cavity by the approximate formula

$$\varepsilon_r'' = \left(\frac{1}{Q_s} - \frac{1}{Q_o}\right) \cdot \frac{a \cdot L}{4 \cdot \pi \cdot r^2} \tag{3}$$

Where Q_o is the Q of the unloaded cavity at the frequency of interest, L is the cavity's axial length, a is the cavity's width, and r is the sample radius.

Again the values obtained from the two methods are in good agreement (see figure 3). As shown the FDTD slightly overestimates the Q, but values obtained are reasonable and the code seems to be sufficiently accurate.

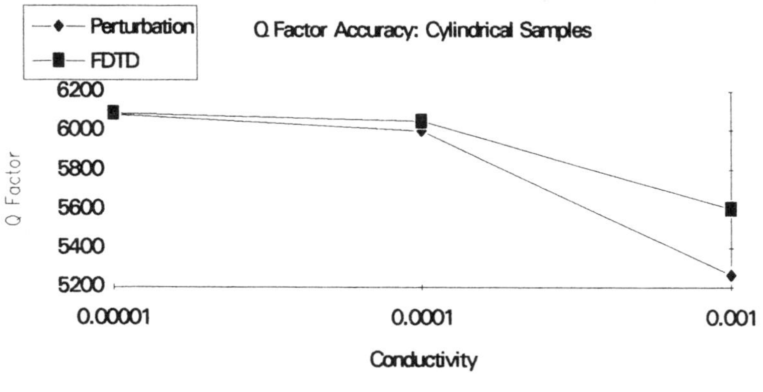

Figure 3. Variation of cavity Q vs. electrical conductivity of samples. Sample has real permittivity of 4.0. The cavity dimensions are 4.6x7.7x9.2cm^3.

FDTD SIMULATION OF SINTERING IN SINGLE-MODE CAVITY

Along with the calculation of resonant frequency and Q of the loaded cavity, electric field distributions and SAR values are calculated for various geometries and types of ceramic samples and insulations. The parameters that we examined include: (1) Effect of size and shape of coupling iris on resonant frequency; (2) Effect of geometry, size, and properties of sample on resonant frequency and Q of cavity; (3) Maximum sample size in a given sintering arrangement.

FREQUENCY SHIFTS WITH IRIS SHAPE

The resonant frequency shift due to the coupling iris is calculated for a slit iris of various height to width ratios as shown in figure 4.

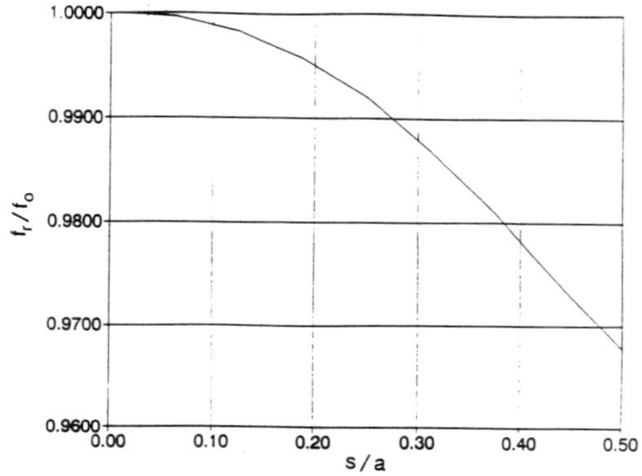

Figure 4 Variation of normalized cavity resonant frequency for various ratios of slit iris' width to width of cavity (s/a).

VARIATION OF RESONANT FREQUENCY FOR DIFFERENT SAMPLE, SIZES, SHAPES, AND ORIENTATIONS

We calculated the resonant frequency shifts due to cubic samples placed in the center of the cavity. Both sample size and real permittivity are varied. The conductivity for all samples is zero. Results are shown in figure 5.

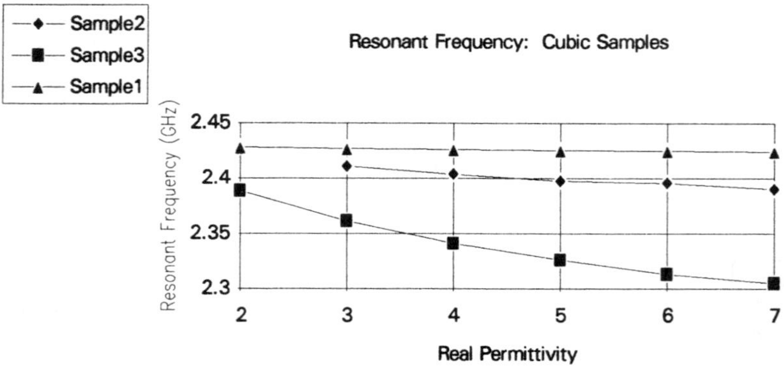

Figure 5 Resonant frequency shifts due to cubic samples centered in cavity. Sample1 .5x.5x.5cm^3. Sample2 1x1x1cm^3. Sample3 1.5x1.5x1.5cm^3. All samples are lossless ($\varepsilon_r''=0$). Cavity dimensions are 4.6x7.7x9.2cm^3

We also examined the change in resonant frequency shifts when parallelepiped samples were placed parallel and perpendicular to the direction of the axial electric field in the cavity (see fig. 6). The samples have a fixed volume of (2 x 1 x 1cm^3). Real permittivity ranges from 2 to 10 and the samples are oriented with the long dimension parallel to the axial electric field (Sample A), and perpendicular to the axial electric field (Sample B and Sample C). It can be seen that when samples are oriented parallel to the axial field the resonant frequency shift is noticeably larger, while the two different perpendicular orientations give identical resonant frequencies.

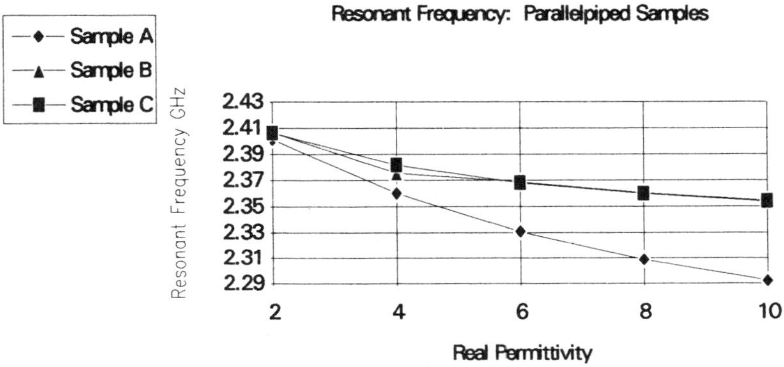

Figure 6 Resonant Frequency of cavity loaded with parallelepiped samples. Sample A is oriented with the axial electric field (y direction). Sample B is oriented perpendicular to the axial electric field (z direction). Sample C is oriented perpendicular to the axial electric field (x direction). All samples are assumed lossless ($\varepsilon_r''=0$) and have volume 2x1x1cm^3.

We also examined the variation of cavity Q versus the variation of the electric conductivity (σ) for cubic samples of different sizes (see figure 7). Note that a noticeable knee exists at approximately $\sigma=.0001$ S/m.

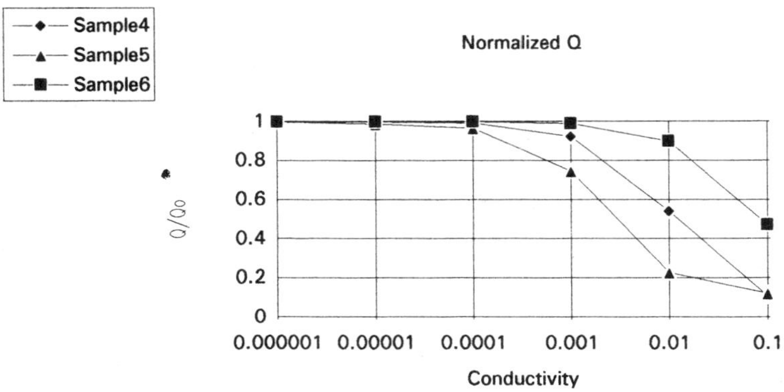

Figure 7 Normalized Q for three cubic samples centered in the cavity. Samples had the following dimensions: Sample4 .5x.5x.5cm^3; Sample5 1x1x1cm^3; Sample6 1.5x1.5x1.5cm^3. All samples have a real permittivity of 4.0.

GUIDELINES FOR SUITABLE SAMPLE SIZES IN SINGLE-MODE CAVITIES

The maximum sample size that can be sintered in a single-mode cavity depends on several parameters. First, the mode which is chosen has a specific frequency which must be satisfied. Second, as shown above, the size, shape, and material being used contribute to the actual resonant frequency of the loaded cavity. Third, the magnetrons have a limited frequency band which falls somewhere between 2.40GHz and 2.50GHz. The increase in the electrical size of the cavity due to the presence of the sample can be compensated for by using a moving short to reduce the size of the cavity. By iteratively running the FDTD code altering either the electrical parameters for a given volume sample, or altering the volume of the sample for a given set of electrical parameters, it is possible to calculate the maximum sample size suitable for a given cavity size. As an example of these calculations we assume that the low frequency end of the magnetron is 2.40GHz and that the cavity was excited at the lowest operating mode. It is further assumed that the moving short decreases the cavity length by one half. We also assume a material with $\varepsilon'=11.3$, $\sigma=0.00041$S/m. For this set of parameters the maximum sample size was found to be 16cm^3 or one eighteenth the original cavity volume.

This parameter is significant since one can begin sintering larger samples which has relatively low ε' at room temperature and maintain good tuning but be unable to tune the cavity at higher temperatures as ε' continues to increase with temperature.

FDTD SIMULATION OF REALISTIC MICROWAVE SINTERING EXPERIMENTS

In addition to the above calculations of resonant frequency and Q of empty and loaded cavities as well as to the determination of maximum sample size in a given cavity, the developed FDTD code was used to simulate and calculate SAR patterns in sample and insulation in typical microwave sintering experiments in a single-mode cavity. The following are examples of these calculations.

In example 1 we investigated the effect of insulation on the sintering experiment. A 1cm^3 sample of $\varepsilon'=4.0, \sigma=64\times10^{-6}$S/m surrounded by 0.75 cm thick insulation of $\varepsilon'=1.557$, $\sigma=15.37\times10^{-6}$ S/m is placed in the center of the cavity. The cavity resonates at 2.2181GHz. When the insulation is removed, the resonant frequency increased to 2.3602GHz and the SAR in the sample increases by 30%. In other words, under the above specified conditions, the presence of insulation not only caused a shift in resonance frequency (as expected) but also a decrease in the SAR value in the sample.

In example 2 we investigated the effect of the sample orientation on the sintering experiment. A 1 x 1 x 2 cm^3 sample without insulation is placed in the cavity. The sample is initially oriented parallel to the axial electric field and then oriented perpendicular to the field. In agreement with figure 5 above, a greater frequency shift is observed when the sample is oriented parallel to the field. More importantly, however, the heating patterns were found to be significantly different, and specifically the SAR decreased by 82% when the sample was oriented perpendicular to the axial electric field.

In example 3 we investigated the effect of SiC rods on the sintering experiment. We found that small, appropriately placed SiC rods can be used in a single-mode cavity while maintaining reasonable Q (~1500) and resonant frequency (~2.38GHz). In these simulations we considered two geometries. First two small (1.0 x .26 x .26 cm^3) SiC rods were placed symmetrically around the axis of the cavity. The rods were separated by 2.5cm and the 1cm^3 sample is placed in the center of the two rods. For this case the cavity resonated at 2.384GHz and Q was found to be approximately 1300.

The distance between the rods was then increased to 3.5cm. The resonant frequency increased to 2.3889GHz, the cavity Q increases to approximately 1500, and the SAR in the sample increased by 21%.

These results illustrate the feasibility of using SiC rods in single-mode cavity sintering. Several trade offs must, however, be considered including the size of these rods, minimizing their effect on the cavity parameters, and at the same time providing an adequate stimulus effect on the ceramic sample

CONCLUSION

We have developed a 3D FDTD code to simulate realistic microwave sintering experiments in single-mode cavities. The code calculates the resonant frequency and the Q of empty and loaded cavities, the field and SAR distributions in the ceramic sample and insulation. We verified the codes accuracy in determining resonant frequency and Q by comparing values obtained using FDTD calculations with those obtained using perturbation formulas.

We used the code to simulate the effects of sample shape, size, orientation, and electrical parameters on the resonant frequency and Q of the cavity, and the SAR distribution. We demonstrated the effect the insulation material has on the resonant frequency and Q of the cavity, and the SAR distribution in the sample. We also examined the feasibility of using small SiC rods in single-mode sintering experiments.

REFERENCES

[1] James Tucker et al., "Dynamic Model for Calculating Heating Patterns During Microwave Sintering", ibid.

[2] Yee, K.S., IEEE Trans. Antennas Propagat.,14,302-307,(1966).

[3] Mur, G., IEEE Trans. Electromag. Compat., 25, 433-440, (1983).

DYNAMIC MODEL FOR CALCULATING HEATING PATTERNS DURING MICROWAVE SINTERING

James Tucker, Ray Smith, Magdy F. Iskander, and Octavio M. Andrade
Dept. of Elect. Eng., Univ. of Utah, Salt Lake City, UT 84112

ABSTRACT

Analysis of dynamic development of heating patterns during microwave sintering provides vital information on the evolution of the heating process and the contributions from the various components in a complex sintering arrangement (picket fence) to the heat-transfer mechanism. Measured heating patterns often provide overall effects, and it is difficult to isolate and control the various contributions. To this end, results from numerical simulation may be significant.

In this paper we describe a thermal model that calculates the temperature distribution in ceramic samples and insulation under realistic sintering conditions. The calculation process involves a two-step procedure. The first step is to calculate the microwave power deposition in the sample and surrounding insulation. 3D FDTD calculations described in a companion paper are used for this purpose [1]. The other step involves calculation of the temperature distribution using a 3D finite-difference heat-transfer program developed in our department.Results illustrating the effect of thickness of insulation and the placement of SiC rods in picket-fence arrangement are presented. Also, the need to measure additional parameters such as thermal conductivity and density of green samples as a function of temperature during sintering is discussed.

INTRODUCTION

Microwave sintering of ceramics appears to proceed differently from conventional sintering. Lower activation energies and lower sintering temperatures resulting in smaller grain size have been reported [2]. Because of the continued interest in microwave sintering there is a need to better model and simulate microwave sintering experiments. Before microwave sintering can be fully commercially viable a thorough understanding of the microwave interactions as a function of frequency, temperature and geometry is needed. As the sample is sintered its properties change as a function of temperature and time. With these changes in properties, the microwave absorption changes thus affecting the heating rate of the sample. In order to accurately model microwave sintering experiments, the temperature distribution pattern must be calculated.

In this paper we show the results from using the Finite-Difference Time-Domain (FDTD) technique to model and simulate realistic microwave sintering experiments. The electric field pattern calculations are described in a companion paper [1]. The electric field pattern is used to generate a power deposition pattern through the sample and surrounding insulation. When SiC rods are used as susceptors, their influence is also accounted for in the generation of the power deposition pattern. The developed heat-transfer program takes the power deposition pattern (SAR) as a heat source and calculates the transient temperature distribution. When the temperatures have changed by a significant amount the heat-transfer program is temporally stopped and the properties of the sample, insulation and SiC rods are then evaluated at their new temperatures. A new SAR pattern is then obtained and the heat-transfer program is restarted with the new SAR (heat-source) pattern. The process (Fig. 1) is continued until sintering is achieved.

NUMERICAL HEAT-TRANSFER METHODS

A heat-transfer code that accepts the output SAR from a preexisting FDTD field code was written. The FDTD method approximates the temperature in an object by calculating the temperature at uniformly spaced points. Each point is at the center of a cell. The temperature of the point represents the temperature of the cell. The calculation for the temperature at each point begins with the Heat Transfer Equation.

$$\rho c \frac{\partial T}{\partial t} - \nabla k \nabla T = W_{EM} \qquad (1)$$

where T = temperature; c = specific heat
ρ = density; k = thermal conductivity
W_{EM} = Power absorbed minus heat lost due to convection and radiation

When two adjacent cells have different thermal conductivities (k) there is a discontinuity in the heat-transfer equation due to taking the derivative of k. The thermal conductivity was approximated by taking the average k between adjacent cells. The central difference method was used to discretize the spatial derivatives. Euler's method was used to discretize the time derivative. This is an explicit method where the new temperature of a cell is calculated based on its previous temperature and the previous temperatures of the cells around it. The Finite difference equation for an updated value of temperature at a point is:

$$T_{cell}^{n+1} = \frac{\Delta t}{rc\Delta h^2} \left\{ \sum (T_{adj}^{n} - T_{cell}^{n}) * k_{ave}^{n}(i,j,k) + W_{EM} * \Delta h^2 \right\} + T_{cell}^{n} \qquad (2)$$

where

T^n is the temperature at time $n(\Delta T)$
T^{n+1} is the temperature at time $(n+1)(\Delta T)$
T_{adj} is the temperature of the adjacent cells
Δh is the cell size
W_{EM} is the absorbed power minus power lost through convection and radiation

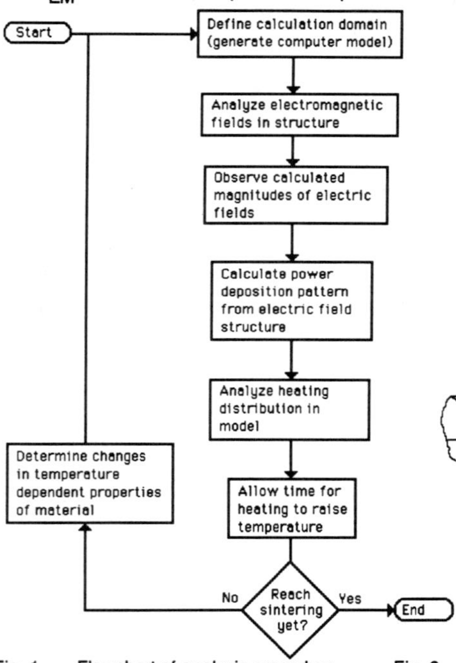

Fig. 1. Flowchart of analysis procedure.

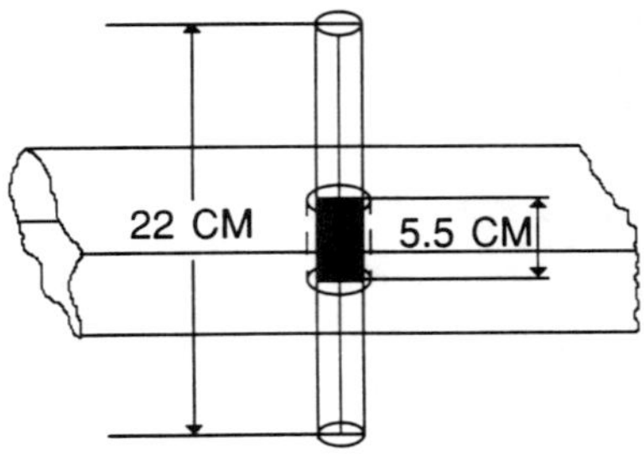

Fig. 2. Model of a ceramic rod being sintered in a wave-guide

The convective heat-loss term was obtained from Newton's law of convection:

$$q = hS(t_{surface} - t_{air}) \tag{3}$$

where

q	is the heat lost	t_{air}	is the ambient air temperature in Kelvin
S	is the surface area	$t_{surface}$	is the surface temperature
h	is the surface convection coefficient		

The radiation heat loss term, on the other hand, was obtained from the equation for grey body radiation:

$$q = \lambda A \sigma T^4 \tag{4}$$

where:

λ	is the emissivity	σ	is the Stefan-Boltzmann constant
A	is the area in meters^2	T	is the surface temperature in Kelvin

Since the implemented finite difference is an explicit method the solution is not unconditionally stable. The condition for stability is to require the coefficient of T^n_{cell} in equation (2) to be greater than zero. This requires that the new temperature of a cell be related to its previous temperature by a positive coefficient. The stability condition is then [3]:

$$\frac{k\Delta t}{rc\Delta h^2}\left[1 + \frac{\Delta h\left(\lambda\sigma T^3 + h\right)}{k}\right] < \frac{1}{6} \tag{5}$$

where:

σ	is the Stefan-Boltzmann constant	k	is the thermal conductivity
T	is the surface temperature in Kelvin	λ	is the emissivity
h	is the surface convection coefficient	Δh	is the cell size

The cell size, Δh is set to the same value used in the FDTD code used for calculating SAR values. The only variable to be set is the time step Δt. The time step is chosen so as to meet the stability condition for the worst case material but as large as possible to minimize computational time.

VERIFICATION

In order to verify the developed code many heat-transfer problems were simulated [4]. In each case one or more aspects of the code was tested and compared with available data. Excellent agreement are found and typically errors of only a few percent were obtained. Some of the test cases involved: steady state convergence, transient response, change in temperature for specified power, convection, conduction, radiation ,etc. In each case the problems involved simple geometries such as plates, infinite walls and semi-infinite slabs. In order to perform a full 3D verification involving heat generation, conduction, convection and radiation, a commercially available program (ANSYS) was used to simulate a test case and the obtained results were compared with those obtained using the developed code.

The verification with ANSYS was performed using the following simulation [5]. A cylinder of radius 8 mm and length of 22 cm is microwave sintered in a waveguide. The center 5.5 cm section is placed in the waveguide while the rest is outside (Fig. 2). The 5.5 cm section being heated is assumed to be uniformly absorbing 100 W. The temperature distribution in the rod is calculated as a function of time. To simulate this test case a rectangular geometry of the ceramic rod was required for the FDTD code. The geometry used for both programs was a rod of length of 22 cm, width = 6 mm, thickness = 6 mm. Figure 3 shows the temperature on the axis of the sample after 200 seconds of heating. The maximum observed difference in the ANSYS and developed code results is 3.4%.

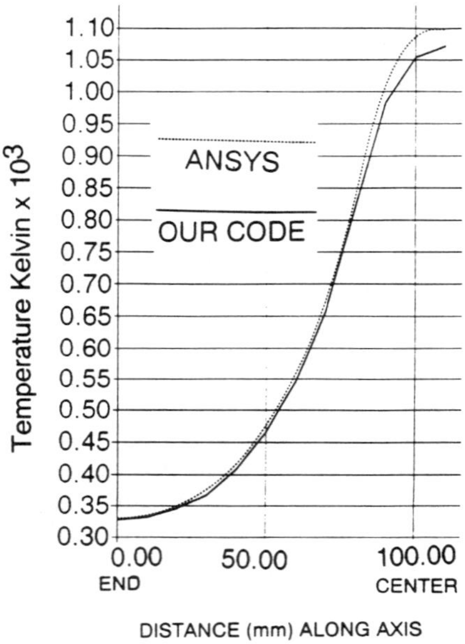

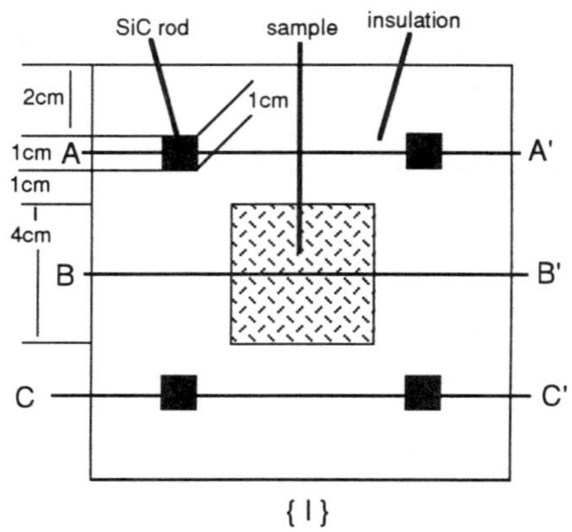

	insulation	ceramic	SiC
k m/K	0.1	30.1	490
C J/kg K	787.3	787.3	675
ε_r	1.537	4.13	29.363
σ S/m	15.37E-6	64E-6	0.9
ρ kg/m3	395	2370	3160

Fig. 3. Temperature distribution along the axis of the sample after 200 seconds of heating.

Fig. 4 Typical model used in the microwave sintering experiments. Top, The placement of the SiC rods and sample in the insulation. Bottom, Material properties used.

RESULTS

A typical arrangement for a sintering experiment is to surround the sample on all sides with insulation and place typically 4 to 6 SiC rods within the insulation as shown in Fig. 4 . The sample and insulation are located in the center of a microwave oven. For this arrangement the electric field pattern calculated using the FDTD code is shown in Fig. 5. The temperature after 33.3 minutes of heating was calculated based on the electric field pattern of Fig. 5. The temperature pattern was then used to update the material properties at this new elevated temperature. These updated parameters were next used in the FDTD EM code to generate a new field pattern. The heat code was then restarted with the temperature pattern calculated at the 33.3 minutes point and the new field pattern. The calculated temperature and field patterns are three dimensional. What is shown is a slice through the center of the sample perpendicular to the electric field. The process was repeated generating the temperature patterns after heating for 50 and 66.6 minutes. The results from the 66.6 minute run are shown in Fig. 6. This shows the samples temperature pattern as it heats to a steady sintering temperature. The SiC rods heat up very fast then stay at a constant temperature. It may be noticed from Fig. 6 that two of the SiC rods heated to a higher temperature than the others. This is not due to a heat transfer process but mainly caused by the uneven electric field pattern with the microwave cavity as can be seen from Fig. 5.

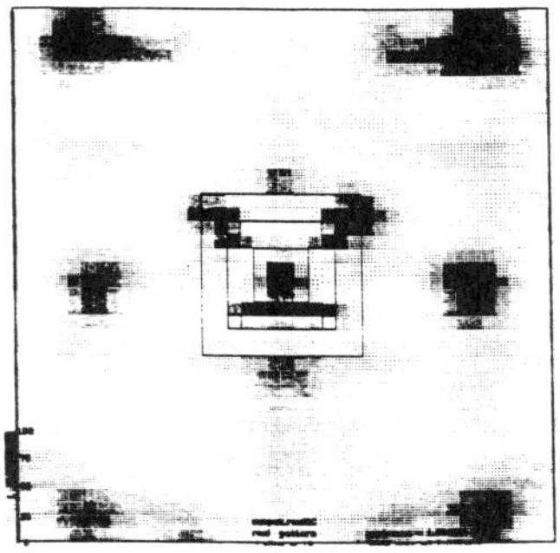

Fig. 5. Electric field pattern using picket-fence arrangement in multimode cavity. This was generated using our FDTD code under the typical sintering arrangement.

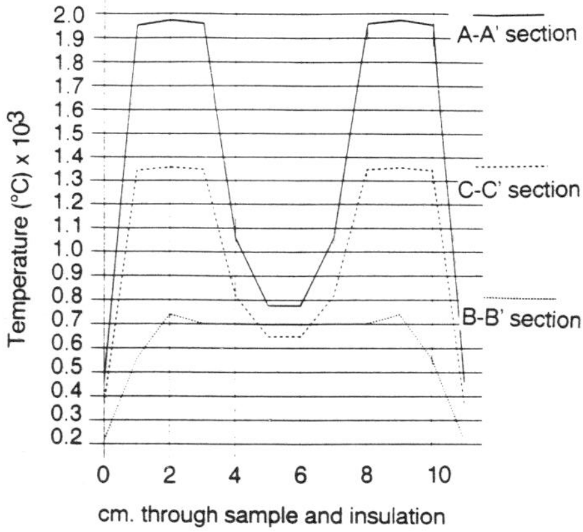

Fig. 6. Temperature distribution pattern in sample and insulation microwave sintered in multimode cavity using "normal arrangement" Time was 66.6 minutes.

This coupled heat transfer and electromagnetic absorption pattern continues as a function of temperature until sintering is achieved. In an attempt to show the temperature pattern at the end of a sintering experiment the heat transfer code was used to calculate temperature distribution based on fully sintered material properties at 1000°C. The obtained results are shown in Fig. 7 where it may be seen that the sample temperature reached approximately 1600°C under these sintering conditions and picket-fence arrangements.

The next simulation experiment was to increase the sample size until its edges touched the SiC rods. All other variables were maintained as in the previous simulation. Insulation was replaced by sample to allow the sample to increase in size (Fig. 8). Figure 9 shows the temperatures along a diagonal through the insulation, SiC rods and sample. Since the rods are touching the sample in this simulation case, the sample was found to heat much faster and the rods generally stay at lower temperatures.

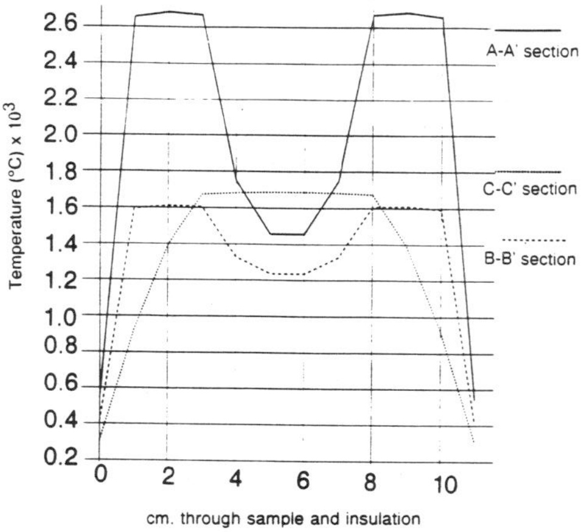

Fig. 7. Steady-state temperature distribution pattern in sample and insulation using "normal arrangement" after microwave sintering is completed.

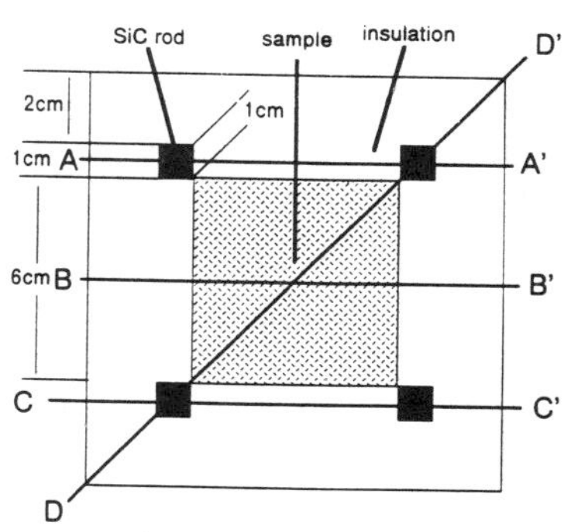

Fig. 8. Model of a larger sample in a typical microwave sintering experiment.

Another simulation experiment was to move the SiC rods closer to the outside surface of the insulation (Fig. 10). All other variables were maintained as in the first simulation. The temperatures along the A, B and C slices are shown after three hours of heating (Fig. 11). Based on Fig. 11 we may conclude that by placing the rods closer to the outside surface of the insulation most of the power absorbed by the SiC rods is lost to the surroundings and hence it was not possible to achieve sintering.

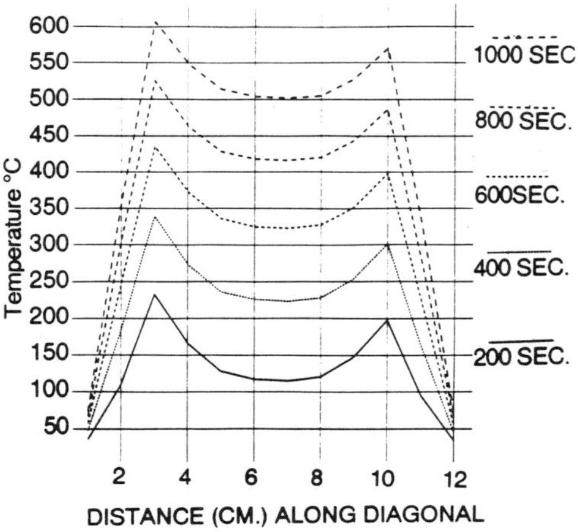

Fig. 9. Temperature distribution pattern in larger sample sintered in multimode cavity using picket-fence arrangement. These temperatures are along a diagonal slice D as shown in Fig. 8.

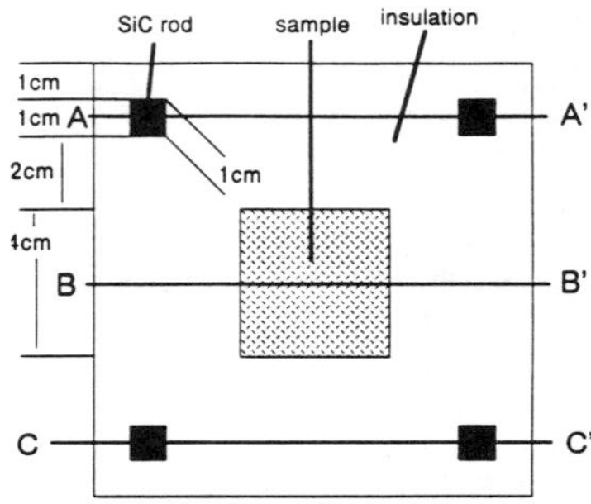

Fig. 10. Model used to show the affects of moving the SiC rods further away from the sample and closer to the outside surface of the insulation.

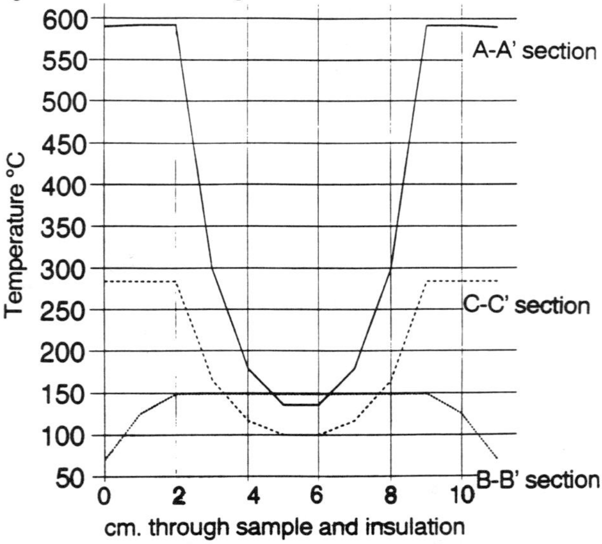

Fig. 11. Steady-state temperature distribution in cross sections of sample and insulation for case when SiC rods are moved towards outside surface of insulation.

CONCLUSIONS

A 3D FDTD heat transfer program was developed to calculate temperature distribution patterns in realistic microwave sintering experiments. The accuracy of the code and specifically its steady-state convergence, convection and radiation terms were verified using simple geometries. The developed code was also tested further by comparing the results with those obtained from the commercial software package ANSYS.

The developed code was used to calculate temperature distributions in typical microwave sintering experiments. In all cases the power deposition patterns were calculated from an FDTD EM code developed by our group. Results from several simulations of typical sintering experiments were shown.

The developed 3D code accounts for the various heat exchange mechanisms and utilizes EM power distribution patterns also calculated based on 3D simulations of realistic sintering experiments. The accuracy of the obtained results, however depend on accurate knowledge of the materials' thermal and dielectric properties as a function of temperature. To this end additional efforts are needed to measure thermal properties, density and dielectric properties as a function of temperature during microwave heating. This will clearly improve the accuracy of the obtained results and hence improve the predictability and optimize the microwave heating process.

Ongoing efforts include integrating the developed heat transfer code with the FDTD code for SAR calculations. The dielectric properties of ceramics, green and sintered, are also being measured at the University of Utah. These properties measurements will also improve our modeling of realistic microwave sintering experiments.

ACKNOWLEDGEMENT

Research sponsored by the U.S. Department of Energy Assistant Secretary for Conservation and Renewable Energy, Office of Industrial Technologies, Advanced Industrial Concepts (AIC) Materials Program, under subcontract 19X-SJ125V and under contract DE-AC05-84OR21400 with Martin Marietta Energy Systems, Inc.

REFERENCES

[1] R. L. Smith, *et al.*, "Finite-Difference Time-Domain (FDTD) Simulation of Microwave Sintering in Multimode Cavities," ibid.

[2] Y.L. Tian, D.L. Johnson, and M. E. Brodwin, "Ultrafine Microstructure of AL_2O_3 Produced by Microwave Sintering, " Ceramic Power Science II, G. L. Messing, E. R. Fuller, and H. Hausner, eds., American Ceramic Society, Westerville, Ohio, 1988, pp. 925-932.

[3] D. G. Watters, An Advanced Study Of Microwave Sintering, Phd. thesis, Evanston, Illinois.

[4] The Heat Transfer Problem Solver, Research and Education Association (1984).

[5] Yong-Lai Tian, Practices Of Ultra-rapid Sintering Of Ceramics Using Single Mode Applicators, Ceramic Transactions, Vol. 21, The American Ceramic Society, Westerville, Ohio.

ELECTROMAGNETIC FIELD MODELING OF A LOADED MICROWAVE CAVITY

P. CHAUSSECOURTE*, J.F. LAMAUDIERE* AND B. MAESTRALI**
* EDF Direction des Etudes et Recherches, Clamart, France
** EDF Direction des Etudes et Recherches, Les Renardières, France

ABSTRACT

The unavoidable way of a better conception of cavities for industrial applications is electromagnetic field numerical computation with the aim of placing the fields E and H into the material, and furthermore to obtain some characteristics like the input impedance of the cavity or the dissipated power in the sample. From Maxwell's equations and using a Finite Element Method with 3D edge elements we get a linear system from which, after resolution, we can obtain several kinds of results: research of eigen modes of an empty or partially dielectric loaded cavity, that is, for a bounded domain, all the frequencies that could be excited, using only dimensionnal (L,l,h) and physical (ε',ε'') characteristics, or response of a loaded cavity to an external excitation, that is, setting a TE_{10} mode in a waveguide feeding a cavity, obtain the distribution of field inside the oven and the dielectric. An original approach allows us, by projection of the excitation on eigen modes, to understand the behavior of the cavity and the interaction wave-product. Our computation is in very good agreement with analytical results we can obtain in some specific configurations (eigenvalues of an empty cavity, or transmission of only the TE_{10} mode in a correctly shaped waveguide).

INTRODUCTION

Conception of multimode cavities used in thermal applications of microwaves suffers nowadays a lack of knowledge in the exact repartition of electromagnetic fields into the cavity and the workload. This lack of understanding does not allow, for example, to have precise values and repartition of the power dissipated in the sample and in the walls of the cavity. Impedance value of the system cavity-product, which allows quantification of the adaptation, seen from the generator, may be obtained from the electromagnetic field. All these quantities being functions of the electromagnetic field are not accessible analytically and thus a numerical modelization is necessary.

METHOD

Maxwell's equations governing the phenomena are:

$$\text{curl } e = - \partial b/\partial t \quad (1)$$

$$\text{curl } h = J + \partial d/\partial t \quad (2)$$

$$\text{div } d = \rho \quad (3)$$

$$\text{div } b = 0 \quad (4)$$

where e and h are the vector electric and magnetic fields, respectively, b and d are the vector magnetic and electric flux densities, J and ρ are the source current and charge densities. Materials are considered linear, isotropic and homogeneous by pieces, which allows us to take constitutive equations:

$$d = \varepsilon e \quad (5)$$

$$b = \mu h \quad (6)$$

Hypothesis on solving domain are:

Three-dimensionnal closed domain
Taking into account Perfect Conductors $\quad n \times e = 0$
External excitation source of imposed frequency and amplitude
Harmonic 'steady-state' $\quad \partial/\partial t = i\omega$
Taking into account lossy dielectric materials $\quad \varepsilon = \varepsilon' - i\varepsilon''$
Ability to consider symmetry planes $\quad n \times h = 0$

The problem can be solved in electric or magnetic field, the two formulations being dual. By combining Maxwell's equations we can find a Helmholtz equation in magnetic field:

$$\text{curl}(\sigma + i\omega\varepsilon)^{-1}\text{curl } h + i\omega\mu h = 0 \qquad (7)$$

We can make a discretization of this equation and get a variational formulation of the problem:

$$\int_\Omega \frac{1}{\varepsilon_k - i\frac{\sigma_k}{\omega\varepsilon_0}} \text{curl } h \text{ curl } h' \, d\Omega - \omega^2 \varepsilon_0\mu_0 \int_\Omega \mu_r \, h \, h' \, d\Omega = -i\omega\varepsilon_0 \int_\Gamma (n, e_s, h') \, d\Gamma \quad \forall h' \in V \qquad (8)$$

with $V = \{ h \in (L^2(\Omega)) \text{ such that curl } h \in (L^2(\Omega)) \}$
e_s source field
(n, e_s, h') mixed product

Domain discretization is done by using edge elements (1st-order Whitney elements); the choice of these elements was induced by two main reasons [1]:

These elements while associating a vector field as unknown of the system, do impose only tangential continuity of this field but allow a normal discontinuity, which is exactly representative of the physic phenomenon.

These elements avoid the problem of spurious modes (i.e non-physical solutions). With this discretization, all these solutions are of a known number and concentrated (as far as concerns their associated eigenvalue) around zero: these are true gradients.

Equation (8), after spatial discretization, gives us a linear system:

$$[K - \omega^2\varepsilon_0\mu_0 M] X = F \qquad (9)$$

with K complex symmetrical (non-hermitian) matrix
M real symmetrical positive definite matrix
F source term vector

RESOLUTION

Two different approaches were used, as much for numerical than physical reasons:

I. A "modal" method, which allows us to project influence of source term (external excitation) on eigen modes of the system cavity-product and thus relative importance of different modes between them in comparison with imposed frequency.

II. A "direct" method, solving linear system (9) by an iterative algorithm, and giving general solution in electric and magnetic fields.

I Modal method

1 Whole spectrum of eigen modes

If there is no source term, system (9) can be reduced to an eigen value problem, which can be written:

$$[K - \omega^2 \varepsilon_0 \mu_0 M]\, X = 0 \qquad (10)$$

with K and M complex symmetrical matrices
K being non-hermitian i.e $^tK\, K^{-1} \neq I$

a) Real eigen modes

° In particular case of cavity filled with perfect dielectric material (i.e without losses) the system is real and matrices are symmetrical, and we can perform resolution by a QL algorithm with Householder reduction (FORTRAN routine issued from NAG scientific database). This gives us the whole spectrum of eigen modes in the limit of spatial discretization, with precision truly proportionned to thickness mapping, and being worse for highest modes.

° Academic cases:

An empty parallelepipedic cavity (non-degenerated, i.e with 3 different dimensions a,b,c) owns eigen values spectrum whose analytical values can be found by the formula

$$\text{wave number } k = \sqrt{\left(\frac{m\pi}{a}\right)^2 + \left(\frac{n\pi}{b}\right)^2 + \left(\frac{p\pi}{c}\right)^2}$$

with $(m,n,p) \in N^3$

Results are given in figure 1.

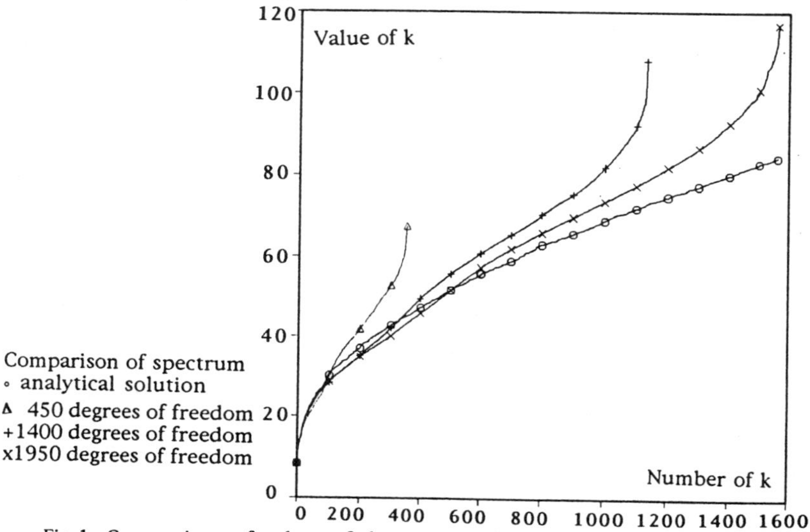

Fig.1 Comparison of values of the wavenumber versus mapping thickness

Two more configurations with analytical solutions were tried, an homogeneous lossless dielectrical filled cavity, and a cavity filled by a slab of perfect dielectric along one of its section.

A fourth configuration, which does not have analytical solution but was tried by some teams [2], composed of a dielectric cube on bottom of a cavity, has been calculated. Results are in table 1.

° Main disadvantage of this solving method is in the way of matrices storage which have to be "decompacted" and full stored, which limits system to about 2500 degrees of freedom. Constraint of this inconvenience is that for a real case, the mesh would not be fine enough for a good representation of associated eigen vectors (typically, less than 6 points by wavelength does not allow a correct representation of modes).

b) Complex eigen modes

If we consider lossy dielectric material filling the cavity, the system is complex, and solving it by a QZ algorithm from NAG does not allow more than 1250 degrees of freedom. However this is possible and has been done in configuration 4 of previous paragraph. It induces mainly a shift in phase but no spatial variation on repartition of electric field.

2 Part of eigen modes spectrum

Find out the whole spectrum is thus limited for discretization reasons and furthermore, if first eigen vectors are understandable, solutions for highest modes give correspondant fields completely false (though wavenumbers are good) because of the too rough mesh .

So we are now interested in searching only a certain part of all the modes spectrum.

a) In the case of a partially (or completely) filled cavity with perfect dielectric, and without external excitation, we can find a sketch of all eigen modes by a generalized power method [3]. This method is giving a finite number of eigen values by a Conjugated Gradient algorithm (iterative process) with introduction of a shift to avoid N first zero eigen values. Comparison between this method and that of 1-a paragraph is good.

b) If we consider that almost all dielectric materials have a loss factor (ε'') smaller than permittivity (ε') and often smaller in an order of magnitude (example of polymers, paper, water and so on) we can think that the perturbation of real solutions by introduction of a dissipative term (ε'', with $\varepsilon''/\varepsilon' <1$) will be a good approximation of complex solutions. We can obtain this by construction of the first terms limited development of matrix R ($R = (K-\omega^2\varepsilon_0\mu_0 M)^{-1}$) similar to system (10), [4], [5]. Thus we have complex eigen values of perturbed problem.

Results for one mode are on figure 2. Surprisingly, we see that up to a value of $\varepsilon''/\varepsilon'$ equal to 0.3 (for the first ten modes) comparison with exact solution (that of 1-b paragraph) is very good.

3 Response to an excitation, by projection of source term on eigen modes

Idea underlying this research is that total electric and magnetic field in a cavity do result of superposition of whole spectrum of modes excited in this cavity, and that it is easier, a priori, to recombine this ensemble of modes to find out electromagnetic field than to proceed on the opposite way.

Furthermore the main influence seems to be the fact of first modes in terms of power, so we can cut the spectrum at a given order without altering the result.

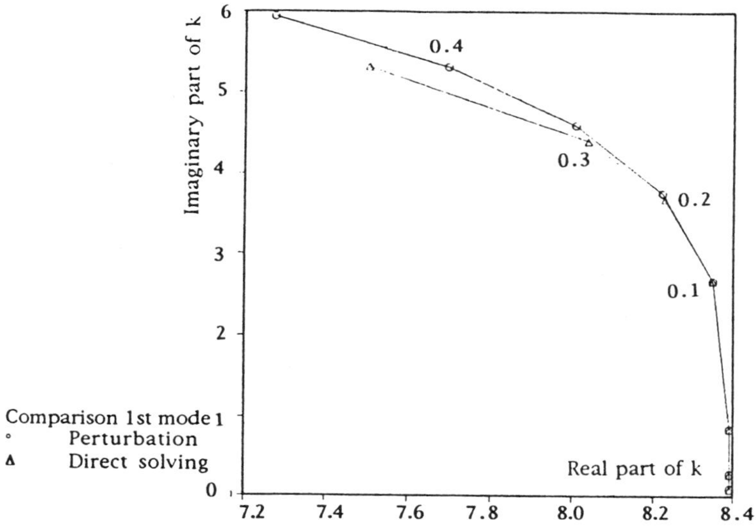

Fig.2 Comparison of direct solving versus perturbation method

Projection of sinusoïdal excitation on subspace generated by eigen values found previously, can give us response of each eigen mode of the cavity as a function of the source term.

We will be able to deduce main modes of all system cavity-product (compared to signal frequency, and facing to the source or sources position), and their relative magnitude.

II Direct method

System to be solved is general system (9) with an external source.

To avoid retroacting coupling terms between imposed source and cavity, excitation is introduced by the means of a waveguide in such transverse dimensions that only TE_{10} mode is able to propagate (in backward or forward way), and at a sufficient distance of the discontinuity waveguide-cavity for evanescent modes to be no more existent.

Discretization of source vector is given by a sinusoïdal function in a transverse plane of the waveguide and has to be calculated as circulation along edges of the elements.

Resolution is made using a Preconditionned BiConjugate Gradient algorithm.

Some examples of what we can obtain are in figures 3 and 4.

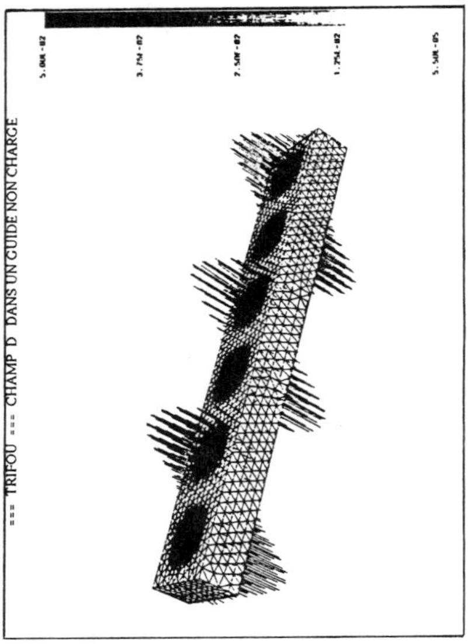

Fig. 3 D field (TE_{10}) in a waveguide

Fig. 4 E field in a TM_{010} cavity feeded by a waveguide via an aperture coupling

CONCLUSION

At this point we have a software able to give us important informations: we can obtain electric and magnetic field in any configuration we want, impedance, reflection coefficient or eigenmodes of the whole system (cavity plus product). Application of this software is first understanding of interaction wave-product and optimization of cavity for a better coupling, but it can be extended to hyperthermia or electromagnetic compatibility. Furthermore this computation is a step toward a coupling with thermal or chemical phenomena.

REFERENCES

1. A. Bossavit, IEEE Trans.MAG-24,1988, pp.74.

2. L. Pichon, A. Razek, SEE journées d'études, Modélisation 3D dans les systèmes électromagnétiques, Mars 1991 pp81-85.

3. C. Hazard," Etude des résonances pour le problème linéaire des mouvements d'un navire sur la houle," Thèse de doctorat, ENSTA, Mai 1991.

4. Perturbation theory for linear operators, 2nd Ed. T. Kato, Springer-Verlag Berlin.

5. L.C. Randrianarivony," Analyse numérique et Micro-ondes," Note EDF, Juin 1991.

PART III

Microwave Processing System Design

MULTIKILOWATT VARIABLE FREQUENCY MICROWAVE FURNACE*

D. W. BIBLE*, R. J. LAUF*, and C. A. EVERLEIGH**
*Oak Ridge National Laboratory, Oak Ridge, TN 37831-6006
**Microwave Laboratories, Inc., 8917 Glenwood Ave., Raleigh, NC 27622

ABSTRACT

We describe a new type of microwave processing furnace in which the frequency can be varied continuously from 4 to 8 GHz and the power level varied from zero up to 2.5 kW. The extraordinary bandwidth of this furnace is achieved by using a traveling wave tube (TWT) amplifier originally developed for electronic warfare applications. The TWT is a linear beam device characterized by a traveling electromagnetic wave that continuously extracts energy longitudinally along the path of an electron beam. The TWT, unlike other microwave tubes such as the magnetron, klystron, gyrotron, and others, does not depend upon resonant RF fields and is therefore capable of wide bandwidth operation.

INTRODUCTION

Microwave processing of materials includes such diverse applications as drying, sintering, melting, and plasma processing [1]. With few exceptions [2-4] most work to date has been conducted at 2.45 GHz because of the ready availability of magnetron sources at this frequency. Like the magnetron, other conventional microwave sources such as the klystron and gyrotron also operate at fixed frequency because they employ a resonant cavity in their design. The desire for a variable-frequency microwave processing system is motivated by several considerations: (1) in an untuned cavity, mechanical mode stirring does not adequately eliminate power nonuniformities, (2) the efficiency of coupling to a loaded cavity is highly frequency-dependent [5] even when the source varies over a 4% bandwidth, (3) plasma processes such as diamond film growth are frequency-dependent and a certain amount of controlled frequency sweeping can help stabilize the plasma, (4) the effects of frequency selection in solid-state sintering are not well-understood, requiring experiments that cover a wide range of frequencies under otherwise identical conditions.

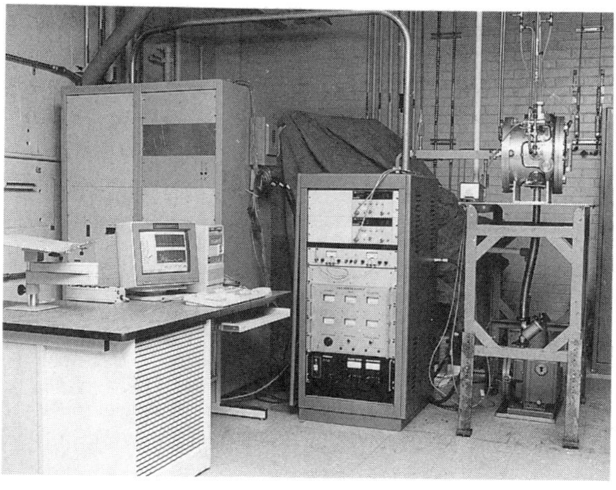

Fig. 1 Variable Frequency Microwave Furnace

* Work supported by the U.S. Department of Energy under contract DE-AC05-84OR21400 with Martin Marietta Energy Systems, Inc.

A microwave processing system has been designed and built using a high-power traveling wave tube (TWT) amplifier capable of supplying up to 2.5 kW over the range from 4 to 8 GHz, Fig. 1. Even at this early stage in its development, the results obtained thus far have shown that this system can be a versatile tool for materials processing research.

FURNACE DESIGN

The furnace layout is shown in Fig. 2. The individual components are described below.

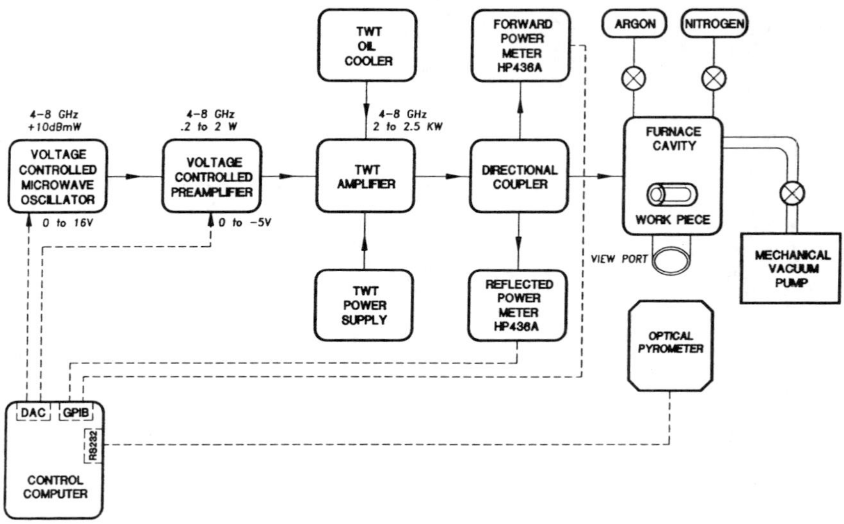

FIG.2 VARIABLE FREQUENCY MICROWAVE FURNACE

<u>Signal Generation And Control System</u>

Microwave signals are generated by a voltage controlled oscillator, Delphi components model VCO-100-XXX, at a power level of +17 dBm. Input control voltages, in the range from 1 to 16 volts, provide output frequencies of 4 to 8 GHz to the furnace cavity. To provide adequate signal power into the TWT, the primary signal level must be increased to approximately 33 dBm. This is accomplished with a National Hybrid model NHI4080 MNW-2 GaAs FET microwave power amp. The TWT typically operates with fixed gain, and power to the cavity is controlled by gain adjustment of the pre-amp. The pre-amp gain can be varied over a 10 dBm range (.2 to 2W) with a provided control voltage of 0 to -5 volts.

Forward power delivered by the TWT, and reflected power from the furnace, are sampled by a Microwave Engineering Corporation broadwall dual directional coupler, model R202-5N and displayed on HP436A power meters.

The data acquisition and control system comprises a '486 PC with a multi-channel D/A card, a GPIB controller, and application software developed using Microsoft "C" and the Lab Windows library from National Instruments.

The system presents a graphical "instrument panel" interface for the operator, which displays measured values (forward and reflected power levels, sample temperature) and allows adjustment of control parameters (center frequency, modulation, power setpoint). A "strip chart" function shows measured values vs. time and also records the data to disk for later analysis.

This system allows the user to slowly scan the frequency range, recording the reflected power ratio at each point in order to determine the points of maximum power absorption by the sample. A PID loop algorithm dynamically adjusts the pre-amp gain to maintain steady forward power as the frequency varies. The "sample profile" is displayed graphically and recorded to disk, Fig. 3. Based on this information the user can then adjust the center frequency, add frequency modulation if desired, and raise the power level to heat the sample.

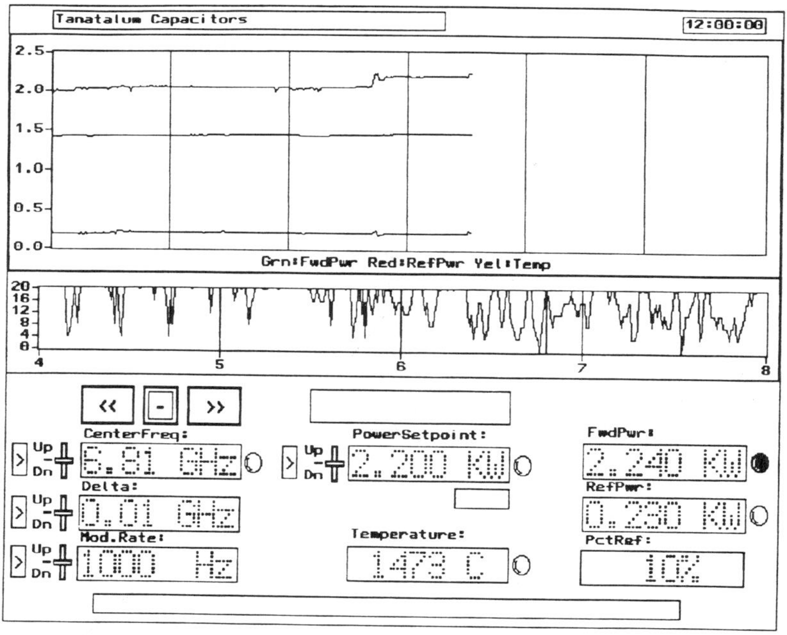

Fig. 3 Control system screen display.

A sample may be heated at a single frequency, modulated about a center frequency, swept through a particular range, or time division frequency shifting can be employed, with or without modulation, where the system shifts among several predetermined frequencies and dwell times. These complex modulations are aimed at improving the heating efficiency and sample temperature uniformity, as well as providing maximum process flexibility.

Also, since the system is both frequency and amplitude agile, the computer can be instructed to effectively notch out those particular frequencies where the reflected power exceeds a preset level.

TWT Amplifier

The TWT amplifier system employs a Microwave Laboratories, Inc., model T-1096 Traveling Wave Tube. The MLI tube is unique within the TWT manufacturing industry with respect to the ability to deliver power in the kilowatt range and withstand relatively large levels of reflected power. Tolerance of reflected power is extremely important for cavity heating applications where a wide range of sample loading and impedance can be expected. The superior performance of the MLI tube is due to the mechanical construction employing a copper helix supported by BeO blocks for efficient heat transfer. A copper backwall with cooling channels surrounds the helix structure. An MLI oil cooler continuously circulates Monsanto Coolanol 35 through the tube to dissipate heat.

Power for operation of the TWT is provided by a Universal Voltronics Model BRE-15-140MM supplying helix voltage from 500 to 15KVDC at 140 mA and collector voltage from 0-10KVDC at 2 A. The solenoid is powered by a Sorensen Model DCR80-33B.

The MLI TWTs have a unique feature that allows modulation from zero to full rated power by adjustment of the anode potential. This feature allows additional tubes to be used with the existing power supplies for increased frequency span. The desired tube is simply biased on while the remaining tubes are held in a standby mode. Thus by a proper bias switching program, a continuous range of frequencies can be swept. A second tube covering the range of 8-16 GHz is under construction for the furnace cavity such that continuous coverage from 4-16 GHz will be available.

Cavity

The furnace cavity is constructed of aluminum with cylindrical dimensions of 12 inches in diam and 10 inches in length. Ten test ports are arranged diametrically around the cavity using standard 2 3/4 inch conflat flanges. A mechanical pump provides a vacuum down to 10 mTorr, while nitrogen and argon are available for purging and inert atmosphere heating. An optical port provides temperature measurement using a Minolta/Land Cyclops 152/152A, interfaced with the control computer.

The microwave waveguide cavity window is constructed of 2 mil. polyimide plastic sheeting. Polyimide is effectively transparent to microwaves, withstands 600oC temperatures, atmospheric pressure, and provides an inexpensive disposable alternative to replacement of expensive quartz windows which often fail due to material deposition and subsequent arcing.

TEST RESULTS AND DISCUSSION

Heating tests were conducted with a variety of materials and casketing strategies [6] as described below. In general, the TWT performed superbly, exceeding specifications and operating stably over a wide range of conditions.

Boron carbide

A pellet of B_4C 1" diam x 1" long containing 8% Varcum (equivalent to 2 1/2% carbon) was pressed to a green density of about 53% and painted with a thin coating of graphite. The pellet was placed in a 3" x 3" boron nitride casket surrounded by granules of BN-coated glassy carbon.

A preliminary heating experiment was conducted in flowing N_2. The frequency scan indicated minimal reflected power at about 6.7 GHz. Approximately 2KW of forward power was applied and the sample heated rapidly. As expected, the optimal frequency changed slightly as the sample heated and by following the changing conditions it was possible to maintain efficient heating.

Sample temperature was not uniform; one end of the pellet had sintered slightly (about 70% dense, estimated from the shrinkage) and the other end had not.

Tantalum

Porous tantalum compacts [7] were distributed in a BN casket surrounded by ZrO_2 fiber. The preliminary frequency scan, Fig. 3, showed minimal reflected power at 6.76 GHz. Applying 2 KW of forward power at this frequency, the sample heated rapidly to ~1450°C. During heatup, the optimal frequency increased from 6.76 to 6.81 GHz.

Tungsten

Several samples of tungsten alloy (W-Ni-Fe), provided by the Hertel Co., were casketed in yttria grit, and heated to 1300°C for 30 minutes for annealing. Previous attempts to heat the samples at 2.45 GHz were unsuccessful. The samples have been returned to Hertel for evaluation.

CONCLUSIONS

1. A variable-frequency microwave processing system has been demonstrated using a high-power traveling wave tube amplifier. The microwave source can be tuned to any frequency from 4 to 8 GHz or swept continuously at rates up to 1 MHz.

2. The frequency range can be expanded by adding TWTs for other bands. Each TWT covers about one octave.

3. Heating efficiency improves dramatically when the frequency can be varied over a wide range. Frequency and power can be controlled in real time to minimize reflected power and/or maintain a preset temperature or heating rate.

ACKNOWLEDGMENTS

C. E. Holcombe and M. S. Morrow assisted in the heating tests. S. R. Maddox developed the control system software. Furnace conception and design were funded by the Office of Industrial Technologies, Advanced Industrial Concepts Materials Program. Equipment and construction were funded by the Office of Defense Programs, Y-12 Advanced Manufacturing Program. Collaborative research is supported by the Office of Energy Research, Laboratory Technology Transfer program and conducted under Cooperative Research and Development Agreement No. ORNL 91-0055 between Martin Marietta Energy Systems, Inc., and Microwave Laboratories, Inc.

REFERENCES

1. W. H. Sutton, "Microwave Processing of Ceramic Materials," Amer. Cer. Soc. Bull. 68 [2], 376-86 (1989).
2. B. Swain, "Microwave Sintering of Ceramics," Met. Prog. 134 [3], 76-82 (1988).
3. T. T. Meek, R. D. Blake, and J. J. Petrovic, "Microwave Sintering of Al_2O_3 and Al_2O_3 - SiC-Whisker Composites," Ceram. Eng. Sci. Proc. 8 [7-8], 861-71 (1987).
4. M. Barmatz, J. L. Watkins, and H. W. Jackson, "Microwave Processing of Materials in Microgravity," AIAA 92-0116, 30th Aerosp. Sci. Meeting, Jan. 6-9, 1992, Reno, NV.
5. A. Mackay, W. R. Tinga, and W. A. G. Voss, "Frequency Agile Sources for Microwave Ovens," J. Microwave Power 1 [1], 63-76 (1979).
6. C. E. Holcombe and N. L. Dykes, "Importance of Casketing for Microwave Sintering of Materials", J. Mat. Sci. Lett. 9, 425-8 (1990).
7. R. J. Lauf, C. Hamby, C. E. Holcombe, and W. F. Vierow, "Microwave Processing of Tantalum Capacitor Anodes," Mat. Res. Soc. Symp. Proc. Vol. 269(1992).

SINGLE FREQUENCY MICROWAVE DIELECTRIC HEATING AND MEASURING

B.Q.TIAN*, W.R.TINGA* AND W.XI*,
* Dept.of Electrical Engineering, University of Alberta, Edmonton, Canada, T6G 2G7

ABSTRACT

Traditional microwave heating and measuring are mostly conducted using two different channels, i.e., a single frequency high power channel for heating and a sweeping frequency low power channel for measuring. Here, a different single frequency system is discussed, by which heating and measuring are conducted simultaneously using one high power source. The theoretical basis emphasizing Q determination at a single frequency is presented. A re-entrant resonator serves as the applicator. Without the normally required swept frequency channel, the experimental system is greatly simplified.

INTRODUCTION

A sophisticated microwave material processing system today should be capable of not only dielectric heating but also of measuring a material's microwave properties. Most methods of dielectric constant and loss factor measurement involve a frequency shift and a Q factor variation measurement, both of which require a swept frequency setup. However, people mostly use a single frequency microwave power source to heat dielectrics. To accommodate the different requirements with respect to frequency and power levels, a two channel system is used, in which the dielectric is heated via a high power single frequency channel while the dielectric constant and the loss factor are measured through a low power swept frequency channel[1,2].

Different from the traditional approach, a single frequency heating and measuring method is presented in this paper. We first discuss the theoretical basis of the method and then implement it using a coaxial reentrant cavity, by which a dielectric is heated and measured simultaneously using one power channel at a single frequency. Instead of measuring Q by the normal bandwidth method, we measure either return loss or the position of a coupling loop at the resonant frequency to determine the loss factor. Then, we verified the method by experiments at ambient temperature and tested the high temperature performance of the system.

PRINCIPLES

When a dielectric is introduced into a cavity, as shown in figure 1, the resonant frequency of the cavity will be altered by Δf, which is a function of the dielectric constant, ε'. Therefore, ε' can be determined by a measurement of Δf. On the other hand, it is equally possible to determine ε' by measuring the variations in cavity dimensions required to maintain the resonant frequency constant.

The loss factor, ε'', can be determined by the measurement of the variation in Q factor, $\Delta(1/Q)$, which itself can be measured at a single frequency using the method suggested in [4], with Q expressed as

$$Q=\frac{2\ n^2 \omega\ S}{Z_o\ R} \tag{1}$$

where n is the transformer ratio of the coupling mechanism, S is the ratio of the energy stored in the cavity to the square of a field integration around the coupling device, ω is the angular resonant frequency, Z_o is the characteristic impedance of the transmission feed line and R is the normalized resonant input resistance. It has been proved that S is a constant if the

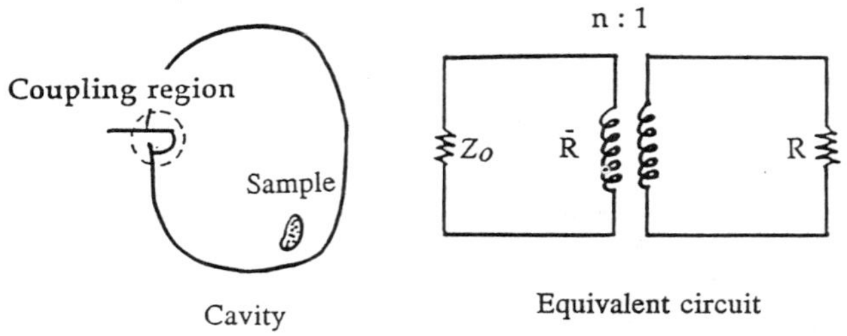

Fig.1 The Cavity Containing a Sample and Its Resonant Equivalent Circuit

perturbation condition is satisfied.[4] Under this condition, consideration of equation (1) suggests two approaches for Q measurement at a single frequency, namely, the constant transformer ratio (CTR) approach and the constant input resistance (CIR) approach.

Constant Transformer Ratio (CTR)

When the coupling device is fixed, n can be considered as a constant. In this case, a simple relation between Q and R can be established as

$$Q=\frac{C_n}{R} \tag{2}$$

where $C_n=(2/Z_o)n^2\omega S$, which is a constant. As a result, one can measure Q variation by measuring the change in the input resistance, or reflection coefficient, $|\Gamma|$, which can be related to R by

$$R=\frac{1\mp|\Gamma|}{1\pm|\Gamma|} \tag{3}$$

Constant Input Resistance (CIR)

A constant input resistance can be maintained for a cavity having a varying load if the coupling transformer ratio is adjustable. In this case, we obtain

$$Q=C_R n^2 \tag{4}$$

where $C_R=(2S\omega/ZoR)$, is now a constant. Using (4), one can measure Q variation by measuring the change in the transformer ratio. The fact that, for a given mode, the transformer ratio is exclusively determined by the geometric structure of the coupling device has made it possible to determine n through the measurement of the coupling loop dimensions and location even though a theoretical solution to an actual structure can be complex. Note that, when R is set equal to unity, the Q measurement can be made under a matched condition.

An understanding of the distinction between these two approaches is of importance in practical applications. In the cases where variation in the input resistance is tolerable, fixing the coupling device is recommended since the measurement of reflection coefficient is relatively convenient from the microwave measurement point of view. On the other hand, when a stable input impedance condition is of primary concern, such as in most microwave material processing systems, where a match at the input of the applicator is desirable for efficient operation, the constant resistance approach is a better choice.

IMPLEMENTATION

In order to implement the foregoing principles, the structure of the cavity has to be given, so that the relationship between ε', ε" and the measurable parameters can be specified.

Cavity Description

The cavity used consists of a coaxial cavity with a gap in its hollow center conductor at the end of the cavity and a movable short circuit plunger, as shown in figure 2. An adjustable coupling loop is mounted on the short circuit plunger. The dielectric of interest is held in the gap and the adjacent hole region.

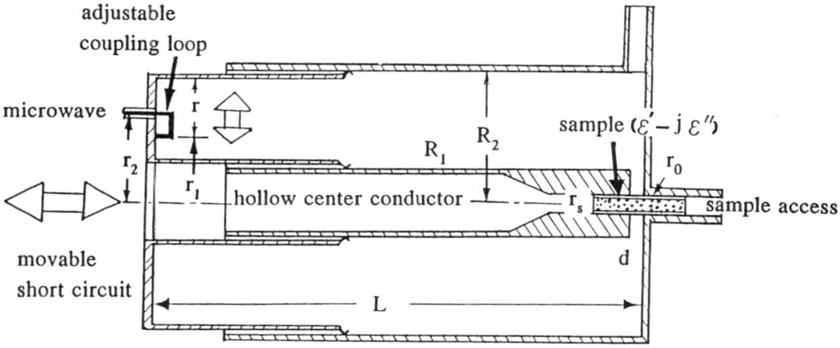

Fig. 2 The Construction of the Applicator
(r_0=0.35, r_s=0.24, R_1=1.27, R_2=5.08, d=0.25, cm)

There are a number of reasons for the choice of the structure. First, it has excellent high temperature performance[2,3]. Secondly, the small filling factor and the location of the sample far away from the coupling device are favorable for satisfaction of the perturbation stipulation[4]. Thirdly, a theoretical model is available to provide the relation between the dielectric properties and the cavity dimensions. In addition, the radially adjustable coupling loop provides the means for altering the transformer ratio, n, by changing the loop location, r, as shown in figure 2.

Dielectric Determination

For a given cavity mode of the coaxial structure chosen, the resonant frequency shift due to the introduction of a sample can be readily calculated by a mode-matching analysis program[2]. Similarly, for a given mode and resonant frequency, the cavity length variation, ΔL, needed to maintain the same resonant frequency can also be calculated by this program. Therefore, a calibration curve can be established for determining the dielectric constant from the measured ΔL. For convenience, this curve is expressed by a least square fit polynomial,

$$\varepsilon'=1.0+4.09245\Delta L+0.46555\Delta L^2+0.05964\Delta L^3. \quad (5)$$

This equation corresponds to the cavity dimensions shown in figure 2 and to the resonant frequency of 915MHz, at which all our measurement and heating experiments were conducted.

It is common practice to determine the loss factor from the drop in Q factor due to the loss of the inserted sample

$$\varepsilon''=\frac{\varepsilon'}{F}\left(\frac{1}{Q}-\frac{1}{Q_o}\right) \tag{6}$$

where Q_o and Q is the cavity Q before and after the sample is loaded respectively. Using the mode-matching program the filling factor, F, can be calculated and expressed by a polynomial

$$\frac{\varepsilon'}{F}*10^{-3}=0.4873+0.02666\varepsilon'-0.73027*10^{-4}\varepsilon'^2 \tag{7}$$

Depending on how Q is determined, the loss factor can be further expressed, for the constant transformer ratio case, as

$$\varepsilon''=C_n\frac{\varepsilon'}{F}\left(\frac{1-|\Gamma|}{1+|\Gamma|}-\frac{1-|\Gamma_o|}{1+|\Gamma_o|}\right) \quad \text{for over coupling} \tag{8}$$

$$\varepsilon''=C_n\frac{\varepsilon'}{F}\left(\frac{1+|\Gamma|}{1-|\Gamma|}-\frac{1+|\Gamma_o|}{1-|\Gamma_o|}\right) \quad \text{for under coupling} \tag{9}$$

where Γ_o and Γ are the reflection coefficients at resonant frequency before and after the sample is loaded. For the constant input resistance case,

$$\varepsilon''=\frac{\varepsilon'}{C_R F}\left(\frac{1}{n^2}-\frac{1}{n_o^2}\right) \tag{10}$$

where n_o and n are the transformer ratios of the loop for a given reflection coefficient before and after the sample is loaded. The transformer ratio, n, of the loop can be approximately expressed as[4]

$$n=\frac{1}{\ln\frac{r_2}{r_1}} \tag{11}$$

where r_1 and r_2 are the two dimensions of the coupling loop shown in fig.2. The constants C_n and C_R in (8) to (10) have to be determined according to the Q value of the unloaded cavity.

EXPERIMENTAL VERIFICATION

In order to verify the proposed method, we measured the dielectric constant and loss factor using both the single frequency method and the swept frequency method. Figure 3 shows the experimental systems used in these two methods. The results for five different samples, ranging from 5-80 in dielectric constant and 0.30-4.2 in loss factor, are presented in table I.

The results of the single frequency method show a maximum difference of 9% from those of the swept frequency method. One error source is the

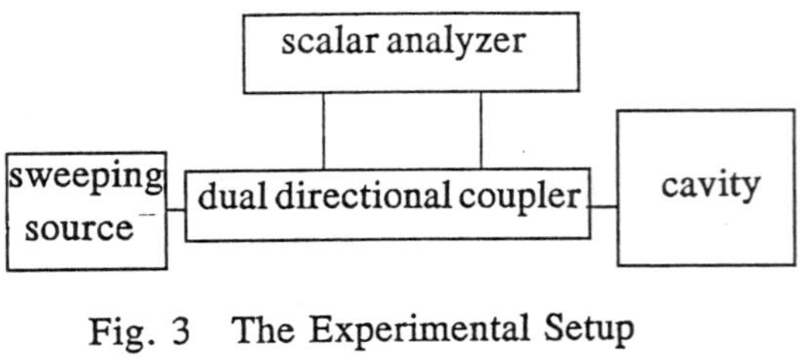

Fig. 3 The Experimental Setup

Table I Comparing Results Between the Single Frequency and Sweeping Methods

SAMPLE NO.	$\varepsilon'-j\varepsilon''$			DIFFERENCE %			
	SINGLE FREQUENCY		SWEEPING	CTR		CIR	
	CTR	CIR		ε'	ε''	ε'	ε''
1	5.3-J0.34	5.3-J0.31	4.9-J0.33	8.2	3.0	8.2	-6.1
2	8.1-J1.33	8.1-J1.29	7.5-J1.19	7.9	12.0	7.9	8.4
3	12.5-J1.58	12.5-J1.58	11.7-J1.42	7.0	11.0	7.0	8.5
4	19.8-J0.51	19.8-J0.49	20.0-J0.43	1.2	19.0	1.2	14.0
5	81.4-J4.2	81.4-J4.47	78.6-J5.3	3.5	-26	3.5	-15.7

CTR — Constant Transformer Ratio   CIR — Constant Input Resistance

deviation of the actual dimensions of the cavity from the ideal dimensions by which the calibration curves were calculated. As we can see, in figure 2, that the short circuit bucket section is obviously smaller in diameter than the rest of the cavity. The frequency drift of the source may also contribute some errors. The resolution of ε' at 915 MHz band is about 0.05, which is restricted by the resolution of the cavity length measurement. The differences in ε'' are mostly due to the differences in Q values obtained by the two different Q measurement methods, the errors of which may be as large as $\pm$ 10% for either method[2,4]. Considering that ε'/F is the same for the swept frequency and the single frequency method, we can estimate that the reflection coefficient errors are of the same order for both methods.

Having verified the single frequency method at room temperature, this method was applied to a high temperature dielectric measurement in conjunction with simultaneous microwave heating. The experimental system, as shown in figure 4, consists of the cavity, as the applicator, a 915 MHz 60 W power source with a reflected power indicator and a thermometer. By observing the reflected power, we adjust the short circuit and the coupling loop to tune and match the cavity, while recording their positions for later determinations of the dielectric constant and the loss factor. Two sets of high temperature curves are presented here to demonstrate the high temperature performance of the system. One is the dielectric constant and loss factor versus temperature for zeolite and the other is that for mullite, as shown in figures 5 and 6.

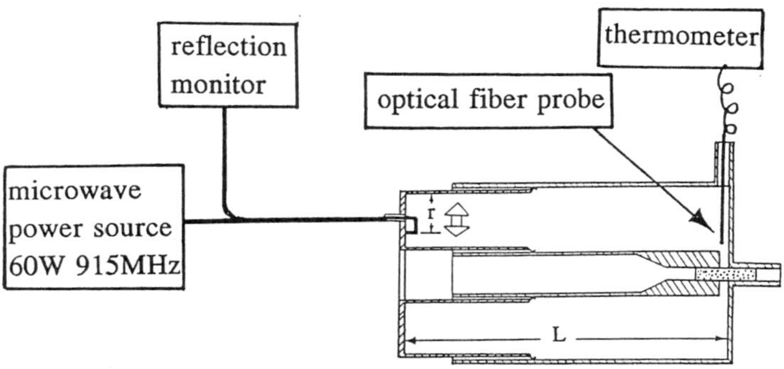

Fig. 4 High Temperature Experiment Setup

Both the zeolite and the mullite data were taken under the matched input impedance condition.

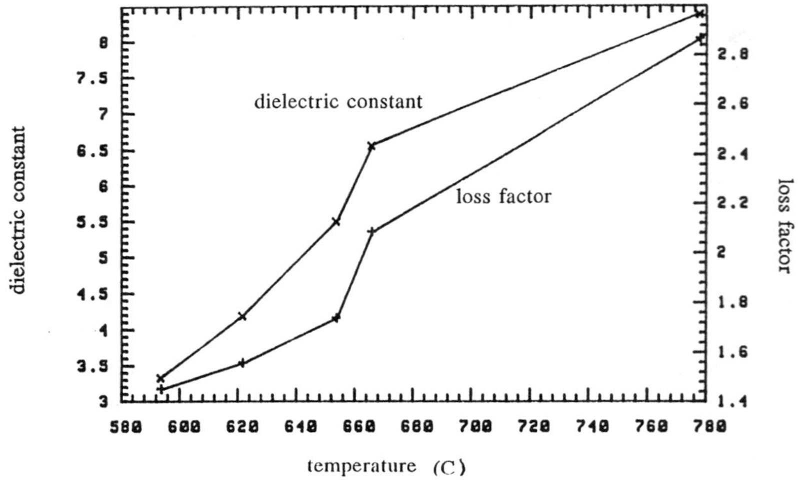

Fig. 5 Dielectric Constant and Loss Factor vs. Temperature (Zeolite)

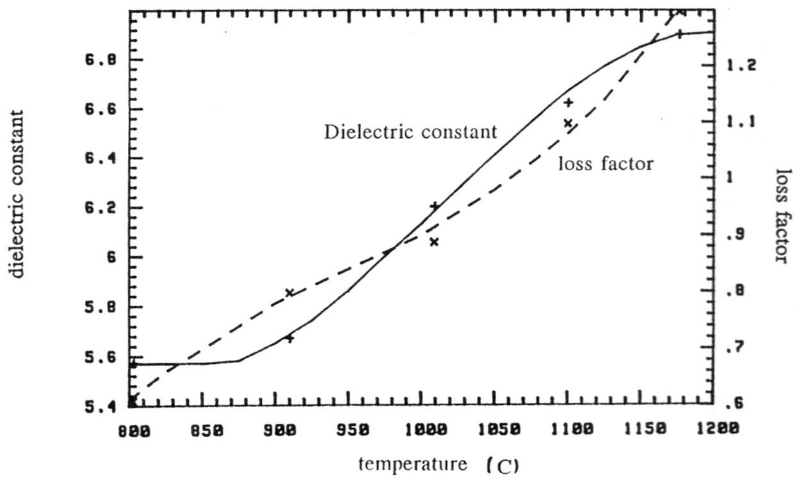

Fig. 6 Dielectric Constant and Loss Factor vs. Temperature (Mullite)

CONCLUSIONS

In this paper, we have presented the single frequency method for the measurement of dielectric constant and loss factor. The theoretical basis of this method has been discussed, followed by the experimental verification. Applying this method using a coaxial reentrant cavity, we have developed the single frequency high temperature microwave dielectrometer, in which a dielectric is heated and measured simultaneously using one power channel. This approach harmonizes the heating and measuring requirements, permitting the whole process to operate under a matched condition. Owing to the removal of the normally required swept frequency channel, the experimental system is greatly simplified.

ACKNOWLEDGEMENT

This work is sponsored by the Natural Sciences and Engineering Research Council of Canada. The financial assistance provided by the Faculty of Graduate Studies and Research and the Alma Mater Fund of the University of Alberta is also acknowledged.

REFERENCES

1. D. Couderc, M. Giroux and R.G. Bosisio, J. of Microwave Power, 11, 69, (1976)
2. Weiguo Xi and W.R. Tinga in Microwaves: Theory and Application in Materials Processing, edited by David E. Clark, Frank D. Gac, and Willard H. Sutton, (Amer. Cer. Soc. Ceramic Tran., 21, 1991) pp. 215-224
3. W.R. Tinga, B.Q. Tian and W.A.G. Voss in Microwave Processing of Materials, edited by William B. Snyder, Jr., Willard H. Sutton, Magdy Iskander and P. Lynn Johnso, (Mater. Res. Soc. Proc. 189, San Francisco, CA 1990) pp. 111-116. Soc. Proc. Symp, L, (1990).
4. B.Q. Tian and W.R. Tinga, Internal Report B11, Dept. of Electrical Engineering, Univ. of Alberta, Edmonton, Canada, T6G 2G7, 1991.

COUPLING WITH AN ADJUSTABLE COMPOUND IRIS IN A SINGLE MODE APPLICATOR

Hussamaldin S.Sa'adaldin*, W.Murray Black*, Iftikhar Ahmad† and Richard Silberglitt†
*Department of Electrical and Computer Engineering, George Mason University, Fairfax, VA 22030
†Technology Assessment and Transfer Inc., Annapolis, MD 21401

ABSTRACT:

Coupling of the input power to a dielectric in a rectangular microwave cavity by using a compound iris having both inductive and capacitive slits, has been investigated. The compound iris is used in combination with a variable tuning plunger to maintain a resonant frequency of 2.45 GHz and minimize the reflected power. Under certain conditions, the compound iris has been observed to achieve an improved coupling of the input power, resulting in higher temperature of cylindrical 0.95 cm rods of silicon carbide (SiC) and alumina (Al_2O_3). By placing the SiC perpendicular to the applied E field (with SiC enclosed in a porous insulation partially filling the applicator), coupling with the compound iris by adjusting both inductive and capacitive slits was achieved for a wide range of values compared to the use of a single pair of inductive slits. The coupling was more sensitive to a precise positioning of the single inductive slit. Such a difference was not observed for the case where the sample was placed parallel to the E field. In temperatures over 1600°C have been reached with a 600 W of power injected into the cavity. In tests with low loss alumina samples (99.8% pure) placed parallel to the E field, the compound iris permitted a greater coupling and higher temperature. The presence of alumina insulation had a significant effect in achieving higher temperatures, and occassionally preventing the initiation of arcing.

INTRODUCTION:

Single mode microwave resonant applicators are often used in the processing of dielectric materials in order to maximize the electric field strength and achieve high temperatures in the material [1]. The amount of incident microwave energy entering the cavity without being reflected depends upon many factors such as, the frequency of the radiation, the cavity dimensions including the aperture size and shape, the type of materials comprising the walls of the cavity, and the material placed within the cavity. In order to optimize the heating process, typically the length of the cavity and the size of the input iris or aperture are adjusted to minimize the amount of reflected power at the input. The problem is compounded by the change in electrical properties (dielectric constant) of the dielectric material as it absorbs microwaves and its temperature increases. Consequently, the process is a dynamic one in which it is desirable to be able to constantly adjust the axial dimension of the cavity and the size of the input iris. Typically for a given experiment the input iris has been fixed in size and a moveable plunger has been used to optimize the power input to the cavity. Fukushima et al. [2] have used a feedback control system to adjust the iris as well as the axial plunger. Metaxas [3] has conducted some initial tests with a compound iris to improve the heating rate of many different materials to a maximum temperature of 1000°C. The purpose of this work is to compare the use of a compound iris to the single inductive iris in coupling power into a single mode cavity with a dielectric material placed in a region of maximum electric field.

EXPERIMENTAL APPARATUS

Figure 1 shows the single mode TE_{103} resonant cavity applicator which consists of a rectangular waveguide type WR284 of length 230 mm with a circular flange supporting the adjustable compound iris on one end and an adjustable short circuit plunger on the other end. The applicator is cooled by a recycling water system. The 0.95 cm cylindrical samples are inserted through one of two pairs of slots, such that they extend out of the waveguide. One pair is located at the center of the wide dimension of the guide and the other at the narrow dimension, both are at an axial position of 157 mm from the iris for maximum electric field strength. Leakage of microwave energy from the two slots in the direction of the H-lines was not detectable while a low leakage was detected at high temperatures with the samples in the direction of E-lines.

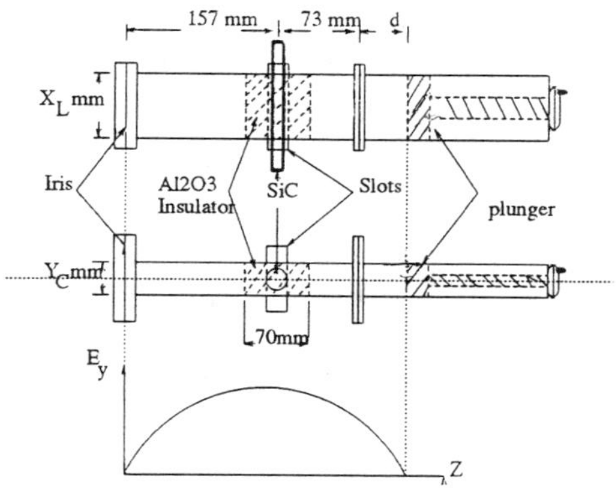

Figure 1. Single mode TE_{103} applicator.

A single mode resonant heating cavity is required to fulfil two criteria, that of operating at the required frequency and that of optimum impedance matching for minimum reflection of energy at the input of the applicator. Figure 2 shows the adjustable compound iris which connects to the opening of the applicator to act as a variable susceptance aperture. This device consist of two discs with two movable rings. One disc controls two vertical plates (forming an inductive aperture) and the other controls the two horizontal plates (forming a capacitive aperture). The ring arrangement has been designed to accommodate a later modification of control by a servo motor. The plates within the discs can independently be made of any size from a full open position to a short circuit condition. The flexibility of the arrangement should permit rapid change of either dimension. This is particularly useful since the dimensions of the resonant aperture for critical coupling depends on the effective loss factor of the inserted material. Consequently, the device can be used to match different ceramic materials as they heat and change their effective loss factor.

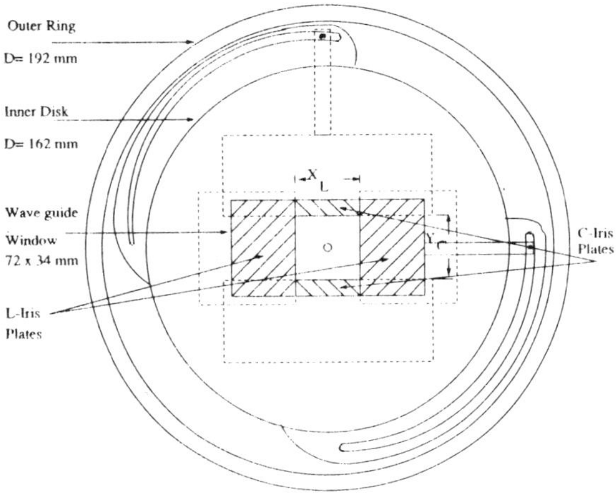

Figure 2. The Adjustable Compound Iris

RESULTS

The investigation was divided into four categories as follows:

(i) *Samples placed without insulation parallel to the E-field lines.*

In this case, using SiC as the dielectric it was difficult to couple enough power, using either L-iris alone or both L and C-iris to achieve high temperature. The sample temperature never exceeded 700°C and hence was not monitored by the optical pyrometer (700°C is the minimum temperature read by the pyrometer installed on the system).

Using alumina samples, the effect of C-iris is shown in figure 3. The coupling improves as the C-iris aperture is reduced.

(ii) *Samples enclosed in porous alumina insulation and placed parallel to the E-field.*

Using SiC as the sample for this configuration, with the C-iris completely open and varying L-iris, good coupling was achieved. Closing the C-iris window added capacitive susceptance and reduced the coupled power. With 600 watts incident power, the max power coupled was 470 watts and the maximum temperature achieved was 1610°C. Note that although the SiC rod was passing through the waveguide the power was absorbed by only the portion of the specimen within the waveguide (34 mm).

When alumina rod was used as the dielectric with 500 watts incident power, it was possible to couple well by means of the L-iris alone with only 50 watts of reflected power. As shown in figure 4, the maximum temperature achieved was 1115°C. Using the compound L- and C-iris the reflected power was only 30 watts and the maximum temperature achieved was 1170°C.

(iii) *Sample placed without insulation perpendicular to the E-field lines.*

In this configuration using SiC as the sample,it was possible to achieve good coupling for a narrow movement range of the L-iris and a fairly wide movement of C-iris. However, since the sample was not insulated high temperature achievement was not possible. The maximum temperature achieved at the best coupling was only close to a 1000°C.

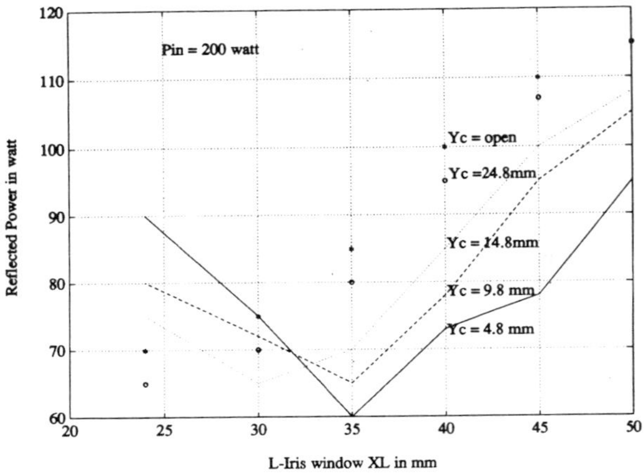

Figure 3. Effect of C-iris window on the reflected power.

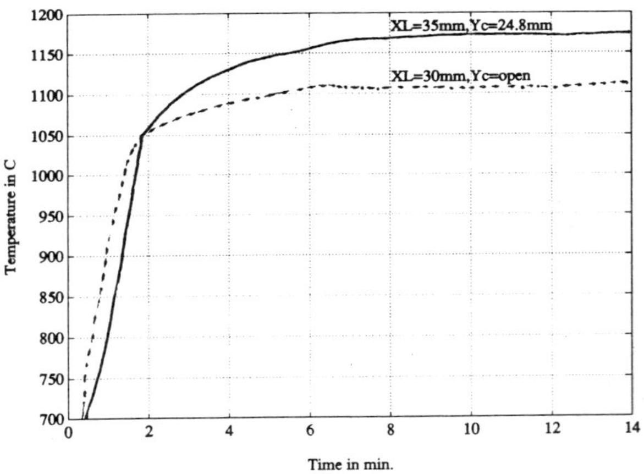

Figure 4. Heating response of alumina for L and compound iris using 500 watts input power.

(iv) *Sample enclosed in insulation and perpendicular to the E-field lines.*

Using SiC sample, this configuration gave excellent coupling. At times the reflected power was practically zero. Typically a few watts of reflected power were observed with 600 watts incident power. It was possible to couple similar power with the C-iris completely open, but the L-iris movement for good coupling was restricted to a narrow range. As shown in figure 5, the presence of the insulation substantially improves the resulting temperature for a wide range of positions of the C-iris. With insulation, although the coupled power decreases as the C-iris is opened, the temperature is still higher than the case without insulation. Figure 6 compares the coupling range of the compound iris to the L-iris. The compound iris has the advantage of providing a wide range of aperture settings to achieve critical coupling.

Using alumina as the specimen, it was not possible to achieve sufficient coupling to reach a temperature high enough for the pyrometer to read. This was true in both cases, using the L-iris alone or the compound iris.

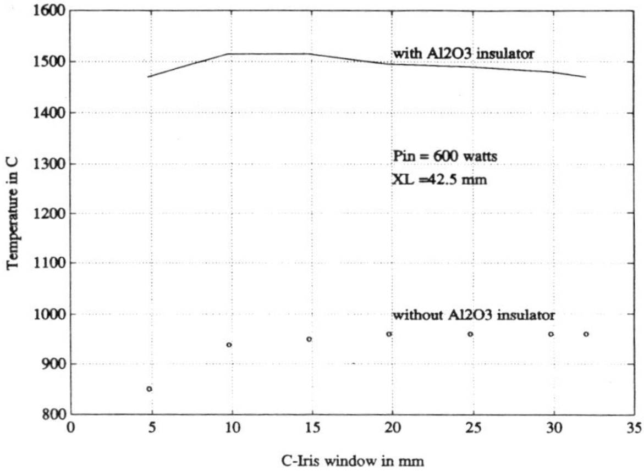

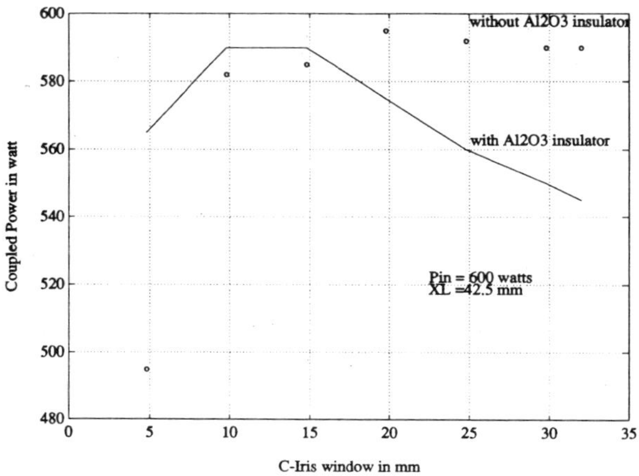

Figure 5. Effect of the insulation on the (a) maximum temperature achieved (b) power absorbed for different C-iris aperture values.

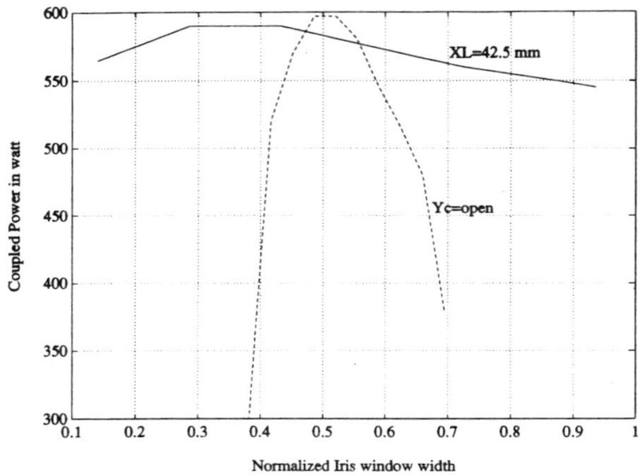

Figure 6. Coupled power versus normalized iris window width for L iris alone and compound iris.

CONCLUSION:

Preliminary experimental results using a compound iris in a single mode cavity suggests that the capability of being able to adjust both irises will provide a wider range of aperture sizes to achieve critical coupling. Under certain conditions, for example, heating an Alumina rod with the field parallel to the rod, the coupling and resulting temperature were improved over the case of using the L-iris alone. The insulation appears to provide several advantages. It increases the volume of the load thereby reducing arcing. Furthermore, the insulation retains the heat generated within the sample allowing achievement of high temperatures. It also helps to increase the dynamic range of the tuning plunger.

ACKNOWLEDGEMENTS

This work was sponsored in part by the Virginia Center for Innovative Technology. The single mode joining apparatus was provided by the Department of Energy through the Los Alamos National Laboratory under Subcontract Number 9-XG1-U0554-1.

REFERENCES:

1 W.H. Sutton, "Microwave Processing of Ceramic Materials," American Ceramic Society Bulletin, 68, (2), 376 (1989).

2 H. Fukushima, T. Yamanaka and M. Matsui, "Microwave Heating of Ceramics and its Application to Joining," J. Mater. Res., 6, (2) 397 (1990).

3 A.C.Metaxas," Rapid feasibility Tests Using a TE10n variable aperture resonant applicator", J.Microwave Power and Electromagnetic Energy, 28, (1), 16 (1990).

STEADY STATE TEMPERATURE PROFILE IN A SPHERE HEATED BY MICROWAVES

M. BARMATZ AND H.W. JACKSON
Jet Propulsion Laboratory, California Institute of Technology, Pasadena, CA 91109

ABSTRACT

A new theory has been developed to calculate the microwave absorption and resultant temperature profile within a sphere positioned in a single mode rectangular cavity. This theory is an extension of a total absorption model based on Mie scattering results. Temperature profiles have been computed for alumina spheres at the center of a rectangular cavity excited in the TM354 mode. Parametric studies reveal significant structure in those profiles under special conditions that are associated with electromagnetic resonances inside the spheres. Anomalous behavior similar to thermal runaway occurs at moderate temperatures when there is enhanced absorption associated with resonant conditions in the sphere.

INTRODUCTION

Prediction and control of microwave induced temperature profiles provides a means for improving the processing of materials. For example, theoretical modeling suggests that under certain controlled conditions, an interior portion of a sample can be melted and subsequently resolidified. This scenario is grounded in the fact that with microwaves, the temperature can be made hotter at the center than at the surface. Improvements in the composition, densification, and grain size characteristics of sintered samples can also be achieved with better temperature control enabled by theoretical modeling.

These potential improvements in materials processing can be aided by a new theory that has been developed to calculate the microwave absorption and resultant equilibrium temperature distributions within a lossy dielectric sphere located in a single mode rectangular cavity. This theory is an extension of a total absorption model [1] based on Mie scattering results. The new extension relies on the observation that when the dimensions of the cavity, the mode of excitation, and the position of the spherical sample are properly selected, the absorbed power density in the sample will be nearly isotropic. This implies that the average value of the power absorbed per unit volume in any thin-shell region concentric with the sphere will be essentially equal to the local power absorption density.

In this new model, the sphere is partitioned into thin shells, and the complex dielectric constant and the thermal conductivity are assumed to be uniform in each shell. The temperature profile depends parametrically on microwave absorption as a source density for heat and the microwave absorption depends parametrically on the temperature profile through the dielectric constants. An iteration procedure is applied to find a self-consistent solution of the microwave absorption problem and the thermal problem. Errors due to the discrete shell structure can be made negligibly small by using a large number of very thin shells. Numerical studies of this model have yielded some remarkable results.

THEORY

Total Absorption Model

The total absorption model [1] utilizes the observation that the resonant modes of an empty rectangular microwave cavity can be represented as a superposition of transversely polarized progressive plane wave fields. In general there will be eight plane waves, but this can be reduced to four if one of the mode indices is zero. When a homogeneous lossy dielectric sphere is located in the cavity, each of these plane wave fields is scattered from the spherical sample. For each plane wave, the electric (or magnetic) field in the region outside the sample will be a superposition of the incident plane wave and an outgoing spherical wave. At the boundary of the sphere, the external fields are matched to a solution of the vector wave equation internal to the sphere. Mie [2] solved this scattering problem for a single incident plane wave early in this century. Because Maxwell's equations and the wave equation that governs the fields are linear in the fields, one can superpose these results to find the fields when an empty cavity resonant mode is the incident wave. These results, found with the aid of Mie's theory, can be used to calculate the Poynting vector and total power absorption by the homogeneous sphere.

The results for fields and absorbed power calculated by this method are only approximately correct because the outgoing spherical waves do not satisfy the correct boundary conditions at the cavity walls. This is due to neglecting multiple scattering events. If the sample is small compared to the cavity dimensions and far from the nearest wall, the effects of multiple scattering should be small. These caveats should be observed not only for the total absorption model, but also for the new shell model.

Microwave Absorption in a Spherical Shell Model

In the spherical shell model, the microwave field incident on the sphere is associated with the empty cavity resonant mode treated as a superposition of plane waves, just as in the total absorption model. Also, the scattered field is a superposition of outgoing spherical waves. The internal fields are treated differently here because the dielectric constant may vary from one shell to another. However, throughout each shell, the fields can be expressed as a superposition of ingoing and outgoing spherical waves with unspecified coefficients. These coefficients are then fixed by a set of equations obtained by requiring continuity of the tangential components of the electric and magnetic fields at the boundaries of the shells. This procedure uniquely determines the fields throughout the sphere and cavity. The Poynting vector can then be constructed and the time-averaged net flux of electromagnetic power into each shell can be calculated. One finds the time- and space-averaged absorbed power density in each shell by dividing the net flux by the shell volume. This quantity is then used as the input for the thermal calculation described next.

Thermal Spherical Shell Model

Under steady state conditions, the total microwave power absorbed by the sphere must equal the net thermal radiation from its surface. This determines the surface temperature within the context of a gray body model. The heat flux density at the surface is just the total microwave power absorbed divided by the surface area of the sphere. For a spherical shell that is heated uniformly throughout its volume, formulas have been derived for the heat flux density and temperature at any point in the shell. The formulas depend parametrically on the heat flux density and the temperature at the shell's outer boundary, and on the thermal power source density which is equal to the microwave absorbed power density in the shell. Using the continuity of the heat flux density and temperature at the shell boundaries, one can start at the outer shell, evaluate the thermal conductivity in a shell, and work inward one shell at a time. This procedure yields expressions for the thermal conductivity, heat flux density, and temperature throughout the sphere.

The temperature at the outer boundary of each shell is then used to find a new value for the dielectric constant in each shell. This updated distribution for the dielectric constant can be used to recalculate the microwave power absorption profile, from which a new thermal profile can be determined. This procedure is then repeated until successive iterations yield the same temperature profile to within some specified accuracy. Such self-consistent, steady state properties have been studied numerically [3] and examples of the resultant microwave heating behavior are presented in the following section.

DISCUSSION

This new spherical shell model was applied to the heating of alumina spheres. Alumina was selected for these calculations because it is an important commercial ceramic and the required thermodynamic parameters have been experimentally measured. The real and imaginary dielectric constants for alumina were measured by Fukushima et. al. [4] to 1800 C and thermal conductivity data have also been compiled [5]. For modeling purposes, best fit functions were determined from the experimental data. A TM354 rectangular mode was chosen for the calculations and cavity dimensions of L_x = 10.25 cm,

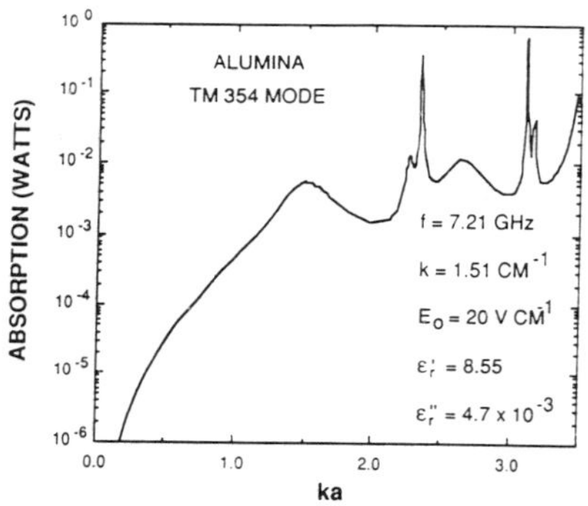

Fig. 1. Total microwave absorption in alumina spheres

L_y/L_x = 1.776, and L_z/L_x = 1.463 were used to produce an essentially isotropic electric field intensity near the center of the cavity. For these experimental conditions, the mode frequency is f = 7.21 GHz and the empty cavity wave number is k = 1.51 cm^{-1}.

The previously developed total absorption model [1] was used to calculate the expected microwave absorption near room temperature. The resultant absorption for alumina is shown in Fig. 1 as a function of ka. The spectrum shows several absorption peaks which are associated with microwave resonances in the sphere. An analysis of the structure of these peaks indicated that, in general, two or more multipole fields are associated with the absorption in each peak. A pure multipole field corresponds to a resonance of the sphere in empty space. Since many of these absorption peaks are very large, a significant enhancement of the microwave heating process may be achieved by choosing spherical samples with radii associated with these peaks.

We have investigated the shift in the lowest absorption peak (ka ≈ 1.5) due to the variation of the dielectric constant as a function of temperature. Figure 2 shows the temperature dependence of the real part of the dielectric constant and the absorption peak radius. The absorption peak radius decreases by ≈ 25% over the range from room temperature to 1500 C. The fact that the absorption peak shifts with temperature suggests that unusual heating phenomena will occur during processing of samples with sizes near the peak. The new thermal spherical shell model was used to investigate these phenomena. The equilibrium temperature at the sphere center was calculated as a function of increasing electric field strength for a set of sphere sizes near the lowest absorption peak. For all the calculations presented in this paper, each sphere was divided into 20 equally spaced spherical shells. Figure 3 shows the results of these calculations for alumina spheres located at the center of a rectangular cavity excited in the TM354 mode. For a sample radius above the room temperature absorption peak, such as a = 1.25 cm, the equilibrium temperature increases only slowly with field strength. For a sample radius equal to the room temperature absorption peak, a = 0.99 cm, the equilibrium temperature increases more strongly with field strength leading ultimately to thermal runaway at a high temperature (T ≈ 1700 C). When the sample radius is slightly below the room temperature peak, such as a = 0.925 cm or 0.900 cm, the equilibrium temperature becomes enhanced for temperatures corresponding to a near equivalence between the sample radius and absorption peak (T ≈ 400 - 800 C). Thermal runaway occurs again at high temperatures (T ≈ 1800 C).

For even smaller samples, such as a = 0.875 cm and 0.850 cm, higher field strengths are required to attain the higher temperatures needed for correspondence

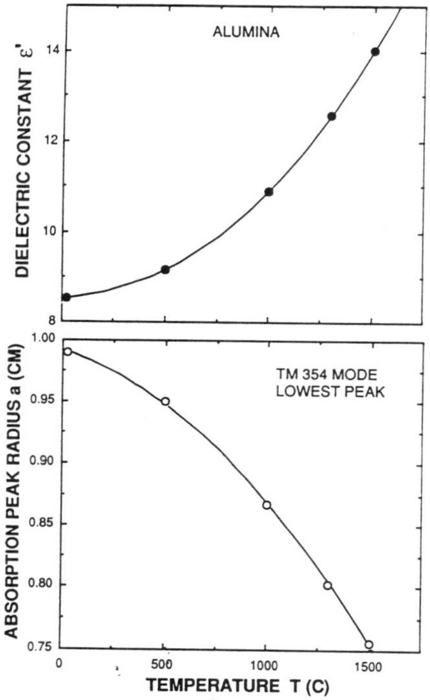

Fig.2. Temperature dependence of the real part of the dielectric constant and the absorption peak radius

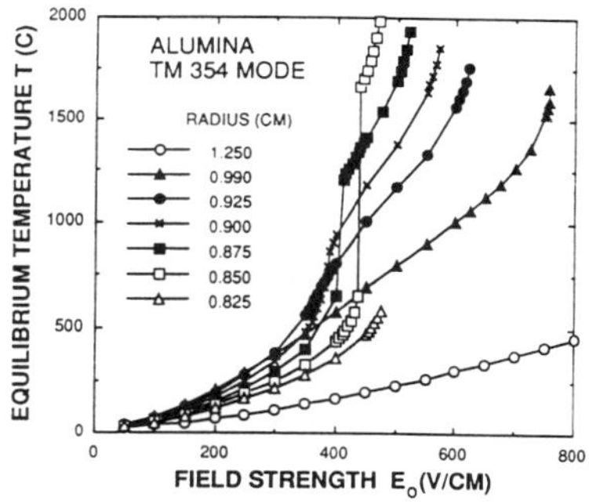

Fig. 3. Equilibrium temperature at center of alumina sphere

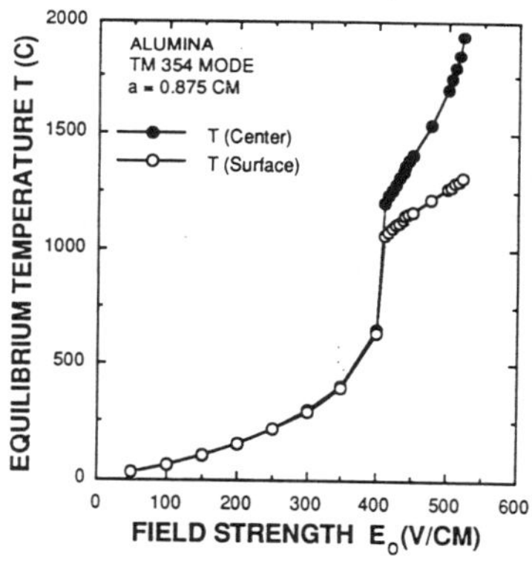

Fig. 4. Equilibrium temperature at center and surface of sphere

between the radius and absorption peak. At these higher field strengths, energy is being absorbed at such a high rate and the imaginary part of the dielectric constant is also increasing with temperature in such a manner that thermal runaway conditions are initiated at a rather low temperature (T ≈ 600 C). However, as the sample temperature increases, the absorption peak moves to smaller radii and the rate of microwave absorption decreases sufficiently to stop the thermal runaway! In this higher temperature region, the sample still heats up until the runaway condition is again established (T ≈ 1950 - 2000 C). For slightly smaller radii, such as a = 0.825 cm, runaway is again initiated at a low temperature (T ≈ 600 C) and is present until the sample melts. Finally at much smaller radii, the absorption peak radius never reaches that of the sample and much higher field strengths are required to initiate high temperature thermal runaway.

A significant temperature gradient can occur within the sphere at higher sample temperatures. To illustrate this effect, Fig. 4 shows the field strength

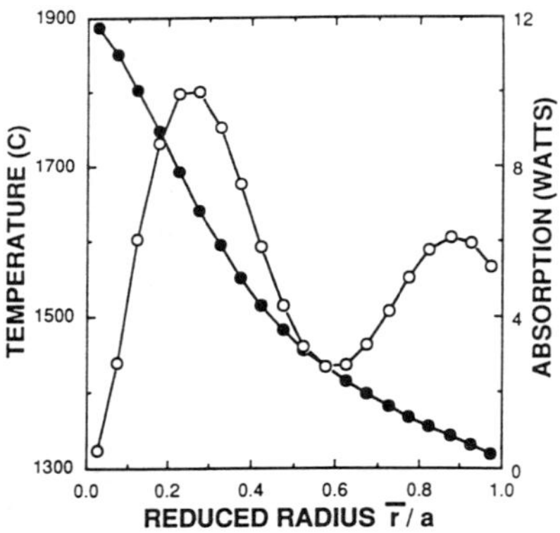

Fig. 5. Temperature and absorption profiles within a sphere

dependence of the equilibrium temperature at the center and surface of the a = 0.875 cm radius sphere. The temperature gradient is small for temperatures below 600 C which corresponds to the onset of thermal runaway. For temperatures above the suppression of the initial runaway, the gradient becomes very large, T_C - T_S being ≈ 600 C just before the sample melts due to a new runaway condition.

Figure 5 gives the actual temperature distribution and the microwave absorption profile within the sphere, based on absorption in each of the evenly spaced shells. These data are for a field strength near the final thermal runaway state shown in Fig. 4. $\bar{r}$ corresponds to the average radius of a given shell. It is seen that the radial distribution of microwave absorption has maxima and minima with very little absorption occurring at the center of the sphere. On the other hand, the temperature profile is monotonically increasing from the surface to the center. This type of inverse temperature profile is being used in sintering and annealing studies of Yttrium-barium-copper oxide high temperature superconductors [6] to produce bulk samples with uniform oxygen content. It is clear that large gradients such as shown here may also be used to induce sample melting which is initiated at the center of the sphere [7] - [9]. The ability to internally melt and resolidify a region within a homogeneous material could reduce volatility and improve stoichiometry during crystal growth. This would be especially important for processing II - VI and III - V compounds. For certain experimental conditions, this internal melting process may also lead to the preparation of new high pressure phases of commercial interest.

CONCLUSIONS

We have developed an advanced spherical shell absorption model that predicts the absorption and temperature profiles within a sphere heated by microwaves. This model was applied to the processing of alumina spheres with radii corresponding to the lowest absorption peak. The analysis predicts runaway behavior at intermediate temperatures and large temperature gradients at higher temperatures. At even higher temperatures, we also observe the normal runaway phenomenon. We expect that the resonant absorption peaks found for alumina spheres will also be present in the processing of other materials of various shapes and thus could account for unusual heating effects reported during processing [9]. This spherical shell model is also ideally suited for calculating the time dependent temperature profiles of a sphere heated by microwaves. We have already obtained preliminary results for such a model [10]. The calculation of the transient heating behavior of the sphere during processing will provide even greater insight into microwave/material interactions and permit more precise control of the processing parameters required to develop improved material properties.

ACKNOWLEDGMENT

We wish to thank Dr. J. L. Watkins for informative discussions and P. Wagner for computer support. The research described in this article was carried out at the Jet Propulsion Laboratory, California Institute of Technology, under contract with the National Aeronautics and Space Administration.

REFERENCES

1. H. W. Jackson and M. Barmatz, J. Appl. Phys. 70, 5193 (1991).
2. G. Mie, Ann. Phys. 25, 377 (1908).
3. H. W. Jackson and M. Barmatz (unpublished).
4. H. Fukushima, T. Yamanaka and M. Matsui, J. of Japan Soc. of Prec. Eng. 53, 743 (1987).
5. Y. S. Touloukian, Thermophysical Properties of Matter, (IFI/Plenum, New York- Washington,1972), Vol. 8,.p. 98.
6. J. Binner, Advanced Ceramics Report, (Elsevier Science Publishers Ltd., February, 1992), p. 8.
7. B. J. Pangrle, K. G. Ayappa, H.T. Davis, E. A. Davis, and J. Gordon, AlChE J. 37, 1799 (1991).
8. Y. L. Tian, Microwaves: Theory and Application in Materials Processing Edited by D. E. Clark, F. D. Gac, and W. H. Sutton, (Ceramic Transactions 21, 1991), p. 283.
9. Y. L. Tian, H. S. Dewan, M. E. Brodwin, and D. L. Johnson, Sintering of Advanced Ceramics, Advances in Ceramics, (Am. Ceram. Soc., 1988), p. 391.
10. H. W. Jackson, M. Barmatz and P. Wagner (unpublished).

PART IV

Microwave/Plasma Processing

PROCESSING OF REFRACTORY ORES IN A MICROWAVE INDUCED "COLD" PLASMA

D.E. BULLARD and D.C. LYNCH
Both at Arizona Materials Laboratory, Department of Materials Science and Engineering, University of Arizona, Tucson, Arizona 85721

Abstract

A Microwave Induced Plasma has been used to investigate the processing of metallurgically significant refractory oxides in a non-equilibrium or "cold" plasma. The main emphasis of this work has centered on the reduction of TiO_2 in a hydrogen plasma. The plasma was maintained using microwave energy (2.45 GHz) in a single mode resonant cavity. The reaction was monitored for volatile species with a quadrupole mass spectrometer, while the extent of reaction was determined by an external standard X-Ray diffraction technique. The effect of process variables (absorbed power by the plasma, chamber pressure, and time) on the extent of reaction was investigated. Hydrogen reduction in a plasma has produced up to 60% conversion of TiO_2 to Ti_2O_3 in 11 minutes with the sample on a grounded support rod, while only an 8% conversion is observed for a sample on an insulated support rod. Sample temperatures were below 735 K for most plasma conditions, based upon thermodynamic considerations.

Introduction

Conventional processing of refractory oxide ores is performed at high temperatures (T>2000°C) in order to make the thermodynamics of the desired reaction favorable. One method of obtaining these temperatures is the use of a thermal plasma. In this application the plasma acts as a high enthalpy source capable of generating gas temperatures of 3000 to 6000 K for the heating of entrained particulates or underlying melts [1]. However, high bulk sample temperatures are not the only method available to make the change in chemical potential of the desired reactions negative. "Cold" plasmas can produce the same chemically reactive species that are generated conventionally by high temperatures simply by applying an electromagnetic field to a low pressure gas. "Cold" plasmas have been used to enhance chemical reactions for years in the semiconductor industry. One of the primary uses of this technology is the dry etching of unwanted material from a substrate. These unwanted materials consist of metals and oxide materials. The latter are very stable and resistive to chemical attack (eg. SiO_2, WO_2, photoresist) and are removed while the substrate remains at relatively low temperatures [2].

The applied electromagnetic field generates the highly reactive plasma. Free electrons (present in small quantities in all gases) gain energy from the applied field. These high energy electrons produce the reactive ions, free radicals (neutral particles with an unpaired bonding electron, eg. H or CH_3) and more free electrons through inelastic collisions with gas molecules. The reactive nature of the plasma can provide new reaction paths that are not available in conventional processing.

Along with these reactive species, voltage gradients in a plasma can be used to enhance chemical reactions. A potential gradient is formed around all insulating objects in the plasma as a result of the difference in the mobilities of the various charged plasma species. This voltage drop is called the Debye sheath and has a magnitude of 1 to 100 Volts. This voltage gradient will cause positive ions to be accelerated across the sheath toward the sample surface, resulting in ballistic impact [3]. These ions can gain enough energy crossing the sheath to break bonds on the surface of the sample (7 eV can break a Ti-O bond). These broken bonds can then easily react with radicals or ions to form volatile products such as water.

For more information on the fundamentals of plasma processing the reader is referred to one of many excellent references that exist on the basics of processing materials in a non-equilibrium plasma [4][5][6], as this subject is beyond the scope of this paper.

Experimental

An experimental program was developed to investigate the effects that power absorbed by the plasma, plasma pressure and time have on the reduction of TiO_2. The experiments were conducted in the pressure range of 6 to 36 Torr and at applied powers of 0.4 to 1.2 kW. These experiments were conducted with the sample holder and support rod either conducting (copper) or insulating (quartz). A ground was maintained in the plasma when the conducting holder and support rod was used.

Figure 1 shows a schematic of the experimental set-up. A Gerling Laboratories model 102-A microwave generator (2.45 GHz) was used to generate and maintain the plasma in the quartz reaction chamber. Power meters allowed the amount of power being absorbed by the plasma to be determined. An Extrel C-50 quadrupole mass spectrometer was used to continuously monitor the plasma gas for volatile reaction products. The hydrogen used for the plasma was research purity, and all gas flows were controlled through rotameters.

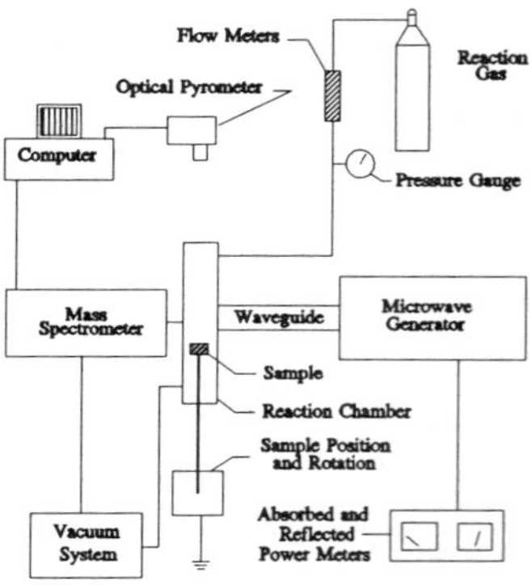

Figure 1. Schematic of plasma processing system.

The three nines TiO_2 (anatase) used in these experiments was obtained from Aldrich. This powder was dried at 400°C for 24 hours to remove any adsorbed H_2O. Upon removal from the furnace the TiO_2 was placed in a desiccator (while still warm) to minimize water pick up from the atmosphere during cooling. The powder was placed in a sample holder and lightly packed to ensure a smooth, uniform surface. This sample was placed on the support rod in the reaction chamber and the reaction chamber evacuated. The system was then flushed with hydrogen until the signal from the mass spectrometer for the impurity gases (ie. N_2, O_2, etc.) was negligible. After flushing the system, the gas flow was adjusted in order to achieve the desired pressure and the plasma was ignited. The sample was raised up into the plasma and allowed to remain in contact with the plasma for the desired amount of time. The sample was continuously rotated while in contact with the plasma to maintain a uniform specimen temperature.

Upon completion of an experiment, the sample was allowed to cool in the chamber under a continuous flow of hydrogen. The chamber was then flushed and back filled with dry nitrogen and the sample was removed. The surface product layer was analyzed with a Scintag XDS 2000 X-Ray Diffractometer (XRD) to determine what crystalline products were formed during contact with the plasma. Quantitative analysis of this product layer was performed using an external standard technique with the XRD spectra [7].

Results and Discussion

The stability diagram in Figure 2 reveals that the reduction of TiO_2 should follow different progressions depending upon the temperature of the sample:

$$TiO_2 \rightarrow Ti_3O_5 \rightarrow Ti_2O_3 \rightarrow TiO \rightarrow Ti \qquad T > 735K$$

$$TiO_2 \rightarrow Ti_2O_3 \rightarrow TiO \rightarrow Ti \qquad T < 735K$$

Therefore, sample temperatures can be inferred from this diagram based upon which phases are present. For most samples in this investigation, no Ti_3O_5 was found which indicates that the sample surface remained below 735 K. Ti_3O_5 was observed in one sample that was reacted in a grounded sample holder at 26 Torr and 1.2 kW. Conditions of high pressures and high power absorption by the plasma correspond to elevated temperatures. The conditions noted correspond to some of the highest specimen temperatures employed in this investigation. The presence of Ti_3O_5 at conditions of high temperature is confirmed by the stability diagram.

Typical mass spectrometer results are shown in Figure 3. This is a plot of the intensity of the H_2O signal as a function of time. A background signal was collected over the first four minutes so that any water that was already in the system could be accounted for. This background signal has been subtracted out of the data presented in Figure 3. The plasma was ignited at 4 minutes and was shut off at 15 minutes. The points in the "peak" (open circles) were averaged over a 6 second interval in order to increase the resolution of the data. The rest of the data was averaged over a 1 minute interval. The sharp increase in the amount of water, followed by a rapid decrease, is typical of a surface reaction. The major portion of the reaction appears to be over after approximately 5 minutes of reaction time. The peak of this graph represents a water fraction on the order of 10 mole percent, which is a 6 order of magnitude increase over that predicted by thermodynamics for conventional thermal processing performed at 735 K.

The steady state production of water after 11 minutes shown in Figure 3 is assumed to be due to the formation of a non-stoichiometric oxide, TiO_{2-X}. This oxide has been shown to form by direct heating of TiO_2 in diatomic hydrogen [8]. The production of

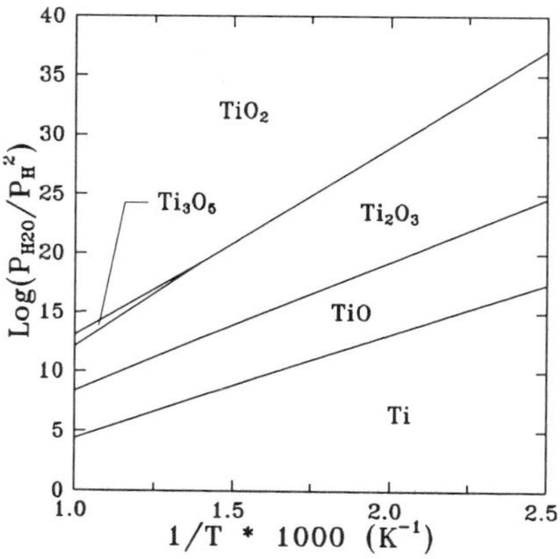

Figure 2. The stability diagram for the Ti-O system in the presence of monatomic hydrogen.

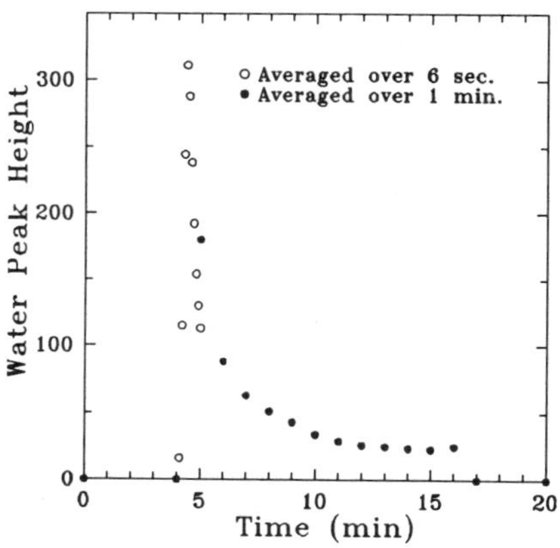

Figure 3. Plot of a typical H_2O signal from the mass spectrometer as a function of time.

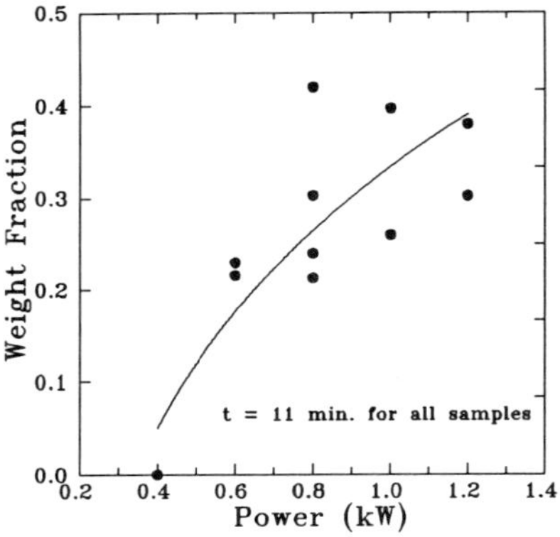

Figure 4. A minimum power is required in order for the conversion of TiO_2 to Ti_2O_3 to occur. All samples reacted at a Pressure of 16 torr.

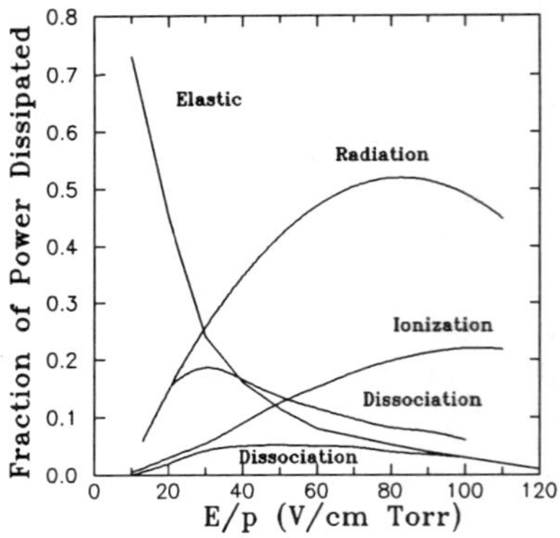

Figure 5. The distribution of power between the various modes of power dissipation in a hydrogen plasma (After [8],[9],[10]).

water below the surface layer where the plasma does not participate in the reaction can cause the re-oxidation of the Ti_2O_3 product layer as the water moves up to the surface. The presence of other higher oxides of titanium (ie. Ti_8O_{15} and Ti_9O_{17}) have been found in the XRD spectra of samples run for long periods of time.

At the same time that the mass spectrometer is showing the large signal for water, the surface of the sample changes from white (TiO_2) to black. This black product layer extended down into the sample to a depth of approximately 1 to 2 mm. Analysis of this surface product was performed by XRD, and showed the surface to be principally a mixture of Ti_2O_3 and residual TiO_2. The XRD spectra for both of these phases showed excellent agreement with the standard spectra. The XRD analysis of the surface also indicated the presence of an amorphous phase, such as TiO_{2-X}, as the weight fractions determined from the quantitative analysis never summed to 1. This would be in agreement with the steady state production of water as mentioned above.

Since a plasma and solid must be in contact for a plasma-solid reaction to take place, the sharp increase, and rapid decrease in the water peak (Figure 3) indicates that the reaction is limited by the amount of surface exposed to the plasma. As the Ti_2O_3 forms, there is less TiO_2 in contact with the plasma, therefore, the plasma must penetrate down into the sample to find TiO_2 to react with. Unfortunately, the solid particles in contact with the plasma become charged and the Debye sheath forms over the sample. The sheath bridges between the particles preventing the plasma from penetrating deep within the sample. This will shut the plasma-solid reaction off below the sample surface.

The influence of the amount of power absorbed by the plasma on the extent of TiO_2 reduction is shown in Figure 4. Quantitative analysis revealed that up to 40% of the sample surface, to a depth of 5 μm, was converted to Ti_2O_3. The data in Figure 4 have a large amount of scatter, but indicate that a minimum amount of power is required in order for the reduction to proceed. There is also a definite trend of increasing amounts of reduction with increasing the amount of power absorbed by the plasma.

The enhancement of reactions in a plasma is closely related to the energy of the free electrons in the plasma. The average kinetic energy of the electrons in a plasma can be approximated as:

$$\overline{\epsilon_e} \propto \frac{E}{p} \tag{1}$$

where E is the electric field strength and p is the pressure. In general, the amount of power absorbed by the plasma is proportional to the electric field strength. Therefore, from Equation 1, increasing the absorbed power or decreasing the pressure of the plasma will increase the energy of the electrons, and thus the ability of the electrons to break bonds during inelastic collisions.

Figure 5 (after [9],[10],[11]) shows the fraction of power dissipated through the various modes (elastic collisions, radiation, ionization, etc.) as a function of the average electron kinetic energy for a hydrogen plasma. It can be seen that for low electron energies elastic collisions predominate whereas at high electron energies the formation of reactive particles (ions and radicals) will consume a large portion of the applied power. With the proper conditions over 20% of the total power dissipated in the plasma can produce monatomic radicals.

Figure 6 indicates that the reduction proceeds better at the extremes of pressure. A minimum amount of transformation occurs at 9 torr. This does not agree with Figure 5, which shows that a maximum number of reactive species will be produced at some "ideal" pressure ($E/p \approx 30$ V/cm torr). Increasing the pressure from this "ideal" point (at constant field strength) would increase the amount of power lost through elastic collisions at the expense of the production of reactive species. This would result in heating of the sample, due to the increased number of elastic collisions, and enhance the reaction kinetics. On the other hand, decreasing the pressure from the "ideal" pressure would increase the amount of power lost through radiation and ionization while decreasing the amount of dissociation in the plasma. Therefore, a decrease in pressure should result in a decrease in the extent of reaction observed. This is the opposite of what is observed.

The increase in the amount of reduction at lower pressures as shown in Figure 6 may be due to the increased involvement of ions in the surface reaction. Ion bombardment in conjunction with chemical etching has been shown to enhance the etch rates of materials in the electronics industry [12]. This combined effect is known as Reactive Ion Etching (RIE). As the pressure decreases, the fraction of ions (primarily with positive charge) in the system will increase, along with increasing the mean free path of the plasma species. The potential drop across the sheath will also increase sharply with decreasing pressure. This will result in an increased flux of high energy particles to the sample surface. Ions that are drawn down through the Debye sheath can cause ballistic

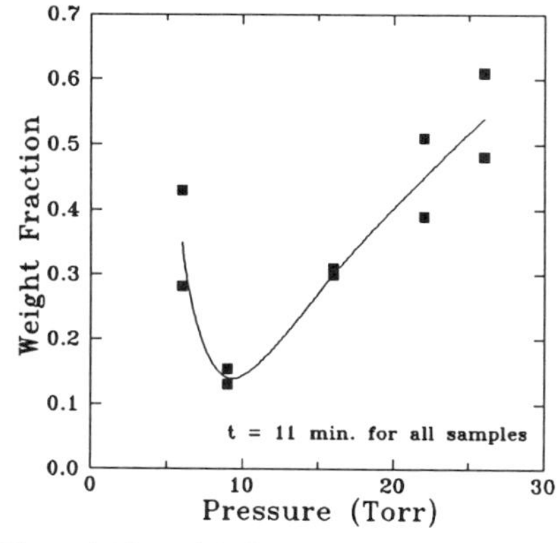

Figure 6. The weight fraction of Ti_2O_3 produced as a function of pressure has a minimum. All samples run at 0.8 kW of absorbed power.

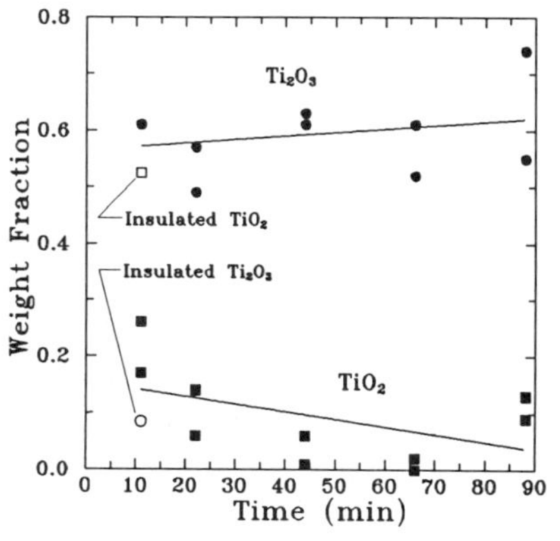

Figure 7. The extent of reduction of TiO_2 as a function of time. All samples were run at 0.8 kW and 26 Torr.

damage to the sample surface. This damage can result in broken and dangling bonds which can react easily with any radicals that are on or near the surface. This increased involvement of ions in the surface reaction could therefore help compensate for the decrease in the amount of radicals produced at lower pressures, resulting in an increase in the extent of reduction.

The effect of time on the transformation is shown in Figure 7. This data gives an indication of the repeatability of these experiments, as each data point represents a new sample. Both the weight fraction of TiO_2 and Ti_2O_3 are shown in this figure. The amount of Ti_2O_3 produced as a function of time shows a slight increase with time, while the weight fraction of TiO_2 decreases with time. This is consistent with the mass spectrometer data which shows the production of H_2O from the surface reduction of TiO_2 is over in approximately 5 minutes of reaction time. The continued depletion in the amount of TiO_2 with time is due to the production of TiO_{2-X}. Again, it should also be noted that the sum of the volume fractions do not add up to 1 for any give time, the difference being associated with TiO_{2-X}.

There is also a single experiment at 11 minutes in Figure 7 (open symbols) comparing the effect of having the support rod and sample holder grounded and insulating. Using the insulating support system appears to inhibit the reduction of TiO_2. XRD analysis shows that only 8% of the TiO_2 was reduced to Ti_2O_3 while the major portion of the sample is residual TiO_2 (56%). The amount of TiO_{2-X} that formed is approximately the same as that produced (in the same reaction time) with the ground in the plasma.

The reason for increased extent of reaction with the ground present is not well understood at this time. One possibility is that the presence of the ground increases the induced voltage on the sample (ie. makes the floating potential less negative). This would effectively decrease the voltage drop across the Debye sheath, increasing the number of electrons with sufficient energy to reach the sample surface. This increase in electrons reaching the surface would result in the generation of more reactive species near the surface by inelastic collisions with neutral molecules that have passed through the sheath region. These particles will be generated close enough to the surface to react with the surface before being removed from the system (ie. recombination with other gas species or free electrons).

The idea of modifying the voltage drop across the Debye sheath was taken one step further. An external DC bias (±100 V, 5 mA max. current) was applied to the plasma in order to try to increase the extent of the reduction by increasing the flux of charged particles to the sample surface. However, XRD analysis of the sample showed no increase in the extent of reaction over the samples run with only the ground maintained in the chamber.

The ineffectiveness of the DC bias to improve the extent of reaction could be due to one of the following reasons. First, neutrals (ie. radicals) may be the predominant species involved in the surface reaction for these conditions, as indicated by Flamm and Herb[2]. These neutral species would not be affected by the voltage gradient developed across the Debye sheath. Therefore, applying the DC bias to the system would not show any increase in the extent of the surface reaction. A second possibility is that the reaction times investigated were too long. The Figure 3 shows the reaction is effectively over after the first five minutes of plasma-solid contact. Increasing the charged particle flux to the surface and increasing the amount of damage on the surface would make the reaction proceed faster. This would not necessarily make the reaction proceed to a larger extent. Figure 7 shows that the extent of reaction is not strongly time dependent. Shorter periods of contact times between the plasma and the solid are required to investigate exactly what is occurring at the sample surface. Shorter times have not been investigated at the time of this manuscript.

Conclusions

The experimental data indicate that plasma processing of refractory oxides offers advantages over conventional thermal processing (ie. increased conversion at lower temperatures). Powdered TiO_2 samples have been reacted in a hydrogen plasma show a 60% conversion of the TiO_2 to Ti_2O_3. This conversion has been achieved with sample temperatures below 735 K. The reduction to Ti_2O_3 has been enhanced by maintaining a ground in the plasma. However, from the above discussion it is apparent that more work is required to understand exactly what is occurring at the plasma-surface interface when the ground is present. The application of the ground to the plasma may decrease the voltage gradient across the Debye sheath. This will allow more electrons to reach the sample surface to generate reactive particles through inelastic collisions.

Acknowledgements

The authors are grateful to the U.S. Bureau of Mines' Pyrometallurgy Center at the University of Missouri at Rolla (contract number MU-USDI-G1175129 and MU-USDI-G1105129) for financial support. In addition the authors would like to express their appreciation to the University of Arizona's Space Engineering Research Center, a NASA sponsored program under contract number NAGW-1332, for financial support for the equipment used during this research. One of the authors, DEB, would like to acknowledge financial support from the U.S. Department of Interior's Mineral Institutes program administered by the Bureau of Mines under allotment grant number G1164104.

References

1. J. Feinman and D. R. MacRae, "*Plasma Technology in Process Metallurgy*," Proceedings of the Elliot Symposium on Chemical Process Metallurgy, ed. P. J. Koros and G. R. Pierre, Iron and Steel Soc., (The Iron and Steel Society, Warrendale, PA, 1991).

2. D. L Flamm and G. K. Herb, "*Plasma Etching Technology - An Overview*," Plasma Etching an Introduction, ed. D. M. Manos and D. L . Flamm, (Academic Press, San Diego, California, 1989), 1-89.

3. G. S. Oehrlein, *Reactive Ion Etching*, Handbook of Plasma Processing Technology: Fundamentals, Etching, Deposition, and Surface Interactions, ed. S. M. Rossnagal, J. J. Cuomo, and W. D. Westwood (Noyes Publications, New Jersey, 1990), p. 196.

4. A. T. Bell, "An Introduction to Plasma Processing," Solid State Tech., April 1978, 89-94.

5. Plasma Etching an Introduction, ed. D. M. Manos and D. L. Flamm (Academic Press, San Diego, 1989), 1-183.

6. Handbook of Plasma Processing Technology: Fundamentals, Etching, Deposition and Surface Interactions, ed. S.M. Rossnagal, J.J. Cuomo, W.D. Westwood (Noyes Publications, New Jersey, 1990).

7. B. D. Cullity, Elements of X-Ray Diffraction, 2nd ed., (Addison - Wesely, Massachusetts (1978), p. 409-411.

8. F. Shadman and Y. Zhao, NASA Space Engineering Research Center 1990-91 Annual Progress Report, (University of Arizona, Tucson, AZ, 1991) p. 1A-26.

9. R.W. Lunt and C.A. Meek, Proc. Roy. Soc. A, 157, 146 (1936).

10. K.G. Emeleus, R.W. Lunt and C.A. Meek, Proc. Roy. Soc. A, 156, 394 (1936).

11. A.T. Bell, Chemical Reactions in Radiofrequency Discharges. The Oxidation of Hydrogen Chloride., Sc.D. Thesis, Massachusetts Institute of Technology, Cambridge, Massachusetts, 1967.

12. J. W. Coburn and H. F. Winters, J. Vac. Sci. Technol., 16, 391 (1979).

UNIFORM LOW-TEMPERATURE DEPOSITION OF OPTICAL LAYERS BY PLASMA-ACTIVATED CVD

Hans Lydtin and Arnd Ritz
Philips Research Laboratories,
P.O. Box 1980, D-5100 Aachen, Germany

ABSTRACT

Plasma activated CVD in a remote operation mode is applied to the deposition of dielectric layers on extended substrates. Layer thickness uniformity of ±1% over areas of 80mm diameter is demonstrated. The microwave applicator sustaining the plasma source of condensable species is compared in its deposition characteristic to conventional evaporation sources. Deposition efficiencies up to 70% are reached. The layer materials used are SiO_2 and TiO_2. Multilayer structures are prepared and optically characterized.

INTRODUCTION

Thin film coatings on flat or slightly curved substrates are desirable in many fields of applications. Examples are optical layers reducing annoying reflections from display screens, colour filters for projection-TV tubes or (semi-)conducting, dielectric, and passivation layers for large area electronics in novel display systems.

Many of these applications demand very high accuracies in layer thickness uniformity coupled with processing of large substrate areas. Furthermore the substrates are often sensitive to high temperatures and set limits to the process.

The plasma-activated chemical vapour deposition (PCVD) process [1] seems to be a very suitable technique to solve this problem. The plasma excitation obviates elevated substrate temperatures in order to start reactions. In particular techniques with a separation of the activating plasma and the substrate surface [2,3] enable the processing of sensitive materials, because the surface is subject to thermal and radiation loads only to a minor degree.

Our objective is the development of a deposition technique that complies with these needs and is capable of uniformly coating extended substrate areas at low processing temperatures.

First results for the preparation of SiO_2 and TiO_2 layers are reported in this paper.

EXPERIMENTAL

The concept for coating larger areas described here utilizes a localized deposition zone with dimensions small compared to those of the substrate which is scanned over the whole substrate. This method allows adjustment of temperature and adatom energy in the growing layer to a certain degree by the choice of plasma power and residence time of the deposition zone on the surface.

The layer materials prepared are SiO_2 and TiO_2 as low and high refractive index materials, respectively. The overall reaction chemistry for the layer formation is as follows:

$$SiCl_4 + O_2 \rightarrow SiO_2 + 2Cl_2 \quad \text{and}$$
$$TiCl_4 + O_2 \rightarrow TiO_2 + 2Cl_2.$$

The experimental set-up is shown schematically in fig. 1. The plasma for the activation of reactions is sustained in the cavity of a microwave applicator (2.45 GHz). The mixture of precursor gases $SiCl_4$ or $TiCl_4$, O_2 and Ar delivered from a gas supply system is fed into the discharge region. After passing the plasma the gases flow through an aperture of the cavity into the substrate chamber. Due to the pressure gradient between plasma region and substrate chamber a gas beam directed to the substrate evolves. Upon reaching the surface the condensable species of the gas beam are built into the growing layer. Process parameters typically used in the experiments are reactor pressure 1 - 10 mbar, total gas flow 3 - 200 sccm, plasma power 2 - 100 W and substrate temperatures from room temperature (unheated substrate) to 260°C, as measured with a thermocouple.

This method is a remote-PCVD technique in the sense that the plasma region is separate from the growing surface. Here all the gases are fed into the plasma to have a high activation efficiency and a well defined deposition gas beam. Some effort has to be given to reduce unwanted deposition on the walls of the plasma reactor itself.

During deposition the gas beam results in a relative narrow deposition zone. To realize a coating on the whole substrate surface, the substrate holder is equipped with a scanning mechanism enabling movement of the substrate relative to the deposition zone in three dimensions.

The substrates used are slices of quartz glass, float glass or plastics with diameters up to 100mm. The substrate surfaces were cleaned in alcohol and subjected to an in-situ remote plasma clean prior to deposition.

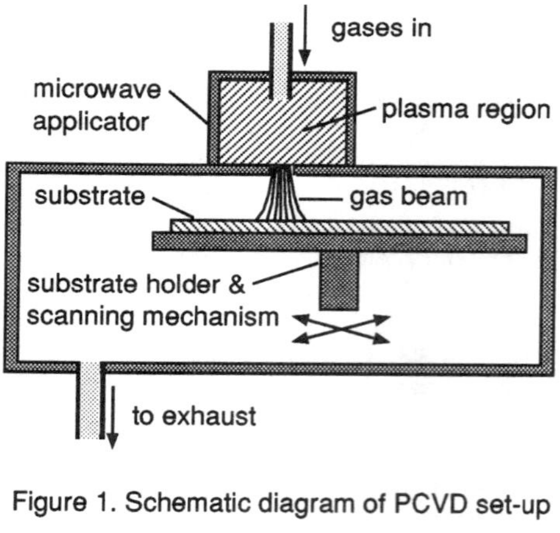

Figure 1. Schematic diagram of PCVD set-up

RESULTS AND DISCUSSION

Properties of the plasma effusion cell

The microwave cavity in combination with the gas inlet and aperture to the substrate chamber comprise a deposition source for thin film formation, referred to in the following as plasma effusion cell (PEC). The deposition characteristic of this PEC has been examined in some detail. With a stationary substrate vis-á-vis at some distance from the aperture a deposition thickness profile as that given in fig. 2 is found. The static profiles have been mechanically measured using a Talysurf 5-120. Under the center of the aperture the maximum of the profile is observed with the layer tapering off to the sides. The profile can be made very narrow with a full width at half maximum (FWHM) of less than 6mm, depending on the deposition geometry, e.g. the distance between aperture and substrate. The form of the profile is rotationally highly symmetric and is the same for the deposition of different materials as seen in the figure for a comparison of SiO_2 and TiO_2 layers.

A deposition source with a directional intensity distribution of effused particles of $\cos^n\alpha$, where α is the angle made by the direction of observation with the normal to the aperture of the source, generates on a flat substrate, placed at a distance h above and oriented parallel to that aperture, a layer profile according to the formula:

$$d(r) / d_o = [\, 1 + (r/h)^2]^{-(n+3)/2} \qquad (1)$$

where r is the distance from the point of the profile maximum on the substrate. A comparison of formula (1) for n=4 and a measured profile for the PEC is given in fig. 3. It should be noted here, that in this comparison no further free parameter for adjustment exists. From the high degree of congruence it can be concluded that the PEC has a deposition characteristic analogous to conventional vacuum deposition sources, such as e-beam evaporators, which mostly show n-values of 3 to 5 [4].

The deposition efficiency, i.e. the portion of e.g. the Ti-atoms found in the layer on the substrate as compared to the number of Ti-atoms brought into the plasma zone by the gas supply, is a complicated function of the geometry of the plasma chamber, the microwave power applied, the amount of total gas flow, and the gas phase composition. Some trends have been found in the experiments, but much of the dependancies are, as yet, unexplained. Deposition efficiencies up to 70% have been reached for TiO_2-layers.

Deposition on extended substrates

With the deposition characteristic of the PEC known, we tried to prepare uniform coatings on extended flat substrates by a suitable movement of the substrate relative to the deposition source. The thickness profile for the coating of a 100mm diameter substrate with a TiO_2 layer as measured by ellipsometry is given in fig. 4. At the edges of the substrate there are taper effects due to the fact that our deposition setup prevents movement of the deposition zone beyond a radius of 50mm, the edge of this substrate. So a taper region of about 10mm (one half of the deposition profile from fig. 2) is to be expected. Within this a substrate area of about 80mm diameter with a layer thickness uniformity of about ±1% is found.

Properties of the layer materials

Most of the experiments were done with TiO_2 layers because of the high difference in refractive index as compared to the substrate materials. This makes visualization and optical measurement of the layer thickness profiles much more convenient.

A Rutherford backscattering (RBS) spectrum of an TiO_2 layer on a quartz substrate is given in fig. 5. Besides the measured (scattering) curve a simulation for a TiO_2 layer on quartz is included (smooth curve). Also indicated are the positions on the energy scale where the signals for the respective elements are to be expected provided that they are present at the surface. The first one is a peak from Ti starting at the energy corresponding to a surface layer. The simulated spectrum assumes stoichiometric TiO_2 and fits very well to the measurement in peak hight and width. This indicates that within the accuracy limits of the RBS-method the layer material is really TiO_2 and not a suboxide. The Si peak is shifted to lower energies due to the coverage by the TiO_2 layer. The O signal starts at the surface position due to the oxygen in the thin film. An important point is that the Cl contained in the precursor material is not found back in the layer. No signal is seen in the spectrum and a more detailed inspection of the measurement revealed no chlorine in the layer within the detection limit of the method (2×10^{15} at/cm²).

The crystallographic modification of TiO_2 found here under the indicated conditions of plasma deposition is anatase, a modification stable at temperatures below 650°C. In fig. 6 a measured x-ray diffraction spectrum of a TiO_2 layer is given along with the positions and relative intensities of lines of powder specimen of anatase, indicated by small bars. From a comparison it can be concluded that the layers are of anatase with no significant preferential orientation.

An important property of the layers with respect to their use in optical applications is the index of refraction. For the TiO_2 layers the index of refraction is given in fig. 7 in the wavelength range of 300 to 1500 nm, as deduced from optical transmittance measurements. A comparison is made between films prepared with the remote-PCVD technique (full line) and films prepared with a PCVD technique where the substrate is immersed in the plasma and at temperatures of 300 to 600°C [5]. As seen from the figure the immersing technique gives indices very close to those of single crystal data, where the layers reach near theoretical densities. In contrast the layers from the remote-PCVD method reported here have indices roughly 0.3 lower, indicating the formation of less dense material.

The microstructure and accordingly the density and refractive index of a deposited layer is determined by the energies available to the species building up the layer to increase their mobility on the surface. The experiments reported here with the PEC up to now have mostly been done at very low microwave powers in the range of several watts and substrates at low temperatures of less than 260°C. Applying higher microwave powers is expected to deliver more energy to the growing layer resulting in a high effective temperature of this layer as seen by the incoming species. The bulk of the substrate will be nearly unaffected according to its heat capacity and the short residence time in the deposition zone due to the substrate movement. So this technique is believed to enable the preparation of films with a range of refractive indices by proper adjustment of the deposition parameters. More experiments will be carried out to examine these aspects.

The feasibility of low temperature processing with this technique is shown by the successful deposition of TiO_2 layers on unheated plastic substrates.

Preparation of multilayer stacks

The experiments reported so far have shown that it is possible to prepare SiO_2 and TiO_2 layers and to achieve a good layer uniformity over a certain area of a planar substrate. A first application of this technique is the preparation of optical multilayer interference filters. These consist of a number of dielectric films with little or no absorption and have alternate high and low refractive indices. The thickness of each film and the number of films can be choosen such that a wide variety of transmittance spectra for different applications can be generated.

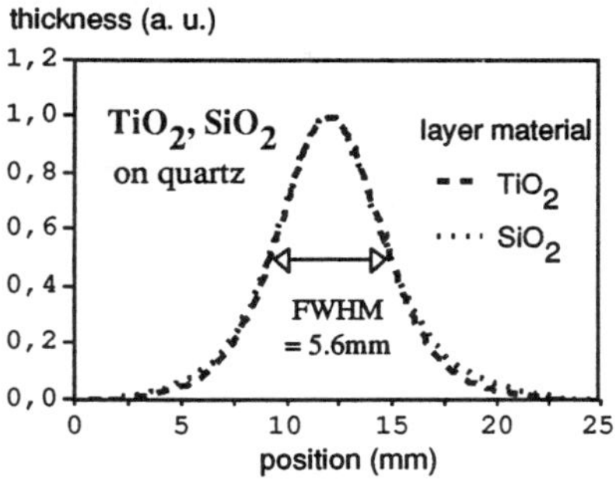

Figure 2. Static deposition profiles of SiO_2 and TiO_2 layers.

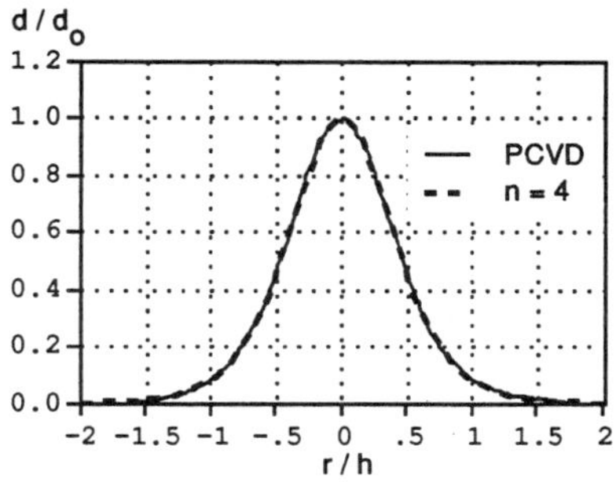

Figure 3. Comparison of static deposition profile from PCVD experiment with theoretical $\cos^n\alpha$ characteristic.

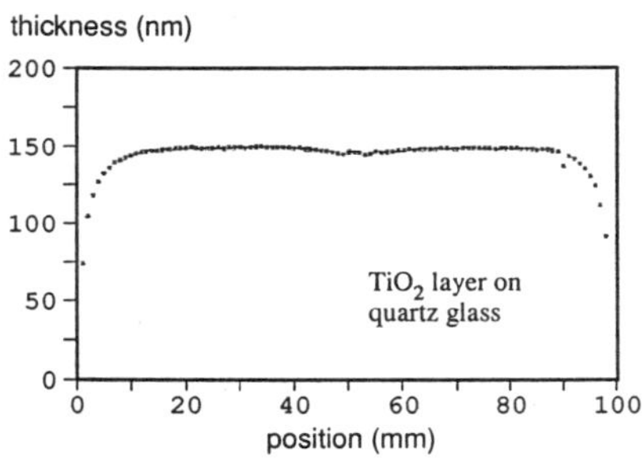

Figure 4. Linescan of the thickness profile of a TiO_2 film on a substrate of 100mm diameter.

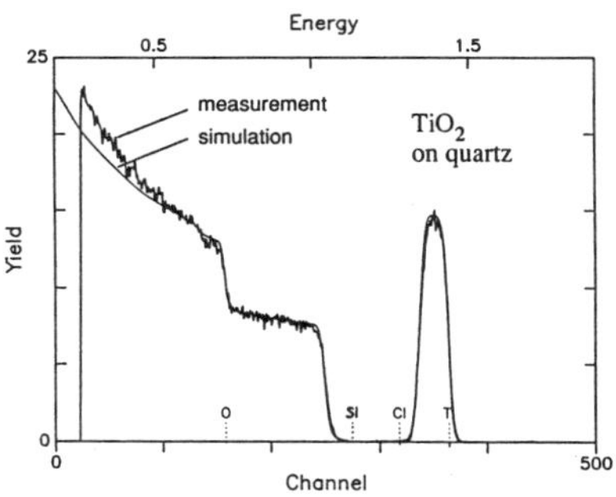

Figure 5. RBS spectrum of a TiO_2 film on a quartz glass substrate.

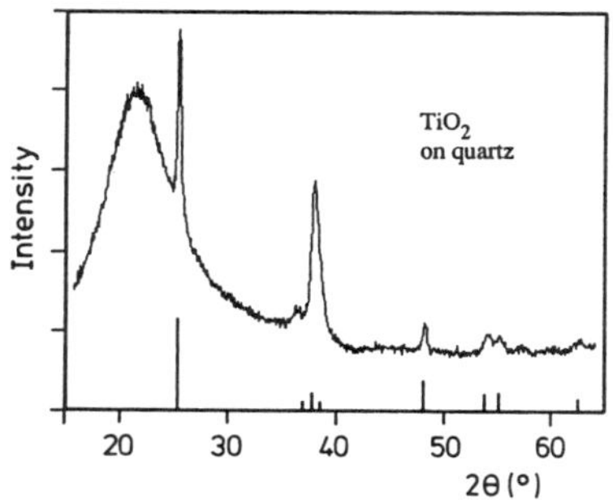

Figure 6. X-ray diffraction spectrum of a TiO_2 film on a quartz glass substrate.

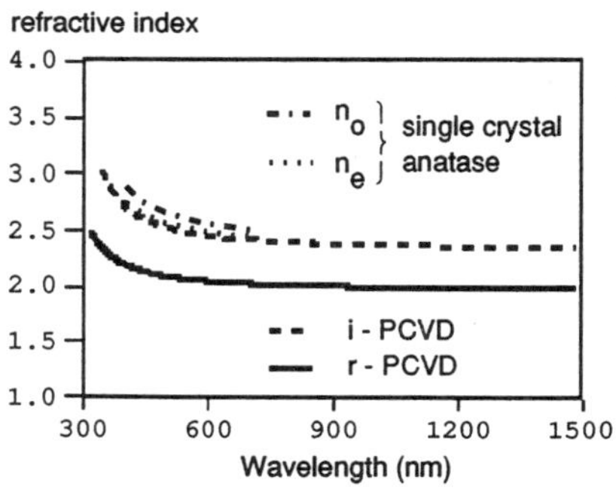

Figure 7. Refractive index of TiO_2 prepared by remote (r) and immersed (i) PCVD and of single crystal anatase (from ref. 6).

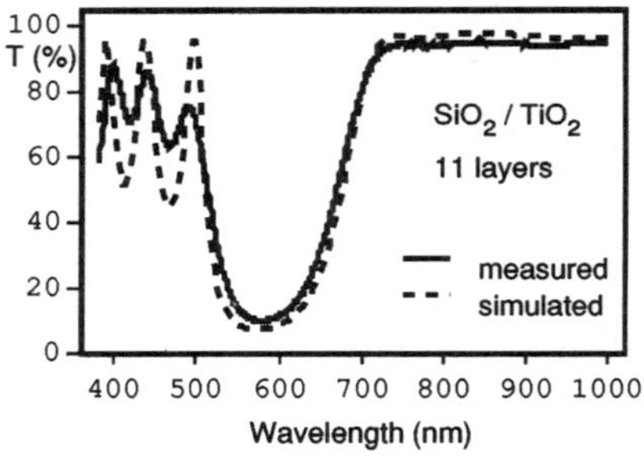

Figure 8. Transmittance spectrum of an 11-layer filter compared to theory.

For the experiment a simple eleven layer design of sequence H/2 $(LH)^4$ L H/2 S is selected, where S refers to the substrate, H and L to the high and low refractive index $\lambda/4$ layers, respectively. H/2 denotes a $\lambda/8$ layer. The total stack thickness is about 1µm. In fig. 8 the measured transmittance spectrum (full line) is given in comparison with the calculated transmittance expected for the given design (dashed line). As seen in the figure it is possible to realize multilayer stacks with this deposition technique. The filter prepared is a simple cold light mirror, reflecting part of the visible range of the wavelength spectrum and transmitting the heat radiation.

CONCLUSION

The combination of the plasma effusion cell and the substrate manipulation system described here has been shown to be suited for the deposition of uniform thin films on extended substrate areas. Thickness uniformities of about ±1% are achieved over areas of 80mm diameter. The plasma effusion cell shows a deposition characteristic analogous to conventional evaporation sources.

With the PCVD technique described, low temperature processing is possible, enabling even the coating of plastic substrates.

The preparation of multilayer stacks has been demonstrated for an interference filter based on SiO_2 and TiO_2.

Matters of concern for further work are process reproducibility, process parameter optimization, deposition rate and enlargement of possible substrate areas.

ACKNOWLEDGEMENTS

The authors gratefully acknowledge the assistance of M. Huppertz during the course of this study. We wish to thank E. Maburgh for the RBS measurement, H.D. Bausen for the x-ray diffraction and D.U. Wiechert for the ellipsometric measurement.

REFERENCES

1. P. Geittner, J. El. Chem. Soc., 84-86, 479(1984).
2. L.G. Meiners, J. Vac. Sci. Technol., 21, 655(1982).
3. P.D. Richard, R.J. Markunas, G. Lucovsky, G.G. Fountain, A.N. Mansour, and D.V. Tsu, J. Vac. Sci. Technol., A3, 867(1985).
4. G. Deppisch, Vakuum-Technik, 30, 67(1981).
5. H. Lydtin and A. Ritz, Rivista della Staz. Sper. Vetro, 1, 53(1989).
6. S. Koritnig, in Landolt-Börnstein "Zahlenwerte und Funktionen", 6th ed., edited by J. Bartels and H. Borchers (Springer-Verlag, Berlin, 1962), II/8, pp. 43-464.

CHARACTERIZATION OF AN ECR MICROWAVE PLASMA ETCHING SYSTEM

MING JIN AND KWAN C. KAO

Department of Electrical and Computer Engineering, University of Manitoba, Winnipeg, Manitoba, CANADA R3T 2N2.

ABSTRACT

An ECR microwave plasma etching system and an alternating bias voltage source with variable frequencies (10kHz to 10 MHz) have been built for the study of the effects of the substrate bias on the submicron feature etching. A quadrupole mass spectrometer was used for the end–point monitoring and the plasma chemistry analysis. A Langmuir probe was used to characterize the etching plasma . The system can generate plasmas with densities ranging from $1x10^{15}m^{-3}$ to $5x10^{16}m^{-3}$ corresponding to gas pressures from $1x10^{-4}$ torr to $6x10^{-3}$torr at a fairly low microwave power. This condition results in a stable electron temperature. The magnetic field profile can be adjusted to obtain a constant plasma density of $2.8x10^{16}m^{-3}$ and an electron temperature in kT_e of 4 eV to 5 eV. The plasma beam is of 75mm in diameter with the ion current intensity fluctuation of about ±5% at the etching position for the gas pressures ranging from $4x10^{-4}$ torr to $3x10^{-3}$ torr at an absorbed power of 70Watts. The plasma density, the electron temperature, the breakdown power, and the sustaining power have been studied as functions of gas pressure in order to optimize the etching performance.

INTRODUCTION

Electron cyclotron resonance (ECR) microwave plasma technique has been successfully used in the semiconductor industry for processing of thin films, especially, for deposition and pattern etching[1–3]. ECR plasma technique can generate plasma with high density and low energy ions at low pressures . The additional bias technique can be used to control the ion bombardment independently[4], thus making it easy to control the radiation damage to the devices . In this paper, we present the characterizations of the ECR microwave plasma system used for submicron feature etching.

EXPERIMENTAL

The ECR microwave plasma etching system and the relevant diagnostic tools are shown schematically in Fig.1a. The microwave power is fed into the plasma chamber through a quartz microwave window. The plasma chamber is a stainless–steel cylindrical chamber with 10cm in diameter and 23cm in length. These dimensions are used in order to excite the TE^{o}_{11} mode. Two electromagnets provide the magnetic field profile required for the ECR condition with the highest plasma density at the etching table surface position. The magnetic field profile along the axis of the plasma chamber is shown in Fig.1b. The process chamber is pumped to a base pressure of less than

$1x10^{-6}$torr. The etching table is connected to a bias voltage source with variable frequencies ranging from 10KHz to 10 MHz.

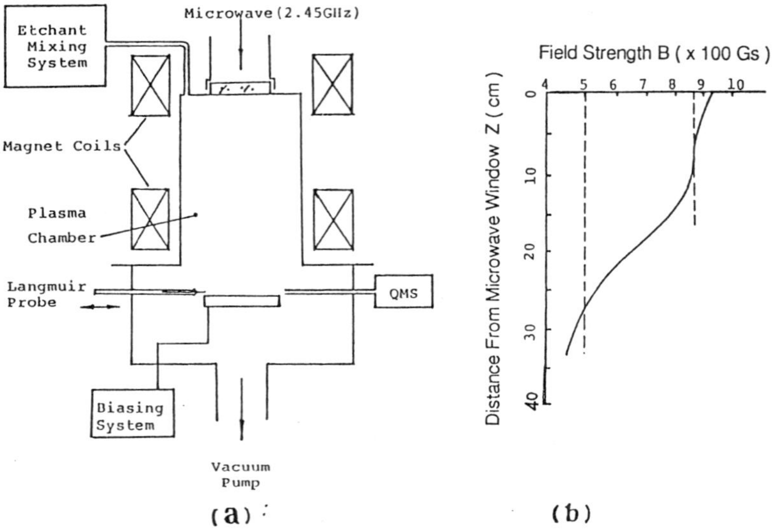

Fig.1 a) The schematic diagram of the ECR microwave plasma etching system; b) the magnetic field strength profile along the z-axis of the plasma chamber.

A Langmuir probe was used to measure the plasma parameters. The probe was located at a distance of 3cm below the ECR plasma chamber and could measure the parameters at a radial distance of 5cm from the center. The resulting I–V characteristics of the probe were fed to a data acquisition system and stored for further analysis.

A quadrupole mass spectrometer was used to measure the partial pressures in the region close to the edge of the etching table in order to achieve the highest sensitivity. It was also used for end–point monitoring the etching process.

RESULTS AND DISCUSSION

The axial profile of the magnetic field with the currents of 270A in the first coil and 180A in the second coil is shown in Fig.1b. This combination of coil currents was chosen for producing the maximum plasma density at the etching table position . The plasma density and the electron temperature as functions of nitrogen gas pressure were measured using the Langmuir probe. The results are shown in Figs.2 and 3. The pressure was varied by adjusting the nitrogen flow rate from 1 sccm to 12 sccm. The microwave plasma power was set at 70watts. The plasma density increases with increasing pressure, but levels off at $2.8 \times 10^{16} m^{-3}$ for the pressures ranging from 3.0×10^{-4} to 1.5×10^{-3} torr. The electron temperature has a fairly constant value of 4 eV to 5 eV in the same pressure region. In this pressure region the plasma is very stable

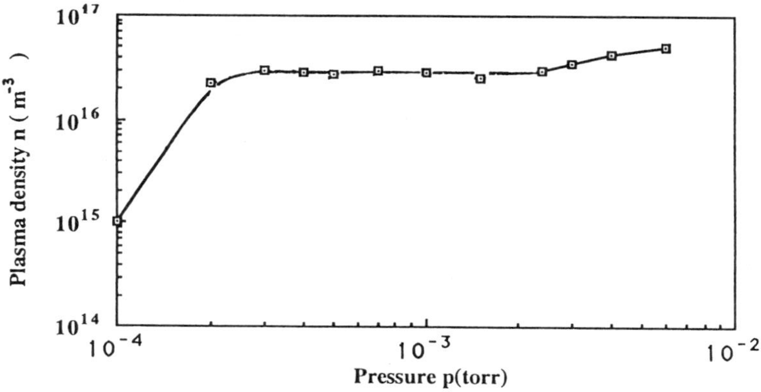

Fig.2 Plasma density as a function of nitrogen pressure

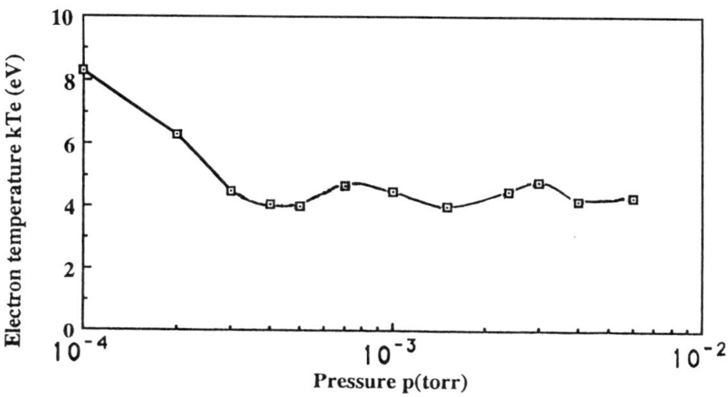

Fig.3 Electron temperature as a function of nitrogen pressure

The plasma density and the electron temperature as functions of applied microwave power are shown in Figs.4 and 5. The results were obtained at the pressure of 4.0×10^{-4} torr. For the microwave power ranging from 10 to 100 watts, the plasma density increases with the power. The electron temperature increases for the power less than about 30watts and then decreases for the power greater than 50watts. This dependance may indicate that the increase in microwave power mainly contributes to the increase in plasma density for P > 50watts.

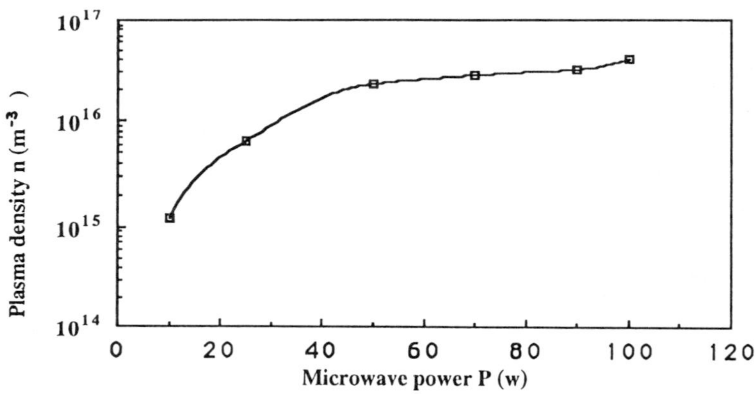

Fig.4 Plasma density as a function of microwave power for the pressure of 4×10^{-4} torr

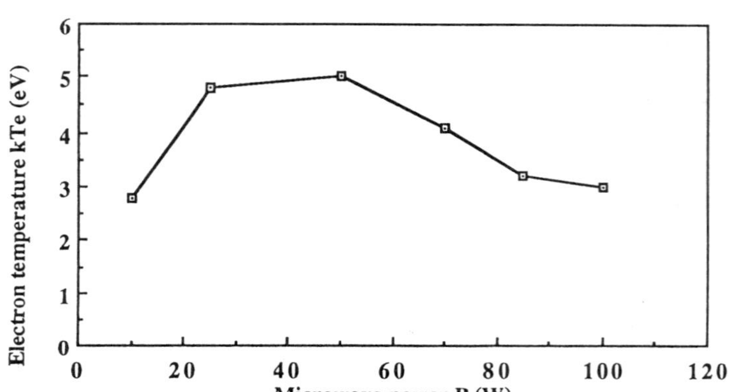

Fig.5 Electron temperature as a function of microwave power for the pressure of 4×10^{-4} torr

The uniformity of the plasma beam at the etching table position located at 3 cm below the ECR plasma chamber was determined by measuring the ion saturation current density at P = 70 watts and p = 4.0×10^{-4} torr. The result is shown in Fig.6 . In a radial range of 3.75cm, the ion current fluctuation is $\pm 5\%$. This uniformity of ion current is just enough for the processing wafers with diameter of 7.5cm.

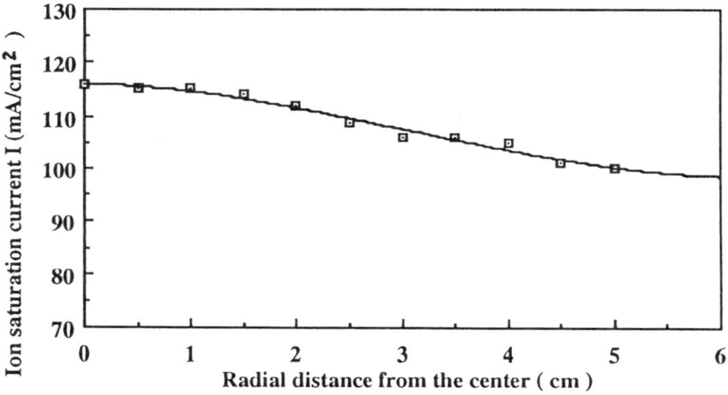

Fig.6 Radial profile of ion saturation current

Figure 7 shows the breakdown power and the sustaining power as functions of gas pressure. The breakdown microwave powers for p > 3.0×10^{-4} torr are less than 20 watts. The minimum sustaining microwave powers are less than 5 watts. The above data indicate that the plasma chamber as a load of the microwave power supply was matched very well.

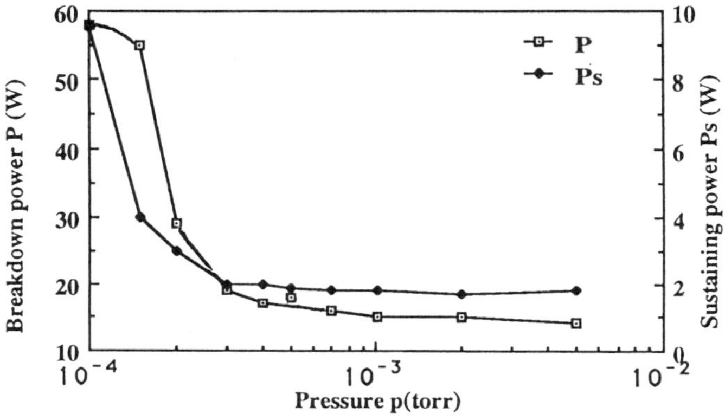

Fig.7 Breakdown and sustaining powers as functions of pressure

To characterize the ECR microwave plasma etching system, we have carried out the etching of Si_3N_4 films using CF_4+5%O_2 etching gas at $p = 4.0 \times 10^{-4}$ torr and P = 70 watts. The bias voltage chosen was 0–20 V at 10 MHz. The etching rate increases with the increasing bias voltage. The etching rate at $V_a = 20$ V was about 500A/min with a uniformity of better than ±5% variation in etching rate across a 3–in wafer. The details of this work has also been presented at the 1992 MRS Spring Meeting[4].

CONCLUSION

The ECR microwave plasma etching system can generate a plasma with a density of $2.8 \times 10^{16} m^{-3}$ and electron temperature of 4–5eV at 70 watts and at pressures ranging from 3.0×10^{-4} to 1.5×10^{-3} torr. The plasma is very stable. The low electron temperatures may result in a low radiation damage to the devices during etching. The plasma beam is uniform, which results in a uniform etching rate for Si_3N_4 films deposited on 3–in silicon wafers.

ACKNOWLEDGMENT

We thank the National Sciences and Engineering Research Council of Canada (NSERC) for supporting this research.

REFERENCES

1. H.H.Sawin, Solid State Technology, March issue, 47 (1991) .
2. S.Samukawa, M.Sasaki, Y.Suzuki, and S.Mori, Digest of Technical Papers–Symposium on VLSI Technology, (IEEE Electron Devices Society, The Japan Society of Applied Physics, Honolulu, Hawaii, 1991) , p.1 .
3. T.D.Mantei and S.Dhole, J. Vac. Sci. Technol. B9 , 26 (1991) .
4. M.Jin and K.C.Kao, presented at the 1992 MRS Spring Meeting, San Francisco, CA, 1992 (unpublished) .

PART V

Microwave/Materials Interactions

THE EFFECT OF PARTICLE SIZE ON MICROWAVE HEATED CARBON AND THE SUBSEQUENT CRYSTALLITE GROWTH

CLAUDE P. LORENSON, MARK C.L. PATTERSON, GAIL RISTO, ROBERT KIMBER
Alcan International Limited, Kingston Research and Development Centre, P.O. Box 8400, Kingston, Ontario, Canada K7L 5L9

ABSTRACT

The dielectric properties of carbon imply that heating with microwave energy should not be significant at elevated temperatures. It has been demonstrated, however, that temperatures in excess of 1900°C can be achieved with low powers. The effect of particle size on the heating rate will be presented and a heat absorption mechanism is postulated to explain the high temperatures observed.

Under controlled conditions, conventional and microwave heating has been used to observe growth of the carbon crystallite size up to a temperature of 1400°C. Under these conditions, no enhanced crystal growth was observed for microwave heating.

INTRODUCTION

The work presented here summarizes preliminary results to evaluate the potential benefits of microwave processing of carbon. Previous studies [1] showed that the penetration depth for carbon was small and that in theory it would be impossible to volumetrically heat this material to high temperatures. However, in the form of powder, this material exhibits a good microwave absorption. The properties of the treated material and its heating history are partially defined by the length of the crystallite (L_c) as measured by X-Ray diffraction. L_c is some sort of a measure of the heating mechanism. This work summarizes the measurements of the dielectric properties on carbon powder and the growth of L_c as a function of temperature for both conventional and microwave processing.

METHOD AND EXPERIMENTAL SET-UP

Carbon Powder

The powder we used had a large size distribution. In order to have homogeneous material, we have sieved the powder to keep the particles between 103 μm and 150 μm in most of the tests. For exploring the particle size effect on the heating rate, the starting powder was sieved in 5 different fractions.

Dielectric Properties Measurements

To have an idea of the penetration depth of the material we have measured the dielectric properties of the powder with a co-axial probe. The probe is made from an air filled transmission line fitted to a GPC-7 connector. The termination of the probe consists in an impedance matched dielectric window and a ground plane. The probe is connected to a vector network analyzer (Wiltron 360) which has been calibrated in the reflection mode. In using this technique to measure dielectric properties, one makes the assumption that the sample to be measured behaves as a linear system. The relation between ε^* of the material, which is the complex dielectric constant ($\varepsilon^*=\varepsilon'-j\varepsilon''$), and Γ, which is the complex reflection coefficient of the probe can be found in [2] and [3]. Some intrinsic parameters of the probe are inferred from the measurements of liquids with well known dielectric properties like water and certain solvents. This measuring technique is especially well suited for liquid samples and semi-solid samples. For powder samples, the finer the particles, the more accurate are the results. This technique does not distinguish between the different components that make up ε'', which in theory can consist of a dipolar, electronic, atomic, Maxwell-Wagner contribution and a term due to conductivity.

Temperature Measurements

The temperature of a physical process is a critical parameter which is directly related to many kinetic and thermodynamical effects. The ability to measure temperature with accuracy is therefore of great importance. Generally, the measurement of temperature by contact is a result of a volumetric average of the temperature near the measuring probe. Most probes do disturb the thermal field during measurement. Optical pyrometry needs access to the surface of the sample which results in heat loss from the surface being measured. Under normal equilibrium conditions, thermocouples can be adequate measuring devices and are used in many applications. In a microwave environment, the electromagnetic field present can change the behavior of a metallic thermocouple especially at the tip. For this reason, the ideal probe for measuring the temperature in a comparative study on the heating effects in microwave and conventional heating should have the following requirements: relatively non-perturbing in a microwave field, should be as close as possible to region of interest (preferably immersed) and the same technique should be used in both conventional and microwave process. In our studies, we have used a sapphire probe connected to fiber optic from the Accufibre company in identical conditions for both types of processing. K and S types of thermocouples were also used for samples treated conventionally. The thermocouples were not used in the microwave cavity because previous work [4] demonstrated abnormal behavior for this measuring technique.

HEATING OF CARBON

Figure 1 shows the experimental set-up where the Accufibre probe is immersed in a small amount (0.6 g) of carbon powder inside an alumina crucible. Preliminary tests demonstrated that even such a small mass would produce a significant temperature increase. Additional heating effects at high temperature can be caused by the crucible itself, but these are secondary. The powder containing crucible was placed inside another alumina crucible separated by a layer of boron nitride. Boron nitride is almost transparent to microwaves and is also a good thermal conductor, which helps the heating uniformity of the carbon powder. The assembly was then isolated with a layer of "Saffil" fiber (alumina-silicate fiber).

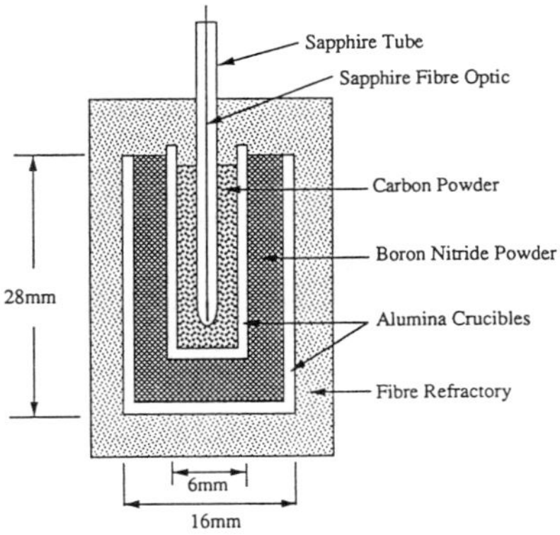

FIGURE 1

Microwave Heating

The sample was placed in the centre of a monomode cavity TM_{010} and was surrounded by more "Saffil" fibers. Microwave energy at 2.45 GHz was supplied by a 1 kW source and the reflected power and the forward power was monitored by directional couplers (Gerling) and crystal detectors manufactured by Hewlett-Packard. The reflected power was minimized at the beginning of each experiment with the use of a screw tuner. A heating rate of 10°C/min (for the heating experiment) was maintained up to slightly below the desired temperature. The processing temperature was reached and held for times of 15, 30, and 45 minutes. Power was turned off and the sample was left in the oven for cooling.

Particle Size Effect

In order to explore the effect of particles sizes on the microwave heating rate, the starting material was sieved into 5 different sizes, 500-850 μm, 300-500 μm, 150-300 μm, 50-150 μm. A similar crucible arrangement, as previously described, was used for these tests with a similar temperature measuring configuration. The heating rates presented here were achieved with samples of constant mass for each size, the volumes were therefore different for each of samples. Two temperature regimes were explored, room temperature to 400°C and 400°C to 1200°C.

Conventional Heating

The crucible is placed in the centre of a small resistance heated furnace ($MoSi_2$). The fiber optic probe used to measure the temperature was placed through the top of the furnace, which is programmed to keep a heating rate of 7°C/min until slightly below the desired temperature. The same holding times were used as with the microwave experiments and the samples were then cooled with the furnace door open.

Analysis of L_c

Once cooled, the carbon powder was crushed to form particles less then 75 μm. The average length of the crystallite (L_c) was measured by X-ray diffraction from peak broadening of the 002 reflection peak. The total width at half-maximum [5] was used to determine L_c from: $L_c=K\lambda/\beta\cos\theta$, where K is a constant (0.89), λ is the wavelength of the incident radiation in A, θ is the Bragg reflection angle, β is the width of the peak in radians.

RESULTS

Dielectric Properties

The reflection coefficient as a function of frequency was measured using the network analyzer between 300 MHz and 3 GHz. The results given in Table I are all at 2.55 GHz, which are the closest to our treatment frequency. The penetration depth (δ) given is calculated by assuming the absorption of a plane wave in a solid material [6]. As the surface of our material is quite active and absorbs moisture, we have also made measurements on the same powders, but after drying at 110°C for one hour.

TABLE I

Sample	ε'	ε''	δ(meter)	σ(siemens/m)
as is	23	65	.002	9
dry	3	.7	.05	.1

With such a small penetration depth, it becomes difficult to understand how we can achieve high temperatures (>1250°C) measured in the centre of a cylindrical sample of 10 cm in diameter and 20 cm in length. Nelson, et al. [7] recently published an article on the relation between the permittivity of a solid material and its particulate form. The air-particle mixture effectively has lower dielectric properties than the solid material, which can explain, in part, the larger penetration depth observed during our experiments. In order to gain a more realistic idea of the penetration of the electric field in the powder, we also measured the absorption coefficient of the field as a function of the thickness of the sample. Rectangular sample holders of different thicknesses made out of polycarbonate were made to fit wr-284 waveguide. The ends of the waveguide were connected to a co-axial/waveguide adaptor and these are connected to the two ports of the network analyzer. A calibration in reflection and transmission mode was necessary to determine S_{11} and S_{12} from which was calculated the percentage of reflected field (R) and transmitted (T) field. The absorption coefficient (A) is calculated from A=100%-T-R. The results presented in Figure 2 were obtained at 2.542 GHz for non-dry samples. Even for a thickness of 3.0 cm only 54% of the field is absorbed, which again helps to clarify the observations made during our heating tests. The powder materials do not behave at all like the bulk, the penetration depth of the field is much larger then expected from the dielectric measurements.

Value of L_c

We can estimate that the temperature reached during microwave heating can be in excess of 1900°C (L_c=84.9 Å). This estimate comes from the formation of Al_3C_4 as a reactive product between the carbon and the alumina crucible. For the tests described previously, L_c was calculated for the samples heated conventionally and by microwaves, and the results are shown in Table II. Figure 3 shows L_c as a function of temperature for

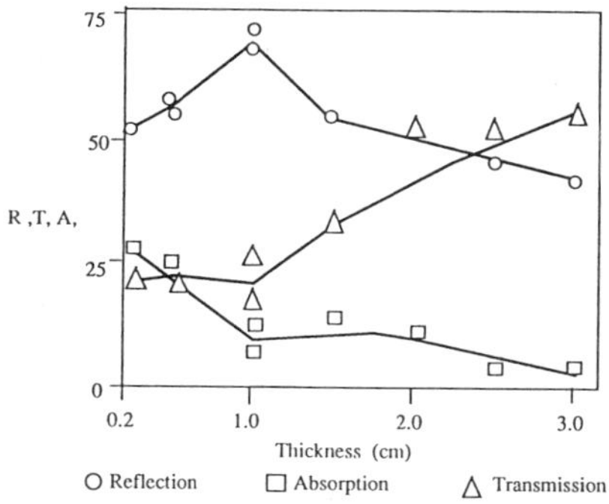

FIGURE 2

samples at a holding time of 45 minutes. In this graph, the temperature for microwave heating was measured with an Accufibre probe and for conventional heating it was measured with thermocouple. A summary of the change in L_c for different holding times is shown in Figure 4. In this graph all the temperatures were measured with an Accufibre probe. We can see that L_c increases with temperature, but that within the experimental error there appears to be no enhanced growth of L_c during microwave processing.

TABLE II: X-RAY DATA FOR CARBON 002 PEAK

45 Minutes at Temperature (Accufibre + Thermocouple)			
Microwaves (Accufibre)		Conventional (Thermocouple)	
Temp. (°C)	L_c (Å)	Temp. (°C)	L_c (Å)
1100	21.34	1100	20.01
1200	27.55	1189	21.86
1300	30.70	1278	24.88
1400	40.88	1390	33.88
Accufibre Data for Temperatures up to 45 Minutes			
Microwaves		Conventional	
	Low - High		
1100	19.64 - 21.34	1125	22.02
1200	21.50 - 27.55	1181	23.31
1300	26.04 - 30.70	1337	36.72
1400	40.88 - 48.31	1391	42.30
		1395	46.85

Particle Size Effect on Heating Rate

Figure 5 shows the heating rate from 400 to 1200°C for 4 different particle sizes for samples of equal mass. The power used in each case was maintained between 90 and 100 Watts. It can be seen that the particle size can have a dramatic effect on the heating rate. In order to explain the shape of this curve, one has to consider the penetration depth (we will call it "d") of the field for an individual particle. This quantity differs from δ in the sense that no neighboring particles are taken into account. If we call the diameter of the particle D, we can understand that for d>D/2, we can expect "complete" heating. As the particle size D is increased the heating rate should be constant to about a particle size of D/2. For d<D/2, the penetration of the field is not complete and since only a "layer" of the

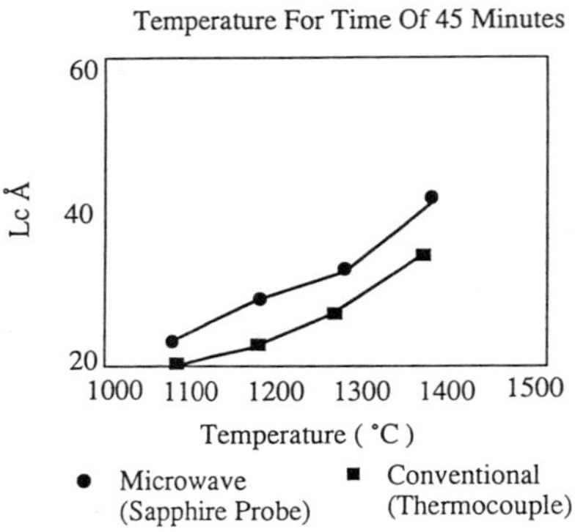

FIGURE 3

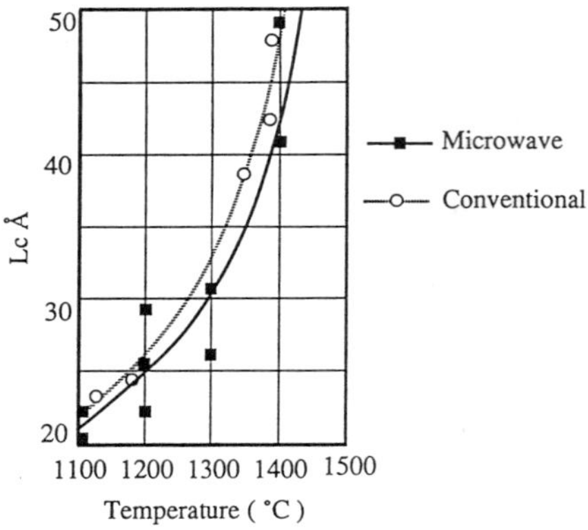

FIGURE 4

material gets heated we can expect a decrease in the heating rate as D increases. This would produce the "heating rate" curve on Figure 5. Another factor to take into account is the fact that the heat loss depends on surface area and is a quadratic function of D, this gives rise to the "heat loss" curve shown on Figure 5. Depending on the magnitude of these two effects, it is possible to construct a curve like the one measured for the heating rate.

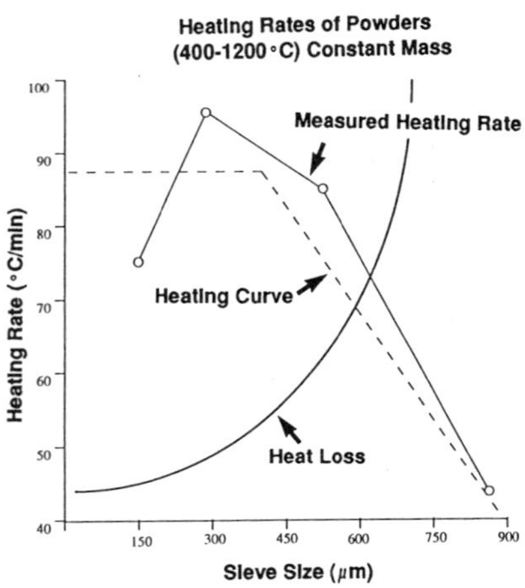

FIGURE 5

DISCUSSION AND CONCLUSION

It was shown that it is possible to use microwave energy at 2.45 GHz to heat carbon to a temperature of over 1900°C with a L_c of 84.9 Å. A penetration depth of only 2 mm is calculated from the measured ε' and ε'' of the powder. In order to explain the high temperatures reached in some samples, we also measured the absorption of the electric field in the powder directly and found that 54% of the field is absorbed in a layer of 3 cm. This is in contradiction with the calculated δ, but emphasizes the fact that we need to consider the porous nature of the material when evaluating the heating efficiency.

The comparison between microwave heating and the conventional heating can be summarized in three points:

- Comparison between different measuring systems for temperature;

- Comparison between two processes using the same probe for temperature;
- Effect of holding time at temperature on L_c.

The data obtained for the samples treated up to 1400°C for 45 minutes used an optical fibre in the microwave cavity and a S type thermocouple in the conventional treatment. This set of data seems to indicate an enhanced growth for L_c at a given temperature for the microwave process. When similar results obtained for holding times of 15 and 30 minutes are compared, this enhanced growth is much less evident. For all data taken with the same measuring device (optical fibre) in both processes, it is clear that within the measuring uncertainty, the growth of L_c is comparable for both techniques. The holding times effect on L_c are weaker than the effect due to treatment temperature.

The heating rate of this material is dependent on particle size and this can be explained by balancing the heating effect due to microwave absorption and the heat loss, which is proportional to surface area.

The main conclusion from this work is the fact that there is no enhanced growth of the carbon crystallite as measured by L_c that can be attributed to the microwave treatment. This implies that absorption mechanism of the microwave energy is not preferential to the crystal boundary (since that would accelerate the growth) and that the heat transfer rate at the absorption site is comparable to the rate at which energy is transferred resulting in only a small temperature difference at the boundary. It seems more probable that the microwave energy is absorbed uniformly by the carbon particle or that the energy is absorbed at the surface of the particle possibly by induced current and that the interior of the powder sample is heated by thermal conduction similarly to conventional heating. This last hypothesis can explain why there is no enhanced growth observed in the microwave treatment because there would not be enough crystallites affected in the whole sample to make a difference in the L_c measured.

REFERENCES

1. Internal communication.

2. Stuchly, M.A., Stuchly, S.S., "Coaxial Line Reflection Method for Measuring Dielectric Properties of Materials at Radio and Microwave Frequencies-A Review", IEEE Trans., Vol. IM-29(3), 176, 1980.

3. Buideve, E.C., Cain, F.L., Seal, J., "Measurement of Radio Frequency Permittivity of Biological Tissues with an Open Ended Coaxial Line: Part 1", IEEE Trans., Microwave Theory Tech., MTT-30, 82, 1980.

4. Patterson, M.C.L., Apte, P.S., Kimber, R.M., and Roy, R., Enersearch Report #600049, "Microwave Heating for Ceramic Joining".

5. Klug, H.P., Alexandre, L.E., "X-Ray Diffraction Procedures for Polycrystalline and Amorphous Materials", John Wiley & Sons, 2nd Ed., pp. 700-704, 1975.

6. Metaxas, A.C., Meredith, R.J., "Industrial Microwave Heating", Peter Peregrimus Ltd., p. 80, 1983

7. Nelson, S., Kraszewoki, A., You, T., "Solid and Particulate Material Permittivity Relationships", Journal of Microwave Power and Electromagnetic Energy, Vol. 26, #1, p. 45, 1991.

STUDIES OF NONTHERMAL EFFECTS DURING INTENSE MICROWAVE HEATING OF CRYSTALLINE SOLIDS

JOHN H. BOOSKE,* REID F. COOPER,** LEON MCCAUGHAN,* SAM FREEMAN,*** AND BINSHEN MENG*
*Electrical and Computer Engineering Dept., University of Wisconsin, Madison, WI 53706
**Materials Science and Engineering Dept., University of Wisconsin, Madison, WI 53706
***Materials Science Program, University of Wisconsin, Madison, WI 53706

ABSTRACT

Reports of enhanced sintering rates associated with microwave heating may be due to nonthermal lattice fluctuation statistics. Recent theoretical analyses reviewed in this paper confirm the feasibility of this phenomenon for a wide variety of situations involving very different microwave absorption mechanisms. For materials with weak microwave absorption coefficients, the effect is expected to be uniformly distributed throughout the volume. For strongly absorbing materials, however, the effect is expected to be concentrated near the material surface, with a characteristic exponential penetration depth of $L_{nt} \sim 10$ - $100\ \mu m$. An "observable" nonthermal effect depends on the relative magnitude of the microwave electric field strength $|E|$ and the lattice ion energy relaxation rate γ with the most pronounced effects occurring for larger values of $|E|$ and smaller values of γ.

INTRODUCTION

A growing body of experimental data [1-7] supports the observation that microwave heating of ionic crystalline solids (e.g., ceramics) leads to enhanced diffusion and solid state reaction rates when compared with conventional heating at the same "temperature." Specific observations include evidence of enhanced sintering rates of alumina powder compacts heated at 28 GHz,[1] different pore evolution for microwave and conventionally fired alumina,[2] enhanced tracer ion diffusion (^{18}O) in microwave heated single crystal sapphire,[3] finer final grain size in microwave sintered ceramics,[4] enhanced diffusion rates across grain boundaries in microwave-annealed dense alumina,[5] evidence of enhanced inter-specie diffusion in microwave heated alumina-chromia solid solution,[6] and evidence of enhanced cation exchange reaction rates in glass surface modification experiments.[7]

Two possible ways in which microwave sintering may macroscopically involve an athermal steady state are: (1) the deposition of microwave energy is spatially nonuniform on microscopic dimensions (i.e., on the scale of grain sizes), or (2) microwave heating induces an athermal distribution of lattice ion energies or lattice phonons. The first possibility was proposed in Ref. 8, but remains in doubt in view of other recent experimental [5] and theoretical [9] studies. The second scenario has been suggested in previous publications.[10,11] More recently, our studies have focused on quantitatively estimating the magnitude of athermal lattice fluctuations under various conditions. This article will summarize the results of these recent investigations.

MICROWAVE ABSORPTION MECHANISMS

Before continuing, we note that analysis of intense microwave field effects on ion fluctuation statistics in crystalline solids require *a priori* models for microwave-solid coupling. We address this by classifying microwave absorption into two basic categories, depending on the type of crystalline material involved:

a) **microwave absorption in insulating ionic crystalline solids.** In the absence of free conduction electrons, microwave absorption in insulating ceramics must occur through direct coupling to lattice normal modes (i.e., photon-to phonon conversion). Of greatest interest are those lattice modes that dominate ionic diffusion, typically having frequencies in the infrared or far-infrared regime. Due to the large disparity in both frequency and phase velocity of these elastic waves and the electromagnetic microwaves, it is not realistic to assume a direct coupling between the microwave fields and a single normal mode (in this high frequency regime). Thus, any significant microwave absorption resulting in crystalline heating of ionic insulators must involve multi-wave processes. An example of this is life-broadened-two-phonon-difference absorption which has been convincingly proposed as the dominant microwave absorption mechanism in alkali halides.[12] It is probable that this mechanism or similar ones are responsible for the majority of microwave absorption in ionic insulating materials.

b) **microwave absorption in conducting and semi-conducting crystalline solids.** At sufficiently high densities of free conduction charges, conventional ohmic dissipation dominates the microwave absorption. The microwave electric field harmonically accelerates the conduction charges and this ordered motion is converted to random "thermal" fluctuations by irreversible collisions with other conduction charges and with the crystal lattice.

The microscopic details of microwave energy absorption depend on the type of material, such as ionic insulator (e.g., alumina, magnesia), covalent semiconductor (e.g., Si), metallic conductor, and possibly whether the material is single or poly-crystalline. Nevertheless, in most materials the microwave absorption is expected to involve either one (or a mixture of both) of the two basic mechanisms described above. In the following sections, we will quantitatively examine the relationships between the various absorption mechanisms and the generation of athermal lattice fluctuation distributions. Here, we will briefly outline the underlying physical principles in the context of different material types.

Ionic Insulators

In the case of coupling to polarized insulators (coupling to ionic dipoles), the lattice phonon distribution is distorted from thermal equilibrium due to the preferential absorption of microwave photons (by a two-phonon difference process) into a narrow band of high frequency transverse optical lattice vibration modes at the edge of the first Brillioun zone.[12] Thermalization only occurs after the relaxation of this "pumped" phonon mode into all of the other possible lattice vibration modes. For single crystal ionic insulators, this microwave absorption mechanism predicts an absorption coefficient that increases with the square of the microwave frequency (in the regime 1-300 GHz). Such a frequency scaling has been experimentally observed in both alkali halides[12] and single crystal sapphire.[13]

Conductors and Semi-conductors

In conductors (e.g., metals) and covalent semi-conductors (e.g., silicon), microwave absorption is an ohmic process dominated by acceleration of the "free" conduction electrons, since these electrons have much less inertia than their corresponding charged ions (i.e., the "holes" in semi-conductors). This results in a nonthermal "high energy tail" on the electron energy distribution (for strong microwave field intensities). These same nonthermal distribution features are transferred to the lattice ions by collisions between the conduction electrons and the lattice. If the conduction electron density is modest, such that the microwave electric field penetrates in a relatively uniform way throughout the medium, then total power absorption[14]

$$P_{diss} = \frac{1}{2}\,\mathrm{Re}(\hat{E}\,\hat{J}^*) = \frac{\sigma_{dc}}{(1+\frac{\omega^2}{\nu_e^2})}\,\frac{|E|^2}{2}$$

is expected to be relatively insensitive to frequency, since $\omega/\nu_e << 1$ for temperatures above room temperature. In the above equation, E is the microwave electric field, J is the conduction current density, ω is the microwave frequency, ν_e is the electron collision frequency, σ_{dc} is the dc conductivity, and '^' represents a complex (Fourier transform) amplitude. As suggested elsewhere,[11] microwave ohmic absorption in semi-conductors is expected to vary nonlinearly with temperature, due to the exponential increase in conduction electrons with increasing bulk crystal temperature.

On the other hand, in highly absorbing materials (such as good conductors), the microwave field will *not* be uniform throughout the medium. Instead, the fields and absorption will be confined to a thin "skin depth" layer at the surface. This, too, can yield nonthermal energy distributions, but only in that narrow region near the material surface where absorption of the microwave energy takes place. As the heat conducts inward, the ion energy distribution rapidly thermalizes, and thus a "microwave effect" is only expected for thin film (0.1 - 1.0 mm) applications when the material is a good conductor.

Hybrid Effects

An important class of materials which require a modification of the two basic microwave absorption models (pure photon-to-phonon absorption and basic ohmic absorption) is the polycrystalline form of ionic insulators. Impurities are impossible to avoid in these ceramics (typically transition metal impurities such as iron) and their segregation at grain boundaries is expected to lead to a "hybridization" of the two absorption mechanisms. For example, in polycrystalline alumina, beryllia, or magnesia, it is expected that microwave absorption in the grain (crystallite) interior occurs through ionic polarization (specifically, the two-phonon difference absorption of Ref. 12), while microwave absorption at the grain boundaries occurs ohmically, as in an impurity-doped (transition metal) semiconductor with a small bandgap. As a result, the microwave-regime behavior of the absorption coefficient is expected to be somewhere between the weak frequency dependence of a good conductor (or semi-conductor) and the frequency-squared dependence of a pure, single-crystal, ionic insulator. This is consistent with measurements on polycrystalline alumina, in which the microwave absorption coefficient was observed to vary linearly with frequency (rather than frequency squared, as in single crystal sapphire).[13]

NONTHERMAL LATTICE FLUCTUATION DISTRIBUTIONS DUE TO INTENSE MICROWAVE HEATING

Insulating ionic crystalline compounds

An athermal ion energy distribution and an athermal phonon distribution are simultaneous phenomena (i.e., when the ion fluctuation energies deviate from Maxwell-Boltzmann statistics, the lattice phonon spectrum will deviate from Bose-Einstein statistics, and *vice-versa*). Hence, to establish the feasibility of obtaining an athermal ion energy distribution (and enhanced ion transport) during microwave heating, it is sufficient to establish the feasibility of obtaining either an athermal phonon distribution, *or* an athermal energy distribution. The existence of one implies the existence of the other. We will choose which distribution to evaluate based on computational convenience.

In this section, we concentrate on the interaction of an intense microwave field with an insulating ionic crystalline compound. To segregate the probable multiplicity of microwave absorption mechanisms present in polycrystalline materials (cf. discussion above) we only consider single crystal structures here. The fact that all the electrons are tightly bound to valence shells in these insulating materials makes it improbable that microwave absorption is due to a conventional "ohmic" process. In Ref. 12, careful comparisons of data and theoretical predictions convincingly indicate that a "line-broadened two phonon difference" process is the dominant microwave absorption mechanism in alkali halides. Due to structural similarities, it is probable that the same mechanism also accounts for microwave absorption in insulating ionic crystalline materials such as Al_2O_3, ZrO_2, MgO, and BeO, especially in single crystal form (this particularly true for MgO and BeO, which have NaCl structure). For illustration, we will investigate the feasibility of sustaining an athermal phonon distribution during intense microwave heating of NaCl. The calculation parameters will be characteristic of the other insulating ionic compounds mentioned above. Furthermore, since the absorption mechanism consists of microwave radiation fields (photons) coupling to lattice phonons, it is more convenient to compute athermal features of phonon distributions, rather than ion energy distributions.

"Line-broadened two phonon difference" absorption actually involves three normal modes of the crystal lattice. The microwave field initially couples into the lattice via the fundamental reststrahlen transverse-optical phonon.[12] Ordinarily, this coupling should not lead to absorption--i.e., it should be a reactive, reversible coupling--due to the large disparity between the frequencies of the microwave field and this "far-infrared" mode. However, the microwave frequency is "quasi-resonant" with a "lifetime-broadened" phonon state transition between a transverse acoustic and a transverse optical mode which "straddle" the forbidden region and are located at the large-wavenumber-edge of the first Brillioun zone. Furthermore, the geometric properties of the transverse reststrahlen mode have the necessary force vectors to simultaneously couple with the Brillioun-zone-edge-acoustic (BZEA) and Brillioun-zone-edge-optical (BZEO) modes mentioned above. As a result, instead of reversibly slipping in and out of phase with the reststrahlen mode, the microwave field induces a microwave frequency excitation of the reststrahlen mode followed by an irreversible absorption associated with a phonon "raising" transition from the BZEA mode (lower state) to the BZEO mode (upper state). We can see how this process can result in a (nonthermal) population inversion between the BZEO and BZEA states whenever the raising transition rate exceeds the relaxation rate back to thermal equilibrium between these two states. Incidentally, the coupling should be more effective as the microwave frequency increases into the millimeter-wave band (specifically increasing as frequency-squared) as observed in measurements with sapphire,[13] as well as the alkali halides.[12]

We now proceed to investigate the feasibility of a microwave-sustained athermal phonon excitation in NaCl--a characteristic insulating ionic crystalline compound. Specifically, we will estimate the degree to which the energy density in the BZEO transverse optical (TO) mode deviates from thermal equilibrium. The differential equation governing the instantaneous phonon number density, n, in the TO-mode can be written as

$$\frac{dn_{TO}}{dt} = -\gamma_{TO}(n - n_o) + S \tag{1}$$

where the first term on the right hand side represents relaxation or decay to a thermal equilibrium level, and the second term symbolizes the repopulation or pump rate of the TO-mode by the microwave photons. In Eq. (1), n is the instantaneous phonon number density of the TO mode, n_o is the thermal equilibrium value, and γ_{TO} is the spontaneous lifetime of the TO-mode. An athermal, or driven steady state solution to Eq.(1) occurs when the pump rate equals the decay rate. Hence, we can calculate a fractional deviation from thermal equilibrium for the TO-mode as

$$\frac{\delta n}{n_o} = \frac{n - n_o}{n_o} = \frac{1}{n_o}\left(\frac{S}{\gamma_{TO}}\right) \tag{2}$$

The repopulation rate S (in phonons-per-unit-time-per-unit-volume) can be estimated from the absorption rate of microwave photons. Using the formulae and parameter values of Ref.12, one can obtain the order-of-magnitude estimate[15]

$$\frac{\delta n}{n_o} \sim \frac{10^{21}\,|E|^2}{n_o\,\gamma_{TO}}. \tag{3}$$

for microwave absorption in NaCl at $\omega_\mu/2\pi \approx 30$ GHz and T = 1000 ^oK. As before, $|E|$ is the amplitude of the microwave electric field (in V/m). Since phonons are bosons, we can estimate the thermal population level from a Bose-Einstein distribution

$$n_o = \eta_{BE}(\omega TO)\, d\omega \approx \eta_{BE}(\omega TO)\, \gamma TO \tag{4}$$

where

$$\eta_{BE}(\omega TO) \approx \frac{3\omega_{TO}^2}{2\pi^2 v_o^3 (e^{\hbar\omega_{TO}/k_BT} - 1)} \tag{5}$$

and v_o is the acoustic velocity, k_B is Boltzmann's constant, and $\hbar$ is Planck's constant (divided by 2π). Again, for 1000 ^oK in NaCl we can estimate[12] that $\omega_{TO} \sim 10^{13}$ sec^{-1}, and $v_o \sim 10^5$ m/sec, yielding $n_o \sim 10^{10}$ (γ_{TO}) m^{-3}. Substituting this into Eq. (13), we obtain the fractional excess energy in the TO mode due to microwave excitation:

$$\frac{\delta n}{n_o} \sim \frac{(10^{11})\,|E|^2}{(\gamma_{TO})^2} \tag{6}$$

From Eq.(6) apparently the fractional deviation of the BZEO mode energy from a thermal distribution level depends on how hard we "shake the lattice", i.e., the magnitude of the microwave electric field strength squared. The other parameter we require is an estimate for γ_{TO}, the spontaneous lifetime of the transverse BZEO mode. Consistent with the lifetime-broadened two-phonon-difference absorption model as well as reported microwave absorption data,[12,13,16] we use $\gamma_{TO} \sim 10^{11}$ sec^{-1} to obtain

$$\frac{\delta n}{n_o} \sim 10^{-11}\,|E|^2 \quad . \tag{7}$$

To compare with experiments, one can estimate the electric field strength in studies at Oak Ridge National Laboratory[2,3,5] to be of the order $|E| \sim 10^5 - 10^6$ V/m.[9] Using similar values in Eq. (7) gives

$$\frac{\delta n}{n_o} \sim 0.1 - 10 \tag{8}$$

To summarize, Eq.(18) represents an estimate of the fractional deviation of transverse optical (Brillioun zone boundary) mode energy from thermal equilibrium levels during microwave heating of NaCl at 1000 ^oK and 30 GHz. The calculation indicates that a significant deviation of the phonon distribution from thermal equilibrium is feasible during intense microwave heating of insulating ionic crystalline compounds. It should be kept in mind that relatively small deviations from thermal distributions (i.e., 0.1 - 1.0 % in the ion fluctuation statistics) are sufficient to significantly enhance ion jump rates (and thus ion transport) during microwave sintering.[10,15]

Conducting or semiconducting crystalline solids

In this section, we investigate the feasibility of sustaining a nonthermal ion energy distribution in conducting or semiconducting crystalline solids during intense microwave heating. In contrast to the insulating materials, the dominant microwave absorption mechanism in conducting crystalline materials is ohmic dissipation. In other words, the microwave electric field couples to and accelerates the free conduction electrons which irreversibly transfer this extra kinetic energy to the crystal through electron-ion collisions. For this process, it is more convenient to compute the effect of the microwave fields on energy distributions, rather than phonon distributions.

To proceed, we will use the "Lorentz gas" model of the Boltzmann transport equations for the free conduction electron and the bound lattice ion velocity distributions, where the Boltzmann collision operator has been approximated by a "Krook" collision term[17]

$$\frac{\partial f_e}{\partial t} + \mathbf{v}\cdot\frac{\partial f_e}{\partial \mathbf{x}} + \mathbf{a_e}\cdot\frac{\partial f_e}{\partial \mathbf{v}} = -\nu_{ei}(f_e - f_i) - \nu_{ee}(f_e - f_{eo}) \tag{9a}$$

for the electrons and

$$\frac{\partial f_i}{\partial t} + \mathbf{v}\cdot\frac{\partial f_i}{\partial \mathbf{x}} + \mathbf{a_i}\cdot\frac{\partial f_i}{\partial \mathbf{v}} = -\nu_{ei}(f_i - f_e) - \gamma(f_i - f_{io}) \tag{9b}$$

for the ions. In Eqs. (9), ν_{ei} is the (constant) electron-ion collision frequency, $f_e(v)$ and $f_i(v)$ are the instantaneous electron and ion velocity distributions, respectively, and γ is the rate at which the ion velocity distribution relaxes to the thermal equilibrium (Maxwell-Boltzmann) distribution, $f_{io}(v)$. Note that this model treats electron-phonon collisions as effective electron-ion collisions. This is justified by the fact that any irreversible transfer of electron momentum to phonons will ultimately show up as an increase in ion momentum. We now employ the following simplifying assumptions: both distribution functions are spatially uniform, the acceleration $\mathbf{a_e}$ is due to the microwave electric field, direct coupling between the microwave field and the ions is negligible (due to the higher inertia of the latter), and we concentrate on a one-dimensional calculation. Writing $f_e = f_{eo} + \delta f_e$ and $f_i = f_{io} + \delta f_i$, where $f_{eo}=f_{io}=f_o$ [$f_o \sim \exp(-\varepsilon/T_b)$, a thermal Maxwellian with bulk temperature T_b], we can obtain the steady state solution (i.e., $df_e/dt = df_i/dt = 0$): [15]

$$f_i(\varepsilon) = f_0(\varepsilon) + \delta f_i(\varepsilon) = \left\{1 + \frac{e\sqrt{2m_i}}{m_e}\frac{E\sqrt{\varepsilon}}{T_b\gamma}\frac{1}{\left(1+\frac{\nu_{ee}}{\nu_{ei}}+\frac{\nu_{ee}}{\gamma}\right)}\right\} f_0(\varepsilon). \tag{10}$$

How significant is the δf_i term? As an illustration, we consider the following parameter values (consistent with our previous calculations for insulating materials): $m_i \sim$ 20-40 amu's, $T_b \sim 2 \times 10^{-20}$ J (approximately 1300 ^oK), $|E| = 10^5$ V/m, $\gamma \sim 10^{12}$ sec^{-1}, and typical of many crystalline conductors,[18] $\nu_{ee}/\gamma \sim \nu_{ei}/\gamma \sim 1$. For these values, we obtain

$$\frac{\delta f_i}{f_0} \sim 10\sqrt{\frac{\varepsilon}{T_b}} \tag{11}$$

Again, this represents a very significant perturbation from thermal equilibrium in the energy range of the *mobile* ion fraction, $\varepsilon/T_b \sim$ 5-10.

Finite penetration depths associated with highly absorbing materials

In this section, we evaluate the possibilities and penetration depths of nonthermal effects associated with *surface heating* methods. This would include microwave heating with highly absorbing (small skin depth) materials, as well as plasma sintering.[19] Here, we assume that the microwave energy does not penetrate directly into the bulk, but rather gets absorbed at the material surface. The sample interior becomes heated by inward conduction of the surface heat. Thus, we envision a model system in which the lattice ions near the surface of the material are characterized by nonthermal fluctuations associated with the microwave driven effect discussed in the previous section.

Again, we simplify by considering a one-dimensional problem. The ion velocity distribution is assumed to be a function of penetration distance "x" into the sample interior. Presumably it may be nonthermal near the material surface, but due to lattice relaxation effects it is expected to be thermalized (a "Maxwellian") at distances far from the surface, i.e.,

$$f_\infty(\infty,v) = A_\infty e^{-mv^2/2T_b} \tag{12}$$

where the bulk temperature is specified in energy units and is measured far from the surface (x >> 0). To qualitatively illustrate the expected physical phenomenon, we characterize the expected nonthermal ion distribution near the sample surface by a "bi-Maxwellian":[10]

$$f(x=0,v) = A_0\{\Delta_0 e^{-mv^2/2T_{hot}} + [1-\Delta_0]e^{-mv^2/2T_b}\} \tag{13}$$

where Δ_0 represents a small fraction of ions in a suprathermal "high energy tail" and T_{hot} represents the characteristic average energy of these suprathermal ions. Since these suprathermal ions transfer their energy to the bulk by relaxation processes we generally expect an x-dependent ion velocity distribution of the form:

$$f(x,v) = A\,\{\Delta(x,v)\,e^{-mv^2/2T_{hot}} + [1-\Delta(x,v)]e^{-mv^2/2T_b}\}. \tag{14}$$

The coefficients A_0, A_∞, and A are determined by the normalization condition

$$1 = \int_{-\infty}^{+\infty} dv\, f(x,v). \tag{15}$$

In the field free interior of the sample bulk, f(x,v) is determined by the Boltzmann equation which for our one-dimensional problem is just

$$\frac{df}{dt} = \frac{\partial f}{\partial t} + v\frac{\partial f}{\partial x} = -\gamma\,\{f(x,v) - f_\infty(v)\} \tag{16}$$

where γ again refers to the lattice-ion relaxation rate due to anharmonic lattice potentials, grain boundaries and dislocations, and other imperfections. For steady state conditions, Eqs.(14) and (16) are equivalent to a differential equation describing the spatial variation of the energetic ion fraction $\Delta(x)$:

$$\frac{\partial\Delta(x,v)}{\partial x} = -\{\frac{\gamma}{v}\frac{1}{[1-\exp(-\frac{mv^2}{2k_B}(\frac{1}{T_\infty}-\frac{1}{T_s}))]}\}\,\Delta(x,v) \tag{17}$$

with solution[19]

$$\Delta(x,v) = \Delta_0 e^{-x/\lambda(v)} \quad ; \quad \lambda(v) = \frac{v}{\gamma}\,[1-\exp(-\frac{mv^2}{2}(\frac{1}{T_b}-\frac{1}{T_{hot}}))]. \tag{18}$$

The parameter $\lambda(v)$ of Eq.(18) governs the spatial rate at which the suprathermal ions are relaxed into the thermal Maxwellian bulk distribution. As a rough measure, when $\Delta_0 << 1$ and for $T_{hot} > T_b$, we can estimate $\lambda \approx v_b/\gamma = \sqrt{2T_b/m_i}\,/\,\gamma$. Then, the nonthermal ion fraction decays with distance into the sample as approximately

$$\Delta(x) \approx \Delta_0\, e^{-x/L_{nt}} \tag{19a}$$

where

$$L_{nt} = \frac{\sqrt{2T_b/m_i}}{\gamma} = \frac{v_b}{\gamma} \tag{19b}$$

Typically, $\gamma \sim 10^{11} - 10^{12}\ sec^{-1}$ and for temperatures between 100 and 1000 ^oK, $v_b \sim 10^7$ m/s. Hence, we can expect $L_{nt} \sim 10 - 100\ \mu m$.[19]

SUMMARY

If reports of enhanced sintering rates associated with microwave heating are accurate, then they may be due to nonthermal lattice fluctuation statistics. Recent theoretical analyses reviewed in this paper confirm the feasibility of this phenomenon for a wide variety of situations involving very different microwave absorption mechanisms. For materials with weak microwave absorption coefficients, the effect should be uniformly distributed throughout the volume. For strongly absorbing materials, however, the effect is expected to be concentrated near the material surface, with a characteristic exponential penetration depth of $L_{nt} \sim 10 - 100\ \mu m$. The observation of a "resolvable" nonthermal effect depends on the relative magnitude of two relatively ubiquitous parameters--the microwave electric field strength $|E|$ and the lattice ion energy relaxation rate γ. Nonthermal effects should be most pronounced for large values of $|E|$ and smaller values of γ. This corresponds to high microwave power levels and/or lower bulk temperatures. Nevertheless, parameter values relevant to some realistic sintering conditions are consistent with significant nonthermal effects.

ACKNOWLEDGEMENTS

The authors appreciate valuable discussions with Drs. I. Goldberg, W. Ho, M. Janney, R. Joynt, N. Tabatabaie, L. Shohet, and N. Hitchon. An anonymous conference attendee to the 1991 Microwave Processing Symposium in Cincinnati, Ohio, is also thanked for his suggestion to investigate the microwave absorption theory of Sparks and Mills. This research has been supported by the National Science Foundation Presidential Young Investigator Program and by a grant from the Electric Power Research Institute.

REFERENCES

1. B. Swain, "Microwave Sintering of Ceramics," in *Adv. Materials and Proc. incorporating Metal Progress* magazine, September issue (1988).
2. M.A. Janney and H.D. Kimrey, "Microstructure Evolution in Microwave Sintered Alumina," in *Advances in Sintering*, Ed. J. Bleninger and C. Handwerker, (Amer. Cer. Soc., Westerville, OH, 1990).
3. M.S. Janney, ""Microwave Processing of Ceramic Materials," seminar presented at the Materials Science Seminar Series, University of Wisconsin-Madison, (1989).
4. Y.L. Tien, D.L. Johnson, and M.E. Brodwin, pp. 925-32 in Ceramic Transactions, Vol. 1B., *Ceramic Powder Science II*, Ed. G.L. Messing, E.R. Fuller, and H. Hausner, (Amer. Cer. Soc., Westerville, OH, 1988).
5. M.A. Janney, H.D. Kimrey, M.A. Schmidt, and J.O. Kiggans, J. Amer. Cer. Soc. 74 1675 (1991).
6. J.D. Katz, Proc. 93rd Ann. Mtg. Amer. Cer. Soc., (Cincinnati, OH, Apr. 1991), paper 4-SXI-91.
7. Z. Fathi and D.E. Clark, Proc. 93rd Ann. Mtg. Amer. Cer. Soc., (Cincinnati, OH, April, 1991), paper 11-SXIP-91.
8. T.T. Meek, J. Mater. Sci. Lett. 6, 638 (1987).
9. D.L. Johnson, J. Amer. Ceram. Soc. 74, 849 (1991).
10. J.H. Booske, R.F. Cooper, I. Dobson, and L. McCaughan, in Microwaves: Theory and Application in Materials Processing; Ceramics Transactions, Vol. 21, Eds. D.E. Clark, F.D. Gac, and W.H. Sutton, pp. 185-192 (American Ceramic Society, Westerville, OH, 1991).
11. J.H. Booske, R.F. Cooper, and I. Dobson, J. Mater. Res. 7, 495 (1992).
12. M.Sparks, D.F. King, and D.L. Mills, Phys. Rev. B 26 6987 (1982); K. R. Subbaswamy and D.L. Mills, Phys. Rev. B 33, 4213 (1986).
13. W. Ho and I. Goldberg, (private communication, 1992).
14. S. Ramo, J.R. Whinnery, and T. Van Duzer, Fields and Waves in Communication Electronics, 2nd Ed., (Wiley and Sons, New York, 1984).
15. J.H. Booske, "Feasibility of Athermal Ion Energy and Phonon Distributions During Microwave Heating of Crystalline Solids," submitted to J. Appl. Phys. (1992).
16. J.C. Owens, Phys. Rev. 181, 1228 (1969).
17. N. Krall and A. Trivelpiece, Principles of Plasma Physics (McGraw-Hill, New York, 1973).
18. Kruezer, Nonequilibrium Thermodynamics and its Statistical Foundations, (Oxford University Press, New York, 1981).
19. J.H. Booske, "Penetration Depth of Nonequilibrium Kinetics During Plasma and Microwave Ceramic Sintering, " to be submitted for publication (1992).

THEORY OF SUBLINEAR POWER DEPENDENCE OF MICROWAVE HEATING IN SOME CERAMIC MATERIALS

V. M. KENKRE*, M. KUS* and J. D. KATZ**
* Department of Physics and Center for MicroEngineered Ceramics, University of New Mexico, Albuquerque, NM 87131
** Los Alamos National Laboratory, Los Alamos, NM 87475

ABSTRACT

A sublinear dependence of the microwave heating rate on the incident microwave power implying a saturation effect appears to have been observed recently. We present simple model calculations to address this observation on the basis of an idea of spatial confinement of the absorbing charges in grain boundary regions. Two natural lengths exist in this model: the spatial extent of the confining region, and the maximum distance an absorbing charge travels under the combined action of damping and of the oscillating microwave field. We suggest that a mismatch of these lengths results in the observed saturation, more generally, in the observed decrease in absorption efficiency.

INTRODUCTION

Standard calculations of the heating rate of materials subjected to electromagnetic waves, such as microwave radiation, predict a linear dependence on the incident power: the heating rate, equivalently the power absorbed, depends quadratically on the electric field E of the microwave field. This behavior is indicative of standard Joule heating. Even recent theories [1,2] of the pheonomenon of thermal runaway, which explain the runaway on the basis of nonlinear feedback of the energy, predict a linear dependence of the heating rate on the incident power. Such a linear dependence is surely observed in many materials such as silica [3,4]. However, a saturation effect, more generally a sublinear dependence of the rate on the microwave power, appears to have been observed in other materials, typified by ferric oxide [4]. No theoretical starting point in the literature known to us, is currently able to address this effect. In the present paper, we describe our recent effort in this direction.

The essential idea behind the calculation we present here is one of *spatial confinement* of absorbing charges in grain boundary regions in a ceramic material. The charges absorb power from an applied oscillating electric field by following the field oscillations and dissipating the absorbed power to a damping reservoir such as the solid lattice. The instantaneous power absorbed is proportional to the square of the velocity of the charge, i.e., to the square of the applied electric field. The time average yields the average power absorbed. This is the situation in a spatially unrestricted medium. However, if the charges are constrained to move in a finite region of space, they may find themselves unable to follow the time dependence of the applied field simply because they may reach the walls of the confining region before the electric field has reversed its direction. The encounters with the wall could result in velocity reversal and, thus, work against efficient absorption of power from the field. Surely, this effect will be negligible in small fields or large regions. However, for a given extent of the confining region, if the input power is increased, the electric field will eventually reach a magnitude consonant with a traversal distance which is larger than the confining length. The consequence will be a sublinear dependence of the absorbed power on the incident power.

Spatial confinement of charges can occur in materials in the proximity of grain boundaries. In many ceramics, particularly those which are clearly ionic in character, atomic diffusion is thought of as occurring in two parts: diffusion of the cations through the bulk, and the diffusion of larger anions primarily through the grain boundary regions. Let us focus attention on the anions. The grain boundary may be considered as made up of a number of segments along which the anion motion takes place. These linear segments of the boundary provide a possible model for our spatially constraining regions. Another possibility arises from the space charge accumulation region *across* a grain boundary: electrostatic interactions are known [5,6] to lead to the accumulation of space charges on the two sides of a grain boundary and to potentials which are felt by charged interstitials or vacancies. The confining length in the former case could be of the order of microns or tenths of a micron, while in the latter case it could be much smaller, viz. of the order of 200 A. The following is a brief description of our basic theory and of a recent application to observations on the dependence of the saturation temperature on input microwave power.

BASIC THEORY AND APPLICATION TO EXPERIMENT

For the sake of simplicity, the nature of charge motion in the confined region will be assumed to be free except for damping and the action of the electric field. The equation of motion is

$$\frac{dv}{dt} + \gamma v = \frac{qE}{m}\sin \omega t. \tag{1}$$

where v is the velocity of the charge q with mass m, E is the electric field, and γ is the damping constant. With the notation $\varepsilon = \frac{qE}{m}$, the solution of (1) is

$$v(t) = e^{-\gamma t}\left[v(0) + \frac{\omega\varepsilon}{\gamma^2+\omega^2} \right] + \frac{\varepsilon}{\sqrt{\gamma^2+\omega^2}}\sin(\omega t - \varphi). \tag{2}$$

where the phase lag angle φ is given by $tan^{-1}(\omega / \gamma)$. In an unconfined region, the steady-state heating rate R is given by

$$R = \frac{1}{2}\frac{q^2}{m}\frac{\gamma}{\gamma^2+\omega^2}E^2. \tag{3}$$

Since the incident power is proportional to E^2, the rate as given in (3) depends linearly on the incident power. Under the assumption of very large damping, known to be valid for microwave interactions with ceramics, the solution given in (2) simplifies significantly. The result, which can also be obtained directly by taking the large damping limit of (1), is a complete *slaving* of the velocity by the electric field:

$$v(t) = \frac{qE}{m\gamma}\sin \omega t. \tag{4}$$

Results such as (2), (3) and (4) are valid when there are no walls to confine the motion of the charges. Let us now consider the case of a spatially restricted region. The charge is accelerated but hits a confining wall before the oscillating electric field has reversed sign. The charge bounces back, with its velocity reversed in sign. If the collision with the wall is elastic, the magnitude of the velocity is unchanged. The velocity now decreases in magnitude because the electric field acts against the motion. The charge may return to the wall one or more times before the electric field reverses and forces it to move towards the other restraining wall.

It is possible [7] to show that, if the charge is at rest at one of the walls initially, the average heating rate is given by

$$R = \frac{1}{2}\frac{q^2}{m\gamma}E^2, \qquad \frac{L}{L_E} \geq 1 \tag{5}$$

$$R = \frac{1}{2}\frac{q^2}{m\gamma\pi}E^2\left[\cos^{-1}\left(1-\frac{2L}{L_E}\right)-2\left(1-\frac{2L}{L_E}\right)\sqrt{\left(1-\frac{L}{L_E}\right)\frac{L}{L_E}}\right], \qquad \frac{L}{L_E} \leq 1 \tag{6}$$

Equation (5) is simply the large damping limit of (3). However, (6) has novel content and involves the quantity $L_E = 2qE/m\gamma\omega$ in addition to the distance L between the walls. The rate depends on the electric field not only through the E^2 term as in (5) but also through the dependence of L_E on the electric field. The overall effect is to lower the heating efficiency. We present Fig. 1 to make this point clear.

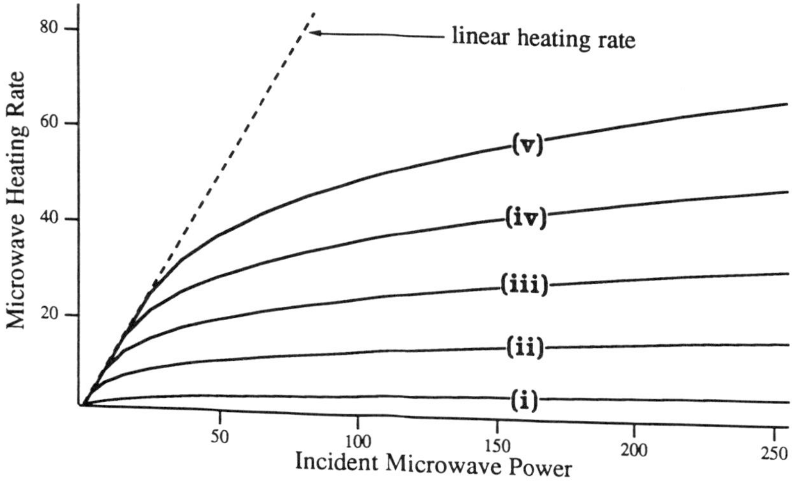

FIG.1: Theoretical prediction of a sublinear dependence of the heating rate on the incident microwave power, for several values of the confining length L. Values of L for curves (i), (ii), (iii), (iv) and (v) are respectively in the ratio 1:2:3:4:5. The limit of infinite L is approached by all finite-L curves for appropriately small values of the incident power. Units are arbitrary.

An application of these ideas results in a possible understanding of the observed sublinear power dependence in ferric oxide [4]. Thermal runaway in a variety of materials has been explained recently in terms of a microscopic theory [1,2] which results in the following temperature-time dependence:

$$\frac{dT}{dt} = [k_A + k_M f(T)]P - \sigma_1 T^4 \tag{7}$$

where k_A and k_M are related to the absorption coefficients of two kinds of species of absorbers present in the material and $f(T)$ denotes the fraction of the M-absorbers which are free to absorb [8], P is the incident power, and σ_1 is proportional to the Stephan-Boltzmann constant. Equation (7), which has provided a successful description [1,2] of the temperature-time evolution in a large number of materials, when combined with the spatial confinement ideas presented above, results in

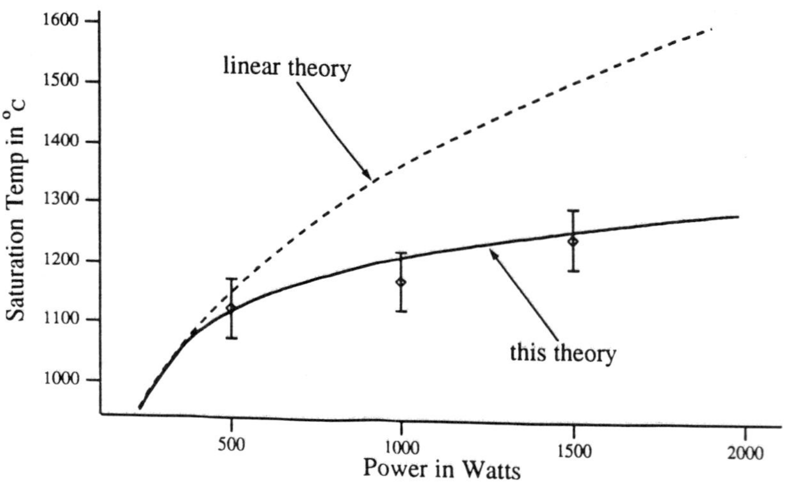

FIG.2: Saturation temperature during microwave heating of ferric oxide plotted as a function of the incident microwave power, along with the results of our theory. Error bars show the extent of uncertainty introduced by the procedure used to estimate values of the saturation temperature from the time-temperature curves reported in [4].

the following relation between the *saturation temperature* T_s and the input microwave power P:

$$\frac{\sigma_1 T_s^4}{k_A + k_M f(T_s)} = P, \qquad \frac{L}{L_E} \geq 1 \tag{8}$$

$$\frac{\sigma_1 T_s^4}{k_A + k_M f(T_s)} = \frac{P}{\pi}\left[\cos^{-1}\left(1-\frac{2L}{L_E}\right) - 2\left(1-\frac{2L}{L_E}\right)\sqrt{\left(1-\frac{L}{L_E}\right)\frac{L}{L_E}}\right], \qquad \frac{L}{L_E} \leq 1 \tag{9}$$

It is important to notice that , in (9), L_E is a function of the field and, consequently, of P.

In Fig. 2 we have shown the saturation temperature T_s as a function of input microwave power with and without the spatial confinement effect along with data deduced from the observations in ref. [4]. There is a probable error of ± 50 deg. C in the saturation temperature, as depicted in Fig. 2. The confining length which corresponds to the solid curve is of the order of a tenth of a micron. At this stage, the comparison of theory and experiment is relatively crude, but we can certainly conclude from Fig. 2 that the spatial confinement mechanism we have proposed is a possible source of the observed sublinear power dependence of the heating rate in ferric oxide.

REMARKS

We have assumed in the analysis presented above that the damping is infinitely large so that the velocity is *completely* slaved by the applied field as in (4). If the damping is large but not infinite, we may follow a method that has been recently used in the context of nonlinear quantum transport in presence of dissipation [9]. The Laplace transform of (1) can be written in the form

$$\tilde{\mathrm{v}}(\mathrm{s}) = \frac{\tilde{\mathbf{E}}(\mathrm{s})}{\mathrm{s}+\gamma} = \frac{\tilde{\mathbf{E}}(\mathrm{s})}{\gamma}\left(\frac{1}{1+(s/\gamma)}\right) \approx \frac{\tilde{\mathbf{E}}(\mathrm{s})}{\gamma}\left(1-\frac{s}{\gamma}\right) \approx \frac{\tilde{\mathbf{E}}(\mathrm{s})}{\gamma}\exp\left(-\frac{s}{\gamma}\right) \tag{10}$$

where $\mathbf{E}(t)$ is the time-dependent electric field, tildes denote Laplace transforms, and s is the Laplace variable. The expansion in powers of (s/γ) that we have made in (10) is obviously valid for large damping and for long times. Using the time-displacement property of Laplace transforms, we can invert (10) into the simple form

$$\mathrm{v}(t) = (1/\gamma)\mathbf{E}\left(t-\frac{1}{\gamma}\right) = \frac{qE}{m\gamma}\sin\omega\left(t-\frac{1}{\gamma}\right) \tag{11}$$

Finite damping corrections to the theory presented above can be derived easily with (11) as a point of departure. For want of space, we do not show the details here.

We have presented a simple possible explanation of the sublinear power dependence of the microwave heating rate on the basis of spatial confinement ideas. Many of the assumptions made here can be relaxed. These include the complete confinement of the moving charges to segments of the grain boundary, and the assumption that the nature of the motion in the confining region is free. For a fuller discussion of these issues we refer the reader to ref. [7].

ACKNOWLEDGEMENTS

This work was carried out in part with the support of the UNM-NSF Center for Microengineered Ceramics, a collaborative effort of the National Science Foundation (CDR-8800352), the Los Alamos and Sandia National Laboratories, the New Mexico Research and Development Institute, and the Ceramics Industry.

REFERENCES

1. V. M. Kenkre, L. Skala, M. Weiser and J. D. Katz, J. Materials Science, **26**, 2483 (1991).
2. V. M. Kenkre, Ceramics Transactions **21**, 69 (1991).
3. L. M. Sheppard, Ceramic Bulletin, **67** (1988) 1656.
4. S. L. McGill, J. W. Walkiewicz and G. A. Smyres,in Mat. Res. Soc. Symp. **124**, Materials Research Society, (Pittsburgh 1988), p. 247.
5. M. F. Yan, R. M. Cannon, H. K. Bowen and R. L. Coble, J. American Ceramic Soc., **60**, 120 (1977).
6. S. K. Tiku and F. A. Kröger, J. American Ceramic Soc., **63**, 183 (1980).
7. V. M. Kenkre, M. Kus and J. D. Katz, phys. stat. solidi (b), **171**, scheduled for publication.
8. M. Kus and V. M. Kenkre, Phys. Rev. B, scheduled for publication.
9. V. M. Kenkre and H.-L. Wu, Phys. Lett. A **135**, 120 (1989).

MEASUREMENT OF MICROWAVE INDUCED FORCES

JOHN L. WATKINS, H. W. JACKSON, AND M. BARMATZ
Jet Propulsion Laboratory, California Institute of Technology, Pasadena, CA 91109.

ABSTRACT

Electrically polarizable materials in an inhomogeneous electric field experience a dielectrophoretic force which is proportional to the gradient of the square of the electric field. For high frequency radiation, the time averaged electromagnetic field provides a dc force. The force exerted on a sapphire and a metal sphere by the microwave field in a single mode resonant cavity has been measured. The force measurements have been made at equally spaced points along three orthogonal axes centered in a cylindrical cavity operating in the TE_{111} mode using 20 watts of microwave power at about 4.84 GHz. These data are compared to the simple theory of dielectrophoresis on a small sphere and verify that, for this mode and in the absence of other forces, there is a unique and stable equilibrium position at the center of the cavity. Measurements of the quality factor of the cavity provide a means of making an absolute check to the theory. The TE_{111} mode is one of an infinite set of modes predicted to have stable equilibrium positions that exist away from the cavity walls. It is proposed that dielectrophoresis in a microwave resonant cavity can be used to position samples for material processing applications in a microgravity environment. Knowledge of these forces may also be important for interpreting thermogravimetric measurements made in a microwave oven.

INTRODUCTION

As part of NASA's Microgravity Science and Applications program, the Jet Propulsion Laboratory is exploring technologies to support materials processing in space. Part of this effort involves investigating ways to position materials containerlessly in a microgravity environment. Acoustic, electrostatic, and electromagnetic forces are some of the methods that have been considered for this use. It is advantageous to have multiple techniques since each positioning method has its own set of operating constraints. For example, a material being positioned acoustically needs to be immersed in an atmosphere of some sort and electrostatic positioning requires active positioning feedback. In order to be able to accommodate the broadest range of materials science needs, new techniques that provide a means of remote manipulation need to be identified and demonstrated. It has been proposed that the forces induced by microwave fields in a resonant cavity may be exploited for positioning neutral dielectric materials.[1,2] Simple dielectrophoretic theory predicts that the forces induced by certain resonant modes can stably position an object away from the cavity walls. This paper describes experiments carried out to measure these forces and compare them with the theory.

THEORY

The force on a small electrically polarizable sphere in an electric field $\mathbf{E}(\mathbf{r},t)$ is given by

$$\mathbf{F}(\mathbf{r},t) = \frac{3}{2}\frac{(\varepsilon_2 - \varepsilon_1)}{(\varepsilon_2 + 2\varepsilon_1)} V \nabla \mathcal{E} \tag{1}$$

where

$$\mathcal{E} = \frac{1}{2}\varepsilon_1 \, |\mathbf{E}(\mathbf{r},t)|^2 \tag{2}$$

is the energy density of the electric field, V and ε_2 are the volume and dielectric constant of the sphere and ε_1 is the dielectric constant of the surrounding medium.[3] (The bold symbols are used to represent vector quantities.) In this approximation, **E** is the electric field intensity when the sphere is not present. The quantity $-\mathcal{E}$ acts as a potential and the sphere will tend to move toward the point of the highest electric energy density. When equation (1) is applied to a sphere in a microwave cavity, the ac component of the force due to the high frequency time variation of the electric field is averaged out since the sphere's inertial response is very slow. The resulting force is just half of the maximum instantaneous force.

Figures 1 and 2 show the time average of the potential, $-\langle\mathcal{E}\rangle$, for the TE_{111} mode on two orthogonal surfaces through the center of a cylindrical cavity. In this particular mode, since there is a minimum in the potential located at the center of the cavity, the center is a stable equilibrium point. The angular dependence of the electromagnetic field of this mode causes the behavior seen in Fig. 2.

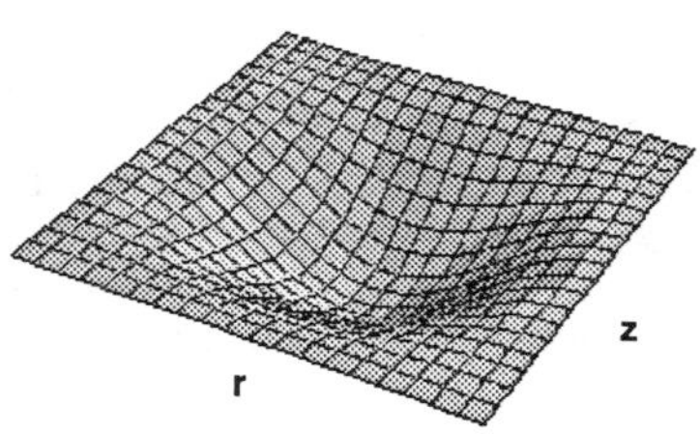

FIGURE 1. POTENTIAL SURFACE
$\theta = 0, \pi$

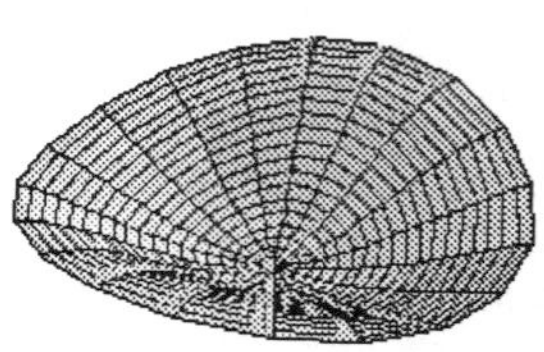

FIGURE 2. POTENTIAL SURFACE
Z = L / 2

In order to calculate the magnitude of the force, the absolute value of $\langle\mathcal{E}\rangle$ must be known. It can be determined from directly measurable quantities using the definition of the quality factor Q

$$Q = \frac{\omega \int \langle E \rangle \, dV}{P} \tag{3}$$

where ω is the resonant frequency, P is the power into the cavity, and the integral is over the volume of the cavity. By measuring Q and the power into the cavity, the maximum of $\langle E \rangle$ can be calculated for any mode where the field configuration is known.

EXPERIMENT

A schematic diagram of the experimental apparatus is shown in Fig. 3. The microwave resonator was a right circular cylindrical cavity with an interior diameter a and length L both equal to 47.8 mm. The cavity was mounted on a dovetail slide which allowed it to be moved and positioned linearly with an accuracy of 25 μm. Force measurements were made with the sample on a 1 mm diameter quartz rod which in turn rested on an electronic microbalance. For axial force measurements, the cavity was placed with its axis vertical and the support rod extended inside it through a 6 mm hole centered on the lower end plate. The sample temperature was monitored using a remote sensing IR thermometer through another identical hole centered on the upper end plate. The sample and thermometer remained fixed while the cavity was translated. Radial force measurements were made in the same manner with the cavity placed horizontally. Opposing holes in the side wall at the axial midpoint allowed extension of the support rod into the cavity and permitted viewing of the sample by the thermometer.

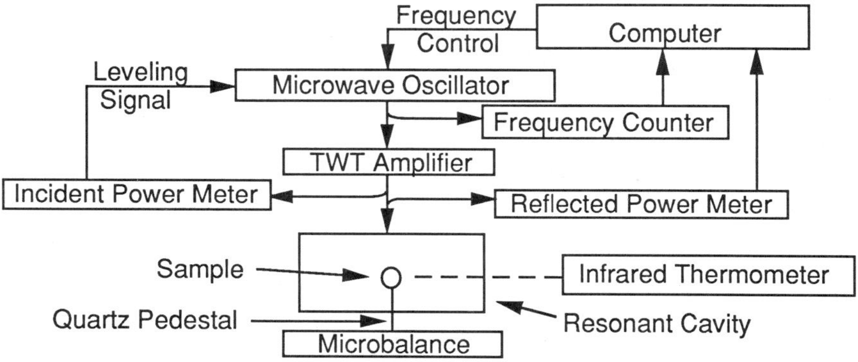

FIGURE 3. EXPERIMENTAL APPARATUS

For this resonator, the TE_{111} mode has a frequency of about 4.84 GHz. The mode was excited by an adjustable inductive loop. A voltage controllable oscillator provided the input to a 20 W TWT amplifier. Both the incident and reflected cavity power were monitored. Tuning of the cavity was done by minimizing the reflected power with respect to the frequency. The cavity Q was determined from the shape of the reflected power. With the cavity critically coupled, a computer was used to step the frequency through the TE_{111} resonance. The reflected power was measured at each frequency and the resulting power curve was fit by a theoretical lorentzian shape. A Q of 6100 was determined as one of the fit parameters. This value was taken with the sphere located at the center of the cavity and was used to calculate the theoretical force curve. More accurate force calculations would use the Q measured for each position of the sample. However, since the sapphire sample has a very low microwave loss tangent (< 0.0001), changes in the quality factor were very small and were neglected in this work.

The sample on which the force measurements were made was a 9.5 mm sapphire sphere. The force was determined by measuring the change in weight of the sphere. This method measured only the force in the vertical direction; reorientation of the microwave cavity as described above permitted both axial and radial force measurements. At the same time, the temperature of the sapphire was monitored to ensure that convection caused by microwave heating of the sphere did not interfere with the measurement. Because of sapphire's low loss tangent and the low power levels used in this experiment, the sample never reached more than about 5 C above room temperature. This almost negligible amount of heating was not enough to cause appreciable convection. An additional indication that convection was not affecting the force measurement was that the observed force was almost instantaneous with the application of the microwaves.

Figures 4 and 5 show the measured force in dynes compared against the force calculated with equation 1 for positions along three orthogonal axes in the cavity. The axial force was measured along the resonator axis and the radial forces were measured on two diameters 90 degrees apart at the midpoint of the cavity. The difference in the two sets of radial force data is due to the angular dependence of U for this mode. There is excellent agreement between the measurements and the predictions of the simple dielectrophoretic theory.

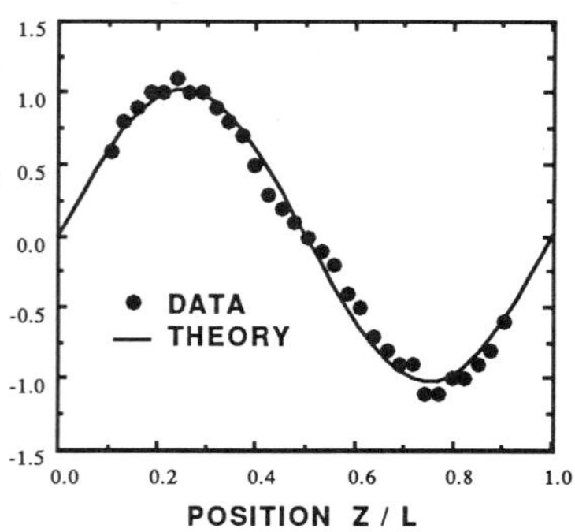

FIGURE 4

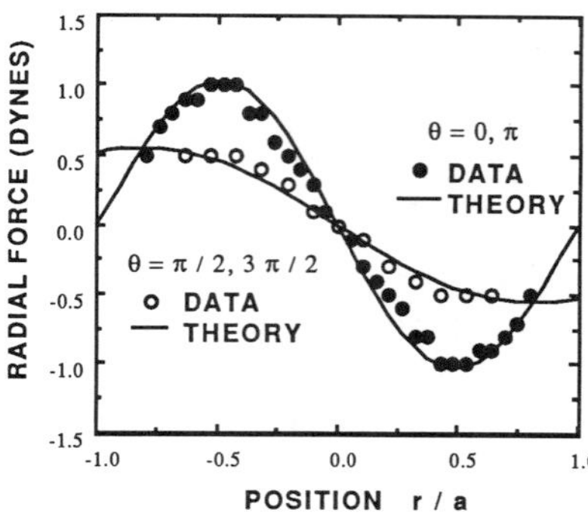

FIGURE 5

Microwave induced axial forces were also measured on brass and aluminum spheres at several positions on the axis of the cavity. Metals are not dielectrics and application of the dielectrophoretic theory is not valid. However, forces were measured and their magnitude was slightly larger than those for sapphire. This is what would be expected if the metals were treated as having an infinite dielectric constant. Both the brass and aluminum samples experienced forces of the same strength.

DISCUSSION

The TE_{111} mode was used for these experiments because it has a single stable position away from the cavity walls. Provided that externally induced accelerations were small, material located at this point would remain there with no active control required. This is a desirable feature for many containerless processing applications. However, all cavity modes will exert forces on a material and by choosing a specific mode or a combination of modes, complex positioning and manipulation of a sample could be done. For example, higher order modes in general have more than one stable position and so multiple samples could be contained at different places in one cavity simultaneously. By switching modes, material could be translated from one position to another. In addition to linear forces, properly phased modes should exert torques and thus provide a means for controlling rotation of levitated materials. Microwave positioning may also be used in conjunction with other positioning techniques as an additional translation or rotation control. There are also electrostrictive forces acting on materials in a microwave cavity. These forces, along with the induced torques, may be exploited to change the bulk shape of materials or affect the internal structure.

The actual magnitude of the force depends on the cavity geometry, the sample size, the mode, the electric field strength, and the dielectric properties of the sample and the background medium. In most cases, a number of these parameters can be chosen to optimize the force for a particular application.

In these experiments, heating of the sapphire was negligible since its microwave losses are small. In general, application of microwaves to position a sample is expected to induce some heating. In the case of most microwave absorbing materials, the point of stable levitation will be at the position of highest heating. For some materials such as certain semiconductors and metals, it may be possible to position a sample with a poor heating mode and then use a second mode to perform controlled containerless heating studies.[4] These aspects may be exploited for high temperature processing of materials.

Finally, microwave ovens are beginning to be used for thermogravimetric analysis.[5] Because microwaves induce forces on materials, it is important to know their magnitude when doing mass measurements. Although the effect is more pronounced in single mode ovens, it needs to be considered in any oven where there may be gradients in the electric energy density.

CONCLUSION

Microwave induced forces have been measured on a sapphire sphere. The magnitude and direction of these forces agree with the predictions of the simple model of dielectrophoresis on a sphere. For modest power levels, these forces can be large enough to be useful in a microgravity environment.

ACKNOWLEDGEMENT

We wish to acknowledge Richard Zanteson for the design and fabrication of the experimental apparatus. The research described in this article was carried out at the Jet Propulsion Laboratory, California Institute of Technology, under contract with the National Aeronautics and Space Administration.

REFERENCES

1. J. L. Watkins and H. W. Jackson, NASA Tech Briefs, **15** (9), 109 (1991).

2. M. Barmatz, J. L. Watkins, and H. W. Jackson, "Microwave Processing of Materials in Microgravity," 30th Aerospace Sciences Meeting and Exhibit, AIAA Paper #92-0116, January, 1992.

3. H. A. Pohl, Dielectrophoresis (Cambridge University Press, New York, 1978), p.34.

4. H. W. Jackson and M. Barmatz, J. Appl. Phys. **70** (10), 5193 (1991).

5. E. H. Moore, I. Ahmad, and D. E. Clark in Microwaves:Theory and Application in Materials Processing, edited by D. E. Clark (Ceramics. Transactions 1991) pp. 675-681.

MICROWAVE ABSORPTION BY LOSSY CERAMIC MATERIALS *

Isidoro E. Campisi, Lynda K. Summers, Keith E. Finger and Anne M. Johnson
Continuous Electron Beam Accelerator Facility
12000 Jefferson Avenue
Newport News, VA 23606

ABSTRACT

Microwave energy generated by the electron beam in the superconducting cavities of the Continuous Electron Beam Accelerator Facility (CEBAF) is absorbed by special loads fabricated with a novel lossy ceramic material (AlN-glassy carbon) developed especially for this application. Strict environmental constraints (ultra-high vacuum compatibility, operation at 2 K, brazeability, etc.) are imposed on the materials which can be used. Several other ceramics were sintered with AlN and various minority conductive powders to obtain the desired electrical properties according to the "artificial dielectric" model. Dielectric permittivity data and results of low temperature measurements are reported.

INTRODUCTION

Materials capable of absorbing microwaves or, in general, electromagnetic radiation, can rely on several loss mechanisms: the description of the losses is complicated, and the materials used for absorption are often selected based on empirical criteria. It is possible, however, to produce materials in which one loss mechanism dominates over the others, and for which the microwave losses can be controlled over a wide range of temperatures and over a chosen frequency band. The possibility of controlling such properties is particularly useful in a number of accelerator applications.

In many particle accelerators being built or presently being designed (SSC, LEP, high current storage rings etc.) microwave power generated by the charged particle beams (in some cases from below 1 GHz to several hundred GHz and with power ranging from a few watts to several kilowatts per load) must be suppressed and absorbed to prevent beam instabilities. In all these cases materials are required which must be compatible with ultra-high vacuum (less than 10^{-10} torr), and which must be capable of handling large power dissipation (this implies good thermal conductivity and the possibility of brazing them to effectively remove the energy deposited by the microwaves). The materials to be employed cannot contain organic compounds and must withstand large temperature excursions during medium-temperature bakeout of the vacuum structures. Ceramic materials meet all of the conditions imposed by the accelerator environment, but most available materials' dielectric properties are poorly characterized or controlled.

As in other accelerators, absorbers at CEBAF are used to damp higher-order modes (HOM) generated by the electron beam [1]. But at CEBAF the accelerator consists of 338 superconducting cavities and, in addition to the strict environmental constraints common to most accelerators, the microwave absorbers are located in a cryogenic environment at 2 K, the only such arrangement in any accelerator. This additional constraint led to a thorough examination of materials capable of absorbing microwaves and compatible with the severe conditions of the superconducting accelerator. No material was found available which met all of the requirements. Some of the existing materials (e.g., AlN-SiC) which are often used in microwave absorption at room temperature showed temperature dependences related to their semiconducting properties. Most microwave loss mechanisms

* This work was supported by the U.S. Department of Energy under contract DE-AC05-84ER40150.

become ineffective at low temperatures because they are related to thermally activated phenomena; an analysis of the possible loss mechanisms reveals that a special class of properly designed materials exists which can provide the necessary interaction with the microwave fields, irrespective of the low temperatures.

LOSS MECHANISMS IN SOLIDS

Electromagnetic losses in solids can be effected by two basic mechanisms: direct photon-phonon conversion in polar dielectric insulators, or through the mediation of electrons from the conduction band in non-insulating materials. The electrons acquire energy from the electromagnetic wave and then deposit it into the lattice in an electron-phonon interaction [2],[3].

The direct photon-phonon conversion phenomenon occurs in insulating dielectrics; the polarization induced in crystal lattices by incoming electromagnetic fields relaxes into normal modes of oscillation, modes which eventually redistribute the electromagnetic energy into thermal phonons. This type of loss mechanism is, in general, not very effective, since it depends on the number of available normal modes of the ions in the lattice, which is a strong function of temperature and of frequency, and it certainly is ineffective at low temperatures, where, for instance, crystalline solids exhibit very low attenuation constants. This process becomes more effective at high temperatures, up to the point where, together with the increasing semiconduction losses, it can lead to thermal runaway phenomena [4],[5]. In a more controlled way, the same loss mechanisms are employed to effect ceramic sintering.

In conducting and semiconducting materials the loss processes involve the presence in the conduction band of electrons which are scattered off ions, impurities, dislocations, grain boundaries etc., exchanging kinetic energy with the lattice. The basic phenomenon is common to many materials, but the exact scattering phenomenon leads to drastically different absorption behaviors.

In semiconducting materials the number of electrons available in the conduction band decreases with temperature, thus leading to substantially different absorption properties as a function of temperature. Moreover, in semiconductors the inherent losses are a strong function of the doping, and the intrinsic resistivity can change by many orders of magnitude by small changes in impurities' composition or concentration and due to vastly different mobilities of electrons and holes.

The loss phenomena based on semiconduction are often difficult to control and might lead to unreliable behavior from sample to sample. This is the case, for instance, of sintered SiC, the resisitivity of which can span over ten orders of magnitude [6].

Both conductors and semiconductors can have loss mechanisms which rely on grain boundary resistance due to hopping conductivity across powder grains: these loss phenomena are often used in microwave systems at room temperature and when small power densities are present. Lossy foams, or even the losses in carbon resistors, are typical examples of these mechanisms which do not depend very strongly on the material's conductivity, but which rely on grain boundary or surface effects: these are mostly thermally excited loss mechanisms and are therefore ineffective at low temperatures.

Metal conductors have electrons in the conduction band at any temperature. These electrons are always available to mediate the photon-electron-phonon interaction, but at microwave frequencies the skin effect prevents absorption on the part of bulk metal materials. However, controlled and effective microwave absorption can be obtained by making use of the metallic behavior of conductive particles suitably dispersed into a dielectric insulating material. This composite is known as an "artificial dielectric".

ARTIFICIAL DIELECTRICS

These materials and their properties have been known for a long time and have been the subject of extensive theoretical and experimental studies [7],[8], [9],[10].

The basic loss mechanism is here given by scattering of electrons within each elementary grain due to the current induced in each grain by the incoming electromagnetic wave. Depending on the size and resistivity of the grains, the losses in each grain can be electric or magnetic, and a particular frequency can be best attenuated [11].

Typically, radiation is best absorbed if the grain diameter is of the order of the skin depth in the grain material. By choosing a proper conducting material in powder form, and by knowing its resistivity, its grain size (or size distribution), its concentration and the dielectric properties of the host material, it is in principle possible to obtain specific combinations of dielectric constant and losses.

The concentration of the conducting particles alters the effective dielectric permittivity of the host material, but it also has a secondary effect on practical materials: if the volume concentration of the conductive phase is larger than 15 percent, then aggregation of particles can take place. In this case, loss mechanisms other than pure ohmic losses within the elementary grains can occur, and these additional losses will be temperature dependent [12].

If the conductor's concentration approaches or exceeds 50 percent, then a metallic behavior will screen the currents from the bulk of the material and absorption will be inhibited (at or above the percolation threshold).

To avoid temperature dependent absorption (at least from 0 K to several hundred degrees C) one must rely solely on the metallic behavior of the minority powders in concentrations below 15 percent by volume and choose powders with the proper sizes and resistivity.

In two limiting cases the resistivity of elementary grains is described by phenomena which are essentially independent of temperature. If the electron mean free path is much smaller than the grain size the conductivity is independent of temperature because the impurity scattering dominates (dirty limit). If the mean free path, conversely, is much larger than the grain size, then scattering at the grain boundary is responsible for the losses, which will be independent of temperature from 0 K to several hundred C. Intermediate cases have in principle a mild temperature dependence, due to the varying skin depth as a function of resistivity (square root law), but this dependence eventually saturates due to the anomalous skin effect. This effect limits the surface resistance variations to a factor of six between room temperature and 0 K. Due to this limited change in microwave losses in elementary grains as a function of temperature, the overall losses in an artificial dielectric will be limited. Because most artificial dielectrics contain a distribution of particle sizes, the changes in losses due to the changing skin depth in a specific grain size will be compensated for by the increased losses in a contiguous grain size [11].

The actual choice of materials that will yield specific electric and magnetic properties is difficult in practice, because of the poor control of powder properties, and becomes even more difficult if the artificial dielectric to be manufactured is in a ceramic form. However, artificial dielectrics present properties which are a much milder function of the constituents' properties than other types of lossy materials, and for this reason ceramic artificial dielectrics can in many cases provide the required loss characteristics.

CERAMIC ARTIFICIAL DIELECTRICS

Lossy ceramic materials have been used in several applications where thermal conductivity, good vacuum properties or high dielectric constant are important. Among these are terminations for traveling wave tubes (TWT's) and lumped resonators in planar microwave circuits. Most of these applications have relied on the semiconducting losses of materials such as SiC, properly sintered by itself or within a ceramic host. In many cases the properties of these materials are not well understood or reproduced, because the very process of sintering can drastically alter the semiconducting properties of the lossy material.

In order to obtain a well-behaved ceramic version of the artificial dielectric materials, the following conditions must be satisfied: a) The conducting material (not semiconducting) cannot melt at the sintering temperature (in the range of 1500 to 2000 C for typical host materials). b) The conducting material cannot react chemically with the host ceramic at the sintering temperature. (Chemical reactions can alter the conducting properties of the original powders beyond the point of prediction of properties.)

The conductivity of the powders can be altered by the high temperature treatment since, for instance, recrystallization at high temperatures tends to lower the resistivity, whereas impurity diffusion tends to increase it. Often unpredictable conductivities can result from these competing mechanisms.

Among the materials which satisfy the above requirements there are just a handful of metals, semimetals, alloys and compounds: W, Mo, Ta, Nb, C and a few other rare metals and their alloys. Most of these materials and compounds can be obtained in a powder form, for instance through plasma spraying techniques [13].

For this work several materials were sintered [14] in a ceramic form, in order to produce specific dielectric properties which would meet required absorption for CEBAF [1]. Among the materials that were tested, we report here on mixtures of ceramics composed of AlN as a matrix, and with conductive powders of Mo, TiC and glassy carbon (an amorphous form of carbon obtained by pyrolysis of phenolic resins) [15].

The relative dielectric constant of AlN is 8.5. The data in Table I show the increase in dielectric constant due to the presence of the conductive powders. The data are consistent with the artificial dielectric model predictions for the given concentrations.

At these frequencies the loss tangent of AlN is negligible compared to the loss tangents of the composite materials. Therefore most of the losses occur within the conductive grains.

Table I. Physical properties of conducting powders

Mat'l	T(melt) [C]	Avg. Grain Size [μm]	Vol. [%]	ϵ_r [@ 2 GHz]	tg δ	ρ [$\mu\Omega$·cm]	δ[@ 2GHz] [μm]
Mo	2617	2.7	12	10	0.02	5.7[16]	2.7
Mo	2617	2.7	15	19.5	0.01	5.7	2.7
Mo	2617	2.7	16	27.5	0.02	5.7	2.7
TiC	3140	15	16	67.5	0.18	170[18]	14.7
TiC	3140	1.3–3	16	18	0.01	170	14.7
TiC	3140	1.3–3	18	23	0.03	170	14.7
TiC	3140	1.3–3	20	32	0.07	170	14.7
C	3550	3–12	15	22	0.16	1k–5k	35–80
C	3550	20–50	15	33	0.22	1k–5k	35–80

The minority conducting phase is always held near or below 15 percent in order to avoid temperature dependent phenomena. In the case of glassy carbon this choice of concentration also provides structural strength even though glassy carbon does not sinter with the dielectric ceramic: larger concentrations would make the structure unstable. Structural integrity considerations not only limit the powder concentration, but also the size of the powders that can be used. Whereas for general artificial dielectrics any powder size can be used, as long as it satisfies the electrical properties criteria, in sintering ceramic structures only a limited range of powder sizes is permissible in order to maintain structural integrity of the composite ceramic. Typical sizes range from less than a micron to several tens of microns. This size limitation has also an impact on the range of grain resistivities which can be chosen to more effectively damp the microwave frequency band.

Table I also gives the expected skin depth in the grains at 2 GHz, a critical frequency to be damped in the CEBAF cavities. For most powders the skin depth and the grain sizes are chosen to be very close at this frequency. The theoretical skin depth of the various materials is calculated at 2 GHz for published values of conductivity of bulk materials [16], except for glassy carbon, the conductivity of which is derived from data of Guillot *et al.* [17].

Measurements of permittivity and loss tangent were performed with an HP85070A dielectric probe kit with an HP8753C network analyzer.

Samples of these materials have been tested for rf properties at room temperature to obtain baseline dielectric data, and then absorbers have been cooled to 2 K to determine the presence or absence of thermally induced absorption phenomena. All of the materials presented here showed no modifications of the dielectric properties with temperature, unlike other samples containing SiC, which show low temperature changes of dielectric properties.

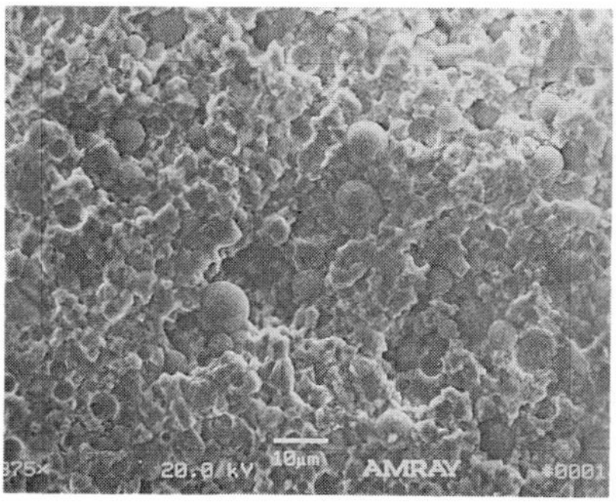

Figure 1. SEM image of an AlN-glassy carbon composite ceramic. The carbon particles are easily identifiable by their spherical shape.

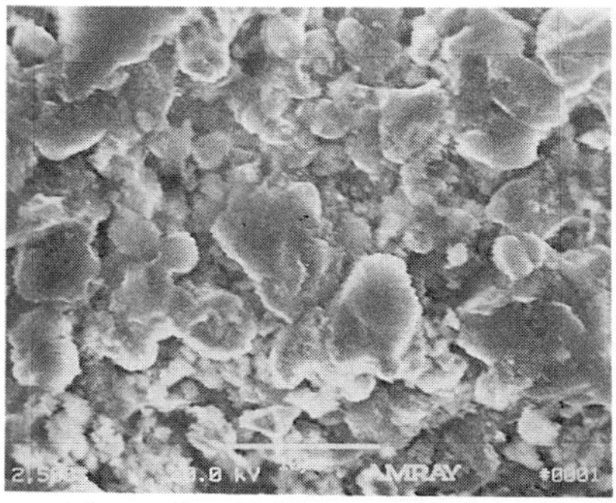

Figure 2. SEM image of the AlN-Mo composite.

For the production of absorbers in the accelerator the AlN-glassy carbon ceramic was chosen, having a reproducibly high loss tangent. The perfectly spherical particles of this material (Figure 1) make a possible theoretical modeling simpler.

The vacuum properties of this fully dense ceramic have been tested and upper limits to its outgassing rates have been set at about 1×10^{-10} (torr liter)/(cm^2 sec).

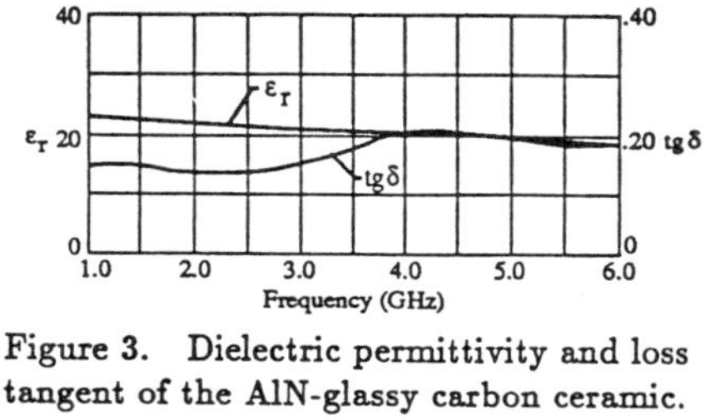

Figure 3. Dielectric permittivity and loss tangent of the AlN-glassy carbon ceramic.

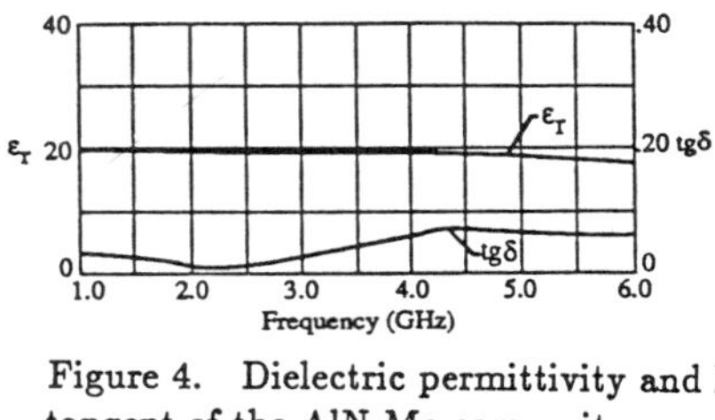

Figure 4. Dielectric permittivity and loss tangent of the AlN-Mo composite.

CONCLUSIONS

Various ceramic materials have been studied and produced for practical applications where microwave absorbers must coexist with very stringent environmental conditions (ultra-high vacuum, low or high temperatures, high power densities etc.). Specific dielectric constants and loss tangents can be achieved and reproduced by using the artificial dielectric model to design the ceramic composition. The additional low temperature studies help in discriminating the basic loss mechanisms which take place in these lossy ceramics. The resulting understanding can benefit both the development of methods to sinter ceramics through microwave absorption and the development of ceramics sintered for microwave absorption applications.

ACKOWLEDGMENTS

The authors would like to thank Dr. L. R. Doolittle of CEBAF for kindly providing the SEM pictures of the ceramic samples.

REFERENCES

1. I. E. Campisi, K. E. Finger, L. K. Summers, and A. M. Johnson, "Higher-Order-Mode Damping and Microwave Absorption at 2 K," presented at the Third European Particle Accelerator Conference, Berlin, March 24-28, 1992.
2. J. M. Ziman, *Electrons and Phonons* (Oxford, 1960), Chap. 9.
3. F. Seitz, *The Physics of Metals* (McGraw-Hill, 1943), Chap. 21.
4. R. E. Newnham, S. J. Jang, Ming Xu, and F. Jones, "Fundamental interaction mechanisms between microwaves and matter," *Ceramics Transactions* **21**, 51–67 (1991).
5. V. M. Kenkre, "Theory of microwave interactions with ceramic materials," *ibid.*, 69–80.
6. Y. Takeda, "Development of high-thermal-conductive SiC ceramics," *Ceramic Bull.* **67** (12), 1961–1963 (1988).
7. M. M. Z. Kharadly and W. Jackson, "The properties of artificial dielectrics comprising array of conducting elements," *Proc.* Institute El. Eng. **9100**, 199–212 (1953).
8. R. E. Collin, *Field Theory of Guided Waves*, 2nd ed. (New York, 1991), Chap. 12.
9. W. T. Doyle and I. S. Jacobs, "Effective cluster model of dielectric enhancement in metal-insulator composites," *Phys. Rev. B* **42**, (15), 9319–9327 (1990).
10. N. E. Russell, J. C. Garland, and D. B. Tanner, "Absorption of far-infrared radiation by random metal particle composites," *Phys. Rev. B* **23**, 632–639 (1981).
11. I. S. Jacobs, "Advanced artificial dielectric materials for millimeter wavelength applications," Report 90-SRD-001, GE CRD, Schenectady, 1990.
12. S. Kirkpatrick, "Percolation and conduction," *Rev. Mod. Phys.* **45**, (4) 574–588 (1973).
13. R. F. Cheney, R. L. Daga, R. M. German, A. Bose, and J. W. Burlingame, "Heavy alloys from rapidly solidified powders," *Proceedings of the 1988 International Powder Metallurgy Conference and Exhibition*, Orlando, Florida, June 1988.
14. Sintering done by Ceradyne, Inc., Costa Mesa, Ca.
15. Manufactured by SIGRI, Inc.
16. *CRC Handbook of Chemistry and Physics*, 62nd ed.
17. T. Guillot and G. Bobillot, "Mesure en hyperfrequence de la conductivite' electrique d'un grain elementaire d'une poudre conductrice," presented at SEE Conference, Limoges, 1991.
18. *Handbook of Thermophysical Properties of Solid Materials*, Vol. IV, (New York, 1961).

INTERACTION BETWEEN GYROTRON RADIATION AND POWDER MATERIALS

V. SKLYAREVICH [*,**], A. DETKOV [*], M. SHEVELEV [*,**], R. DECKER [**]
* E. O. Paton Electric Welding Institute, Kiev, Ukraine
** Paton Tek, Inc., 717 East Huron, Ann Arbor, Michigan, (313)995-5550, USA

ABSTRACT

To capitilize on the special advantages of gyrotron radiation for fast and continuous materials processing, surface and interface layers are "alloyed" with lossy chemicals and fine-disperse powders. The principles and practical uses are described herein.

Since 1986 the E. O. Paton Electric Welding Institute has been involved in research and development on application of microwave radiation for technological purposes. A gyrotron, the simplest and most efficient generator of millimetric waves, is used in the research. The gyrotron generates waves in the form of Gaussian beams having the power of tens of kilowatts along with high (more than 30%) efficiency. By means of simple metal mirrors this radiation can be given practically any shapeincluding strip, circle and focused spot, etc.

The gyrotron radiation generated at frequencies from 35 to 100 GHz has some special features, as compared to the widely used radiation of the centimetric range generated at frequencies of 915 MHz and 2.45 GHz. First of all, the millimetric radiation is by 3 orders of magnitude more efficient in interaction with nonmetallic materials than centrimetric radiation. Besides, it is characterized by the high selectivity of heating an irradiated material. All this determines the fields of the most rational application of the gyrotron radiation. In our opinion, it is better to use it for fast processes, when a process is based on short-time heating. In such cases, no insulation is needed opening the way for continuous processes.

When developing such technologies one has to solve two main problems. The first problem consists of insuring that sufficiently high rate of physico-chemical or mechanical processes occur in a material exposed to the gyrotron radiation so that this high rate leads to the desirable final result, e.g. polymerization, melting and diffusion. It had already been proven that centimetric radiation could reduce autoclave times of hours to microwave curing times of minutes. Now our new practice opens the way for engineering design and development of new compositions, primers and solders which can respond adequately to the faster heating afforded by the gyrotron. The end result is technological process cycles in seconds rather than minutes. As an example, this has been observed in burning of pastes in microcircuits.

The second problem is the need to concentrate heating in only those layers or volumes that need thermal processing, without overheating other sensitive components. Again, this can be solved by the increase in absorbing capacity of the specific material of the layers being processed. It can be achieved by different methods: e.g. by adding absorbing agents, by creating resonant electromagnetic systems or by selecting the workpiece components. An example of the latter is the technology we developed for deposition of anticorrosion coatings on oil and gas pipelines. According to existing technologies the two-layer film "polymer-primer" is glued to the metal pipe surface by gas heating. The temperature of heating is 120 - 160 °C, thus it does not exceed the temperature of destruction of a polymer. However, with such coating deposition technology, there is drastic decrease in adhesion strength under the elevated temperature and higher humidity service conditions. This is associated with the chemical nature of adhesion which, in this case, is caused by a bond forming between active radicals and hydroxyl OH present on the metal pipe surface. It was expected that in case of the drastic elevation of the primer heating temperature the mechanism of adhesion would change for the better, but then it was important to prevent destruction of the protective polymeric film.

Therefore, we sought to increase the absorbing capacities of the primer relative to the film. To solve this problem we made calculations, where it was assumed that radiation in the form of the plane electromagnetic wave fell on the three-layer medium in the free space and consisting of the successive layers of polyethylene, the absorbing layer (primer) and the metal layer. The equations for the amplitudes of electric field, E, and temperature, T, :

$$\frac{d^2E}{dx^2} + \frac{\omega^2}{c^2}\left[\varepsilon(\chi,T) - \frac{i4\pi\sigma(\chi,T)}{\omega}\right]E = 0$$

$$\frac{\partial T}{\partial t} - \frac{\partial}{\partial x}\left[a(\chi,T)\frac{\partial T}{\partial x}\right] = \frac{\sigma(x,T)E^2}{2C_p(\chi,T)\rho(\chi,T)}$$

where: ω is the frequency of the microwave radiation, ε is the effective apart of dielectric permeability, σ is the conductivity, $\mathbf{C_p}$, ρ , **a** are the heat capacity, the density and the thermal conductivity of the heterogeneous medium, **c** is the velocity of light in vacuum, were solved numerically for the following conditions: losses of the microwave radiation in the polyethylene layer are small, according to our measurements, $\sigma = 1.1 \cdot 10^{-13}$ (0hm cm)$^{-1}$. For metal $-\sigma = \infty$. Conductivity of the primer layer was set equal to $\sigma = 1.8 \cdot 10^{-3}$ (0hm cm)$^{-1}$ and its dependence on the temperature could be described by equation $\sigma = \sigma_0 \exp[\beta(T - T_0)]$.

It can be seen from distribution of the temperatures in Fig. 1 that the primer can be heated for a short time up to the temperatures of 600 °C without any destruction of polyethylene since it remains at low temperature.

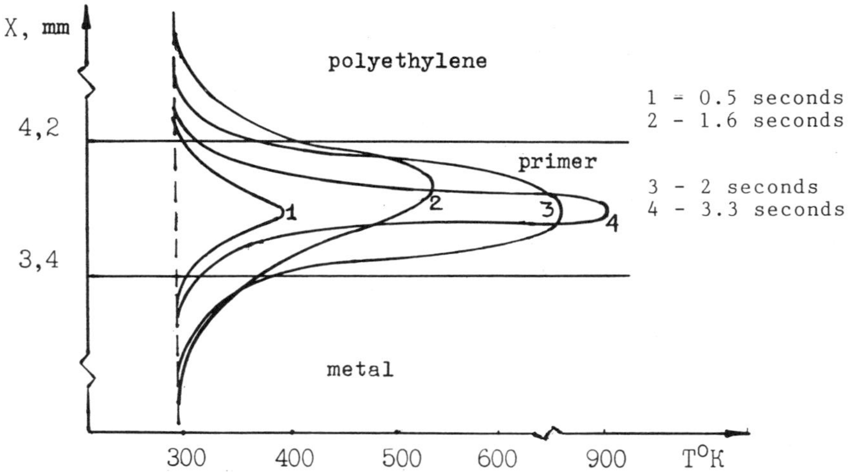

Fig. 1. Distribution of temperature in three-layer system at different moments of time

In accordance with the calculations performed, we developed a special primer having a factor of absorption of the gyrotron radiation of 0.33 cm^{-1} provided by selection of resins, polyolefins and other components. As a result of irradiation of the film with this primer, adhesion of a coating under the elevated temperature and higher humidity does not practically change with time in service, and this allows to considerably increase the life of pipelines. This technology was used to deposit coatings on the 1420 mm dia. main pipelines.

The problem of deposition of ceramic coatings on refractory items was also solved by selecting the components. We succeeded in providing preliminary melting of the underlying surface of an item coated before melting of the coating. This resulted in formation of a wide transient zone and, hence, a thick heat-resistant ceramic coating.

In a similar way, we managed to provide high-temperature brazing of elements of the structural ceramics items by developing a brazing alloy with high absorbing capacity. Melting point of the alloy can reach 1600 - 1800 °C, thus insuring the layer strength required for the high-temperature operation.

However, in a majority of cases we can not realize the advantages of the gyrotron radiation just simply by selecting materials for the treated item components. For this we often have to use absorbing agents, mainly in the form of powders, added to a material irradiated. In fact fine-disperse metal powder turns out to be most efficient for this purpose.

It is a known fact that microwave radiation interacts but slightly with metal, since its reflection factor exceeds 99%. But, with powder, due to multiple reflection from each particle, the total absorbing capacity increases. A pronounced enhancement is observed with the mixture of metal and dielectric powders. Finding the dependence of absorbing capacity of such mixtures on a) size of the particles, b) their electrophysical characteristics, c) their concentration and d) the radiation wavelength is of a great interest for practical purposes.

In a general case, the task is to find a solution for the electrodynamic problem in order to determine absorption of the microwave radiation by a multi-component medium with a pronounced frequency dispersion. Another issue in solution of such problems is frequency dispersion. The main issue in solution of such problems is the correct description of the multi-component system by so-called effective characteristics of the medium. For this it is necessary to take into account the multiple dissipation of the microwave radiation at particles and to statistically average the result of sizes, shapes and orientation of the particles. The most correct calculation of the effective electrodynamic parameters is possible only within the theory of multiple dissipation of waves [1,2]. However, in this case the result will not reflect the entire picture of interaction between the microwave radiation and the fine-disperse medium depending on the radiation frequency, the mean sizes of the particles and their concentration. It is associated with the fact that the statements of this theory about peaks (narrow lines) of the resonance absorption related both to the plasma oscillations of electrons and the resonance transitions [3, 4, 5] are not observed in the mixtures used in practice. Therefore, we solved the problem within the scope of classical electrodynamics and for the systems "metal-dielectric" we chose the solution based on the method of the effective medium [6]. At the high degrees of filling the medium with the metal component (of the order of 35%) it is necessary to take into consideration the dipole-dipole interaction between the particles, this being possible when using the approximated diagram of self-consistent mean field (method of coherent potential [7]).

In a general case for the spherical particles, on the basis of the methods and the assumptions made, we obtain for $\tilde{\varepsilon}$ and $\tilde{\mu}$:

$$\tilde{\varepsilon} = \varepsilon_m + \frac{3f\tilde{\varepsilon}(\varepsilon - \varepsilon_m)}{2(\tilde{\varepsilon} + \varepsilon) - f(\varepsilon - \tilde{\varepsilon})};$$

$$\mu = \mu_m + \frac{3f\tilde{\mu}(\mu - \mu_m)}{2(\tilde{\mu} + \mu) - f(\mu - \tilde{\mu})};$$

where: ε , μ are the dielectric and magnetic permeabilities of the dielectric, ε_m, μ_m are the dielectric and magnetic permeabilities of the metal, determined on the basis of Mie theory [9].

The accurate solution for the problem of dissipation of the plane electromagnetic wave at the spherical conducting particle was suggested by Van de Hulst [8]. Using this solution we determine ε_m and μ_m as functions of the particle sizes (**r**), the frequency of the incident radiation (ω) and the conductivity (σ).

Assume the particular parameters of the two-component system polymeric composition -copper with the following values: $\varepsilon_{p.c.} = 3.7$, $\mu_{p.c.} = 1$, $\sigma_{Cu} = 5 \cdot 10^{17} s^{-1}$, $\varepsilon_{Cu} = 2$ and the variable values of sizes of the fine particles (**a**) and the volume concentration (**f**), as well as consider the law of absorption of the plane-polarized microwave which is known to have the form of $\bar{I}(x) = \bar{I}_o \exp(-\alpha \cdot x)$, where $\bar{I}(\mathbf{x})$ is the intensity along the line of the wave propagation; $\alpha = 2\,\omega / \mathbf{c} \cdot x$, where ω is the radiation frequency, $\chi = I_m \sqrt{\tilde{\mu}\,\tilde{\varepsilon}}$. As a result, we have the possibility to calculate the dependencies of the absorbing capacity, P, on the radius of the copper particles for various frequencies (Fig. 2a) and on the concentration of the copper particles in the system (Fig. 2b). It can be seen that there is the clearly defined maximum of the absorbing capacity of the mixture, depending on the particle sizes, which varies with the frequency of the radiation and the concentration of the copper particles.

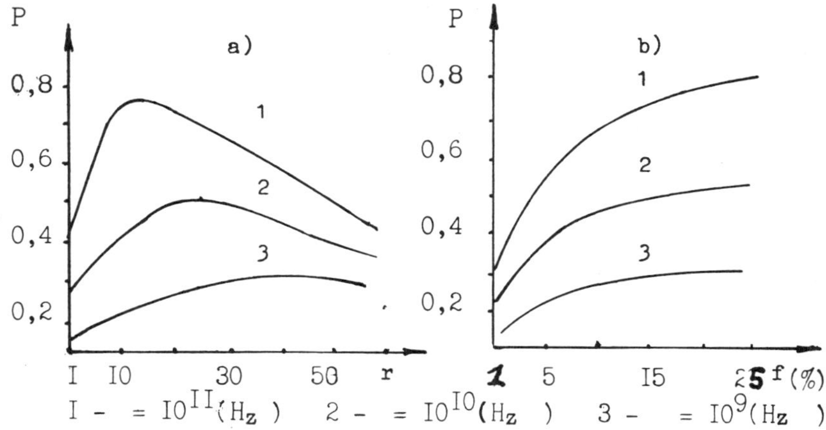

Fig. 2. Dependence of absorption on sizes of particles (a) and concentration (b)

The research conducted allowed optimization of a number of technologies associated with heating of a mixture of the metal-dielectric powders. First of all, this is the process for manufacture of printed circuit boards following the so-called thick-film technology. The main point of the process is selective heating of the copper paste deposited on the surface of glass-reinforced plastic. By selecting the sizes of the copper powder particles and the radiation wavelength we achieved the excess of the paste absorbing capacity over the substrate material capacity by more than an order of magnitude. This, in its turn, allowed selective heating of the paste up to 240 °C. Here, the temperature of the glass-reinforced plastic did not exceed 60 - 80°C. The technology is fast and continuous and provides the required properties of the conductors.

Similarly, we have taken into account the research results for other technologies associated with soldering using the metal solders. By selecting the sizes of the solder particles, as well as the fluxes with certain electrophysical properties, we can achieve local melting of the solder and avoid inadmissible heating of other elements of an item. For example, this soldering was realized in compacting synthetic diamonds and in soldering radio components to printed circuit boards.

The results of the researches conducted are equally suitable for powders of inorganic materials having high conductivity. Examples are carbides. We used them to develop a few interesting processes. One process is heating of silicon or boron carbide in nitrogen-oxidation environment by the gyrotron radiation. At the temperature of 1400 - 1600 °C carbon is oxidized and remaining boron or silicon form nitride. The peculiar feature of this process is that as soon as the nitride particle is formed, its heating automatically stops, because tangent of the loss angle of nitride in the millimetric range is by two orders of magnitude smaller than that of carbide. As a result, the thermal process will last until transformation of all carbide powders into nitride ones is completed. Another example concerns the field of surface alloying treatment of metals. The technology has been developed which consists of heating a mixture of boron carbide powders and various oxides at the surface of a metal workpiece. During a few seconds the mixture can be heated up to the melting temperature of the treated metal surface, thus accelerating the boron diffusion into its structure by an order of magnitude. Here, the workpiece is not subjected to excessive heating, hence, its properties are not deteriorated.

Of interest is annealing of fine atomized iron powder used for manufacture of flux-cored welding wire. The short-time effect of the gyrotron radiation leads to a 5 time decrease in the content of the OH groups at the particle surfaces. Conventional multi-hour annealing of such powder in a furnace at the 300 - 350 °C temperature leads to decrease in OH groups only by 25 - 30 %. At the same time the decrease in the hydroxyl content reduces the hydrogen diffusion in welding of critical structures of high-strength and high-alloy steels.

Annealing of the semiconducting crystals also leads to the unexpected results. During the short-time effect of the gyrotron radiation we can stimulate the appearance of new impurity lines (Fig. 3). Here, no degradation of physical properties of the crystals is observed, since there are no deformation stresses usually formed in application of other heat sources.

Conclusions

Gyrotron microwave energy is highly efficient and affords fast, continuous processing of materials without encumbrance by insulation.

Appropriate materials design can enhance the benefit of super-high frequency. Localized heating at surfaces and interfaces is offered by "alloying" at those locations with lossy chemicals and fine-disperse powders of optimum size and volume percent. Then, surrounding materials are not subject to harmful overheating.

Some useful materials design and processing techniques were demonstrated as follows:

a. Alloying of primer layers to adhere polymer to metal
b. Alloying of powders for surface glazing of ceramics
c. Brazing of ceramics and diamonds
d. Fine-disperse powders for curing thick films
e. Fine-disperse powders for soldering of circuit boards
f. Synthesis of and cleaning of powders
g. Treatment of semi-conductor crystals

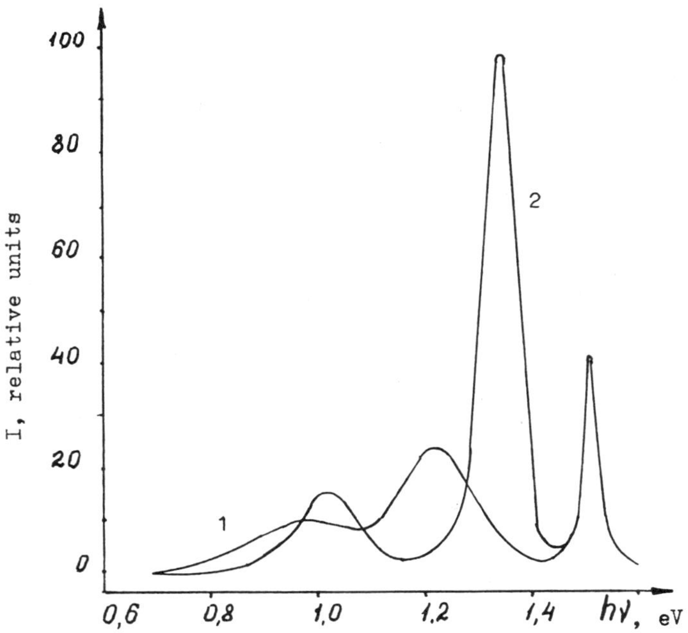

Fig. 3. Spectra of non-irradiated (1) and irradiated (2) crystals of n-GaAs

References

1. B. Twersky. J. Opt. Soc. Amer. 52, (2), 145-171 (1981).

2. A.P. Prishivalko, V.A. Babenko, V.N. Kuzmin. Dissipation and Absorption of Light by Heterogeneous and Anisotropic Spherical Particles. (Nauka i Tekhnika, Minsk, 1980), p. 192.

3. L.Gensel, T.P. Martin. Surf. Sci. 34, 33-49 (1973).

4. C.F. Borren, N.C. Wickramasingle. Astrophys. Space Sci. 50, 461-472 (1977).

5. C.F. Borren, D. Hafmen. Absorption and Dissipation of Light by Small Particles. (Mir, Moscow, 1986), p. 660.

6. D. Stroud. Phys. Rev. B, 12, 3368 (1975).

7. G. Erenreih, L. Shvarts. Electron Structure of Alloys. (Mir, Moscow, 1979), p.200.

8. H.C. Van de Hulst. Light Scattering by Small Particles. (Dover, New York, 1981), p.150.

9. C. Mie. Ann. Phys. 25, 377 (1908).

PART VI

Microwave Processing of Ceramics

MICROWAVE PROCESSING OF CERAMICS: Guidelines Used at the Oak Ridge National Laboratory

MARK A. JANNEY, HAL D. KIMREY, AND JAMES O. KIGGANS
Oak Ridge National Laboratory*
P.O. Box 2008
Oak Ridge, Tennessee 37831-6087

ABSTRACT

To make meaningful comparisons between conventional and microwave processing of materials, one must conduct experiments that are as similar as possible in the two environments. Particular attention must be given to thermal conditions, sample parameters, and furnace environment. Under thermal conditions, one must consider temperature measurement (pyrometer or thermocouple, sheath type, and arcing of thermocouples), thermal history (heating and cooling rates, thermal gradients), and exothermic reactions. Regarding sample parameters, one must consider sample size, and packing powders and insulation systems. With respect to furnaces, one must consider differences in atmosphere, impurities, and uniformity of heating. Examples will be drawn from diffusion, grain growth, sintering, nitridation, and drying experiments conducted at the Oak Ridge National Laboratory (ORNL) over the past six years.

INTRODUCTION

There is a critical need to compare the results obtained from microwave processing with those from conventional processing when conducting an investigation of the effects of microwave heating on materials. Side-by-side comparisons are required for at least three reasons. First, literature data from similar investigations are not sufficient. For example, in the study of diffusion, the scatterband of data from various investigators using different materials is typically an order of magnitude or greater. Second, there are lot-to-lot variations in the composition and processing of materials that greatly affect their properties. If the same material is not used in both the conventional and the microwave studies, meaningful comparisons might not be made. Finally, the claims in the microwave processing literature such as lowered sintering temperatures [1,2], accelerated kinetics in ceramic, polymeric, and organic systems [3-6], and reduced activation energies [1,3-5] are suspect to many researchers both inside and outside the microwave processing community.

STANDARD ASSUMPTIONS

The experimental procedures described in this paper are built upon a few standard assumptions that are common to all of the experiments cited herein. Multimode cavities were used in all of the work reported here. Thus, there are certain details that might not be strictly applicable to work in single-mode cavities. Also, most of the experiments reported here were conducted at 2.45 GHz, the most common microwave processing frequency. However, the general principles involved are the same irrespective of cavity mode structure or frequency. It is further assumed that processing parameters such as heating rate, soak temperature, processing atmosphere and total pressure, and calibration of instruments such as pyrometers and thermocouples are the same for both conventional and microwave processing.

---

*Operated by Martin Marietta Energy Systems, Inc., under contract DE-AC05-84OR21400 for the U.S. Department of Energy.

THERMAL EFFECTS

The ability to heat a material

There are major differences between microwave and conventional thermal processing. To heat a part conventionally only requires putting the part into a furnace and heating the furnace to the required temperature. If enough time is taken to permit heat to flow into the part, it must, by necessity, heat to the required temperature. Such is not the case when heating a material in a microwave furnace. Whereas heat is supplied externally to the part being processed conventionally, it is generated within the part being processed by microwaves. Depending on the dielectric properties of the part, several different phenomena may occur: (1) it may heat quickly; (2) it may not heat at all; (3) it may heat uniformly; or (4) it may generate hot spots. How the part heats depends critically on the dielectric loss properties of the part, how those properties change with temperature, and on the microwave field distribution in the furnace. It is important to remember that microwave furnaces are not like conventional furnaces.

In addition to chemical and thermal considerations, dielectric properties must be considered in microwave heating. This includes the material being processed, the insulation surrounding the part, and auxiliary materials such as thermocouples. As an example, consider the parts shown in Fig. 1. The ZrO_2 - 12 mol% CeO_2 part shown in Fig. 1a has a conical shape from top to bottom because the temperature was much higher on the top. The temperature gradient led to severe cracking because of differential sintering. The insulation that was used around this part was pelletized ZrO_2 - 12 mol% CeO_2, which heated very nonuniformly. The sample shown in Fig. 1b sintered very uniformly because uniform heating was achieved. In this case, zirconia fiber insulation of composition ZrO_2 - 8 mol% Y_2O_3 was used, which heated much more uniformly than the pelletized ZrO_2 - 12 mol% CeO_2.

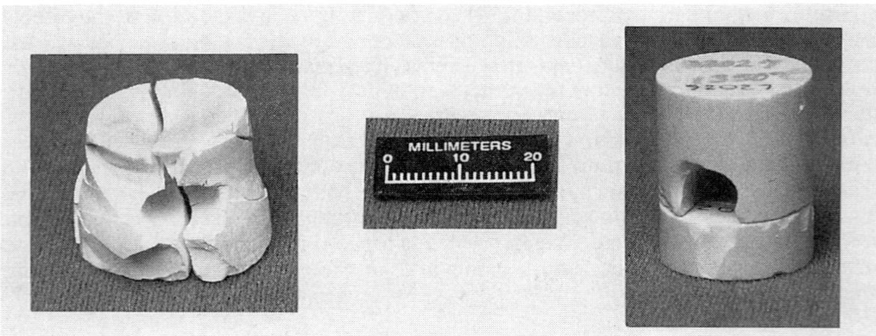

ZrCe 92006
Sample Hotter on Top
Differential Sintering
Cracking
Conical Shape
Pelletized TZ12CE Matched Insulation

ZrCe 92027
Uniform Heating
Good Sample
ZrO_2 (8% Y_2O_3) Bulk Fiber Insulation

Figure 1. ZrO_2 - 12 mol % CeO_2 was challenging to process at 2.45 GHz.

Some materials cannot be processed at all. For example, carbon fiber preforms arc continuously upon exposure to microwave energy. The arcing is a result of the high electrical conductivity of the fibers combined with their small radii, which tend to concentrate the electrical field around them.

The dielectric loss factor of materials being microwave processed must be within certain ranges to be heated successfully. The microwave conductivity of bulk metals is much too high to permit their heating by microwaves; they simply reflect the energy back into the cavity. However, finely divided metals have been heated in many cases [7-9]. The increased surface area provided by powders accentuates the role of the "skin depth" of the metal. The skin depth is that near-surface region (on the order of 1 to 100 μm) which absorbs a small amount of the microwave energy incident on the metal surface owing to its finite conductivity. In a metal powder, a significant fraction of the material is in the near-surface region, and hence can play a role in microwave absorption. At the other end of the spectrum are low loss insulators, which are difficult to heat because they absorb little of the microwave energy incident upon them. Materials such as high purity alumina, high purity magnesia, and high purity silica can be heated at 2.45 GHz only at high field strengths (e.g., in a tuned cavity at high power density), at high frequencies (e.g., above ~20 GHz), or with auxiliary preheating. Table 1 lists several materials that have been successfully processed in multimode cavities at 2.45 GHz at ORNL.

The use of indirect heating may be required (1) to preheat a material to a temperature at which it will couple to 2.45 GHz microwaves or (2) to mitigate problems associated with thermal runaway. Examples of indirect heating techniques include the "picket fence" [2,14,15] coatings [16], and lossy second phases [8,17]. All of these have been used successfully in materials processing.

Temperature measurement

The ability to measure temperature accurately is crucial to conducting meaningful microwave and conventional experiments. In fact, the first question asked by observers from outside the microwave community is often - "how do you know that you are measuring a meaningful temperature in the microwave furnace?" Even after standard temperature measurement procedures have been established in a laboratory, it is important that the need for accurate temperature measurement remain in the forefront of the researcher's concerns.

Table 1. Various ceramics have been processed at 2.45 GHz at ORNL.

Material	Maximum Temperature Achieved (°C)	Auxiliary Heating Required	Comments
SiC	1600	No	Parts as large as 2.5 kg have been heated to 1600°C at rates up to 2000°C/h.
$LaCrO_3$	1700	No	Arcs to bare thermocouple [10].
NiO	900	No	Could not be heated above 900°C due to increase in reflectivity of NiO at high temperature
Fe_3O_4	600	No	Oxidation to Fe_2O_3 prevented further heating.
B_4C	2500	No	Heating must be done quickly [11].
TiB_2	2200	No	Yttria insulation [12].
Silicon	1400	No	Melting prevents further heating. [13].
Al_2O_3	1600	Yes	Conducting particles added. [8].
Al_2O_3	1600	Yes	Picket fence used [14].
ZrO_2	1600	Yes	Picket fence used [2].
Al_2O_3-ZrO_2	1500	Yes	Picket fence used [15].

At ORNL, both thermocouples and radiation pyrometers are used extensively for temperature measurement. Several guidelines can be formulated regarding the preferred techniques at ORNL, based on our experience. Thermocouples are most often used for direct measurement of sample temperature and control of the heating program for materials processing. Thermocouples must be shielded from the microwave field to prevent arcing of the fine thermocouple wires. Molybdenum, Inconel, stainless steel, and platinum sheaths have been used successfully on Type C (tungsten-rhenium), Type K (chromel-alumel), and Type S (platinum-rhodium) thermocouples. The sheath diameter is typically 1/8 inch (3mm). This diameter represents a reasonable compromise between response time of the thermocouple (one prefers a small diameter sheath to reduce response time) and the tendency of the sheath to arc in the furnace (arcing is reduced by using a large diameter sheath). Figure 2 shows the variation in time of both the electric field and the sample temperature (as indicated by a Type-C thermocouple embedded in an alumina part) in a 28-GHz microwave furnace[3]. As the microwave electric field is cycled off and on, the thermocouple signal responds with decreasing, then increasing, temperature. However, there are no discontinuities in the temperature measurement when the power is turned off or on; this shows that the microwave field has no effect on the temperature measurement as made by the thermocouple. Only normal heating and cooling effects are observed, which are similar to those that would be observed in a conventional furnace if the power were turned off and on. Additional support that thermocouple temperature measurements are correct is given by the following example. For the highest microwave annealing temperatures that were used in the alumina heating study, i.e., 1600°C, no degradation, densification, or embrittlement of the alumina bulk fiber insulation used was observed. When the bulk alumina fiber insulation is heat treated conventionally, it degrades rapidly at temperatures above 1600°C. If the temperature measurement in the microwave furnace had been significantly incorrect, say that 1500°C was recorded when the temperature was actually 1700°C, then densification and embrittlement of the insulation would have been observed. Since no such observations were made, the indicated temperature is likely the correct temperature. As a further example, it has been shown [10] that there is excellent agreement (~5°C) between the temperature measured by an embedded thermocouple and the temperature measured by a two-color infrared pyrometer when a sample of $LaCrO_3$ was heated in the microwave furnace. $LaCrO_3$ approximates a black body and represents an excellent material for temperature measurement using a two-color IR pyrometer.

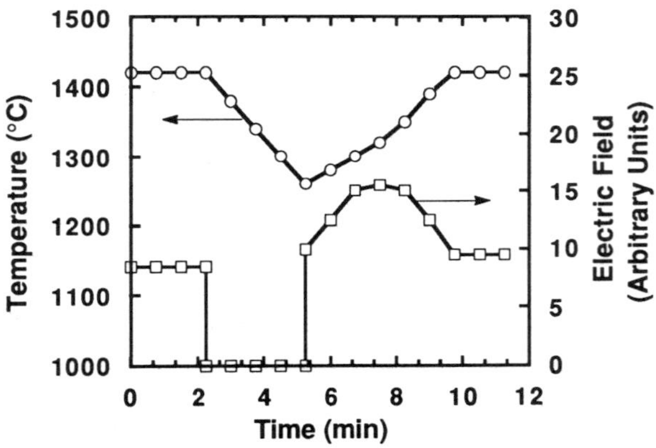

Figure 2. Turning the electric field on and off had no effect on the thermocouple readings during annealing of alumina at 28 GHz [3].

In conventional furnaces, accurate measurement of temperature is often obtained by using a sensing thermocouple in addition to the control thermocouple. The sensing thermocouple is placed directly on, or within a few millimeters of, the sample. In the procedures normally used at ORNL, the sensing thermocouple normally is the control thermocouple. To assure that good temperature measurements are made, particular attention must be made to placement of the thermocouple. Figure 3 shows two examples of thermocouple placement that have been found to give reproducible results. For a cylindrical specimen, a thermocouple well is drilled about 3/4 of the way through the sample. The thermocouple is then inserted until it abuts the back wall of the well. For plate specimens, the thermocouple is inserted between two plates. This ensures that the thermocouple sees the same thermal environment in both directions. Simply placing the thermocouple in contact with the surface of a cylinder, or on the top or bottom of a single plate, guarantees that the temperature that is measured will be too low. Instead of measuring the temperature of the part, the thermocouple will measure some average of the part temperature and the temperature of the surrounding insulation.

Temperature uniformity

Temperature uniformity within a sample is a concern on several scales. Johnson [18] has shown that there can be essentially no microscopic gradient within a sample caused by preferential heating of grain boundaries. His calculations were based on a model in which all of the microwave energy is deposited in a region close to the grain boundary. Even in grains as large as 100 μm, the worst case temperature differential was only ~0.1°C. In contrast, significant gradients can exist on a macroscopic scale in both microwave and conventional processing. Conventional heating deposits energy on the surface of a sample and depends upon thermal diffusion to transfer the heat to the center of the sample. Microwave heating deposits energy throughout the volume of the sample (assuming the penetration depth of the microwaves is larger than the thickness of the sample) with cooling of the sample from the surface. Therefore, the normal temperature gradient in a conventionally heated sample is opposite to that in a microwave heated sample, Fig. 4. The difference in surface and centerline temperature can be rather large. Figure 5 shows the surface and centerline temperatures for microwave and conventional drying of a castable insulating refractory, K3300 (Thermal Ceramics, Augusta, Ga.) The centerline temperature of the microwave dried sample is 30 to 40°C above the surface temperature. The centerline temperature of the conventionally dried sample is 20 to 30°C below the surface temperature. In this case, a one-to-one comparison of the microwave and conventional processes is difficult because the physics of the situation are so different in the two cases.

Thermal gradients often arise from improper insulating systems. During microwave processing, such gradients often lead to differential sintering as is shown in Fig. 1a. Development of a workable insulating system may require extensive effort. For example, the sample shown in Fig. 1b was the first "good" sample processed in the ZrO_2-12 CeO_2 system at ORNL. This was the 27th sample of the series. Numerous other insulating schemes were tested to try to alleviate the thermal gradient problems. The first 26 attempts were unsuccessful. Unfortunately, such efforts are not unusual. Development of a workable, repeatable insulating system remains one of the most challenging tasks in the high temperature microwave processing of ceramics. To determine whether the insulating system is working appropriately normally requires investigation of the microstructure of the sample after firing. If the microstructure is uniform, one can be assured that the thermal history of all parts of the sample was the same. Microstructural development is the most sensitive probe of thermal history available.

Sometimes the gradients developed by microwave processing can be used to advantage. Clark, et al [19] have shown advantages in solid state high temperature synthesis (SHS) reactions from microwave heating. The reaction front is initiated at the center of the pellet, and then proceeds toward the periphery, which results in more complete reaction of the pellet than conventional heating. Likewise, Kiggans, et al [13] showed that nitridation of silicon proceeds from the inside of a powder compact to the surface under microwave processing. In this way, the material was converted from silicon to Si_3N_4 more completely than in conventional processing.

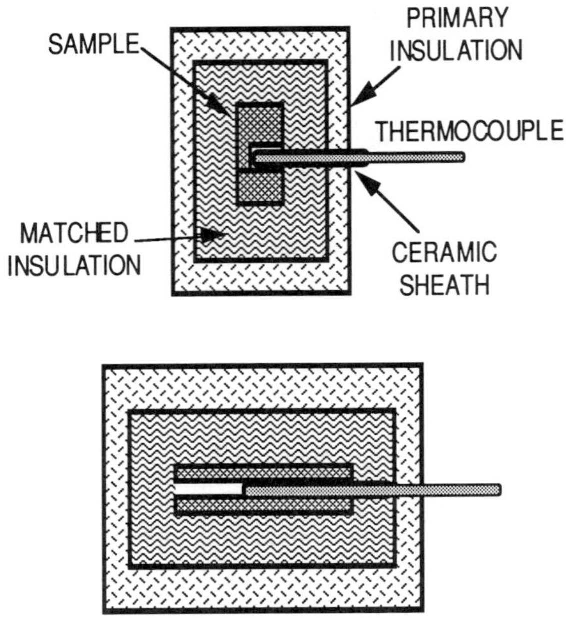

Figure 3. Examples of thermocouple placement for accurate temperature measurement in a microwave furnace for cylinder and plate sample geometries.

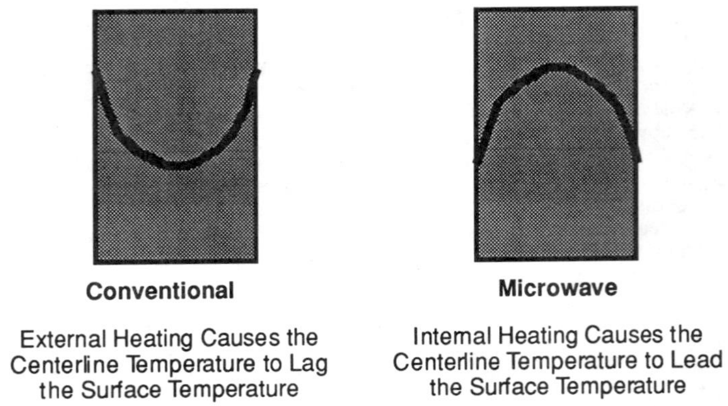

Figure 4. Temperature gradients generated during microwave and conventional heating of samples tend to be different.

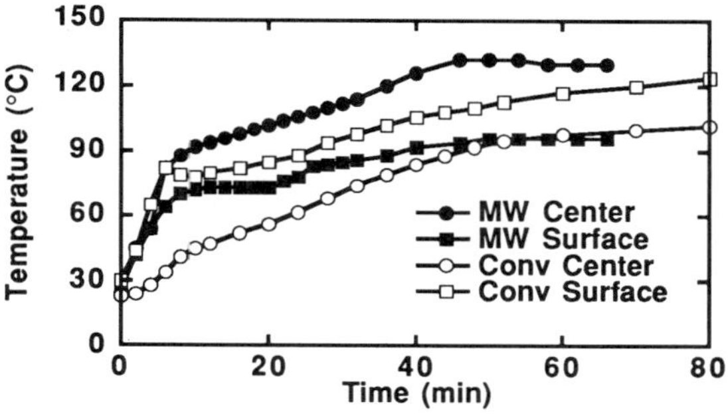

Figure 5. Microwave drying of a castable refractory (Thermal Ceramics K3300) raised the internal temperature well above the surface temperature.

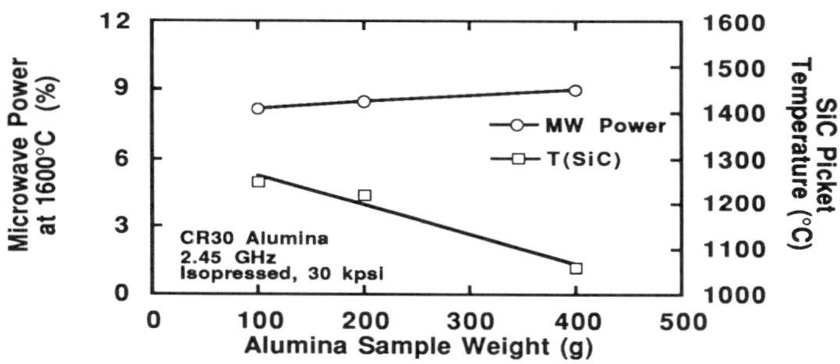

Figure 6. Increasing the size of an alumina sample reduced the SiC picket temperature and raised the total power indicating that more of the power in the furnace was absorbed by the alumina sample.

SAMPLE CONSIDERATIONS

Size

In general, samples used in microwave processing studies at ORNL are larger than those used in conventional studies. For example, conventional sintering studies are typically performed using 1 to 2g samples. Microwave sintering studies at Oak Ridge have typically used 30 to 150g samples [1,2,4,10,11,14,15]. The use of large samples in the microwave furnace is related to coupling power into the sample and to temperature measurement. Very small samples do not couple well to the microwave field. Larger samples couple better. (A demonstration of this is easily done with popcorn in a home microwave oven. A single popcorn kernel cannot be heated; a cupful of popcorn is easily popped.) Temperature measurement using small samples is difficult if thermocouples are used. The thermal mass of the sample relative to that of the thermocouple is small; hence, the thermocouple reads a temperature that is indicative of both the sample and the surrounding insulation, which is normally at a lower temperature than the sample, rather than just the sample temperature. If a radiation pyrometer is used (IR or optical), a small sample may experience significant radiative cooling. The pyrometer sight hole can represent a large fraction of the total solid angle through which the sample radiates. Again, a temperature lower than the true temperature will be recorded.

Another effect of size has been observed for the firing of samples using the "picket fence" indirect heating arrangement. The picket fence [2,14] consists of a ring of silicon carbide rods surrounding the sample and its insulation and enclosed by an outer insulating cylinder. The picket fence is used to preheat the sample to a temperature high enough that the sample will couple directly to the microwaves. This arrangement has worked well for microwave sintering of zirconia [2], alumina [14], and alumina-zirconia composites [15]. Figure 6 shows the total microwave power used and the temperature of the silicon carbide pickets for firing various alumina samples at 1600°C. As the alumina sample size increases, the temperature of the silicon carbide rods decreases. This indicates that the amount of power absorbed by the silicon carbide rods decreases as the alumina sample size increases. This observation is consistent with Kimrey's model [14] of microwave absorption in multimode cavities. It also indicates that the microwave power level in the alumina increases as the alumina sample size increases.

In the study of some materials phenomena, small samples must be used. Examples include grain growth studies [3] and diffusion experiments [4]. In both of these cases, a ballast/crucible [3] was constructed to permit the processing of the small samples. The basic premise behind the ballast/crucible is to increase the dielectric load in the microwave furnace with a material having dielectric properties either identical or very similar to the sample being studied. Figure 7 shows a schematic of a ballast/crucible used for the alumina grain growth and diffusion experiments. In the case of the grain growth experiments, 40 test coupons 0.5 cm^3 in volume were cut from a single hot-pressed sample for annealing in both conventional and microwave furnaces. A single hot-pressed sample was used to assure that the initial materials conditions for grain growth were identical for all of the microwave and conventional annealing studies. The ballast/crucible was made by isopressing and sintering powder from the same processing batch as the hot-pressed samples. In the case of the diffusion experiments, single crystal sapphire coupons 1 x 1 x 0.1 cm were used. The ballast/crucible for these experiments was made by isopressing and sintering ultrahigh purity (undoped) alumina powder, which gave as close a match as possible to the dielectric properties of the sapphire. To confirm that the ballast/crucible concept was a reasonable approach, it was tested by sintering small (3g) samples of alumina. The sintering behavior of the small alumina samples in the ballast/crucible was identical to that of large (150g) samples.

Insulation and packing powders

As was mentioned above, development of an insulating system that gives reproducible results is perhaps the most difficult task in the high temperature processing of ceramics using microwaves. A few guidelines are in order regarding how one proceeds. As is shown in Fig. 3, the insulation packages used for microwave processing at ORNL are typically multilayered. The primary thermal insulation, i.e., the outer envelope, should be microwave transparent to allow penetration of the microwaves into the sample. The inner layer of insulation should be matched both chemically and dielectrically to the sample as closely as possible. If done properly, this approach results in essentially isothermal conditions within the sample. This is observed experimentally as a uniform microstructure in the final sintered part [1]. All of the thermal gradient is produced in the insulation. In some cases, the inner insulation will actually be a packing powder of the same composition as the sample. This is typical for silicon nitride, for example [20].

FIELD EFFECTS

Microwave field effects are important to consider in the processing of materials in a microwave furnace. Three phenomena that are important are uniformity of the field in the furnace, concentration of the field around the tip of a thermocouple, and arcing. Nonuniformity of the microwave field is a serious concern because it leads to the creation of hot-spots in parts that are being processed. Figure 8 shows a representation of the variation of the electric field in a typical small microwave cavity (similar to a home microwave oven) [21]. Field concentrations of this type are ubiquitous in cavities operating at 2.45 GHz [2,21]. Figure 9 shows the effect of field concentration around the tip of a thermocouple in the sintering of ZrO_2 - 12 mol% CeO_2. The microwave field was higher at the tip of the thermocouple than along its length. This produced more heating at the right side of the sample than at the left side. Differential sintering is evidenced by the "banana" shape of the sample and the crack on the left-hand side that developed during sintering. Figure 10 shows an example of arcing at 2.45 GHz between a thermocouple

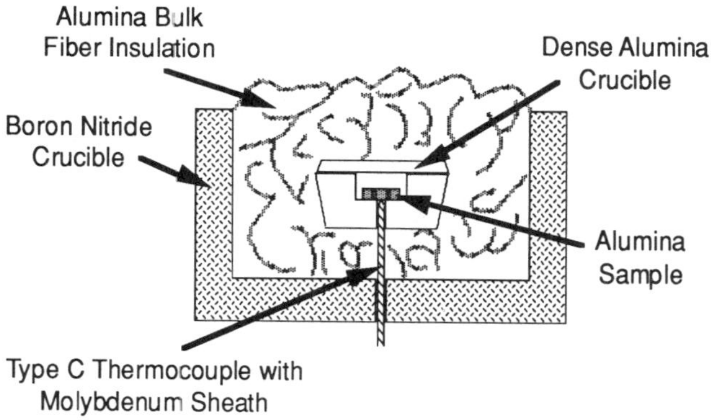

Figure 7. A ballast-crucible arrangement was developed for processing alumina samples for grain growth and diffusion experiments [3].

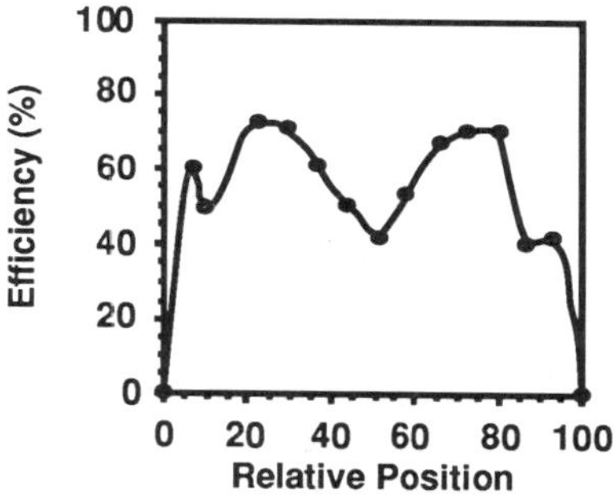

Figure 8. Microwave field variation is sizable in a 40x40x50 cm microwave furnace operating at 2.45 GHz [21]

Figure 9. Concentration of the microwave field around the thermocouple sheath caused differential heating of ZrO_2 - 12 mol % CeO_2.

and a $LaCrO_3$ sample. Arcing caused the tip of the thermocouple to heat excessively, which produced a series of upward spikes in the temperature profile. Because the rapid heating pushed the thermocouple temperature well above the controller setpoint, the furnace controller shut off the power, the arcing stopped, and the temperature of the thermocouple tip dropped rapidly. Since the thermocouple temperature was now below the setpoint temperature of the furnace controller, power was reapplied and arcing resumed. Thus a rapid periodic temperature fluctuation was established, Fig. 10. To prevent arcing, a dielectric barrier of calcium aluminate cement (Alundum, Norton Co., Worcester, MA) was placed between the thermocouple sheath and the sample, which reduced the electric field strength in that region, and hence the propensity to arcing. The resulting furnace run using the Alundum-coated Pt-sheathed thermocouple is shown as the top trace in Fig. 10, which shows no evidence of arcing. The microwave power-time traces (not shown) mimicked the temperature-time curves in both cases; i. e., the trace was irregular for the arcing case and smooth for the non-arcing case. For other applications, alumina and boron nitride thermocouple protection tubes have been used to accomplish the same task of mitigating electric field concentration.

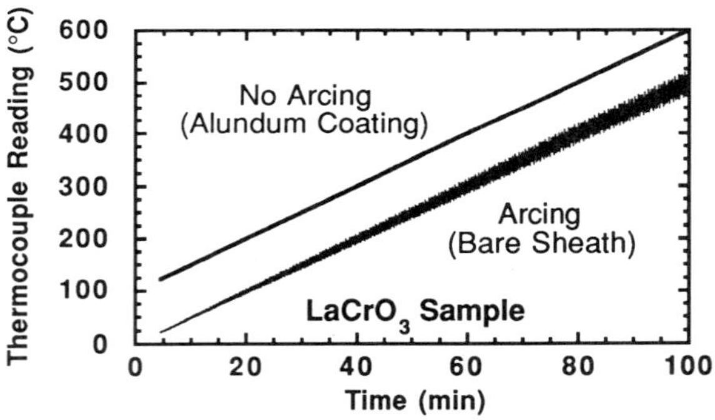

Figure 10. Arcing from a platinum thermocouple sheath to a $LaCrO_3$ sample at 2.45 GHz was eliminated by coating the sheath with calcium aluminate (Alundum®) cement [10].

ATMOSPHERE EFFECTS

Gas composition in conventional processing of ceramics is normally used to fix the partial pressure of oxygen in the furnace. Therefore, neutral atmospheres of argon or nitrogen or vacuum are often considered to be equivalent. In microwave processing, an additional factor must be considered; that is the breakdown potential of the gases under the influence of the microwave field [22]. Figure 11 shows the effect of pressure, gas composition, and frequency on thebreakdown field strength. At all pressures, air (or nitrogen) has a much higher breakdown potential than does argon. This difference becomes pronounced at high pressures approaching atmospheric pressure. Nitrogen is often used instead of argon if a non-oxidizing atmosphere is required to prevent problems associated with arcing. At intermediate pressures, say between 0.01 and 10 torr, plasma formation is a real possibility. Unless specifically desired, plasmas are best avoided during microwave processing. At very low pressure, e.g. below 10^{-4} torr, neither arcing nor plasma formation is a problem; there just are not enough gas molecules around to cause either phenomenon to occur. Figure 11 also shows that the higher the frequency, the higher the breakdown field strength required to cause arcing or plasma formation. Therefore, operation at 28 GHz is less sensitive to plasma formation than is operation at 2.45 GHz.

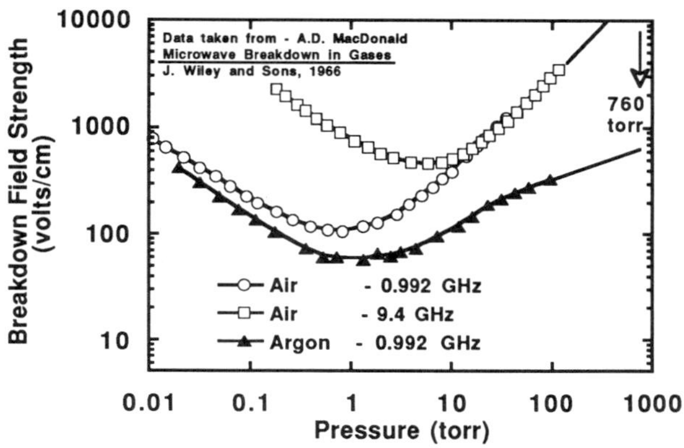

Figure 11. Dielectric breakdown in a microwave field is sensitive to total pressure, gas composition, and frequency.[22]

SUMMARY AND CONCLUSIONS

Guidelines have been developed at ORNL for conducting meaningful side-by-side conventional and microwave thermal processing experiments on ceramics. These guidelines may be summarized as follows:

1. Microwave and conventional furnaces are inherently different in terms of materials processing.
2. Dielectric properties must be explicitly considered in planning microwave experiments, including both temperature and frequency effects.
3. Temperature measurement in a microwave furnace is always an issue. In particular, each new material or experimental setup brings with it new problems in temperature measurement.
4. Thermal gradients may be inherently different for conventional and microwave processing. Conventional samples tend to be cooler in the middle than on the surface; microwave samples tend to be warmer in the middle.
5. Sample size may have to be adapted to microwave processing conditions; in general, a minimum sample size of 50 to 100g is required in the microwave furnace unless indirect heating or a ballast/crucible is used.
6. Field effects must be managed in a microwave furnace including arcing of thermocouples to samples and field concentration around thermocouples. Also, the field uniformity in the microwave cavity (or lack thereof) can have a decided effect on how materials can be processed
7. It must be recognized that the processing atmosphere interacts with the microwave field. Plasma formation and arcing can be a problem, especially when operating in argon or other gases with low breakdown potential.

REFERENCES

1. M. A. Janney and H. D. Kimrey, "Microwave Sintering of Alumina at 28 GHz," **Ceramic Powder Science, II**, pp 919-924, G. L. Messing, E. R. Fuller, and H. Hausner, eds., American Ceramic Society, Westerville, Ohio, 1988.

2. M. A. Janney, C. D. Calhoun, and H. D. Kimrey, "Microwave Sintering of Solid Oxide Fuel Cell Materials, I: Zirconia-8 mol% Yttria," *J. Am. Ceram. Soc.,* **75** [2]341-6 (1992).

3. M. A. Janney, H. D. Kimrey, M. A. Schmidt, and J. O. Kiggans, "Alumina Grain Growth in a Microwave Field," *J. Am. Ceram. Soc.,* **74** [7] 1675-81 (1991).

4. M. A. Janney and H. D. Kimrey, "Diffusion-Controlled Processes in Microwave-fired Ceramics," ***Microwave Processing of Materials, II***, *Proc. Mater. Res. Soc.,* Vol. 189, 215-27 (1991).

5. D. Lewis, "Review of Polymer Processing with Microwaves," this proceedings.

6. R. Giguerre, "Review of Organic Synthesis Using Microwaves," this proceedings.

7. P. Marquardt, G. Nimitz, G. Heite, and H. Peters, "Microwave Evidence for a Size-Induced Metal-Insulator Transition in Mesoscopic Conductors," ***Microwave Processing of Materials, I***, *Proc. Mater. Res. Soc.,* Vol. 124, pp. 155-160 (1989)

8. T. Nishitani, **Method for Sintering Refractories and an Apparatus Therefore**, U. S. Patent 4,147,911; filed Aug. 10, 1976; issued Apr. 3, 1979.

9. C. E. Holcombe; personal communication.

10. M. A. Janney and H. D. Kimrey, "Microwave Sintering of Solid Oxide Fuel Cell Materials, II: Lanthanum Chromite," submitted to *J. Am. Ceram. Soc.*

11. C. E. Holcombe and N. L. Dykes, "Importance of 'Casketing' for Microwave Sintering of Materials," *J. Matl. Sci. Lett.*, **9**, 425-8 (1990).

12. C. E. Holcombe and N. L. Dykes, ""Ultra" High-Temperature Microwave Sintering," *Ceramic Trans.*, Vol. 21, 375-85, 1991.

13. J. O. Kiggans and T. N. Tiegs, "Characterization of Silicon Nitride Synthesized by Microwave Heating," *Ceramic Trans.*, Vol 21, 403-410, 1991.

14. H. D. Kimrey, "A Quasi-Optical Model for Microwave Processing," this proceedings.

15. H. D. Kimrey, J. O. Kiggans, M. A. Janney, and R. L. Beatty, "Microwave Sintering of Zirconia-Toughened Alumina Composites," ***Microwave Processing of Materials, II***, *Proc. Mater. Res. Soc.,* Vol. 189, 243-55 (1991).

16. R. A. Peterson, **Microwave Heating Temperature Control**, U. S. Patent 3,859,493; filed Nov. 21, 1973; issued Jan. 7, 1975.

17. W. H. Sutton and W. E. Johnson, **Method of Improving the Susceptibility of a Material to Microwave Energy Heating**, U. S. Patent 4,219,361; filed Jun. 9, 1978; issued Aug. 26, 1980.

18. D. L. Johnson, "Microwave Heating of Grain Boundaries in Ceramics," *J. Am. Ceram. Soc.*, **74**[4]849-50(1991).

19. D. E. Clark and D.C. Foltz, "Microwave Processing Activities at the University of Florida," *Ceramic Trans.*, Vol. 21, 29-34, 1991.

20. J. O. Kiggans and T.N. Tiegs, "Microwave Processing of RBSN and SRBSN Silicon Nitrides," this proceedings.

21. C.R. James, W.R. Tinga, and W.A.G. Voss, "Energy Conversion in Closed Microwave Cavities," in E. C. Okress, ed., **Microwave Power Engineering, Vol. 2**, Academic Press, N.Y., 1973

22. A.D. McDonald, Microwave Breakdown in Gases, J. Wiley and Sons, 1966.

PULSED MICROWAVE PROCESSING OF HIGH-T_C SUPERCONDUCTING FILMS

R. B. JAMES*, R. A. ALVAREZ**, A. K. STAMPER***, X. J. BAO***, T. E. SCHLESINGER***, D. S. GINLEY****, K. F. McCARTY*, T. A. FRIEDMANN*, and R. H. STULEN*
*Sandia National Laboratories, Advanced Materials Research Department, Livermore, CA 94550
**Lawrence Livermore National Laboratory, Livermore, CA 94550
***Department of Electrical and Computer Engineering, Carnegie Mellon University, Pittsburgh, PA 15213
****Sandia National Laboratories, Advanced Materials and Devices Department, Albuquerque, NM 87123

ABSTRACT

We have used 2.0-μsec microwave pulses at a frequency of 2.856 GHz to rapidly heat thin amorphous yttrium-barium-copper-oxide (YBCO) films deposited onto silicon substrates. The samples were irradiated inside a WR-284 waveguide by single-pass TE_{10} pulses in a traveling wave geometry. X-ray diffractometry studies show that an amorphous-to-crystalline phase transition occurs for incident pulse powers exceeding about 6 MW, in which case the amorphous YBCO layer is converted to Y_2BaCuO_5. Microscopy of the irradiated film reveals that the phase transition is brought about by melting of the YBCO precursor film and crystallization of the molten layer upon solidification. Time-resolved in situ experiments of the microwave reflectivity (R) and transmissivity (T) show that there is an abrupt change in R for microwave pulse powers exceeding the melt threshold, so that measurements of R and T can be used to monitor the onset of surface melting.

Introduction

High-temperature superconductors (HTS) are soon expected to make an impact in electronic applications, such as circuitry interconnects and passive microwave devices. These applications will be primarily based on the use of HTS thin films. Processing of the HTS films are currently performed by ex-situ or in-situ techniques. The structural properties of the HTS films depend on several factors and are sensitive to the thermal treatments during and after film deposition. Ex-situ processing typically involves extended high-temperature (~1100 K) sintering and annealing steps, where the original non-superconducting, partially oxidized stoichiometric metals are converted to superconducting phases through solid state or liquid phase sintering. The primary obstacle facing the use of ex-situ processing of HTS films in most electronics applications is the interdiffusion of the substrate material into the film during the high-temperature thermal anneal. Silicon is presently the substrate material of choice for most of these applications. Unfortunately, silicon diffuses readily into the HTS during the ex-situ furnace treatment, and this diffusion destroys the film's superconducting properties.

Thermal processing with microwaves has been successfully applied to numerous solids.[1] Most of these investigations have relied on using continuous microwave sources to induce heating of ceramics, rubber, composite materials, polymers, and metal oxides [1]; although a few investigations have been conducted with short high-power microwave pulses [2]. Heating with pulses is preferable in circumstances where rapid thermal processing is required, and localized heating of only the near-surface region (within a distance of few times the skin depth) is desired. The primary objective of this work is to determine if high-power pulsed microwave radiation can be used to replace a conventional furnace for heating yttrium-barium-copper-oxide (YBCO) HTS precursor films to the temperature required for crystallization of the amorphous film. Ideally, the microwave-induced heating technique can be optimized to achieve improved mechanical and electrical properties of the film, compared to those achieved with thermal processing in an oven. Microwave heating offers some potential advantages, since the heating is localized to the near-surface region and the deleterious effects associated with heating the silicon substrate and diffusion of the silicon into the film can be greatly reduced. In addition, the short pulse time ($t < 2$ μsec) allows for liquid phase sintering of the precursor film, which results in improved intergrannular properties.

Experiment

A high-power pulsed klystron was used to generate microwave pulses having a frequency of 2.856 GHz and a maximum peak power of 12 MW. By gating the rf drive, the pulse duration τ_p was variable between about 135 ns and 2 μsec, with a risetime of approximately 70 ns. A schematic diagram of the experimental setup is shown in Fig. 1. The output of a highly stable cw master oscillator is gated by a PIN diode, which provides a variable-length 500-mW pulse to a driver klystron. The driver klystron produces a 135-2000 ns pulse that is further amplified by a high-power klystron. The pulse from the high-power klystron passes through a circulator, which protects the microwave source from large pulses reflected from the test cell. The output from the circulator then feeds into a WR-284 copper waveguide. The waveguide is terminated with an impedance-matched, water-cooled load to minimize reflections and ensure that the sample is irradiated by a single-pass, TE_{10} traveling-wave pulse.

The test cell consists of a copper waveguide section that allows insertion of a sample into the microwave field without venting of the waveguide system. The samples are inserted into the test cell with a linear motion feedthrough and are suspended in the center of the WR-284 waveguide by an alumina holder. The waveguide assembly, not including the test cell, is pressurized to 30 psig with Freon-12. The test cell was evacuated to a pressure of about 5 x 10^{-6} torr, so that any problems associated with dielectric gas breakdown could be avoided. Bi-directional couplers and calibrated fast diodes are placed both upstream and downstream from the test section. The diodes, which have a time resolution of 2 ns, are used to measure the shape and amplitude of the incident, transmitted, and reflected pulse envelopes.

The moderately doped (100) silicon substrates used in the experiment have a thickness of about 300 μm. A thin native oxide of SiO_2 is present at the substrate surface. YBCO films were sputtered on-axis onto the silicon substrates to a layer thickness of about 1 μm. The sputtering gas was 90 mTorr of a 90:10 argon:oxygen mixture, and the target was $YBa_{3.4}Cu_{4.1}O_x$. The films produced with this sputtering gas, target, and selected target-to-substrate spacing yielded films that were within 5 % of the desired $YBa_2Cu_3O_x$ composition. The silicon substrates were not heated during the film deposition, so the precursor high-temperature superconducting films are initially amorphous. The film resistivity is about .003 Ω-cm at room temperature. After coating the silicon wafers with a YBCO film, the wafers were cut into smaller samples having a size of 5x5 mm, so that the cross-sectional area of the microwave-irradiated specimens (~0.25 cm^2) would be much smaller than the cross-sectional area of the test cell (~24.5 cm^2).

The specimens were inserted into the test cell through a thin (~0.4 cm) slit in the narrow wall of the WR-284 waveguide. The slit is oriented parallel to the wall currents. Each sample is held at the center of the test cell, where the **E**-field is at a maximum and uniform in the unloaded waveguide, and it is oriented so that the YBCO-deposited side faces upstream and normal to the traveling-wave pulses. Some specimens were irradiated by a single microwave pulse, while others were irradiated by multiple pulses. Whenever multiple-shot experiments were conducted, a 1-Hz repetition rate was selected. Time-resolved measurements of microwave reflectivity and transmissivity were performed to monitor changes in the dielectric properties of the specimens during each pulse. Post-irradiation studies were also conducted to determine the structure of the microwave-heated films, electrical and optical properties, surface topography, and presence of inhomogeneous heating.

Results and Discussion

The coupling of the 2.856-GHz microwave radiation to the YBCO-coated Si samples is predominantly by heating of the free carriers. In a plane-wave approximation, the electromagnetic waves in a conductive media are exponentially damped with the absorption coefficient, α, given approximately by [3]

$$\alpha = (2\mu\omega\sigma)^{1/2} . \qquad (1)$$

Here, α is the magnetic permeability, ω is the angular frequency associated with the incident microwave field, and σ is the electrical conductivity. For a microwave frequency of 2.856 GHz,

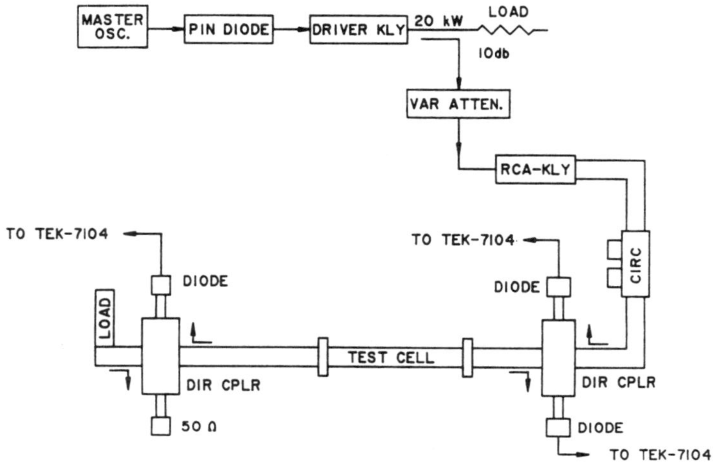

Fig. 1 Schematic of the microwave source and experimental setup.

the calculated linear value of α at room temperature is about 200 cm^{-1} for the YBCO film and about 50 cm^{-1} for the Si substrate. The dielectric losses in both the YBCO and Si are expected to be accelerated as the samples are heated.

Samples were irradiated over a range of excitations conditions, and x-ray diffractometry was used to determine if the degree of film was sufficient to induce crystallization of the amorphous YBCO film. Figure 2 shows the x-ray diffraction pattern of a YBCO film on a silicon substrate after irradiation by 1000 microwave pulses, each having an incident pulse power of 9.1 MW and pulse duration of 2 μsec. (In this paper values of the incident microwave power are quoted rather than the power density, because our measurements suggest that the presence of the sample modifies the TE_{10} traveling wave pulse, especially for pulse powers exceeding the melt threshold of the film. If we assume that the presence of the sample does not significantly perturb the TE_{10} guided mode in the test chamber, a pulse power of 9.1 MW corresponds to a power density of about 750 kW/cm^2 at the location of specimen.) No peaks associated with the film were observed prior to microwave heating, since the film was amorphous. The x-ray diffraction data shown in Fig. 2 confirm for the first time that YBCO films can be transformed from the amorphous to crystalline state by the absorption of high-power microwave pulses. The location of the diffraction peaks indicates nucleation of Y_2BCuO_5 (i.e, the 2-1-1-5 YBCO compound), which is commonly referred to as the green phase of the material. Diffractometry measurements on different specimens indicated that, in general, crystalline material was present after microwave irradiation for pulse powers exceeding approximately 6 MW.

Field ion microscopy was used to study the topography of the microwave-irradiated specimens. The top photo of Figure 3 shows an unirradiated sample. The surface appears somewhat rough at this magnification with the roughness on the order of a few thousand angstroms. The bottom photo shows a sample after irradiation by 1000 microwave pulses, each having a pulse power of 11.5 MW and pulse duration of 2 μsec. Examination of the irradiated surface reveals strong evidence for melting of the film, indicating that the amorphous-to-crystalline phase transition is most likely associated the liquid phase. Surface damage is sometimes observed at the corners of the rectangular specimens, suggesting that the **E**-field of the TE_{01} wave is enhanced at (or near) the edges.

In order to obtain information on the pulse-power conditions required to induce melting of the YBCO film, it is useful to perform in situ measurements to monitor the degree of sample heating. The microwave-induced melting of the sample surface causes an abrupt change in the complex conductivity, which produces changes in the sample impedance and reflectivity of the microwave radiation from the test cell. We have performed time-resolved measurements of the reflectivity (R) and transmissivity (T) of the specimens. As the pulse power was progressively increased, an

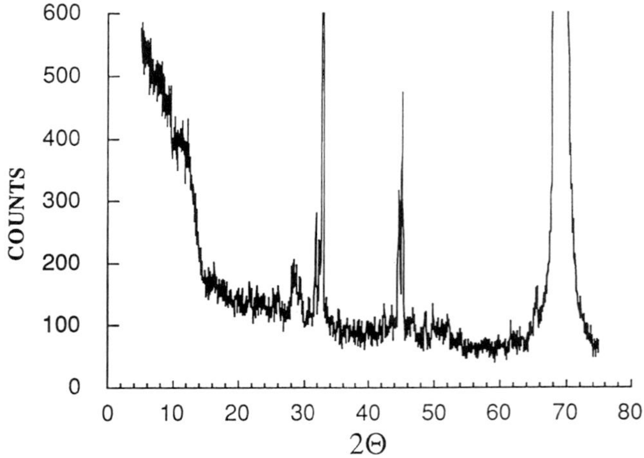

Fig. 2 X-ray diffraction pattern of a microwave-heated YBCO film after irradiation by 1000 pulses at a repetition rate of 1 Hz. Each pulse had a duration of 2 μsec and power of 9.1 MW.

abrupt increase in the reflected microwave power from the test cell and a corresponding drop in the transmitted power were observed. When the sample was removed from the test cell, no changes in R and T were observed in the empty waveguide over the same range of incident microwave powers. The reflectivity (top traces) and transmissivity (lower traces) are shown in Fig. 4 for nine 2-μsec pulses, each at an incident pulse power of 11.5 MW. At the beginning of the pulse, the reflectivity of the test cell is only a few percent. After a time of about 0.7 μsec, the reflectivity of the test cell changes abruptly and persists at a higher value throughout the remainder of the pulse. During these times of increased microwave reflectivity, there is a corresponding drop in the transmissivity of the test cell, as shown by the lower traces in Fig. 4. For lower intensity microwave pulses the onset of the increased reflectivity occurs at later times. For excitation by 2-μs pulses, the threshold for observing the increased microwave reflectivity was found to be about 5.6 MW, which corresponds to about 0.9 J/cm^2 incident on the sample surface. Post-irradiation studies on the samples revealed that surface melting occurred whenever the incident microwave power exceeded the threshold for the abrupt change in R. Thus, monitoring of the microwave reflectivity from the test cell was found to provide a measure of film melting and crystallization.

Summary

Silicon samples with amorphous YBCO superconducting precursor films were irradiated in a WR-284 waveguide by high-power microwave pulses. The results of in-situ and post-irradiation studies established that sufficient heating was achieved to induce crystallization of the amorphous YBCO film. The microstructure of the film was found to be Y_2BaCuO_5, instead of the desired superconducting phase, $YBa_2Cu_3O_{7-\delta}$. In order to obtain the (123) superconducting phase of YBCO, it will be necessary to perform the irradiations in an oxygen atmosphere, instead of a vacuum. The exposure of the microwave-heated specimen to an oxygen environment will allow for a reduction in the loss of oxygen from the film, and by controlling the O_2 partial pressure in the test cell, one should obtain an increased control of oxygen states in the microwave-heated YBCO film. Although this work has clearly demonstrated that a sufficient degree of heating is possible using microwave pulses, the usefulness of the technique for ex-situ thermal processing of amorphous YBCO layers requires that the crystallized material be a high-T_c superconducting phase.

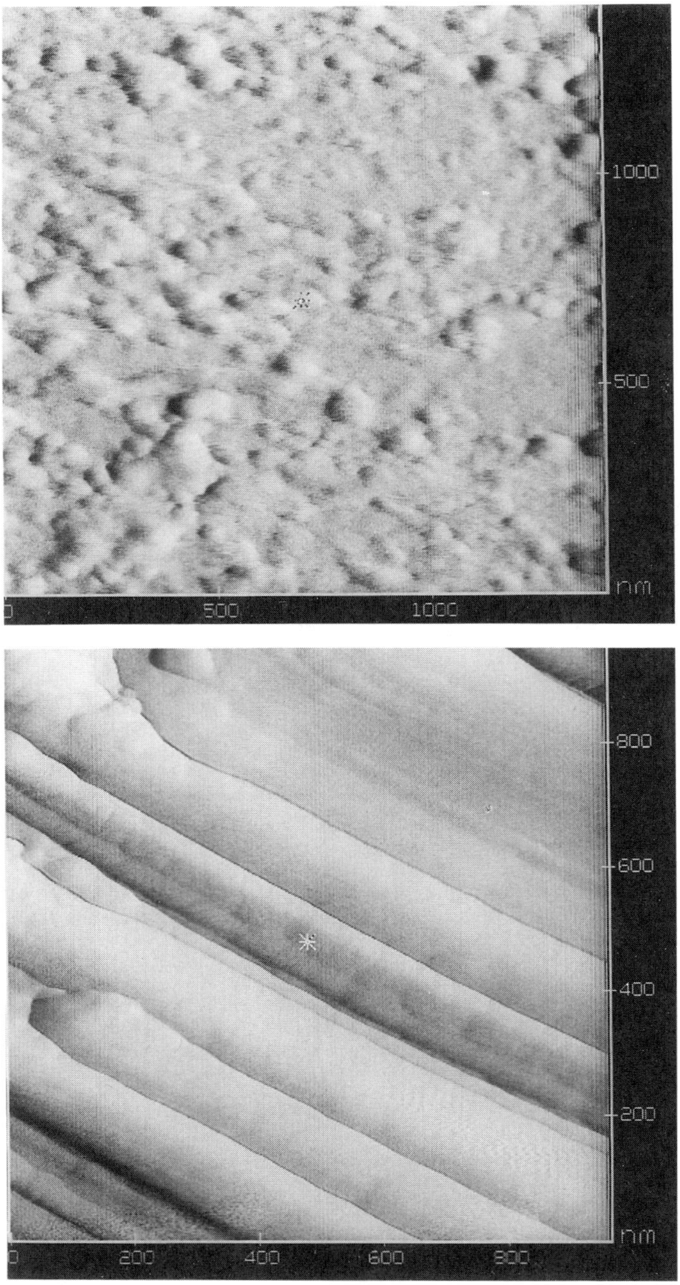

Fig. 3 The top photo shows an unirradiated YBCO-coated Si sample. The bottom photo shows a sample after irradiation by 1000 pulses, each having a power of 11.5 MW and duration of 2 μsec.

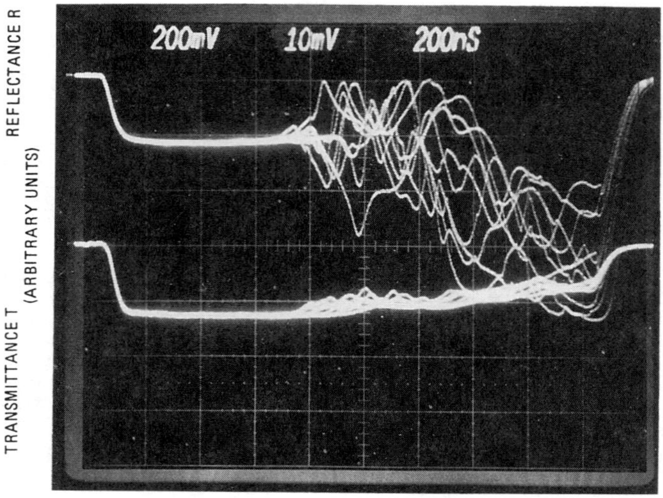

Fig. 4 Oscilloscope traces of the reflectivity R (top traces) and transmissivity T (bottom traces) for nine 2 .0 μsec pulses, each having an incident power of 11.5 MW. The two detectors used to measure R and T had different levels of attenuation; thus, the vertical scale in the figure is in arbitrary units. Increases in the reflectivity and transmissivity are shown by going downward on the vertical scale. The horizontal scale is 200 nsec/division.

Acknowledgments

This work was supported by the U. S. Department of Energy. The authors would like to acknowledge A. J. Antolak and R. Q. Hwang for useful discussions and B. Patterson for technical assistance.

References

1. See, for example, Microwave Processing of Materials, Vol. 124, edited by M. H. Brooks, I. J. Chabinsky and W. H. Sutton (Materials Research Society, Pittsburgh, PA, 1988); and Microwave Processing of Materials II, Vol. 189, edited by W. B. Snyder, Jr., W. H. Sutton, M. F. Iskander, and D. L. Johnson (Materials Research Society, Pittsburgh, PA, 1991).
2. See, for example, R. B. James, P. R. Bolton, R. A. Alvarez, and W. H. Christie, in Mikrowellen and HF Magazine, Vol. 16, 332 (1990); and R. B. James, P. R. Bolton, R. A. Alvarez, W. H. Christie, and R. E. Valiga, J. Appl. Phys. 64, 2049 (1988).
3. J. D. Jackson, in Classical Electrodynamics, 2nd Edition (Wiley, New York, 1975), p. 298.

METALLORGANIC- AND MICROWAVE-PROCESSING OF MONOLITHIC AND POLYPHASIC CERAMICS.

M. WILLERT-PORADA, University Dortmund, Dept. Chem. Eng., 4600 Dortmund 50, Emil Figge Str. 66, Germany.

ABSTRACT

A method for microstructural modelling of ceramic materials utilising precursor compounds and heating by microwave radiation to introduce a second phase as well as to control grain growth is described. The precursor compounds are introduced into the ceramic material by infiltration of a porous compact or by impregnating a classical powder with the precursor prior to compaction. Microwave radiation is used for decomposition of the precursor and for sintering. Decomposition rate controlled pyrolysis of the precursor is only possible by microwave heating. The resulting microwave processed ceramics display a finer and in some cases more bimodal microstructure as compared to conventionally processed samples with the same density.

INTRODUCTION

Microwave heating offers the unique potential for selective heating a strongly MW-absorbing material within a non absorbing matrix, providing the mixture is homogeneous and the absorbing phase is diluted enough to retain a penetration depth sufficient for volumetric heating. The requirement to disperse a reinforcing second phase within a ceramic matrix is difficult to fullfill when two fine components with usually a substantial degree of aggregation have to be combined [1]. Furthermore some of the most promising reinforcing particles like SiC-, Al_2O_3- or Si_3N_4-whiskers are difficult to process because of health hazards [2]. The work presented in this paper focuses on the development of a method to achieve homogeneous dispersions of potentially reinforcing particles within a ceramic matrix without excessive mixing, by "in situ" formation of the reinforcement from a precursor using microwave heating (MW). Two cases have to be considered in this processing route: a) MW-absorbing ceramic precursor compound / non absorbing matrix b) non absorbing precursor / absorbing matrix. An absorbing ceramic precursor should exhibit permanent dipols or at least easily polarizable electrons within a chemical bond. Furthermore occurrence of self dissociation into ions would be advantageous to dissipation of MW-radiation into heat. Alcoholates as well as acetylacetonates meet this requirement due to a substantial degree of covalent-ionic resonance as well as self dissociation into $[M^x(OR)_{x-1}]^+$ $[M^x(OR)_{x+1}]^-$ in the case of alcoholates [3].

EXPERIMENTAL

Two matrix-materials as well as precursors with different coupling behaviour to MW (2.45 GHz) were used. Alumina (99.6 % Al_2O_3) served as low-absorption- and SiC as high absorption-matrix. The precursors are the alcoholates $Al(OiPr)_3$ and $Zr(OiPr)_4$ (abr. ATIP, ZTP) as good absorbing sources of Al_2O_3 and ZrO_2 and a mixture of tar and polysilane $-(SiR(OR)-)_x$ as a good absorbing source for SiC. A poorly absorbing precursor is the silane. Acetylacetonates of Zr and Al were only used for preliminary investigations because they sublime out of the matrix upon MW-heating. Two different methods were employed to introduce the precursor : 1. infiltration of a porous compact by the liquid precursor or a solution of a solid precursor and 2. impregnation of a ceramic powder with the liquid precursor [4] prior to green body processing . The processing scheme is shown in Fig. 1. Commercial powders (CT8000 SG, A16 SG, Alcoa) were used for the alumina matrix. The CT8000 powder was mainly used for infiltration experiments whereas the A16

powder due to its lower aggregation tendency was employed in the impregnation route. The SiC-matrix for infiltration was a coatmix-material (provided by KFA, Jülich) with 30 % open porosity. For impregnation experiments a SiC powder (ESK, 5 m^2g^{-1}) was used. The alcoholates and acetylacetonates were commercial products (Fluka, Hereaus). ATIP was vacuum distilled prior to use. The silane (Wacker Chemie) and ZTP were used without further purification.
The porous compacts were infiltrated under Ar or in vacuum. Decomposition of the alcoholate precursor within the porous matrix occured under weight loss-control in a multi mode MW-cavity and for comparison in a conventional furnace as shown in Fig.2. The silane as well as the silane/tar-mixtures were decomposed in case of the SiC-matrix in a MW-cavity under Ar. With Al_2O_3-matrix a graphite vacuum furnace was used due to insufficient coupling. All powder-impregnation experiments were performed in a rotating, MW-heated apparatus, described in detail elsewhere [4].
Sintering was performed in a multi-mode MW cavity under air or in argon-atmosphere in case of SiC-matrix and -dispersion materials. For comparison, sintering was also performed in a conventional high temperature furnace with $MoSi_2$-heating elements (air) or in a graphite vacuum furnace.

RESULTS

As shown by Fig. 3. for an Al_2O_3/ATIP infiltrated sample, control of the decomposition rate can be maintained by MW-heating but not by conventional heating. In case of MW-pyrolysis ATIP-decomposition is not a function of the measured macroscopic temperature. Rather the macroscopic temperature is a function of the ATIP-amount left and the applied MW power as indicated in Fig. 4a,b. Decomposition occurs at any macroscopic temperature. Compared to this, in case of conventional heating no decomposition is observed at temperatures below 200°C.

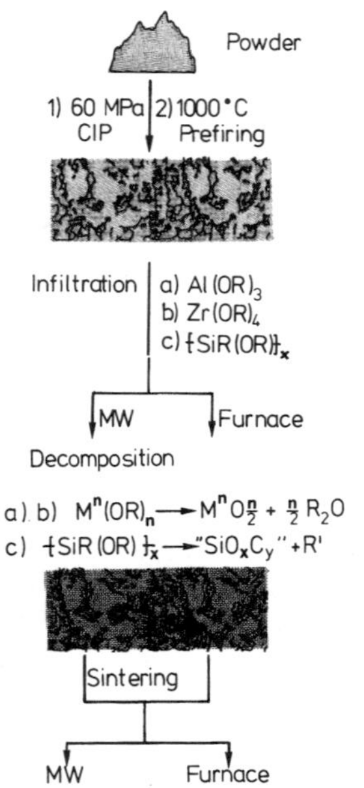

Fig.1.: General processing scheme.

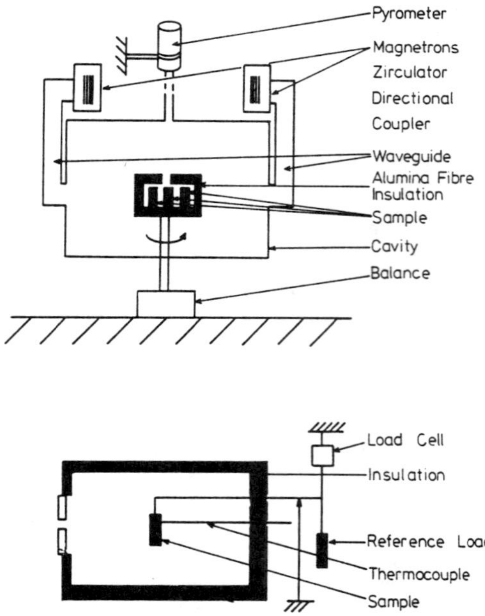

Fig.2.: Balance equipped MW and conventional furnace.

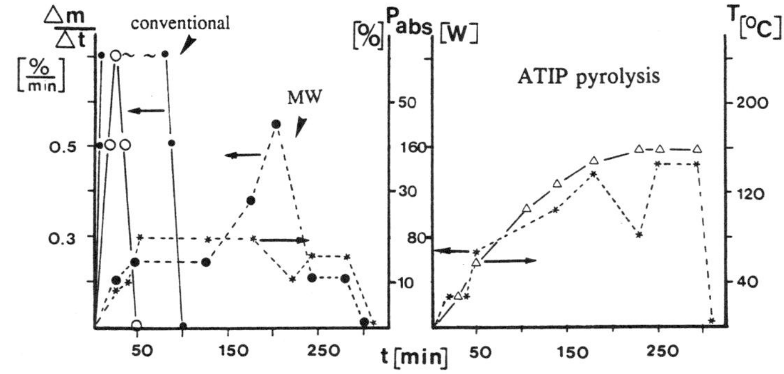

Fig.3.: Decomposition rate and surface temperature versus absorbed MW-power.

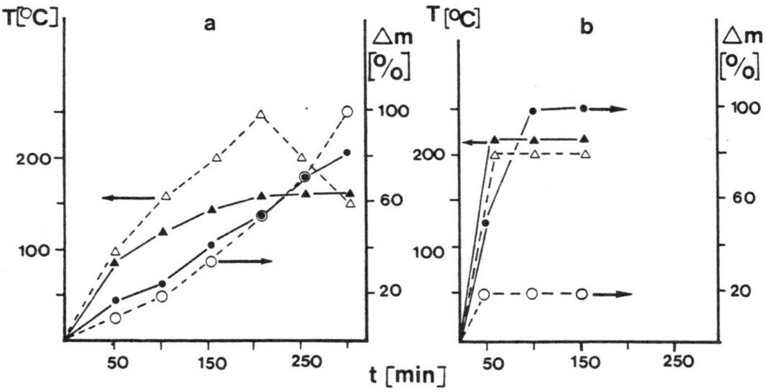

Fig.4.: ATIP decomposition rate as function of temperature a) MW, b) conventional heating.

Starting with 200°C an incomplete decomposition takes place as illustrated in Fig. 3. Increasing the temperature to 220°C causes the ATIP to decompose completely within a short period of time. However, a constant decomposition rate could not be maintained in any of the conventional heating experiments.

Sintering has been performed with different heating profiles, as shown in Fig. 5. for MW and conventional heating. Density data obtained in these experiments are summarized in Table 1. When higher MW-heating rates were employed as those in Fig. 5, thermal runaway occurred. To obtain comparable materials, no effort was made to sinter to full density, but rather 1450°C was maintained as soak temperature in the case of alumina- and 1700°C in the case of SiC-matrix materials. In previous work alumina ceramics with a bimodal microstructure were obtained by annealing CT 8000 samples at 1450°C using MW-hybrid heating followed by conventional sintering or by infiltrating CT 8000 with ATIP and conventional sintering [5]. In this work the influence of MW-heating on infiltrated samples was investigated. The influence of MW-processing on grain growth of CT8000 infiltrated with ATIP as compared to a not infiltrated sample with comparable density is qualitatively illustrated in Fig. 6a and c. The infiltrated material has a smaller matrix grain size and a more bimodal microstructure than CT 8000 itself. The comparison of metallorganically modified conventional and microwave processed samples is given in Fig. 6a,b; d,e and f,g. A smaller grain size is found due to MW-sintering for ATIP infiltrated and ZTP impregnated Al_2O_3 as well as for ZTP impregnated SiC. The composition of SiC/ZrO_2-materials also depends upon the heating method. Whereas after the MW-heating experiments ZrO_2/SiC is still present, it is converted to ZrC/SiC in the conventionally heated furnace . In the case of silane infiltrated SiC-matrix materials a slight

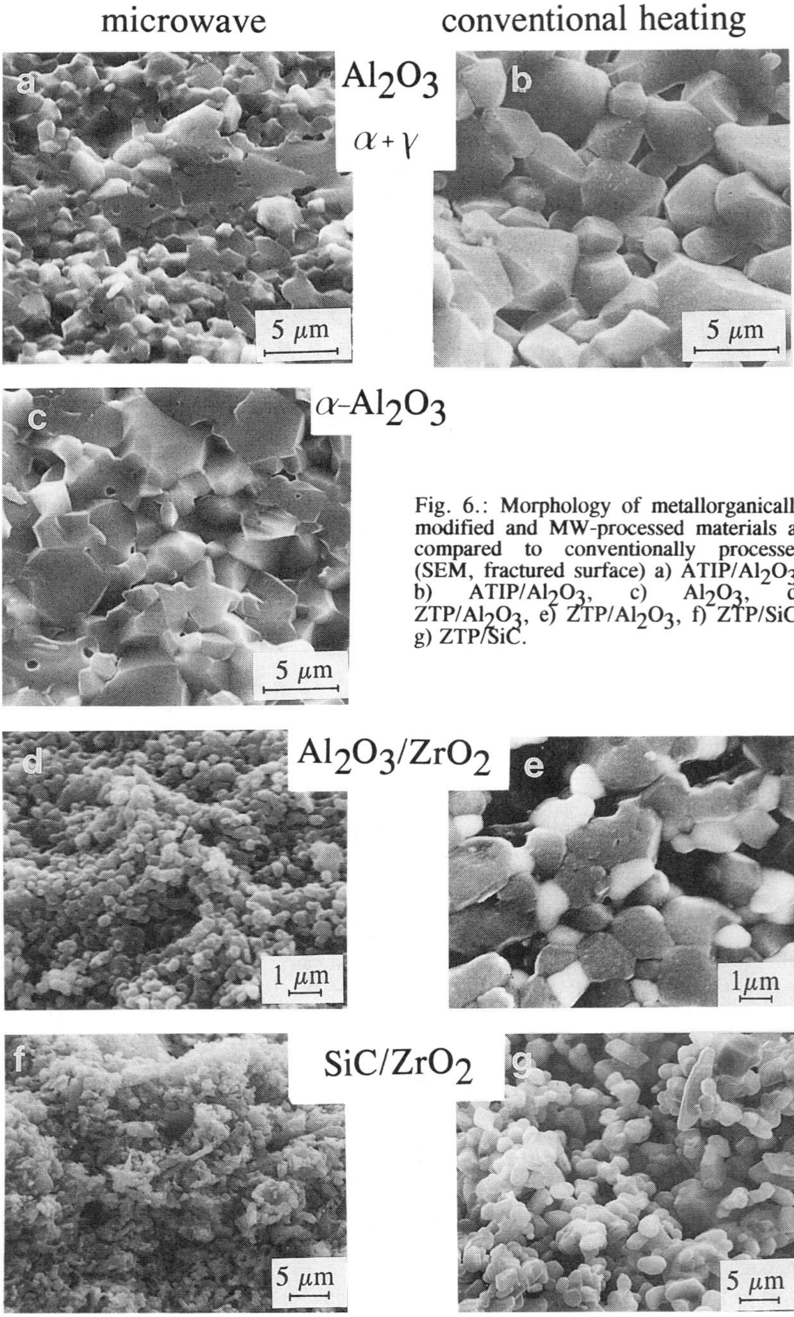

Fig. 6.: Morphology of metallorganically modified and MW-processed materials as compared to conventionally processed (SEM, fractured surface) a) ATIP/Al_2O_3, b) ATIP/Al_2O_3, c) Al_2O_3, d) ZTP/Al_2O_3, e) ZTP/Al_2O_3, f) ZTP/SiC, g) ZTP/SiC.

densification is observed only for the material infiltrated with polysilane/tar but no densification occurs for silane/SiC as shown in Table 1. The influence of infiltration or impregnation on mechanical properties of MW- processed Al_2O_3-matrix samples is shown in Fig. 7.

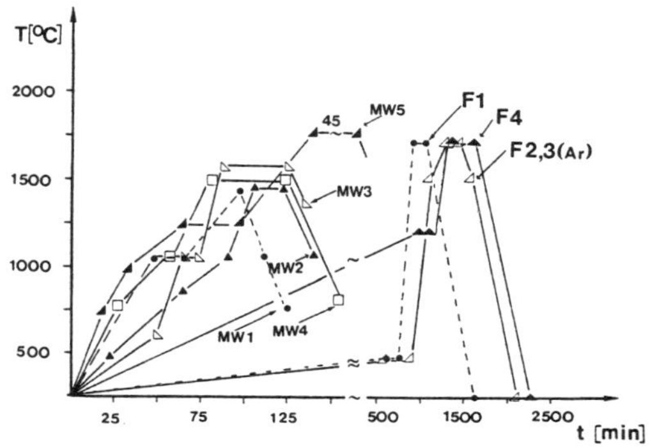

Fig.5.: Sintering profiles: MW and conventional (F) .

Table.1.: Sintering and density data of materials from Fig.6, 7.

Sample	Sintering profile	Density gcm^{-3}
CT8000/ATIP	F1	3.75
CT8000/ATIP	MW1	3.75
CT8000	MW1	3.83
A16/ZTP(10%)	F2	4.09
A16/ZTP(10%)	MW2	3.79
A16	MW3	3.92
CT8000/Silane	MW4	3.30
CT8000/Silane	F3	3.09
CT8000	MW4	3.80
SiC/ZTP	F4	1.39
SiC/ZTP	MW5	1.65
SiC/tar/Silane	MW4	1.79
SiC/Silane	MW4	1.54
SiC	MW4	1.54

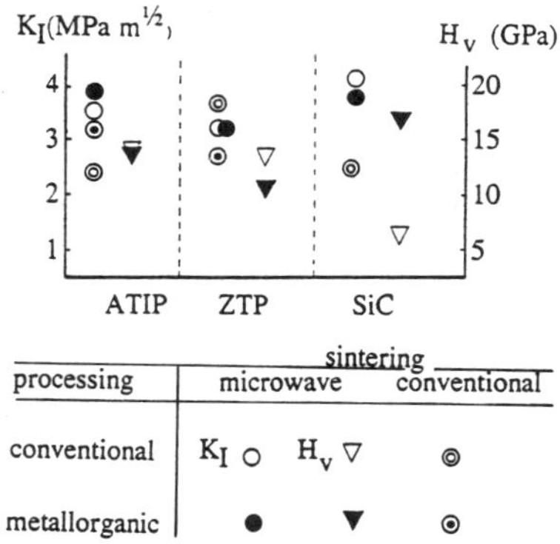

Fig.7.: Mechanical data of infiltrated/impregnated MW processed Al_2O_3-matrix materials as compared to conventionally processed.

DISCUSSION

A pronounced difference between MW-heating and conventional processing is observed for the pyrolysis of metallorganic compounds. The fact that decomposition of ATIP occurs by MW-heating relatively independently of the macroscopic temperature measured at the sample surface indicates real volumetric heating. The additional "exotherm" observed in the temperature profile can not be identified yet. ATIP-decomposition is not expected to be strongly exothermic [3]. However, heating by MW could give different results, compared to conventional heating, because H_2O and C_3H_7OH formed during the reaction could couple with the MW-radiation in addition to the ATIP left in the sample. A completely different decomposition chemistry is not expected because no significant differences were found among the decomposition products from the two pyrolysis routes. Together with water and propanol small amounts of numerous esters and acids were found by GC, but not explicitly identified yet.

A comparison of MW- and conventionally sintered infiltrated or impregnated samples with not infiltrated/impregnated ones shows in both cases inhibition of grain growth due to a second phase. For conventionally sintered samples this was already observed [6,7,8] and attributed to the different sintering activity of powders generated by MW-pyrolysis [5] as compared to classical powders. In addition to this effect MW sintering of infiltrated or impregnated samples promotes densification at reasonably low temperatures. This is very important for SiC/Al_2O_3-materials due to their sensitivity towards oxidation of the fine dispersed SiC to SiO_2, followed by mullite formation [9]. In the case of ZrO_2/Al_2O_3 the ZrO_2 grains can be kept small due to a low sintering temperature. In the case of alumina the bimodal character of the microstructure, which is interesting only for a very fine grained matrix [5], is further enhanced in MW sintering. At higher MW-sintering temperatures the difference in densification between MW- and conventional heating disappears [5]. Preliminary results on mechanical properties summarized in Fig. 7. indicate for the infiltrated/impregnated materials in the case of ATIP sligthly better values for both MW and conventional sintering and also better values for MW sintered single powder as compared to conventionally sintered single powder [7]. In the case of ZTP/Al_2O_3 conventionally sintered samples are better than the MW sintered ones [8]. However, MW sintering of the ZTP material is not optimized yet. The SiC-infiltrated Al_2O_3 materials show no improvement in mechanical properties compared to the dummy materials [9]. Because SiC could not be detected by XRD, despite the fact that the samples displayed a greenish colour typical for SiC/Al_2O_3-materials, we assume that the amount of SiC is probably too low to significantly influence the mechanical properties. For SiC-matrix materials a slightly increased sintering activity at 1700°C was observed only in MW processing of the tar/silane infiltrated material, probably due to free carbon. In the case of ZrO_2/SiC the observed ZrC formation agrees with the reactivity of this materials reported in literature [10]. We attribute the compositional difference between MW and conventional heating partially to the reducing atmosphere in the graphite furnace. However, gravimetric analysis shows that the reducing agent to a large extent is SiC itself. Because of the different temperature dependence of the loss tangent of SiC as compared to ZrO_2 MW-heating could influence the reactivity of SiC towards ZrO_2 by reversing the temperature gradient at the SiC/ZrO_2 interface. To verify this, further experiments are necessary.

CONCLUSION

Metallorganic and microwave processing offers the opportunity for microstructural and compositional modelling of ceramic materials. All second phases introduced by infiltration or impregnation of precursor compounds act as extremly well dispersed fine inclusions with "inert" character, inhibiting grain growth. Due to MW-sintering a high density is achieved at low temperature, preserving the "inert" character of the inclusions. Controlled, very slow decomposition of precursor compounds is easy to achieve using MW-heating but very difficult by conventional heating. For alumina-matrix ceramics slightly improved mechanical properties were found. Some potential for pressureless sintering of infiltrated SiC-matrix materials is expected from preliminary results using the metallorganic and microwave processing route.

LITERATURE

[1] M. Yang, R. Stevens, J. Mat. Sci. 25, 4658 (1990)
[2] T.F. Cooke, J. Amer. Ceram. Soc. 74, 2959 (1991)
[3] F.A. Cotton, in Progr. Inorg. Chem. , Vol. II, (Interscience Publishers, New York, 1960), p. 331 ff.
[4] M.Willert-Porada, T. Krummel, D. Moormann, B. Rohde,this volume (/)
[5] M. Willert-Porada, S. Vodegel, Ceram. Trans. 21, 631 (1991)
[6] F.Schmidt, M. sc.thesis, University Dortmund, 1990
[7] H. Liebertz, M. sc.thesis, University Dortmund, 1991
[8] T. Gerdes, M. sc. thesis, University Dortmund, 1991
[9] T. Krummel, M.sc. thesis, University Dortmund, 1991
[10]N. Claussen, K.L. Weisskopf, M. Rühle, J. Amer. Ceram.Soc, 69, 288 (1986)

Financial support of Land NRW (IV A 6 214 501 89) is greatfully acknowledged. We are indebted to Dr. Dias, KFA Jülich for providing a coatmix-SiC sample.

CERAMIC POWDERS GENERATED BY METALLORGANIC AND MICROWAVE PROCESSING

M. WILLERT-PORADA, T. KRUMMEL, B. ROHDE, D. MOORMANN, Universität Dortmund, Dept. Chem. Eng., D-4600 Dortmund 50, Emil Figge Str. 66

ABSTRACT

A novel approach to synthetic ceramic powders based on microwave heating of metallorganic ceramic precursor compounds or mixtures of such compounds with a appropriate matrix material is described.
Numerous oxides as well as nonoxide powders with high specific surface area were obtained. Pyrolysis as well as enhanced reactivity is observed with microwave heating at lower temperatures than in conventional heating.

INTRODUCTION

Synthetic powders with "tailored" morphology, agglomerate structure and composition are necassary to improve the reliability of ceramic materials. Extremely fine, monosized powders can be obtained by controlled reactions using sol-gel processing, gas phase synthesis following the CVD-methodology or by low temperature oxidation of metal vapor [1-3]. A further route to monophase and polyphase fine powders is given by the EDS (Evaporative Decomposition of Solutions) and related spray drying processes [4]. In this synthetic route powder properties are controlled by, e.g., the temperature gradient, contact time on a hot surface and evaporation of solvents. A serious disadvantage of all methods mentioned above comes from the large amount of solvents which have to be evaporated as well as from the problem of separating a small amount of a fine powder from a large carrier gas or solvent vapor stream. Our approach to prepare fine ceramic powders from ceramic precursors by utilizing microwave heating is therefore focused on avoiding a solvent and achieving control of powder properties by certain decomposition profiles, the use of a particular applicator as well as by chemical modification of the reaction mixture [5].
Microwave heating has already been introduced into an EDS-related process [6], to heat and evaporate the solvent and to calcine the residual solid. In contrast the main objective of our investigation is to heat the precursor compound directly in order to obtain a ceramic powder due to simultaneous decomposition of the precursor within the whole volume of the reaction mixture. Besides a simple pyrolysis of a single precursor compound, decomposition of precursor-mixtures is also possible, eventually followed by further reaction of the components. Furthermore an inert or reactive powdery matrix-material can be impregnated with the precursor and than converted to "alloyed" ceramic powder. Inert powders, Al_2O_3 and BN, were introduced into the investigation as model powders to simulate ZrO_2-dispersion ceramics, e.g., ZTA and ZrO_2-bonded boron nitride. Selective heating using microwave radiation offers the possibility of pyrolysing a precursor as well as the opportunity for high temperature solid/solid-reactions, when an appropriate powder mixture is utilized. These types of reactions were investigated for a system with a strongly microwave absorbing solid combined with a weakly absorbing one, e.g., CuO/Al_2O_3, C/SiO_2 as well as for two strongly absorbing components, e.g., SiC/ZrO_2. An overview of the methods is given in Fig. 1 and Table 1.

EXPERIMENTAL

As ceramic precursors alcoholates and/or acetylacetonates of Al, Zr, Ti, Si, Cu, Mg were used, as shown in Table 1. Abbreviations used thruout this paper are summarized in Table 2. As reactive powders, activated carbon (Fluka, 75% $\leq$ 40μm) and commercial SiC

(ESK, 5 m^2/g) were used. As inert powders, different aluminas (CT 8000 SG and A 16 SG, both Alcoa) and a BN powder (Aldrich, 99% hex-BN) were used.
Solid precursor materials for pyrolysis and copyrolysis experiments, e.g., ATIP, MgDE, $Cu(acac)_2$, were ground under an Ar blanket and sieved (160 μm) prior to reaction. Liquid precursor compounds or solvates of them, e.g., TEOS, PMMS, TTP, ZTP were used as wet solids in the case of impregnation of Al_2O_3, BN, SiC, C followed by pyrolysis. Because of substantial losses due to evaporation of the precursor in the case of TEOS-, ZTP- and TTP-impregnated carbon, a small amount of acetic acid was added to the mixture prior to microwave heating. In this case the slurry was dried at a low microwave power level before the solid state reaction. PMMS/C- mixtures were directly decomposed.
Most of the microwave heating experiments were performed in a rotating reaction vessel shown in Fig. 2. In this case the process was controlled by adjusting the microwave power level and the vapor pressure during a complete pyrolysis reaction to approximate a constant pyrolysis rate. The power and pressure data for each compound were evaluated from a number of experiments with prolonged reaction time and increasing power level in which the weight loss was controlled. Temperature measurement in the rotating vessel is difficult. Therefore some heating profiles were registered in a horizontally rotating system under weight control, as described in [7]. For high temperature experiments heating and power profiles were evaluated in a static system from step-wise experiments with increasing reaction time. In this case the yield of carbide was used as process control. Due to pressure changes in the course of reaction, evaluation of a "weight controlled" or reaction rate controlled heating profile was impossible.

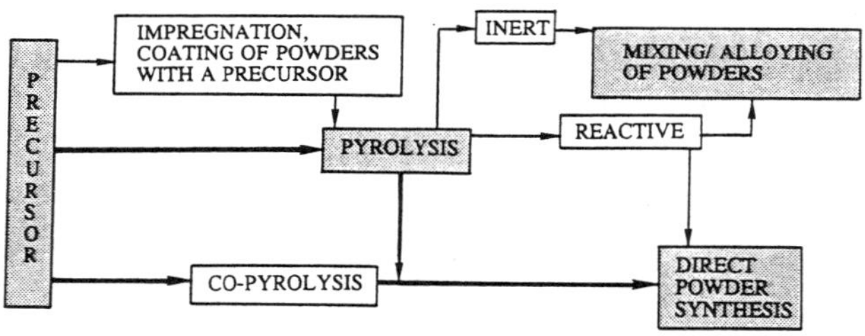

Fig. 1.: Processing scheme for powder generation by microwave pyrolysis.

Table 1.: Methods and materials utilized in powder generation by metallorganic and microwave processing.

METHOD	REACTANDS	PRODUCTS	surface area m^2g^{-1}
Pyrolysis	ATIP	Al_2O_3	≤ 350
Pyrolysis	ZTP	ZrO_2	≤ 30
Copyrolysis	ATIP + ZTP	Al_2O_3, ZrO_2	≤ 100
Copyrolysis	ATIP + $Cu(acac)_2$	Al_2O_3, CuO, or $CuAlO_2$	≤ 50
Copyrolysis	ATIP + MgDE	Al_2O_3, MgO or $MgAl_2O_4$	≤ 50
Impreg./reactiv	TEOS + C_{act}	SiO_2,C_{act}/SiC	≤500/10
Impreg./reactiv	PMMS + C_{act}	SiC	n.est.
Impreg./reactiv	TTP + C_{act}	TiO_2,C_{act}/TiC	≤ 700/-
Impreg./reactiv	ZTP + C_{act}	ZrO_2,C_{act}/ZrC	≤ 550/100
Impreg./reactiv	ZTP+ SiC	SiC,ZrC	≤ 1
Impreg./inert	ZTP+ Al_2O_3	Al_2O_3,ZrO_2	≥ 20
Impreg./inert	ZTP+ BN	BN,ZrO_2	n.est.
Impreg./inert	ZTP+ SiC	SiC/ZrO_2	≤ 30

Table 2: Abbreviations used for metallorganic compounds.

COMPOUND	ABBREVIATION
$Al(O\text{-}iC_3H_7)_3$ Al-triisopropanolate	ATIP
$Zr(O\text{-}nC_3H_7)_4$ Zr-tetrapropylate	ZTP
$Mg(OC_2H_5)_2$ Mg-diethanolate	MgDA
$Si(OC_2H_5)_4$ Tetraethoxysilane	TEOS
$Ti(O\text{-}iC_3H_7)_4$ Ti-tetrapropylate	TTP
$Cu(C_5H_7O_2)_4$ Cu-acetylacetonate	$Cu(acac)_2$
$(\text{-}SiR(OR)\text{-})_x$ Polymethylmethoxysilane	PMMS

RESULTS

- Pyrolysis, Co-pyrolysis

With alcoholates and ATIP, pyrolysis as well as co-pyrolysis does occur due to microwave heating, as shown from gravimetric measurements summarized in Fig. 3a. The corresponding surface area data, obtained from N_2-BET measurements, are shown in Fig. 3b. MW-pyrolysis yields products with some organic residue present, as indicated by IR-spectra of the "after pyrolysis", 400°C- and 800°C-annealed powders [8]. From XRD-measurements no conclusion can be drawn about the phase composition of the "as pyrolysed" powders. TEM and electron diffraction results of some of the powders reveal crystallinity of the ATIP- and ATIP/$Cu(acac)_2$-pyrolysis product. The "agglomerates" shown in Fig. 4a-d are built up of particles with apr. 0.1 μm size. The material surrounding the crystalline particles could be amorphous or even contain organics. A closer characterization of this material has not been done yet. Comparison of microwave-decomposed with furnace- decomposed powders shows in the case of CuO/Al_2O_3 an interesting difference in spinel-formation. Whereas the $CuAlO_2$-spinel is detectable in the MW-decomposed powder after annealing at 400°C, in the furnace decomposed powder spinel is first formed at 800°C, as shown in Fig. 5. The lowest, to our knowledge, reported $CuAlO_2$-formation temperature is 800°C for a highly active citrate-process derived powder [9]. In the case of Al_2O_3/MgO no difference in temperature of $MgAl_2O_4$-formation is found between MW- and conventionally-decomposed powders [8].

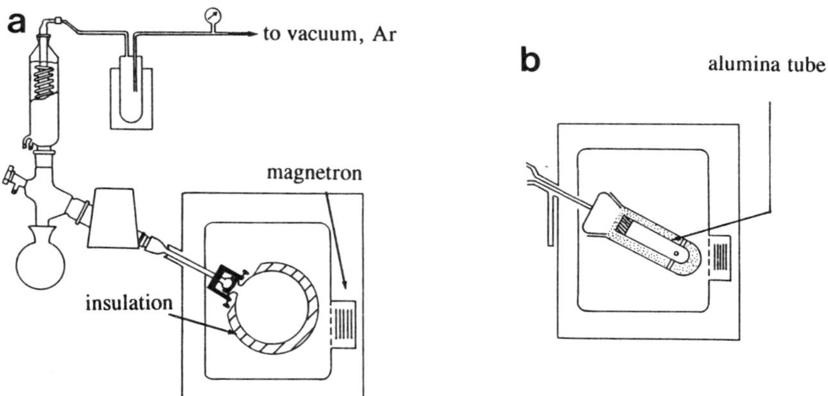

Fig. 2.: Microwave pyrolysis apparatus a) T ≤ 500°C, b) T ≥ 500°C

-Inert impregnation/coating

For all ZrO_2-composite powders, e.g., Al_2O_3-, BN- and SiC-matrix, direct MW-decomposition of the ZTP was achieved. According to XRD measurements after subsequent annealing up to 800°C MW-pyrolysis of ZTP yields t-ZrO_2 first. Upon heating to 1200°C only apr. 4% of the t-ZrO_2 is not transformed to m-ZrO_2 [10]. Conventional as well as MW sintering yields ceramics with very uniform ZrO_2 distribution [7, 10]. With SiC/ZrO_2 no ZrC-formation was observed in MW sintering experiments. However, graphite furnace sintering yielded a SiC/ZrC composite with no ZrO_2 left [7]. The density of the MW-"sintered" SiC/ZrO_2 is low, (apr. 50 %TD) because heating above 1700°C was avoided due to contact reactions with the alumina fibre insulation.

-Reactive impregnation/coating

When activated carbon is impregnated with ZTP, TTP, PMMS or TEOS a reactive mixture results, which upon further MW heating up to apr. 1500°C (surface temperature

Fig. 3.: Pyrolysis and copyrolysis of ATIP-based materials a) weight loss of the reaction mixture due to microwave pyrolysis and subsequent annealing, b) specific surface area (N_2-BET) of the resulting powders at subsequent processing steps, with comparison data of some conventionally decomposed samples.

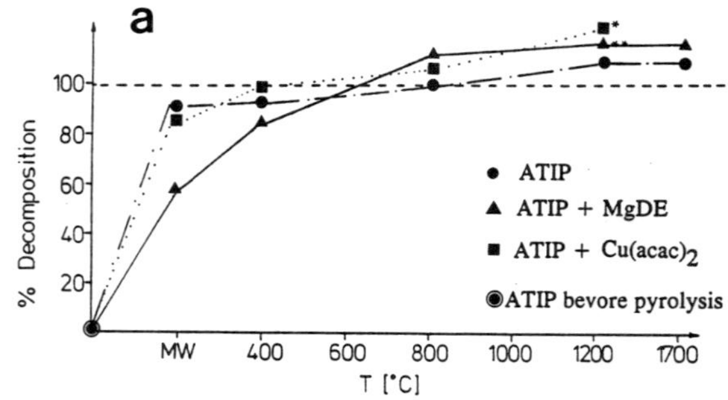

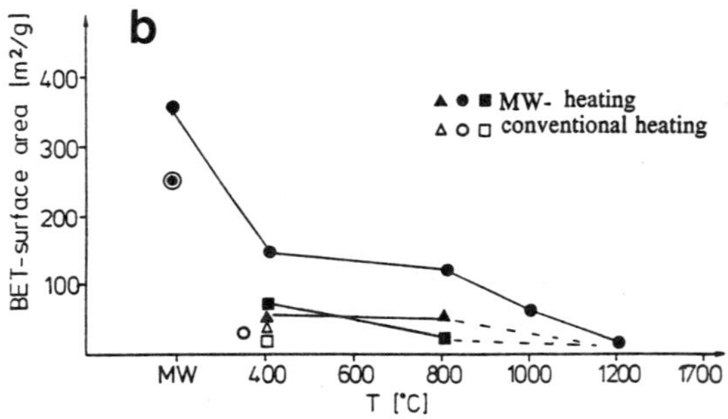

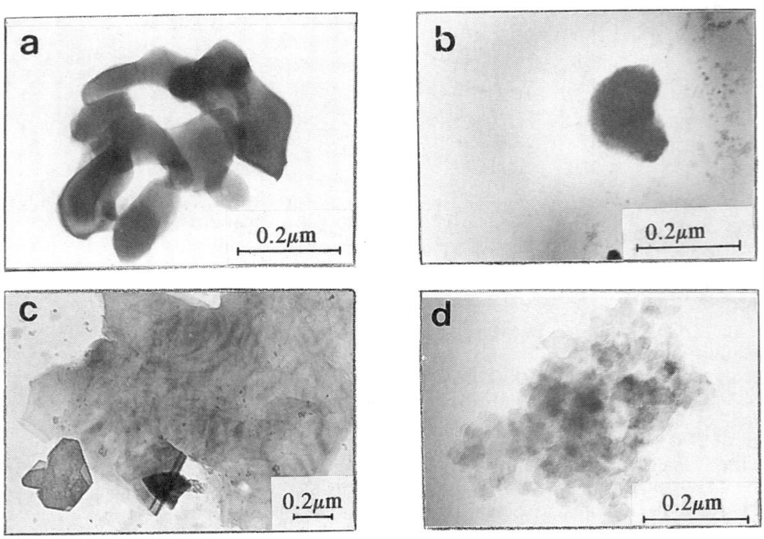

Fig. 4.: TEM of powders from pyrolysis of metallorganic precursor compounds a) MW-pyrolysis of ATIP, b) conventional ATIP-pyrolysis, c) ATIP/Cu(acac)$_2$ MW-pyrolysis, d) ATIP/Cu(acac)$_2$ conventional.

of the alumina reactor) is converted to the corresponding carbides. The conversion rate achieved can be quantified only in the case of SiC, because the carbon as well as the SiO_2-residue can be separated from the product [11]. The best overall yield of apr. 80%mol of the SiO_2 used was achieved with a TEOS/C-ratio of 1:11 [11]. As shown in Fig. 6a,b for SiO_2/C no grain growth is observed when going from the activated carbon to the coated/impregnated carbon. The resulting SiC is finer than the starting carbon due to the reaction mechanism of carbothermal synthesis [12]. The morphology of the SiC obtained varies in the rough mixture from whiskerc to small crystallites, as visible in Fig. 6c. XRD indicates ß-SiC and in some cases SiO_2. After HF-washing the SiO_2 disappears.

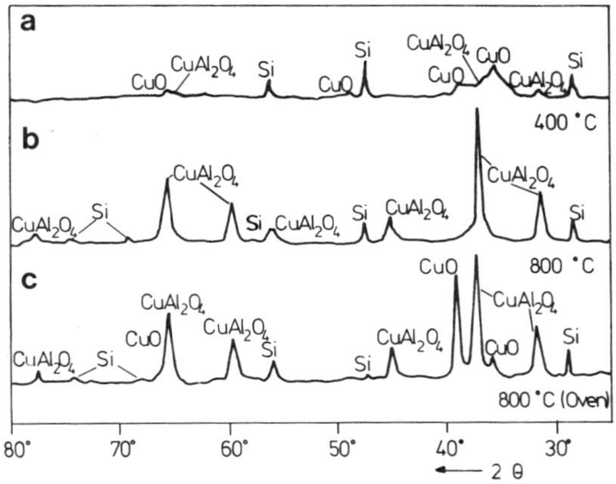

Fig. 5.: a), b) XRD of powder from microwave pyrolysis of ATIP and $Cu(acac)_2$ 1:1 mixture after annealing at 400°C and 800°C to allow for crystallite growth, c) conventional pyrolysis (400°C) of a mixture with the same composition; Al_2O_3 first detectable at 1200°C (Si added as standard).

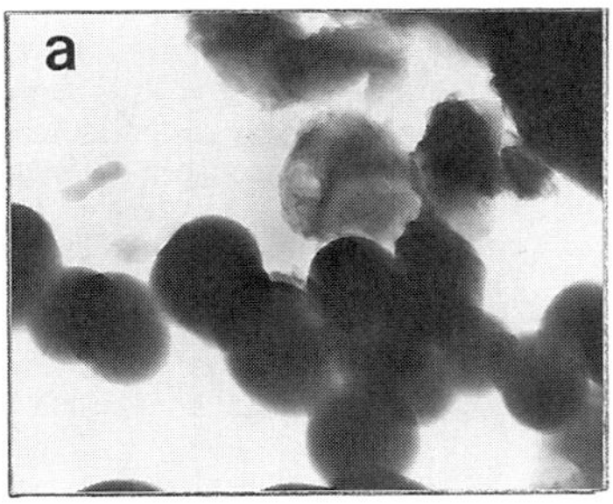

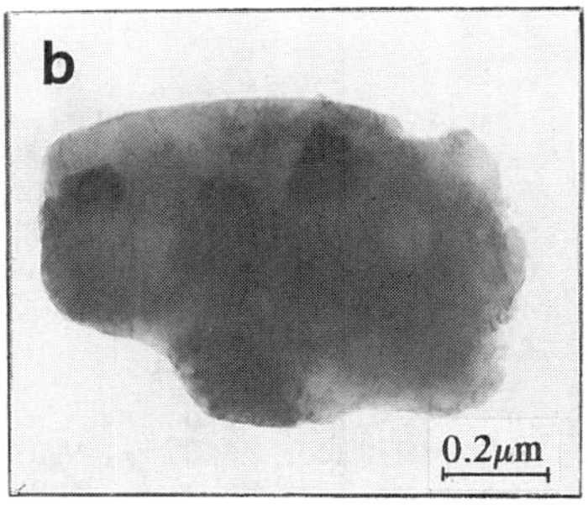

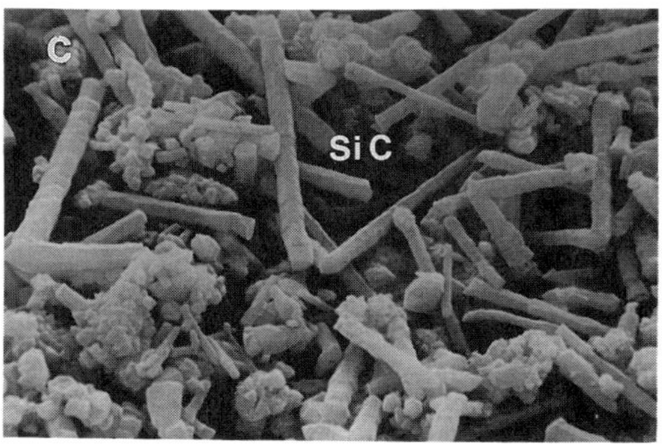

Fig. 6.: MW processing of a reactive mixture composed of SiO_2/C a) act. carbon (TEM), b) after impregnation with TEOS / pyrolysis (TEM), c) after MW-heating to 1450°C / 0.5 h (SEM).

DISCUSSION

Dissipation of MW radiation into heat readily occurs with alcoholates and acetylacetonates causing decomposition reactions as shown by gravimetric analysis. Depending upon the precursor, differences in decomposition tendency and/or reactivity are detectable. The decomposition temperature profile is indicative of enhanced MW-absorption and fast, complete reaction in the case of the CuO/Al_2O_3-mixture as compared to MgDE/ATIP. Because of carbonate impurities often present in MgDE, incomplete decomposition alone would not be indicative of diminished MW-absorbtion. But in contrast to the ATIP/Cu(acac)$_2$ case, no enhanced spinel formation as compared to conventional decomposition is detectable in ATIP/MgDE mixtures. An enhanced MW-absorbtion and decomposition reaction is found also in other precursor compounds for MW-absorbing oxides , e.g., Cr_2O_3/, NiO/, CoO/ and TiO_2/Al_2O_3 [8].
Specific area measurements show the powders derived by MW-pyrolysis or copyrolysis to be sinterable. The expected increase in specific area due to pyrolysis is found only in MW pyrolysis of ATIP, not in conventional heating. Preliminary results on sintering yield $\leq$ 1μm grain size materials. However, compaction without pressing aids is difficult, and with no further annealing shrinkage is too high to yield crack free products [10].
In the case of inert coating/impregnation powders are obtained with increased surface area and better sieving properties. Compaction is sometimes more difficult as with the matrix powder alone. The method is proved to be able to achieve second phase dispersions better or at least comparable to commercial ZTP ceramics [10], without time- and energy consuming milling.
The reactive mixtures can be used to generate single phase nonoxide powders as well as mixed nonoxide or oxide/nonoxide powders at reasonably low temperatures, thus preserving the small grain size of the starting mixture. Further optimization of the synthesis, particularly the conversion rate, is necassary.

CONCLUSION

With MW-heating using metallorganic precursor compound decomposition, spinel formation and red-ox reactions take place at much lower macroscopically measured temperatures as compared to conventional thermal processing. Due to the mild reaction conditions a wide selection of mono- and polyphasic powders with reasonably high surface area can be synthesized by this method.

LITERATURE

[1] T. Ogihara, H. Nakajima, T. Yanagawa, N. Ogata, K. Yoshida, N. Matsushita: J. Am. Ceram. Soc., 74, 2263 (1991).
[2] S. Hori, Y. Shigaki, Y. Hirata, M. Yoshimura, S. Somiya: Mat. Res. Soc. Symp. Proc., Vol. 155, 3 (1989).
[3] J.A. Eastman, Y.X.Liao, A. Narayanasamy, R. W. Siegel: Mat. Res. Soc. Symp. Proc., Vol. 155, 255 (1989).
[4] T.J. Gardener, G.L. Messing: Am. Ceram. Soc. Bull., 63, 1498 (1984).
[5] M. Willert-Porada, T. Gerdes, T. Krummel, B. Rohde: Ger. Pat. Appl. P4213832.9 and P 4214631.3.
[6] W.F. Kladnig, J.E. Horn: Ceram.Internat., 16, 99 (1990)
[7] M. Willert-Porada, this volume.
[8] M. Willert-Porada, B. Rohde, unpubl. results.
[9] C. Marcilly, P. Courty, B. Delmon: J. Am. Ceram. Soc. 53, 56 (1970).
[10] T. Gerdes, M. Willert-Porada, unpubl. results.
[11] T. Krummel, D. Moormann, M. Willert-Porada, unpubl. results.
[12] G.C. Wei, C.R. Kennedy, L.A. Harris: Am. Ceram. Soc. Bull. 63, 1054 (1984)

Financial support of Land NRW, contract IVA6-214 501 89, is greatfully acknowledged. We are indebted to Prof. Köster and his coworkers for helpful discussion of the TEM results.

METALLORGANIC AND MICROWAVE PROCESSING OF CERMETS.

M. WILLERT-PORADA, T. GERDES, S. VODEGEL; University Dortmund, Dept. Chem. Eng., 4600 Dortmund 50, Emil Figge Str. 66, Germany

ABSTRACT

Preliminary results are decribed on synthesis of ceramic-metal composites with a functional gradient material character utilizing infiltration of a porous metallic matrix with ceramic precursors and thermal processing by MW-heating. Infiltration of the metal matrix with a ceramic precursor facilitates bonding between metal and ceramic. By repeated infiltration and decomposition, complete filling of the pores with a ceramic phase is achieved.
Using SiC as MW absorbing material in a sandwich-like configuration, heating of the metal-ceramic composite in a strong MW field is possible. Temperatures of apr. 1000°C were achieved with this method.

INTRODUCTION

Ceramic metal composites play an important role as thermomechanically and corrosion resistant materials. For thin ceramic coatings CVD or PVD are the methods of choice. However, for heat and corrosion resistant applications, demand exists for thick ceramic layers of > 1000 μm. Such composites naturally suffer from high internal stresses due the thermal and mechanical properties mismatch within a ceramic-metal laminate structure.
One way to potentially overcome this difficulty is the "functional gradient material", abr. FGM, approach [1]. FGM`s are composed of a ceramic substrate with gradually increasing metal content, ending up with a pure metal or vice versa, as illustrated in Fig. 1. Such materials are usually processed by powder metallurgical (PM) methods, using gradient mixtures of metal and ceramic powders [2] or by infiltrating a ceramic matrix with a metal melt [3]. However, in the PM route the usually very high processing temperature causes the metal powder to melt and infiltrate the ceramic, so microstructure control of the metal as well as ceramic part within the composite is difficult if not impossible. The infiltration route into a ceramic with controlled microstructure, on the other hand, often suffers from poor wettability of ceramics by metal melts [4]. Thus, none of the methods mentioned above offers the opportunity for microstructural control of both the ceramic and the metal component. Infiltration of molten ceramics into a metal matrix isn`t practical. However, infiltration of a porous metal by ceramic precursor compounds is practicable, in analogy to the CVI-process of ceramic composites [5].
The key step of such a processing route to FGM with controlled metal microstructure is the development of a selective heating method which combines a high penetration depth and a high heating rate within the ceramic part with a low penetration depth and low heating rate in the metal part of the composite. Because of the high penetration depth in ceramic materials and high reflection at metal parts microwave radiation seems to be advantageous over IR or visible radiation for this type of application.
However, MW heating of metal-ceramic parts has some inherent problems, illustrated in Fig. 2. Particularly in a laminate composite, arcing due to E-field concentration at metal edges as well as the E-field reflection pattern at a metal surface have to be considered.
In this preliminary investigation we attempted to facilitate coupling of MW radiation into the ceramic part of a FGM by using a sandwich-like configuration with a supporting SiC central layer of varying thickness as indicated in Fig. 2.. The removable SiC layer also virtually increases the ceramic to metal ratio.
Zirconia was chosen as ceramic part of the FGM due to its high- temperature resistance and low thermal conductivity as well as its good MW absorbing properties at higher

temperature. Furthermore, metallorganic Zr-compounds have been reported in the literature [6] as precursors for ZrO_2 coatings in conventional processing.
As metal matrix, porous Cu-, Fe- based alloys as well as Ni were used.

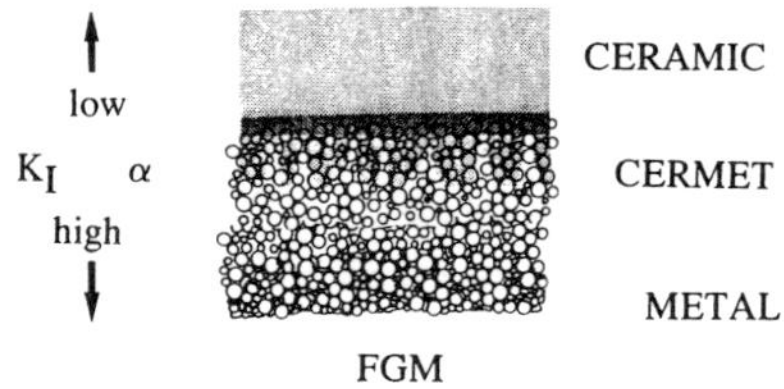

Fig.1.: Functional Gradient Material (FGM) concept [1]

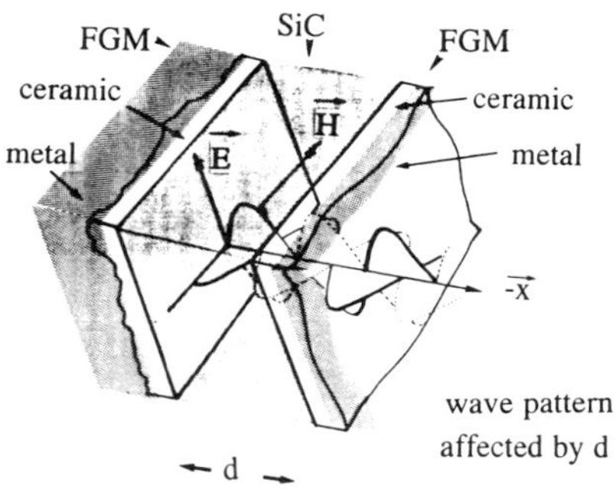

- absorption of MW in ceramic as function of E and λ
- reflection of MW at metal surface:
- arcing at metal edges
  → sites of high E-field density
- composition-, size- and shape-dependent surface currents (skin effect)

$E = 0$ at $-x = n \cdot \lambda/2$,
$E = E_{max}$ at $-x = m \cdot \lambda/2$

E: electric field, H: magnetic field

Fig. 2.: Microwave (MW)- material interaction in a FGM.

EXPERIMENTAL

The matrix materials used were commercial filters produced by conventional powder metallurgy. Ni-matrix material was obtained by axial pressing of commercial powders. The mechanical strength of the as-pressed pellets was sufficient to directly infiltrate them after surface finishing. Properties of the matrix materials are listed in Table 1. The matrix metals were ultrasonicated in acetone and dryed in vacuum prior to infiltration. For infiltration 70%vol $Zr(OC_3H_7)_4$, Zirconiumtetrapropylate, in propanol, abr. ZTP, was used. For slip casting, commercial m-ZrO_2 powder as slurry in dry ethanol was employed. The infiltration procedure was performed by pouring the metallorganic compound onto the metal plate through a nozzle with gentle pressure under an Ar-blanket. Additives such as Mg- and Y-nitrate were added as ethanolic solutions (0.25 M in metal-ion) to the precursor solution prior to infiltration.
Heating by MW was performed in a 2.5 kW multi mode cavity (Püschner Mikrowellentechnik) on a rotating support plate within a quartz-cylinder under Ar or in a 2.5 kW single mode cavity (Puls Plasma Technik, Dortmund) in vacuum. Curing was performed using an alumina-casket to insulate the samples and keep the temperature at a medium high level during the decomposition. Recrystallized silicon carbide, R-SiC, (Schunk) was used to facilitate MW-coupling. Heating to higher temperatures was performed with no insulation, by cooling the metal with Ar.

Symbol	Composition.	Density (gcm^{-3})	Porosity (%)	Pore size/ Grain size (µm)
R 3*	1.4404	5,2 - 5,6	26 - 30	3 / -
R 20*	1.4404	5,1 - 5,5	32 - 36	20 / -
R 80*	1.4404	4,3 - 4,7	32 - 38	80 / -
B 8*	CuSn10	5,8 - 6,2	27 - 30	8 / -
B 40*	CuSn10	5,0 - 5,4	33 - 36	40 / -
B 150*	CuSn10	4,9 - 5,3	32 - 38	150 / -
Ni 3	99.7Ni	6,36	28,6	- / 3
Ni 150	99.0 Ni	6,35	28,7	- / 150

* Thyssen Siperm Tradename of Thyssen

Table 1.: Properties of matrix materials used for infiltration with ZTP.

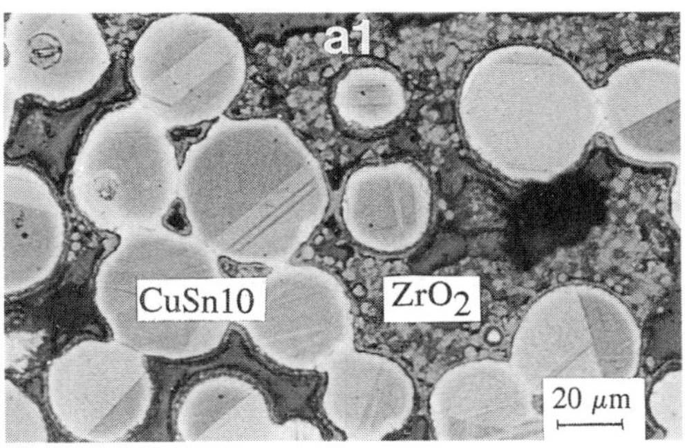

Fig. 3.: Bonding between metal and ZrO_2 derived from ZTP as function of pore size (a1,a2), pore shape (a,b,c) and matrix-material (a,b,c); a) bronze, b) steel, c) Ni.

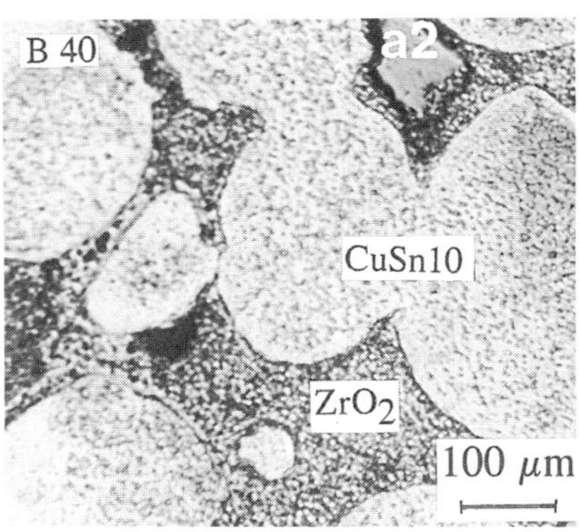

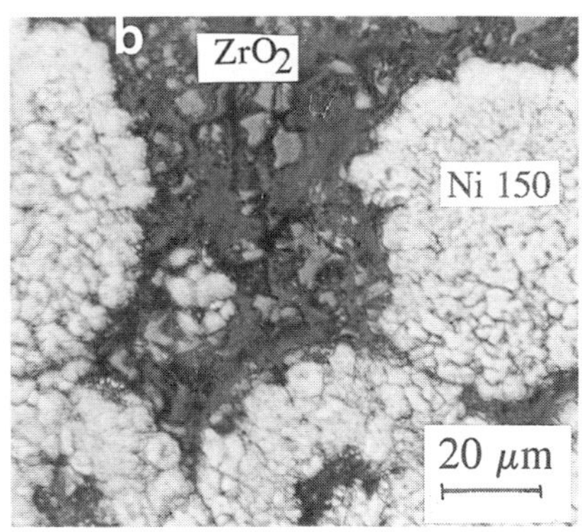

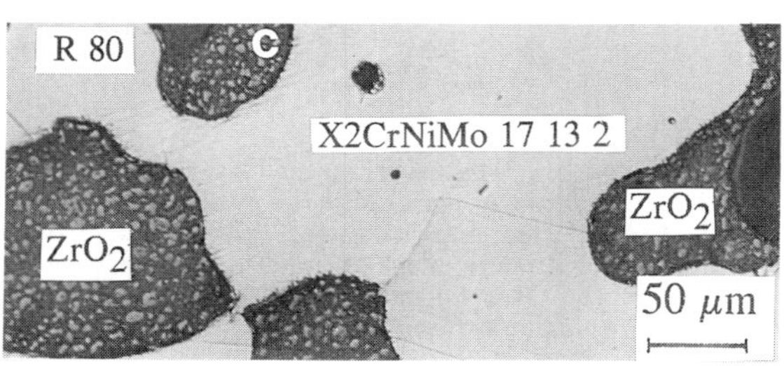

1.4404 :
X2CrNiMo 17 13 2

RESULTS

Infiltration with ZTP provides a good bonding of the ceramic powder generated from the precursor to the metal, as visible from the close contact betwen the ceramic and the metal shown in Fig. 3. a-c. The pore size and shape strongly influence the effectiveness of the infiltration procedure. Good results were obtained with the B 40, R 80 and R 20 material. With Ni 150 some infiltration occurs, though not nearly as effectively as with the steel and bronze material. With Ni 3 no infiltration was possible. With B 8, single infiltration is possible with the 70%-ZTP solution but multiple infiltrations are not very effective. Also, chemical incompatibility seems to play a role. The worst bonding is found with Ni and ZTP Fig. 3c.).
Heating profiles were investigated using the same configuration of SiC with and without the metal-ceramic composite. As visible from Fig. 4a,b only the reflected power is affected by the metal present, but not the temperature profile. The higher power level necassary to achieve apr. 1000°C could be accounted for as due to the higher heat capacity and the higher heat losses with the metal present.
Some influence of the SiC-layer thicknes on the heating profile is found as illustrated by Fig. 5 . Also the SiC to metal ratio affects the heating profile as shown in Fig. 5b. However, the maximum temperature which could be reached is not affected by this parameter. Mainly the tendency for arcing is greatly increased, when the SiC layer is less than 1 cm thick.
The sintering atmosphere also influences the bonding between the ceramic and the metal. Heating in air causes separation of the ceramic from the metal as illustrated by Fig. 6a. Slip casting of a ethanolic ZrO_2-slurry from a commercial ZrO_2 powder onto an already ZTP-infiltrated and decomposed cermet gives after drying originally good bonding between this two materials. However, repeated MW-heating of this composite causes the slip cast layer to separate from the cermet layer as shown in Fig. 7.

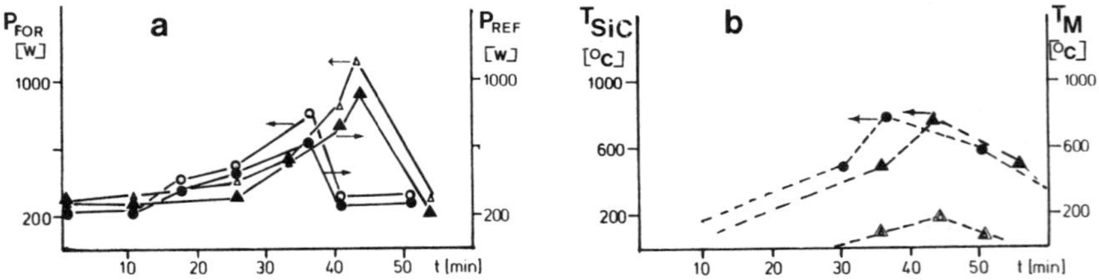

Fig. 4.: Heating profile as function of metal-presence: a) reflected and forwarded power, b) surface temperature of the SiC and the metal (M) part; ●○without metal,▲△with metal;

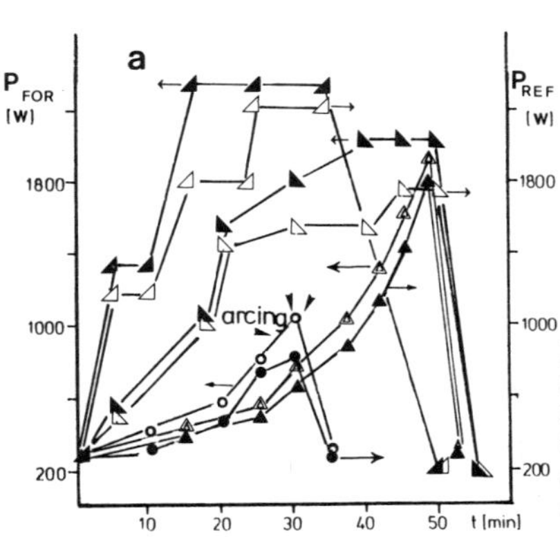

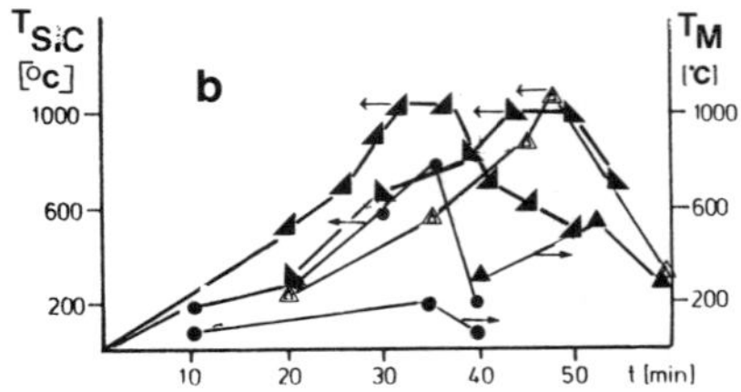

Fig. 5.: Influence of SiC layer thicknes and ceramic to metal-ratio on heating profile; a) reflected and forwarded power, b) surface temperature of the SiC and the metal (M) part; SiC layer thicknes: ●○ ≤ 1cm, ▲◭ ≥ 1cm; ceramic to metal ratio: ◢◿= SiC:M 3:1, ◣◺= SiC:M 1:3.

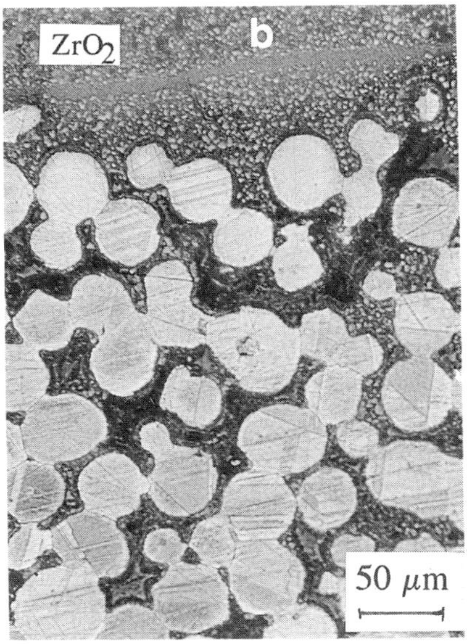

Fig. 6.: Influence of atmosphere on bonding between ceramic and metal; a) air, b) vacuum

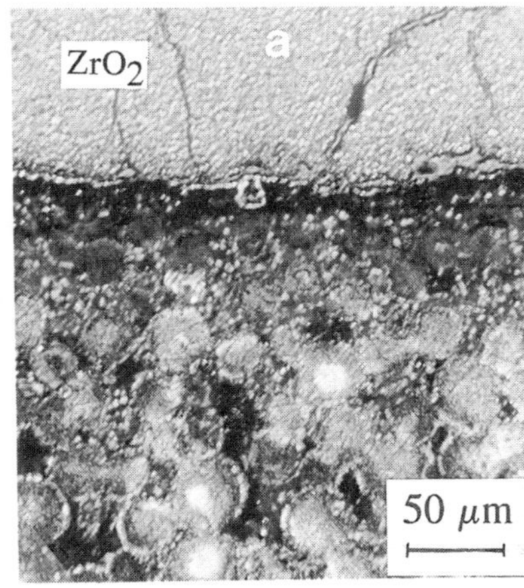

Fig. 7.: Separation of a slip cast m-ZrO_2-layer from the ZrO_2/metal cermet after repeated MW-heating.

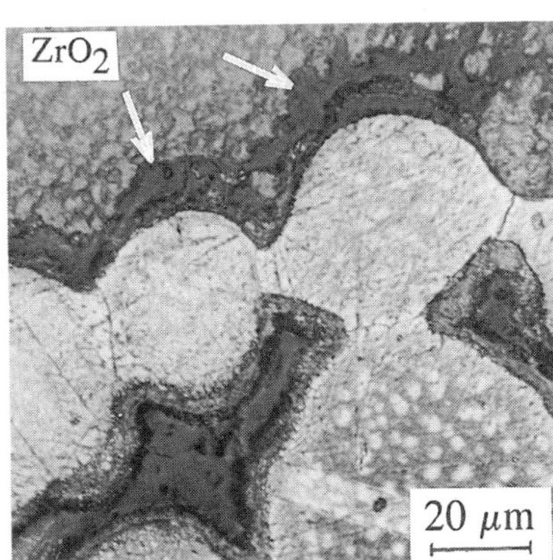

DISCUSSION

Metallorganic infiltration as well as MW-heating of the composite appears to be a practicable method for the synthesis of cermets with a functional gradient. The readily occuring oxidation shows that the ceramic coating is possibly not gastight, but at this stage of the experiments the high oxygen diffusivity of ZrO_2 cannot be excluded as also facilitating the oxidation.
A further increase in the ceramic layer thickness by slip casting seems to need additives to overcome the large difference in particle size and density between the precursor derived and commercial ZrO_2 powder. The main problem in this unstabilized ZrO_2-powder is probably the m→t ZrO_2 transformation. For this purpose slip casting with yttrium or magnesium stabilized precursor derived powder will be tried in future work. Direct slip casting a ZrO_2-powder slip onto the porous matrix did not give good results. Only a low penetration depth was achieved. The close bonding between the metal and ceramic shown for ZTP-derived materials is not observable with a slip cast commercial powder. However, mixing the commercial powder with ZTP and slip casting gives some bonding between these two components. In order to sinter the ZTP-derived ZrO_2 onto the metal part higher temperatures could be necessary than those achieved with SiC. Furthermore the decoupling of SiC at apr. 1000°C has to be overcome. For this purpose a SiC/ZrO_2-composite [7] will be used in the sandwich configuration as well as some insulation to decrease the radiation losses.

CONCLUSION

By using ZTP as precursor and multiple infiltration into a porous metal matrix, a cermet can be produced. Pore size and morphology of the substrate strongly influence the effectiveness of this process. The precursor as well as slip cast thicker ceramic layers together with the metallic substrate can be heated by MW in a multi mode cavity when a SiC-sandwich configuration is used.

LITERATURE

[1] T. Hirano, T. Yamada, J. Teraki, M. Niino, A. Kumakawa, Proc. 16th Int. Symposium on Space technology and Science, Edited by K. Kondo, AGNE, Tokyo, 1988, pp. 375-380.
[2] Y. Miyamoto, Ceram. Bull., 69, 686 (1990).
[3] C. Toy, W.D. Scott, J. Am. Ceram. Soc., 73, 97 (1990).
[4] K. Suganuma, Y. Miyamoto, M. Koizumi, Ann. Rev. Mater. Sci. , 18, 47 (1988).
[5] W. B. Hillig, J. Am. Ceram. Soc., 71, C-96 (1988).
[6] K. Izumi, M. Murakami, T. Deguchi, A. Morita, N. Tohge, T. Minami, J. Am. Ceram. Soc., 72, 465 (1989)
[7] M. Willert-Porada, T. Gerdes, R. Makowski, unpubl. results.

Financial support of DFG, SFB 316, A6 is greatfully acknowledged.

REACTION SINTERING OF CERAMIC MATERIALS BY MICROWAVE HEATING

Ph. BOCH, N. LEQUEUX and P.PILUSO
Ecole Supérieure Physique Chimie Paris, 10 rue Vauquelin, 75005-Paris, France

ABSTRACT

Microwave heating was used to reaction sinter Al_2O_3 + SiO_2 and Al_2O_3 + TiO_2 powder mixtures to prepare mullite and aluminum titanate ceramics, respectively. In comparison with conventional heating, microwave heating leads to a decrease in the temperatures of treatment of 50 to 100°C.

INTRODUCTION

There is a growing evidence that microwave heating can decrease the sintering temperature of ceramic materials [1,2]. Sintering involves four phenomena (consolidation, densification, grain growth, and eventually reactions between constituents) but most of the microwave experiments were related to the study of consolidation and densification. Grain growth was rarely examined [3] and very few data exist about the effect of microwaves on reactions between constituents [4]. The aim of the present study was to bring information on the microwave reaction sintering of two ceramic materials: mullite and aluminum titanate.

REACTION SINTERING OF MULLITE AND ALUMINUM TITANATE

Mullite is the only binary compound in the Al_2O_3-SiO_2 system [5]. Stoichiometric "3:2 mullite" ($Al_6Si_2O_{13}$) melts (congruently or incongruently, depending on conditions) at temperatures superior to 1820°C. The preparation of mullite ceramics by reaction sintering [6] must take into account three characteristics, namely: i) silica generally transforms into cristobalite before its reacts with alumina to give mullite, ii) diffusion phenomena are slow in mullite, which suggests to select conditions which favor densification before reaction, and iii) the formation of mullite is accompanied by an expansion (about 4% in volume, when silica is cristobalite).

Aluminum titanate (ß-Al_2TiO_5) is the only binary compound in the Al_2O_3-TiO_2 system [7]. It melts at 1850°C but decomposes into Al_2O_3 + TiO_2 below about 1200°C. The formation of aluminum titanate is accompanied by an expansion (11% in volume). The study of the reaction sintering of alumina + rutile mixtures [8] has shown two main points. First, densification develops before reaction. And secondly, reaction occurs according to a two-stage process. The first stage corresponds to the nucleation and fast growth of Al_2TiO_5 cells which trap alumina and rutile inclusions, and pores. The second stage corresponds to the slow growth of Al_2TiO_5 cells with elimination of non-reacted residues and slow densification.

EXPERIMENTAL

Pre-sintering processes

For the mullitic system, the starting powders were a 99.99% pure alumina (*Baikowski*) and a 99% pure cristobalite (*Sifraco*). Powders were mixed in the ratio of 3 moles of alumina to 2 moles of silica, which corresponds to 71.8 wt% Al_2O_3 + 28.2 wt% SiO_2. For the aluminum titanate system, the starting powders were a 99.8% pure alumina (*Alcoa*) and a 99.8% pure rutile (*Kronos Titan*). Powders were mixed in the ratio of 1 mole of alumina with 1 mole of titania, which corresponds to 56.1 wt% Al_2O_3 + 43.9 wt% TiO_2. The characteristics of powders are shown in table I.

Ceramic powders were mixed with a binder (3 wt% polyvinylbutyral) and a deflocculant (0.3 wt% phosphate ester), then attrition milled for 5 h (with $ZrSiO_4$ balls, in alcohol) and finally granulated through a 200 μm sieve. Disks of 13 mm in diameter and 4 mm in height were uniaxially pressed (150 MPa). Subsequently, non-ceramic compounds were burnt (700°C, for 1 h) and a thermocouple hole (1.6 mm diameter, 6 mm long) was drilled into the pre-sintered samples. After pre-sintering, the relative density ($d/d_{theoretical}$) was 0.45 for Al_2O_3 + SiO_2 materials and 0.56 for Al_2O_3 + TiO_2 materials.

Table I : Characteristics of powders

Commercial source	Average grain size (μm)	Purity (%)	Composition
Baikowski CR30	<0.5	99.99	65% α Al_2O_3 35% sub-α Al_2O_3
Sifraco M5000	1.8	99	100% cristobalite
Alcoa A16	0.3-0.5	99.8	100% α Al_2O_3
Kronos Titan R1053	1.5	99.8	100% rutile

Reaction-sintering

Microwave treatments were always compared with conventional ones, using identical temperature-to-time cycles.

The conventional furnace was an electrical, fast-heating one, using $MoSi_2$ resistors (T_{Max} ≤ 1700°C).

The laboratory-assembled microwave furnace was composed of seven sub-systems, namely: i) a generator, with a 2.45 GHz, 1200 W magnetron and a control of incident and reflected powers, ii) a circulator, which diverts the reflected power from the applicator, iii) a brass waveguide (WR-340 standard), iv) a rectangular coupling iris, v) a cavity, vi) an alumina crucible, and vii) a temperature measurement system.

The rectangular cavity is made of brass. It resonates on a TE_{103} single mode and is adjustable by a short circuit. The characteristics of generator and cavity are critical parameters, which must be optimized when the goal is to heat low-loss dielectrics, such as compounds based on high-purity alumina (tn $\delta \approx 0.005$ at 3 GHz, R.T.). In particular, the quality factor must be put at a high value to lead to a high electric field inside the cavity.

The disk shaped sample is put into a crucible made of alumina fibers which allow thermal insulation. The sample is positioned in the zone of the cavity where the electric field reaches a maximum value. Several shapes and orientations were studied and it was found that a disk shape, with its axis parallel to the electric field, decreases thermal gradients and limits the occurrence of hot spots in the sample.

The temperature measurement system uses both a thermocouple and an IR pyrometer. The shielded Pt-PtRh thermocouple is aligned perpendicularly to the electric field and its tip is introduced inside the hole drilled into the sample. It was verified that the microwave field does not disturb temperature measurements. The pyrometer is focused on the fiber crucible. The accuracy on the temperature measurements is thought to be better than ± 10°C, from R.T. to 1600°C. Figure 1 shows an example of temperature to time cycle (alumina + silica sample).

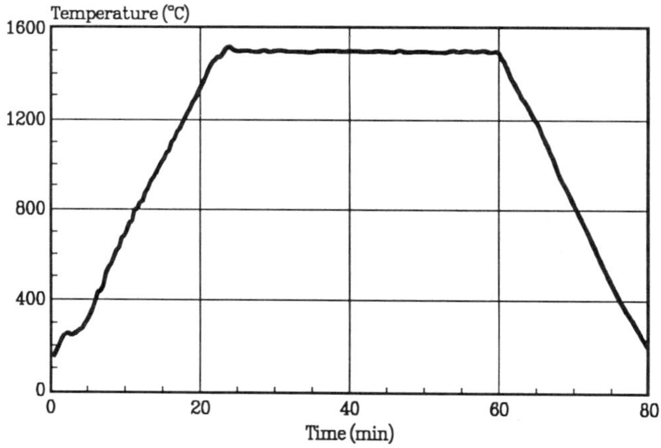

Figure 1 Temperature to time cycle (alumina + silica sample)

Materials characterization

Density measurements were done by Archimedes'method. Microstructures were observed by optical and scanning electron microscopy on etched samples (thermal etching at 50°C below sintering temperature or ion etching). XRD was used for a quantitative determination of phases, using reference mixtures of known composition. The relevant ratios for XRD peaks intensities were $I(220)_{mullite}$ / $I(204)_{alumina}$ and $I(121)_{mullite}$ / $I(204)_{alumina}$ for mullite, and $I(023)_{titanate}$ / $I(101)_{rutile}$ and $I(023)_{titanate}$ / $I(104)_{alumina}$ for aluminum titanate. The accuracy on phase content was better than 5%.

RESULTS AND DISCUSSION

$Al_2O_3 + SiO_2$ system

Three thermal treatments were used (1400°C, 1500°C, and 1580°C, for 35 minutes), in conventional and microwave furnaces.

Figure 2 shows densification and Figure 3 shows reaction. Microwave heating accelerates both phenomena, in a similar manner. For a treatment of 35 min at 1580°C, for instance, conventional heating leads to a densification of 68% and a mullite content of 84% whereas microwave heating leads to a densification of 81% and a fully mullitized material.

A problem for microwave heated materials is temperature homogeneity. Microstructure is very sensitive to thermal gradients and, therefore, can be considered as a probe to show their occurence. In the present case, all reaction sintered samples exhibited an identical microstructure (grain size and shape, porosity) at the surface and at the center. For example, micrographs of materials treated at 1580°C (not represented here) show a homogeneous grain size of about 3-4 μm.

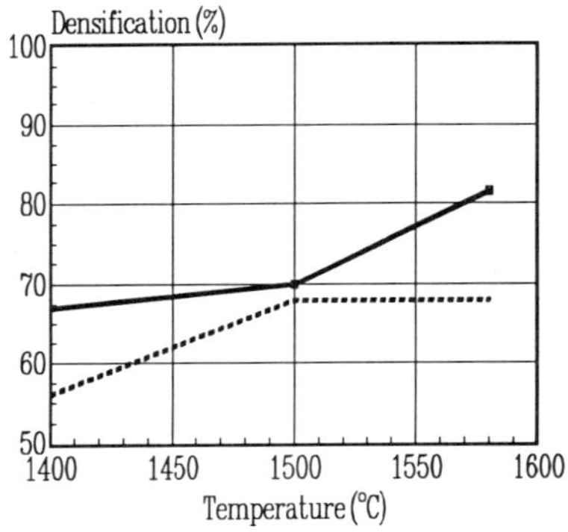

Figure 2 Densification of $Al_2O_3 + SiO_2$ mixtures

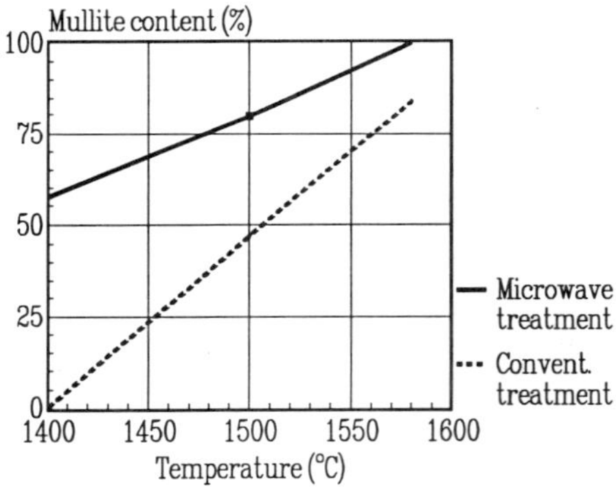

Figure 3 Reaction in $Al_2O_3 + SiO_2$ mixtures

$Al_2O_3 + TiO_2$ system

Four thermal treatments were used (1250°C, 1290°C, 1330°C, and 1370°C, for 60 min), in conventional and microwave furnaces.

Figure 4 shows densification and Figure 5 shows reaction. Microwave heating accelerates both phenomena but its influence is more pronounced on reaction. For a low temperature of treatment (1250°C, which is close to the equilibrium temperature for the titanate formation), conventional heating leads to a densification of 74% and no titanate formation whereas microwave heating leads to a densification of 85% and a beginning of reaction. For a higher temperature (1290°C), conventional heating leads to a densification of 86% and still no titanate whereas microwave heating leads to a densification of 92% and a large titanate content (78%).

Optical micrographs (not represented here) confirm the other data. In particular, the materials treated by microwaves at 1250°C contain elongated aluminum titanate cells (50 x 10 μm), similar to those observed in ref. [8], whereas the materials treated by conventional heating do not contain any cell.

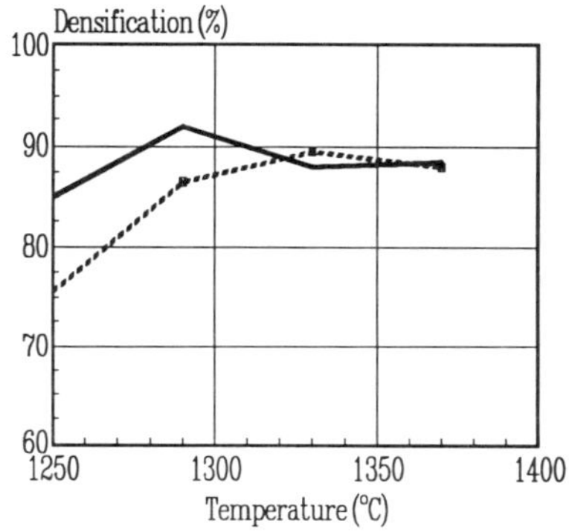

Figure 4 Densification of Al_2O_3 + TiO_2 mixtures

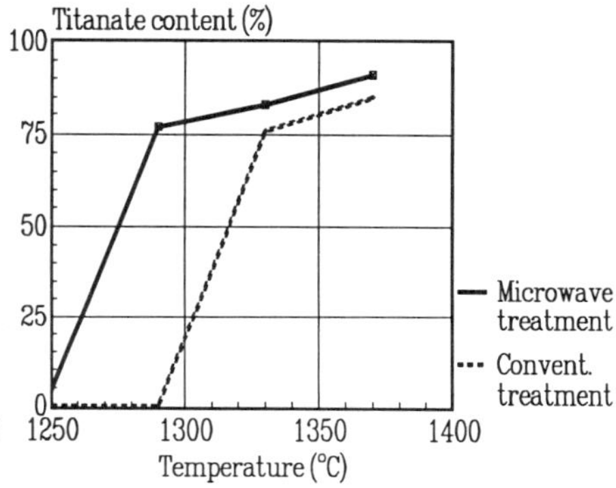

Figure 5 Reaction in Al_2O_3 + TiO_2 mixtures

CONCLUSIONS

Microwave heating facilitates the reaction sintering of Al_2O_3 + SiO_2 and Al_2O_3 + TiO_2 powder mixtures. For the mullitic system, densification (at a given reaction rate) and reaction are accelerated in a similar manner. Both the temperature of reaction and that of densification decrease about 100°C in comparison with conventional heating. For the aluminum titanate system, the microwave effect is more pronounced on reaction than densification. The temperature of reaction decreases about 100°C whereas the temperature of densification (in the pre-reacted state) decreases about 50°C. A study is presently under way to apply microwave reaction sintering to the joining of ceramic parts.

Acknowledgments

The authors gratefully acknowledge *Electricité de France* and *Ministère de la Recherche et de la Technologie* for supporting this research program.

References

[1] Microwave Processing of Materials, edited by W.H. Sutton, M.H. Brooks, and I.J. Chabinsky (Mat. Res. Proc. 124, Pittsburgh, PA 1988).
[2] Microwave Processing of Materials II, edited by W.B. Snyder, W.H. Sutton, M.H. Iskander, and D.L. Johnson (Mat. Res. Proc. 189, Pittsburgh, PA 1991).
[3] M.A. Janney, H.D. Kimrey, M.A. Schmidt, and J.O. Kiggans, J. Am. Ceram. Soc., 74, 1675 (1991).
[4] D. Patil, B. Mutsuddy, and R. Garard, J. Microwave Power and Electromagnetic Energy, 27, 49 (1992).
[5] I.A. Aksay and J.A. Pask, J. Am. Ceram. Soc., 58, 507 (1975).
[6] P.D.D. Rodrigo and P. Boch, Int. J. of High Tech. Ceramics, 1, 3 (1985).
[7] E. Kato, K. Daimon, and J. Takahashi, J. Am. Ceram. Soc., 63, 355 (1980).
[8] B. Freudenberg and A. Mocellin, J. Am. Ceram. Soc., 70, 33 (1987).

MICROWAVE PROCESSING OF TANTALUM CAPACITOR ANODES*

R. J. LAUF*, C. HAMBY*, C. E. HOLCOMBE*, AND W. F. VIEROW**

*Martin Marietta Energy Systems, Inc., Oak Ridge, Tennessee 37831-6087
**AVX Tantalum Corp., 69 Landry St., Biddeford, ME 04005

ABSTRACT

Porous tantalum anodes were sintered at temperatures from 1600 to 1900°C using a conventional high-vacuum furnace as well as both 2.45 GHz fixed-frequency and 4-8 GHz variable-frequency microwave furnaces. Various insulation and casketing techniques were used to couple the microwave power to the tantalum compacts. Several types of tantalum powder were used to assess the effect of microwave processing on sintered surface area and impurity levels.

Some microwave sintered anodes have an unusual surface rippling not seen on conventionally fired parts. The rippling suggests that a microscopic arcing or plasma might have been generated. Two important effects could be exploited if this phenomenon can be controlled. First, the effective tantalum surface area could be increased, yielding higher capacitance per volume. Second, surface impurities might be cleaned away, allowing the formation of a better dielectric film during the anodization process and, ultimately, higher working voltage.

INTRODUCTION

Tantalum capacitors are widely used in electronics where high reliability is needed, such as cardiac pacemakers and other implants, hearing aids, pagers, diagnostic equipment, and military products. Tantalum capacitors are manufactured from a powder of pure Ta metal pressed to form a pellet and vacuum sintered. The spongelike pellet, although of high mechanical strength and density, is still highly porous giving a large internal surface area. This forms the positive "plate" of the capacitor.

A dielectric layer of tantalum pentoxide is formed on the porous tantalum by anodizing, covering all available surface area. The cathode is formed by infiltrating the anodized pellet with layers of manganese dioxide. Electrical contact is established by deposition of carbon and silver onto the surface of the pellet. Cathode connection is then made by means of conductive contact to a lead frame. Packaging is carried out to meet individual specification and customer requirements.

Two types of Ta powder are used commercially depending on specific requirements. Sodium reduced powder has the highest surface area but somewhat lower purity. Electron beam refined powder has the highest purity for higher working voltages but at the expense of somewhat lower surface area. It is interesting to note that because the dielectric is an anodic oxide film grown directly on the Ta, damage can, to some extent, be self-healing. Conversely, failure can occur at weak spots in the oxide, possibly where impurities have accumulated.

The long term goal of this work is to explore several possible aspects of microwave processing: 1. Microwave-generated plasmas or arcing at particle surfaces might clean away surface impurities such as sulfur that are deleterious to the formation of a good dielectric oxide film. 2. Lower processing temperatures and/or shorter cycles could improve process economics and provide tighter control over the microstructure. 3. Observation of surface chemistry and morphology can give fundamental insights into the actual mechanisms of microwave sintering, which are still poorly understood.

EXPERIMENTAL

Four batches of Ta anodes were pressed from representative commercial Ta powders, two each of sodium reduced and electron beam purified. The pressed compacts contained a small amount of organic binder, which was removed by holding 30 min at

---

* Work supported by the U.S. Department of Energy, Office of Energy Research, under contract DE-AC05-84OR21400 with Martin Marietta Energy Systems, Inc.

480°C in vacuum. The parts were sintered in either a conventional Brew vacuum furnace, a 2.45 GHz fixed-frequency microwave furnace [1], or a 4-8 GHz variable-frequency microwave furnace [2]. In order to initiate microwave heating as well as provide thermal insulation at high temperatures, various casketing arrangements [3] were used in the microwave furnaces. Sintering conditions are summarized in Table 1.

Table I. Sintering Conditions for Tantalum Capacitors.

Run	Powder*	Furnace**	T, °C	Atm.	Casketing
1	Na-1	C	1600	Vac.	
2	Na-2	C	1600	Vac.	
3	EB-1	C	1600	Vac.	
4	EB-2	C	1600	Vac.	
5	Na-1	C	1600	Vac.	
6	Na-2	C	1700	Vac.	
7	EB-1	C	1700	Vac.	
8	EB-2	C	1700	Vac.	
9	Na-1	C	1800	Vac.	
10	Na-2	C	1800	Vac.	
11	EB-1	C	1800	Vac.	
12	EB-2	C	1800	Vac.	
13	Na-1	C	1900	Vac.	
14	Na-2	C	1900	Vac.	
15	EB-1	C	1900	Vac.	
16	EB-2	C	1900	Vac.	
17	Na-2	M	1700	Vac.	Y_2O_3 grit
18	Na-2	M	1700	Ar	Y_2O_3 grit
19	Na-2	M	1700	Ar	Y_2O_3 grit, parts coated with BN
20	Na-2	M	1500-1600	Vac.	ZrO_2 fiber
21	Na-2	M	1700	Vac.	ZrO_2 fiber: Y_2O_3 grit 50:50
22	Na-2	M	1600	Vac.	ZrO2 bubbles, BN casket
23	Na-2	M	1700	Vac.	ZrO2 bubbles, BN casket
24	Na-2	V	~1450	N_2	ZrO2 fiber, BN casket
25	Na-2	M	1800	Vac.	ZrO_2 bubbles, BN casket

**C = Conventional, M = 2.45 GHz, V = Variable frequency
*Na = Na reduced, EB = electron beam purified
All samples sintered 30 min. except #24 which was sintered 90 min.

RESULTS AND DISCUSSION

The structures of sintered anodes were characterized by fracturing the pellets and examining the fractured surface by SEM. The results are given in Table 2. In general, all samples were sufficiently sintered to form strong necks between adjacent grains.

Table II. Sintering Results for Tantalum Capacitors.

Run	Surface Reactions	Ripples	Comments
1-16	No	No	Conventional process
17	Yes	No	Well sintered
18	Yes	Yes	Well-sintered without coarsening
19	Yes	Yes	Similar to #18
20	Yes	Yes	
22	Yes	Few	Thick surface crust
23	Yes	Few	
24	Yes	Few	Less sintering than others
25	Yes	--	Reacted w/ZrO_2 bubbles

*All capacitors showed particle-to-particle sintering.

Microstructural development in the controls (Samples, 2, 6, 10, 14, 4, 8, 12, and 16, Fig. 1) illustrates the appearance of "typical" anodes made from the two types of powder. The difference in powder size and surface area is clear because sintering is not carried out to full densification.

Sodium Reduced | Electron Beam Refined

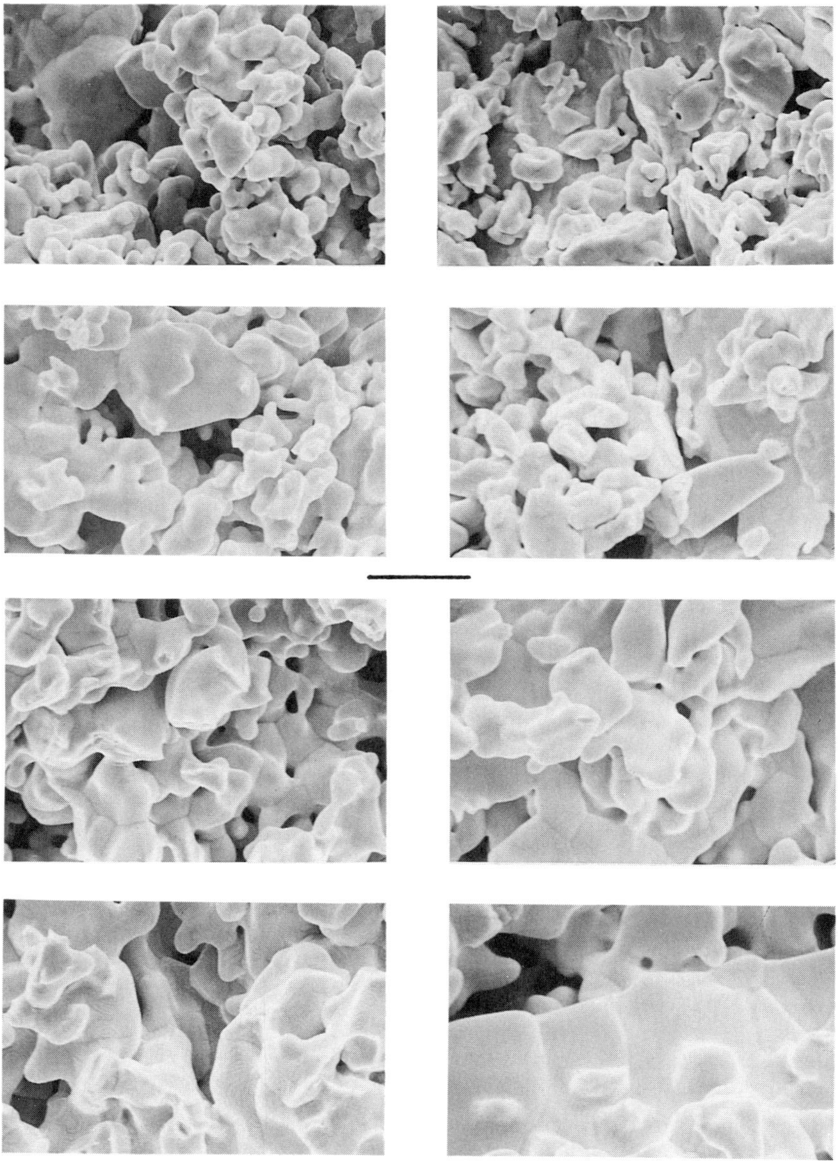

Fig. 1. Ta anodes from two powders conventionally sintered 30 min. in vacuum at (top to bottom): 1600, 1700, 1800, and 1900°C. Bar = 10 μm.

Figure 2 shows a similar progression for three microwave sintered specimens (samples 22, 23, and 25). It is clear from the high degree of sintering that the third sample experienced a much higher temperature than its nominal 1800°C sinter.

Many microwave sintered anodes show unusual surface ripples on their individual Ta grains, Fig. 3. The origin of this effect is presently unknown, but it suggests the possible existence of microscopic arcing or plasma formation within the pore spaces. Such phenomena could be helpful by encouraging rapid sintering without coarsening and/or by cleaning away surface impurities. Diagnostic tools are presently under development with which we hope to detect microscopic discharge effects as they occur [4].

Fig. 2. Ta anodes from Na reduced powder microwave sintered 30 min in vacuum at (top to bottom): 1600, 1700, and 1800°C (nominal). Bar = 10 μm.

All microwave processed samples had some degree of surface reaction such as a thin (presumably oxide) layer, Fig. 4, or a more catastrophic reaction with the insulation medium, Fig. 5. Because cleanliness of the anode is crucial to the fabrication of a good tantalum capacitor, it is clear that casketing techniques must be improved to minimize material interactions and temperature nonuniformities as well as to facilitate the attainment of the best possible vacuum.

Sample 24 was sintered at 1450°C in nitrogen at 6.7-6.8 GHz in a preliminary test of the variable-frequency furnace [2]. Despite the low temperature, some sintering was observed in the specimens.

Fig. 3. Surface ripples on Ta particles after microwave sintering.

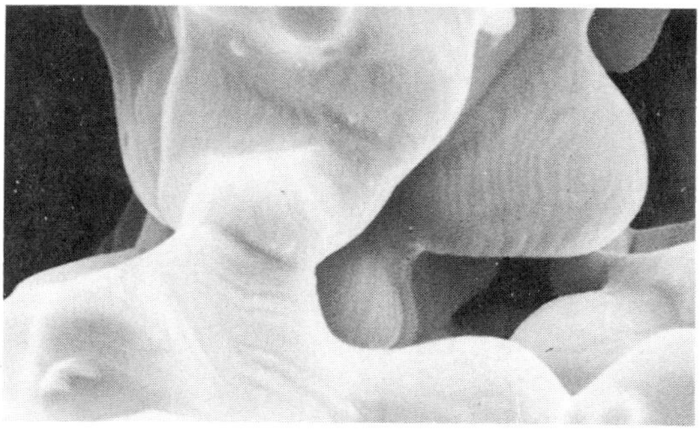

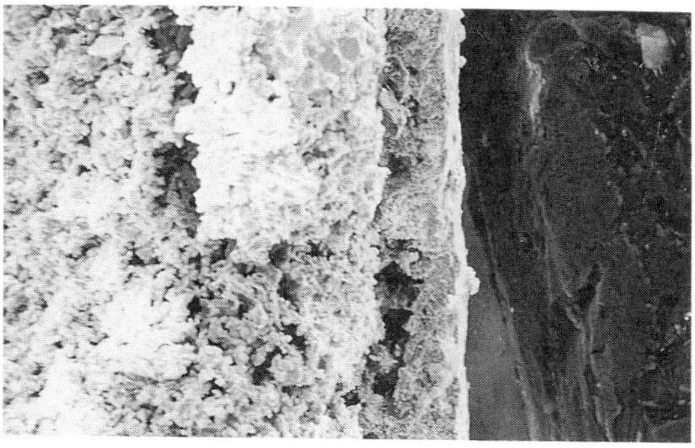

Fig. 4. Layer on outside of microwave sintered pellet.

Fig. 5. Pellet fired at 1800°C (nominal) showing reaction with ZrO_2 bubbles on surface.

CONCLUSIONS

1. Tantalum capacitor anodes can be sintered using microwave heating provided the casket contains lossy materials to initiate heating from ambient temperatures.

2. Microwave sintered anodes often show unusual surface ripples on the individual Ta particles; their exact origin is presently unknown.

3. Casketing techniques must be improved in order to maintain high vacuum, cleanliness, and uniform temperatures within the working volume.

ACKNOWLEDGMENT

This project was supported by the U.S. Department of Energy, Office of Energy Research, Laboratory Technology Transfer Program, and conducted under Cooperative Research and Development Agreement No. ORNL 91-0056 between Martin Marietta Energy Systems, Inc., and AVX Tantalum Corporation.

REFERENCES

1. C. E. Holcombe and N. L. Dykes, "Microwave Sintering of Titanium Diboride," J. Mat. Sci. 26, 3730-38 (1991).

2. D. W. Bible, R. J. Lauf, and C. A. Everleigh, "Multikilowatt Variable Frequency Microwave Furnace," Mat. Res. Soc. Symp. Proc., Vol. 269 (1992).

3. C. E. Holcombe and N. L. Dykes, "Importance of 'Casketing' for Microwave Sintering of Materials," J. Mat. Sci. Lett. 9, 425-8 (1990).

4. D. W. Bible, Oak Ridge National Laboratory, personal communication (1992).

MICROWAVE SINTERING OF MULTILAYER CERAMIC CAPACITORS*

R. J. LAUF*, C. E. HOLCOMBE*, AND C. HAMBY*
*Martin Marietta Energy Systems, Oak Ridge, TN 37831-6087

ABSTRACT

Multilayer ceramic capacitors of several compositions (both Z5U and NPO types) were sintered in air in both conventional and microwave furnaces. Several casketing and insulation techniques were used to improve temperature uniformity and minimize dopant losses. Zirconia and yttria are suitable packing materials as either powders or fibers. After sintering, the packing material remains friable and can be recycled. The presence of metallization layers did not interfere with the microwave process and no arcing was observed at surfaces where the internal electrodes emerged. In some runs, however, local thermal runaway was observed, which could severely decrease process yields and must be eliminated in a commercial process.

INTRODUCTION

Multilayer ceramic devices have become a multibillion-dollar family of products in the modern electronics industry [1]. Multilayer capacitors are used extensively in computers, telephones, and other applications where size is critical. Multilayer packages increase the interconnection density and speed of microprocessors and enhance their reliability. Transducers and actuators relying on a multilayer approach are being developed for a variety of applications.

All of these multilayer devices are composite materials containing ceramic phases and metallizations. Cost-effective compositions and processing strategies to optimize electrical properties must therefore also consider materials compatibility at processing temperatures and, later, in service.

To manufacture ceramic multilayer capacitors (MLCs), ceramic dielectric powders are mixed with an organic vehicle and tape-cast to form thin layers. Electrodes are screen printed onto the dielectric layers, which are then carefully stacked, laminated, and cut to form green multilayer chips. The green chips are then sintered at high temperatures to form dense monolithic bodies with internal metallization layers. End terminations, lead attachment, and packaging complete the process.

Many types of MLCs exist to satisfy different performance and cost requirements. For example, a $BaTiO_3$-based dielectric with a K of about 5200 requires high-Pd electrodes in order to withstand the high sintering temperature. A rare-earth titanate with a K of about 80 is fired at lower temperature, allowing a less costly electrode composition (Table I).

This work was motivated by several considerations. First, many studies have been reported on microwave sintering of ceramics [2] and in some cases microwaves are thought to lower sintering temperatures by as much as several hundred degrees [3]. This "microwave effect", if real, could possibly allow the use of low-cost silver metallizations with $BaTiO_3$ dielectrics for huge cost savings. Second, the behavior of ceramic/metal composite structures in a microwave processing environment is of fundamental interest, and the internal metallizations can directly indicate true local temperatures. Thus, a successfully densified $BaTiO_3$ capacitor with intact Ag electrodes would be unambiguous evidence of a "microwave effect."

---

* Research sponsored by the U.S. Department of Energy, Assistant Secretary for Conservation and Renewable Energy, Office of Industrial Technologies, Advanced Industrial Concepts (AIC) Materials Program, under contract DE-AC05-84OR21400 with Martin Marietta Energy Systems, Inc.

Table I. MLC Characteristics

EIA type*	Z5U	NPO
Dielectric constant (K)	~8000	~80
Dielectric composition	$BaTiO_3$	$NdTiO_3$
Electrode composition	90Pd-10Ag	30Pd-70Ag
Firing temperature	1375°C	1121°C

*Temperature characteristics are designated according to EIA Specification RS198-C. The designations Z5U and NPO refer to the temperature dependence of the dielectric constant and do not specify dielectric composition as such.

EXPERIMENTAL

Green MLCs of several sizes and compositions representing both Z5U and NPO types were obtained from a manufacturer. The green parts contained about 10 wt % organics, which were removed by first heating to ~600°C in N_2 followed by brief oxidation in air to remove residual carbon. All sintering runs were done in air with the parts buried in refractory powder to minimize loss of dopants. Conventional sintering was done in a horizontal tube furnace using Y_2O_3 powder and ZrO_2 crucible. Microwave sintering was done in a 2.45 GHz mixed-mode cavity [4] using a variety of casketing arrangements [5]. Sintering conditions are summarized in Table II.

RESULTS AND DISCUSSION

Sintering results are summarized in Table III. In the best cases, $BaTiO_3$ parts were sintered to adequate density and the microstructures of both dielectric and metallization layers looked good. In the worst cases, parts were heated catastrophically, sometimes to the point of melting. There was often a large disagreement between the thermocouple and pyrometer as well as other evidence of large temperature inhomogeneities within the crucible.

Fig. 1. Multilayer capacitors partially removed from yttria fiber insulation after microwave sintering in air at 1285°C. Parts are approximately 5 × 6 × 1 mm.

Table II. MLC Sintering Conditions

Sample	Type	Mode *	Temp. (°C)	Time (h)	Casketing
1	Z5U	M	1375	0.5	Y_2O_3 grit
2	Z5U	M	1285	0.5	Y_2O_3 fiber
3	Z5U	M	1285-1300	.75	Y_2O_3 fiber, Y_2O_3 powder
4	Z5U	M	1225-1267	1.0	ZrO_2 fiber
5	NPO	M	1225-1400	0.5	Y_2O_3 fiber
6	Z5U	C	1130	2.0	Y_2O_3 powder
7	NPO	C	1130	2.0	Y_2O_3 powder
8	NPO	M	1130	2.0	Y_2O_3 powder, ZrO_2 bubbles
9	Z5U	M	1130	2.0	Y_2O_3 powder, ZrO_2 bubbles
10	Z5U	M	---------		Y_2O_3 powder
11	NPO	M	1130-1523	2.0	Y_2O_3 powder, no crucible
12	NPO	M	1130	2.0	Y_2O_3 powder
13	NPO	M	1130	1.7	Y_2O_3 powder, SiC bars
14	NPO	M	--------		Y_2O_3 powder, SiC bars
15	NPO	M	1294	2.0	Y_2O_3 powder, SiC grit
16	NPO	M	1000	2.0	Y_2O_3 powder, SiC grit
17	NPO	M	1130	2.0	Y_2O_3+ZrO_2 powder, SiC bars
18	Z5U	M	1130	2.0	Y_2O_3+ZrO_2 powder, SiC bars
19	Z5U	M	1130	2.0	Y_2O_3+ ZrO_2 powder, SiC bars

All sintering was done in air. Typical heating rates were 5°C/min.
*C = conventional, M = microwave

Table III. MLC Sintering Results

Sample	Results
1	Parts sintered but not uniformly
2	Most parts (~80%) well-sintered; some underfired
3	Very inhomogeneous sintering
4	Five good parts; three overfired; one underfired
5	Parts warped badly; temperatures too high
6	Good parts; control samples
7	Good parts; control samples
8	Good parts; similar to #7
9	Some parts heavily overfired, others underfired
10	Sample melted
11	Run aborted; sample did not heat
12	Sample melted
13	Catastrophically overheated
14	Run aborted; sample did not heat; arcing at high power
15	Catastrophically overheated
16	Fairly good parts, similar to #7
17	All parts heavily overfired
18	Heavily overfired, some worse than others
19	Some parts heavily overfired, others practically melted

An example of a successful run is shown in Fig. 1 with some of the sintered Z5U capacitors removed from the packing. At these firing temperatures the packing material remains friable and is reusable. (Recycling is desirable because "used" powder more effectively prevents dopant loss as it does in the conventional process.)

The microstructures of NPO capacitors fired at 1130°C conventionally, Fig. 2a, and by microwaves, Fig. 2b, are quite similar indicating that the internal metallizations do not interfere with the heating process. There is no evidence of arcing or other surface effects at the edges where the electrodes emerge.

In some runs, large part-to-part variations were observed, Fig. 3. In extreme cases, some pieces were virtually unfired and others less than 1 cm away nearly melted. Thermal runaway in this system is no doubt exaggerated by the large temperature coefficient of dielectric loss and by the highly nonuniform power density observed in the 2.45 GHz furnaces. It is possible that more sophisticated furnace designs and mode stirring concepts [6] will help to control thermal runaway, which must be eliminated for a commercial process to become viable.

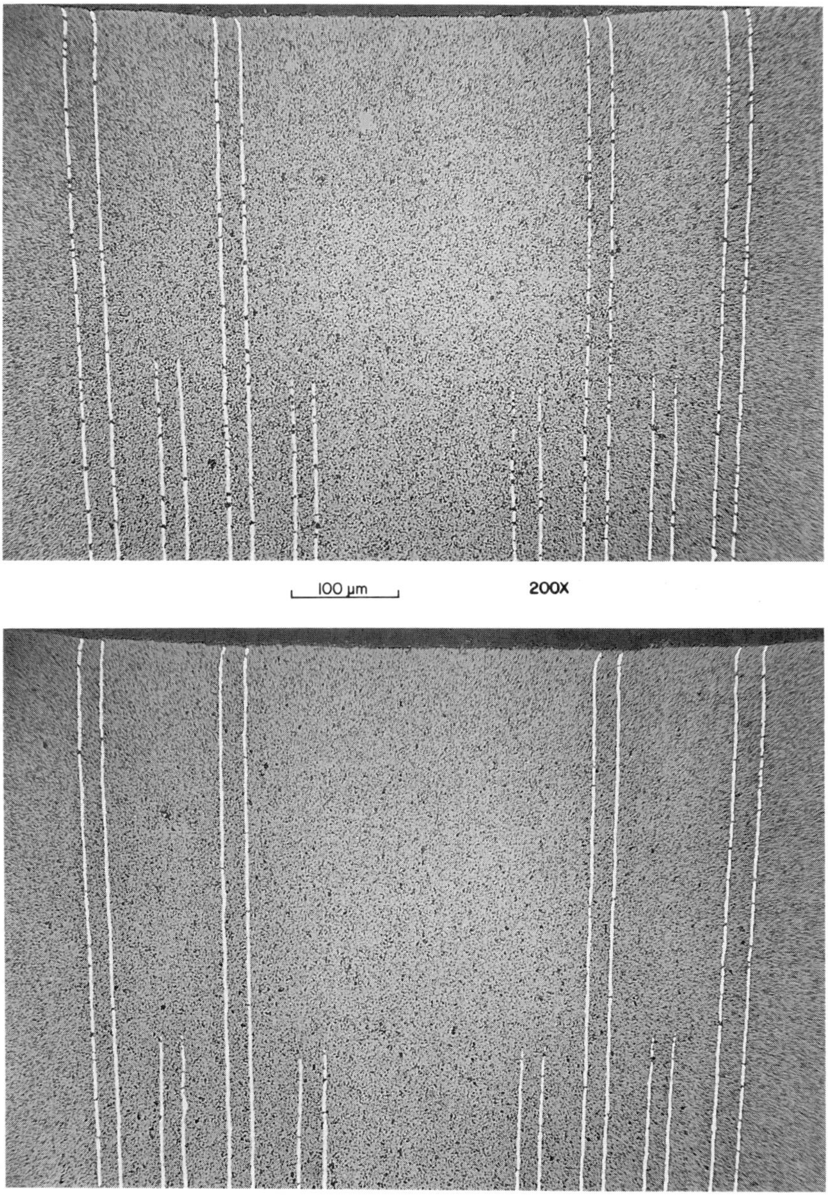

Fig. 2. Optical micrographs of NPO capacitors sintered 2 h at 1130°C conventionally (top) and by microwave heating (bottom). Plane of polish is normal to the internal metallization layers (lighter phase).

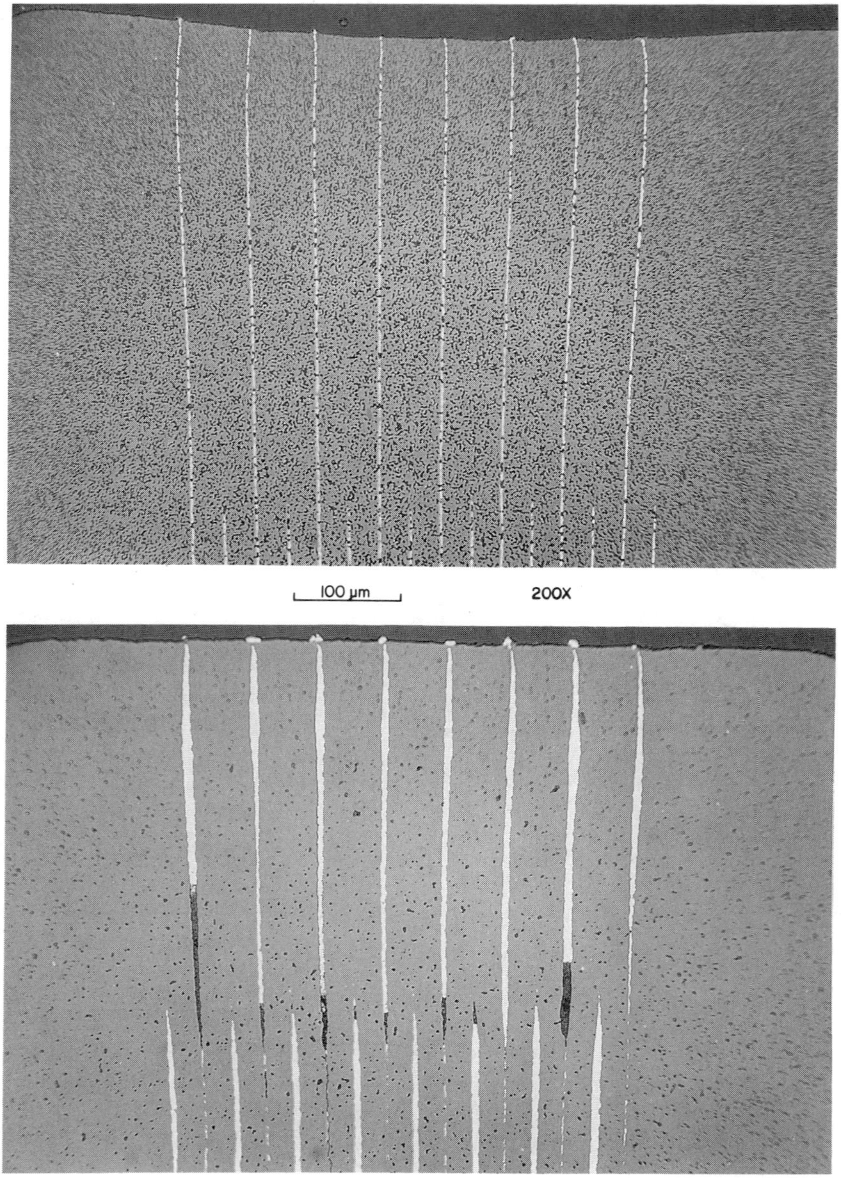

Fig. 3. Optical micrographs of two Z5U capacitors from a single batch microwave sintered 2 h at 1130°C (Run 9). Comparison of a "good" part (top) and a heavily overfired part (bottom) indicates large differences in thermal history within the casket.

CONCLUSIONS

1. Multilayer ceramic capacitors can be sintered in a 2.45 GHz microwave furnace. Zirconia and yttria fibers or powders are suitable packing materials and can be reused.
2. Internal metallization layers do not seem to interfere with the sintering process and there is no evidence of arcing at the edges where they emerge.
3. General thermal runaway as well as local temperature inhomogeneities were sometimes observed, possibly aggravated by poor field uniformity in the furnace cavity.

ACKNOWLEDGMENT

Metallographic specimens and photomicrographs were prepared by M. J. Gardner and C. P. Haltom.

REFERENCES

1. J. R. Black, "Technology and Market Trends in Multilayer Ceramic Devices," pp. 3-11 in Adv. in Ceramics, Vol. 19, ed. J. B. Blum and W. R. Cannon, Am. Ceram. Soc., Columbus, OH (1986).

2. W. H. Sutton, "Microwave Processing of Ceramic Materials," Am. Ceram. Soc. Bull. 68[2], 376-86 (1989).

3. V. K. Varadan, Y. Ma, A. Lakhtakia, and V. V. Varadan, "Microwave Sintering of Ceramics," Mat. Res. Soc. Symp. Proc., Vol. 124, 45-57 (1988).

4. C. E. Holcombe and N. L. Dykes, "Microwave Sintering of Titanium Diboride," J. Mat. Sci. 26, 3730-38 (1991).

5. C. E. Holcombe and N. L. Dykes, "Importance of 'Casketing' for Microwave Sintering of Materials," J. Mat. Sci. Lett. 9, 425-8 (1990).

6. D. W. Bible, R. J. Lauf, and C. A. Everleigh, "Multikilowatt Variable Frequency Microwave Furnace," Mat. Res. Soc. Symp. Proc., Vol. 269 (1992).

FABRICATION OF HYDROXYAPATITE CERAMICS BY MICROWAVE PROCESSING

DINESH. K. AGRAWAL, Y. FANG, D. M. ROY, AND R. ROY
Materials Research Laboratory, Pennsylvania State University, University Park, PA 16802

ABSTRACT

Various hydroxyapatite (HAp) ceramics were fabricated by microwave processing. By carefully selecting the starting material and controlling the processing conditions, regular HAp ceramics with density up to 97% of the theoretical density, porous HAp ceramics with porosity up to 76%, and fully densified transparent HAp ceramics were obtained by using a 2.45 GHz, 500W microwave oven. These ceramics were characterized for density, microstructure, strength, and phase composition, and the results were compared with the conventionally processed ceramics.

INTRODUCTION

That microwave transparent ceramics such as Al_2O_3 and SiO_2 could be melted and sintered by a microwave field was a surprising finding. Using gels of such materials, Roy, Komarneni and Yang[1,2] first showed that extraordinary rates of temperature increase (hundreds of degrees per minute) could be attained by microwave heating of such materials, since then, microwave sintering has become a new technique to process ceramics. In microwave sintering, heat is generated from inside of materials instead of being transferred from outside as in conventional sintering. Sutton[3] has shown how microwave sintering is basically a volumetric sintering process. This fundamental feature of microwave sintering offers one a unique technique to process a variety of materials, and besides saving energy and time, also produce improved physical and mechanical properties.

Hydroxyapatite [$Ca_{10}(PO_4)_6(OH)_2$, HAp] ceramics have been investigated for years for biomedical applications, but the microwave sintering of these materials was reported only recently[4-6]. This paper summarizes the results of our recent study in microwave sintering of HAp ceramics, in which we have been able to produce various kinds of HAp ceramics by varying the processing conditions and selecting suitable precursor powders.

EXPERIMENTAL

Regular white HAp ceramics

HAp powder was prepared by the hydrolysis of brushite, $CaHPO_4{\cdot}2H_2O$, followed by ripening with $CaCl_2{\cdot}2H_2O$[7]. The mixture was dried at 90°C, ground to -325 mesh and then heated at 500°C for 24 hours before compaction. Figure 1(a) shows the morphology of this powder. Pellets of 12.7 mm in diameter were made at various pressures from 35 to 351 MPa to get various green densities ranging from 32.63 to 51.4 % of theoretical.

Porous HAp ceramics

Two kinds of HAp-precursor powders were used: (a) Whiskers of pure HAp, with typical size 0.2 x 5 μm (Fig. 1(b)), synthesized by the hydrolysis of tricalcium phosphate [α-$Ca_3(PO_4)_2$, or α-TCP] at 60°C, and dried at 165°C; (b) Pure HAp, agglomerations of very fine crystals as used above for white ceramics. This precursor powder was dried at 500°C, and ground to -325 mesh. $(NH_4)_2CO_3{\cdot}H_2O$ was added in some samples to introduce artificial porosity. The pressed samples were heated at 500°C to remove ammonium carbonate before microwave processing. The experimental conditions are presented in Table 1.

Table 1. Experimental conditions to fabricate porous HAp by MW.

Batch →	I[1]	II[2]	III[2]	IV[3]	V[3]
Wt. of HAP, g	1.5	2.8	3.0	2.8	2.8
Sintering Temp., °C	1150	1150	1200	1150	1200
Sintering Time, Min.	1	1	1	3	5

1. Powder from hydrolysis of α-TCP, 2. Powder from hydrolysis of brushite, 3. Powder 2 + Amm. carbonate

Transparent HAp ceramics

HAp powder was synthesized by hydrothermally treating a precipitate of $Ca(NO_3)_2.4H_2O$ and $NH_4H_2PO_4$ at 200°C, 1.5 MPa, for 24 h, then washed with deionized water, dried at 150°C, and then dehydrated at 500°C after grinding. The specific surface area (BET) of HAp powder was 40.7 m^2/g with average particle size of 0.05 μm. Thermal stability study showed that this HAp remains stable up to 1370°C in air. The morphology of the starting powder is shown in Fig. 1(c). Thin, circular pellets of 1.27 cm in diameter were compacted at 70 - 350 MPa.

Figure 1. Morphologies of starting hydroxyapatite powders synthesized by (a) hydrolysis of brushite, (b) hydrolysis of α-TCP, and (c) hydrothermal method.

Sintering

A 500W, 2.45 GHz commercial microwave oven was used for the sintering experiments. Pellets were placed on a platinum foil located in the center of a porous zirconia insulating cylinder of 25.4 mm ID, 51 mm OD and 30 mm in height. A Pt - Pt/10%Rh thermocouple, was used to record the temperature inside the cylinder. Fiberfrax wool was used for insulation. Sintering of various samples was carried out between 1100 and 1300°C, for 1 to 20 minutes. After sintering, the system was allowed to cool naturally to room temperature. For comparison, conventional sintering of the pellets without microwaves was also carried out. The sintered samples were characterized for their density, grain size, microstructure and strength. Density was determined by geometric means. Grain sizes and microstructures were determined by scanning electron microscopy (SEM). Phase compositions were identified by powder X-ray diffraction (XRD) using a Scintag diffractometor. Diametral tensile strength[8] of regular and porous ceramics was measured by diametral-compression (Brizilian) test in a MTS 810 with a crosshead speed of 0.051 mm per minute.

RESULTS AND DISCUSSION

White ceramics

Figure 2 shows the relative densities of the sintered pellets vs the corresponding green densities. It is clear from these plots that microwave sintered samples at 1200 and 1300°C for 10 minutes have shown higher densities than conventionally sintered samples at the same temperature for 120 minutes. The highest sintered density obtained was about 95% at 1300°C with 10 min. soaking time. The results also indicate that the green density of 45% or over is more favorable to the microwave sintering than to the conventional sintering. Although, in both cases the sintered density increases with increasing green density.

The SEM photographs of as-sintered surfaces of some pellets are shown in Fig. 3(a). Higher densification efficiency of microwave sintering is further confirmed from the fact that in the samples microwave-processed at 1200°C, as-sintered surface is very smooth with no visible porosity while the same features are not obvious in the conventionally sintered samples. Table 2 shows the average grain size of the microwave-sintered specimens compared with those of conventionally sintered samples. The grain size at 1200°C for conventionally sintered samples is close to that of microwave sintered. For those sintered at 1300°C, the grain size of both microwave and conventional sintering increased substantially but the grains in conventional sintering are much bigger. The abnormal grain growth was obvious in the conventional process but not so obvious in the microwave process.

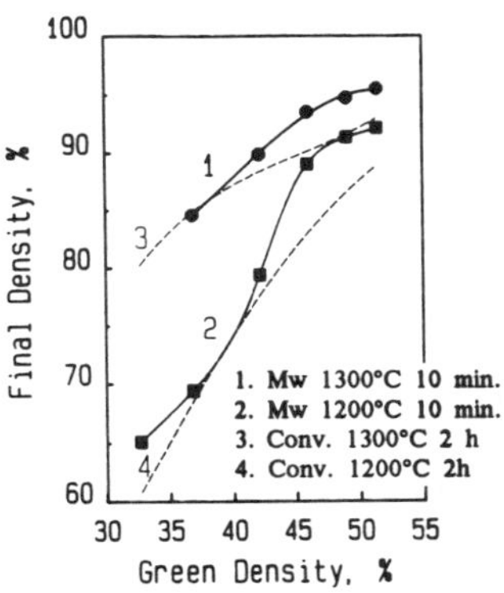

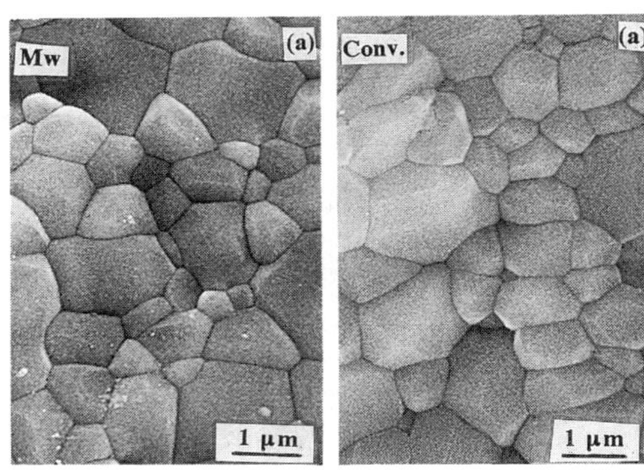

Figure 2. Relative density of mw sintered HAp ceramics.

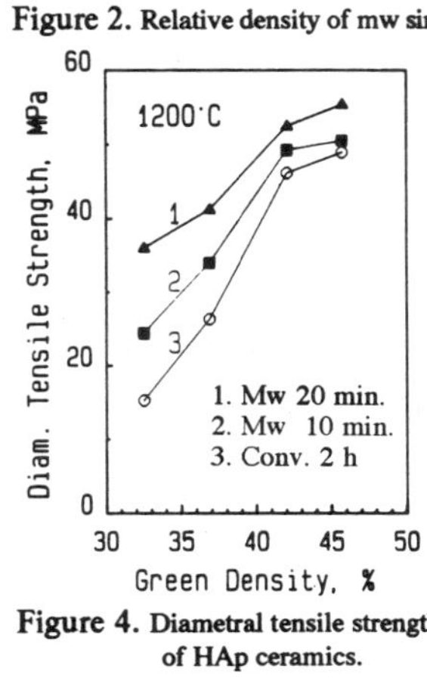

Figure 4. Diametral tensile strength of HAp ceramics.

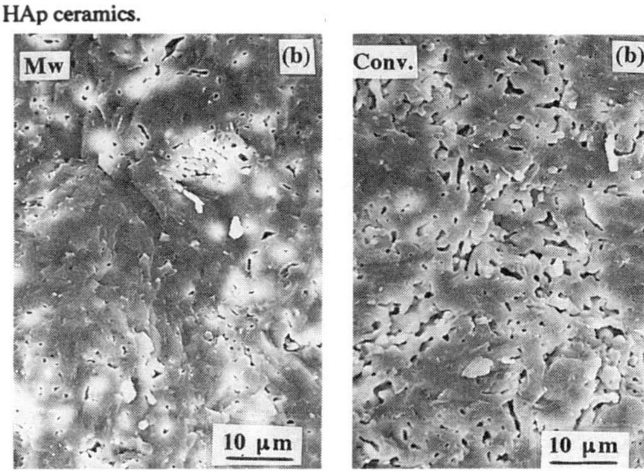

Figure 3. Microstructures of HAp ceramics, (a) as-sintered surfaces, (b) fracture surface.

Table 2. Grain size (μm) of sintered HAp samples under different experimental conditions.

Temperature →	1200°C			1300°C			
	MW		Conv.	MW		Conv.	
Time, min.	10	20	120	5	10	20	120
Grain Size, μm	1.05	1.48	1.12	2.06	3.16	3.39	6.52

Figure 3(b) illustrates the microstructure of the fracture surfaces, indicating that, at the same temperature, the microwave sintering resulted not only in lower porosity, but also finer pore size. The difference in the morphologies could indicate that in the microwave sintering, the process of densification, boundary diffusion and grain growth are all accelerated. However, since the soaking time is very short in microwave sintering, the rate of grain growth is limited. Therefore at the same temperature, in microwave processing smaller soaking times often result in higher densification and a finer grain size. This also prevents secondary recrystallization. Figure 4 shows the tensile strength vs green density of the sintered samples. Again the microwave processed specimens demonstrate much higher strength than the conventionally sintered samples due to a higher density.

Porous HAp ceramics

The sintered samples were porous with varying porosity ranging from 10 to 75%, depending upon the starting material, green density, sintering temperature and duration. All the processed specimens are perfect in shape without any visual microcracks. Table 3 gives the green and corresponding sintered densities of the porous HAp ceramics fabricated in this study. In all cases the final density increases with an increase in the green density. Also the starting materials influence the final porosity due to the difference in sinterability and microstructure. The samples which had ammonium carbonate had obviously developed high porosity (batch IV and V). To evaluate the open porosity, a porosimetry experiment was carried out on three representative samples (apparent density around 50%).The pellet made of whiskers showed total open porosity, and those of powders (b) and (b) + amm. carbonate, had 82% and 84% open porosity respectively. The difference is again due to the sinterability of the starting materials.

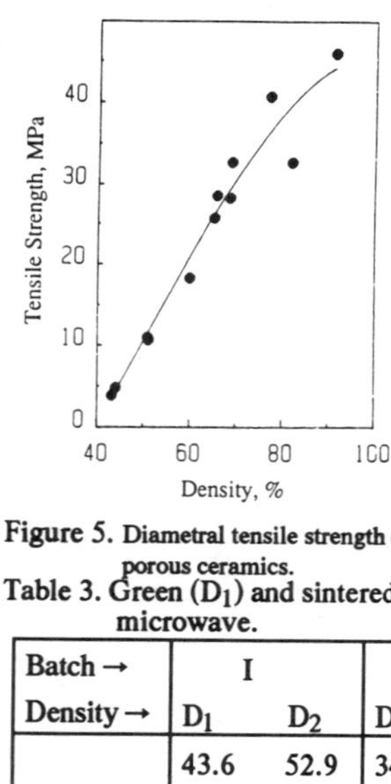

Figure 5. Diametral tensile strength of HAp porous ceramics.

Table 3. Green (D_1) and sintered (D_2) densities (%) of porous HAp ceramics processed by microwave.

Batch →	I		II		III		IV		V	
Density →	D_1	D_2	D_1	D_2	D_1	D_2	D_1	D_2	D_1	D_2
	43.6	52.9	34.2	51.4	34.1	58.0	16.3	26.7	15.5	30.3
	49.3	55.8	39.0	65.8	39.0	77.1	25.5	43.3	23.2	44.5
	51.8	68.8	44.5	68.7	44.5	81.9	27.3	51.2	27.3	52.5
	54.3	72.0	48.3	82.4	48.7	91.2	32.2	56.6	30.4	58.4
							37.6	65.3	36.3	69.6
							40.4	69.0	38.8	75.5

XRD analysis conducted on the specimens after microwave processing shows that the samples of powder b (the ripened HAp) contain pure HAp, but those made of the whiskers (powder a) show small amount of α-TCP besides HAp, which is attributed to the partial decomposition of the starting HAp during microwave processing.

The tensile strength measured by the diametral-compression method is shown in Figure 5. It can be seen that most porous HAp ceramics fabricated in the current research have shown "reasonable" strength values, while, of course, there is a trade-off between the porosity and the strength of the HAp ceramics.

Transparent ceramics

Only the samples made of hydrothermally synthesized HAp powder show transparency, and those made of HAp powder synthesized by the hydrolysis of brushite were opaque (Table 4). Total transparency was achieved in the pellets compacted at 350 MPa, sintered for 5 min. at 1100 to 1150°C. The pellets compacted at lower pressures, although sintered under the similar conditions, showed lower degree of transparency. The XRD analysis of the transparent ceramics shows single phase crystalline HAp.

Table 4. Processing parameters and characterization of Transparent HAp ceramics.

Starting HAp Powder	Cold Pressure MPa	Sintering Temp.°C	Sintering Time, min.	Density %	Sample condition
Hydrothermal	70	1100±5	5	99.82	Transparent
same	140	1175±25	10	100	Transparent
same	140	1150±10	5	100	Transparent
same	175	1150±10	5	99.70	Transparent
same	210	1100±5	5	98.99	Transparent
same	350	1125±25	5	99.73	Transparent
Hydrolysis of	140	1150±10	5	95.71	Opaque
Brushite	210	1100±5	5	93.0	Opaque

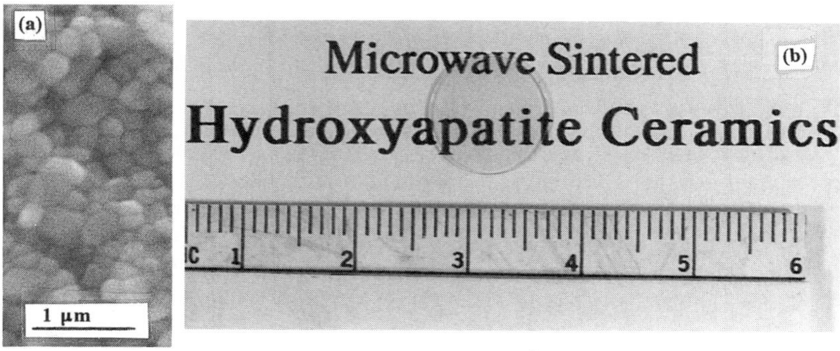

Figure 6. (a) Microstructure of as sintered surface of transparent HAp ceramic, (b) Transparent HAp ceramics.

Figure 6 (a) shows a typical microstructure of the as sintered surface of the transparent HAp ceramics prepared in this study. The grains are nearly spherical with the average size of 0.25 μm. Figure 6 (b) shows a transparent sample prepared in this study. A brief comparison between the HIP method and microwave processing in the processing conditions for fabricating transparent HAp ceramics is made in Table 5; from this the superiority of the microwave process is obvious. We also used a conventional method of heating to prepare transparent HAp ceramics using the same hydrothermal powder but we could not achieve complete transparency. The success of microwave sintering of the transparent HAp ceramics lies in the uniformly ultrafine, pure and stoichiometric HAp powder which offered high packing efficiency, good sinterability, and excellent thermal stability, as well as rapid and uniform densification with very limited grain-growth during sintering. Since the microwave sintering takes place in air without applying any pressure, it is much simpler in operation, and also much more economic. At the same time, relatively higher green density is obviously helpful to achieve full density and transparency in a short period of time.

Table 5. Comparison of method and processing conditions for fabrication of transparent HAp ceramics.

Method	Processing conditions	Reference
Microwave sintering	1100-1200°C, 5-10 min,in air, pressureless	Current Study
Filter cake + HIP	800-1000°C, 100 MPa, 2 h	[9]
Conv. Sint. + HIP	1050°C, 3 h + HIP at 1000°C, 200 MPa, Ar	[10]
CIP + HIP	1275°C, 200 MPa, 1 h	[11]

CONCLUSIONS

By careful control of processing conditions in microwave sintering, it was shown that various kinds of hydroxyapatite ceramics can be prepared. Samples ranging in density from fully dense and transparent to highly porous were prepared. The nature of the starting powder of HAp plays an important role in the sinterability of the material. Also a comparison with conventionally processed HAp ceramics showed that that microwave processed samples had better properties.

Acknowledgement: This research was supported by a grant from National Science foundation under the contract No.88-12824, and Consortium on Chemically Bonded Ceramics.

REFERENCES

1. R. Roy, S. Komarneni, and L.J. Yang, *J. Am. Ceram. Soc.*, 68, 392-95 (1985).
2. S. Komarneni and R. Roy, *Materials Letters*, 4, 107-110 (1986).
3. W. H. Sutton, *Am. Ceram. Soc. Bull.*, 68, 376-86 (1989).
4. Y. Fang, D.K. Agrawal, D.M. Roy, and R. Roy, Preparation and Microwave Sintering of Hydroxyapatite Ceramics, Presentation in Cements Division, abstract No. 31-T-90, 92nd Annual Meeting of Am. Ceram. Soc., Dallas, Texas, April 24, 1990.
5. Y. Fang, D.K. Agrawal. D. M. Roy, and R. Roy, Ceramic Transactions:Microwaves, Vol.21, Eds. D. E. Clark, F. D. Gac and W. H. Sutton, The Am. Ceram. Soc. Publ., Westerville, Ohio, 1991.
6. Y. Fang, D.K. Agrawal. D. M. Roy, and R. Roy, J. Mater. Res., 7, 490-93 (1992).
7. H. Monma and T. Kamiya, J. Mat. Sci., 22, 4247-50 (1987).
8. A. Rudanick, A.R. Hunter, and F.C. Holden, "An Analysis of the Diametral-Compression Test", Materials Research Standards, pp. 283, April, 1963.
9. K. Uematsu, M. Takagi, T. Honda, N. Uchida, and K. Saito, J. Am. Ceram. Soc., 71, 1476 (1989).
10. J. Li and L. Hermansson, Interceram, vol. 39, No.2 (1990).
11. M. Yoshimura, K. Ioku, and S. Somiya, Euro-Ceramics, Vol.3, Eds. G.de With, R. A. Terpstra, and R. Metselaar (Elsevier Applied Science, New York, 1989), p. 316.

MICROWAVE SINTERING OF OXIDE CERAMICS

J.R. BRANDON, J. SAMUELS AND W.R. HODGKINS

EA Technology Ltd., Capenhurst, Chester, CH1 6ES, UK

ABSTRACT

The activities of EA Technology in microwave processing of oxide ceramics are reviewed. Comparative sintering rate measurement techniques have been used to demonstrate microwave enhanced densification of oxide ceramics. Temperature distributions within microwave and conventionally heated ceramics are simulated. Based on these simulations a method for firing large components has been developed which achieves uniform microstructures and low thermal stresses by careful control of temperature distribution.

INTRODUCTION

Microwave sintering of ceramics has been the focus of attention of numerous researchers in recent years. This interest has been centred on the potential benefits of the technique which include improved energy efficiency, more uniform and refined microstructures, reduced thermal mismatch stresses [1] and the possibility of achieving densification at lower firing temperatures [2]. Results of experimental investigations indicate that these claims are indeed valid for a range of materials including oxide and non-oxide ceramics [1]. Despite this fact a number of technical and economic problems remain to be solved before widespread commercialization of microwave furnaces can be achieved. Of the many studies reported in literature only a few have detailed attempts to scale up the process to a realistic level [3-5].

Following a feasibility study which confirmed the benefits of microwave sintering of ceramics, a comprehensive R&D programme has been established at EA Technology. Included in this programme are investigations of microwave interactions with a range of materials and the development of new microwave furnace technology relevant to the ceramics processing industries.

COMPUTER SIMULATION OF MICROWAVE HEATING OF CERAMICS

Sintering rate measurements derived from small test specimens give important information on sintering mechanisms. However, attention must also focus on the practical aspects of firing realistically sized components in volume quantities. At EA Technology the challenge was, therefore, to sinter a number of large specimens simultaneously in microwave furnaces while achieving uniform microstructures and low thermal stresses. To this end, blocks of alumina and alumina based materials, weighing from 250g to several kilograms were fired using a 2450MHz, 5kW microwave sintering furnace with microwave susceptors to preheat the component. This early furnace design, however, proved unsatisfactory, with poor field uniformity within the cavity and thermal break-away effects causing hot spots and thermal mismatch cracks.

From these tests the importance of correct furnace design was soon realized. In an attempt to characterize the important design parameters of a microwave furnace a numerical model was constructed which examines the interaction of microwaves with the furnace contents and the role of microwave susceptors as preheaters. In choosing a suitable model to simulate the microwave heating of ceramics, it has to be borne in mind that few reliable data exist on the dielectric properties of ceramics, particularly at high temperatures. Furthermore, the complex geometry of an actual furnace makes calculation of electromagnetic field distribution extremely difficult. Recognizing these difficulties this model uses simple, easily characterized, furnace and specimen geometries. In addition, all calculations are made assuming an even electromagnetic field distribution which can be varied infinitely to dissipate the required power within the material. In practice the field within a multi-mode cavity will certainly be non-uniform and will have a relatively low Q factor which will limit the maximum field strength attainable. Thermal conductivity, specific heat and dielectric loss behaviour of specimen, susceptor and thermal insulation are defined as a function of temperature only. With a simple modification the model can also be used to predict the temperature profile of a conventionally heated specimen for comparison. The heat flow, which is described by the solution to a non-linear partial differential equation, is calculated using a Crank-Nicholson implicit finite difference scheme with a uniform spatial mesh.

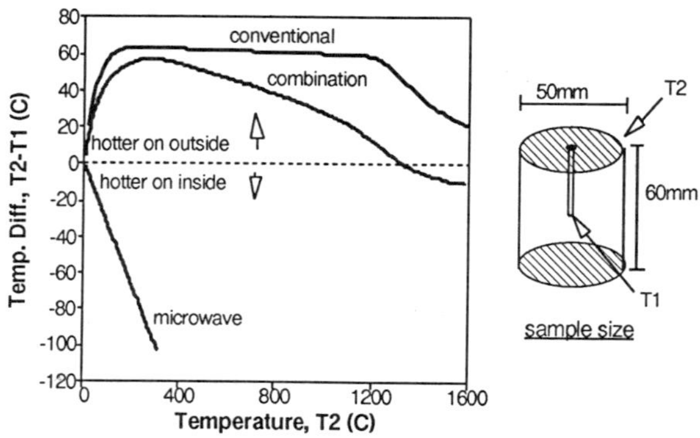

Figure 1. Theoretical data for alumina specimen heated to 1600°C at 5°C/min in conventional, microwave and combination microwave furnaces.

Results from a series of simulations are summarized in figure 1. In this example a block of alumina of radius 25mm and height 60mm is placed within a typical ceramic fibre insulation lining and heated at 5°C/min (measured at the surface of the specimen) conventionally, with microwaves or with a combination of microwaves and susceptor. The dielectric loss properties of the alumina, which are defined as the product of the loss tangent, $\tan\delta$, and the dielectric permittivity, $\varepsilon_0\varepsilon'$, are described as increasing exponentially as a function of temperature starting at an initially low value. Equivalent data for the susceptor are described as being much higher at the start temperature but increasing more slowly with temperature. In figure 2 temperature uniformity is indicated by the difference between the interior, T_1, and surface temperature, T_2, of the specimen, and this is plotted against T_2. A positive difference denotes hotter on the surface of the specimen while a negative difference denotes the opposite. Interpretation of these data is as follows. During conventional heating the surface is necessarily hotter than the inside of the specimen at all stages of the heating cycle. The maximum difference observed is a function of the specimen geometry, the heating rate and the thermal conductivity of the material. The large temperature difference below 1100°C reflects the poor thermal conductivity of unsintered alumina powder (<1W/m.K) and the relatively large radius chosen. Conversely, microwave heating of the same specimen sets up an inverse temperature distribution as the specimen loses heat to the surroundings. For materials such as alumina which display an exponential increase in dielectric loss as a function of temperature, this profile is accentuated and thermal break-away results. Temperature differences in excess of 1000°C are predicted by this model under these conditions and have been observed experimentally during microwave heating of YSZ doped alumina in non-ideal furnace configurations. For the combination heating arrangement temperature difference is initially similar to that of the conventionally heated specimen. As the temperature increases, however, the increasing absorption of microwave energy progressively raises the temperature of the interior of the specimen to the point where it is actually losing heat to the surroundings and an inverse temperature profile is finally generated.

Despite the simplifications used in this model, these types of prediction have been used to gain an insight into the likely behaviour of different microwave furnace designs. From this insight a new approach to microwave furnace design has been developed in which a combination of susceptor types of known characteristics are carefully matched to the furnace and its contents in order to maintain as closely as possible a uniform temperature profile throughout the entire heating cycle.

EXPERIMENTAL

Sintering Rate Studies

The first sintering rate studies at EA Technology were carried out using modified 650W domestic microwave ovens fitted with optical temperature measurement devices and capable of heating to 1700°C with a semi-manual control method. Comparative measurements made in conventional furnaces confirmed the enhancement of sintering rates in microwave fields. The need to measure densification rate as a function of temperature, in-situ, led to the development of a high temperature dilatometer apparatus capable of rapid heating to temperatures in excess of 1700°C [6,7]. In this instrument a recrystallized alumina push-rod assembly is used in conjunction with an LVDT displacement transducer to measure linear shrinkage. Heating of this assembly is with a purpose built, fully controllable, 1.3kW microwave furnace or with a rapid heating, $MoSi_2$ resistive heating element furnace. Temperature measurement is by means of an Accufiber optical fibre thermometer (OFT) in close proximity with the specimen, which consists of a pellet up to 15mm in diameter. Using a spreadsheet programme, a series of calibration routines correct for the linear thermal expansion of the alumina dilatometer and the specimen, and the differential expansion between the inner and outer tubes of the dilatometer. Corrected data are then used to calculate linear shrinkage and linear shrinkage rate with changing temperature.

Volumetric Heating Studies

Specimens were prepared by isostatic pressing of alumina (ALCOA A1000SG) and alumina doped with 20wt% yttria stabilized zirconia (YSZ) (TOSOH TZ-8Y) using 2wt% organic binder. Yttria stabilized zirconia was chosen as being a suitably microwave absorbent additive to improve the ability of the alumina to be self-heated with microwaves. Non-transformable, stabilized zirconia does not provide a toughening mechanism for the alumina.

Cylindrical specimens of 50mm diameter by 60mm length were chosen and a concentric hole of 30mm depth and 5mm diameter was formed to accommodate a temperature probe. Firing was conducted in a 2.5kW, 2450MHz microwave furnace in which measures were taken to ensure a uniform field distribution. The specimens were placed within an insulated lining of approximately 1.5 litre volume, and surrounded by a quantity of tubular microwave susceptible heating elements chosen to match the characteristics of the furnace and its contents. Tubular susceptors have the convenience of being easily added to or removed from the furnace as required, although they have largely been superseded by microwave susceptible paints applied directly to the furnace insulation [8]. Temperature measurement was made using a pair of calibrated Accufiber light-pipe sensors placed within the concentric hole and in close proximity to the surface of the specimen. Conventional firing was carried out on identical specimens using a rapid heating, electric resistance furnace and with temperature measurement by thermocouple. In this way the effects of heating rate, susceptor configuration and specimen composition were studied and compared with data taken from conventionally fired specimens.

RESULTS AND DISCUSSION

An example of sintering rate data obtained from 99.5% pure alumina (Alcoa, A1000SG), heated at 5°C/min, is shown in figure 2. These results indicate a microwave enhancement of the sintering process. The peak sintering rates of both microwave and conventionally fired alumina are similar, although sintering is approximately 100°C more advanced in the microwave furnace at a given temperature. The well defined peaks in the microwave shrinkage rate data compared to the single broad peak of the conventional data suggest a modification to the sintering mechanisms caused by the microwaves. Measurements performed on higher purity alumina (>99.95%) are characterized by having a single peak in sintering rate during both microwave and conventional firing. The mechanism from which this effect results is not yet known.

Shrinkage rate data have also been used to calculate the activation energy of the sintering process [9] following the method described by Wang and Raj [10]. At present no conclusive results are available for alumina based materials, although measurements performed on 12wt% yttria stabilized zirconia have shown a clear reduction in activation energy from approximately 535kJ/mol to 255kJ/mol. These energies are in general agreement with other published work [2].

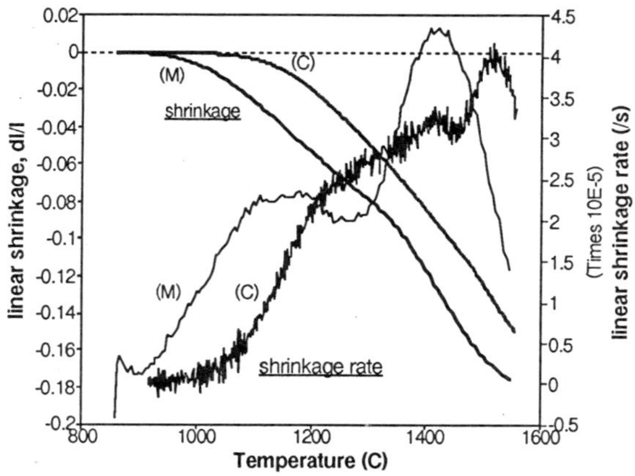

Figure 2. Sintering data for 99.5% pure alumina heated at 10°C/min to 1550°C in microwave (M) and conventional (C) sintering dilatometers.

Results are presented in figures 3a and 3b for specimens of alumina and alumina + 20wt%YSZ heated at 5°C/min to 700°C, at 10°C/min to 1500°C and then held for 30 minutes before cooling. In the first graph the temperature difference between the centre and surface of the specimen is plotted against the surface temperature as in figure 1. Due to the limitations imposed by the short wavelength optical temperature probes, measurements were not possible below 600°C and data are presented for 700°C upwards.

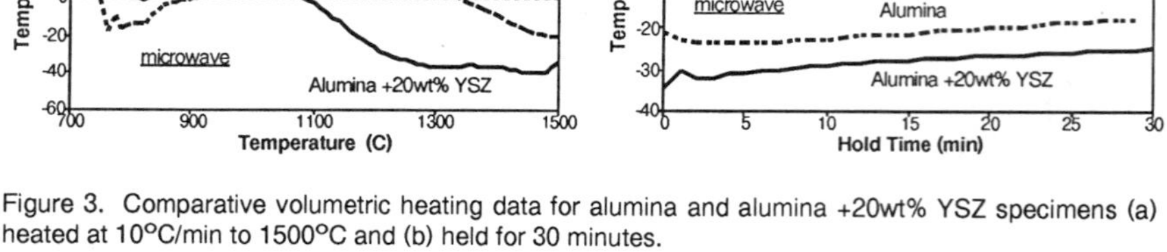

Figure 3. Comparative volumetric heating data for alumina and alumina +20wt% YSZ specimens (a) heated at 10°C/min to 1500°C and (b) held for 30 minutes.

Although not strictly comparable, results from the conventionally fired specimen show a similar pattern to those predicted in figure 1. The temperature difference increases to over 100°C and is maintained near this value until the onset of sintering where it begins to fall to a final value of 40°C hotter at the surface when 1500°C is reached. For the microwave fired specimens, however, the temperature difference for both sets of data falls from an initial value of 40°C hotter on the surface to a value close to zero near the start of the 10°C/min heating rate. Then after becoming hotter on the surface, the specimens once more become significantly hotter internally. In the case of the YSZ doped specimen this change occurs at approximately 1100°C and reaches a maximum value of over 40°C, whereas for alumina the change takes place just above 1300°C and reaches 20°C. The difference in behaviour observed between these two materials is accounted for by the addition to the alumina of stabilized zirconia which is a good absorber of microwaves. The higher microwave absorption of the YSZ doped specimen causes it to heat more rapidly at lower temperatures than the alumina specimen. This behaviour is maintained to 1500°C and is responsible for producing the more pronounced inverted temperature distribution in YSZ doped alumina.

In Figure 3b the temperature difference is plotted as a function of time during the 30 minute hold at 1500°C. The temperature difference observed for the conventionally fired specimen reduces fairly quickly initially and then more slowly until the specimen is at a uniform temperature. The microwave heated samples, on the other hand, remain hotter on the inside throughout the period of the dwell. The addition of YSZ to alumina appears to increase the steady state temperature difference slightly from 20°C to over 25°C. This difference is significant and can be expected for two reasons. First the presence of a lossy material such as zirconia in alumina will mean that the specimen will absorb a greater proportion of the power being supplied to the microwave furnace during the hold than the undoped specimen. This in turn means that more power will be lost from the surface of the specimen to maintain thermal equilibrium and a correspondingly larger temperature gradient will be set up to maintain this heat flow. Second, the thermal conductivity of the zirconia doped alumina will be lower than for alumina, partly through the presence of zirconia, which has a lower intrinsic thermal conductivity than alumina, and partly through the effect the zirconia has on inhibiting the densification of alumina (cf density measurements). The thermal conductivity of a material is highly sensitive to the presence of porosity within its structure. Despite the limited absolute accuracy of the optical temperature sensors, which may incur a 5°C error even when calibrated, the consistent experimental layout and regular calibrations made throughout the series of experiments should ensure good repeatability of temperature measurements.

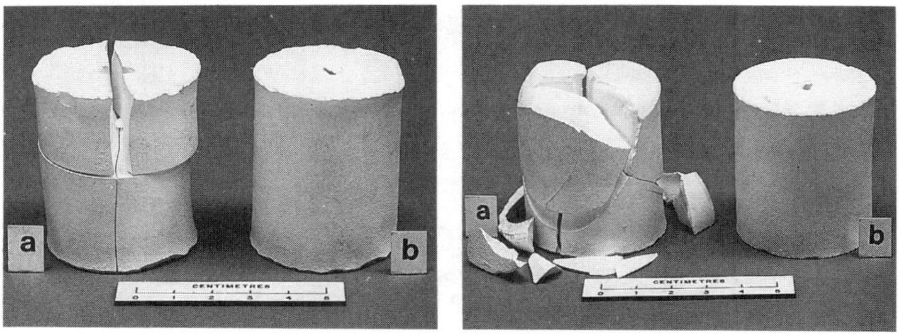

(4a) (4b)

Figure 4. Examples of alumina (4a) and alumina + 20wt% YSZ (4b) test specimens fired conventionally (a) and microwave fired (b) at 1500°C with a heating rate of 10°C/min.

The appearance of the specimens at the end of the firing cycle is illustrated in figures 4a and 4b. All the specimens fired in the conventional furnace were found to be cracked (even with a 5°C/min heating rate) while those heated in the microwave furnace were not.

A visual examination of the microstructures at the centre and surface of polished and thermally etched test specimens was made using SEM. In figure 5 representative micrographs are shown for the alumina specimen. It can be seen that the grain size of the conventionally fired specimens is larger on the surface than in the centre. The microwave fired specimen, however, does not show such a clear variation. Overall the grain sizes of the microwave fired specimen are larger that those of the conventionally fired specimen. A higher density was measured for the microwave fired alumina (95% theoretical density) than for the conventionally fired alumina which had a density of 90% theoretical.

Similarly a value of 93% t.d was measured for microwave fired alumina doped with YSZ, compared to 88.5% t.d. for the conventional firing. The YSZ doped samples also showed a similar pattern of grain size variations to that observed for alumina.

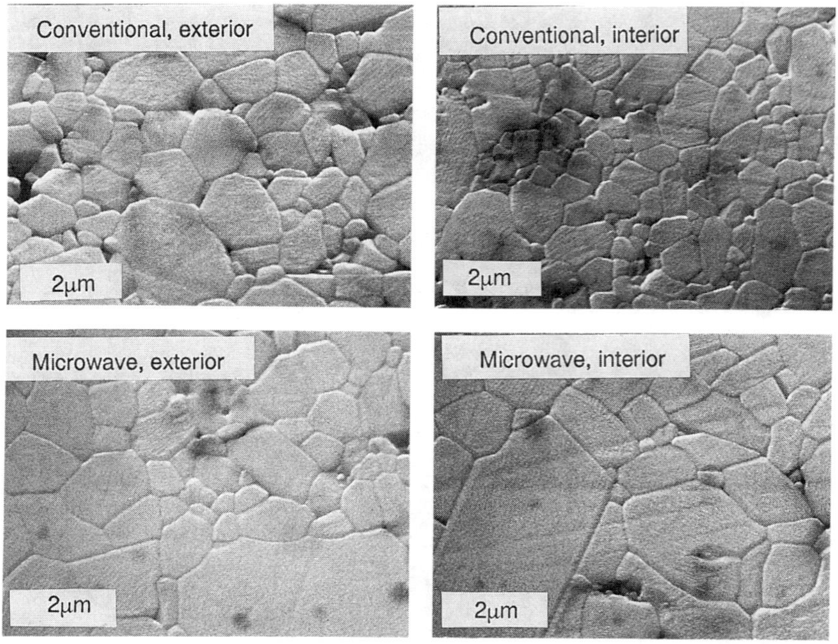

Figure 5. Microstructures of alumina specimens fired in conventional and microwave furnaces at 1500°C for 30 minutes.

The temperature difference data measured for conventionally fired specimens are consistent with the microstructural differences and thermal stress fracture observed. Higher temperatures at the surface of the specimen appear to have caused more grain growth than was observed in the centre of the specimen. In contrast, the reduced temperature differences measured for the microwave fired materials is consistent with the more uniform microstructure observed and the absence of any thermal mismatch cracking. Increasing the heating rate to 15°C/min in later tests was found to be insufficient to cause any thermal stress fracture in these specimens since the temperature difference remained low. The microwave enhancement of the sintering process detected using sintering dilatometry is also manifest in the larger test specimens in the form of higher densities and correspondingly larger grain sizes.

Overall the results from these tests are promising. A single furnace configuration has been demonstrated to be effective in the uniform sintering of large specimens of two different materials having different microwave absorption characteristics. Conventional firing of identical sample was unsuccessful even at heating rates one third of those possible in microwave furnaces. Possibilities for this approach to microwave sintering include the manipulation of temperature profiles to achieve a desired microstructure as well as controlling stress distributions in difficult specimen geometries. Future developments of this measurement technique include extending the temperature measurement to below 500°C and the elimination of possible sources of experimental error associated with optical temperature probes, as well as further refinements to the use of susceptors.

CONCLUSIONS

Comparative measurement of sintering rates using sintering dilatometry has shown an apparent enhancement by microwaves with greater shrinkage being observed at lower temperatures for alumina.

The use of microwave heated susceptors as a means of controlling temperature distributions within larger specimens has been demonstrated and a single furnace design has been shown to be capable of successfully processing two materials with different microwave absorption behaviour.

Microwave furnaces have been found to be capable of outperforming conventional furnaces during difficult firing operations while being reliable and simple to operate in a laboratory environment.

ACKNOWLEDGEMENTS

The authors would like to thank the Director of EA Technology for permission to publish this work. The authors would also like to thank Dr R M Perkin for his advice on applicator design and F A Fellows for carrying out much of the experimental work.

REFERENCES

1. W.H. Sutton, Ceramic Bulletin. 68, 376 (1989).

2. M.A. Janney and H.D. Kimrey in Microwave Processing of Materials II, edited by W.B. Snyder, W.H. Sutton, M.F. Iskander and D. Lyn-Johnson (Mater. Res. Soc. Proc. 189, Pittsburgh, PA 1990) pp. 215-227.

3. W.H. Sutton in Microwave Processing of Materials, edited by W.H. Sutton, M.H. Brooks and I.J. Chabinsky (Mater. Res. Soc. Proc. 124, Pittsburgh, PA 1988) pp. 287-295.

4. M.C.L. Patterson, R.M. Kimber and P.S. Apte in Microwave Processing of Materials II, edited by W.B. Snyder, W.H. Sutton, M.F. Iskander and D. Lyn-Johnson (Mater. Res. Soc. Proc. 189, Pittsburgh, PA 1990) pp. 257-266.

5. J.D. Katz and R.D. Blake, Ceramic Bulletin, 70, 1304 (1991).

6. J.R. Brandon, Report No. ERDC/R2569, 1991.

7. J. Samuels and J.R. Brandon, J. Mat. Sci., to be published (1992)

8. J.R. Brandon, patent application, (Feb. 1992).

9. J. Samuels and J.R. Brandon, presented at the 2nd European Ceramics Society Symposium, Augsburg, 11-14 September 1991 (unpublished).

10. J. Wang and R. Raj, J. Am. Ceram. Soc. 73, 1172 (1990).

PROCESSING AEROSOLS AND FILAMENTS IN A TM_{010} MICROWAVE CAVITY AT 2.45 GHz

Gerald J. Vogt and Wesley P. Unruh
Los Alamos National Laboratory, Los Alamos, NM 87545

ABSTRACT

As part of the development of generic microwave processes for spray-drying of homogeneous complex metal oxide powders and for inorganic fiber processing, we have investigated the use of 2.45 GHz microwaves in a high-Q single-mode TM_{010} cavity coupled directly to aerosols and fibers. Aqueous and ethanol aerosols of ferric nitrate solutions have been successfully dried at 1.8 kW of cavity power for a loaded Q greater than 6000 in flowing nitrogen gas. Similarly, we have observed extremely rapid heating rates in the TM_{010} cavity for small-diameter confined cylinders of water and lossy inorganic fibers. These observations suggest using 2.45 GHz microwave power for drying, calcining, and sintering extruded ceramic filaments. Droplet modeling indicates that the large dielectric shielding for spherical droplets can significantly limit the coupling of 2.45 GHz microwave with spherical aerosols, but not with fibers. Experimental observations on the microwave interactions with ferric nitrate aerosols and with ceramic filaments in the TM_{010} cavity are described.

INTRODUCTION

Microwave-driven spray drying and fiber processing are developing alternate technologies to conventional thermal processing of aerosols and ceramic filaments. In thermal processing, thermal energy is transported from a hot gas or solid through the exterior surface toward the center of the aerosol or filament. In microwave processing, microwave energy is directly coupled to the entire body of the aerosol particle or filament by volumetric absorption of the microwave electric field. A major advantage of microwave processing is therefore the direct rapid heating by volumetric coupling to the aerosol particles and filaments.

Microwave spray drying and fiber processing have been conducted with a stable and versatile microwave system at 2.45 GHz with a single-mode TM_{010} cavity for microwave processing. Microwave interactions with aqueous and organic ferric nitrate aerosols and with inorganic filaments are described in this paper.

EXPERIMENTAL METHOD

The microwave supply and cavity system, shown in Figure 1, have been described earlier in detail [1]. The 3 kW power generator (Gerling Laboratories, Modesto, CA, Model GL119) is designed for low ripple and stable frequency and power output. A system of irises and cavity couplers of the beyond-cutoff type allows us to externally reduce the operating loaded Q of the cavity from a maximum of ~ 7000 and to critically couple microwave power from the source into the cavity. This flexible system makes it possible to fully couple any power level into the cavity and thus achieve any microwave electric field up to, and including, breakdown in dry nitrogen. Directional couplers allow us to measure the power bled from the cavity (in adjusting its Q) and to measure incident and reflected power, thus providing accountability of all power in the system. The high isolation between the microwave power source and the cavity-coupling system provides great flexibility in setting up the appropriate parameters. We can attain the electric field intensity desired with a high degree of control and stability of the system. After a few minutes warm up, the magnetron source can be run open loop, without feedback control, maintaining its tuning on the cavity resonant frequency for long periods of time without adjustment.

The TM_{010} mode is a cylindrical mode, producing a uniform axial electric field parallel to the aerosol flow direction and to the fiber axis. The resonant frequency of

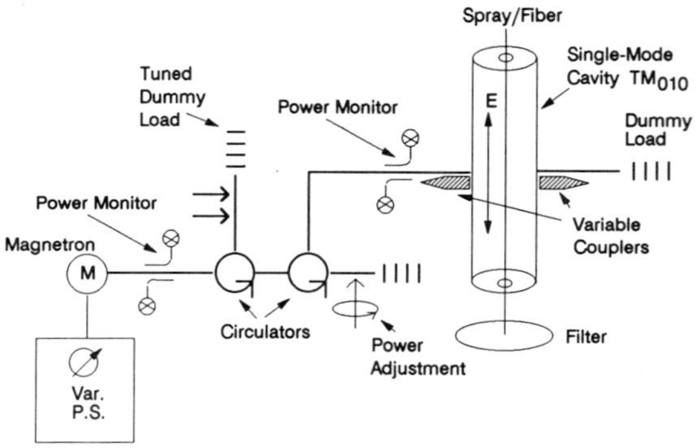

Figure 1. Schematic diagram of the microwave and cavity system.

this mode depends only upon the diameter of the cavity cylinder, which is 9.36 cm for 2.45 GHz. The length of the cavity was set at 42 cm for spray drying and at 11.5 cm for fiber processing. The cavity modules were machined out of 6061-T6 aluminum and the end plates out of brass. The end plates have a 1-cm centered hole to provide an inlet for the aerosol stream or fiber. The exiting orifice for the spray dryer was a meshed hole with a ~ 3-cm diameter.

A resonant TE_{102} rectangular cavity was also used for microwave processing of fibers. This cavity consisted of a stainless steel waveguide section with a coupling iris for the microwaves and an adjustable short to tune the cavity frequency. A variable-stub tuner was located before the iris to minimize reflected power from the cavity. The waveguide section was equipped with a pair of opposing orifices aligned parallel with the electric field vector. These fiber orifices were located near the electric field maximum of the microwave standing wave. The loaded Q of this cavity was typically near 1000.

Ferric nitrate aerosols were generated with an ultrasonically driven nozzle (Sono-Tek Corp., Model 8700-120MS) which produces a polydisperse low-velocity spray. Typical throughput in the drying experiments was 1-2 g/min of ferric nitrate solution. The ultrasonic nozzle was mounted atop an auxiliary spray chamber rather than spraying directly into the microwave cavity. A fraction of the nitrate aerosol was swept into the microwave cavity with a dry nitrogen carrier gas at 1 standard liter per minute (slpm).

The aerosol stream was axially injected into the top of the cavity and was swept through the cavity with an additional 3 slpm of dry nitrogen carrier. The processed particles were collected on a polycarbonate filter at the cavity exit, just below the mesh screen which forms the bottom end of the cavity. A mild underpressure, several Torr below ambient, was applied to the cavity and filter by means of a small vacuum pump at the exit of the filter collector to enhance the flow through the filter and make particle collection on the filter more efficient.

Solution droplets were swept into the cavity without any significant change in cavity coupling. The total power applied to the cavity was 1.8-1.9 kW. After collecting particles for 10-15 minutes, the trapped particles on the filter were examined by scanning electron microscopy (SEM) for the degree of drying. Comparative experiments were also run under identical conditions without microwave power to verify the effect of the microwave field.

Microwave interactions with commercial inorganic fibers were investigated in both the TM_{010} and TE_{102} cavities. Fiber samples were tested by first mounting a length of fiber through the cavity parallel with the electric field vector and then applying microwave power. Longer fiber lengths were slowly pulled through the cavity at ~ 4 cm/min by means of a slow-speed DC motor. The fibers were heated in ambient air at ambient temperature (25°-30° C).

Microwave interaction experiments were conducted with commercial fibers of NICALON and Nextel™. The NICALON silicon carbide fibers were a woven continuous carbon-coated fiber with a 500-tow and 1800 denier with PVA sizing. The Nextel™ fibers were continuous polycrystalline oxide fiber with an alumina, boria, and silica composition.

RESULTS AND DISCUSSION

Spray Drying

Microwave spray drying has been successfully demonstrated at 2.45 GHz for aqueous and alcoholic ferric nitrate aerosols. Drying was attained with 6 w% and 30 w% aqueous solutions of ferric nitrate and with 15 w% solutions of ethanol and methanol. Results for a controlled pair of experiments, where the same drying experiment was performed both with and without microwave power, are shown in Figure 2 for aerosol droplets from 30 w% aqueous ferric nitrate solution. Particle drying with a microwave field appears to be quite complete, as judged from the spherical morphology of the particles. At these low aerosol feed rates, the injection of the droplets into the cavity did not significantly disturb the cavity frequency and Q.

The submicron-sized particles in Figure 2 dried at ~ 25° C with a microwave field and, as well, without microwave power. However, the larger micron-sized particles required microwave energy to complete drying before filter collection. Without a microwave field the larger aerosol particles were not dried, as indicated by splatting on contact with the collection filter surface, as seen in Figure 2. The microwave dried particles retained their spherical shape without any sign of flowing upon impact with the filter surface.

For the alcohol solutions the microwave energy was essential to produce any drying for these aerosols. Dried particles for methanol and ethanol solutions are shown in Figure 3. Without microwave energy the alcohol aerosols deposited a continuous film of ferric nitrate on the collection filter. Note that the methanol droplets were not completely dried, but even this degree of drying was solely due to microwave heating.

Microscopic observations showed that microwave drying was achieved in our TM_{010} cavity when the loaded Q was 6000 or higher. For lower values of loaded Q the morphology of the collected droplets was essentially identical to the particles in Figure 2b without microwave heating. At a loaded Q of 5200 drying was evident by the particle morphology, although a few of the larger droplets had not completely dried before impact. At a loaded Q of 3000 the collected particles did not show any significant drying for the larger droplets, similar to those particles in Figure 2b. This observation indicates that microwave heating of gas-borne droplets is significantly less effective than expected from experience with heating bulk water in beakers and fine water streams in capillaries [1].

Microwave Droplet Interactions

The experimental difficulty in drying aerosol particles with microwave heating at 2.45 GHz is due to the basic electromagnetic boundary conditions governing the interaction of applied microwave energy with spherical droplets. The microwave energy dissipation per unit volume, P_{abs}, within a droplet is given by

$$P_{abs} = 1/2\, \varepsilon_o\, k''\, \omega\, E_i^2,$$

where E_i is the rms microwave electric field strength at frequency ω in the interior of the droplet and k'' is the relative imaginary dielectric constant (or, loss factor) for the

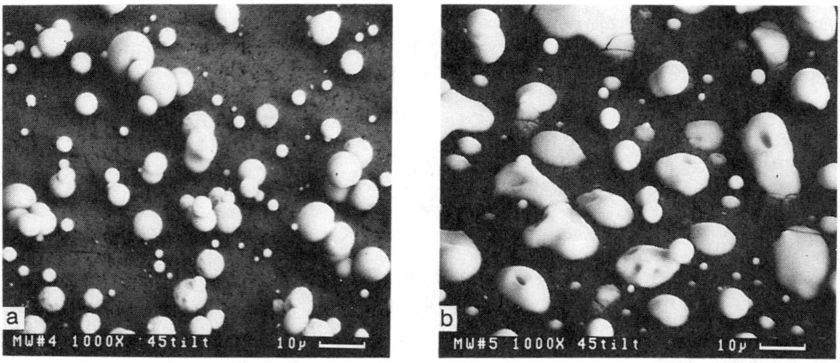

Figure 2. SEM images of ferric nitrate particles from a 30 w% aqueous solution dried with microwave cavity power at (a) 1.8 kW and (b) 0.0 kW.

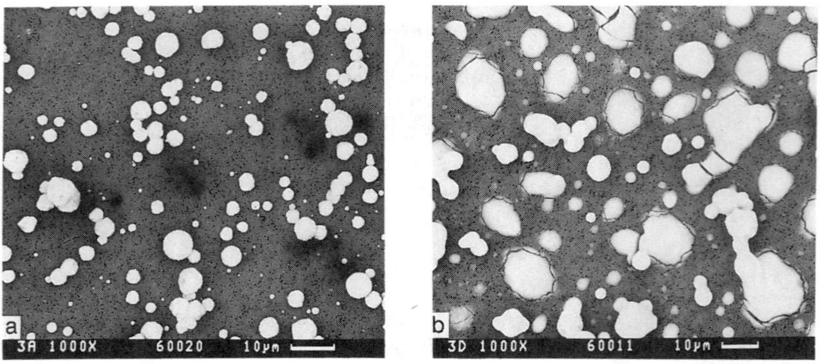

Figure 3. SEM images of ferric nitrate particles from a 15 w% solution in (a) ethanol and (b) methanol dried with microwave cavity power at 1.8 kW.

solution. For water at 2.45 GHz the relative real dielectric constant at 27° C is k′ ~ 80, although at higher microwave frequencies k′ is substantially less than this value. In spherical droplets the standard solution for the E_i with electromagnetic boundary conditions shows that E_i is less than the applied external field by the factor of 3/(k′ +2). Consequently, at 2.45 GHz the internal electric field is reduced by ~ 27 and the microwave energy dissipation is substantially decreased by ~ 750. Thus the applied cavity electric field strength must be near breakdown to affect droplet drying at this frequency. The volumetric power dissipation in the droplet is directly proportional to cavity Q, all other parameters being constant. In our TM_{010} cavity an applied power of 2.2 kW for a loaded Q of ~ 7000 produced field strengths sufficient to cause electrical breakdown in dry ambient air, ~ 1 MV/m at 2.45 GHz. If the loaded Q was reduced below ~ 6000, even at power levels of ~ 2 kW the internal electric field strength in the droplets was insufficient to produce complete drying of the droplet spray.

A phenomenological numerical model of the drying process has been developed to provide some qualitative understanding of the evaporation of small droplets and to provide some insight into the effectiveness of microwave heating [1]. In general, the model suggests that microwave energy deposition rates at 2.45 GHz are too severely limited by dielectric self-shielding due to a high real dielectric constant and the spherical geometry of droplets. For droplet diameters less than 50 μm, with very high surface area to volume ratios, the heat losses to the surrounding atmosphere are quite significant, requiring much higher microwave heating rates than possible at 2.45 GHz to offset these thermal losses and to affect the drying rate by raising the droplet temperature. However, for droplet diameters greater than 50 μm microwave heating is far more effective due to the greater volumetric microwave heating compared to the heat transfer to the surrounding gas. Therefore, our drying model predicts that at 2.45 GHz microwave-driven spray drying has potential for the larger droplet size range(>50 μm) only.

This model prediction is generally consistent with our experimental observation for aqueous solution aerosols. As noted earlier for the dried particles in Figure 2, the finer aerosols generated from aqueous ferric nitrate solutions appeared to dry independently of the applied microwave energy. Whereas, microwave heating was truly needed to dry the larger aerosol particles shown in Figure 2a.

These limitations would be removed by operating at microwave frequencies 10 GHz and higher. Increasing the applied microwave frequency makes a dramatic improvement in the the model evaporation rates, even for very small droplets. With increasing frequency the relative real dielectric constant k′ decreases substantially [1,2], allowing greater penetration of the applied microwave electric field. Fortuitously, the dielectric loss factor k″ increases with frequency[1,2], providing for a greater volumetric energy dissipation in the droplet. In general, our model indicates that it is possible to heat and dry quite small droplets (diameters < 50 μm) at frequencies of 10 GHz and above, at microwave electric fields substantially below breakdown in air.

Fiber Processing

Microwave-driven fiber processing possesses a geometrical advantage over spray drying. Because the internal microwave electric field in a long, thin cylindrical dielectric oriented parallel to an applied field is equal to the applied field, far greater microwave heating rates can be obtained for filaments and fibers than for aerosol droplets. As described above, for water at 27° C the heating rate at 2.45 GHz is expected to be ~ 750 times greater in filaments than in droplets. This dramatic increase in energy dissipation was qualitatively verified by experiments on heating a cylindrical stream of water. Aggressive heating of a flowing water cylinder contained in Teflon tubing easily generated reactive steam at 1 gm/min feed rate. This efficient heating with 2.45 GHz microwaves indicates a great potential for processing fibers through drying, calcining, and sintering ceramic fiber.

Initial microwave heating experiments were conducted with commercially-available NICALON carbon-coated silicon carbide fibers in air. This fiber was readily heated in both the TM_{010} and TE_{102} cavities to radiant temperatures above 600°C with relatively small power levels, only 10 to 40 watts. A heated zone was easily maintained as the fiber was continuously pulled through a cavity. The intensity of the thermal glow over the heated segment was continuously controlled by directly increasing or decreasing the incident microwave power.

Within the 11.5-cm high TM_{010} cavity only a small centimeter-long segment of the NICALON fiber would heat in the aligned microwave field. The heated segment became quite lossy, mixing higher-order cavity modes and preventing other segments along the fiber within the cavity from heating up. The lossy segment was readily translated back and forth between the cavity orifices without initiating heating along a new segment of the fiber. As the lossy segment was drawn out of the cavity, the adjoining fiber segment behind the lossy segment would begin to absorb microwaves and become radiant. The heated zone on a moving fiber was maintained just inside the cavity at the exit orifice, as the fiber was pulled through the cavity at linear velocities up to 37 cm/min at low microwave power levels of less than 100 W.

The heating characteristics of NICALON in the TE_{102} were identical to those observed in the TM_{010} with the exception of the length of the heated zone. Within the 3.5-cm high cavity the NICALON fiber would heat over a 3-cm long segment.

In microwave test heating, oxide ceramic fibers proved to be as difficult to heat as bulk oxide ceramics due to their negligible dielectric loss factor for temperatures below $\sim 600°$ C. Heating low loss fibers typically resulted in thermal runaway with the fibers breaking due to rapid melting or rapid grain growth. This phenomenon was observed while heating Saphikon sapphire monofilaments and Nextel™ alumina-based filaments. We were not able to controllably heat these materials in either cavity in a neat form. Special strategies must be developed to heat low loss dielectric fibers in a controllable, reproducible manner.

CONCLUSIONS

Aerosols of aqueous and alcoholic ferric nitrate solution have been successfully dried by direct microwave heating of the gas-borne droplets. However, because of the unfavorable dielectric properties of water in conjunction with the spherical geometry of droplets, the coupling of 2.45 GHz microwave power to micron-sized droplets appears to be far weaker than is convenient for spray drying, certainly when compared to that for a cylinder of water.

Microwaves at 2.45 GHz are strongly coupled to lossy ceramic filaments. Microwave processing of low loss fibers, like the oxide ceramics, requires further development to effect controlled heating without thermal runaway.

ACKNOWLEDGMENTS

Support for this was provided by the Department of Energy, Office of Industrial Technologies, Advanced Industrial Concepts Division, Materials Program. We gratefully acknowledge discussions with and the technical assistance of Joel Katz, Frank Gac, Brian Rusnak, and James Kennedy. We give our special thanks to Allan Holtz of the 3M Company for his generous samples of experimental Nextel™ fibers.

REFERENCES

1. D.E. Christiansen and W.P. Unruh in Microwaves: Theory and Application in Materials Processing, edited by D.E. Clark, F.D. Gac, and W.H. Sutton (Ceramic Transactions, Vol. 21, American Ceramics Society, Westerville, OH, 1991) pp. 597-604.

2. A. von Hippel, Dielectric Materials and Applications, (Wiley, New York, 1954), p. 361.

AN EXPERIMENTAL INVESTIGATION OF HYBRID CONVECTIVE/MICROWAVE DRYING OF ELECTRICAL PORCELAIN

Walter A. Hendrix* and Tom Martin**

**microGas* Corporation, 127 Bellevue Way SE, Suite 200, Bellevue, WA 98004

**Lapp Insulator Company, Gilbert Street, LeRoy, New York 14482

ABSTRACT

The drying of electrical porcelain using a combination of convection and microwave heating has been experimentally studied. To date, a total of 58 individual tests have been conducted to evaluate the effect of a range of microwave and convection conditions on the drying behavior of both solid and cavity core bodies. Preliminary indications are that order of magnitude reduction in drying time may be achieved. Furthermore, improved control over quality problems appears to be possible because of the ability of microwave heating to minimize the amount of convective surface heating.

INTRODUCTION

The drying of extruded bodies used to manufacture electrical porcelain insulators is especially challenging because of their sensitivity to in-process cracking when shrinkage is taking place. The drying rate is initially dependent on the need to minimize stresses due to thermal and moisture gradients. Once the body is sufficiently dried and shrinkage is complete, these stresses are no longer important, but the drying rate is then subject to heat and mass transfer limitations. Because of these considerations, the bodies may require as much as several hundred hours to dry, depending on size.

Conventional drying processes for electrical porcelain involve a trade-off between time, energy usage and product quality. These materials can be dried relatively defect-free at ambient conditions, but require very long time periods. The drying time can be reduced through the use of a thermal dryer, but only to a point where energy consumption is not excessive or unacceptable quality results.

The energy for conventional ceramic drying is supplied by external heating accompanied by conduction to the point where liquid water resides. Since green ceramics are usually very good thermal insulators, conduction is a relatively ineffective heat transfer mechanism, particularly as moisture content decreases. This circumstance serves to increase drying times, especially during the latter portion of the drying cycle when mass diffusion rates are low.

Combining microwave with convection seeks to overcome the deficiencies of conventional drying. Since the microwaves heat the water in the body directly, the internal and surface heating can be uncoupled. This circumstance allows the surface heating to be minimized and results in more uniform moisture and temperature gradients. The net effect anticipated is an increase in drying rates, at reduced energy consumption, without problems caused by uneven volume changes and case hardening (overdrying of the surface which reduces its permeability to diffusion of water).

This paper describes the results of a test program to evaluate the use of combined convection and microwave heating to dry electrical porcelain bodies at increased speeds and improved quality. The following discussion provides a description of the experimental equipment and procedures and the results observed thus far.

EXPERIMENTAL EQUIPMENT AND PROCEDURES

Three different types of porcelain electrical bodies were used in the test program: two representative of smaller production bodies and one of larger production bodies. All of the test bodies were cylindrical in shape with the smaller ones having dimensions of 6.75 inches diameter and 10 inches length. One type of the small test bodies was solid and weighed about 25 pounds (referred to as a solid post body). The other small test body contained a 2 inch diameter blind hole extending about two-thirds the way down its longitudinal axis (referred to as a cavity core post body). The weight of the cavity core test samples was slightly less than the solid samples at just under 24 pounds. The conventional drying time is around 150 hours for the solid bodies and around 130 hours for the cavity core bodies.

The larger test body was of the cavity core type with a diameter of 14.75 inches and length of 16 inches. The corresponding weight was much greater than the smaller test samples at over 200 pounds. The composition also differs in that alumina is used as the filler material rather than silica, to increase the strength of the final fired product (i.e., resulting in a so-called high strength insulator). The conventional drying time for this body can approach 300 hours. **Figure 1** provides a comparison of the size difference for the small and large test bodies.

Figure 1. Cavity Core Test Bodies

The drying trials were performed in a 915-mHz batch microwave oven capable of continuously varying microwave power in a range of 0-6 kW. Uniform microwave field distribution is provided by a waveguide splitter which introduces energy into the cavity at two locations, 180° out of phase. An electrically-driven mode stirrer and carousel are also contained in the cavity to assist in promoting even distribution of the microwave heating.

The microwave oven has been modified to allow for simultaneous or sequential convective heating with humidity control. Sensible heating of the convection air up to 250°F is accomplished using a bank of four individually-controlled 5-kW electric heating elements. The air temperature can be controlled to within +/-5°F at steady-state conditions. The convection air may flow upward or downward in the microwave cavity with impingement velocities up to several thousand feet per minute.

Moisture may be added to the oven environment for humidity control using a free-standing 20 lb/hr, 6-kW electric boiler equipped with an accumulator tank. The humidity is controlled manually based on readings from a hygrometer mounted in the oven cavity.

The major process parameters measured during testing were: (1) net microwave power input, (2) convection conditions (temperature, volume flow rate, relative humidity), (3) sample weight and moisture content, (4) surface and internal sample temperatures, and (5) sample appearance. Microwave power level was observed using a Hewlett-Packard Model 432A bolometer. Convection conditions (air temperature, volume flow rate and relative humidity) were monitored using hand-held instruments or from instrumentation in the control panel of the oven. The weight of the test article was tracked using one of several laboratory balances available or from reading a 0-500 lb load cell installed in the carousel assembly. Sample moisture content was determined using an Ohaus Model MB200 Moisture Determination Balance or, in the case of multiple samples, from weight change after oven drying.

The experimental investigation addressed the effect of a range of microwave and convection conditions on the drying behavior of the test samples. Measurements were made of relevant process parameters to establish temperature and moisture profiles, drying rates and impact on quality. The first set of trials established the handling and treatment procedures and response to various microwave and convection conditions using the small solid test bodies. Next, a series of testing parametrically evaluated combined microwave and convection drying of these same bodies. A third group of trials involved parametric evaluation of the small cavity core test bodies. In experimental work currently underway, drying of the large cavity core bodies is being investigated. To date, 58 individual experimental tests have been conducted.

RESULTS

The temperature gradients developed in a ceramic piece heated by microwaves differs substantially from one heated by conventional means. **Figures 2** and **3** present the longitudinal and radial temperature profiles, respectively, for a cavity core test sample at about 4 hours into the drying cycle. Note that the radial profile is reversed from that seen in surface heating where interior temperatures are always less than surface temperatures. The longitudinal profiles illustrate the fact that the bottom surface of the test sample is not cooled; i.e., the temperatures decrease with distance from the bottom of the sample. This behavior, observed in both of the figures, is a direct result of internal heating by the microwaves.

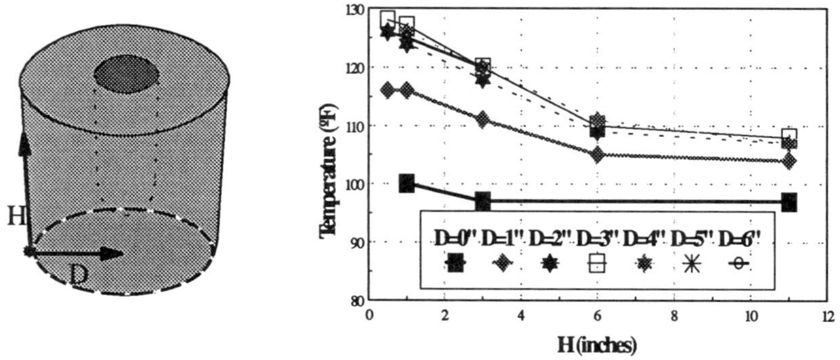

Figure 2. Longitudinal Temperature Profiles in Cavity Core Body

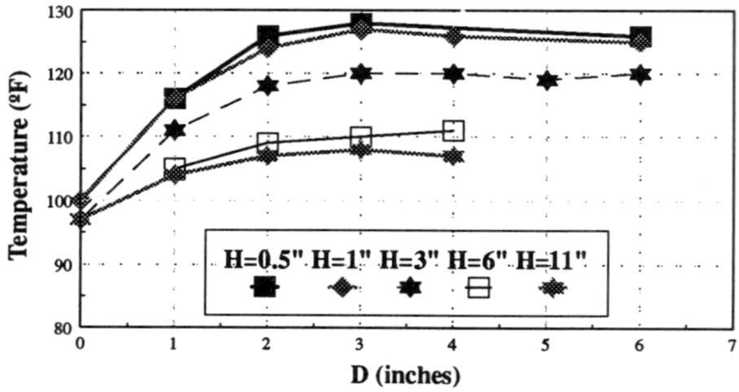

Figure 3. Radial Temperature Profiles in Cavity Core Body

The longitudinal and radial moisture profiles, rendered dimensionless by dividing the current moisture content by its initial value, for the same test body are presented in **Figures 4** and **5**. The longitudinal profiles shown in **Figure 4** are somewhat similar to those experienced in conventional drying with differences of as much as several percent developed along the body. As shown in **Figure 5**, the magnitude of the radial moisture gradients increase near the bottom of the body and could be as much as 1.5% to 2% along the bottom surface.

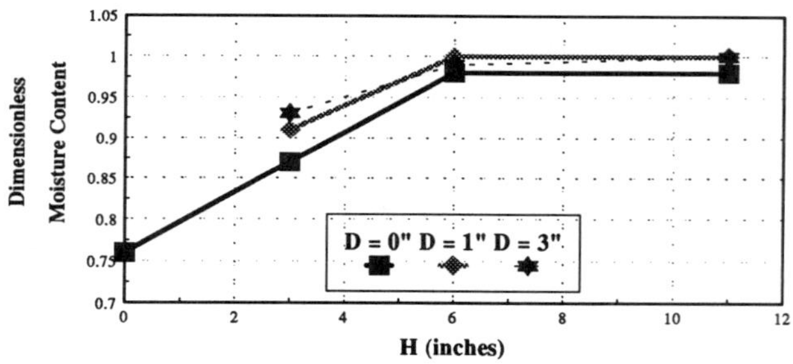

Figure 4. Longitudinal Moisture Profiles in Cavity Core Body

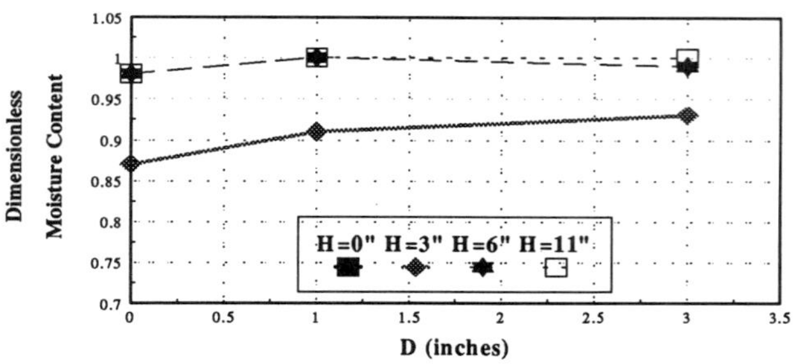

Figure 5. Radial Moisture Profiles in Cavity Core Body

A problem unique to microwave heating arises as a result of the temperature and moisture profiles that are developed. The edges of a ceramic body dried by conventional means are in compression when shrinkage is taking place. By contrast, when these ceramic pieces are dried using microwave heating their edges and surfaces are in tension. In this case, excessive temperature or moisture gradients lead to stress relief cracking. Furthermore, this problem is amplified by the presence of surface and internal imperfections (e.g., dents, scratches, occlusions, anisotropies, etc.) that provide an initiating point for stress relief. **Figure 6** shows a test body with a tension crack on the bottom edge resulting from an excessive temperature gradient.

Figure 6. Tension Crack in Test Body

This behavior during body shrinkage tends to reduce the maximum drying rates and requires careful control of the drying conditions. A series of more than 20 experiments was required to establish suitable conditions for drying the smaller test bodies through the shrinkage period without cracking. Efforts are still proceeding to establish such conditions for the large test bodies.

Figure 7 shows the complete drying rate curves for two small cavity core bodies. The curves are very similar to each other and somewhat characteristic of those observed for conventional drying. The drying rate increases initially to a constant rate period during the shrinkage portion of the drying cycle, then exhibits falling rate behavior thereafter.

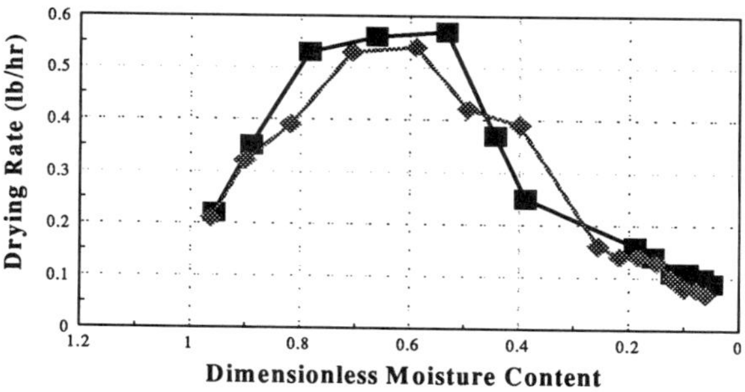

Figure 7. Drying Rate Curves for Small Cavity Core Bodies

Figure 8 shows the drying curves corresponding to the drying rate curves shown in **Figure 7**. The total drying time is observed to be around 15 hours, which represents an order of magnitude improvement over the conventional drying cycle. Furthermore, the drying time appears to be quite repeatable with equal amounts associated with shrinkage and non-shrinkage portions of the cycle.

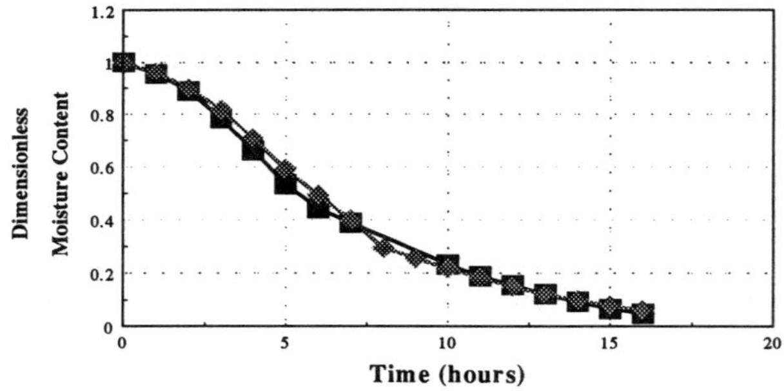

Figure 8. **Drying Curves for Small Cavity Core Bodies**

CONCLUSIONS

Extensive experimental testing of combined convection and microwave drying of electrical porcelain bodies has resulted in reduction in drying times of 80-90% over that required for conventional equipment. This success has been achieved while avoiding stress relief cracking associated with uneven shrinkage of the bodies. Experimental activity is continuing to further optimize the combined convection and microwave drying conditions and to extend them to the drying of much larger bodies. Successful completion of this work would be more representative of an entire insulator production operation. Implementation of this new drying process would greatly impact the manufacturing process, reducing energy consumption, in-process inventory, factory space and labor requirements.

ACKNOWLEDGMENTS

The authors gratefully acknowledge the support of the Lapp Insulator Company for the performance of the experimental work described here. Mr. Dennis Moraski and Mr. John Keder, both of *microGas* Corporation, are also recognized for their involvement in the testing and very helpful discussions of the test results.

Thermal Runaway and its Control in Microwave Heated Ceramics

Gregory A. Kriegsmann
Department of Mathematics
Center for Applied Mathematics and Statistics
New Jersey Institute of Technology
Newark, NJ 07102

Abstract.

The heating of a ceramic slab under TEM illumination is modeled and analyzed in the small Biot number regime. The temperature distribution is almost spatially uniform in this limit and its evolution in time is governed by a first order nonlinear amplitude equation. This equation admits a time independent solution which is a multivalued function of the microwave power. The dynamics of the heating process are deduced, from the amplitude equation and the multivalued response, and are dependent upon the microwave power and initial conditions. The results of this analysis give a plausible explanation of certain difficulties arising in sintering experiments, such as thermal runaway. A simple control process is presented and analyzed which mitigates against these deleterious effects. Abbreviated parameter studies are performed showing trends in the controlled heating process.

A quasi-three-dimensional problem modeling the heating process of a thin cylindrical sample in a waveguide applicator is also presented. For certain excitations the physical phenomenon deduced from this model and the required control process are the same as those obtained for the slab. For other excitations there is a strong spatial structure along the axis of the sample. For a certain choice of parameters the middle portion of the sample is at an elevated temperature while the remaining portion is at a much lower temperature. This phenomenon may be useful in joining applications.

Introduction.

In many microwave heating experiments the ratio of convective heat losses to conductive heat transfer (i.e., the Biot number) is quite small. We have exploited this fact and have developed an asymptotic analysis of the dynamic microwave heating of a ceramic slab under TEM illumination [1,2]. This analysis has produced a theory which takes into account both convective and radiative losses at the ceramic surface and predicts, to leading order, that the temperature distribution is essentially uniform in space. The physical reason for this uniformity is that the convective and radiative losses are weak in a sense that the slab boundaries are nearly insulated in this limit. The analysis has also produced higher order corrections that are relatively small compared to the leading order approximation and have spatial structure which describes nonuniform heating.

This work was supported by the Air Force Office of Scientific Research Under Grant No. AFOSR 91-0252.

According to the theory referenced above the leading order approximation to the temperature satisfies the first order nonlinear initial value problem

$$\rho C_p \frac{dT}{dt} = -2[\frac{h}{d}(T - T_A) + \frac{se}{d}(T^4 - T_A^4)] + \frac{\sigma(T)}{2}||U^2||\sqrt{\mu_0/\epsilon_0}P \quad (1a)$$

$$T(0) = T_A \quad (1b)$$

where T_A is the ambient temperature, μ_0 and ϵ_0 are respectively the permeability and permittivity of free space, ρ and C_p are respectively the density and heat capacity of the ceramic, d is its thickness, and $\sigma(T)$ is its effective electrical conductivity. The parameters h and e are respectively the heat transfer coefficient and the emissivity of the ceramic surface, s is the radiation heat constant, and P is the microwave power defined by $P = \sqrt{\epsilon_0/\mu_0}E_0^2$ where E_0 is the amplitude of the impinging microwave. The electric field within the ceramic is given by $E = E_0U$ where U is the electric field produced there by an incident plane wave of unit amplitude. Finally, the factor $||U^2||$ in (1b) is defined by

$$||U^2|| = \frac{1}{d}\int_0^d |U|^2\, dx \quad (2)$$

so that $\frac{\sigma(T)}{2}||U^2||\sqrt{\mu_0/\epsilon_0}P$ is the power generated by microwave heating.

This equation admits a steady state solution which is a multivalued function of the microwave power. The dependence of this function on some of the myriad of physical parameters has been described elsewhere [3-4]. We exhibit here in Figure 1, for the sake of discussion and motivation, a typical graph of the steady temperature as a function of microwave power which is qualitatively the same for either an exponential or Arrhenius model of the electrical conductivity. We shall describe below how this s-shaped response curve gives a plausible description of some difficulties experienced in microwave sintering experiments and in particular the phenomenon of thermal runaway.

By observing the sign of the right hand side of (1a) we deduce that for power levels below the threshold P_1 the temperature evolves along the vertical line, beginning on the horizontal axis where $T = T_A$ and reaching its steady state value on the lower branch of the s-shaped response curve. As the power P is increased the temperature will evolve to a new steady state further up the lower branch. If the maximum value of the temperature, T_1 on the lower branch is less than that required for sintering, then the ceramic will not sinter. If the power is increased slightly beyond P_1 in an attempt to reach the sintering temperature, then the temperature will evolve along the corresponding vertical line to its new value on the upper branch, T_2. If the temperature T_2 is greater than the melting temperature of the ceramic, then the sample is destroyed. This scenario seems to be the case in a number of important practical situations [5-8].

The Control Process.

In the context of the present physical model we have proposed a simple linear control process which has the potential to mitigate against thermal runaway and to insure sintering [1,2]. It is given by

$$\frac{dP}{dt} = -\gamma(P - P_s) - \beta(T - T_s) \quad (3a)$$

$$P(0) = P_0 \tag{3b}$$

where γ and β are constants and P_0 is the initial value of the power. The rate of change of the power is on a very large time scale compared to a microwave period so that γ and β must be suitably restricted. The point (P_s, T_s) lies on the s-shaped response derived where T_s is the sintering temperature and P_s is its corresponding power. Finally, we observe that the description of the heating process is now contained in the solutions of equations (1) and (3) and their properties are deducible by phase plane techniques [9].

We note that (3) may be considered the mathematical generalization of the control process suggested in Reference 10. That process is essentially given by (3) with $\beta = 0$ so that the power is an exponential function of time increasing from its initial value P_0 to its final value P_s.

There are two interesting and inclusive possibilities which will be described separately. In the first, we assume the sintering temperature lies on the upper branch below T_2 as shown in Figure 2. Then we observe that the steady state solutions to the **system** of equations (1) and (3) are geometrically given by the intersection of the control line $\gamma(P - P_s) + \beta(T - T_s) = 0$ and the s-shaped response curve. By design the point (P_s, T_s) is a steady state solution. Then depending on the slope of this line there are either two more steady state solutions or none. We assume here that there are none and that the slope $m = \gamma/\beta$ is suitably restricted. This case is also shown in Figure 2. A phase plane analysis of this situation shows that all solutions of (1) and (3) evolve in time from their initial value (P_0, T_A) to the final state (P_s, T_s). We have shown our results for two numerical calculations in Figure 2 for $P_0 = 0$. In both cases the slope m was the same but the solid curve had larger values of γ and β than the dashed curve which experiences a large overshoot which might drive the ceramic above its melting temperature. This was a general trend and suggests that large feedback coefficients result in a well controlled process.

In the second case we assume that (P_s, T_s) lies on the middle branch of the s-shaped response curve. Again we observe that the steady state solutions of the system (1) and (3) correspond to the points of intersection of the control line and the s-shaped response curve. Once again there are two possibilities and we assume that the slope m is restricted so that there is only one point of intersection as shown in Figure 3 for $P_0 = 0$. A phase plane analysis for this case shows that if γ and β are large enough, then all the solutions of (1) and (3) evolve in time from their initial value (P_0, T_A) to the final state (P_s, T_s). A typical trajectory converging in a spiral to the final point is also shown in Figure 3. On the other hand, if γ and β are small enough, then the system experiences a Hopf bifurcation [9] and the point becomes s center around which the system oscillates. This case is shown in Figure 4. Thus, the ceramic is periodically heated and cooled and the final sintering outcome is dubious. The important trend here is similar to the first case discussed above: The feedback parameters must be sufficiently large for the process to work.

Before proceeding to the next section we would like to give a few points of explanation about the numerical simulation of (1) and (3). First, we note that the term $||U^2||$ in (1a) requires that the electric field is known in the slab. This can be written explicitly for a given function $\sigma(T)$. Secondly, we observe that there is a natural time scale associated with (1a) which physically is the time required for the sample temperature to decay to $1/e$ of its original value in the absence of microwave power and radiative losses. It is given by $\theta = \rho C_p d/h$ and for ceramic slabs on the order of a few millimeters thick $\theta \sim 300 sec$.

RESPONSE CURVE AND FEEDBACK LOAD LINE

Figure 1

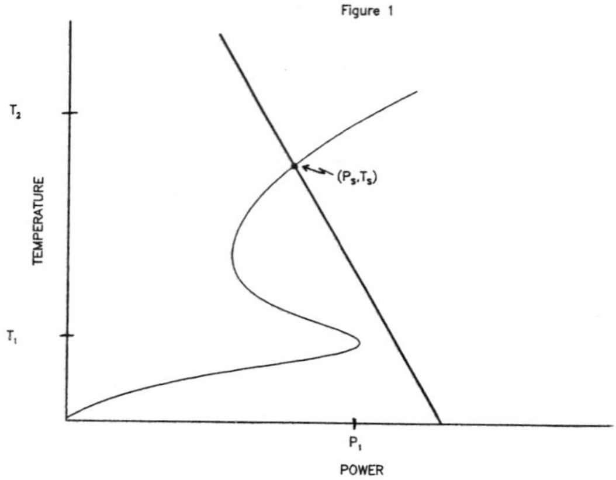

RESPONSE CURVE AND TWO TRAJECTORIES

Figure 2

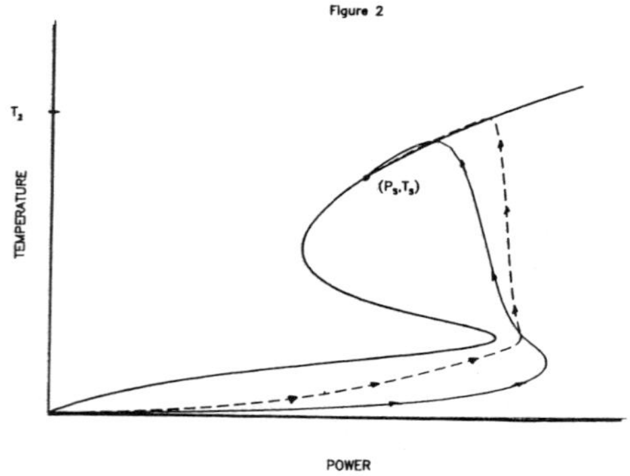

RESPONSE CURVE AND ONE TRAJECTORY

Figure 3

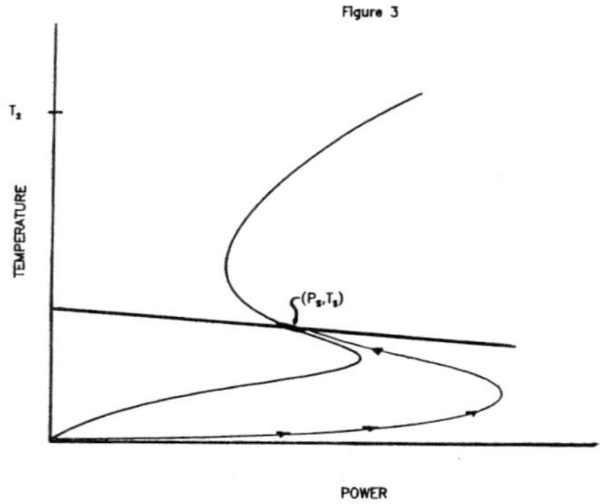

RESPONSE CURVE AND ONE TRAJECTORY

Figure 4

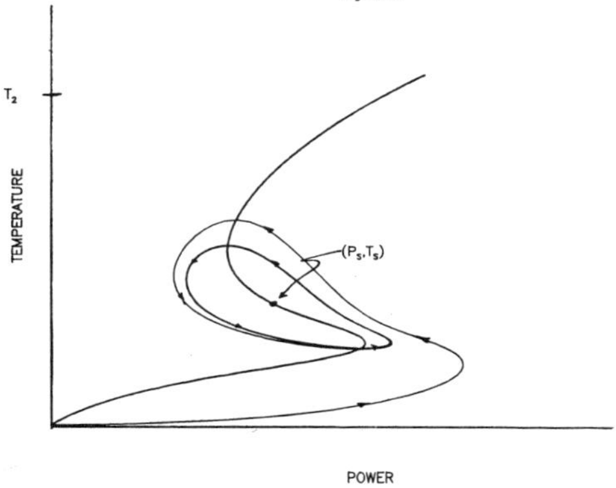

TEMPERATURE PROFILE FOR A THIN ROD

Figure 5

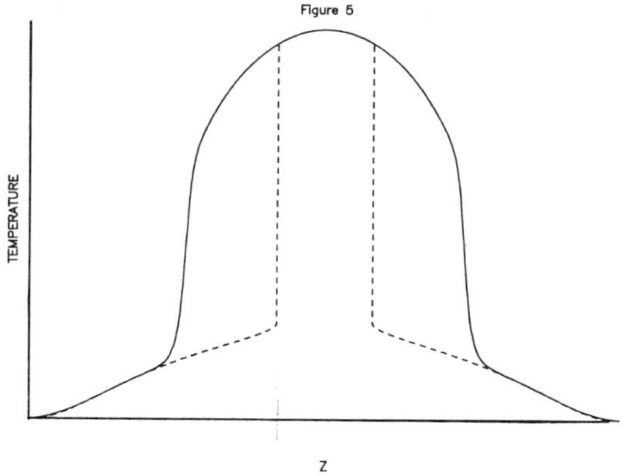

We require that the constant γ is of the same order as $1/\theta$ and that β is chosen similarly. Then both terms on the right hand side of (6c) are of comparable size and both P and T_0 change on the same time scale. Thus, θ^{-1} can be thought of as a measure of the heating rate for the process.

Three Dimensional Considerations.

We shall now analyze a "quasi-three-dimensional" problem modeling the heating of a very thin cylindrical sample in a waveguide applicator. The cylinder has a length L and a radius d which is so small compared to both the skin depth and wavelength within the ceramic that the electric field is uniform across its cross section. We call the problem quasi-three-dimensional because the electric field is assumed to be known along the axis of the cylinder, $0 < z < L$, as a function of z. We also assume that the L is comparable to a wavelength and accordingly the fineness ratio $d/L << 1$. Finally, we not only require that the Biot number is small but that it is now of the same order as the square of the fineness ratio. Then using similar asymptotic methods to those described in References 1 and 3 which incorporate this new scaling we find to leading order that the temperature now satisfies

$$\rho C_p \frac{\partial T}{\partial t} = \frac{\partial}{\partial z}(K \frac{\partial}{\partial z} T) - 2[\frac{h}{d}(T - T_A) + \frac{se}{d}(T^4 - T_A^4)] + \frac{\sigma(T)}{2}|E(z)^2| \tag{4a}$$

$$T(z, 0) = T_A \tag{4b}$$

where K is the thermal conductivity of the ceramic and $|E|$ is the amplitude of the electric field within the cylinder. The appropriate boundary conditions at the ends of the cylinder, $z = 0$ and $z = L$, depend upon how the sample is held in the applicator. If the sample is held in place by thermally insulated push rods, then the boundary conditions are

$$\frac{\partial}{\partial z} T(o, t) = \frac{\partial}{\partial z} T(L, t) = 0. \tag{4c}$$

These boundary conditions are also applicable if the ends are allowed to convect and radiate because of the scaling relationship between the Biot number and the fineness ratio.

The mathematical analysis of the nonlinear, initial boundary value problem (4) is a far more difficult task than the corresponding analysis for the (1) and (3). However, if the sample is being heated by an incident TE_{10} mode, then the electric field is independent of z. Then (4) admits a special class of solutions that are also independent of z and that satisfy equation (1) with $||U^2|| = 1$ and with $E = E_0$. Thus, the analysis and conclusions of the previous section are valid. However, we are still in the process of proving that these are the only solutions.

If on the other hand the sample is being heated by an incident TM_{11} mode, then $E = E_0 f(z)$ where $f(z)$ is a $2\pi/L$ periodic function with a maximum value at $z = L/2$ and a minimum at $z = 0$. The solution of (4) must be determined by numerical simulation. Even the determination of the steady state solution(s) of (4) require numerical simulation because of the z dependence of the $|E^2|$ term. A bit of insight can be gleaned in this case by assuming the thermal conductivity K is small and by consequently dropping the diffusion term. Then, for each point z in the cylinder the temperature evolves according to (1). We thus have a very interesting possibility: For those points on the cylinder where $P(z) = \sqrt{\epsilon_0/\mu_0}|E(z)^2|$ is less than P_1 the temperature will evolve to a point on the lower branch of the s-shaped curve and for those points where $P > P_1$ the temperature will evolve to the upper branch. This scenario is shown in Figure 5. as a the discontinuous dashed curve and a numerical simulation of case where K is small but finite is shown as the solid curve. Although the nonzero thermal conductivity smooths out the jump in temperature, the center of the rod is at a much elevated temperature which may cause melting. On the other hand, if the elevated temperature at the center can be controlled and limited to the sintering temperature, then such an arrangement may be useful in joining processes.

Conclusions.

We have reviewed the heating process for a ceramic slab under TEM illumination in the small Biot number limit and have given a plausible explanation for the phenomenon of thermal runaway. We have suggested and analyzed a linear control process which mitigates against runaway. We have shown through several calculations that when the feedback parameters are properly chosen the process will insure that the slab safely reaches its sintering temperature.

We have also described a quasi-three-dimensional problem which models the heating of a very thin cylindrical sample in a waveguide applicator. We have derived a nonlinear heat equation in the longitudinal spatial variable in which the electric field is assumed known. If the ends of the cylinder are insulated and the electric field is uniform along its length, then we have shown that the controlled heating process for this scenario is the same as the slab case. On the other hand, if the electric field varies sinusoidally along the length of the cylinder, then we have demonstrated that the center of the ceramic sample may be hot enough to melt.

Bibliography.

1. Kriegsmann, G. A., "Thermal Runaway in Microwave Heated Ceramics: A One-Dimensional Model", **Journal of Applied Physics**, 71 (1992), pp. 1960-1966.
2. Kriegsmann, G. A., "Feedback Stabilization of Thermal Runaway in Microwave Heated Ceramics", **Journal of American Ceramic Society**, submitted.
3. Kriegsmann, G. A., "Microwave Heating of Ceramics", in Ordinary and Partial Differential Equations 3, ed. B. Sleeman and R. Jarvis, Pitman Research Notes in Mathematics Series, No. 254, Longman Scientific, Essex, 1991, pp. 126-145.
4. Kriegsmann, G. A., "Microwave Heating of Ceramics: A Mathematical Theory", in Microwaves: Theory and Applications in Materials Processing, ed., D.E. Clark, F.D. Gac, and W.H. Sutton, Ceramic Transactions 21, American Ceramic Society 1991, pp. 177-183.
5. Brodwin, M. E. and Johnson, D. L., "Microwave Sintering of Ceramics", **MIT-S**, K-5 (1988), pp. 287-288.
6. Tian, Y. L., Johnson, D. L. , and Brodwin, M. E., "Ultrafine Structure of Al_2O_3 Produced by Microwave Sintering". Proceedings of the First International Conference on Ceramic Powder Processing Science, Orlando Florida, Nov. 1987.
7. Araneta, J. C., Brodwin, M. E., and G. A. Kriegsmann, "High Temperature Char-

acterization of Dielectric Rods", **IEEE MTT**, 32 (1984), pp. 1328- 1334.

8. Bertrand, A. J. and Badot, J. C., "High Temperature Microwave Heating in Refractory Materials", **Journal of Microwave Power**, 11 (1976), pp. 315- 320.
9. Stoker, J. J., "The Theory of Nonlinear Oscillations", Wiley Interscience, N.Y., 1969.
10. Tian, Y. L., "Practices of Ultra-Rapid Sintering of Ceramics Using Single Mode Applicators", in Microwaves: Theory and Applications in Materials Processing, ed., D.E. Clark, F.D. Gac, and W.H. Sutton, Ceramic Transactions 21, American Ceramic Society 1991, pp. 283-300.

AUTOMATIC CONTROL TO PREVENT THERMAL RUNAWAY DURING MICROWAVE JOINING OF CERAMICS

Guy O. Beale and Francisco J. Arteaga
Department of Electrical and Computer Engineering, George Mason University, Fairfax, VA 22030

ABSTRACT

A temperature control system is designed and simulated for microwave joining of ceramics. The objective is to develop a control system for reliable, high quality joints to be produced on a variety of ceramics. The described procedure permits the derivation of a closed-loop configuration for controlling temperature. The ceramics (alumina) heating model equation is linearized around an operating point and represented as a transfer function. Conditions for closed-loop stability with any linearized model are presented and mathematically proven. A computer simulation is implemented for the closed-loop control and numerical integration is used for solving the non-linear heating equations. Results are very good and show that the temperature can be regulated to a set point value and thermal runaway in the ceramics can be prevented.

INTRODUCTION

Advanced ceramics and composites based upon them can survive hostile chemical environments and temperatures as high as 1500-2000°C. There is great interest in using these materials in many diverse applications. Technical barriers have prevented the commercialization of ceramics to date. These all stem from the brittleness of the materials. Inexpensive and reliable methods to join ceramics are needed so that more complex components could be fabricated from smaller and simpler shapes. Microwave heating of the interface, unlike other joining methods, provides the heat necessary to effect the joint very rapidly, and for non-conducting ceramics, of heating the interior directly and inmediately through the effect of the electromagnetic field [1],[2]. Results have demonstrated that alumina, mullite, and silicon nitride can be joined directly by microwave heating [3],[4],[5]. Research efforts in automatically controlling the joining process have also been reported [5],[6],[7]. The microwave loss factor of ceramics increases rapidly with temperature, causing positive feedback and thermal runaway in the temperature region of interest for forming strong chemical bonds. The objective of the temperature control system requires development of methods to prevent thermal runaway and regulate the interface temperature throughout the heating process.

MICROWAVE HEATING PROCESS

A model for the heating process has been developed [3],[8],[9]. The time rate of change of temperature is determined from the power incident on the cavity, P_{IN}, and the materials properties described in Table I. Thermal equilibrium is established when the input power is balanced by the thermal radiation from the material.

Table I
Parameters for Alumina (Al_2O_3)

Name	Symbol	Value	Units
Real Part of Dielect.Const.	k'	9.8	Dimensionless
Sample volume	V	$2.714 \cdot 10^{-6}$	m^3
Sample height	h	$3.81 \cdot 10^{-2}$	m
Sample radius	r	$4.762 \cdot 10^{-3}$	m
Surface emissitivity	ϵ	$5.99 \cdot 10^{-1}$	$watts/m^2$
Mass density	ρ	$3.97 \cdot 10^3$	Kg/m^3
Specific heat capacity	c_p	774.9	$J/Kg \cdot^\circ C$
Thermal Conductivity	k	29.7	$watts/m \cdot^\circ C$

The rate of change of temperature ($^\circ K$) can be approximated by:

$$\frac{dT}{dt} = f(T, P_{IN}) = k_1 Q(T) k''(T) P_{IN} - k_2(T^4 - (296)^4) - k_3(T - 296)^{5/4} - k_4(T - 296) \tag{1}$$

where $Q(T)$ is the loaded quality factor for the resonant cavity including the sample, $k''(T)$ is the loss factor, and the constants k_1, k_2, k_3, k_4 are function of the materials properties (Alumina) [8],[9]:

$$k_1 = \frac{8Q(T)}{\rho c_p k' V} \; ; \; k_2 = \frac{\epsilon\sigma}{\rho c_p} \; ; \; k_3 = \frac{1.42}{L^{1/4} \rho c_p} \; ; \; k_4 = \frac{2\pi k L}{\rho c_p ln(2L/r)}$$

In the right side of equation (1) the first term represents the absorption by the material due to the microwave power and to the reflected heat of the walls, and the second, third, and forth term are the losses due to radiation, convection, and conduction, respectively. The expression for the loss factor $k''(T)$ can be derived based on experimental work with alumina [5],[10]:

$$k''(T) = 3.7276 \cdot 10^{-3} \cdot 10^{(T-273)/700} \tag{2}$$

Since k'' increases rapidly with T, a runaway temperature can quickly result. A factor which partially offsets the effects of the increase in k'' is that Q decreases; $k'' \cdot Q$ approaches an approximate value of 44, assuming that the resonant frequency is held constant. The relationship between Q and k'' for our experimental conditions can be approximated by [12]:

$$Q = \frac{1500(\frac{V_{cavity}}{4V_{sample}} + k')}{3000k'' + \frac{V_{cavity}}{4V_{sample}} + k'} \tag{3}$$

The solution of equation (1) and (2) is shown in Figure 1. The input power is held constant at 400 watts (open-loop control), all relationships are assumed to remain valid, and critical coupling is assumed.

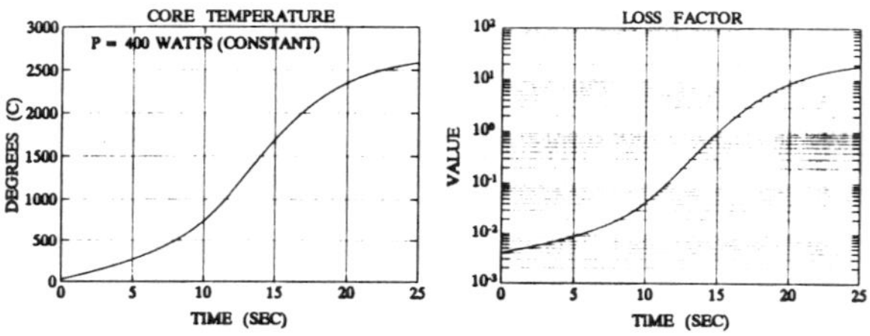

Figure 1. Core Temperature T_o and Loss Factor k'' vs. Time.

Assuming that the desired temperature was in the normal range for microwave joining, it is seen that feedback control is required to regulate temperature. Equation (1), (2), and (3) can be linearized about a particular operating point:

$$\frac{\delta T}{dt} = \frac{\partial f(T, P_{IN})}{\partial T} \cdot \delta T + \frac{\partial f(T, P_{IN})}{\partial P_{IN}} \cdot \delta P_{IN} = A_1 \cdot \delta T + B_1 \cdot \delta P_{IN} \tag{4}$$

where the control input= P_{IN} =microwave power, δT, δP_{IN} represent deviations from the nominal operating point, and $\partial f/\partial T$, $\partial f/\partial P_{IN}$ are evaluated at the operating point (T_{op}, P_{op}). From (1)

$$A_1 = \frac{\partial f}{\partial T} = k_1 P k''(T) \cdot \frac{\partial Q(T)}{\partial T} + k_1 P Q(T) \cdot \frac{\partial k''(t)}{\partial T} - 4k_2 T^3 - 1.25 k_3 (T - 296)^{1/4} - k_4 \tag{5}$$

$$B_1 = \frac{\partial f}{\partial P} = k_1 Q(T) k''(T) \tag{6}$$

Figure (2) shows the effect of temperature on the coefficients A_1 and B_1 for different power levels. With $A_1 > 0$ in the linearized model, except for extremely low power levels or extremely high T, equation (1) is an unstable function of time over much of its operating range. This leads to the thermal runaway problem unless proper control of input power is achieved.

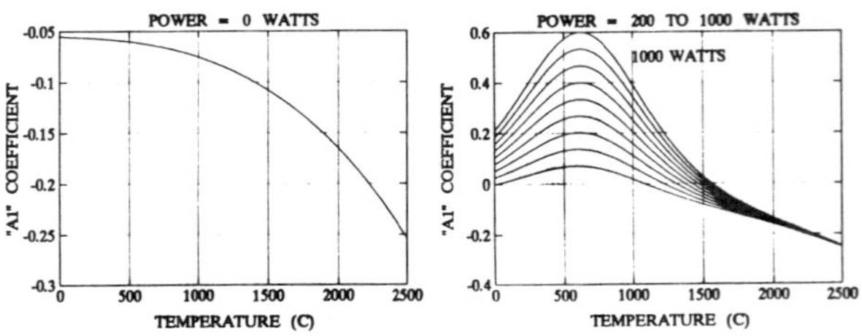

Figure 2. Effect of Temperature on Linearized Coefficient A_1.

As Figures 2a-2b show, the degree of instability increases with increasing power level. If an upper bound on power level can be defined, a "most unstable" linear model results. This model and all "less unstable" models must be stabilized by the same compensator.

CLOSED-LOOP CONFIGURATION FOR TEMPERATURE CONTROL

A system configuration for temperature control is shown in Figure 3.

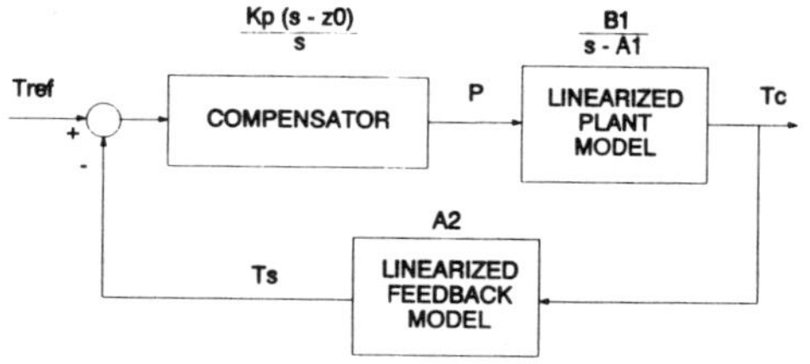

Figure 3. Linearized System Model for Temperature Control.

The plant and compensator transfer functions are represented by $G_p(s)$ and $G_c(s)$. The transfer function $H(s)$ represents the relationship between the surface temperature T_s and the core temperature T_o. Initially, $H(s)$ is assumed to be 1. From the linearization in (4), $G_p(s)$ becomes:

$$G_p(s) = \frac{\delta T(s)}{\delta P_{IN}(s)} = \frac{B_1}{s - A_1} \tag{7}$$

where A_1 and B_1 are the linearized coefficients for an operating point (T_{op}, P_{op}). With $H(s) = 1$, the structure of the control system assumes that T_o can be measured exactly and instantaneously. The actual value T_o is compared with the desired value T_{ref} for the experiment. The difference between the actual and desired temperatures is the system error which is the input to the compensator. The compensator relates microwave power to temperature error. At a given instant of time, the commanded microwave power (compensator output), is a function of the time history of the temperature error. This power command is applied to the magnetron. The compensator, by measuring the difference between the actual and desired temperatures, accounts for the nonlinear nature of the equations and is designed to force the temperature error to go to zero. Once the desired temperature value has been reached, the compensator will maintain the temperature at that value. Realistically, only the surface temperature can be directly measured. The estimated thermal gradient can be expressed by:

$$\Delta T = 2.9 \cdot 10^5 \cdot P_o \cdot \frac{k''}{10k'' + 1} \cdot \frac{1}{k} \cdot r^2 = T_o - T_s \tag{8}$$

where k is the thermal conductivity assumed constant. The quadratic radial dependence of (8) is slowly varying enough to provide for accurate process control. Equation (8) can be linearized about a particular operating point and expressed by

$$\delta T_s = \left[1 - \frac{\partial}{\partial T_o}(\Delta T)\right] \cdot \delta T_o \tag{9}$$

and a linear model for $H(s)$ is given by

$$H(s) = \frac{\partial T_s}{\partial T_o} = 1 - \frac{\partial}{\partial T_o}(\Delta T) = A_2 = 1 - \frac{3.556 \cdot P \cdot r^2 \cdot 10^{T_o/700}}{(10k'' + 1) \cdot k} \tag{10}$$

For T very small and T higher than $2000°C$, the coefficient A_2 is almost 1, and T_s and T_o are practically the same. When $P = 0$ watts, $A_2 = 1$ exactly, regardless of the temperature. In general, $A_2 \approx 1$, and $T_s \approx T_o$ for our sample material.

DESIGN OF THE COMPENSATOR G_c

A compensator will be designed based on a "worst case" linearized model of equations (1), (2), (3), and (8). The design of the compensator provides the stabilization of all linearized plant models and also the capability that the actual material temperature will equal a constant desired value in equilibrium. In

Table II three set of values for A_1, B_1, and A_2 are presented for three operating points (T_{op}, P_{op}). The corresponding transfer functions $G_{p_i}(s)$ and $H_i(s)$ are:

$$G_{p_i} = \frac{B_{1i}}{s - A_{1i}} \quad , \quad H_i(s) = A_{2i} \quad ; \quad i = 1, 2, 3 \tag{11}$$

In the first case, G_{p_1} represents the linearized plant for the "most unstable" model (largest positive A_1). In the second case, G_{p_2}, represents the "worst case" for stability. The third case represents an equilibrium condition. In Figure 2a, for $P = 0$W, as T increases A_1 becomes more negative and $G_p(s)$ is a stable function. From Figure 2b, for $P > 0$, $A_1 > 0$ and it increases with T up to a maximum value around 620°C; then A_1 decreases to a very small value at high T. For the interval [0,2500]°C, A_1 is always positive, except when $P \approx 0$, and $G_p(s)$ is unstable.

Table II

A_1, B_1, A_2 For 3 Operating Points

No.	Oper.Point	A_1	B_1	A_2
1)	$T = 620°C$, P=1000W	0.5995	0.3976	0.9874
2)	$T = 80°C$, P=1000W	0.2665	0.1141	0.9968
3)	$T = 2000°C$, P=200W	-0.1595	0.7964	0.9995

It is necessary to add a compensator $G_c(s)$ in the configuration so that the closed-loop system becomes stable with: i) Stability at all operating points, ii) Steady-state error=0 for T_{ref}=constant, and iii) Small overshoot to a step response for any given linearized plant model $G_p(s)$ (stable or unstable). Since $G_p(s)$ is type 0, choosing a compensator with an integrator will provide a system response with zero steady-state error to a constant T_{ref}, assuming $A_2 = 1$ in equilibrium [10]. The proposed PI compensator is represented by

$$G_c(s) = \frac{K \cdot (s - z_o)}{s} \tag{12}$$

where K is the gain and z_o is a compensator zero. Figure 4 is a block diagram for a discrete-time implementation of the compensator.

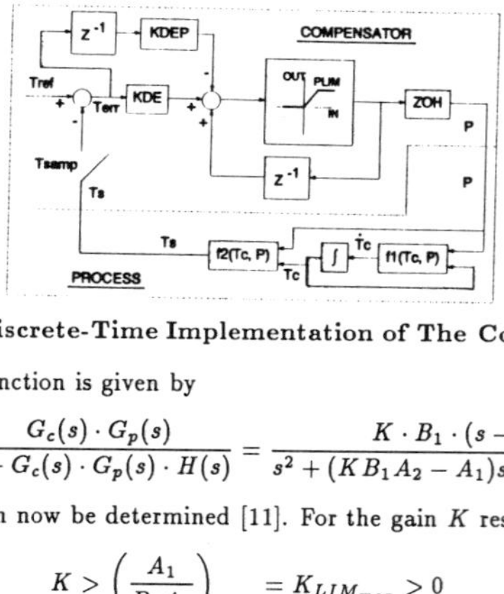

Figure 4. Discrete-Time Implementation of The Compensator.

The closed-loop transfer function is given by

$$G_{CL}(s) = \frac{G_c(s) \cdot G_p(s)}{1 + G_c(s) \cdot G_p(s) \cdot H(s)} = \frac{K \cdot B_1 \cdot (s - z_o)}{s^2 + (KB_1A_2 - A_1)s - KB_1A_2z_o} \tag{13}$$

The conditions for stability can now be determined [11]. For the gain K results in

$$K > \left(\frac{A_1}{B_1 A_2}\right)_{max} = K_{LIM_{max}} > 0 \tag{14}$$

where A_2, and B_1 are always positive and A_1 positive gives a tighter constraint than any negative A_1. The condition for the zero z_o is given by

$$sgn(z_0) = -sgn(K) \tag{15}$$

Thus, the zero z_o is required to be in the LHP. Expressions (14) and (15) are necessary and sufficient conditions for closed-loop stability in the case of any linearized model $G_p(s)$ in the region of operating points (T_{op}, P_{op}). The expression for the standard $2^n d$ order system with a zero is:

$$G_{CL}(s) = \frac{(\omega_n^2/A_2|z_o|)(s - z_o)}{s^2 + 2\zeta\omega_n s + \omega_n^2} \tag{16}$$

Comparing expression for $G_{CL}(s)$ given by equation (13) with equation (16) yields

$$\zeta = \frac{KB_1A_2 - A_1}{2\sqrt{KB_1A_2|z_o|}} \quad ; \quad \omega_n = \sqrt{KB_1A_2|z_o|} \tag{17}$$

where ζ and ω_n are the damping ratio and natural frequency. From (17) and $\zeta > 1$:

$$|z_o| < \left(\frac{B_1A_2}{4}\right)\frac{(K - K_{LIM})^2}{|K|} = z_{max} \tag{18}$$

and the step response $Y_{step}(s) = T_o(s)$ is expressed as

$$Y_{step}(s) = \frac{KB_1(s - z_o)}{s(s - p_1)(s - p_2)} = \frac{a_o}{s} + \frac{a_1}{s - p_1} + \frac{a_2}{s - p_2} \tag{19}$$

where p_1 and p_2 are the closed-loop poles, given by

$$p_{1,2} = -\zeta\omega_n \pm \omega_n\sqrt{\zeta^2 - 1} \tag{20}$$

and a_0, a_1, and a_2 are expressed by

$$a_o = \frac{1}{A_2} \quad ; \quad a_1 = \frac{KB_1(p_1 - z_o)}{p_1(p_1 - p_2)} \quad ; \quad a_2 = \frac{KB_1(p_2 - z_o)}{p_2(p_2 - p_1)}$$

The peak time t_p and overshoot for the compensated system are given by:

$$t_p = \left[\frac{1}{p_1 - p_2}\right] ln\left[\left|\frac{p_2a_2}{p_1a_1}\right|\right] \quad ; \quad P.O. = \left(\frac{a_1e^{p_1t_p} + a_2e^{p_2t_p}}{a_o}\right)100\% \tag{21}$$

The selection of K and z_o is based on: (1) the minimum possible value for the overshoot in the step response of the linearized plant model and (2) the stability conditions $K > K_{LIM}$ and $|z| < z_{max}$. In Figures 5a and 5b, the overshoot and the location of z_o are plotted as functions of K.

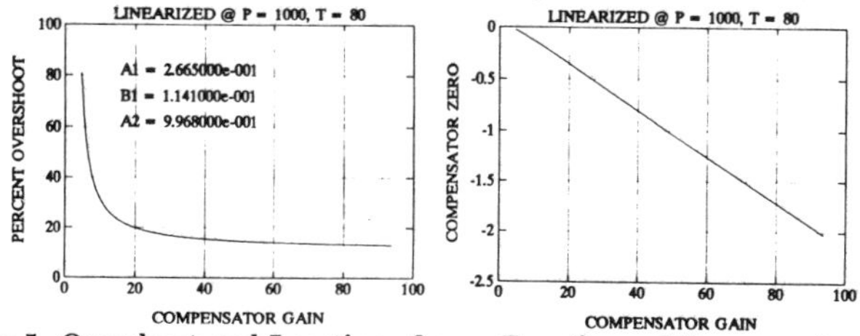

Figure 5. Overshoot and Location of z_o as Functions of Compensator Gain K.

The coefficients are $A_1 = 0.2665$, $B_1 = 0.1141$, and $A_2 = 0.9968$, for the "worst case" linearized model $G_p(s)$, $T = 80°C$, $P = 1000W$. The minimum possible overshoot is approximately 14% for the closed-loop linearized system with a gain $K \geq 60$. For K=60, a value of $z_o = -1.75$ satisfies the condition (18) for z_o.

RESULTS

The control of temperature for the process with alumina is simulated in the computer. Equations (1), (2), (3), and (8) are evaluated to obtain values for T_o and T_s. Results are obtained for three set of initial conditions for T and P_{IN}, with the power limited to $250W$. Case 1: T_{init}=80°C, P_{init}=1000W, Case 2: T_{init}=25°C, P_{init}=0W, Case 3: T_{init}=620°C, P_{init}=1000W. Case 1 starts the process at the initial conditions of the "worst case" operating point for closed-loop stability. The power is assumed 1000 watts, and then the closed-loop control system is activated. Case 2 represents the most typical scenario. The material is initially at room temperature, and the microwave power is at 0 watts. Case 3 starts the process at the initial conditions for the "most unstable" linearized open-loop model. The power is set at 1000 watts and the temperature is forced to 620 degrees prior to the activation of the closed-loop control. In all three cases, the maximum value for microwave power allowed by the compensator is 250 watts. Any calculated power greater than 250 watts is limited to that value. In cases 1 and 3, the compensator inmediately drops the power from 1000 watts to 250 watts as soon as the closed-loop system is activated. The plot for case 2 is shown in Figure 6. The compensator yields good control even when an upper limit is imposed on P_{IN}.

At the desired $T_{ref} = 1900°C$, the equilibrium T_{error} is 0, and the equilibrium $P_{IN} \approx 0$. Once T_{ref} is reached no further change in T occurs since the control system forces an equilibrium condition to exist.

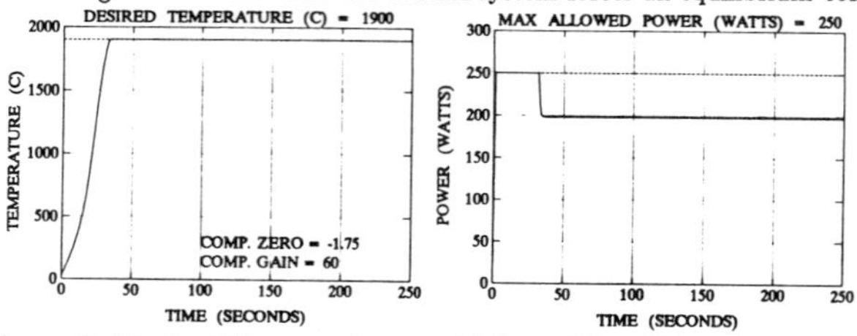

Figure 6. Desired Temperature and Max. Allowed Power vs. time.

CONCLUSIONS

A control algorithm has been designed and evaluated with the complete nonlinear model representation of temperatures at the surface and material center and the temperature dependence of the loss factor and cavity factor Q. The performance of the compensator have been illustrated with an example for small samples of alumina. The simulation yielded very good results for the desired values in the material temperature, with zero steady-state error, closed-loop stability, and no overshoot of the desired temperature. The compensator stabilized all linear models resulting from linearizing the nonlinear system at different operating points. The results have demonstrated the ability to overcome thermal runaway and to maintain good temperature control. The first part in the research is intended to design, simulate, and implement the control system for the process. The objective of the system will be to maintain constant sample temperature by: (1) maintaining the cavity at resonance; (2) maintaining critical coupling of microwave energy into the cavity in order to maximize the power incident on the material; and (3) controlling the generated microwave power. The interaction of these requirements must be accounted for in order to prevent instability.

ACKNOWLEDGEMENTS

This research has been supported in part by Virginia Center for Innovative Technology and is a joint effort with Technology Assessment and Transfer, Inc., MD. The authors sincerely thank Dr. W. Black, Dr. R. Silberglitt, Dr. D. Palaith, and Dr. I. Ahmad for their advice and suggestions for this paper.

REFERENCES

1. M. M. Schwartz, Ceramic Joining, ASM International, Materials Park, OH, 1990.
2. W. H. Sutton, Amer. Ceram. Soc. Bull., vol. 68, no. 2, p. 376, 1989.
3. D. Palaith, R. Silberglitt, C. C. M. Wu, R. Kleiner, and E. L. Libelo, Mat. Res. Soc. Symp. Proc. vol. 124, p. 255, 1988.
4. D. Palaith and R. Silberglitt, Amer. Ceram. Soc. Bull., vol. 68, no. 9, p. 1601, 1989.
5. H. Fukushima, T. Yamanaka, and M. Matsui, J. Mater. Res., vol. 6, no. 2, p. 397, 1990.
6. X. D. Yu, V. V. Varandan, V. K. Varandan, and D. Ghodgaonkar, Ceramic Transactions, The American Ceramic Society, vol. 21, pp. 251-259, 1991.
7. X. D. Yu, V. V. Varandan, V. K. Varandan, Ceramic Transactions, The American Ceramic Society, vol. 21, pp. 497-505, 1991.
8. Yong-Lai Tian, Ceramics Transactions, The American Ceramic Society, vol. 21, pp. 283-300, 1991.
9. J. P. Holman, Heat Transfer. New York, McGraw-Hill, 1972.
10. Guy O. Beale, Francisco J. Arteaga, and W. Murray Black, to be published in IEEE Transactions on Industrial Electronics.
11. J. D. Jackson, Classical Electrodynamics, New York, Wiley, 1962.
12. R. E. Collins, Foundations for Microwave Engineering, New York, McGraw-Hill, 1966.
13. John J. D'Azzo and C. H. Houpis. Linear Control System Analysis and Design, New York, McGraw-Hill, 1988.
14. G. Thaler, Automatic Control Systems, St.Paul, MN, West Publishing, 1989.

MICROWAVE JOINING OF SILICON CARBIDE USING SEVERAL DIFFERENT APPROACHES

IFTIKHAR AHMAD*, RICHARD SILBERGLITT*, W. MURRAY BLACK†, HUSSAMALDIN S. SA'ADALDIN† and JOEL D. KATZ‡
*Technology Assessment & Transfer, Inc., Annapolis, MD 21401
†Department of Electrical and Computer Engineering, George Mason University, Fairfax, VA 22030
‡Los Alamos National Laboratory, Los Alamos, NM 87544

ABSTRACT

Microwave joining of sintered silicon carbide, both to itself and reaction bonded silicon carbide, has been accomplished in a single mode rectangular resonant cavity. Several approaches using different interlayer materials were employed to join sintered silicon carbide. Effective joining of reaction bonded silicon carbide to itself and sintered silicon carbide was accomplished without the use of any interlayer material in the single mode resonant cavity as well as in a multi-mode oven. Specimens cut from 0.95 cm diameter rods were joined in the single mode cavity, whereas a variety of arbitrary shapes and larger specimens (of reaction bonded silicon carbide) were joined in the multi-mode oven.

INTRODUCTION

The combination of high thermal conductivity, excellent thermal shock resistance and demonstrated corrosion resistance in high temperature oxidative environments makes silicon carbide (SiC) an excellent candidate material for numerous applications. These applications include structural components for advanced heat engines, heat exchanger tubing, radiant burner tubes and pump components. The cost and difficulty of manufacturing components of complex shapes has limited the use of SiC in these applications.

A simple and cost effective silicon carbide joining technique would allow complex components to be fabricated from smaller and simpler shapes, which would be easier to reliably manufacture. A significant amount of work has been performed on conventional joining of silicon carbide [1]. Joining was achieved using either a metallic or glassy interlayer. Some of the joints exhibited good mechanical properties at low temperatures. Performance of these joints at high temperatures is expected to be unsatisfactory because of oxidation of the metal alloy and creep, both in the metallic and glassy interlayers. Direct joining of SiC has also been attempted by polishing the pieces to mirror-like finish before heating. The joined specimens of SiC exhibited lower than desired strengths [1].

The use of microwave energy to efficiently join oxides and silicon nitride having some intergranular material has been demonstrated [2,3]. Silicon carbide material has no intergranular phase, which poses difficulties in conventional and microwave joining. To surmount these difficulties, one may interpose a material compatible with silicon carbide or use some form of silicon carbide having an intergranular phase. Reaction bonded silicon carbide is one such material. This paper discusses the results of joining sintered SiC using various interlayer materials as well as the direct joining of reaction bonded silicon carbide to itself and sintered SiC.

Reaction bonded silicon carbide (RBSC)

Reaction bonded silicon carbide (RBSC) is manufactured by infiltrating molten silicon into shaped bodies of SiC-carbon. The SiC-carbon mix contains an organic polymer and is well suited to near-net-shape forming by extrusion or injection molding. Capillary action provides the driving force for the infiltration of silicon, which reacts with carbon to form reaction bonded silicon carbide. The 5 - 10 % excess of silicon present in RBSC reduces its strength and service temperature only slightly as compared to sintered SiC.

Conventional joining of reaction bonded silicon carbide was reported by Iseki et al. [4] using aluminum metal as the interlayer. No reaction layer was detected because aluminum and silicon in the reaction bonded SiC formed a continuous liquid phase at high temperature. However, some cracks were found at the front of the aluminum-penetrated region. These cracks are believed to be formed because of the difference in thermal expansion of aluminum, SiC and silicon [4].

Microwave joining of reaction bonded silicon carbide with aluminum as the interlayer was reported by Yiin et al. [5]. A single mode resonant cavity was used to heat the reaction bonded silicon carbide. Aluminum penetrated into Si regions between the SiC grains. Room temperature 3-point and 4-point bend tests demonstrated that the joined specimen was stronger than the as-received material, similar to the earlier microwave joints in oxides and nitrides [2,3].

Microwave heating has been observed to result in enhanced diffusion and reaction rates [6]. Earlier microwave joining times have been much less than conventional methods [2,3,5]. Thus, microwave heating of RBSC should result in enhanced mass transport across the interface, leading to a diffusion bonding of the pieces to be joined. This should eliminate the need to use any interlayer material, thus increasing the service temperature of the joined specimens.

MATERIALS AND METHODS

Materials used in the joining experiments include specimens of Carborundum Hexoloy™ (sintered SiC) and reaction bonded silicon carbide from Coors Ceramics and Laing Ceramics Corporation.

Hexoloy™ SiC specimens were joined using several interlayer materials. These include blended powders of Si, C and Ti to form in situ SiC/TiC and a plasma-sprayed thin layer of Si on one of the SiC specimens [7].

For direct joining of RBSC to itself and sintered SiC, specimens of 0.95 cm diameter and 0.5 cm height were cut with a low speed saw using a diamond wafering blade. Without any further surface preparation, the specimens were joined in a single mode TE_{103} rectangular cavity. Microwaves from a 1 kW (max) 2.45 GHz power source were directed through standard waveguide to the sample. The specimens were enclosed in alumina insulation (with a viewing hole to monitor temperature) and placed at the maximum electric field position in the cavity. Microwave power was coupled to the specimen by adjusting a plunger and an adjustable iris such that the reflected power was minimum. In a few minutes the specimens turned red hot and the surface temperature was monitored with an optical pyrometer. Typical time to reach 1450 - 1500°C was 5 - 8 minutes with 400 - 600 watts of power. The specimens were held in this temperature range for 10 - 15 minutes and then cooled by turning off the microwave power.

Specimens identical to the ones described above and of arbitrary shapes, such as faucet disks furnished by Laing Ceramics and small portions (5 cm diameter x 2.5 cm length) of RBSC tubes furnished by Coors, were joined in a commercial 900 watt multi-mode oven with hybrid heating. The specimens were placed in alumina insulation lined on the inside with a thin layer of silicon carbide. When microwave power is turned on, the lining of this insulation starts heating rapidly and the heat generated by the lining and the sample is retained within the insulation. Such microwave (hybrid) heating has been reported to provide uniform heating and very high heating rates even in a low loss ceramic like alumina [8,9]. Typical heating schedules were about 20 - 30 minutes to reach 1450 - 1500°C (depending on the size of the sample) and a holding time of 30 - 40 minutes at these temperatures.

RESULTS AND DISCUSSIONS

The joined specimens were characterized by sectioning the interfacial region and examining it with the scanning electron microscope. Specimens of sintered silicon carbide were joined with either a blend of Si, C and Ti, or only a Si interlayer. In the case when a blend of Si, C & Ti, or Si & C was used, although the powders did undergo a reaction to form SiC/TiC, the reaction was not uniform over the cross section of the specimen. Exposure of the sides of the joint interface to air resulted in some oxidization of the metals before the reaction occurred. Moreover, the reaction product was porous in the interfacial region. However, when Si was used as the interlayer, the interfacial region was dense and homogeneous. The width of the joint was dependent on the method employed for applying the interfacial material [7]. The joint made with a plasma-sprayed Si layer on one SiC specimen, was less than 5 microns in width.

Successful use of Si as the interlayer material for joining of SiC suggested the direct joining of reaction bonded silicon carbide. Specimens of RBSC were joined (to themselves) in the single mode cavity as well as the multi-mode microwave oven. Some results of pressureless direct joining of RBSC were presented earlier [10]. The joints were sectioned and polished to further examine the interfacial region. Figure 1 shows an example of the micrograph of Coors RBSC specimens joined in a single mode cavity.

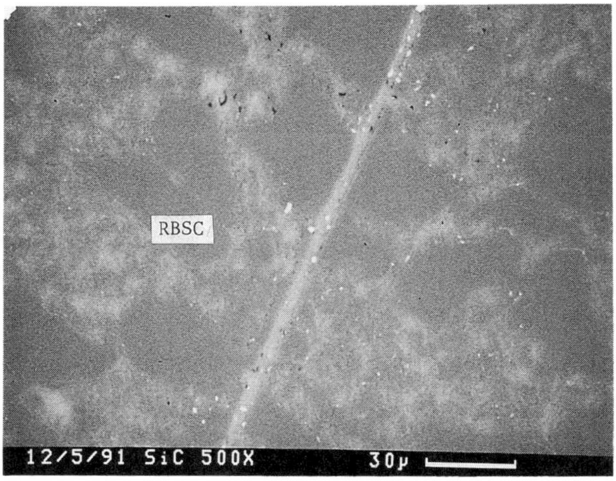

Figure 1. SEM of Coors RBSC-RBSC joint.

The interfacial region is identical to the bulk of the specimens. A silicon rich band of about 5 micron width is observed at the interfacial region. Since similar patches of silicon are seen throughout the RBSC material, the presence of this band is not expected to negatively impact the mechanical properties of the joint. This argument is strongly supported by the essentially identical appearance of micrographs of sectioned joints that had greater strength than as received RBSC [Yiin et al. Ref. 5].

Short lengths (2.5 cm) of small scale (5 cm diameter) RBSC tubes intended for radiant burner tube applications were supplied by Coors. These tube sections were joined directly without any polishing, using the multi-mode oven with hybrid heating. The bleeding of silicon filled the void space making the joint leak tight. The joined tube was installed on a vacuum line which held the vacuum (12×10^{-3} torr) as well as a single tube. The tube was cut into test bend bars (by Coors) for mechanical testing. The poor surface match of the rough cut specimens resulted in flexural strengths ranging from 1/3 to 1/2 of the actual value (about 300 MPa) for the Coors RBSC. Similar joining experiments have been performed with surfaces polished (by LANL) to a 5 micron finish. Figure 2 is a picture of these joined RBSC tubes. Joining is apparent all around the circumference with substantial bleeding of silicon. Mechanical property measurements of this joint are in progress.

Figure 2. Photograph of Coors RBSC tube sections joined in a multi-mode oven.

Joining of RBSC to Hexoloy™ (sintered SiC) specimens was carried out in a single mode cavity. The joined specimens were sectioned and polished for electron microscopy. Figure 3 shows the micrograph of a joint between RBSC and Hexoloy™. The joint is not visible in the interior of the sample (region A) but is identified close to the edge of the sample (region B) because of the poor matching of the two surfaces. A crack in the RBSC (region C) where the joint is good indicates that the joint is stronger than RBSC. Figure 4 shows a slightly higher magnification of region B close to the edge. Although the micrograph exhibits substantial cracking at the interfacial region, good wetting and bonding of silicon is apparent to both RBSC and Hexoloy™. The crack in Hexoloy™ at the top of the micrograph and not at the interface indicates that the bonding is probably as strong as the sintered SiC material.

Figure 3. SEM micrograph of RBSC to SiC joining. Note a crack in RBSC where joining appears to be good.

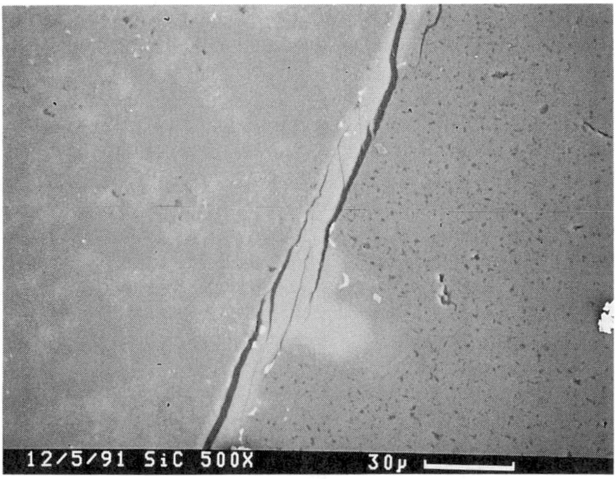

Figure 4. RBSC to SiC joint showing crack in sintered SiC.

CONCLUSIONS

Joining of RBSC to itself and sintered SiC has been demonstrated to be feasible by the use of microwave energy. External pressure may not be necessary but the surfaces to be joined should be flat and smooth. RBSC may be used as the interlayer material for the joining of sintered SiC. For industrial applications mechanical design and joint geometry will be critical factor in obtaining good mechanical properties of the joints.

ACKNOWLEDGEMENTS

The authors are grateful to Laing Ceramics Corporation for providing RBSC faucet disks and to Coors Ceramics for furnishing RBSC tubes and performing mechanical testing on joined tubes.

This work was sponsored in part by the U.S Department of Energy Advanced Industrial Concepts (AIC) Materials Program, through Subcontract 9-XG1-U0554 with LANL. Work at GMU was performed under contract to TA&T, with additional support provided by the Virginia Center for Innovative Technology.

REFERENCES

1 C.H. Bates, M.R. Foley, G.A. Rossi, G.J. Sundberg and F.J. Wu, "Joining of Non-oxide Ceramics for High-Temperature Applications," Ceramic Bulletin, 69, (3), 350 (1990).

2 D. Palaith and R. Silberglitt, "Microwave Joining of Ceramics," Ceramic Bulletin, 68, (9), 1601 (1989).

3 H. Fukushima, T. Yamanaka and M. Matsui, "Microwave Heating of Ceramics and its Application to Joining," J. Mater. Res., 6, (2) 397 (1990).

4 T. Iseki, T. Kameda and T. Maruyama, Interfacial Reactions between SiC and Aluminum during Joining, Journal of Materials Science, 19, 1692 (1984).

5 T-Y. Yiin, V.V. Varadan, V.K. Varadan and J.C. Conway, Microwave Joining of Si-SiC, Ceramic Transactions, Microwaves: Theory and Application in Materials Processing, 21, 507 (1991).

6 I. Ahmad, "Effect of Microwave Heating on the Solid State Reactions and Mass Transport in Ceramics," Ph.D. dissertation, University of Florida, (1991).

7 R. Silberglitt, D. Palaith, W.M. Black, H.S. Sa'adaldin, J.D. Katz and R.D. Blake, "Investigation of Interlayer Materials for the Microwave Joining of SiC," Ceramic Transactions, Microwaves: Theory and Application in Materials Processing, 21, 487 (1991).

8 A. De, I. Ahmad, E.D. Whitney and D.E. Clark, "Microwave (Hybrid) Heating of Alumina at 2.45 GHz: I. Microstructural Uniformity and Homogeneity," Ceramic Transactions, Microwaves: Theory and Application in Materials Processing, 21, 319 (1991).

9 A. De, I. Ahmad, E.D. Whitney and D.E. Clark, "Microwave (Hybrid) Heating of Alumina at 2.45 GHz: II. Effect of Processing Variables, Heating Rates and Particle Size," Ceramic Transactions, Microwaves: Theory and Application in Materials Processing, 21, 329 (1991).

10 I. Ahmad, W.M. Black and R. Silberglitt, "Direct Microwave Joining of Reaction Bonded Silicon Carbide," Presented at the 16th Annual Conference on Composites and Advanced Ceramics, Jan. 1992, Cocoa Beach, Florida, Ceramic Engineering and Science Proceedings (in press).

NITRIDATION OF NON-ISOTHERMAL SILICON COMPACTS

JEFFREY J. THOMAS, RENEE R. JESSE, D. LYNN JOHNSON, AND HAMLIN M. JENNINGS
Depts. of Materials Science and Engineering and Civil Engineering
Northwestern University, Evanston, IL 60208

ABSTRACT

The use of microwave heating for the processing of reaction-bonded silicon nitride offers at least two advantages over conventional heating methods. First, the reaction can be made to proceed preferentially in the interior of the reacting compact by maximizing the temperature gradient resulting from microwave heating. This helps maintain the flow of nitrogen into the compact during the later stages of the reaction, when the overall porosity is reduced. Second, the reaction rate can be controlled by changing the power, which gives better control over the processing than can be obtained using conventional heating.

INTRODUCTION

Reaction-bonded silicon nitride (RBSN) has better high temperature properties than other types of silicon nitride because it is made without the addition of oxide sintering aids that form a glassy grain boundary phase. RBSN has lower room temperature strength than other types of silicon nitride because of its lower density. However, because it maintains its mechanical properties at temperatures above 1400 C, it is an attractive candidate for use in high-temperature ceramic engines.

RBSN is formed by pressing silicon powder into a billet which is then heated to reaction temperatures (1150-1450 C) in an atmospheric nitrogen source. Silicon nitride forms as a mat of fine-grained needles which grows into the porosity of the compact[1]. If the starting porosity is too low, nitrogen will eventually be unable to diffuse into the interior of the specimen, and the reaction will not go to completion[2]. For this reason, the final product must always have significant porosity.

There has not been extensive research involving microwave processing of Si_3N_4 to date. Tiegs et al.[3] did some microwave RBSN experiments involving insulated (near isothermal) silicon compacts, and found that the nitridation reaction appeared to be enhanced by the microwaves. Other studies have involved sintering of Si_3N_4 powder and annealing rather than RBSN.[4],[5]

Microwave heating may offer significant advantages for the processing of RBSN. Because microwave energy heats the volume of the material, and because the surface radiates energy, a temperature gradient can be maintained in specimens heated with microwaves. Thus, a silicon compact will tend to react faster in the interior than near the surface. This effect is especially significant if the heat loss from the surface is maximized by not using insulation around the specimen. As a result, nitrogen will diffuse into the compact during the later stages of the reaction more easily than with isothermal heating, because the porosity at the surface remains open longer. This means that compacts with a lower initial porosity could be more fully reacted than is currently possible using conventional heating, thus increasing the maximum strength and toughness of the final product.

Another possible advantage of microwave processing of RBSN is the ability to quickly and accurately change the specimen temperature by changing the input power. This unique aspect of microwave processing, which results from the direct heating by absorption of microwave energy by the material, allows the reaction rate to be controlled by utilizing the weight gain of the specimen to control the microwave power.

This paper discusses experiments that show that microwave heating does, indeed, have the above advantages over conventional heating. Also, initial results of fracture toughness comparisons between microwave heated and conventionally heated specimens are reported.

EXPERIMENTAL

The silicon powder used (HQ/10 grade, Elkem Metals Co, Pittsburgh, PA) was 99.6% pure and had particles $< 10\ \mu m$. Silicon compacts were formed by pressing in hardened steel dies. For the processing experiments, a 19 mm diameter cylindrical die was used, and specimens were typically 7 mm thick. For the fracture toughness measurements, a 25 mm diameter cylindrical die was used to make 2.5 mm thick disks, and a rectangular die was used to make bars with dimensions 3x4x50 mm. In all cases the densities were 52-54% of theoretical. In all experiments the reactant gas was a 90% nitrogen/10% hydrogen mixture. A hot copper getter and molecular sieves were used to remove oxygen and water vapor before the gas entered the reaction zone.

Microwave processing was done using a 2.45 GHz microwave generator connected to a tunable cylindrical cavity (Wavemat Inc, Plymouth, MI). In order to maximize radiative heat loss, the specimens were put into a basket of alumina fiber insulation which was cut so that much of the outside edge of the specimen was exposed (see Fig. 1). The specimen was sandwiched between silicon nitride powder compacts of the same diameter (SN-9FW, Denka, Tokyo, Japan) to prevent reaction with the insulation. The basket was suspended inside the cavity from an electronic analytical balance as shown in Fig. 2. An optical pyrometer (Model 100, Accufiber, Beaverton, OR) was focussed onto the sample to monitor the surface temperature.

Conventionally nitrided samples were made in an alumina muffle tube furnace with a SiC heating element. These samples provided isothermally nitrided controls for comparison with the microwave heated samples.

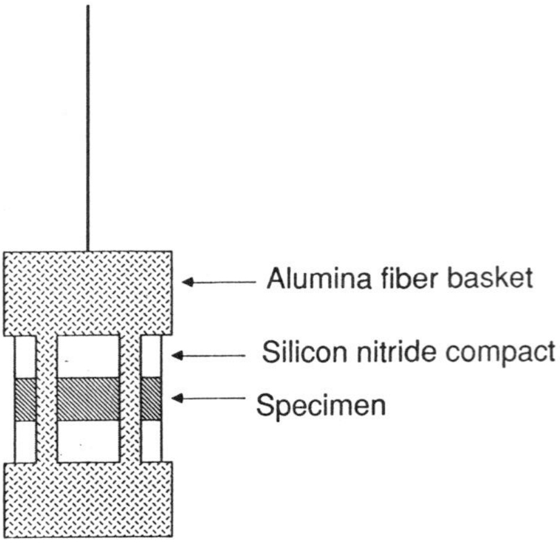

Figure 1. Diagram of cylindrical basket used to hold the silicon compacts during microwave nitriding. Most of the outer edge of the specimen was exposed in order to maximize radiative heat loss and create a large radial temperature gradient.

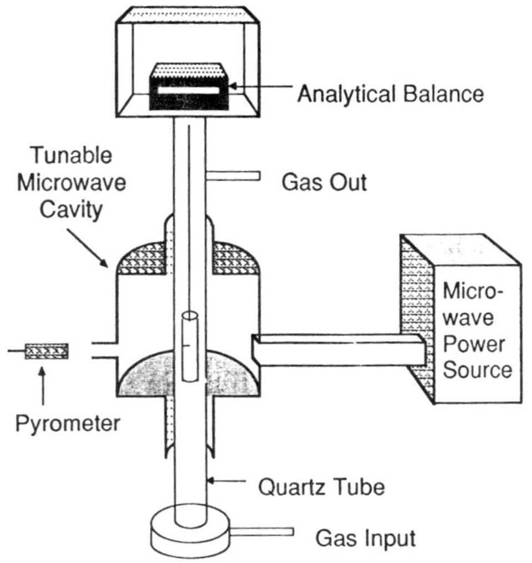

Figure 2. Schematic of the microwave processing apparatus used to nitride silicon compacts. The alumina fiber basket which holds the specimen was suspended from the balance in order to measure the weight gain.

To see if the temperature gradient was causing a composition profile, the microhardness was measured across the diameter of polished sections of partially reacted samples using the Vickers indent method.

Mechanical testing was done using specimens with identical green states reacted in both the microwave apparatus and the tube furnace. After nitriding these specimens had densities in the range of 61-63% of theoretical, as measured by the Archimedes method. A commercial RBSN (Cercom, Inc., Vista, CA) which contained no sintering aids was also purchased to use as a baseline comparison. This material, which was 68% dense, was received in rectangular slabs and was machined into 3x4x50 mm bars.

The indentation strength-in-bending was measured for these specimens. After making an indent using a 29.4 N load, the bars were tested in four-point bending and the disks were tested using the ball-on-ring method.[6]

RESULTS AND DISCUSSION

Controlling the processing by maintaining a desired reaction rate rather than a constant temperature has two advantages. First, the problem of melting during the initial stages, which is caused by the reaction exotherm, can be avoided by keeping the initial reaction slow. This is particularly important for silicon powder with a high surface area, which reacts very quickly.[7] Second, the nitriding rate can be prevented from dropping off in the later

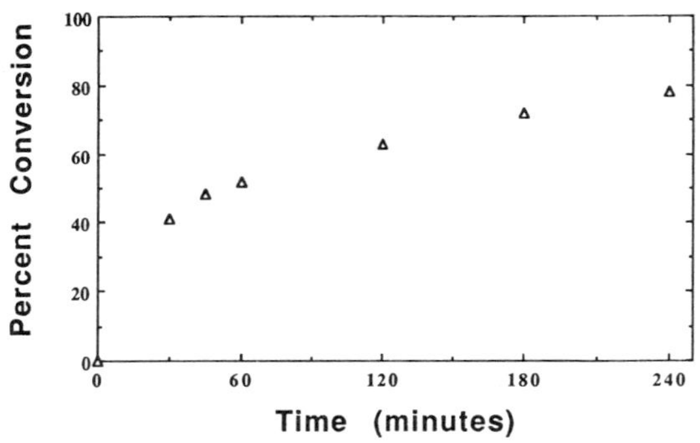

Figure 3. Isothermal reaction kinetics for silicon compacts nitrided in a conventional furnace.

stages of the reaction by increasing the temperature as needed, which results in a shorter overall processing time.

In this study, the overall reaction rate of the specimen was known *in situ* from the weight gain. This rate was controlled by a closed-loop computer interface which adjusted the microwave power to keep the reaction proceeding along a predetermined program. The microwave-reacted samples used for this study were nitrided using a constant weight gain rate of 10-15% of the total per hour. This linear profile is quite different from the continuously decreasing reaction kinetics which occur with constant temperature processing.[8],[9] Figure 3 shows the reaction kinetics for specimens nitrided in the tube furnace at 1350 C. In order to maintain linear kinetics in the microwave system, the computer feedback system gradually increased the input power over time.

In order to test the hypothesis of a temperature gradient causing preferential reaction in the interior of silicon compacts, 19 mm diameter disks were reacted in the microwave system to 50-80% of completion. A relative measure of the degree of reaction was obtained using hardness testing, since the hardness increases as the density increases. In all cases, microwave processed specimens were significantly harder in the center than near the edges, which conclusively shows that the interior was more fully reacted than the outside. The hardness profile for a specimen reacted to 65% of completion is shown in Fig. 4(a). This disk had a hardness of over 3 GPa at the center, which is near the maximum value of 3.3 GPa for fully reacted disks.

To make a comparison with the isothermal case, identical disks were reacted using the tube furnace. Figure 4(b) shows the hardness profile for a disk reacted to 65%: the reaction has occurred preferentially near the surface. This composition profile, which is typical for RBSN processing, is an indication that the availability of nitrogen in the interior of the compact was rate-controlling.

The fracture toughness of the disks and bars described earlier was calculated from the indentation strength-in-bending using the method of Chantikul et. al.,[10] given as:

$$K_{IC} = 0.59\left(\frac{E}{H}\right)^{1/8}(\sigma P^{1/3})^{3/4}$$

where E is the elastic modulus, H is the hardness from mean contact pressure, σ is the indentation bend strength, and P is the indentation load.

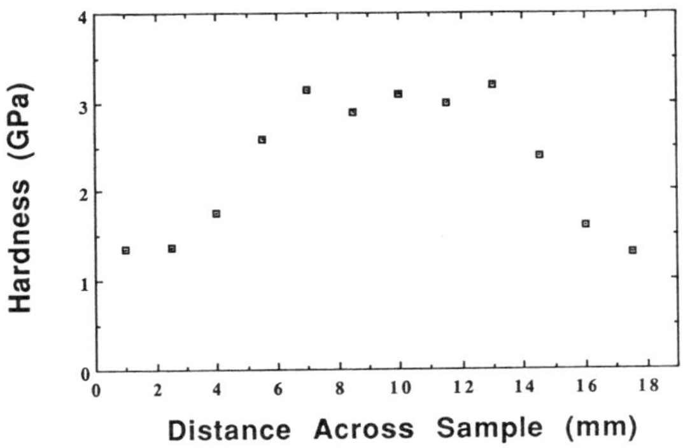

Figure 4(a). Hardness profile for a silicon compact nitrided to 65% conversion using microwaves.

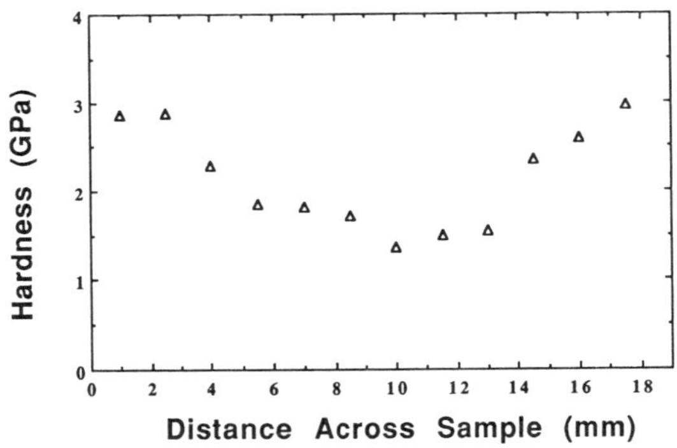

Figure 4(b). Hardness profile for a silicon compact nitrided to 65% conversion using conventional (isothermal) heating.

TABLE I. Fracture Toughness Results

Material	% Theoretical Density	KIC (MPa√m)	# of Specimens
Commercial RBSN	68	2.09 ± 0.12	31
Conventional RBSN	61-63	1.68 ± 0.13	7
M-RBSN	61-63	1.90 ± 0.12	3

The fracture toughness results are shown in Table 1. The toughness value for the commercial RBSN was the largest, as was expected from its higher density. The toughness of the microwave heated material is higher than that of the conventionally heated material with the same density. This seems to indicate that microwave processing improves the fracture toughness. More testing needs to be performed, as this is only a preliminary evaluation.

SUMMARY

Microwave processing appears to be superior to conventional processing for RBSN. The ability to change the reaction rate quickly by changing the input power allows for more controlled and efficient processing. The specimens in this study were nitrided using a constant weight gain.

The temperature gradient which exists in microwave-heated compacts causes the nitridation reaction to proceed more quickly near the center of the specimen, leaving the surface porosity open. This composition gradient, which is opposite to the composition gradient that forms during conventional processing, should provide a significant advantage for the nitridation of silicon compacts with a low starting porosity. The results of this work show that significant composition gradients can be created in compacts with moderate starting porosity.

Initial results indicate that the fracture toughness of fully reacted RBSN may be enhanced by the use of microwaves, compared with conventionally processed specimens.

ACKNOWLEDGEMENTS

This work is supported by the National Institute of Standards and Technology under Contract SC91209.

REFERENCES

1. H. M. Jennings, B. J. Dalgleish, and P. L. Pratt, "Reactions Between Silicon and Nitrogen, Part 2: Microstructure," *J. Mat. Sci*, **23**, 2573-2583 (1988).

2. I. Amato, D. Martorana, and M. Rossi, "The Nitriding of Silicon Powder Compacts," *Powder Met.*, **18** [36], 339-48.

3. T. N. Tiegs, J. O. Kiggans, Jr, and H. D. Kimrey, Jr, "Microwave Processing of Silicon Nitride," Proceedings of MRS Symposium "Microwave Processing of Materials", (1990).

4. T. N. Tiegs, J. O. Kiggans, H. D. Kimrey, "Microwave sintering of Silicon Nitride," *Ceram. Eng. Soc. Proc.*, **12** [9-10], 1981-92 (1991).

5. M. K. Ferber, T. N. Tiegs, and M. G. Jenkins, "Effect of Post-Sintering Treatment on the Mechanical Performance of Silicon Nitride," *Ceram. Eng. Soc. Proc.*, **12** [9-10], 1993-2004 (1991).

6. G. de With and H. H. M. Wagemans, "Ball-On-Ring Test Revisited," *J. Am. Ceram. Soc.*, **72** [8], 1538-1541 (1989).

7. O. J. Gregory, S. B. Lee, and R. C. Flagan, "Reaction Sintering of Sub-micrometer Silicon Powder," *J. Amer. Ceram. Soc.*, **70**, C-52-C-54 (1987).

8. A. Atkinson, A. J. Moulson, and E. W. Roberts, "Nitridation of High Purity Silicon," *J. Am. Ceram. Soc.*, **59** [7-8] 285-289 (1976).

9. H. M. Jennings and M. H. Richman, "Structure, Formation Mechanisms and Kinetics of Reaction Bonded Silicon Nitride," *J. Mat. Sci.*, **11**, 2087-98 (1976).

10. P. Chantikul, G. R. Anstis, B. R. Lawn, and D. B. Marshall, "A Critical Evaluation of Indentation Techniques for Measuring Fracture Toughness: II, Strength Method," *J. Amer. Ceram. Soc.*, **64** [9], 539-543 (1981).

CHARACTERIZATION OF SINTERED REACTION-BONDED SILICON NITRIDE PROCESSED BY MICROWAVE HEATING

J. O. KIGGANS, JR., AND T. N. TIEGS
Oak Ridge National Laboratory, P.O. Box 2008, Oak Ridge, Tennessee 37831-6087

ABSTRACT

Sintered reaction-bonded silicon nitride (SRBSN) tiles were fabricated using microwave and conventional heating. Materials from both processes were analyzed at various stages in their fabrication. Microwave processing resulted in a SRBSN material of higher density and strength than the conventionally processed material.

INTRODUCTION

Silicon nitride-based ceramics are being considered for numerous applications in advanced heat engines. Sintered reaction-bonded silicon nitride (SRBSN) is an attractive candidate for the material of choice for these parts [1,2]. Normally, SRBSN is produced in a two-step process. The first step, referred to as nitridation, is the reaction of a mixture of silicon, α-Si_3N_4, and sintering aid powders with N_2 containing gas at between 1200 to 1400°C. The second step is the densification of this reaction-bonded silicon nitride (RBSN) at temperatures from 1750 to 1800°C to effect liquid-phase sintering.

Traditionally, the production of RBSN and SRBSN has been accomplished using resistance-heated furnaces [1,2]. In recent years, microwave heating has been introduced as an alternative process for producing RBSN materials [3,4,5]. This work is an extension of previous work using microwave processing to produce SRBSN materials [6].

EXPERIMENTAL PROCEDURES

The starting materials for the SRBSN consisted of silicon[a], α–Si_3N_4[b], Al_2O_3[c],and La_2O_3[d]. Additionally, 1.3 wt.% PVP K-15[e] and 1 wt.% Darvan 821A[f] were added as dispersants. The final composition after nitridation of the silicon was calculated to be Si_3N_4 - 11.5 wt. % La_2O_3 - 3 wt. % Al_2O_3. However, this composition is considered nominal because of several uncertainties in the processing, such as (1) hydration of the La_2O_3 during storage, (2) oxidation of the Si and Si_3N_4 during milling, and, (3) volatilization of Si during heating to elevated temperatures. Silicon and α–Si_3N_4 were mixed to a solids loading of 35% in water containing the dispersants and aged 3 days prior to further processing. Al_2O_3 and La_2O_3 were added to the slurry which was then milled for several hours in a turbomill using high density ZrO_2 milling media, and finally sieved through a 325-mesh screen. This material was rolled on a ball mill for several days with frequent stops to allow H_2 gas to be vented from the bottle. The slurry was then degassed under vacuum and slip cast into 100 × 100 × 6.3 mm plaster-of-paris molds. The resulting green tiles were slow dried over a period of 1 week by passage through controlled humidity chambers and a final drying in a forced-air furnace. The dried green tiles

---

[a]Elkem Metals Co., Buffalo, NY; Grade Si-HQ.

[b]Starck, Berlin, Germany; Grade LC-10N Si_3N_4.

[c]Reynolds, Malakoff, Tx.; Grade RC-HP DBM.

[d]Molycorp, White Plains, NY; Grade 5205, >99.9% La_2O_3.

[e]GAF Chemicals Corporation, Wayne, N. J.

[f]R. T. Vanderbilt Company, Inc., Norwalk, C. T.

were packed in Si_3N_4 powder in a graphite crucible and heated to 1200°C in flowing argon in a graphite resistance-heated furnace. These tiles, referred to as heat-treated tiles, were green-machined to produce flat plates.

Microwave processing of the heat-treated tiles was conducted in a 500 L cylindrical microwave furnace with a 2.45 GHz microwave generator. Two tiles were processed in each furnace run. The tiles were packed in Si_3N_4 powder containing 4% Y_2O_3 and 4% SiC, inside of a 150 × 150 × 125 mm alumina fiberboard box. A molybdenum sheathed thermocouple with a boron nitride sleeve was inserted between the tiles to measure the sample temperature. Prior to microwave processing, the furnace was purged and backfilled with N_2- 4 vol. % H_2 to a final pressure of ~0.1 MPa (16 psia). Runs were made with the pressure maintained by an automatic pressure-controlled make-up valve that resulted in the input of fresh N_2 gas. Furnace runs were terminated after a 0.5-h hold at 1480°C to recover RBSN samples or continued to 1750 or 1800°C with a 1-h hold for SRBSN samples. Figure 1 shows the temperature and microwave forward power of a typical nitridation and sintering cycle in the microwave furnace. In this run, the tiles were heated to a final temperature of 1750°C for 1 h. A gradual drop in microwave power occurred at approximately 1200°C during the run, which was the start of the silicon nitridation. At approximately 1300°C, the microwave power rose sharply, indicating that most of the silicon had been converted to Si_3N_4. A second drop in power occurred at approximately 1400°C. This dielectric relaxation corresponds to the formation of the intergranular glass phase during the sintering of the SRBSN. An entire heating cycle to 1750°C required approximately 25 h.

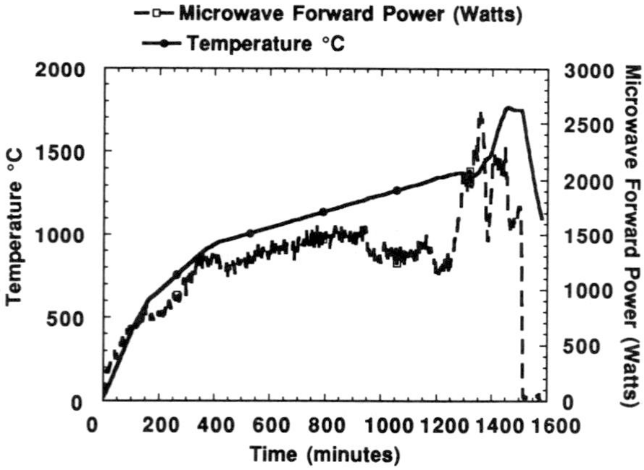

Fig. 1. Time-temperature-power curves for microwave processing of SRBSN tiles.

The conventional processing was carried out in a graphite-resistance heated furnace in a two-stage process. In stage one, the heat-treated tiles were placed on a bed of Si_3N_4 powder in a graphite crucible. The furnace was purged with N_2-4 vol. % H_2 and gas flow was continuous throughout the run. Stage one was terminated after a 0.5-h hold at 1480°C. Some of the samples were removed for analysis and others were then completely covered with the Si_3N_4 packing powder inside the graphite crucible and heated in N_2 gas at 5°C/min. to the 1750 or 1800°C end point. Note that, in this conventional process, a two step process was necessary since tiles which were completely packed in powder in stage one experienced partial melting during this exothermic nitridation reaction stage.

Tiles from conventional and microwave processing at 1480 and 1750°C were measured and weighed. The densities were determined by the Archimedes method. Tiles from the 1480°C treatment were sliced for visual inspection of the interior of the tiles and to obtain samples for X-ray diffraction and SEM analysis. Tiles from the 1750 and 1800°C runs were machined into bend bar specimens with nominal dimensions of 3 mm x 4 mm x 40 mm. The fracture surfaces of some of the bars were analyzed with the SEM. Also, bend bars were cut to sizes suitable for X-ray diffraction analysis.

RESULTS

The densities of the green slip cast tiles ranged from 1.40 to 1.55 g/cm3. Following the heat treatment of 1200°C in argon, the densities ranged from 1.50 to 1.55 g/cm3. Tiles treated to 1480°C by conventional and microwave heating had a 59-60 % weight gain with respect to the initial weight of silicon in the sample. This represented essentially complete nitridation of the tiles. The theoretical weight gain for converting Si to Si_3N_4 is 66.6%. However, due to the oxidation of the Si during milling and Si volatilization during nitridation, the entire gain is not realized [7]. This data was confirmed by X-ray diffraction analysis which showed no residual silicon. Tiles processed by either heating method showed negligible size changes during nitridation.

Densities of the SRBSN materials processed by both methods are summarized in Fig. 2. Tiles sintered by conventional heating at 1750°C had an average density of 3.03 g/cm3 or 91.7% T.D. (3.35 g/cm3) and at 1800°C, a density of 3.23 or 96.6 % T.D. The tiles sintered to 1750°C by microwave heating reached a density of 3.34 or 99.7%. Microwave heating to 1800°C produced tiles with 100% T.D. Microwave sintering of the SRBSN was more effective than conventional sintering. This has also been shown to be true for previous comparisons of microwave and conventional sintering with Si_3N_4 powder compacts [6]. The SRBSN tiles sintered by microwave heating showed a linear shrinkage of approximately 10%.

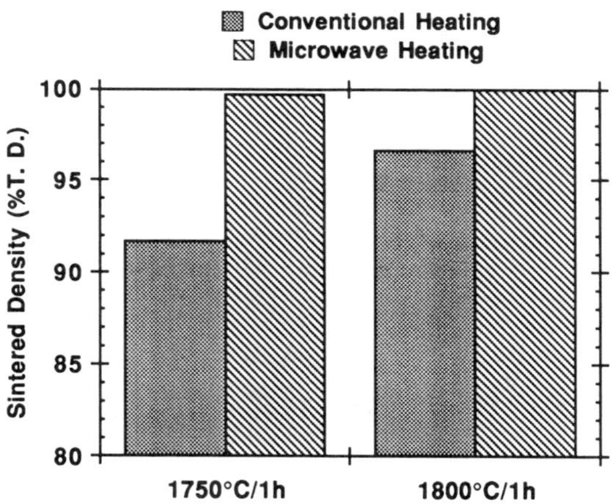

Fig. 2. Densities of SRBSN Tiles Sintered by Conventional and Microwave Heating to 1750°C/1h and 1800°C/1h.

X-ray diffraction analysis was performed on samples processed by conventional and microwave heating to temperatures of 1480, 1750, and 1800°C. As shown in Fig. 3a, the tiles processed to 1480°C by conventional and microwave heating both showed the major α-Si_3N_4 (102) and (210) peaks as well as the β-Si_3N_4 (101) and (210) peaks. Following sintering (Fig. 3b), only the major β(101) and β(210) were present. These findings were expected, since sintering aids were present in the material to cause the formation of β-Si_3N_4.

The flexural strength of bend bars from the conventional 1750°C sintered tile was quite low at 250 MPa. This was expected due to the low 91% density. The flexural strength of the bend bars from the tile processed by microwave heating to 1800°C was ~480 MPa. This strength is higher than for SRBSN materials processed at temperatures ≤1800°C by conventional heating [8]. An SEM examination of a fracture surface of one of the latter samples (Fig.4a) clearly showed the fracture origin to be an inclusion. Inclusions likely entered the materials during the slip-aging process when small dried flakes of the same composition are reintroduced into the slurry. Flaws of this type are believed to be easily eliminated with more careful handling of the slurries. Another SEM photograph (Fig. 4b) of the 1800°C microwave-processed sample shows a predominantly equiaxed grain morphology that was seen in both the microwave and conventional SRBSN materials.

Additional flexural strength testing was performed on bend bars from the 1800°C microwave sintered tiles (Fig. 5) as a function of temperature in an oxidizing environment. The flexural strength was relatively constant up to 800°C at which point there was a rapid decrease in strength with increasing temperature. The large standard deviation is due to the processing flaws present in these samples. The decrease at temperatures ≥ 800°C is caused by the nature of the intergranular phases when La_2O_3 and Al_2O_3 are used as sintering aids. Similar trends have been observed in sintered samples fabricated with high purity α-Si_3N_4 powders using the La_2O_3 and Al_2O_3 sintering aids [9].

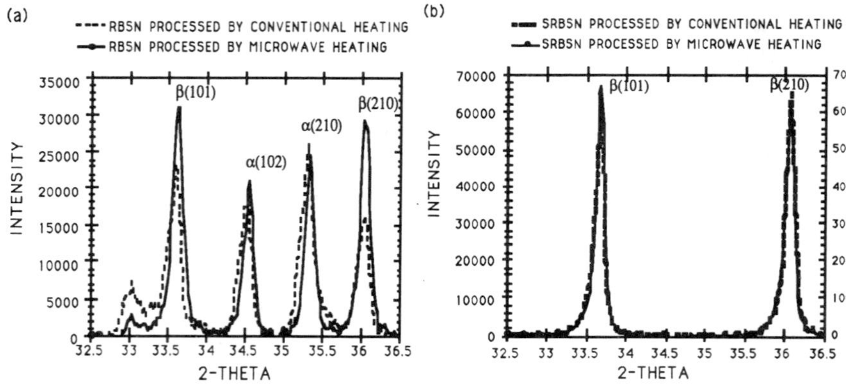

Fig. 3. X-ray diffraction patterns showing selected α– and β–Si_3N_4 peaks of (a) RBSN material processed to 1480°C/0.5 h by conventional heating or microwave heating, and (b) SRBSN materials processed to 1750°C/1h by conventional heating or microwave heating.

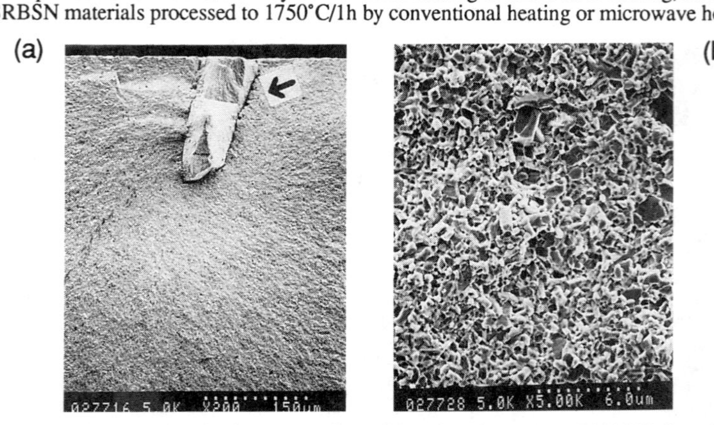

Fig. 4. SEM photographs of a fracture surface of SRBSN sintered to 1800°C/1h by microwave heating showing (a) a 200X magnification of an area with a processing flaw (see arrow) and (b) a 5000X magnification of a typical field of view.

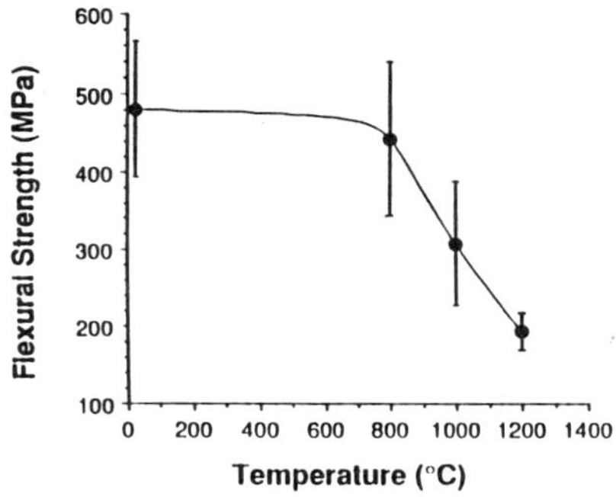

Fig. 5. Plot of flexural strength versus temperature in air for a SRBSN sample sintered to 1800°C/1h by microwave heating.

CONCLUSIONS

There were several important observations made in this comparison of SRBSN production by microwave and conventional processing:

1. SRBSN tiles could be produced in a single microwave heating cycle, whereas two separate runs were required to produce SRBSN tiles by the conventional heating process.

2. Higher densities were attained in SRBSN tiles produced by microwave heating than by the conventional process.

3. Flexural strength of the microwave sintered SRBSN was higher than for the conventional material. However, processing flaws reduced the ultimate strengths of both sets of tiles.

REFERENCES

1. J. A. Mangels and G. J. Tennenhouse, "Densification of Reaction-Bonded Silicon Nitride," Am. Ceram. Soc. Bull., 59 [12] 1216-1222 (1980).
2. R. M. Williams and A. Ezis, "Slip Casting of Silicon Shapes and Their Nitriding," Am. Ceram. Soc. Bull., 62 [5] 607-619 (1983).
3. T. N. Tiegs, J. O. Kiggans, Jr., and H. D. Kimrey, Jr.," Microwave Processing of Silicon Nitride," Mat. Res. Soc. Symp. Proc., 189, 267-272 (1991).
4. J. O. Kiggans, C. R. Hubbard, R. R. Steele, H. D. Kimrey, C. E. Holcombe, and T. N. Tiegs, "Characterization of Silicon Nitride Synthesized By Microwave Heating," in Ceramic Transactions, Microwaves, Theory, and Applications in Materials Processing, edited by D. E. Clark, F. D. Gac, and W. H. Sutton (Am. Cer. Soc., Westerville, Ohio) 267-272 (1991).
5. C. E. Holcombe, Oak Ridge National Laboratory, Oak Ridge, Tenn., unpublished results (1988).
6. T. N. Tiegs and J. O. Kiggans, "Fabrication of Silicon Nitride Ceramics by Microwave Heating," Proc. 4th International Symp. Ceram. Mater. & Compon. for Engines, Swedish Ceram. (1991).
7. A. Ezis, "The Fabrication and Properties of Slip-Cast Silicon Nitride," in Ceramics for High Performance Applications, edited by J. J. Burke, A. E. Gorum, and R. N. Katz (Brook Hill Publishing Co.) 207-222 (1974).
8. J. A. Mangels, "Sintered Reaction-Bonded Silicon Nitride," Proc. of the 5th Annual Conference on Composites and Advanced Ceramic Materials, 589-603 (1981).
9. T. N. Tiegs, Oak Ridge National Laboratory, Oak Ridge, Tenn., unpublished data (1991).

ACKNOWLEDGEMENTS

Research sponsored by the U.S. Department of Energy, Assistant Secretary for Conservation and Renewable Energy, Office of Transportation Systems, as part of the Ceramic Technology Project of the Materials Development Program, under contract DE-AC05-84OR21400 with Martin Marietta Energy Systems, Inc. The authors would also like to thank K. Ploetz, O. B. Cavin, and D. W. Coffey for their valuable contributions in the mechanical testing, the X-ray analysis, and the SEM analysis, respectively, and to Billie Russell for the preparation of the manuscript.

BATCH PROCESS FOR MICROWAVE SINTERING OF Si_3N_4

M.C.L. PATTERSON, P.S.*APTE, R.M. KIMBER and R. ROY
Alcan International Limited, Kingston Research and Development Centre, P.O. Box 8400, Kingston, Ontario, Canada K7L 5L9
* Present Address : Sherritt Gordon Ltd, Fort Saskatchewan, Alberta.

ABSTRACT

A method for sintering silicon nitride using microwave energy at 2.45 GHz is described. Sintering takes place in air, in times of between 30 and 120 minutes and has been scaled up to give isothermal conditions over a batch size of 150 mm diameter by 200 mm in height and weighs approximately 1.0 kg. Additions of 5% alumina and 5% yttria result in a sintered product density of approximately 97% of theoretical, with a density variation better than ±0.5% throughout the batch.

It has been estimated that a load of 7.0 kg can be conventionally sintered using a 12 hour cycle and an energy consumption of 19.7 kWh/kg. In contrast with microwave energy, a batch of 540 g can be sintered in 120 minutes with an energy consumption of approximately 3.1 kWh/kg. This results in a possible energy savings of up to 78% for microwave heating.

INTRODUCTION

Silicon nitride has been identified as an important material for both structural and high temperature applications [1]. Due to its covalent bonding and its lack of solid state mobility at high temperatures it is difficult to sinter without sintering aids [2]. The type and amount of sintering aids used, define not only the temperature and method by which sintering can take place [3], but also the final material properties. Conventionally, Si_3N_4 is sintered in a powder bed of Si_3N_4, in carbon resistance furnaces at a temperature in excess of 1800°C. This temperature is typically above the degradation temperature for Si_3N_4 and sintering must be carried out under an increased nitrogen pressure in order to suppress this reaction. By using a powder bed technique [4], Si_3N_4 with additives described in detail elsewhere [5], has been sintered using microwave energy at 2.45 GHz in an atmosphere of air, without detrimental effects from oxidation.

This process was developed for sintering Si_3N_4 tool-bits in an effort to demonstrate the feasibility of using microwave energy for a commercial sintering operation. Silicon nitride tool-bits were chosen as a product on which to focus for two main reasons;

i) The powder bed technique is particularly suited to the sintering of silicon nitride.
ii) Tool-bits are one of the few high performance ceramic products in use today and information concerning quality and performance of the product can be obtained in a relatively short period of time.

One of the inherent problems which has frequently been encountered when microwave sintering ceramics is that of "hot-spots" caused by the non-uniform deposition of energy within the ceramic which results in thermal runaway [6]. In theory, the time averaged field strength from a travelling wave is uniform, and it should be possible therefore to uniformly heat an infinitely small component. In practice, however, this is not the case and what is meant by "ceramics can be heated uniformly", should be compared with the way in which they are conventionally heated. With conventional sintering, the deposition of energy occurs strictly on the surface of the ceramic component and the process relies on thermal conduction to transport heat throughout the component. Due to the low dielectric loss factor which many ceramics of interest exhibit, microwave energy can penetrate several 10's of cms into the ceramic and therefore appears to heat "uniformly" throughout the entire volume.

Because of the low dielectric factor, however, only a small amount of energy is absorbed according to the equation relating field and properties to the power density;

$$P_d = (2\pi f \varepsilon')(E^2/2)\tan\delta$$

where -

P_d is the power deposited (watts/cm^3)
f is the frequency (Hz)
E is the electric field strength (V/cm)
ε' is the dielectric constant
$\tan\delta$ is the loss tangent

In order to improve the efficiency of energy absorption, it is necessary to place the dielectric material in a cavity where multiple reflections occur. It is this necessity which typically causes the onset of hot-spots, which are brought about by constructive interference between the reflected and forward waves of the electromagnetic field, which is anything but uniform. Compounding this problem of hot-spots are three other factors:

i) The dielectric loss factor for many of these ceramic materials increases with temperature and once a hot-spot is initiated it will tend to absorb energy preferentially.
ii) The thermal conduction properties of many ceramics are poor so hot-spots which do form do not dissipate quickly.
iii) A third factor which appears less significant, but which can still lead to non-uniformity is a change in the relative bulk density of the material being processed. The dielectric loss factor ε'' is a material characteristic, but it is the effective ε'' of a given volume which determines the amount of heat generated in that volume [7,8]. During sintering the powder compact can increase in density from approximately 50% to 100% and is accompanied by an increase in the effective ε'' throughout that volume. If this transition does not occur uniformly throughout the component, there is an additional mechanism for preferrential absorption of energy and therefore non-uniform processing.

Many methods have been investigated to minimise the effects of non-uniform heating. The spatial occurrence of the regions of constructive interference depends upon the half wavelength distance. At 2.45 GHz this approximately 6 cm in air. There is an advantage in going to higher frequencies as demonstrated at 28 GHz [9] and at 60 GHz [10] since these regions of constructive interference occur closer together and the distances for thermal conduction are therefore much shorter. In addition, at these higher frequencies, the dielectric loss of many ceramics is higher [9] making the ceramics easier to heat. However, the cost of energy in the microwave region increases with frequency and is unfortunately thought to be prohibitively expensive for industrial applications at higher frequencies.

An additional advantage in going to higher frequencies which was first adopted by Oak Ridge [11] is the better field uniformity which is achieved in multimode cavities where the cavity dimension is greater than 100 λ. To obtain this advantage at a frequency of 2.45 GHz, which from the cost point of view is far more attractive, this would require a cavity 12.2 m in size and would result in a very low energy density. An additional problem here is that in order to achieve the required energy density, a large power supply would be necessary which would again increase the capital cost.

If microwave sintering of ceramics is be both technically and economically feasible on an industrial scale, there is a great advantage in using 2.45 GHz. This paper decribes a generic process at 2.45 GHz, which has been developed to sinter ceramics and which has been demonstrated for silicon nitride based materials. The process overcomes the problems of non-uniformity at this frequency by the use of field modification devices [12] and allows thermal conduction to play an important role [4].

EXPERIMENTAL

The process described in this paper has been developed for the pressureless sintering of Si_3N_4 based ceramics to achieve closed porosity, such that following hipping they can be used as ceramic tool-bits. In the future, it is hoped that these two densification steps can be carried out in a single step microwave heated hipper [13]. The demonstration process is based on a fully automated batch operation capable of sintering loads of 0.54 kg of material in a 120 minute cycle. One of two cavities can be fed from a single 5 kW magnetron via a switch, which enables continuous operation at a rate of 0.30 kg/hour. In each load it is possible to process up to 90 tool-bits. The material properties, quality and performance are described in a separate paper [5].

Equipment

An illustration of the sintering process is shown in Figure 1. It shows two multimode cavities attached via a switch to a 5 kW magnetron. The magnetron is protected from reflected power by a circulator with an integral directional coupler. An automated three stub tuner and associated IBM controller developed by Astex are used to continuously minimise the reflected power which is typically less than 5% of the forward power. The predetermined sintering schedule is controlled by a Eurotherm programmable controller which controls an Astex A 5000 power supply, and process control is achieved by monitoring the process temperature. Temperature is measured at a reference position in the load using a sapphire fibre-optic probe produced by Accufibre. Information regarding temperature, time, forward, reflected and net power, has been continuously collected using a Kaye data logger.

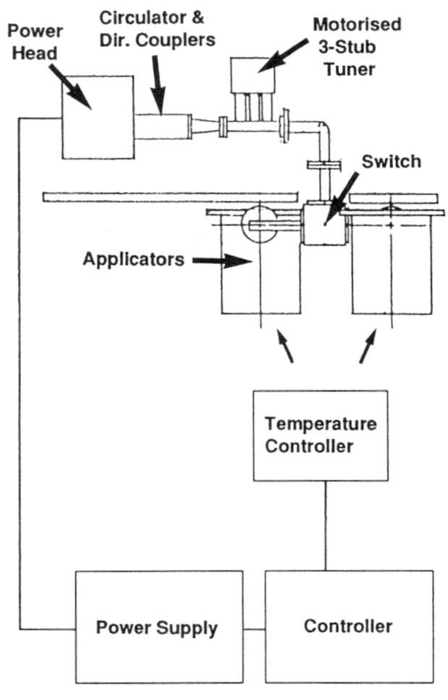

FIGURE 1: Illustration of the automated Microwave sintering process showing the set-up of the equipment.

Cavity and Load Design

Each of the multimode cavities is cylindrical in shape, approximately 12" in diameter by 16" in height. Microwaves enter the cavity at the top and from the side, and are directed onto a mode stirrer. The cavity lid is removed and the insulated load is placed centrally within the cavity. A conductive braid is fixed to the lid to seal against microwave leakage. Construction of the load is an important factor in ensuring that uniform thermal conditions are obtained and any bias in mass, either across a layer or in the depth will tend to result in enhanced heating in that region. In general, the load is arranged symmetrically about a central axis in an alumina crucible up to a size of 15 cm in diameter by 20 cm high, and each component is separated by the powder bed mixture [4] as shown in Figure 2. The bed is a mixed powder consisting of 40% SiC, with 30% BN and 30% Si_3N_4 by weight, which has multiple functions throughout the sintering cycle. Following sintering, the powder bed is removed from the sintered components and after mixing can be reused. Conductive rings are used to modify the field in the region around the load [12] and an additional symmetry should be considered between the rings and the individual components of the load. The load is insulated with a low mass zirconia refractory adjacent to the load, and saffil refractory in the cooler regions adjacent to the zirconia, Figure 2.

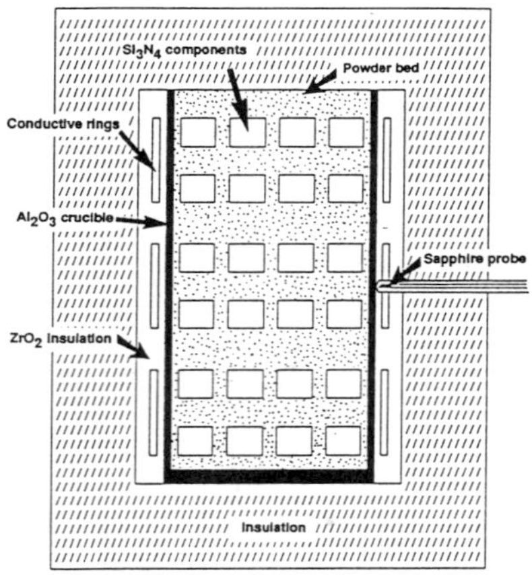

FIGURE 2: Illustration showing the arrangement of components within the load, field modifying rings and the insulation.

For the purpose of these tests, up to 6 temperature measurements adjacent to the load have been made in order to observe the temperature distribution across the load. In practice, it is only necessary to monitor one of these reference temperatures. Temperature is measured using sheathed, sapphire fibre optic probes in contact with the alumina crucible. The probes protrude through the side of the cavity and the insulated load as shown, and are attached via matched fibre-optic cables to an Accufibre model 100 mainframe, capable of monitoring up to 8 channels simultaneously.

Sintering Cycle

Once the load has been positioned and the lid replaced, the mode stirrer is activated and power is increased from 0.5 kW (at a rate of approximately 40 W/min) to 2 kW. At this setting, power is maintained for 15 minutes and then increased at a rate of approximately 10 W/min up to 2.2 kW, when the control temperature is reached. Power is gradually increased and then decreased in order to maintain the control setting, reaching a maximum value of approximately 2.7 kW. The control temperature is maintained for 30 minutes before the power is turned off. The reflected power is approximately 50% of the forward power for the first 20 minutes of the process, following which it can be tuned to below 5% for the remainder of the process.

When sintering a larger load size and mass, the sintering cycle is increased in time to approximately 3 hours in length. A similar heating rate is maintained up to a maximum power of approximately 4.4 kW and the load is again held at the control temperature for 30 minutes.

Following sintering, the load, still insulated, is removed from the cavity and allowed to cool naturally.

Product Analysis

The Si_3N_4 components are arranged in layers within the load and each sample is identified. When cold, the powder bed is removed and the components are inspected for discolouration and visual defects. Quality control is carried out using ultrasonic c-scan and density measurements [5]. The results are described showing the average density as a percentage of theoretical (3.27 g/cm^3) and standard deviation across each layer and throughout the bed. The results described here have been taken from over 50 batch sintering cycles and are representative of typical results.

Five Si_3N_4 mixtures have been evaluated with different amounts of sintering aids and different methods of introduction. All the materials contain Si_3N_4 with 5% Al_2O_3 and 5% Y_2O_3, and are designated by SN55. Some tests have evaluated a 1% addition of ZrO_2 with the usual mixing technique [5] and other tests have evaluated the addition of the sintering aids as a nitrate salt solution. These mixtures have been given the designation SN551 and SN55N or SN551N respectively.

RESULTS AND DISCUSSION

Temperature Uniformity

The measurement of temperature during the sintering cycle was carried out using 6 probes placed at the surface of the inner crucible and the results at certain times throughout the cycle are shown in Table I. These values are illustrated in Figure 3 and show the progression of heating throughout the load. This shows clearly that the load is not heated uniformly but that with time, thermal conduction acts to dissipate the hot-spots and due to the change in dielectric properties of the load with temperature, isothermal conditions can be reached within the bed at the sintering temperature. It is worth stressing that the temperatures shown here are not the sintering temperatures, but merely reference temperatures on the surface of the load.

TABLE I: The progression of temperature with time at various positions on the surface of the load. Numbered positions correspond to Figure 3.

	Time in minutes			
Position	50	60	75	140
Top Right, 1	680°C	1023°C	1584°C	1610°C
" Left , 2	536°C	743°C	1282°C	1581°C
Middle Right, 3	1092°C	1202°C	1331°C	1553°C
" Left , 4	843°C	1003°C	1257°C	1540°C
Bottom Right, 5	1189°C	1206°C	1276°C	1556°C
" Left , 6	957°C	1026°C	1180°C	1538°C

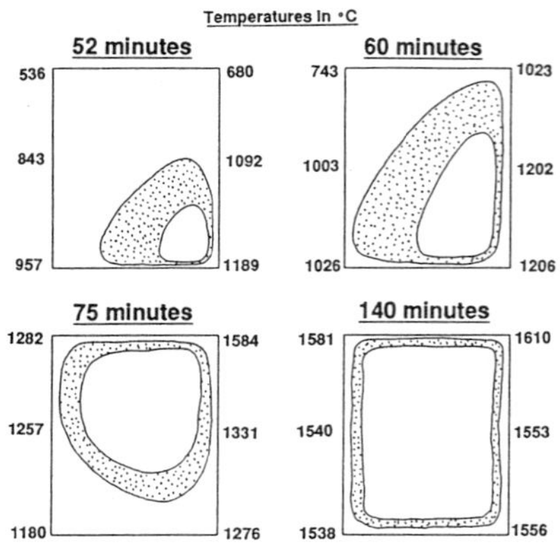

FIGURE 3: The estimated progression of heating throughout the load with time, calculated from surface temperatures.

As the size of the load is increased it is apparent that longer processing times are required in order to achieve isothermal conditions. Since the influence of the field modifying rings has a finite range, it is probable that there is an optimum batch size which can be efficiently processed. Above this optimum size, the processing times will need to be lengthened in order to allow for thermal conduction to dissipate the heat throughout the load, which in turn will result in an inefficiency due to the additional heat loss from the surface of the load. At very much larger loads, the effect of penetration depth will become a more significant issue.

TABLE II: Density variations for sintered Si_3N_4 components with position throughout the load. 72 components in each load.

	Density (%)			
Layer	Test 1	Dev.	Test 2	Dev.
1, top	95.0	±1.1	96.3	±1.1
2	95.5	±1.0	96.8	±0.6
3	95.5	±0.9	95.0	±0.8
4	94.8	±0.7	94.2	±0.2
5	94.0	±0.2	93.7	±0.4
6	93.7	±0.3	93.7	±0.3
7	93.5	±0.3	93.7	±0.2
8, base	93.1	±0.3	93.5	±0.3

Uniformity of Density

The density of components following the sintering of batches of 72 pieces (0.54 kg) is shown in Table II. It shows that there is a bias in temperature towards the top of the load in both of these tests. In each of the 8 layers, the 9 components showed very good uniformity in density except at the top of the load where the variation was much higher. The components at the top of the load were dark grey/black in colour and were darker than the components lower in the bed (light tan). It is thought that the higher temperatures at the top of the load had caused slight dissociation of the Si_3N_4, thereby turning it black and causing the variation in the density.

The region with a slightly higher temperature was not always confined to the top of the load and its position could be influenced by the mass of the load and by the rate at which the load was heated. Also, since no oxidation products were found it was discounted that this was a result of the ingress of oxygen through the top of the powder bed.

Once the problem of the higher temperature either at the top or the bottom of the load was corrected, it was possible to obtain better uniformity of density throughout the load. Table III shows the variation in average density throughout the load for a selection of materials. It shows very good uniformity in density from the top to the bottom of the load as shown in Figure 2 and in addition, shows the effect of a small amount of additional sintering aid and the effect of the manner in which the sintering aids are introduced. The addition of

1% zirconia, only slightly increases the final density, but a significant increase in density is obtained if the additives are introduced in the form of a nitrate solution. Individual pieces have also shown a significant increase in density when 0.1% of carbon is added [14] uniformly throughout the component.

TABLE III: Average density variation throughout the load for various powder compositions.

	Powder Composition				
Layer	SN55	SN551	SN551	SN55N	SN551N
1, top	94.8	94.5	94.8	96.7	96.6
2	94.3	94.8	94.6	96.6	96.4
3	94.9	95.0	94.5	96.0	96.4
4	94.3	94.9	94.5	95.9	96.4
5	95.1	95.4	94.6	95.7	96.5
6, base	95.5	95.7	95.8	96.2	96.4

The final density of the pressureless sintered silicon nitride component is greatly affected by the type and amount of sintering aids and the way in which they are distributed throughout the Si_3N_4. Milling techniques and media have also been shown to influence the final density quite markedly. Since the primary aim of this paper however, is to describe a process for microwave sintering of Si_3N_4 and not to investigate the chemistry of producing high density Si_3N_4 by pressureless sintering, the densities shown here should be taken only as a measure of the ability of this process to uniformly sinter Si_3N_4.

This sintering technique is based on a range of processing conditions within which, Si_3N_4 with 5%Al_2O_3 and 5%Y_2O_3, can be suitably densified in air at atmospheric pressure. At a lower temperature densification does not take place and at a higher temperature, the degradation of Si_3N_4 takes place. It is possible that this technique is made feasible by the preferential absorption of microwave energy by the sintering aids and that with conventional heating, a nitrogen overpressure is essential to inhibit degradation of the Si_3N_4, but this has not been confirmed. The cost advantage in being able to sinter Si_3N_4 in air at atmospheric pressure clearly makes this powder bed technique with microwaves a very attractive process.

Energy Requirements

The energy and times required to sinter Si_3N_4 loads ranging in size from 6 g to 1080 g are shown in Table IV. In each of these multiple component tests, the sintered products showed good uniformity throughout the batch. These results show that although the energy and time required for sintering increase with the mass of the load, the overall energy efficiency is improved from 24.7 kWh/kg to 2.2 kWh/kg as the load size is increased from 6 g to 1.08 kg.

From tests based on a small load, we have estimated [15] that 6.6 kg of Si_3N_4 can be conventionally sintered in a graphite element super VII furnace with the following sintering cycle and energy consumption;

Room temperature to 1100°C:	-	4 hours	
1100°C to 1800°C ramp-up:	-	3 hours	58.8 kWh
Retention at 1800°C	-	2 hours	54.0 kWh
1800°C to 1100°C ramp-down	-	3 hours	5.6 kWh
Natural cooling from 1100°C			

The total energy required is 119 kWh, showing that the energy consumption was approximately 18 kWh/kg. An additional 1.7 kWh/kg has been added to overcome the heat of reaction estimated at 200 kcal/mol which is required to heat the additional mass of silicon nitride above that of the small amount used in this test. This would result in a total energy consumption of approximately 19.7 kWh/kg to conventionally sinter Si_3N_4. Figure 4 shows the energy consumption data for both the conventional and microwave sintered Si_3N_4 and shows that there is a possible energy savings of about 84% and 89% at a load of 0.5 kg and 1.0 kg respectively. Assuming that the magnetrons can work with a 70% efficiency, then for a 0.5 kg load an energy consumption of 4.4 kWh/kg would result in a possible energy savings over conventional sintering of 78%.

TABLE IV: Energy requirement for sintering Si_3N_4 loads of various sizes.

# of Components	Mass (grams)	Time (Mins)	Net Energy (MJ)	Efficiency (kWh/kg)
1	6	30	0.53	24.7
90	540	105	22.80	6.7
12	540	120	12.05	3.1
24	1080	171	24.32	2.2

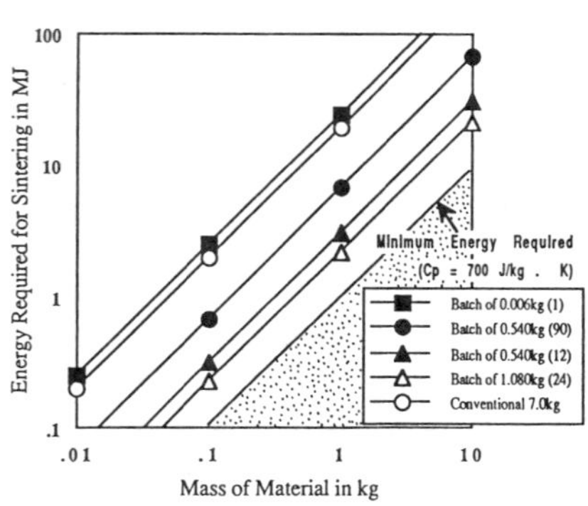

FIGURE 4: Graph showing the energy required to conventionally and microwave sinter Si_3N_4 as a function of the size of the load. The minimum energy requirement is based on the heat capacity for Si_3N_4 to be raised to 1800°C.

CONCLUSIONS

A process based on microwave heating at 2.45 GHz has been developed, capable of sintering batches of up to 1.0 kg of Si_3N_4 with 5% Al_2O_3 and 5% Y_2O_3 additives to a density of approximately 97% of theoretical. Density variations throughout the batch are typically better than ±0.5%.

A combination of field modification devices in the form of metallic rings and a multifunctional powder bed has been successfully used to control the position of the regions of high field and the way in which heat is distributed throughout the load. This process does not initially heat uniformly but results in isothermal conditions throughout the load at the sintering temperature.

For microwave sintering of Si_3N_4 with additives, there is a well defined processing window where heating can be carried out in the powder bed, in air and at atmospheric pressure, with no detrimental effect due to oxidation of the sintered product.

Si_3N_4 loads of 0.54 kg and 1.08 kg can be processed in times of 120 minutes and 171 minutes, with an energy utilisation of 3.1 kWh/kg and 2.2 kWh/kg respectively. Assuming the efficiency of magnetrons to be 70%, this represents a possible energy savings of 78% over conventional sintering of Si_3N_4.

From the commercial viewpoint, this process would be attractive to manufacturers of tool-bits and other small Si_3N_4 components which require the product uniformity and performance which microwave heating offers. The cost of processing at 2.45 GHz is considerably cheaper than higher frequencies and the ability to uniformly sinter 1.0 kg batches in air, at atmospheric pressure compares favourably with conventional sintering. This technology is presently available under license from Alcan.

REFERENCES

1. NTIS Department of Commerce, Ceramic Technology for Advanced Heat Engines Project Semiannual Progress Report for October 1989 through March 1990. Oak Ridge National Laboratory, Sept. 1990.

2. Pickup, H., Ph.D. Thesis, University of Leeds, 1988.

3. Wotting, G., and Ziegler, G., Interceram (2), 32-36, (1986).

4. Apte, P.S., Patterson, M.C.L., Roy, R., Kimber, R.M. and Mitchell, D.N., PCT International publication No. WO 91/05747, 2nd May, 1991.

5. Patterson, M.C.L., Apte, P.S., Kimber, R.M. and Roy, R., "Mechanical and Physical Properties of Microwave Sintered Si_3N_4", this proceedings.

6. Sutton, W., Ceramic Bulletin, 68, (2), 376-386, (1989).

7. Salsman, J.B., in Microwave Processing of Materials II, edited by W.B. Snyder Jr., W.H. Sutton, M.F. Iskander and D. Lynn Johnson, (Mater.Res.Soc.Proc., 189, Pittsburg PA, 1990), pp. 509-515.

8. Patterson, M.C.L., Kimber, R.M., and Apte, P.S., in Microwave Processing of Materials II, edited by W.B. Snyder Jr., W.H. Sutton, M.F. Iskander and D. Lynn Johnson, (Mater.Res.Soc.Proc., 189, Pittsburg, PA, 1990), pp. 257-266.

9. Janney, M.A., and Kimrey, H.D., in Ceramic Transactions, Ceramic Powder Science, II,B, Edited, G.L. Messing, E.R. Fuller Jr., and H. Hausner, (Am. Cer. Soc., Westerville, OH. 1988), pp. 919-924.

10. Meek, T.T., Blake, R.D., and Petrovic, J.J., Ceram. Eng. Sci. Proc., 8, (7-8), pp. 861-871, (1987).

11. Kimrey, H.D., and Janney, M.A., in Microwave Processing of Materials II, edited by W.H. Sutton, M.H. Brooks and I.J. Chabinsky, (Mater. Res. Soc. Proc., 124, Pittsburg PA, 1990), pp. 367-372.

12. Patterson, M.C.L., Roy, R.V., Kimber, R.M., and Apte, P.S., Filed Patent but not Published, PCT.CA92/00031.

13. Apte, P.S., and Pant, A., U.S. Patent No. 5 010 220, (23rd April, 1991).

14. Komatsu, M., and Miyano, T., U.S. Patent No. 4 911 870, (March 27th, 1990).

15. Raghavan, N.S., (private communication).

MECHANICAL AND PHYSICAL PROPERTIES OF MICROWAVE SINTERED Si_3N_4

M.C.L. PATTERSON, P.S.* APTE, R.M. KIMBER and R. ROY
Alcan International Limited, Kingston Research and Development Centre, P.O. Box 8400, Kingston. Ontario, Canada K7L 5L9
* Present Address : Sherritt Gordon Ltd, Fort Saskatchewan, Alberta.

ABSTRACT

Microwave energy at 2.45 GHz has been used to sinter silicon nitride in times of between 30 and 120 minutes to form a dense product with an average grain size of 0.35 μm and substantially improved uniformity of properties over all commercial silicon nitrides and sialons investigated.

A comparison between microwave and conventionally sintered parts of the same density showed the average mechanical properties to be the same for both sintering techniques, but revealed an improvement in the uniformity of mechanical properties of the microwave sintered products.

Silicon nitride components with 5% Al_2O_3 and 5% Y_2O_3 as sintering aids, has been sintered to a density of approximately 97% of theoretical. Hipping of these components at 1800°C for 60 minutes, resulted in the complete removal of residual porosity. Dense components were evaluated in machining tests against commercial ceramic cutting tools. The results show a reduction in wear rate of better than 10% for microwave sintered Si_3N_4 in turning and milling operations on cast iron at a cutting speed of 2000 surface feet per minute.

INTRODUCTION

Due to the unique microstructure of sintered Si_3N_4 it has been studied extensively as an advanced structural ceramic for possible engineering applications. During pressureless sintering, densification is brought about in the presence of a liquid phase, by a solution - diffusion - reprecipitation mechanism [1,2], during which the equiaxed α-Si_3N_4 grains reprecipitate as acicular β-Si_3N_4 grains [3], with preferred growth in the c-dimension. The interlocked acicular microstructure which develops typically determines the room temperature properties, whilst the high temperature properties are determined by the properties of the resulting intergranular phase [4].

The mixed additive system between Si_3N_4, Al_2O_3 and Y_2O_3, has been investigated [5,6,7] due to its good high temperature properties. During sintering of Si_3N_4 with additives of Al_2O_3 and Y_2O_3, Al^{3+} substitutes for Si^{4+} and O^{2-} for N^{3-}, to form a Si_3N_4 - Al_2O_3 solid solution referred to as β'-sialon [8,9]. Sialons have the generic formula $Si_{6-z}Al_zO_zN_{8-z}$, with the z number representing the amount of O and Al in solid solution. The Y remains in the grain boundary phase with SiO_2 and some Al. At relatively low temperatures Y_2O_3-Al_2O_3-SiO_2 forms liquid phases which allow rapid sintering and at high temperatures when densification is complete, these phases react with more Si_3N_4 to give a highly refractory bonding phase [10].

Over the past few years microwave heating has been applied extensively to the sintering of oxide based ceramics and has been shown to offer several advantages over traditional forms of sintering. Conventionally, ceramics must be heated slowly in order to reduce thermal gradients which arise. By microwave sintering heat can be delivered throughout the ceramic volume thereby overcoming the constraint of thermal conduction which is typically slow, allowing thermal gradients to be minimized and the component to be heated rapidly. Microwave sintering of oxide ceramics has been shown to offer only subtle differences [11,12] in the material properties over conventionally processed material, brought about primarily by;

i) small grain size

ii) uniformity of the structure throughout the component.

Microwave heating of silicon nitride is still relatively novel and there exists only limited information on annealing [13,14] and sintering [14]. Successful densification has only been achieved with relatively large volumes of sintering aids or at higher frequencies [14]. This is due in part to the low dielectric absorption of silicon nitride and the inherent problems of atmosphere control and choice of sintering aids. This may also be attributed to the fact that suitable microwave sintering procedures have not yet been perfected.

Although the material advantages produced by microwave heating are relatively minor, they exist in areas of key importance to ceramics and impact directly on the wear behavior and the predictability of the component. As engineering materials, ceramics have generally had poor acceptance due to their unpredictable behaviour and erratic performance, which is often related directly to their method of densification. Since it has been demonstrated that the uniformity of the microstructure can be improved by microwave sintering [11], there is a possibility that in the future we could see an increase in the use of ceramics for engineering applications, as their behaviour becomes more predictable and as the microwave fabrication techniques become better understood and more reliable.

This paper deals with a technique which is described elsewhere [15] which has been developed for pressureless sintering of Si_3N_4 and the relative mechanical and physical properties of this material with 10% by weight additives. A comparison is made with conventional sintering of the same powder composition and with other commercial Si_3N_4 and sialons.

EXPERIMENTAL

A powder blend of Si_3N_4 with 5 wt% Al_2O_3 and 5 wt% Y_2O_3 (Table I), was milled in hexane in an abathane container for 12 hours using Al_2O_3 milling media. The blend was decanted, dried and then remilled for a further 8 hours. Pellets were pressed uniaxially at a pressure of 60 MPa followed by isopressing at 275 MPa.

TABLE I: Source and Properties of the Raw Materials

Material	Supplier	Purity (%)	Average size (μm)
Si_3N_4	Ube - SN 10E	98.60	0.33
Al_2O_3	Baikowski CR30	99.99	0.05
Y_2O_3	Alfa (John.Mat.)	99.90	0.90

A mass of up to 1 kg of pellets was packed into a powder bed [16] and placed centrally in a stirred, top fed, cylindrical multimode cavity. Sintering was carried out at

2.45 GHz, using an Astex 5 kW magnetron and controller and during operation the reflected power was continuously minimized using an automated 3-stub tuner. The process is described in more detail in an additional presentation made at this symposium [15]. The sintering cycle varied between 30 minutes and 180 minutes depending upon the load, but for the work described here, a cycle of 105 minutes was used to sinter a batch of 540 g. Several shapes and sizes of pellets have been sintered but for the purpose of this work we have concentrated on two cylinders with sintered diameters of 15 mm and 21 mm. Following sintering the load was allowed to cool naturally and the pellets were separated from the powder bed, which was re-used. The density of the pellets after pressureless sintering in air was between 96% and 97% and in order to remove the residual porosity, the pellets were hot isostatically pressed. Hipping temperatures between 1700°C and 1850°C were investigated for 1 hour, under an argon and nitrogen atmosphere with an Ar:N ratio of 2:1, at 200 MPa.

The larger diameter cylinders were surface ground to a depth of 1 mm and diced into discs each 1 mm in thickness for measurement of strength from biaxial flexure. The smaller cylinders were surface ground and lapped to a finished size of 12.7 mm in diameter by 7.9 mm thick, (standard tool-bit SNG 45). Ultrasonic evaluation by c-scan at 15 MHz was carried out on all the specimens to observe uniformity. Samples were also sectioned and polished using normal metallographic techniques to reveal their microstructure and in preparation for indentation toughness and hardness measurements. The grain boundaries were enhanced by plasma etching with CF_6 and grain size analysis was carried out using a Contron Ebas image analyzer on images collected directly from a Joel scanning electron microscope (SEM). Multiple areas were analyzed across the sample (17 horizontal and 13 vertical), to establish grain size variation throughout the sample. Vickers indentations were carried out at 1 mm intervals using loads of 10 kg and 15 kg across the polished diameters and values for hardness and toughness were calculated from the crack lengths measured optically in the standard way [17,18]. Microanalysis was also carried out on thin foils using a Hitachi H-9000 transmission electron microscope (TEM) using Link energy dispersive spectroscopy (EDS).

As a comparison similar analyses were carried out on the same powder composition conventionally sintered at 1800°C for 2 hours followed by hipping to full density, and also on a selection of 6 commercial Si_3N_4 and sialon tool-bits.

Density was measured by both helium and water pycnometry and expressed as a percentage of the estimated theoretical density (3.27 g/cm^3)

The microwave sintered β'-sialon was evaluated against four other commercial Si_3N_4 and sialons as a ceramic cutting tool for the machining of cast iron. Standard turning tests were carried out at a constant surface speed of 2000 feet per minute, a feed rate of 0.015" and a cut depth of 0.100". The tool-bit insert was indexed and removed every 2 or 4 passes for measurement of wear on the flank of the insert. Wear was measured using a microscope with a micrometer attachment and is expressed in thousandths of an inch. Evaluations were carried out against 4 commercial tool-bits. Machining tests were repeated 4 times in order to investigate the degree of reproducibility.

RESULTS AND DISCUSSION

Effect of Sintering Time and Load

The sintering time required to obtain uniform density throughout the load using the experimental conditions described [15], depended upon;

i) The mass and size of the load

ii) The distribution of components throughout the load.

With small loads of 100 grams, uniform properties throughout each component could typically be obtained in times of 30 minutes, but for larger loads, longer times were necessary in order to achieve isothermal conditions throughout the load. The test cylinders evaluated in this work were sintered in loads of approximately 540 g (90 small cylinders or 12 large blocks) and a total sintering cycle of 120 minutes was required to achieve isothermal conditions. Under these conditions the sintered densities typically varied between 96% and 97% of theoretical.

Conventionally heated cylinders also gave a density of approximately 97% following pressureless sintering.

Both the microwave and conventionally sintered cylinders were hipped in order to remove the residual porosity. The results indicated that the lowest temperature at which appreciable pore removal took place was 1800°C, and this temperature with a hold time of 60 minutes was used for all further tests.

Ultrasonic Evaluation

Ultrasonic evaluation by c-scan was carried out on all the Si_3N_4 and sialons investigated. This technique highlights areas of attenuation as a signal is passed through a sintered component thereby indicating areas of non-uniformity. The results shown in Figure 1, indicate that microwave sintering followed by hipping result in a very uniform product compared with the same material conventionally sintered and hipped. The cause of the ring structure observed in the conventionally sintered and hipped material, is not fully understood but since there is no visible porosity in the sample it is most probably due to a change in grain size, intergranular phase composition or the overall composition. The commercial tool-bit materials showed various degrees of uniformity, which could be related back to their methods of fabrication. Material which was hot pressed showed good uniformity compared with either the sinter hipped, or glass encapsulated-hipped commercial materials.

ULTRASONIC EVALUATION

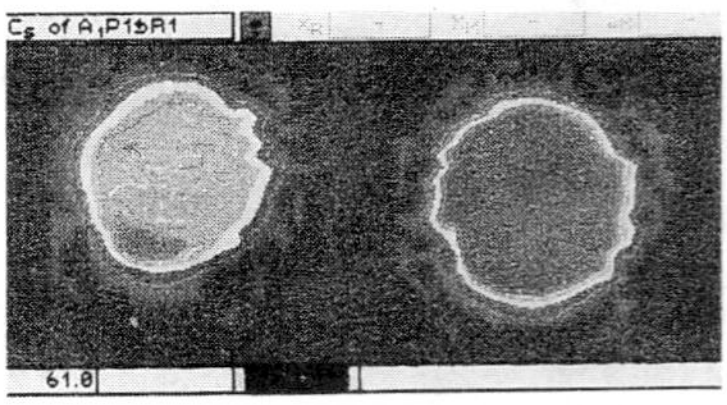

Microwave sintering, 90 minute cycle.

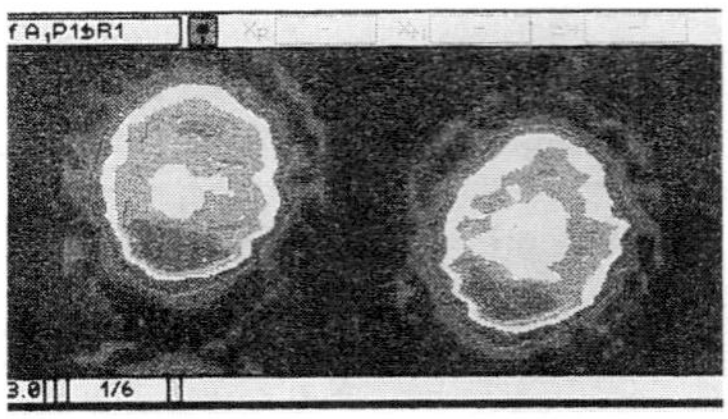

Conventional sintering, 2 hours at 1775°C.

FIGURE 1: Ultrasonic c-scan images at 15 MHz of microwave and conventionally sintered Si_3N_4 showing attenuation in the conventionally sintered material. Both samples have a density of 3.27 g/cm^3.

Shear and logtitudinal ultrasonic transducers were also used to obtain wave velocity data from which Youngs' moduli could be calculated, these values are shown in Table III.

<u>Microstructure</u>

Microstructural observation of the microwave sintered Si_3N_4 in section, reveals fine acicular grains of β'-sialon. Microanalysis of the grains and grain boundary phases carried out by TEM and EDS showed β'-sialon lattice spacings of 7.606 Å for "a" lattice and 2.911 Å for the "c" lattice dimensions. This corresponds to a z number ($Si_{6-z}Al_zO_zN_{8-z}$) of 0.2, indicating that little substitution of Al_2O_3 into the Si_3N_4 structure has taken place. It was also noted that 100% transformation of the α-Si_3N_4 to β'-sialon had taken place. Analysis of the grain boundaries and triple points shown in Figure 2, revealed a Si-Y-Al-O-N glass with no crystallization.

A summary of the average grain size analysis for both horizontal and vertical directions is shown in Table II and shows a mean minimum grain size of approximately 0.35 μm.

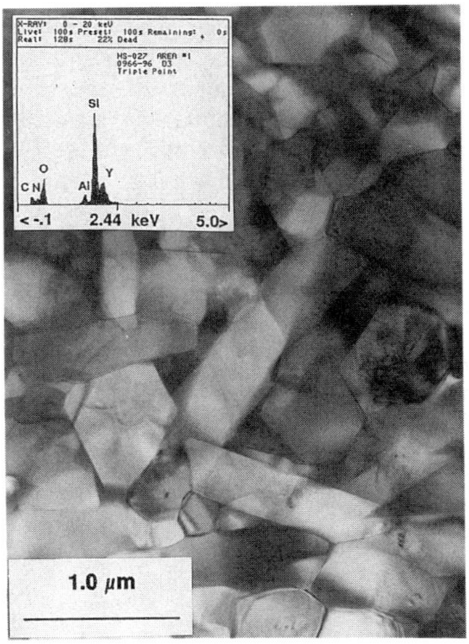

FIGURE 2: Transmission electron micrograph of the microwave sintered structure and associated EDS analysis of the glass phase at a triple point.

TABLE II: Average grain size analysis for microwave sintered and hipped β-Sialon.

Parameter	Mean	Variance	Std. Dev.
Minimum grain size (μm)	V 0.3469 H 0.3631	V 0.0221 H 0.0321	V 0.1487 H 0.1791
Maximum grain size (μm)	V 0.5926 H 0.6210	V 0.1229 H 0.1647	V 0.3505 H 0.4058
Average grain area (μm^2)	V 0.1567 H 0.1778	V 0.0341 H 0.0505	V 0.1847 H 0.2248
Shape factor D.min./D.max.	V 0.6237 H 0.6290	V 0.0203 H 0.0208	V 0.1426 H 0.1441

The standard deviation of the maximum grain size is high since the grain is acicular and as we are viewing it on a planner surface, we will be cutting it at various positions along its true length. (The minimum grain size measurement should always measure the true minimum grain size, irrespective of its orientation). By taking the maximum value recorded for the maximum grain size measurement, referred to as D.max (4.52 μm), this is most likely the dimension for β'sialon cut parallel to its c-axis. If we compare this with the average value for the minimum grain size measurement, referred to as D.min (0.35 μm), we can estimate that the upper bound of the aspect ratio is approximately 13:1. Based on visual analysis of multiple observations in the TEM and SEM we would estimate the average aspect ratio be between 4 and 6:1.

Figure 3, shows the change in the minimum grain size as a function of position across the sintered component and illustrates the existence of very little exaggerated or cannibal grain growth. This tends to indicate that the onset of the α-Si_3N_4 to β'-sialon transformation takes place uniformly throughout the component.

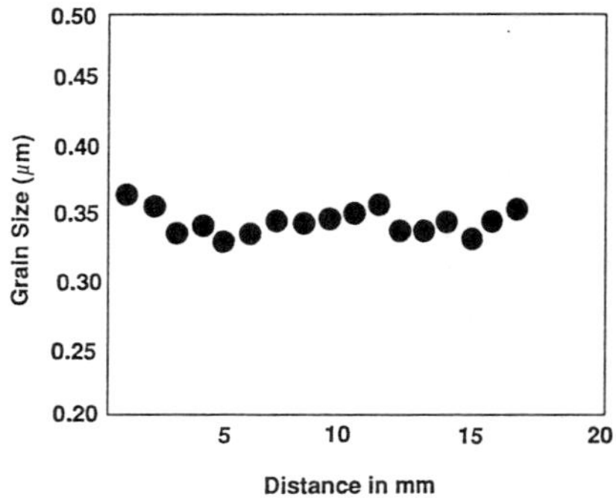

FIGURE 3: Variation in the average grain size across the diameter of microwave sintered Si_3N_4, showing the small variation in grain size.

<u>Hardness, Toughness and Strength</u>

The average hardness, toughness and strength of the microwave sintered and hipped Si_3N_4 ranged from approximately 1670 to 1730 GPa, 5.39 to 5.53 $MPa.m^{-2}$ and from 624 to 1463 MPa respectively and is summarized in Table III. Similar values for conventionally sintered and hipped material are also shown in Table III.

TABLE III: Average Hardness, Toughness and Strength Data for Microwave and Conventionally Sintered and Hipped Si_3N_4

Sintering Technique	Hardness (GPa)		Toughness ($MPa.m^{-2}$)		Strength (MPa)	E (GPa)
Load	10 kg	15 kg	10 kg	15 kg	Weibull 5.5	
Microwave Sint/Hip	17.33	16.74	5.39	5.55	1033	308
	80 tests		80 tests		95 tests	
Conventional Sint/Hip	17.34	17.10	5.25	5.42	-	308
	80 tests		80 tests			

Comparison of the hardness, toughness and strength data for microwave sintered and hipped components, with the conventionally sintered and hipped components, shows no significant difference in any of these average parameters. Likewise, when these values are compared to commercial Si_3N_4 and sialons [19], microwave sintering does not appear to enhance any of these average properties. When the individual values for hardness and toughness are plotted against their position in the component, it becomes obvious that these properties are very uniform throughout the component for microwave sintered material compared to any of the conventionally processed materials. This is shown in Figure 4, as a plot of relative toughness across the diameter, for microwave, conventional and commercial Si_3N_4 and sialons. This uniformity of properties throughout the component has been shown to be typical of microwave sintered components for both oxides [11] and nitrides.

It is expected that the uniformity of properties will be reflected in a higher than normal Weibull value for the strength measurements. The measured Weibull value of 5.5 for nearly 100 tests is not high, but the volume of material which is sampled with the biaxial flexure method (only that directly under the central ball), is very small and would not necessarily reflect this uniformity in the sample. A better way to identify uniformity throughout each component would be to use a technique, such as a four-point bend test, which involves the sampling of a larger volume.

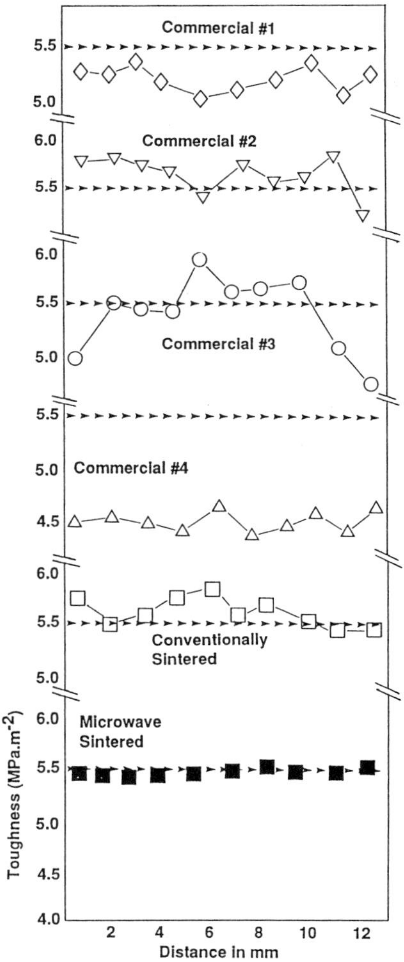

FIGURE 4: Variation in toughness across the diameters of microwave, conventional and commercially sintered Si_3N_4 tool-bits.

Machining Performance

The machining tests carried out under controlled conditions typically showed at least 10% less flank wear for the microwave sintered inserts compared to any of the commercial inserts. Figure 5 shows a summary of the average wear data for 450 inch^3 of metal removed, against the average grain size of the tool-bits, highlighting the clear correlation between the grain size and the flank wear. This type of a correlation typically only exists when the wear mechanism is primarily due to abrasion. Repeated wear trial on commercial insert #2, showed that over 8 identical tests the reproducibility of the wear data obtained from the microwave sintered tool-bit was far higher than that obtained from the commercial insert, as shown in Figure 6. For many machining applications the tool-bit life is defined as 0.025" wear on the flank which would indicate that considerably more metal could be removed with the microwave sintered tool-bit. The sudden jump in flank wear of approximately 0.004" for the microwave sintered tool-bit after approximately 430 inch^3 of metal removal is not real and is a result of a build-up of metal on the wear surface making measurement difficult. Catastrophic failures which were observed with the commercial tool-bits were not observed in any of the tests on the microwave sintered tool-bits. Predictable performance in the machining industry is of great importance and as such the process described in this paper should be attractive to tool-bit manufacturers.

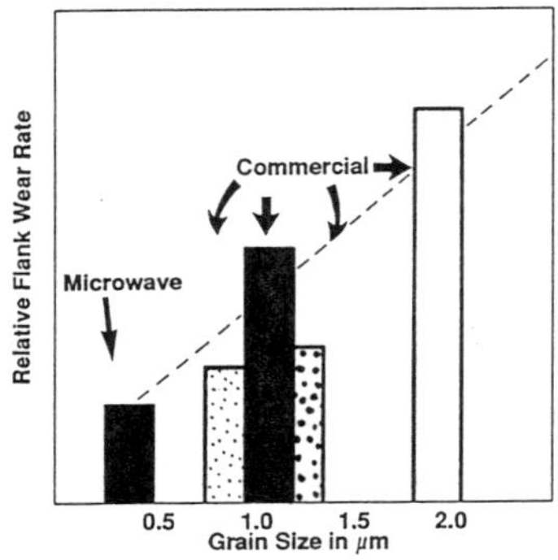

FIGURE 5: Graph showing the relative flank wear and respective grain sizes for microwave sintered and commercial tool-bits. Machining conditions - cast iron at 2000 SFM.

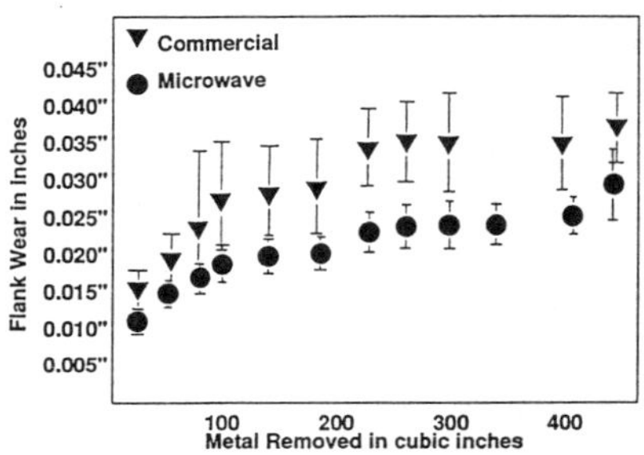

FIGURE 6: Comparison of multiple test, wear data showing the scatter in flank wear for microwave sintered and commercial tool-bits. Machining conditions - cast iron at 2000 SFM.

CONCLUSION

A process has been developed to microwave sinter Si_3N_4 at 2.45 GHz. As the mass is increased from 100 g to 1.0 kg, the sintering time must be increased from 30 minutes to 180 minutes in order to maintain uniformity of product. The final sintered density is approximately 96%, with full density being achieved by Hipping at 1800°C for 60 minutes.

The microstructure of sintered Si_3N_4 with 5% Al_2O_3 and 5% Y_2O_3 consists of fine acicular β'-sialon grains with an average grain size of 0.35 μm and an aspect ratio of approximately 6:1. Substitution of Al_2O_3 results in a sialon z number of 0.2. The uniformity of average grain size across the sintered components is good, indicating that sintering has occurred uniformly throughout the component. No crystallization of the grain boundary phase was observed. Ultrasonic evaluation of the microwave sinter-hipped product shows greater uniformity than conventional sinter-hipped products, with properties similar to those for hot pressed material.

The mechanical properties of the microwave sinter-hipped material are not dissimilar to conventionally sinter-hipped material of the same material composition. The hardness, toughness and strength of the material ranged as follows, 1670 - 1730 GPa, 5.39 -

5.53 $MPa.m^{-2}$ and 624 - 1463 MPa respectively. The uniformity of mechanical properties across each component was shown to be greater than with conventional processing and this appears to be a signature of the microwave sintered materials in general.

In machining tests on cast iron, microwave sintered tool-bits showed at least a 10% reduction in flank wear over conventionally sintered material of the same composition and other commercial inserts. Over repeated tests, microwave sintered tool-bits showed more predictable wear than commercial inserts and over all the tests performed no catastrophic failures were observed for the microwave sintered tool-bits.

ACKNOWLEDGEMENTS

Initial funding for this work was provided by the Ontario Ministry of Energy under the Enersearch Program Project #6041/600048.

We would like to thank the National Aeronautical Establishment and the Materials Technology Centre divisions of the National Research Council of Canada. We should also like to thank Newcomer Products Inc., and Kennametal Ltd. for industrial assistance and technical discussions.

REFERENCES

1. Nighe, J., J. Mater. Sci. 13, pp. 1455, (1978).

2. Lange, F., Int. Metals. Rev., 1, pp. 247, (1980).

3. Lange, F, and Davies, B.I., J. Am. Cer. Soc., 62, pp. 629, (1979).

4. Ekstrom, T, and Nygren, M., J. Am. Ceram. Soc., 75, 2, pp. 259-276. (1992).

5. Stutz, D., Greil, P., and Petzow, G., J. Mater. Sci. Lett., 5, pp. 335-336, (1986).

6. Ekstrom, T., Mater. Sci. Eng., A109, pp. 341-349, (1989).

7. Ekstrom, T., and Persson, J., J. Am. Ceram. Soc., 73, 10, pp. 2834-2838, (1990).

8. Oyama, Y., and Kamigaito, O., Japan J. Appl. Phys. 10, pp. 1637-1642, (1971).

9. Jack, K.H., and Wilson, W.I., Nature Phys. Sci. (London) 238, pp. 28-29, (1972).

10. Jack, K.H., J.Mater.Sci., 11, pp. 1135-1158, (1976).

11. Patterson, M.C.L., Kimber, R.M., and Apté, P.S., in Microwave Processing of Materials II, edited by W.B. Snyder, W.H. Sutton, M.F. Iskander and D. Lynn Johnson. (Mater. Res. Soc. Proc., 189, Pittsburgh, PA, 1990), pp. 257-266.

12. Apté, P.S., Kimber, R.M., and Patterson, M.C.L., Structural Ceramics, (Proc. 11th Risø Int. Symp. Met. Mat. Sci., Denmark, Sept 11th, 1990), pp. 167-174.

13. Tiegs, T.N, Ferber, M.K, Kiggans, J.O, More, K.L, Hubbard, C.M, and Coffey, D.W, in Ceramic Transactions, Microwaves: Thoery and Application in Materials Processing, Vol 21, Edited by E.D. Clark, F.D. Gac and W.H. Sutton, (Am. Cer. Soc., Westerville, OH, 1991), pp. 411-420.

14. Tiegs, T.N., Kiggans, J.O. Jr., and Kimrey, H.D .Jr., in Microwave Processing of Materials II, edited by W.B. Snyder, W.H. Sutton, M.F. Iskander and D. Lynn Johnson, (Mater. Res. Soc. Proc., 189, Pittsburgh, PA, 1990), pp. 267-272.

15. Patterson, M.C.L., Apte, P.S., Kimber, R.M., and Roy, R., "Batch Process for Microwave Sintering of Si_3N_4", this proceedings.

16. Patterson, M.C.L., Roy, R.V., Kimber, R.M., and Apte, P.S., Filed Patent but not Published, PCT.CA92/00031.

17. Evans, A.G., and Charles, E.A., J. Am. Cer. Soc., 59, pp. 371, (1976).

18. Ponton, C.B., and Rawlings, R.D., Mater. Sci. Tech. 5, pp. 865-872, (Sept 1989).

19. Patterson, M.C.L, Apté, P.S, Kimber, R.M and Roy, R., Int. Conf. on Microwaves and High Frequencies, (Comite Francais De L'Electricite, Nice, 8-10 October, 1991), pp. 33-36.

Microwave Sintering of ZnO Varistor Ceramics

Lionel M. Levinson, Holly A. Comanzo, and William N. Schultz
General Electric, Corporate Research and Development, P.O. Box 8,
Schenectady, NY 12301

ABSTRACT

ZnO varistors are electronic ceramics whose electrical behavior is dominated by grain boundary effects. A careful evaluation of the effect of microwave sintering compared to conventional sintering revealed no significant differences between varistor devices processed by these two methods. A slight apparent enhancement of grain growth was observed for microwave processing, but further evaluation of this effect leads us to believe that it arises from a small systematic shift in temperature derived from the different experimental configurations.

1. INTRODUCTION

One of the intriguing aspects of the microwave processing of ceramic materials has been reports[1-4] that microwaves have the ability to enhance process rates such as diffusion and grain growth during sintering. At present, there appears to be no accepted theory to account for these nonthermal phenomena. Enhanced grain boundary heating[5] and microwave-induced changes in the diffusion process[6] have recently been examined as possible causes of this effect. However, the former explanation has been rejected,[5] and microwave-induced nonthermal diffusion processes[6] are not clearly established. Most of the microwave processing studies have concentrated on high-temperature ceramics, such as alumina and zirconia. These materials are insulating at low temperatures and couple well to 2.45 GHz microwave fields only at higher temperatures (>> 1000 °C), where the material conductivity and loss tangent are much greater.

In this paper we discuss the effect of microwave fields upon the sintering behavior of ZnO varistor ceramics. ZnO varistor ceramics provide a good test vehicle to evaluate any nonthermal effects of microwave sintering since the electrical properties of these ceramics are sensitively controlled by grain boundary phenomena. In addition, the ZnO grain resistivity is about 1 Ω-cm[8,9], allowing relatively good coupling and adequate penetration of the microwave field into the ceramic.

Microwave sintering of ZnO varistors, which was recently studied,[10] indicated a varistor breakdown field of a factor of two greater for microwave-processed material than for conventional samples. The higher breakdown field was attributed to smaller grain size in the microwave processed samples, which was in turn attributed to "shorter sintering and annealing times involved with microwave sintering." In general, sintering and annealing times can have profound effects upon ZnO varistor microstructure and electrical behavior. We have therefore chosen to evaluate the effect of microwave processing upon ZnO varistors by comparing microwave and conventional thermal processing under nominally identical conditions.

2. EXPERIMENTAL

A single uniform batch of ZnO varistor powder was prepared for this study. The varistor formulation comprised 1/2 mole percent of each of Bi_2O_3, Co_2O_3, MnO_2, and Cr_2O_3; 1 mole percent each of NiO and Sb_2O_3; and 96 mole percent ZnO. This composition is typical [7,8] of varistor formulations with high nonlinearity coefficient α. The powders were milled, dried, calcined, remilled, and cold-pressed at 4 kpsi. Three different sample sizes were used to ascertain the effects (if any) of sample size upon the study. Samples sintered were cylindrical with the following sizes as pressed: 1.2 in. thick, 1.5 in. diameter; 0.12 in. thick, 1.5 in. diameter; and 0.04 in. thick, 0.75 in. diameter.

The pressed samples were bisque-fired in a conventional furnace at 750 °C to burn out organic binders used to facilitate the pressing. After binder burnout, samples were sintered either conventionally using a resistively heated furnace or by microwaves in a multimode cavity. Samples were placed on edge (on their curved surface) in high purity alumina boats. The firing ambient was air and the furnace atmosphere was exchanged at the rate of a few times per hour.

For both conventional resistive and microwave sintering the furnace temperature was programmed using the following schedule:

Ramp up:	Room Temperature to 950 °C in 20 hours; 950 °C to top temperature at 100 °C/hr.
Top Temperature:	90 minute soak
Ramp Down:	Top temperature to room temperature at 100 °C per hour.

Figure 1 gives a top view of the sample layout for both microwave and conventional firing. Temperature was measured for the conventional firing using a Pt-Rh thermocouple placed a few centimeters from the samples. Tests of temperature homogeneity showed that the temperature variation around the samples in the conventional furnace was less than 1 °C. Top temperature was controlled to within 1 °C.

Microwave sintering was conducted using a 900 W multimode cavity operating at 2.45 GHz. Temperature was measured using a sapphire fiber optic probe which entered a deep (depth = 0.75 in., diameter = 0.14 in.), close fitting hole in the large varistor as shown in Figure 1(b). We note here that ZnO varistor ceramics are black, i.e., opaque to visible and infrared radiation. The fiber optic probe tip is additionally coated with 2 µm of "black" 70% Pt - 30% Rh alloy, and, since this thickness is considerably less than the microwave skin depth for this deposited alloy at high temperatures, the probe will not materially affect the microwave field. The fiber optic probe was calibrated at 1000 °C, and we have confirmed its accuracy (a few degrees celsius) over the temperature range used for this study by comparing the temperature measured in a conventional furnace using both a thermocouple and the fiber optic probe.

After sintering, the samples were electrically contacted using conductive silver epoxy. The thick samples (1.2 in. thick by 1.5 in. diameter) were sectioned into thinner pieces to provide information on any variation of electrical properties with position in the thick sample. Electrical measurements undertaken included dc current-voltage data, pulse current-voltage data up to current densities of 200 A/cm^2, varistor leakage current, and capacitance-dissipation measurements for frequencies f in the range $10^2 < f < 10^7$ Hz.

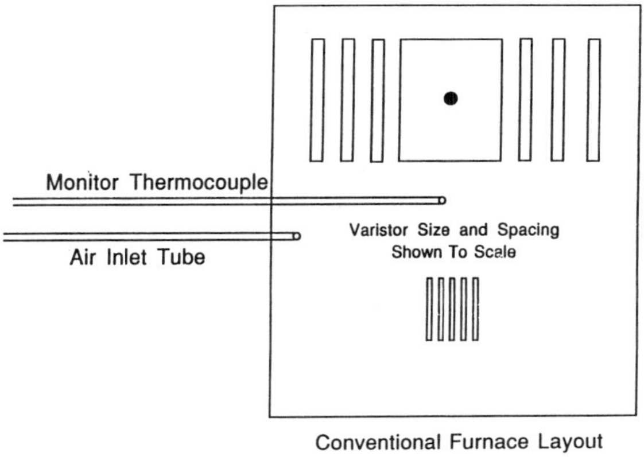

(a)

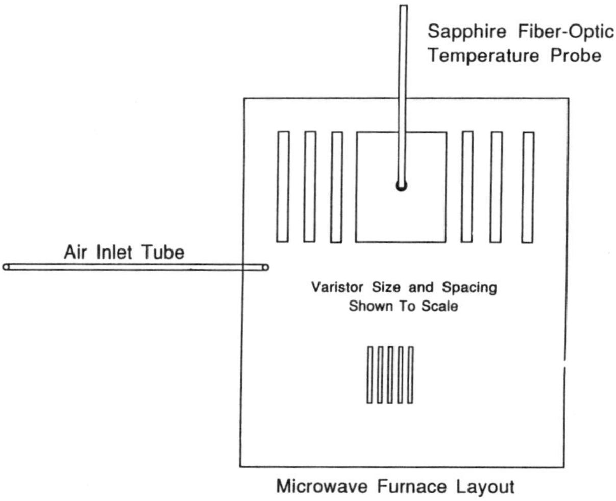

(b)

Figure 1. Experimental arrangement for sintering varistors. (a) Temperature was measured using a Pt-Rh thermocouple for conventional sintering. (b) The temperature for microwave sintering was measured using a sapphire fiber optic probe placed in a 0.14 in. diameter by 0.75-in.-deep hole.

In addition, we measured sample density and microstructure for sintering temperatures T with 950 °C < T < 1200 °C.

3. DATA AND DISCUSSION

In Figure 2 we present current density vs. applied field for varistors sintered at 950, 1000, 1100, and 1200 °C, respectively. Both dc measurements (low currents) and pulse data (high currents) are displayed. As is evident from the figure, at a given current density the breakdown field is lower for the microwave-sintered samples. Since it is generally

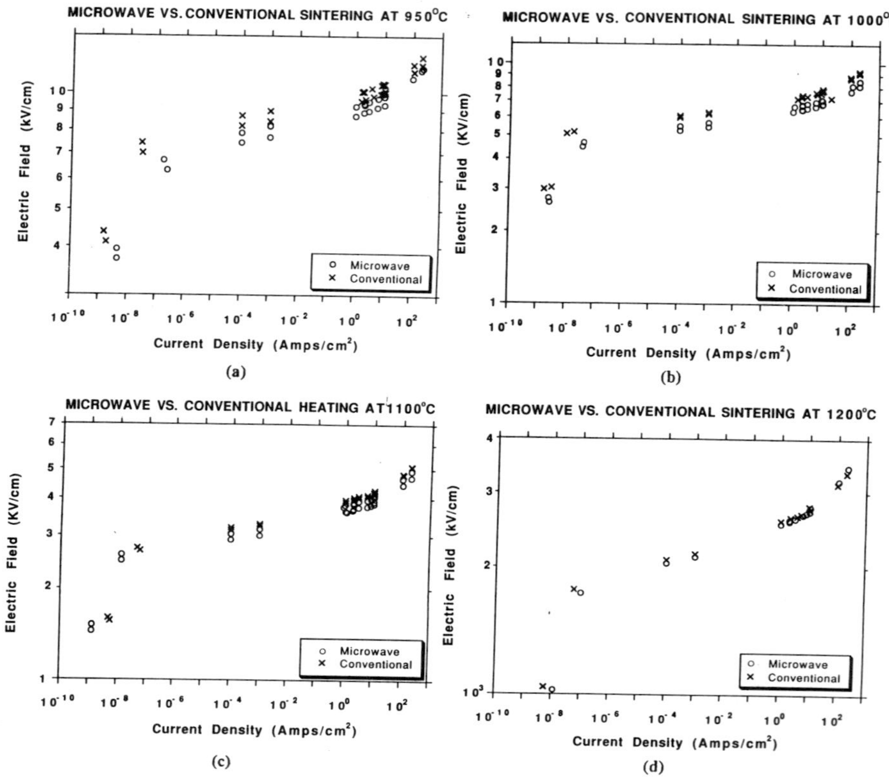

Figure 2. Electric field vs. current density for varistors sintered conventionally and using microwaves at (a) 950 °C, (b) 1000 °C, (c) 1100 °C and (d) 1200 °C.

accepted [7,8] that the breakdown field per grain boundary remains the same as the firing temperature is varied, these results imply that the grain size is larger for the microwave sintered samples, i.e., we have accelerated the grain growth because of microwave effects.

The degree of difference between the breakdown field for microwave-sintered samples and conventionally sintered samples decreases with increasing temperature. This difference is clearer in Figure 3 where we present the breakdown field at low-current density (1 mA/cm^2) vs. temperature. This result is not surprising since grain growth in ZnO varistor ceramics begins around 900 °C and is most rapid for $900 < T < 1100$ °C. At higher temperatures, the grain growth rate decreases and we would therefore expect little variation in grain size (or breakdown field) because of microwave enhancement of the sintering process.

The measured varistor density is consistent with the breakdown field results. Figure 4 gives sample density vs. sintering temperature, and at lower temperatures we again find increased density for the microwave-processed samples.

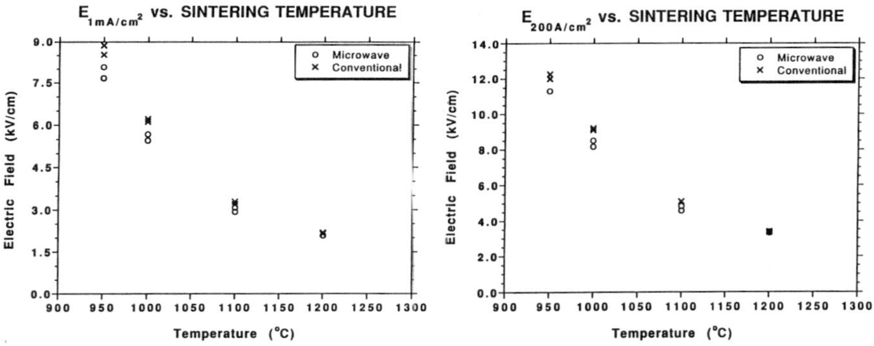

(a) (b)

Figure 3. Electric field vs. sintering temperature for conventional and microwave sintering. The field has been evaluated at a current density of (a) 1 mA/cm^2 and (b) 200 A/cm^2.

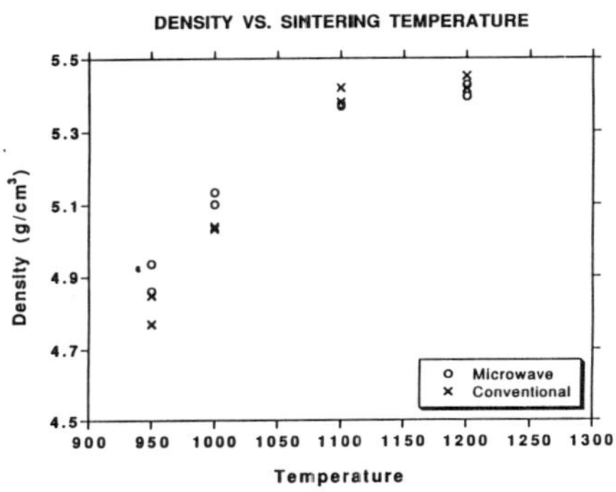

Figure 4. Bulk density vs. sintering temperature for conventionally processed and microwave sintered varistors.

In Figure 5 we present micrographs of polished and etched sections of the varistors. Grain growth and densification is evident as the temperature increases.

Aside from the apparent more rapid densification with microwaves, there appears to be little difference in the electrical behavior between microwave and conventional processing.

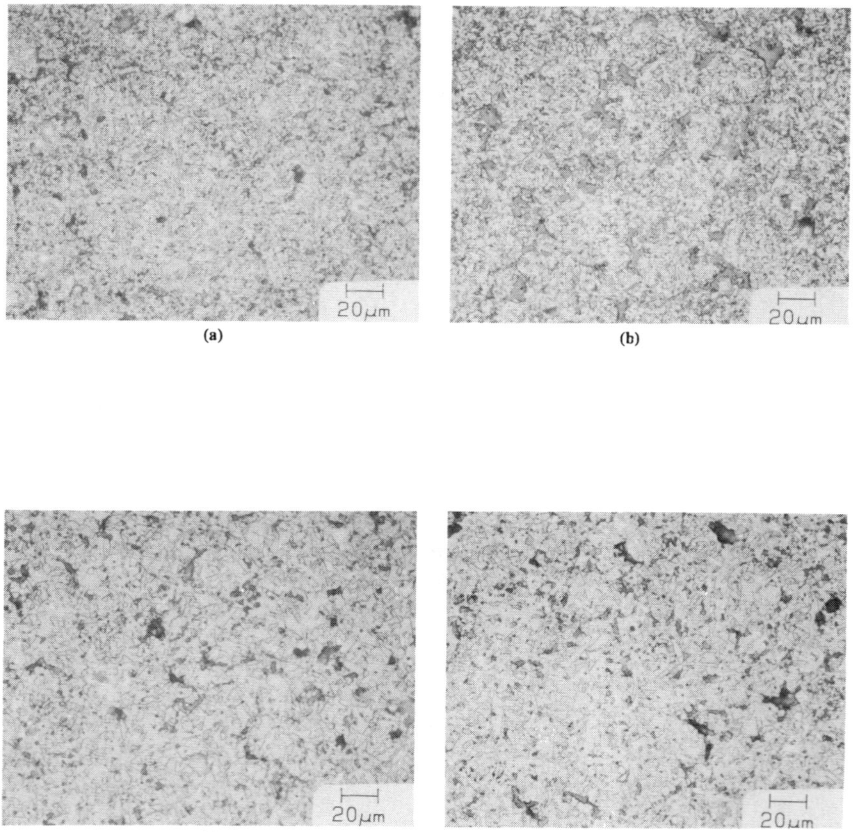

(a) (b) (c) (d)

Figure 5. Micrographs of polished and etched samples of ZnO varistors. The sintering temperatures and conditions are as follows: (a), (c), (e), (g), microwave at 950, 1000, 1100 and 1200 °C, respectively; (b), (d), (f), (h) conventional at 950, 1000, 1100, 1200 °C respectively.

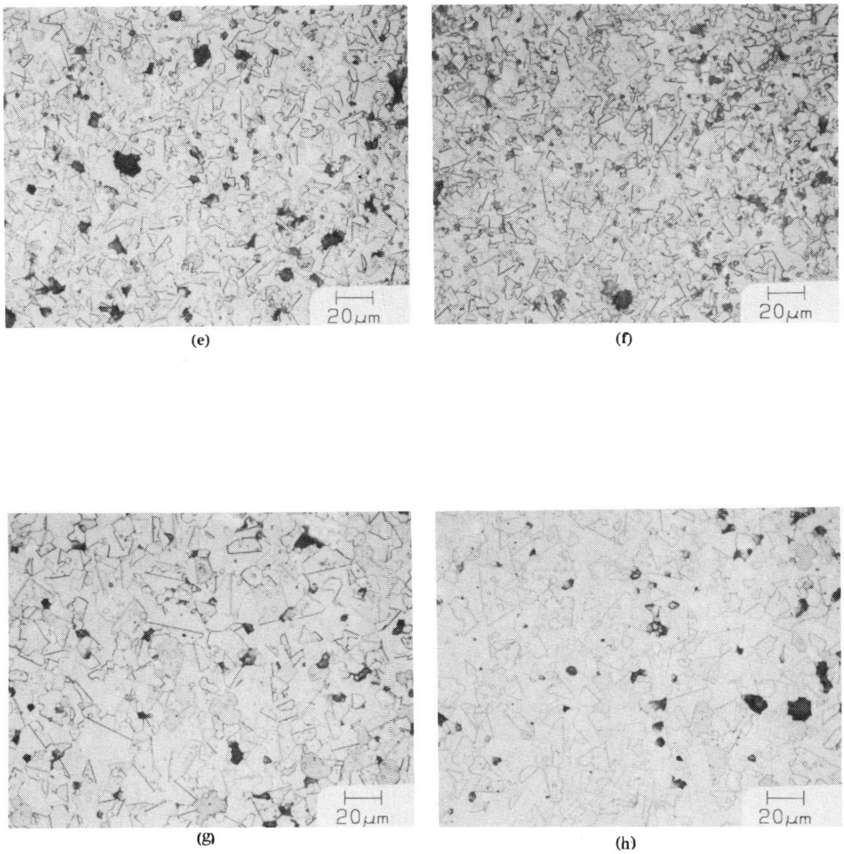

Figure 5. (Continued). Micrographs of polished and etched samples of ZnO varistors. The sintering temperatures and conditions are as follows: (a), (c), (e), (g), microwave at 950, 1000, 1100 and 1200 °C, respectively; (b), (d), (f), (h) conventional at 950, 1000, 1100, 1200 °C respectively.

Figure 6 gives the leakage current density vs. sintering temperature. No clear difference is observed between microwave and conventionally sintered samples. Leakage current is typically a sensitive indicator of varistor process differences and depends on the details of the ZnO grain boundary interface. Thus, the insensitivity of leakage current to microwaves is somewhat surprising if microwaves were to affect the sintering process or grain boundary diffusion in a special way.

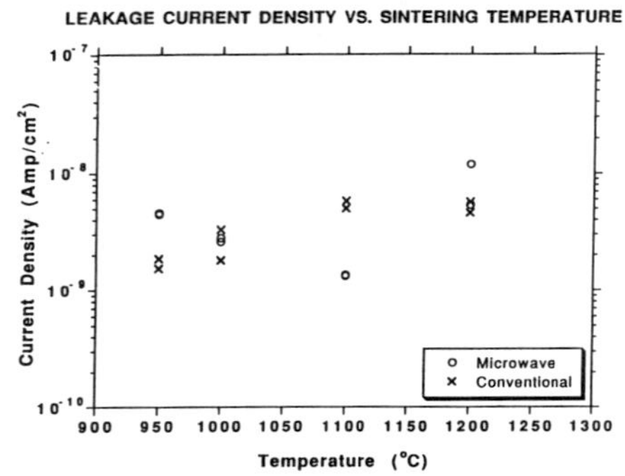

Figure 6. Leakage current density vs. sintering temperature for microwave and conventional processing.

In Figure 7 we plot another relatively sensitive measure of the quality of the varistor samples. The ratios $V_{2\ A/cm^2}/V_{0.1mA/cm^2}$ and $V_{200\ A/cm^2}/V_{2\ A/cm^2}$ give, respectively, low-current and high-current measures of the varistor nonlinearity. If these ratios were equal to unity we would have perfect varistor, i.e., a device with "infinite" nonlinearity. For an ohmic device, the ratio $V_{2\ A/cm^2}/V_{0.1\ mA/cm^2} = 2 \times 10^4$ and $V_{200\ A/cm^2}/V_{2\ A/cm^2} = 100$.

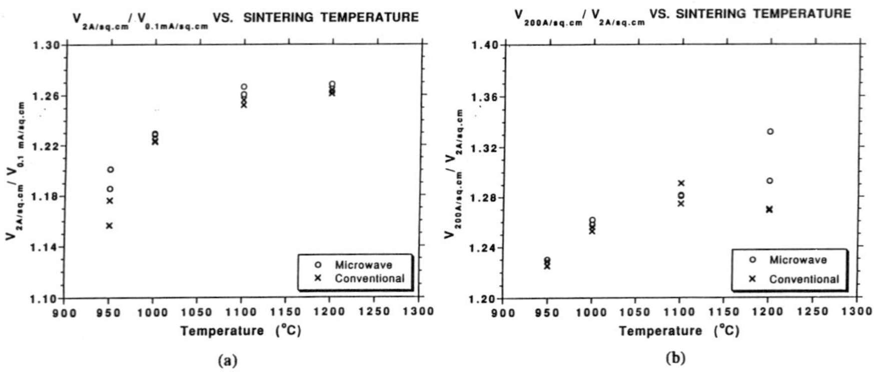

Figure 7. Ratio of varistor breakdown voltage measured at two current densities. The closer the ratio to unity, the more nonlinear the varistor:
(a) Ratio at low current densities, $V_{2\ A/cm^2}/V_{0.1\ mA/cm^2}$
(b) Ratio at high current densities, $V_{200\ A/cm^2}/V_{2\ A/cm^2}$

Figures 7a and 7b show that both the microwave and conventionally sintered devices are good varistors, i.e., have high nonlinearity (note the expanded scale of the graphs). There are no major differences evident between conventional and microwave-processed samples. From Figure 7a one might conclude that conventionally processed samples are slightly superior (ratio closer to unity), but the differences between the sample are close to the scatter in the data.

In Figure 8 we give the dielectric constant and the loss tangent for samples sintered at 950 °C. The shape of these curves has been previously discussed [11]. There is little difference between the curves for conventional and microwave-processed material except for the higher dielectric constant and a slightly higher loss tangent of the microwave samples. The effect on the dielectric constant is to be expected since the dielectric constant is proportional to the grain size, and Figures 2 and 3 show that the microwave-sintered samples have an apparently lower breakdown field and therefore larger grain size.

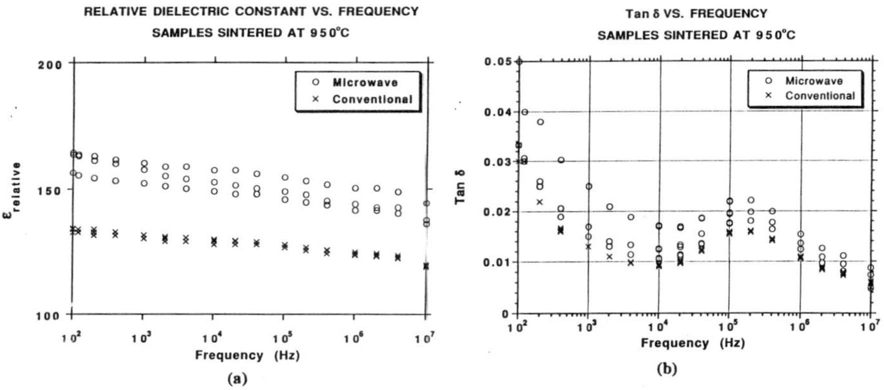

Figure 8. (a) Relative dielectric constant ε and (b) Loss tangent tan d for varistor samples sintered conventionally and using microwaves at 950 °C.

In view of the apparent insensitivity of the varistor leakage and nonlinearity to microwave processing, it was felt appropriate to re-examine with some care the apparent microwave-enhanced grain growth (lower varistor breakdown field) as seen in Figures 2 and 3. Using polynominal fits we have determined the rate of change with temperature of the varistor breakdown field at 1 mA. This data is given in Table 1.

From the data of Table 1 and Figure 3a we estimate that the observed shift in breakdown field between microwave and conventional samples corresponds to a temperature shift of about 14 °C, 14 °C, and 15 °C, for samples sintered at 950 °C, 1000 °C, and 1100 °C, respectively. These are quite small discrepancies in temperature, and it is therefore clear that a systematic error between the microwave and conventional temperature measurements could explain the observed microwave sintering enhancement observed above.

Table 1

Rate of Change of Breakdown Field with Temperature

$$\left(\frac{dE_{1mA/cm^2}}{dT}\right), \text{Vcm}^{-1}\ {}^{\circ}\text{C}^{-1}$$

	TEMPERATURE (°C)			
SAMPLE TYPE	950	1000	1100	1200
Conventional	-59	-42	-18	-6
Microwave	-54	-38	-15	-5

To evaluate potential sources of systematic error in the measured sintering temperatures, we decided to repeat the microwave-sintering experiment at 1000 °C with a slight change in the fiber optic probe configuration. Instead of simply inserting the fiber optic probe in the varistor as shown in Figure 1, we threaded the probe through an additional 0.5 in. of presintered varistor material placed directly above the large (1.2 in. by 1.5 in. diameter) varistor of Figure 1. Since the presintered varistor material (a cylinder 0.5 in. long x 0.625 in. diameter) couples to the microwave field, it has the effect of "heat sinking" the sapphire fiber, i.e., any heat loss along the sapphire is lessened or eliminated. While the magnitude of such an effect is hard to estimate, it is clear that in the absence of such a heat source, heat conduction along the sapphire fiber to the cold microwave cavity will result in the fiber temperature reading being lower than the true varistor temperature.

In Figure 9 we give the measured varistor breakdown field at a current density of 1 mA/cm^2 for samples fired at 1000 °C (a) conventionally, (b) with microwaves, and (c) with microwaves and the "heat source" arrangement described above. Clearly the "heat source" arrangement has resulted in a smaller apparent grain growth enhancement caused by microwave sintering. This is consistent with the view that the apparent enhanced sintering and decreased varistor breakdown field resulting from microwave firing arises from systematic temperature discrepancies between microwave and thermal processing.

4. CONCLUSIONS

ZnO varistors are electronic ceramics whose electrical behavior is dominated by grain boundary effects[7,8]. A careful evaluation of the effect of microwave sintering compared to conventional sintering revealed no significant differences between varistor devices processed by these two methods. A slight apparent enhancement of grain growth was observed for microwave processing but further evaluation of this effect leads us to believe that it arises from a small systematic shift in temperature derived from the different experimental configurations. While the absence of a "microwave effect" in sintering ZnO varistors is not directly relevant to the reported microwave effects in alumina and zirconia, one should perhaps generally consider possible errors caused by the cold walls of the microwave enclosure. In particular, ceramics that are transparent to the IR might produce errors in measurements when pyrometers or thermocouples inadvertently view cold portions of a cavity through a partly transparent ceramic sample.

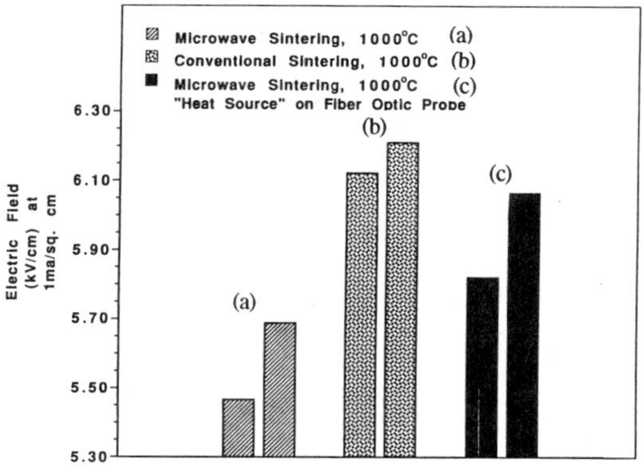

Figure 9. Breakdown field at 1 mA/cm^2 for varistors sintered:
(a) using microwaves at 1000 °C
(b) conventionally at 1000 °C
(c) using microwaves as in Figure 1(b) but with added varistor material placed around the sapphire fiber optic probe in the microwave cavity to act as a "heat source." The breakdown field in case (c) lies between that of (a) and (b)

5. REFERENCES

1. M.A. Janney and H.D. Kimrey, in "Microwave Processing of Materials II", edited by W.B. Snyder, W.H. Sutten, M.F. Iskander and D.L. Johnson (Mater. Res. Soc. Proc. 189, San Francisco, 1990), pp. 215-227.
2. H.D. Kimrey, J.O. Kiggans, M.A. Janney and R.L. Beatty, ibid., 189, p. 243 (1990).
3. M.A. Janney and H.D. Kimrey, in "Ceramic Powder Science II," pp. 919-924, G.L. Messing, E.R. Fuller , and H. Hausner, eds., American Ceramic Society, Westerville, Ohio, 1988.
4. M.A. Janney and H.D. Kimrey, in "Advances in Sintering," J. Bieninger and C. Handwerker, eds., American Ceramic Society, Westerville, Ohio, 1990.
5. D.L. Johnson, J. Am. Ceram. Soc. 74, p. 849 (1991).
6. J.P. Katz, R.D. Blake and V.M. Kenkre, Ceramic Trans. 21, 95 (1991); J.H. Booske, R.F. Cooper, I. Dobson and L. McCaughan, ibid., p. 185.
7. L.M. Levinson and H.R. Philipp, "Application and Characterization of ZnO Varistors" in "Ceramic Materials for Electronics," R.C. Buchanan, Ed., (1991) (in press).
8. L.M. Levinson and H.R. Philipp, Bull. Amer. Ceram. Soc. 65, p. 639 (1986).
9. L.M. Levinson and H.R. Philipp, J. Appl. Phys. 47, p. 3116 (1976).
10. G. McMahon, A. Pant, R. Sood, A. Ahmad and R.T. Holt, in "Microwave Processing of Materials II," edited by W.B. Snyder, W.H. Sutton, M.F. Iskander and D.L. Johnson (Mater. Res. Soc. Proc. 189, San Francisco, 1990), pp. 237-242.
11. L.M. Levinson and H.R. Philipp, J. Appl. Phys. 47, p. 1117 (1976).

DENSIFICATION KINETICS OF ALUMINA DURING MICROWAVE SINTERING

Jiping Cheng, Jinyu Qiu, Jian Zhou, and Neng Ye
Wuhan University of Technology, Advanced Materials Research Institute
Wuhan 430070, China

ABSTRACT

Some kinetics parameters of alumina during microwave sintering were studied and compared with that during conventional sintering. The results demonstrated that the sintering rates for microwave processing were much greater than that for conventional processing, and the grain growth of alumina was rapid with prolonged time at high temperature in a microwave field. It was indicated that the microwave sintering at higher temperatures for a shorter time was favorable for preparing high density and fine-grained alumina ceramics.

INTRODUCTION

Microwave processing is generally characterized by rapid heating and high efficiencies. Numerous investigations have shown that microwave sintering leads to improved ceramic microstructure with better properties[1]. But few studies were undertaken on the sintering kinetics for microwave sintering of ceramics. In present work, some densification kinetics of alumina ceramics during microwave sintering were studied and compared with that during conventional sintering. The results may give some further understanding of microwave processing of ceramics.

EXPERIMENTAL

α-Al_2O_3 powder having a purity of 99.8% and average particle size of o.4~0.5μ doped with 0.2 wt% MgO was used to prepare sample rods. The powder mixture with an added 1 wt% polyvinyltyral as a binder, was wet-milled in water for 48 hours. After drying and sieving through a 60-mesh screen, the rods with size of Φ16x20mm were isostatically pressed at 350MPa. The initial green density was about 52%.

Microwave sintering was performed using a modified TE_{103} single-mode resonator cavity operating at 2.45GHz[2].The sample rod was surrounded by alumina-silica fiber insulator and placed in the maximum electric field region in the cavity. The temperature of the sample was measured by infrared thermometer through a hole in the insulator.

All samples were microwave heated with a rate of 230~250°C per minute by controlling the microwave power. Typically, the power level was controlled in 300~400W in initial heating stage, and gradually increased to 600~800W at the onset sintering temperature then regulated to keep temperature constant. Samples were also sintered by conventional heating using a molybdenum-silicide furnace as contrast.

The shrinkage of sintered samples was measured with a tolerance of 0.02mm and the density was determined by the Archimedes method. The microstructures of samples were observed by TEM and the average grain size was determined by image-analysis method.

RESULTS AND DISCUSSION

1. Densification of Alumina During Microwave Sintering

The density and the length shrinkage of alumina samples with different microwave sintering schedules were shown in Fig.1 and Fig.2, respectively. The densification rate can be expressed

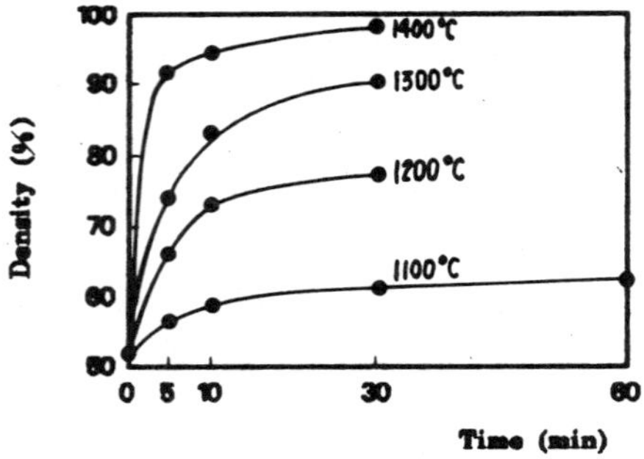

Fig. 1 Densification of alumina during microwave sintering

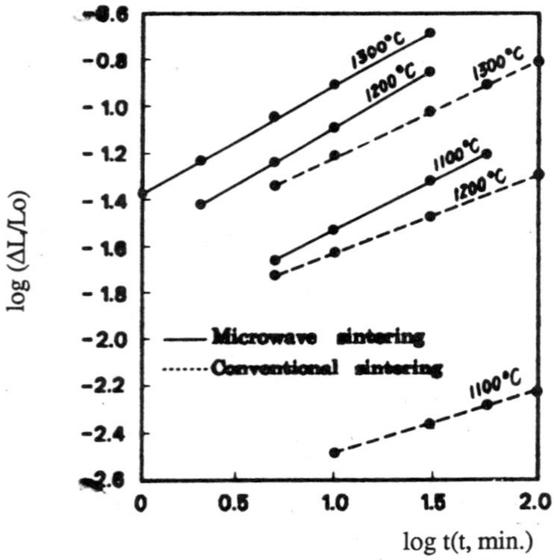

Fig. 2 The shrinkage of alumina versus the sintering time

in form of [3]:

$$\frac{\Delta L}{L_0} = Kt^{1/n} \tag{1}$$

where $\Delta L/L_0$ is the length shrinkage, K is the densification rate constant, t is time and n is correlated to diffusion mechanism during sintering. From the results shown in Fig.2, the densification rates of alumina during microwave sintering at different temperatures were obtained

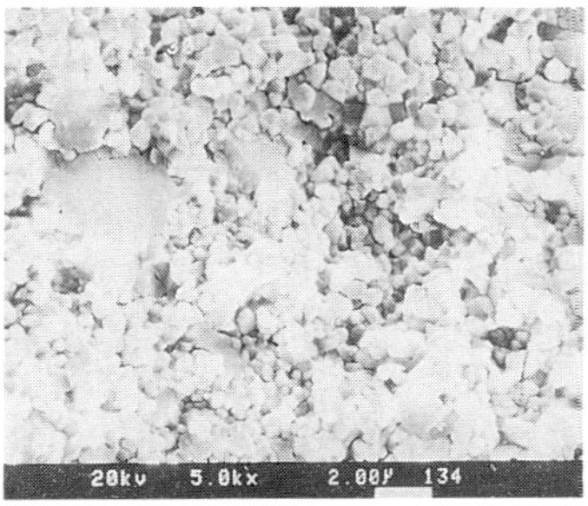

(a) 1350℃ × 10min

(d) 1450℃ × 3min

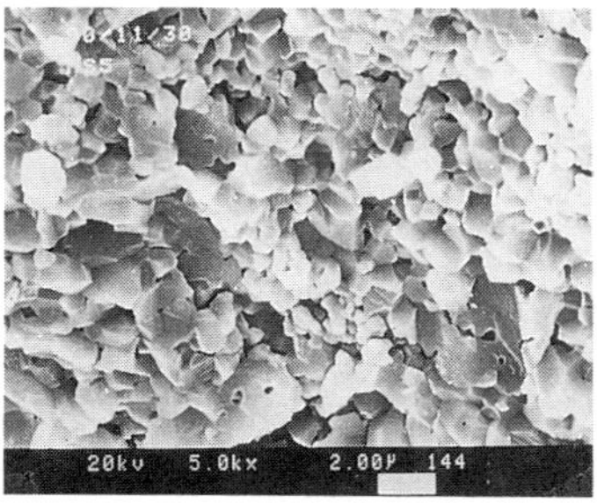

(b) 1450℃ × 10min

(e) 1450℃ × 30min

(c) 1550℃ × 10min

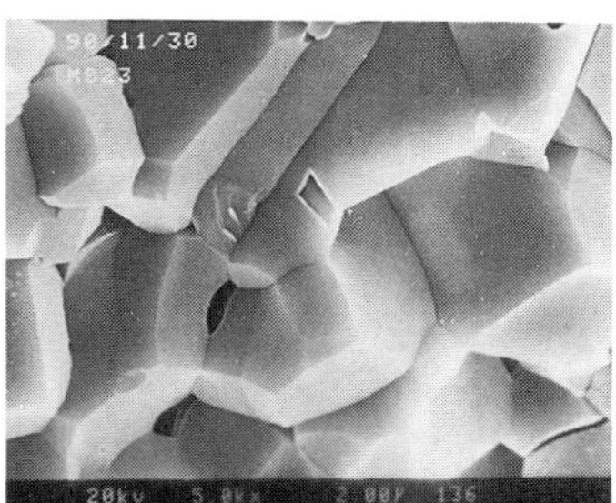

(f) 1450℃ × 60min

Fig. 3 Microstructure of alumina sintered by microwave

as follows:

at 1100°C $\Delta L/L_0 = 1.022\times10^{-2}\, t^{1/2.22}$ (2)

at 1200°C $\Delta L/L_0 = 2.864\times10^{-2}\, t^{1/2.16}$ (3)

at 1300°C $\Delta L/L_0 = 4.221\times10^{-2}\, t^{1/2.20}$ (4)

2. Grain growth of alumina during microwave sintering

The microstructures of the alumina samples sintered by microwave were shown in Fig.3, and the average grain size of the samples was shown in Fig.4.

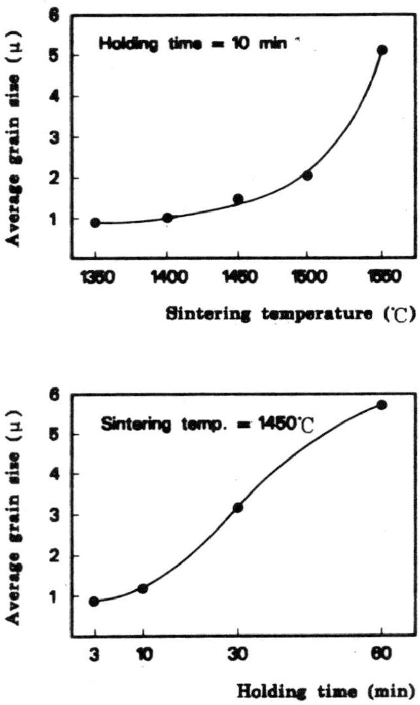

Fig. 4 Grain growth of alumina during microwave sintering

The grain growth rate dG/dt of ceramics during sintering can be expressed as following equation[4]:

$$\frac{dG}{dt} = \frac{CD}{G^n(1-\rho)^m} \qquad (5)$$

where G is the grain size, ρ is the density, D is the diffusion coefficient and C is constant. The indexes n and m are correlated to diffusion mechanism.

By regression analysis on the results shown in Fig.4, the grain growth rate of alumina during microwave sintering was given as

$$\frac{dG}{dt} = \frac{CD}{G(1-\rho)^2} \tag{6}$$

here the values of the indexes n and m were 1 and 2, respectively, which indicated that the grain growth of alumina during microwave sintering fit the model of lattice-diffusion-controlled mechanism[5].

3. Comparison of Kinetics Between Microwave and Conventional Sintering

From the results shown in Fig.2, the densification rates of alumina for conventional sintering were calculated as follows:

at 1100°C $\quad \Delta L/L_0 = 1.181\text{x}10^{-3}\, t^{1/2.72}$ (7)

at 1200°C $\quad \Delta L/L_0 = 9.536\text{x}10^{-3}\, t^{1/2.56}$ (8)

at 1300°C $\quad \Delta L/L_0 = 2.170\text{x}10^{-2}\, t^{1/2.48}$ (9)

In general, the densification rate constant K in equation(1) is related to such parameters[6]:

$$K = f\,(\gamma, a, \Omega, D, T) \tag{10}$$

where γ is the surface energy of the ceramic particles, a is the radius of the powder, Ω is the volume of the vacancy, D is the diffusion coefficient, and T is temperature.

For the same ceramic samples at same temperature, the parameters to effect the value of K such as γ, a, Ω, are identical, then in this condition we can get

$$K = CD \tag{11}$$

where C is a constant.

Therefore, the differences of the values of K between microwave and conventional sintering shown in the above densification rate equations indicated directly the disparity of the diffusion coefficient for two kinds of sintering process. For example, it is obtained at the temperature of 1200°C that

$$\frac{D_w}{D_c} = 3 \tag{12}$$

where Dw and Dc are the diffusion coefficient for microwave and coventiodal sintering, respectively.

4. Preparing of Ultrafine-Grained Alumina by Microwave Sintering

The ultrafine α-Al_2O_3 powder with average particle size of 800Å added with 0.5 wt.% MgO was prepared[7], and sintered by microwave with a very high heating rate (about 600°C/min) and shorter sintering time to prevent a excessive grain growth. Thus an ultrafine microstructure of alumina with grain size less than 0.3μm and density greater than 95% had been obtained(Fig.5.)

Fig. 5 Microstructure of ultrafine-grained alumina made by microwave sintering

CONCLUSIONS

The densification rate of alumina during microwave sintering was much higher compared with conventional sintering. The results indicated that a greater diffusion coefficient in microwave processing may be the main factor to accelerate the densification rate of alumina ceramics. A higher heating rate and shorter sintering time make it possible to prepare ultrafine-grained alumina ceramics.

REFERENCES

1. R.W. Bruce, Microwave Processing of Materials, edited by W.H.Sutton et al.(Mater. Res. Soc. Proc. 124, Pittsburgh, PA 1988), pp. 3-16
2. J.P. Cheng, Ph.D thesis, Wuhan University of Technology, 1991
3. J. Daniels and K.H. Hardfl, Philips Res. Repts., 31, 489-543 (1976)
4. R.J. Brook, in Treatises on Materials Science and Technology, edited by F. F. Wang (Academic Press, New York, 1976), pp. 331-364
5. R.J. Brook, Fabrication Principles for the Production of Ceramics with Superior Mechanical Properties, (Proc. Br. Ceram. Soc., 32, 1982), pp. 7-24
6. H. Salmang and H. Scholze, KERAMIK, (Springer-Verlag Berlin Heidelberg New York, 1982), pp. 213-237
7. W.Y. Zhou, ME thesis, Wuhan University of Technology, 1990

MICROWAVE SINTERING OF SI_3N_4 CERAMICS

Jinsong Zhang,Lihua Cao,Fei Xia
Institute of Metal Research, Academic Sinica, Shenyang 110015, P.R.China

Abstract

The Si_3N_4 ceramics(91wt%Si_3N_4+5wt%Y_2O_3+4wt%Al_2O_3) of room temperature strength 620--760 MPa and fracture toughness 7 MPa·$m^{1/2}$ were prepared by microwave sintering and their microstructure and mechanical properties were studied.The experiment results show that the higher N_2 pressure (>4atm) is very necessary for microwave sintering of Si_3N_4 ceramics; microwave sintering can greatly decrease the sintering temperature to 1500°C, and increase the transformation rate of α to β-Si_3N_4,and reduce the total sintering time to about 1 hour. The mechanical properties of the Si_3N_4 ceramics sintered by microwave are better than that by conventional pressureless sintering.

I.Introduction

Microwave sintering is a new and prospective ceramic manufacturing process , which can greatly cut down the energy loss and processing time compared with conventional process. Since it was proposed in the 1970s, the Al_2O_3, ZrO_2, PTC, YBCO and other oxide ceramics have been sintered successfully by using this technology, and the possibility of microwave sintering SiC and B_4C ceramics also was studied[1-4]. But it is very regrettable that there is less work reported on microwave sintering of Si_3N_4 ceramics. In fact, the excellent room and high temperature mechanical properties have made the silicon nitride ceramics become one of the most important non-oxide ceramics. So, the successful microwave sintering of Si_3N_4 ceramics must be of great value to cutting down its manufacturing cost and enlarging the applied scale of microwave sintering. In this work, the Si_3N_4(5wt%Y_2O_3+4wt%Al_2O_3) ceramics were sintered by 2.45 GHz microwaves with high nitrogen pressure. Mechanical properties and microstructure were studied.

II. Experimental Procedure

The raw composition of the Si_3N_4 ceramic sample prepared was as follows: Si_3N_4 + 5%wt Y_2O_3 + 4%wt Al_2O_3, in which Y_2O_3 and Al_2O_3 were all analytically pure. The chemical composition of the Si_3N_4 powder is listed in table I.

Table I. The chemical composition of Si_3N_4 powder

$Si_3N_4(\alpha)$wt%	Si_f(wt%)	O(wt%)	phase	d(μm)	BET(m^2/g)
90	<1.0	1.4	α,β	1.0	16-18

The Si_3N_4 samples with size of OD20x10(mm) and 45x8x6(mm) were cold isostatically pressed (200MPa), and then sintered by microwave sintering in high N_2 pressure and conventional pressureless sintering respectively.The linear shrinkages of samples sintered at several predetermined temperature were measured.The phase composition and microstructure of sintered Si_3N_4 compacts were studied by means of XRD and TEM respectively.Their bend strength and fracture toughness were determined by the 3--point bending and the indentation microfracture methods.

III. Results and discussion

3.1 Effect of N_2 pressure on the weight loss of Si_3N_4

Fig.1 is the photo of microwave sintering equipment made by our institute, it can work in conditions of vacuum, pressureless, and high atmosphere pressure.

Fig.1 The photo of microwave sintering furnace made by our institute

Fig.2 shows the relationship between the weight loss of Si_3N_4 and the applied N_2 pressure in microwave sintering at 1500°C for 10min.,by which it can be seen that along with the increase of N_2 pressure, the weight loss rapidly decreased from about 15% at 1 atm N_2 to approximate zero at 5 atm N_2. This result significantly shows the importance of N_2 pressure for the successful microwave sintering of Si_3N_4 ceramics.

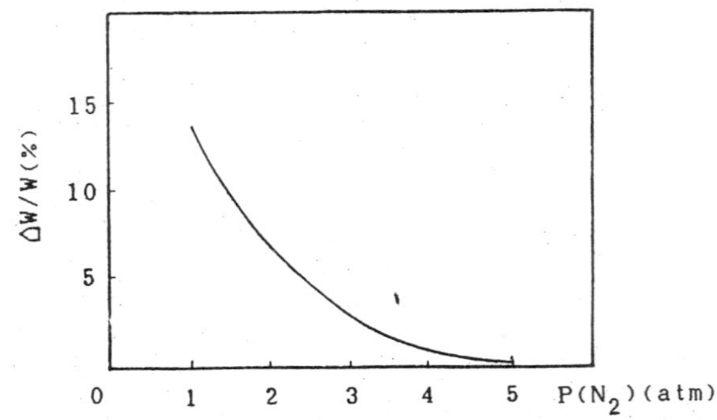

Fig.2 The curve of weight loss of Si_3N_4 ceramics

3.2 Relationship of the linear shrinkage and phase content with sintering temperature

Fig.3 is the ΔL/L--T curves of the Si_3N_4 ceramics sintered by both conventional and microwave sintering. The phase composition of these samples are listed in table II. Fig.4 shows the relationship of β-Si_3N_4 content with sintering temperature for two sintering processes.

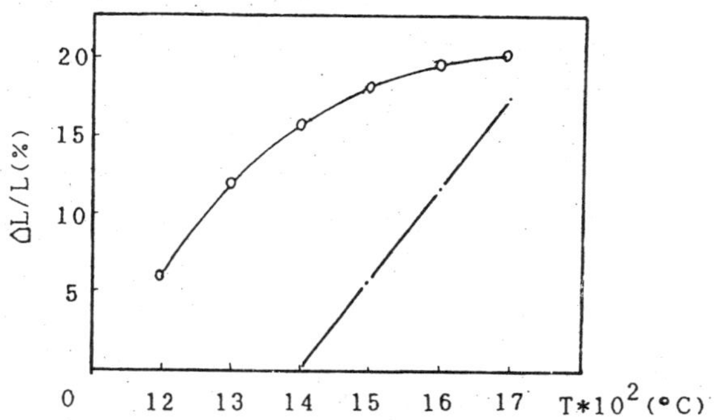

Fig.3 The relationship of ΔL/L with temperature
—o— microwave sintering —·— conventional sintering

Table II. The relationship of phase and temperature

Temp. (°C)	phase	
	microwave sintering	conventional sintering
1200	ß,α,Al_2O_3,$Y_2Si_3N_4O_3$	
1300	ß,α,$Y_2Si_3N_4O_3$,Si_2N_2O	
1400	ß,α,Si_2N_2O,$Y_2Si_3N_4O_3$	Al_2O_3,ß,α,$Y_2Si_3N_4O_3$
1500	ß,α	ß,α,$Y_2Si_3N_4O_3$
1600	ß,Si_2N_2O	ß,α,Si_2N_2O,$Y_2Si_3N_4O_3$
1700	ß	ß,α

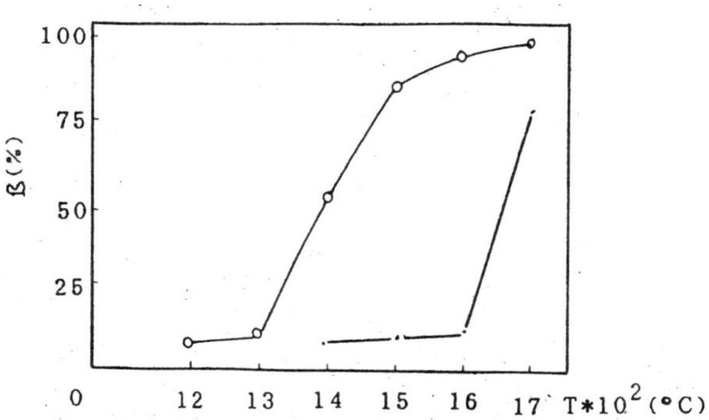

Fig.4 The relationship of ß-Si_3N_4 content with temperature
—o— microwave sintering —·— conventional sintering

The microwave sintering condition is T°C*10min*5atm, this means the Si_3N_4 samples were heated at 40°C/min to the predetermined temperature T and followed by 10min hold under N_2 pressure 5atm, then cooled down without power.The conventional sintering conditions are that the samples were heated at 10°C/min to the predetermined temperature,holding

30 mins. under 1 atm.N_2 pressure. The temperature of all samples were measured by the infrared temperature sensor.

From above curves and tables, it can be seen that by using microwave sintering at 1200 ° C, the liquid phase Y-Si-N-O appeared and the linear shrinkage (▲L/L) reached 6%; at 1300 ° C,α-Si_3N_4 to β-Si_3N_4 transformation started and ▲L/L was 15% ;at 1600 ° C,all α-Si_3N_4 had been transformed to β-Si_3N_4 and ▲L/L was equal to 21% . But by using conventional pressureless sintering, at 1400 ° C,the same liquid phase just turned up and the shrinkage didn't start; at 1600 ° C,▲L/L was about 15%, the transformation of α to β-Si_3N_4 was just starting; at 1700 ° C,about 25% α-Si_3N_4 yet remained and ▲L/L only reached 18% .These results clearly indicate that the densification by both microwave and conventional sintering of Si_3N_4 ceramics involved liquid formation, grain rearrangement and phase transformation: The microwave sintering greatly lowered the temperature of the liquid formation and the phase transformation, and then obviously increased the rate of the grain rearrangement and the phase transformation of α to β-Si_3N_4.

3.3 Mechanical properties and microstructure

The strength and toughness of the Si_3N_4 ceramics are listed in table III. These samples were sintered by microwave at 1500 ° C for 10 min with 5 atm N_2 and conventional sintering at 1750 ° C for 2 hours with 1 atm N_2.

Table III. The mechanical properties

sintering process	bend strength	fracture toughness
microwave sintering	620--760 MPa	7.0 MPa·$m^{1/2}$
conventional sintering	450--540 MPa	6.2 MPa·$m^{1/2}$

Fig.5 ,6 are TEM photos of the sintered Si_3N_4 ceramics, the former microwave sintered, the later conventionally sintered.

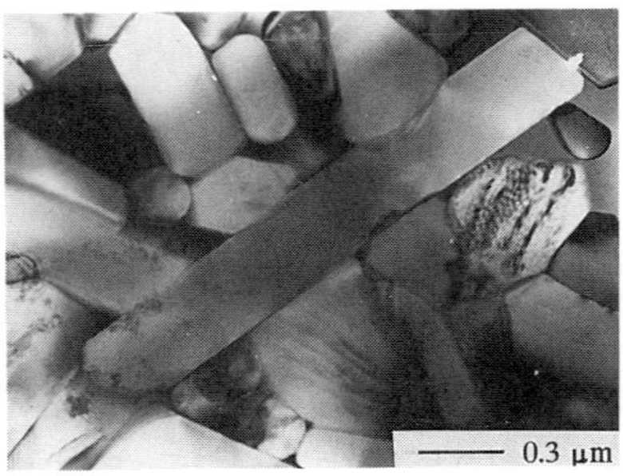

Fig.5 TEM of microstructure of microwave sintered Si_3N_4 ceramic at 1500° C for 10 min.

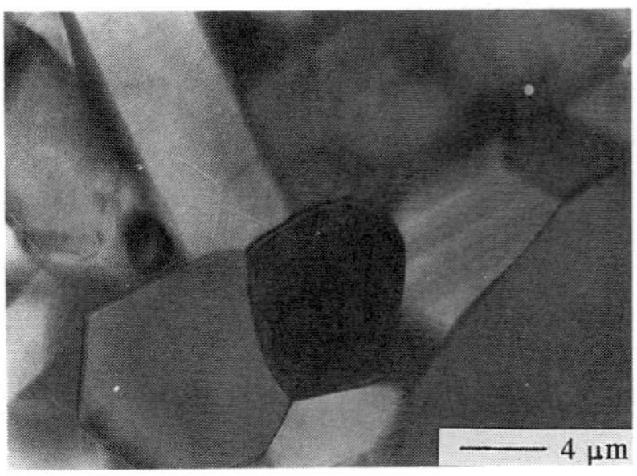

Fig.6 TEM of microstructure of conventionally sintered Si_3N_4 ceramic at 1750° C for 120 min.

By comparing these photos, an obvious difference in microstructure for microwave and conventional sintering is clearly revealed. The grains of the Si_3N_4 sintered by microwave are much finer and better-distributed than that by conventional sintering. The radial size of needle-like β-Si_3N_4 is about 0.3 μm for microwave sintering, but 4--5 μm for conventional sintering. That explains the higher strength of the microwave sintered Si_3N_4. It can be foretold that the optimization of the microwave sintering process and material composition will increase the mechanical properties of the Si_3N_4 ceramics.

3.4 Effect of microwave on liquid phase sintering

Many published results point out that the densification of Si_3N_4 ceramics are mainly influenced by the processes of particle rearrangement caused by formed liquid phase and the transformation of α to β-Si_3N_4. For densification, the particle rearrangement, which is strongly dependent upon the viscosity and quantity of the liquid and its wetting properties, has much more contribution than the phase transformation. For the mechanical properties, the transformation of α to β-Si_3N_4 is one of the most important factors with rate determined by the rates of α-Si_3N_4 dissolving in, diffusing ,and precipitating out of the liquid phase.So all factors, which can decrease the liquid viscosity and increase the rate of the solution-reprecipitation process,will be beneficial for the densification of Si_3N_4 ceramics.

The microwave enhancement of sintering of Si_3N_4 ceramics may be explained by high temperature zone in the sintered compact. Let us first consider several temperatures: the apparent temperature T is the measured temperature of sintered sample;the T_{si} is the temperature of the high temperature zone made up of the interfaces and surfaces of all grains in the sintered compact; and the T_l is the formation temperature of the liquid phase,which is not changed by microwaves. In microwave sintering of the Si_3N_4 ceramics, the T_{si} is postulated to be greater than the T. As the $T_{si} > T_l > T$, the liquid phase forms at the interface of Si_3N_4 grains,and its temperature is also higher than the apparent temperature. Microwave lowering of the transformation temperature of α to β-Si_3N_4 also can be understood through this point of view. Because the liquid viscosity is negatively related to temperature, at the same apparent temperature, microwaves will make the liquid viscosity much lower than in conventional sintering. Besides this, the coupling of microwave with formed liquid will increase the vibration rate and amplitude of the liquid molecules to decrease the liquid viscosity.

The reduced liquid viscosity caused by above two reasons naturally increases the rate of the grain rearrangement to promote the densification process, and the Si and N atoms diffusion in the liquid to accelerate the transformation of α to β-Si_3N_4.

In other words, decreasing the liquid viscosity may be a main reason for microwaves enhancing the liquid phase sintering of Si_3N_4 ceramics.

VI. Summary

1. High nitrogen pressure (>4atm) is a prerequisite for microwaves to sinter Si_3N_4 ceramics successfully.
2. Microwave sintering can greatly decrease the sintering temperature and time of Si_3N_4 ceramics.
3. The Si_3N_4 ceramics sintered by microwave have more perfect microstructure in which the grains are finer and better-distributed, and have better mechanical properties than that sintered by conventional pressureless sintering.
4. Decreasing the liquid viscosity may be the principal mechanism of microwave enhancement of the sintering of Si_3N_4 ceramics.

Acknowledgments

This research was supported by the Chinese high technology plan.

References

1.M.A.Janney,in:Ceramic Transactions,Ceramic Powder Science,II,B, ed. by G.L.Messing,etl.(Amera. Ceram. Soc., 1988)
2.B.Swain, Met.Prog.,134(1988)3
3.Y.L.Tian,in:Microwave processing of Materials,vol.124, ed. by W.H.Sutton(Materials Research Society,1988)
4.J.D.Kate,in:Microwave Processing of Materials, vol.124, ed. by W.H.Sutton (Materials Research Society,1988)

MICROWAVE JOINING OF CERAMICS: A STUDY ON JOINING ALUMINA BOTH DIRECTLY AND WITH ALUMINA GEL

Salwan Al-Assafi* and David E. Clark**
University of Florida, Department of Material Science and Engineering, Gainesville, FL 32611

ABSTRACT

Alumina ceramics of different purities were joined using alumina gel as an interlayer and microwave energy as the heating source. Direct microwave joining of alumina was also accomplished. A comparison between the two joining methods was made. The effect of the gel on joining time and joining temperature was determined. Knoop microhardness was measured across the joint to show the change in the microhardness at the joint region. Joints of hardness as high as that of the bulk were produced when the joining parameters were properly selected. The dielectric properties of the aluminas used in this study were measured. X-ray maps and SEM micrographs were taken for different joints.

INTRODUCTION

Ceramics are generally hard and brittle. These properties cause a lot of difficulty in manufacturing complex shapes. Furthermore, the thermal conductivity of most ceramics is low, and this results in substantial temperature gradients when processing large articles. Joining of ceramics was investigated as a way of overcoming the obstacles associated with processing of large and/or complex shapes. Small and simple pieces can be manufactured separately and then joined together to produce larger shapes.

Microwave processing of ceramics has been investigated recently, due to the unique heating properties that can be achieved with this processing technique. Unlike heating with conventional sources, which normally results in large thermal gradients, heating using microwaves assists in reducing thermal gradients across the material[1]. Microwaves interact with the material, causing the sample itself to generate heat instead of receiving it from the surroundings, as is the case with conventional heating.

Sealing of ceramics using microwave energy was first investigated by Meek and Blake[2]. A glass slurry mixed with a coupling agent was used to join different kinds of ceramics. The microwave oven used was a multi-mode 2.45 GHz microwave with a power of 700 W. Single-mode cavities have also been used for joining of ceramics. H. Fukushima et al. were successful in joining low purity (92%) alumina directly[3]. For joining time of 3 minutes, joining temperature of 1750°C, and pressure of 0.6 MPa, a joint strength was obtained which was equal to the strength of the bulk alumina before joining. High purity alumina was joined using a low purity alumina interlayer. Low purity silicon nitride was also joined directly, and the joint strength was equivalent to the original strength. Different materials were tested as interlayers in joining SiC to SiC in the study carried out by R. Silberglitt et al.[4]. Interlayers composed of Si, C, and Ti were used in joining SiC in both single-mode and multi-mode cavities. In addition, a study on the utilization of combustion synthesis in the interfacial region was also carried out. A cold pressed Ti:C disk was used as an interlayer. The temperature of the SiC disks to be joined rose rapidly to 1500°C after ignition occurred in the interlayer.

In this study, alumina gel was used to join aluminas of different purities. The sol-gel process is an important technique in processing of technical ceramics due to the unique properties that can be achieved. A sol is a dispersion of colloids in a liquid. Colloids are solid particles with diameter in the range of 10-100 nm[5]. The gel is formed by reducing the liquid content of the sol. The formation of gel is normally accompanied by a rapid increase in the viscosity. Microwave energy in a multi-mode cavity was used for the joining process. The controlling parameters were determined by changing joining temperature, joining time, and gel concentration. Direct joining of alumina was also carried out. The advantages of using alumina gel and microwave energy in the joining process were evaluated. In this paper, the term "joint" refers to the interface of two pieces after bonds were formed between them.

EXPERIMENTAL PROCEDURE

Sample Preparation

The alumina sol was prepared according to a process which was first described by Yoldas and was modified by Clark et al.[6] The process consisted of two stages. Hydrolysis of aluminum-sec-butoxide, the precursor, was the first stage:

$$Al(OC_4H_9)_3 + H_2O \rightarrow Al(OC_4H_9)_2(OH) + C_4H_9OH \quad [1]$$

Aluminum-sec -butoxide 90°C Butanol

The second stage was the peptization stage, in which charges were formed on the colloid surfaces:

$$Al(OC_4H_9)_2(OH) + 2H_2O + H^+ \rightarrow \overbrace{AlOOH_2^+ + 2C_4H_9OH + H_2O}^{\text{SOL}} \quad [2]$$

Colloids

The resulting sol had suspended colloid particles in the submicron range. When sufficient liquid is evaporated from the sol, its viscosity will increase. Three different gel concentrations, referred to as gels 1, 2, and 3, were prepared by reducing the liquid content by different amounts. Two of the three gels are actually viscous sols (gel 1 and gel 2). The third, gel 3, was a true gel. Figure 1 shows the particle size for gel 1 and gel 2. Gel 3 would fall to the far right of gel 2. Gel 2 was used in all joining experiments, since it was found that this gel is the best of the three based on ease of processing.

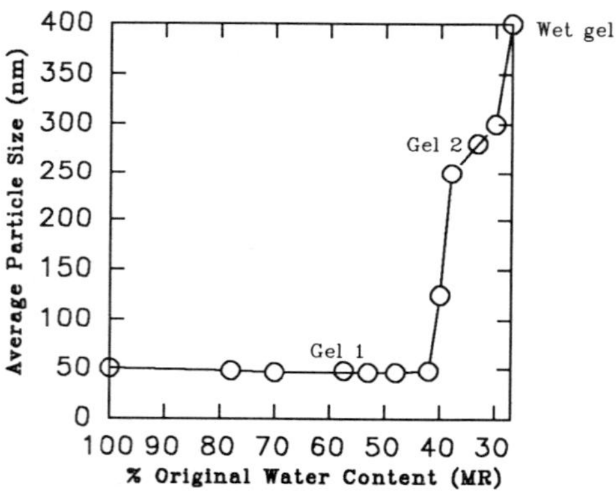

Figure 1. Change of particle size with sol water content[7].

Joining Process

Alumina gel was brushed onto the surfaces to be joined. Since alumina does not absorb microwaves efficiently at room temperature, a suscepting cavity was used to heat samples to a temperature at which they would to absorb microwaves. The suscepting cavity consists of a SALI[1]alumina cylinder, for thermal insulation, with 50% of its inner surface area lined with SiC. SiC absorbs microwaves efficiently at room temperature. After placing the samples in the suscepting cavity, it was placed inside the microwave oven. A pressure of 0.45 MPa was applied on the samples. Three to five magnetrons of 800 W each were used to heat the samples to different joining temperatures and for different joining times. For the purpose of evaluating the effectiveness of the gel in the joining process, alumina samples were joined directly, that is without the use of gel.

Characterization

Dielectric measurements of the samples used in the joining process were done by AECL Research[2]. X-ray maps and SEM micrographs were taken at the joint region for selected samples. Knoop microhardness measurements were taken at and across the joint for different samples.

---

[1]Zircar Products, Inc., New York 10921.

[2]AECL Research, Ontario K0J1J0, Canada.

RESULTS AND DISCUSSION

The behavior in the microwave field of alumina used in the joining process was determined by measuring their dielectric properties. Figure 2 and Figure 3 show the change of tanδ , the loss tangent, with temperature. The power absorbed by the sample per unit volume and transferred into heat is[6]

$$P = 2 \pi f \varepsilon_o \varepsilon_r^{/} \tan\delta \; |E^2| \qquad [3]$$

where f is the frequency, ε_o is the permittivity of free space, $\varepsilon_r^{/}$is the relative dielectric constant, and $|E|$ is the magnitude of the electric field. As seen in this equation, the power absorbed increases as tanδ increases. Figure 2 and Figure 3 show that at a certain temperature, T_c, tanδ starts to increase rapidly with temperature, which results in an increase in the absorbed power and an increase in the temperature in turn. The relatively high tanδ values of 94% alumina is most likely a result of the presence of impurities.These two figures show that 94% alumina interacts with microwaves more efficiently than 99.5% alumina since it has higher tanδ at all temperatures. It can be seen in Figure 4 and Figure 5, that while it is possible to detect the joint line for the 99.5% alumina-gel-94% alumina specimen, the joint line for the 94% alumina-gel-94% alumina is undetectable. The difference in the joint quality may be attributed to the enhanced diffusion of the atomic species in the lower purity alumina.

In determining the effect of the gel on joining time, the following observations were made. At a joining temperature of 1350°C and for joining times of 15 minutes, 30 minutes, 45 minutes, joining occurred for 99.5% alumina-gel-94% alumina specimens, while joining did not occur for the specimens without gel. This shows that the gel assisted in lowering the joining time. At a joining temperature of 1350°C and joining time of 60 minutes, joints were produced for both specimens (with and without gel) but with a significant difference in the quality. Figures 6 and 7 show the joint region for these two specimens. Figures 8 and 9 are x-ray maps for Si at the joint region for the same two specimens. It can be seen that for the 99.5% alumina-94% alumina joint, the joint region is almost depleted of Si. Since Si is present in the aluminas used in the joining process, a region depleted of Si is also depleted of alumina and this corresponds to a poor joint. No depleted region is observed for 99.5% alumina-gel-94% alumina joint indicating a good joint. Figure 10 shows that joint microhardness increases with an increase in joining time.

The effect of the gel on joining temperature was also determined. For joining temperatures lower than 1400°C, 99.5% alumina-gel-94% alumina joints were produced, while no joints without gel were produced. This suggests that the gel assisted in lowering the joining temperature. Joint microhardness was found to increase with joining temperature, as shown in Figure 11. It is seen in this figure that the joint microhardness exceeded the original microhardness at a certain temperature. This may be attributed to a densification process that took place during joining. Joints with microhardness as high as the bulk were produced when the proper parameters were selected. Figure 12 depicts the microhardness across the joint for 94% alumina-gel-94% alumina joined at a temperature of 1550°C and at a time of 30 minutes. As shown in this figure, there is no drop in the microhardness at the joint.

The gel consisted of AlOOH particles in the submicron range. These particles transform into α-alumina at a temperature of 1300°C[6]. Since the surface area of these particles is high, the energy required for sintering is low. These particles have a tendency to bond to each other and to the surfaces to be joined to lower the surface energy. This results in achieving joining at temperatures and times lower than those when there is no gel at the interface.

SUMMARY

Alumina gel and microwave energy can be used to produce strong alumina joints. Several advantages were observed in using the gel as a bonding agent:

1- The use of the gel helped in achieving intimate contact between surfaces to be joined.

2- The gel assisted in lowering the joining time and temperature. This advantage has a significant importance from processing and economic standpoints.

3- The properties were consistent across the joint, since the bonding agent was the same as the bulk.

The use of microwave energy helped in achieving rapid heating rates, short processing times, and uniform heat distribution. We believe that the use of microwaves can be very beneficial in joining large pieces of alumina because microwaves can penetrate through the material and heat the joint as well as the bulk at the same time.

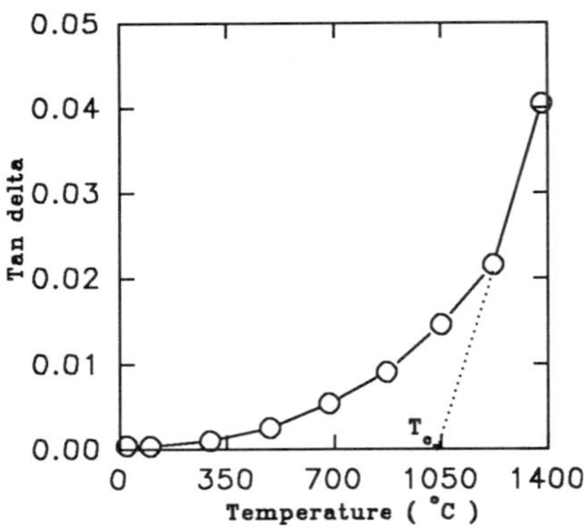

Figure 2. Tan delta change with temperature for 94% alumina; density = 3.76 g/cm^3, frequency = 2.469 GHz.

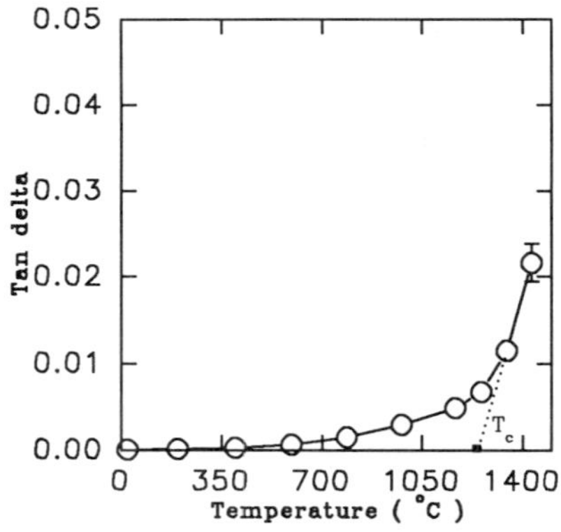

Figure 3. Tan delta change with temperature for 99.5% alumina; density = 3.82 g/cm^3, frequency = 2.469 GHz.

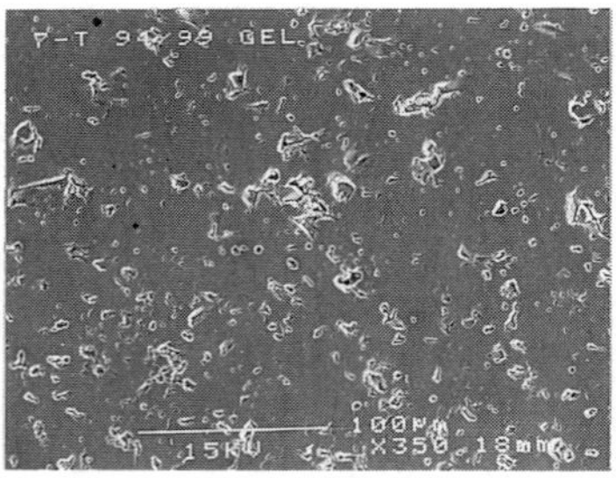

Figure 4. 99.5% alumina (top)-gel-94% alumina (bottom) joint; T = 1550°C, t = 30 min.

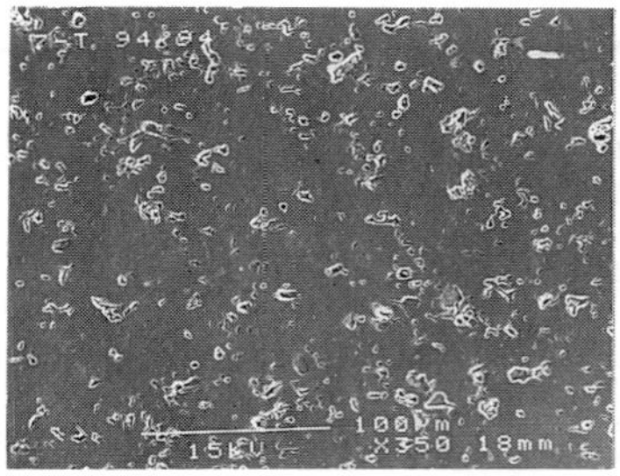

Figure 5. 94% alumina-gel-94% alumina joint; T = 1550°C, t = 30 min.

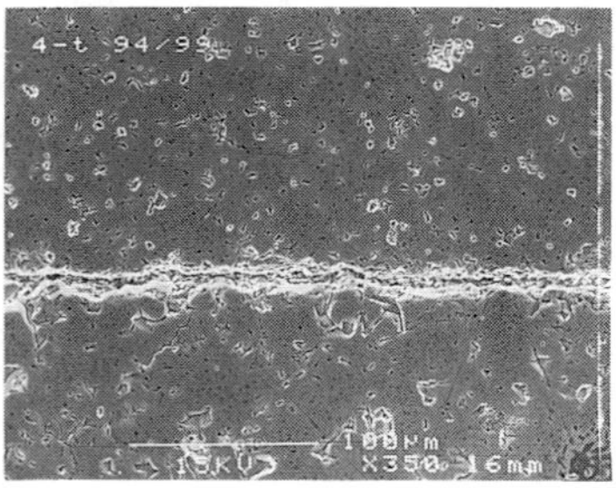

Figure 6. 94% alumina (top)-99.5% alumina (bottom) joint; T = 1350°C, t = 60 min.

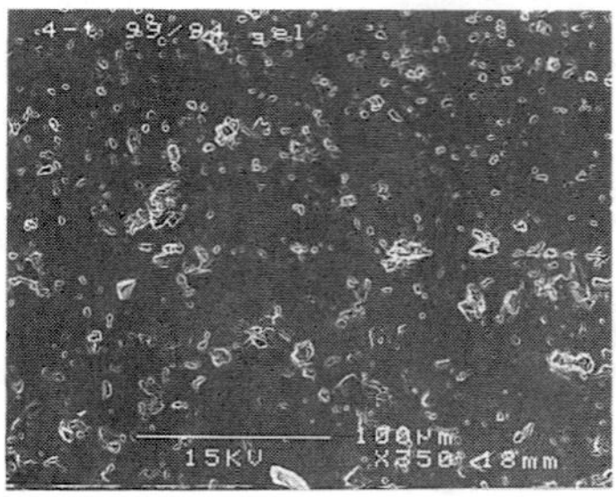

Figure 7. 94% alumina (top)-gel-99.5% alumina (bottom) joint; T = 1350°C, t = 60 min.

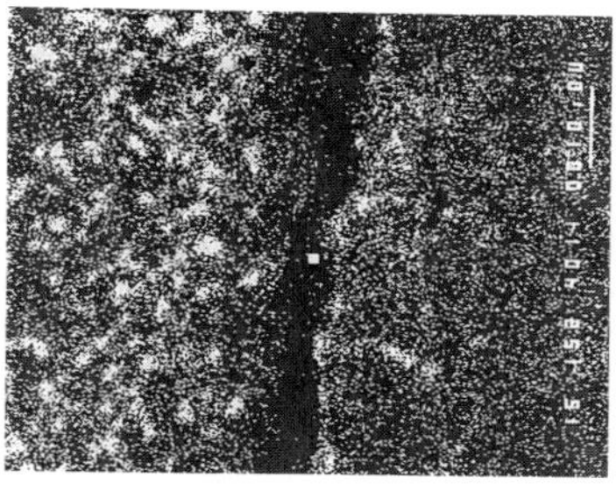

Figure 8. X-ray map for Si at 94% alumina (left) -99.5% (right) joint; T = 1350°C, t = 60 min.

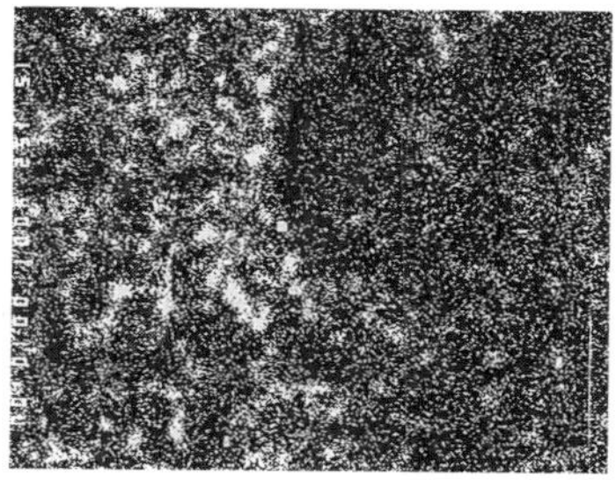

Figure 9. X-ray map for Si at 94% alumina (left) -gel-99.5% (right) joint; T = 1350°C, t = 60 min.

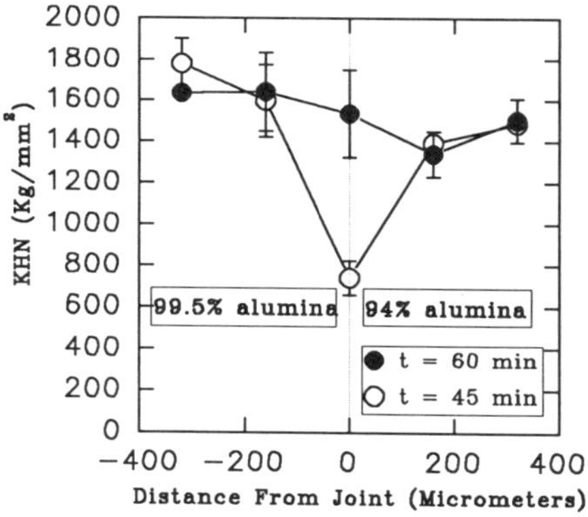

Figure 10. Microhardness across 99.5% alumina-gel-94% alumina joint with two different joining times; T = 1350°C.

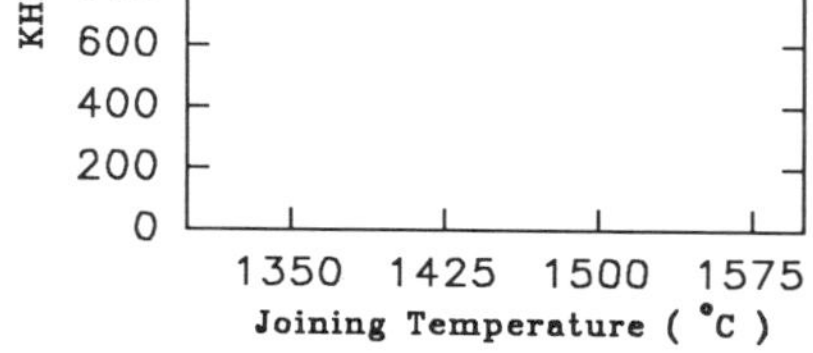

Figure 11. Change of joint microhardness with joining temperature for 94% alumina-gel-94% alumina joint; t = 30 min.

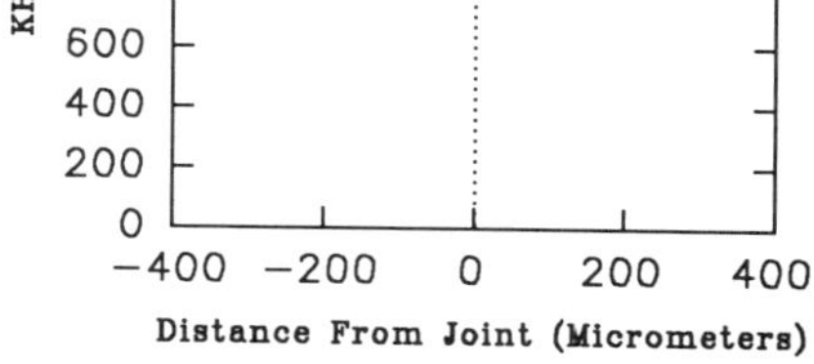

Figure 12. Microhardness across 94% alumina-gel-94% alumina joint; T = 1550°C, t = 30 min.

ACKNOWLEDGMENT

The authors would like to thank R. Hutcheon of AECL for providing the dielectric data necessary for this study.

REFERENCES

1. A. De, Ultra-Rapid Sintering of Alumina With Microwave Energy At 2.45 GHz, a master thesis submitted to the Graduate School of the University of Florida (Gainesville, Florida, 1990).

2. T. T. Meek and R. D. Blake, U.S. Patent No. 4 529 857 (16 July 1985).

3. H. Fukashima, T. Yamanaka, and M. Matsui, Mat. Res. Soc. Symp. Proc. 124, 267 (1988).

4. R. Silberglitt, D. Palaith, W. M. Black, H. S. Sa'adaldin, J. D. Katz, and R. D. Blake, Microwaves: Theory and Application in Materials Processing, edited by D. E. Clark, F. D. Gac, and W. H. Sutton (The American Ceramic Society, Inc., Ohio, 1991), p. 487.

5. J. D. Mackenzie, Ultrastructure Processing of Ceramics, Glasses, and Composites, edited by L. L. Hench and D. R. Ulrich (John Wiley and Sons, Inc., New York, 1984), p. 15.

6. D. E. Clark, W. J. Dalzell Jr., and B. L. Adams, U.S. Patent No. 4 801 399 (31 January 1989).

7. W. J. Dalzell Jr., Alumina Sol-Gel Coatings Formed By Electrophoretic Deposition, a master thesis submitted to the Graduate School of the University of Florida (Gainesville, Florida, 1988).

POLYMETHYL METHACRYLATE BINDER REMOVAL FROM AN ALUMINA COMPACT: MICROWAVE VERSUS CONVENTIONAL HEATING

Edmund H. Moore*, David E. Clark** and Ronald Hutcheon***
*University of Florida, Dept. of Materials Science and Engineering, 136 MAE, Gainesville, FL 32611-2066
**University of Florida, Dept. of Materials Science and Engineering, 136 MAE, Gainesville, FL 32611-2066
***AECL, Chalk River Laboratories, Chalk River, Canada, Ontario K0J1J0.

ABSTRACT

Compact samples of alumina and polymethyl methacrylate have been heated in a 2.45 GHz microwave cavity and by conventional heating in an electric furnace. Various heating schedules have been used to effect the removal of the polymeric binder by thermal decomposition. Dielectric properties, porosity and other physical properties have been investigated in order to better understand the binder removal process in a microwave field. Results of the study emphasize the amount of carbon residuals remaining in the bulk.

INTRODUCTION

Potential advantages of microwave processing include uniform, rapid, volumetric and selective heating of materials [1,2]. These advantages may lead to economic benefits, manifested in lower processing cost via lower energy requirements, increased production and enhanced materials properties.

The polymethyl methacrylate (PMMA) binder and alumina system is presently being used in the electronics industry [3]. Polymers are added to ceramic powders for a variety of reasons, which include acting as a binder, lubricant, plasticizer and dispersant during forming processes (slip casting, extrusion, pressing and injection molding).

The removal of polymeric binder(s) is time consuming and may lead to damage of the ceramic (cracking or bloating). When a very clean ceramic is required, as in the case of a multilayer ceramic (MLC) substrate, the carbon should be removed completely to avoid deleterious consequences such as a lack of strength, low insulation resistance and discoloration [3]. Many factors play a role in the removal of the binder from a ceramic. The overall objective of this study is to understand how microwaves interact with the PMMA/alumina system. However, this paper will emphasize the effects of mass, the type of heating, heating schedules, sample density, dielectric properties, particle size and surface area.

EXPERIMENTAL PROCEDURE

The PMMA/alumina samples were heated in a microwave thermogravimetric analyzer (MTGA) by microwave hybrid heating (MHH) and in an electric furnace by conventional heating. Microwave hybrid heating is a combination of microwave and conventional heating. The MTGA has been discussed by Moore et al [4,5] and has been used to heat compacts of PMMA/alumina.

Apparatus

The MTGA apparatus is a 700 watt, 2.45 GHz microwave oven which allows for the weight of a sample to be recorded as a function of temperature and time. This apparatus contains an insulating cavity with silicon carbide paste as a susceptor. The loss tangent and depth of penetration calculations for the suscepting material used in this insulation cavity indicate that microwaves are penetrating the insulation cavity and interacting with the compacts. Samples were heated by conventional means in a 3500 watt Lindberg oven.

Procedure

Sumitomo alumina powders with a median diameter of 0.38 μm (AKP-30) and 0.67 μm (AKP-15) and DuPont E-2010 PMMA binder were used in all sample preparations. This PMMA is an atactic binder (has random location of side groups along backbone of polymer chain) and has a number average molecular weight of around 50,000 gm/mole. The PMMA was mixed with chloroform in a plastic bottle. Once the PMMA dissolved into the chloroform, alumina powder was mixed into this viscous solution and roll milled for 24 hours. The contents of the plastic bottle were poured onto a sheet of mylar film to air dry under a hood for 24 hours and then dried at 50°C for 24 hours in a drying oven to remove residual moisture and solvent. The mixture was ground to less than 50 mesh (300 μm) with a mortar and pestle and pressed into compacts at 27.57 MPa (4000 psi). All compacts weighed either 8 grams, 12 grams, 16 grams or 24 grams, and were 2.86 cm (1.125 inches) in diameter. The samples used in this study contained either 4 w/o or 8 w/o PMMA. Compacts were processed in the Lindberg oven or MTGA. In all cases a static air atmosphere was used. Usually compacts were processed in the MTGA first. Then the MTGA heating schedule was followed as close as possible with the Lindberg oven. A 5°C per minute heating rate was also used with the Lindberg oven. Selected samples were analyzed for carbon and hydrogen content by an inert gas fusion instrument. A high temperature dielectric properties measurement system, using conventional heating, allowed dielectric information as a function of sample density, temperature, composition, heating rates and frequency to be obtained [6].

RESULTS AND DISCUSSION

Sample Density

The theoretical density of 4 w/o PMMA/alumina compacts was calculated to be ≈3.88 gm/cc using a density for alumina of 4 gm/cc and a specific gravity of 1.2 gm/cc for PMMA. Using the theoretical density as a basis, the compacts pressed at 27.57 MPa were determined to have a density between 2.02 gm/cc to 2.28 gm/cc, which corresponds to a 41-48±2% porosity in these green samples. After the green compacts were heated in the MTGA or Lindberg oven, a sample density of 1.95 gm/cc to 2.27 gm/cc was calculated, which corresponds to a porosity of 43-50±2% in these compacts. The same assumptions were applied to compacts used to collect dielectric information and the green density of those compacts ranged from 1.5 gm/cc to 1.81 gm/cc, which corresponds to a porosity of 50-60±2%. Sample density was measured by dividing the mass by the geometric volume.

Dielectric Properties

The loss tangent versus temperature has been obtained at ≈2450 MHz and ≈915 MHz; only the 2450 MHz data will be shown here. In Figure 1, the loss tangent of 4 w/o and 8 w/o PMMA samples prepared with AKP-15 alumina is shown from room temperature to 1400°C. Likewise, the loss tangent of 4 w/o and 8 w/o PMMA samples prepared with AKP-30 alumina is shown in Figure 2. Figure 3 shows the loss tangent versus temperature for a 99.5% alumina up to 1400°C. A typical heating schedule used to obtain the dielectric data in Figures 1 and 2 is shown in Figure 4. Note that conventional heating was used to obtain this dielectric data and these values may be different if microwave heating was used [7].

The first peak in Figures 1 and 2 corresponds to PMMA interacting with microwaves and decomposing. The decomposition temperature range of PMMA in air is approximately 177°C to 232°C [8] and the first peak in Figures 1 and 2 reach a maximum between 200°C and 240°C. The depolymerization reaction of the PMMA into monomer is shown in equation 1 and it occurs below 500°C.

$$-(C_5H_8O_2)-_n \rightarrow -(C_5H_8O_2)-_{n-1} \quad (1)$$

The PMMA decomposes into monomer, water, benzene and other components. These components are further decomposed into hydrocarbons between 500°C and 1000°C. The carbon reacts with air to form $CO(g)$ or $CO_2(g)$, as shown in equations 2 and 3.

$$C + O_2(g) \rightarrow CO_2(g) \quad (2)$$

$$C + \tfrac{1}{2}O_2(g) \rightarrow CO(g) \quad (3)$$

A second peak is observed in both Figures 1 and 2, and these peaks are believed to be due to residual carbon interacting with the microwaves. The second peaks are centered around 600°C for the samples containing 4 w/o PMMA and around 700°C for samples prepared with 8 w/o PMMA. The shift to the lower temperature for the 4 w/o peak is not fully understood at the present.

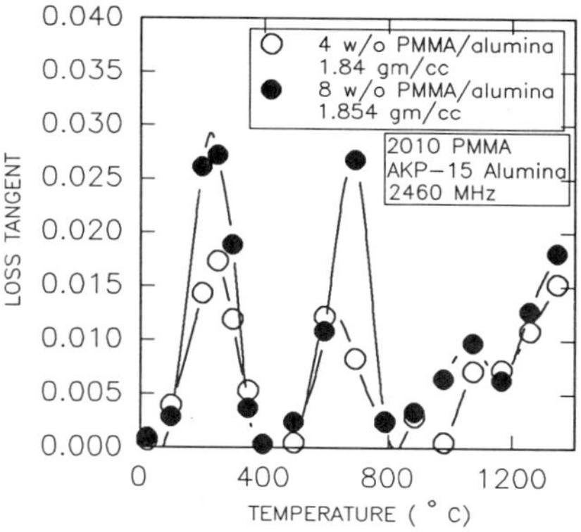

Figure 1. Loss tangent versus temperature for 4 w/o and 8 w/o PMMA/AKP-15 alumina at 2460 MHz.

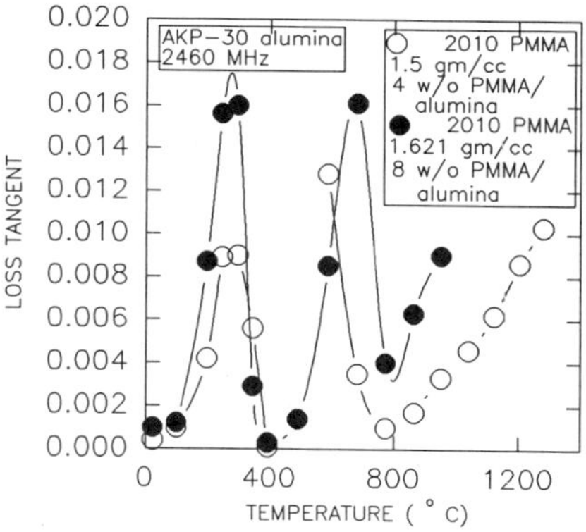

Figure 2. Loss tangent versus temperature for 4 w/o and 8 w/o PMMA/AKP-30 alumina at 2460 MHz.

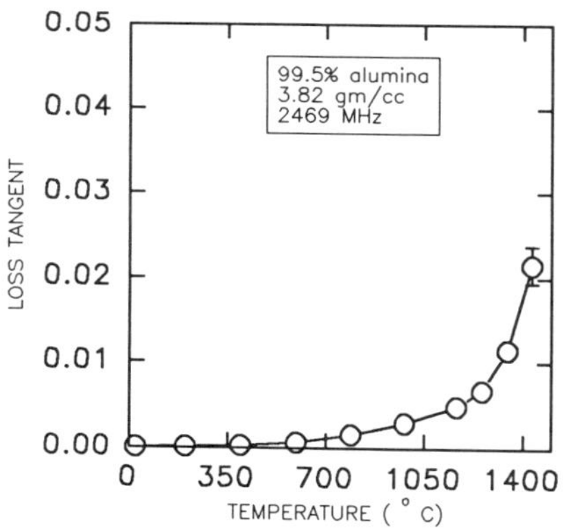

Figure 3. Loss tangent versus temperature for 99.5% alumina at 2469 MHz.

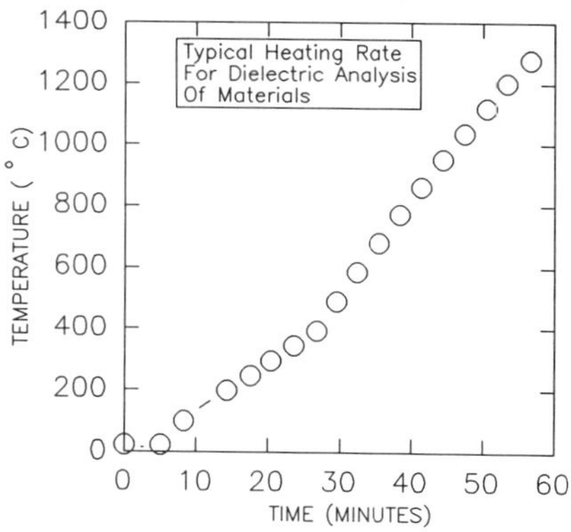

Figure 4. Typical heating schedule used to obtain dielectric data shown in Figures 1 and 2.

The loss tangent in Figures 1 and 2 are dependent on the PMMA concentration. The first peak is due to the PMMA decomposing and the loss tangent is a factor of two times larger for the 8 w/o PMMA samples. This trend is followed for the second peak as well, but not as strongly for the second peak in Figure 2. In Figure 3 the loss tangent of the alumina is less than 0.002 up to 1000°C. This dielectric information suggest that selective heating is occurring in samples containing PMMA, as alumina alone does not absorb microwaves very well in this temperature range. Above 800°C, some residual carbon seems to remain in the samples, as the loss tangent in Figures 1 and 2 are still larger than those of the 99.5% alumina in Figure 3.

The particle size and surface area of the alumina also seem to have an effect on the dielectric properties of these samples. AKP-30 alumina has a smaller particle size and larger specific surface area than AKP-15 alumina. A higher surface area means that more PMMA is absorbed onto the surface of the alumina and less PMMA is free to interact with microwaves. Samples prepared with AKP-15 alumina have a higher loss tangent than those prepared with AKP-30 alumina, as seen in Figures 1 and 2. This suggest that the green microstructure and composition is important in microwave processing.

Binder Removal

Compacts were heated in the MTGA and in the furnace for 90 minutes. In Figures 5 and 6, the MTGA data is shown for the 12 gram and the 24 gram compacts prepared from different alumina powders. It is apparent that as sample mass increases, the PMMA is removed just as efficiently or more efficiently than for the smaller sample mass. After binder removal, in all cases, identical compacts prepared from AKP-15 alumina contained less carbon than those prepared from AKP-30 alumina. The heating schedules of the 12 and 24 gram compacts heated in the MTGA are similar. Error bars used in Figures 5 and 6 were obtained from three samples and reflect the standard deviation.

Compacts were heated in the conventional oven using two heating schedules, as shown in Figure 7. Compacts were heated at 5•C per minute to 400•C and held at temperature for 15 minutes. Compacts were also heated at a rate to mimic the MTGA compacts as close as possible. The compacts loss in excess of 90% of the PMMA by thermal decomposition, as shown in Figure 8. Only 24 gram compacts are shown in Figure 8, but identical prepared 12 gram compacts were also heated in the oven. However, the 12 gram compacts loss binder at a faster rate than the 24 gram compacts. The compacts heated at the MTGA rate lost binder at a much more rapid rate than compacts heated at the 5•C per minute rate.

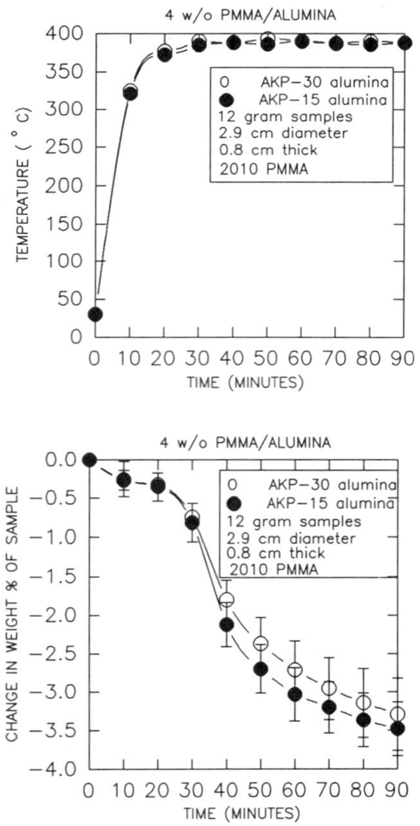

Figure 5. Temperature and change in weight % of of 12 gram samples versus time for the 4 w/o PMMA/alumina (heated in the MTGA).

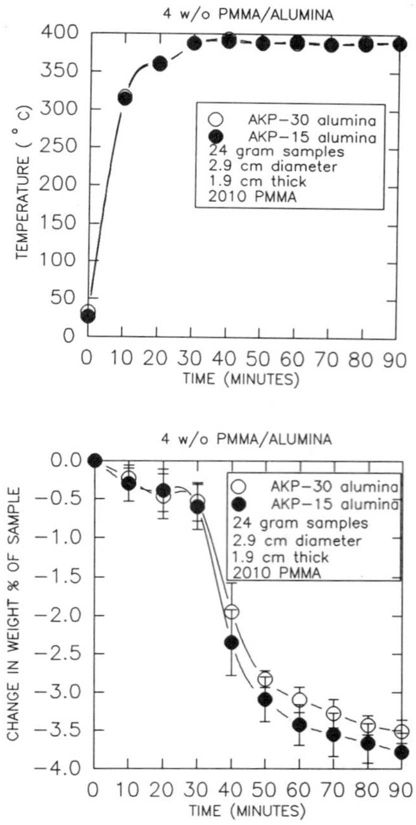

Figure 6. Temperature and change in weight % 24 gram samples versus time for the 4 w/o PMMA/alumina (heated in the MTGA).

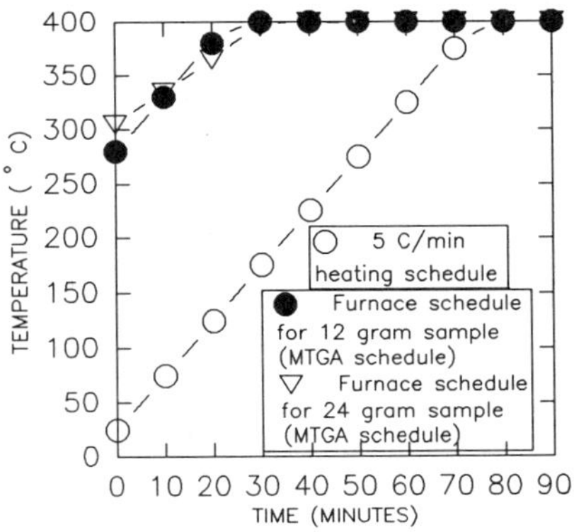

Figure 7. Conventional heating schedules used.

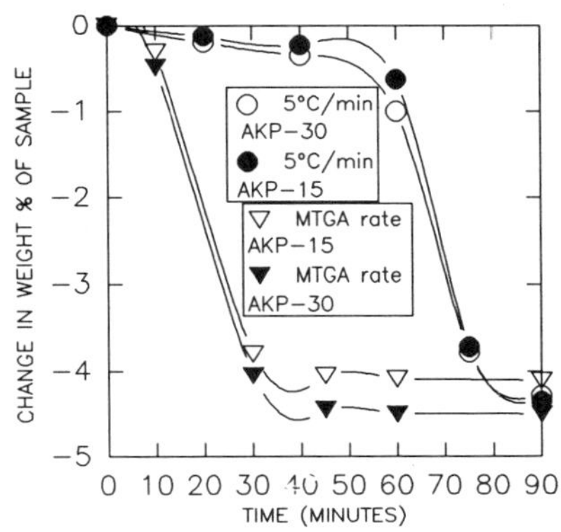

Figure 8. Change in weight % of 24 gram samples versus time for the 4 w/o PMMA/alumina (conventional heated in the furnace).

Carbon Analysis

Compact samples heated in the MTGA and in a conventional furnace were analyzed for carbon across their cross-section. Selected samples were cut in half and powdered samples from four locations across the cross-section were collected and analyzed for carbon and hydrogen. There was not enough hydrogen in these samples to detect. Carbon was detected and the average of the five collected powdered samples is reported in Table I. Table I contains the carbon analysis for alumina compacts containing 4 w/o PMMA heated to 400°C. Sampling data is not significantly influenced by cross-section position for compacts processed in either the MTGA or conventional furnace.

Table I. Carbon Analysis For Alumina Compacts Containing 4 w/o PMMA And Heated To 400°C.

	MTGA	Furnace 5°C/Min. Rate	Furnace MTGA Rate
12 gram, AKP-30	0.84% Carbon	----	----
24 gram, AKP-30	0.61% Carbon	0.31% Carbon	0.06% Carbon
12 gram, AKP-15	0.63% Carbon	----	----
24 gram, AKP-15	0.45% Carbon	0.15% Carbon	0.04% Carbon

In the MTGA, 24 gram compacts contained less carbon than identically prepared 12 gram compacts. Compacts heated in the MTGA, of the same mass, always contained less carbon when prepared from AKP-15 alumina instead of AKP-30 alumina. This result is consistent with the binder removal data shown in Figures 5 and 6 and with the higher loss tangent for samples prepared from AKP-15 alumina.

In the conventional furnace, only the 24 gram compact data is reported. Compacts heated at the MTGA rate contained 4 to 5 times less carbon than those heated at the 5°C per minute rate. Compacts heated in the MTGA contained 10 to 11 times more carbon than those heated in the conventional furnace at the MTGA rate. Compacts heated in the conventional furnace at the 5°C per minute rate contained 2 to 3 times less carbon than those heated in the MTGA.

In all cases, identically prepared compacts heated in the MTGA contained more carbon than compacts heated in the conventional oven, regardless of heating schedule. This is believed to be due to the fact that compacts heated in the MTGA were exposed to an oxygen deficient environment in the enclosed insulation cavity. In all cases, identical compacts prepared with AKP-15 alumina contained less carbon than those prepared with AKP-30 alumina.

SUMMARY

The results of this paper indicates that: (1) As sample mass increases, binder removal in the MTGA becomes more efficient for samples ≥12 grams. The opposite is true for compacts heated by conventional means. (2) Identically prepared compacts heated in the MTGA contained more carbon than their cohorts heated by conventional means. This is regardless of the conventional heating schedule used. These results may be due to the fact that compacts heated in the MTGA were exposed to an oxygen deficient environment while enclosed in an insulation cavity. (3) Samples containing 8 w/o PMMA have a higher loss tangent in the peak regions than those prepared with 4 w/o PMMA. This suggests that the peaks are due to PMMA or its decomposition products, and not due to alumina. (4) Alumina surface area and particle size appear to have a minor role in the binder removal kinetics in the conventional and the MTGA processed compacts. (5) In the MTGA, the binder removal kinetics for the samples prepared with AKP-15 alumina is slightly more rapid. This is believed to be due to the fact that the AKP-15 alumina has a larger particle size and a smaller surface area. Thus, more PMMA is free to interact with the microwaves as there is less PMMA absorbed on the AKP-15 alumina than on the AKP-30 alumina. (6) The samples prepared from AKP-15 alumina have a higher loss tangent in the peak regions than those prepared with the AKP-30 alumina. This is believed to be due to more PMMA being free to interact with the microwaves.

Future work will include conducting research in the area of stand alone microwave heating and processing compacts containing 8 w/o PMMA. The influence of the atmosphere in the cavity will also be investigated.

ACKNOWLEDGEMENTS

The authors thank the United States Air Force for providing funding to E. H. Moore and for performing the chemical analysis. The authors also thank the R. Strohschein for glassblowing services.

REFERENCES

1. L.M. Sheppard, Ceramic Bulletin 67(10), 1656-1662 (1988).
2. W.H. Sutton, Ceramic Bulletin 68(2), 376-386 (1989).
3. A.H. Kumar and R.R. Tummala, The International Journal for Hybrid Microelectronics 14(4), 137-150 (Dec. 1991).
4. E.H. Moore, I. Ahmad and D.E. Clark in Microwaves: Theory and Application in Materials Processing, edited by D.E. Clark, F.D. Gac and W.H. Sutton (The American Ceramic Society, Inc. Ceramic Transactions 21, Westerville, OH 1991) pp. 675-681.
5. E.H. Moore and D.E. Clark, Polymethyl Methacrylate Binder Removal From Alumina By Microwave Heating, presented at the 16th Annual Conference on Composites and Advanced Ceramics, Cocoa Beach, FL, Jan. 1992 (unpublished).
6. R.M. Hutcheon et al, R.F. and Microwave Dielectric Measurement to 1400°C and Dielectric Loss Measurement, to be presented at the 1992 Materials Research Society Spring Meeting, San Francisco, CA, 1992 (unpublished).
7. R. Hutcheon (private communication).
8. Du Pont Company, Brochure on "Elvacite Acrylic Resins, Properties and Uses," Wilmington, DE.

SURFACE MODIFICATION OF CERAMICS USING MICROWAVE ENERGY

Z. Fathi* and D.E. Clark.
Department of Materials Science and Engineering
University of Florida, Gainesville, FL 32611

R. Hutcheon.
AECL Research, Physics Division, Chalk River Lab.
Chalk River, Ontario, Canada KOJ1JO.

ABSTRACT

Sodium alumino-silicate glasses having a constant concentration of alkali ions (15mol%) were prepared. The interdiffusion of potassium ions, K^+, for sodium ions, Na^+, in the glass composition having Al/Na=1 was investigated in the presence and absence of a microwave field. Characterization consisted of dielectric loss measurements, electron microprobe analysis (EMP), and microhardness indentation.

INTRODUCTION

The incorporation of Al_2O_3 in sodium silicate glasses results in structural changes accompanied by a reduction of non-bridging oxygen atoms. The properties of these sodium alumino-silicate (NAS) glasses change when r, the ratio of Al/Na, is taken as a variable. Discontinuous changes of diffusion coefficients, electrical conductivity, and viscosities have been observed when r goes through the equivalency point [1,2,3,4]. Several structural models have been proposed to explain these pronounced property changes [5,6], based on the coordination of the aluminum ion. It is not clear if the Al ion remains tetrahedrally coordinated for all r values between zero and unity, or if there are values at which octahedral coordination occurs. The aluminum coordination for r values greater than one is also in question. However, the models concur on the glass structure when r=1. It has been established that Al goes to tetrahedral coordination, AlO_4^-. It has been shown that the Na/non-bridging oxygen bonds are stronger than the Na/AlO_4 bonds. This explains the extrema in conductivity for the r=1 glass compositions.

In previous work done on surface modification of sodium aluminosilicate glasses using microwave energy [7], a series of glasses were prepared and interdiffusion reactions of K^+ for Na^+ were performed. The sodium aluminosilicate glass compositions investigated were r = 0.2, 0.5, 1.0, and 1.1 where r is the ratio of Al:Na.

The r = 1.0 glass composition exhibited the deepest penetration of K^+ when compared to the other glass compositions. The difference between the penetration depths of microwaved and conventionally prepared samples reached a maximum for the r = 1.0 glass composition. The present work was undertaken to study the dielectric behavior of these glasses and to qualitatively evaluate the influence of microwave energy driving the ion exchange process.

EXPERIMENTAL

Glass Preparation

The glasses were prepared by first milling, for a period of 24 hours high purity silica, sodium carbonate and high purity alumina. The sodium alumino-silicate powders were sintered at 1550°C and subsequently quenched and crushed to glass frits. This operation was performed twice to insure homogeneity of these high viscosity glasses. The glass frits were then melted inside an alumina crucible at 1550°C and were held at that temperature for 6.5 hours after which the furnace temperature was raised to 1610°C for an additional 5 hours. The alumina crucible/melt was annealed at 700°C for 5 hours, and allowed to furnace cool. Only the glass well away from the alumina crucible walls was used in this study.

Glass samples of the r=0.5, 1.0 and 1.1 glass compositions were crushed to powder form for dielectric measurement. Glass samples of the r=1.0 composition were polished to a 1 μm surface finish prior to ion exchange reactions.

Ion Exchange Process

The glass compositions were ion exchanged using a potassium containing medium referred to as Ceramicoat[1]. Once coated with Ceramicoat, the samples were dried for 20 minutes at 120°C. Ion exchange was carried out at 450 and 510°C, each for a period of 30 minutes in both conventional and microwave cavities. After cooling, the samples were sonicated first in water, then in acetone to clean the surfaces. The samples were then cross sectioned and prepared for characterization.

The microwave ion exchange reactions were carried out in a modified industrial Raytheon microwave oven[2]. The samples were placed in an insulating material lined with a microwave susceptor which, in turn, assisted in heating the samples. Changing the thickness of the lining allowed us to process samples at similar temperatures but under different incident microwave powers. Samples subjected to microwave energy were processed at two different power settings, 800W and 1600W. Temperatures were monitored using two Inconel shielded, K-type thermocouples. One thermocouple was positioned at the top of the sample to monitor the gel temperature and one thermocouple was positioned at the side of the sample in direct contact with the bulk glass. Temperature readings between the two thermocouples varied by approximately ± 5°C.

Dielectric loss measurements

Dielectric loss measurements were performed to assist in explaining the observed interdiffusion enhancement for the r=1.0 glass composition. Loss tangent versus temperature were derived at five different frequencies for the r=0.5, 1.0 and 1.1 glass compositions (see figures 1.a, 1.b, and 1.c.).

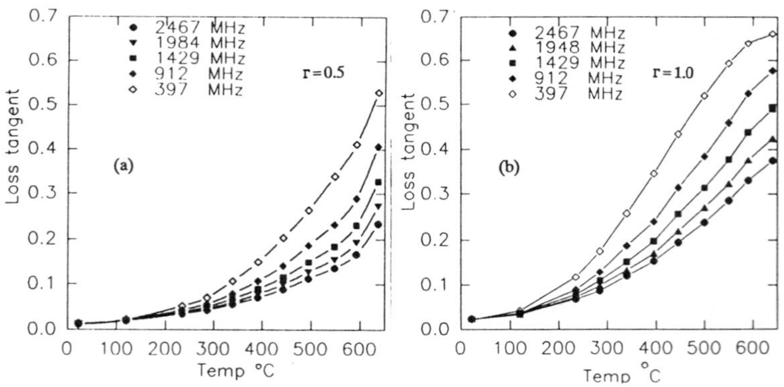

Figure 1. Loss tangent versus temperature at different frequencies for, 1.a r=0.5, 1.b r=1.0, and 1.c r=1.1.

The dielectric losses were found to increase with increasing values of r up to r=1.0 at which the dielectric losses exhibit maximum values.

---

[1]Ceramicoat, G-C International Corp., Scottsdale, AZ 85260

[2]Raytheon Model QMP 2101B-6, Waltham, MA 02254

Figure 1.c

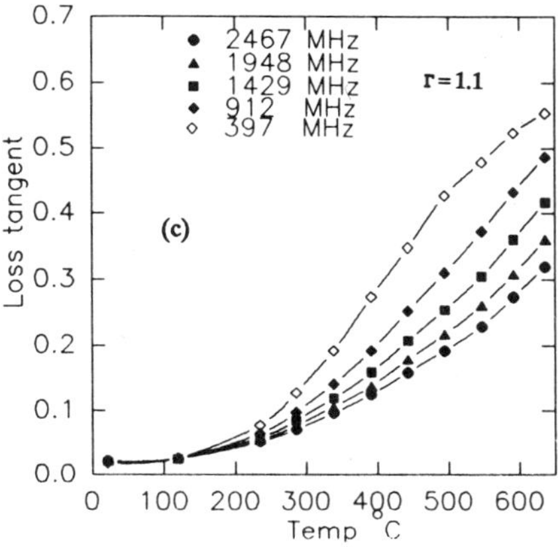

Electron Microprobe (EMP)

EMP was used to evaluate the extent of interdiffusion of potassium for sodium ions. A comparison of the two interdiffusion reactions is shown in Figures 2a and 2b. Conventional and microwave heat treated samples, using an 800W source, are compared on each graph.

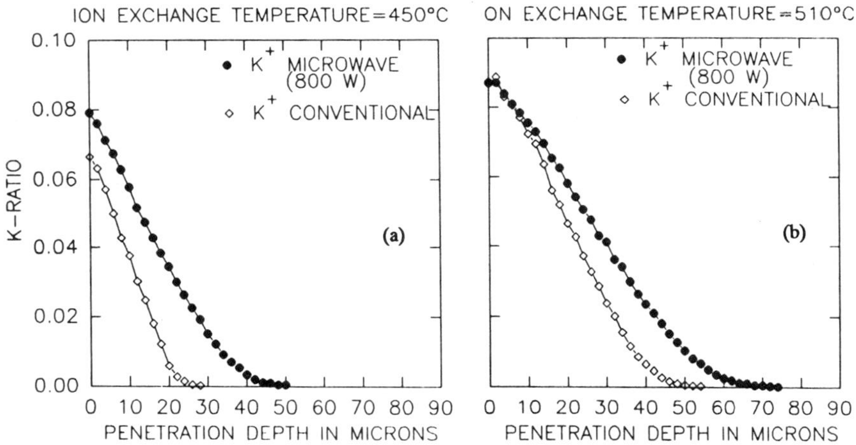

Figure 2. Penetration depths (in μm) of potassium into the r=1.0 glass composition at different temperatures. K-ratios are relative to Tugtupite for potassium.

All microwave treated samples show interdiffusion enhancement when compared to the conventionally processed samples. Furthermore, the samples processed using 1600W microwave energy experienced significantly greater K^+ penetration depths than samples processed conventionally or using 800W of microwave energy. (see Figure 3a and 3b). This suggests that interdiffusion reactions are enhanced under a microwave field and are dependent on the incident power levels.

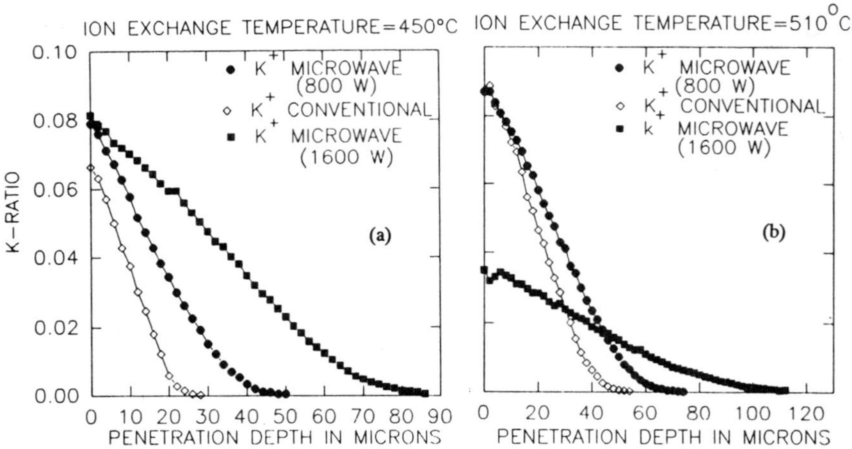

Figure 3. penetration depths (in μm) of potassium into the r = 1.0 glass composition at different temperatures and different microwave incident powers. K-ratios are relative to Tugtupite for potassium.

The results in figure 3.b show a deep K^+ penetration depth but a low surface concentration. The four samples microwave processed under a 1600W incident power were, unlike all the other samples, cleaned in pure ethanol and experienced a surface attack (which is still an unexplained fact).

<u>Microhardness Indentation</u>

Microhardness indentation tests were performed on annealed r = 1.0 glass samples. The Knoop hardness versus load (Figure 4) reveals a load independent region of the glass. A load of 300g was selected to perform all subsequent indentation tests for the microhardness evaluation of the ion exchanged glasses. Knoop hardness data on annealed glass, conventionally and microwave ion exchanged glasses are shown in Figure 5.

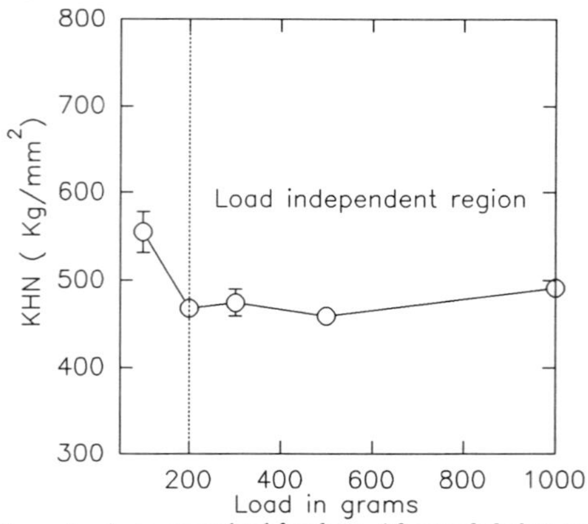

Figure 4. Knoop hardness versus load for the r = 1.0 annealed glass composition.

Microwave heat treated glasses exhibit harder surfaces than the initial annealed glass. However, microwave and conventionally ion exchanged glasses seem to be within the same range of hardness, due to the large standard deviation in the microwaved samples.

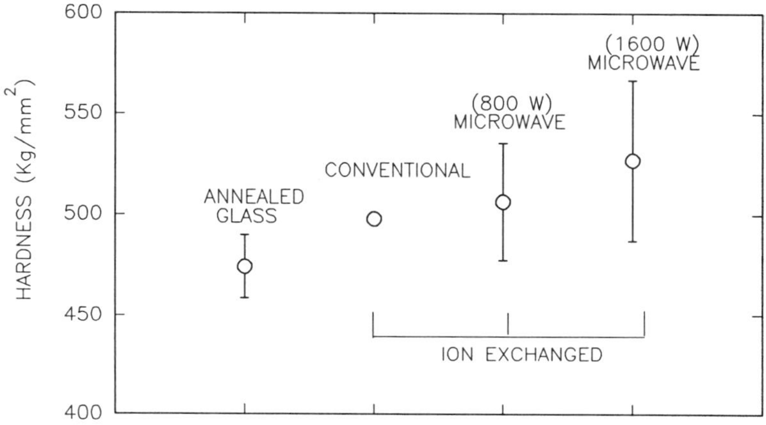

Figure 5. Hardness measurements for the r=1.0 glass composition subjected to different heat treatments. Ion exchange was carried out at 450°C for 30 minutes. A 300 gram load was used for the hardness measurements.

SUMMARY

Dielectric loss measurements revealed a maximum loss tangent at the r=1.0 glass composition. This glass composition also exhibits a pronounced enhancement of $K^+ \leftrightarrow Na^+$ interdiffusion in the presence of a microwave electric field. The extent of $K^+ \leftrightarrow Na^+$ interdiffusion depends on the incident microwave power. Considerable depths of penetration for K^+ have been achieved with a 1600W power source. The glasses ion exchanged in the presence of a microwave field exhibit harder surfaces than the starting annealed glasses.

ACKNOWLEDGMENT

The authors gratefully acknowledge support from the Defense Advanced Research Project Agency (DARPA).

The authors acknowledge W.A. Acree and the Major Analytical Instrumentation Center (MAIC) at the University of Florida for the electron microprobe analysis.

Special acknowledgement for R.L. Schulz for help throughout this study.

REFERENCES

1. D.E. Day and G.E. Rindone, J.Am.Cer.Soc., 45 [10] 489-96 (1962).
2. D.E. Day and G.E. Rindone, J.Am.Cer.Soc., 45 [10] 496-504 (1962).
3. D.E. Day and G.E. Rindone, J.Am.Cer.Soc., 45 [12] 579-580 (1962).
4. J.O. Isard, "Electrical Conductivity in Aluminosilicate Glasses," J.Soc. Glass Technol., 43[211] 113-23 (1959).
5. D.M. Zirl and S.H. Garofalini, J.Am.Cer.Soc., 73 [10] 2848-56 (1990).
6. A. Klonkowski, "The structure of Sodium Aluminosilicate Glass," Phys. Chem. Glasses, 24[6] 166-71 (1983).
7. Z. Fathi and D.E. Clark, Microwaves: Theory and Application in Materials Processing (Ceramic Transactions, 1991) vol. 21, p 623.

MICROWAVE CURING AND DRYING OF CELLULAR CONCRETE

WALTER M. VAN LOOCK
University of Gent, Sint-Pietersnieuwstraat 41, B 9000 Gent, Belgium

ABSTRACT

Traditionally the heating and curing of cellular concrete takes place in an autoclave. This building material is a thermal insulator so the curing and drying processes are energy and time consuming. Volumetric heating is an attractive method for improving the throughputs.

A feasibility study leads to the conclusion that microwave processing is not likely to be applied in a large scale production process.

INTRODUCTION

Cellular concrete is a building material for building blocks which may contain steel reinforcements and for bricks. Basically pulverized sand (~ 20%), CaO (~ 15%) and cement (~ 15%) are mixed into a homogeneous fine powder (roughly 20 μ particles). Then water is added (~ 50%) and a little bit of aluminium powder (< 1%). The components are mixed and then poured into molds. The Al powder reacts to give hydogen gas bubbles to form the cells which cause the mixture to rise like a cake.

Shortly after the rising the cake is demolded and it is sufficiently hard, due to initial curing, for cutting it by wire grids. This results in the building blocks and just as with concrete it takes about 7 days and more for the finished blocks to develop its full strength at normal air temperatures. Steam curing at 150°C in autoclaves reduces the curing time to about 6 hrs. After leaving the autoclave the blocks must be separated as they stick together and an end drying is necessary.

The curing and drying times depend on the thermal conductivity to transmit the heat to the interior. A reduction of these times with the volumetric heating by microwaves was attractive and attention was paid to this application.

DETAILS OF THE SELECTED LINE

A specific production line has been selected for investigating the applicability of microwave power. The line produces bricks of 0.6m x 0.25m x 0.1m at a rate of about 10m^3 per hour. The energy balance for this line is summarized in table I.

TABLE I : Energy balance

raw material, milling and mixing :	+ 3 kWh/1000 kg
$2 Al + 6 H_2O \longrightarrow 2 Al(OH)_3 \downarrow + 3 H_2 \uparrow$:	- 10 kWh
initial curing :	- 10 kWh
steam curing (5 bar, 150°C) :	+ 150 kWh
handling :	+ 4 kWh
TOTAL	165 kWh/1000 kg

Preliminary tests have shown that microwave volumetric heating can assist the cell formation and the initial curing. A compromise between a negligible time reduction but at the cost of larger and inhomogeneous cell formation is possible. However, it was concluded that volumetric heating can be useful for speeding up the curing reducing energy comsumption and the required time, and perhaps also for the end drying. Table II summarizes the relevant properties.

TABLE II : Relevant properties

moisture content at input autoclave :	30	%
time in autoclave :	5	hrs
moisture content out of autoclave :	25	%
moisture content when leaving factory :	15	%
equilibrium m.c. in buildings :	2 - 3	%
diameter of the (open) cells :	1 - 2	mm
compressive strength :	≥ 8	N/mm^2
specific dry mass :	600	kg/m^3
thermal conductivity of dry CC	0.16	W/mK

The gas bricks dry out during storage prior to transportation. This takes place in shelters in the open air and depends on the season. The final drying is naturally obtained in the building construction after a year or so.

The cells are large and open. Although microwave or high frequency drying is a straigh forward technique, there is no industrial interest or benefits to be expected.

MICROWAVE EXPERIMENTS

Permittivity measurements have been performed on standard blocks at 2.45 GHz. The cellular concrete is for the microwaves a homogeneous and isotropic material and lossy due to the water content. Typical averaged results are $\varepsilon^* = 2.7 - j\,0.38$ at a m.c. of 30% and 1.6 - j 0.05 for the dry material. No differences in permittivity have been observed between cured and uncured samples at the same m.c.

Microwave thermal curing was performed with a 1.5 kW laboratory oven equiped with a turntable a mode stirrer and air conditioning for hot and humid air. The samples were cylindrical, $\phi = 78$ mm and 70 mm high for standard test procedures and cut out of the precured material.

Preliminary experiments were performed to determine the range of settings for useful microwave curing. It was decided to load the 1.5 kW oven with 4 and 8 samples for reducing the absorbed power and avoiding high temperatures and deformations.

Various microwave power cycles in humid air (60°C) have been applied in the first 6 hrs for obtaining a similar compressive strength as the autoclaved bricks. The compressive strength tests have been performed on the samples cured for 6 hrs, 1, 7 and 28 days. No optimum microwave processing could be obtained. The general trend is that microwaves tend to increase the initial compressive strength but there is a considerable loss over the 28 days period. Typical averaged results are shown in Table III for 8 sample loadings and trials for diffenent microwave cycles. It should be noted that the samples left in hot humid air cure also. The autoclave processing increases the strength so that the bricks can be handled for storage and packaging. This process clearly degrades also the final compressive strength.

TEST	6 hrs	1d	7d	28d
REF. AT 60°C, 100% m.c.	1.2	3.1	8.1	16
MW 1h, 5 hrs h.h.a.	5.8	6	6.5	7.4
MW 2 hrs, 4 hrs h.h.a.	5	5.7	6.9	7
MW 6 x 15', every hr.	6.1	6.1	6.4	6.8
AUTOCLAVED	6.2	8.2	9.1	9.8

TABLE III, Typical compressive strength, N/mm^2

CONCLUSION

Cellular concrete is an interesting material for volumetric heating experiments. Microwave processing in a conventional industrial production process is not likely due to the costs and the fact that microwaves tend to degrade the product quality more than the autoclave curing. This effect is also known for concrete hardening (1). Also, as the bricks enter the autoclave a temperature rise up to 150°C is necessary. Microwaves may speed up this rising, however the time involved is negligible compared with the total curing time.

REFERENCES

(1) A. Watson, Curing of concrete, Microwave Power Engineering, Vol. 2, edited by E.C. Okres(Academic Press, N.Y. 1968), pp. 108-110.

USE OF THE INVERSE TEMPERATURE PROFILE IN MICROWAVE PROCESSING OF ADVANCED CERAMICS

J.G.P. Binner†, I.A. Al-Dawery†, C. Aneziris† and T.E. Cross‡
† Department of Materials Engineering and Materials Design
‡ Department of Electrical and Electronic Engineering
University of Nottingham, University Park, Nottingham, NG7 2RD, UK

ABSTRACT

Attempts are being made to exploit the inverse temperature profile which can be developed with microwave heating with respect to the processing of certain advanced ceramics. This paper discusses the results obtained to date during the microwave sintering of YBCO high-T_c superconductors and the microwave reaction bonding of silicon nitride.

INTRODUCTION

The development of an inverse temperature profile, that is, a body which has a hotter interior than surface, has been confirmed experimentally by many investigators and explained theoretically [1]. The magnitude of the profile is dependent upon many factors such as power level, electric field intensity and material properties including lossiness, and thermal and electrical conductivity. Understanding the role played by these factors is one of the main research directions currently being taken at the University of Nottingham since control over the power deposition and subsequent temperature profile appears to offer a number of advantages over conventional sintering techniques. These include:

i) The possibility of heating components very rapidly and uniformly due to a reduction in thermal stresses via generation of a uniform temperature profile, and

ii) The potential for sintering ceramics 'from the inside out' via generation of an inverse temperature profile.

The former has been discussed elsewhere [2] whilst the latter has potentially great benefit for situations where a gas phase must be in intimate contact with the ceramic throughout the densification process. Examples of the latter are the high-T_c ceramic superconductors where an oxygen atmosphere is required and during the reaction bonding of silicon nitride. This paper discusses the role of the inverse temperature profile during the microwave sintering of YBCO high-T_c superconductors and the microwave reaction bonding of silicon nitride.

The microwave furnace used operates at 2.45 GHz and is a 0 to 5 kW multimode system with variable atmosphere control. Sample temperatures are sensed by the Accufibre[1] system of optical thermometry, that is a black body tip mounted on a sapphire rod.

---

[1] Accufibre, Beaverton, OR, USA

MICROWAVE SINTERING OF YBCO CERAMIC SUPERCONDUCTORS

Current bulk polycrystalline ceramic superconductors suffer from the major disadvantage that only the surface layers have the correct oxygen stoichiometry and are, therefore, superconducting. This occurs because conventional sintering results in the surface densifying before the interior, cutting off the latter from the oxygen-rich atmosphere required during sintering. This effectively creates a 'shell' of superconducting phases surrounding a central core which is non-superconducting [3].

Yttrium-barium-copper oxide (YBCO)[2] compacts measuring 25 mm in diameter and 3-6 mm thick were formed by die pressing at 20 MPa and heated under a range of conditions using both microwave and conventional sintering techniques [4]. A flowing oxygen atmosphere was used in both cases. The magnitude of the temperature profile was controlled in the microwave furnace by a casketing approach in which the degree of insulation was variable (figure 1). Some of the results obtained are shown in table 1.

The use of microwave energy for sintering YBCO superconductor has resulted in a significantly more uniform oxygen content being present. This is believed to have occurred due to the generation of the inverse temperature profile which permitted oxygen to exist throughout the structure during sintering. The profile has yet to be optimised, however. For example, whilst X-ray diffraction analysis of a disc revealed that it was predominantly the $YBa_2Cu_3O_{7-x}$ phase, the presence of small amounts of CuO, BaO and $Cu_2Y_2O_5$ were detected at the surface and Y_2BaCuO_5 and CuO were found in the centre. In between phase pure $YBa_2Cu_3O_{7-x}$ was found. The presence of these impurity phases indicates [5] that whilst the temperature at the surface was about 890°C it increased to approximately 1000°C at the centre. This observation is supported by a slight increase in grain size towards the centre of the compact (figure 2).

Microwave sintering has occurred at lower temperatures and in significantly shorter times when compared with conventional sintering. The reason for this phenomenon, which has been reported by many investigators working in a range of systems, is still not certain although a number of theories have been put forward. Some of these have been discussed elsewhere [1]. In addition the use of microwave energy permitted heating rates of between 15-20°C min^{-1} to be used, these were approximately five times faster than could be achieved in the conventional furnace if cracking of the samples was to be avoided.

All the samples were compacted and sintered only once leading to low oxygen contents and T_c values with the conventionally sintered pieces. However, the microwave processed samples have T_c values more normally associated with YBCO. This appears to indicate that the microwave sintering reduces the need for multiple calcinations which are both time consuming and introduce impurities from the grinding stages.

A further interesting observation is the existence of YBCO polycrystalline fibres within the microstructure (figure 3). These were obtained when bar-shaped samples were suspended within the microwave furnace during sintering in an attempt to get an aligned microstructure. The samples were sintered at 850°C for 2 hours followed by a further 45 minutes at 900°C for annealing. Further investigations of this phenomenon are underway.

---

[2] SS-ACS YBCO powder, Superconductive Components Inc, Ohio, USA

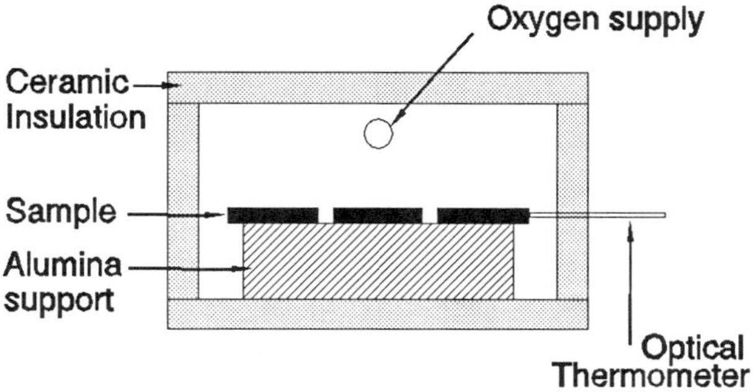

Figure 1: Schematic diagram of the casketing arrangement for YBCO microwave sintering

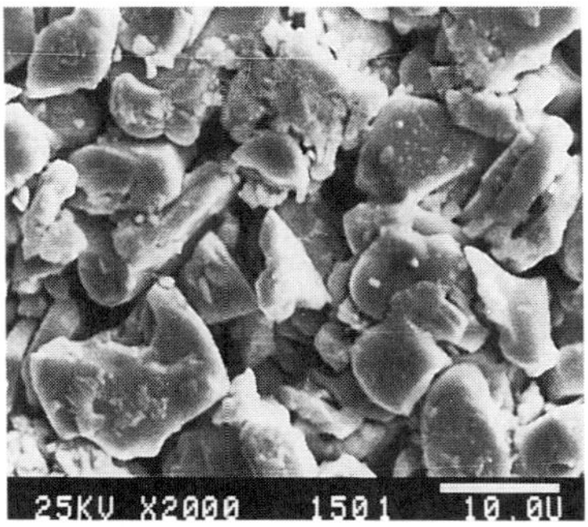

a) Surface of disk

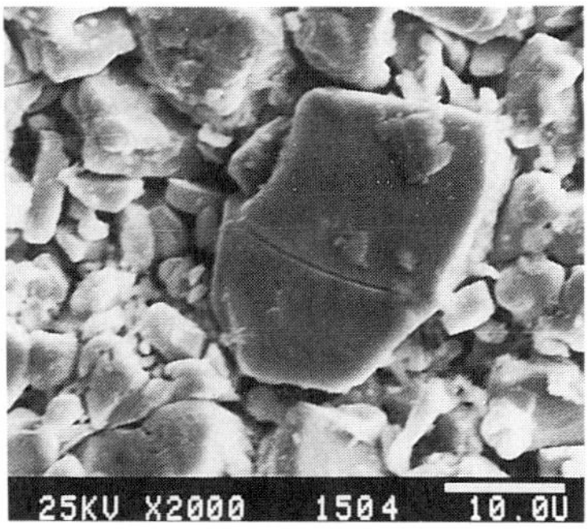

b) Centre of disk

Figure 2: Microwave sintered YBCO disk, 2 hours at 850°C

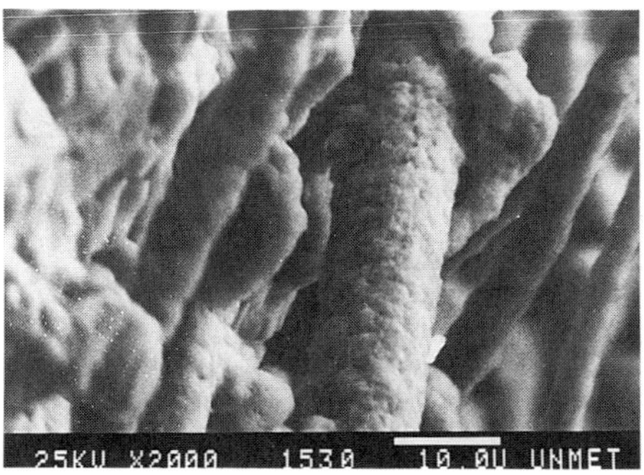

Figure 3: Fibres produced by microwave sintering suspended YBCO disk

Sintering temp/°C	Sintering time/hr	% theor density	T_c K	Oxygen level	
				Surface	Interior
Microwave Processed Samples					
800	1	79	80	6.80	6.80
800	2	84	-	6.80	6.80
855	0.5	78	80	6.80	6.80
855	1	82	-	6.80	6.80
855	1.5	84	-	6.80	6.80
855	2	86	-	6.80	6.80
900	0.5	85	-	6.80	6.80
900	0.75	89	91	7.00	7.00
900	1	90	82	7.00	7.00
900	4	91	-	-	-
950	0.25	92	-	6.56	6.56
Conventionally Processed Samples					
950	12	81	72	6.80	6.56
970	12	85	78	6.80	6.56
970	18	89	86	6.80	6.56

Table 1: Microwave and conventionally sintered YBCO.

Material	Microwave heating temperature/°C	% nitridation by mass
Si	1150	41
Si/20wt% Si_3N_4	1150	42
Si/20wt% SiC	1150	39
Si/10wt% Si_3N_4 /10wt% SiC	1150	34
Si	1200	57
Si/20wt% Si_3N_4	1200	48
Si/20wt% SiC	1200	55
Si/10wt% Si_3N_4 /10wt% SiC	1200	49

Table 2: Microwave reaction bonded silicon nitride (MRBSN).

MICROWAVE REACTION BONDING OF SILICON NITRIDE Si_3N_4

Reaction bonded silicon nitride (RBSN) is a ceramic which retains its strength to high temperatures and displays very low creep rates owing to the fact that glassy grain boundary phases can be avoided. The principle disadvantage is the inability to achieve densities much in excess of about 80% since porosity must be kept open throughout the process to allow nitrogen to permeate the silicon compact. This reduces the strength of the material and also permits internal oxidation at high temperatures if oxidising gases are present during use.

Silicon powder-based compacts[3] measuring 35 mm in diameter and approximately 10 mm thick were formed by die pressing at 20 MPa and have been microwave reaction bonded for one hour in a flowing nitrogen atmosphere. Some of the results obtained are shown in table 2. As expected, the higher the reaction bonding temperature the greater the degree of reaction which has occurred. It is interesting to note, however, that considerably greater nitridation has been found than in the equivalent work performed by Kiggans et al [6]. One explanation for this is that the temperature measured in the current work was the surface of the compacts and, as with the YBCO work described above, the centre of the disc was almost certainly hotter than the surface. In addition, a typical sample mass before reaction bonding was only 12 g compared to 100 g used by Kiggans et al. Iron content may also play a role.

Nitridation of the samples, and thus by implication the temperature profiles, was not uniform. Unlike the YBCO work where there was evidence of a steady increase in temperature towards the centre of the samples, nitridation occurred in a layered fashion with a thick layer just below the surface and the base of the discs being virtually phase-pure silicon nitride whilst the other areas displayed negligible evidence of reaction (figure 4). The addition of silicon carbide and/or nitride powders was intended to reduce this problem however to date this has not occurred and generally the degree of reaction has been slightly impaired.

The percent of free silicon left in some of the samples was calculated using the formula proposed by Gazzara and Messier [7] and the results compared with the weight gains observed, as in the work of Kiggans et al. Very good agreement was obtained, for example in the silicon sample heated at 1200°C 57% nitridation had occurred according to the change in mass whilst Gazzara and Messier's formula indicated 54% nitridation. The nitride formed in all cases was predominantly the α-phase leading to the fibrous microstructure visible in figure 4a.

CONCLUSIONS

The nature of the temperature profile developed within materials heated by microwave energy has been shown to permit rapid heating and to allow ceramics to be sintered from the 'inside out' where appropriate. With yttrium-barium-copper oxide

---

[3] Hermann C Starck, Berlin, Germany, grade B10 Si powder (Fe content: 1400 ppm)

a) Reacted region b) Nonreacted region

Figure 4: Microwave reaction bonded silicon nitride, 1 hour at 1200°C

(YBCO) microwave sintering has yielded uniformly high oxygen contents and high transition temperatures without the need for repeated calcinations.

Significant evidence for lowered sintering temperatures and times has been observed with both the YBCO and reaction bonded silicon nitride systems. However, a greater degree of control over the temperature profile developed within samples is required if advantage is to be taken of the potential offered.

ACKNOWLEDGEMENTS

The authors wish to acknowledge the UK Science and Engineering Research Council and Department of Trade and Industry, and the European Community ERASMUS programme for financial support which enabled the work discussed to be performed.

REFERENCES

1. A.C. Metaxas and J.G.P. Binner, in Advanced Ceramic Processing Technology Vol I, edited by J.G.P. Binner (Noyes Publications, New Jersey, 1990), pp 285-367.

2. J.G.P. Binner and T.E. Cross, to be published in proceedings of Fabrication Science and Technology II, Manchester, England, Dec, 1991.

3. J.W. Ekin, Adv. Ceram. Mat. 2 [3B] 586 (1987).

4. I.A.H. Al-Dawery, J.R. Moon and J.G.P. Binner, to be published.

5. D.R. Clarke and D. Mimos, J. Am. Ceram. Soc. 72 [7] 1103-13 (1989).

6. J.O. Kiggans, C.R. Hubbard, R.R. Steele, H.D. Kimrey, C.E. Holcombe and T.N. Tiegs, in Microwaves: Theory and Application in Materials Processing, Ceramic Transactions Vol 21, edited by D.E. Clark, F.D. Gac and W.H. Sutton, (Am. Ceram. Soc., Ohio, 1991) pp 403-410.

7. C.P. Gazzara and D.R. Messier, Am. Ceram. Soc. Bull. 56 [9] pp 777-81 (1977).

ELECTRON MICROSCOPY STUDY ON ZIRCONIA AND ALUMINA CERAMIC POWDERS SYNTHESIZED BY MICROWAVE PLASMA PYROLYSIS

KURT E. SICKAFUS,** DIETER VOLLATH,* AND RAVI VARMA**
*Kernforschungszentrum Karlsruhe, Karlsruhe, Germany
**Los Alamos National Laboratory, Los Alamos, NM, USA 87545

ABSTRACT

Morphology, chemistry and microstructure of zirconia and alumina based ceramic powders synthesized by microwave plasma pyrolysis were studied by transmission electron microscopy. Electron diffraction was used to determine the phases present, while conventional bright field and dark field imaging were used to characterize grain size and density variations. Materials analyzed in this study were either pure zirconia, pure alumina, or a ternary zirconia-yttria-alumina oxide. Solid spheres, hollow spheres, and fragments of hollow spheres were observed. All three modifications of zirconia were observed in the pure zirconia powder. In alumina, the predominant phase was cubic, gamma alumina. In the ternary oxide, most structures exhibited cubic zirconia diffraction patterns. Energy dispersive x-ray spectroscopy revealed chemical inhomogeneities in the ternary oxide powder.

INTRODUCTION

Pyrolytic synthesis of ceramic powders by plasma processes (energized by DC and RF sources) have been demonstrated previously [1,2]. Temperatures in these processes (>4,000 °C) are sufficient to melt oxide particles so that microstructures produced by these techniques are dependent on melting and resolidification processes. An alternative technique involves powder synthesis in a microwave plasma, where thermal temperatures range from 300 - 800 °C. Gas pressure and plasma temperature can be adjusted to the chemistry of interest, in order to synthesize ceramic powders with diverse microstructures.

EXPERIMENTAL PROCEDURE

Oxide powders were synthesized by a microwave plasma pyrolysis process described elsewhere in these proceedings [3]. For the powders discussed here, nitrate solutions provided the source for the cation constituents of the powders. Specifically, $ZrO(NO_3)_2 \bullet nH_2O$, $Al(NO_3)_3 \bullet nH_2O$, and $Y(NO_3)_3 \bullet nH_2O$, were mixed with water in proportions appropriate to produce three oxide compositions: (1) pure zirconia; (2) pure alumina, or (3) 20 mole % alumina - 3 mole % yttria - zirconia. The latter composition will be referred to as the "ternary oxide" composition.

Powders were collected beneath the plasma zone of the microwave system on platinum foils. Samples for transmission electron microscopy were prepared by scraping holey carbon grids across the platinum foils to dislodge small quantities of deposited powder.

Transmission electron microscopy (TEM) observations were made on either a Philips CM-30 microscope with a twin lens configuration operating at 300 kV, or a Philips CM-30 microscope with a super twin lens configuration operating at 300 kV. Energy dispersive x-ray spectroscopy (EDXS) was performed on the twin lens instrument using a Kevex Delta Series spectrometer, equipped with an ultra-thin window detector, and the Quantex V x-ray analysis system. Scanning transmission electron microscopy (STEM) observations were made on the twin lens instrument. STEM operation on this instrument includes image acquisition in bright field (BF) and annular dark field (ADF) imaging modes with an off-axis detector. In this study, images obtained by the ADF imaging technique utilized a camera length setting of 140 mm.

RESULTS & DISCUSSION

TEM analyses of zirconia, alumina, and ternary oxide powders revealed microstructures that, in each case, consisted primarily of spherical particles (or debris from broken spheres). Sphere diameters ranged from about 25 nm to nearly 4 μm. Larger spheres (around 1μm or larger) were sometimes hollow. TEM analysis is greatly simplified with a hollow sphere because even though the sphere thickness may exceed 1μ m, electron transparency is conserved.

Analyses indicated that most spheres were microcrystalline, though glassy spheres were observed in all compositions studied. Spheres containing nanocrystalline grains were numerous in all powders.

Figures 1a,b, and c show dark-field (DF) TEM images from selected regions of zirconia, alumina, and ternary oxide powders, respectively. The large sphere in Figure 1a from the zirconia powder is hollow and an adjacent sphere (indicated by the arrow) is nanocrystalline. The speckle contrast from the latter sphere suggests an average grain size near 1 nm. Selected-area diffraction patterns (SADP) revealed particles of monoclinic, tetragonal, and cubic zirconia.

The image in Figure 1b from the alumina powder contains the largest diameter hollow sphere observed in this study, approximately 3.7 μm diameter. The grain size within this hollow sphere is nanocrystalline (about 7 nm average diameter). Nanocrystalline grain structure was often observed for hollow spheres. One of the smallest spheres in Figure 1b (indicated by the arrow) appears illuminated in this DF image, indicating that it is a single crystal particle. This was a common observation in the case of small spheres. SADP indicated that the alumina spheres were of the cubic, gamma alumina phase.

Figure 1c contains spheres from the ternary oxide powder. The largest sphere is again a hollow sphere. The sphere indicated by the arrow contains nanocrystalline grains with an average diameter of about 2 nm. The predominant phase observed in SADP patterns from these particles was the cubic zirconia phase.

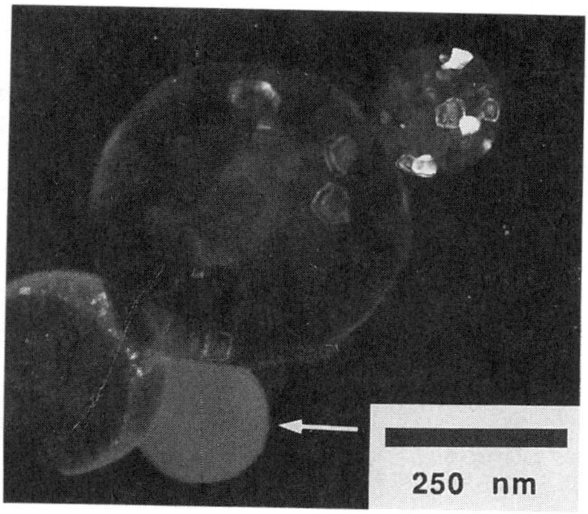

Figure 1a. TEM dark-field micrograph of zirconia particles synthesized by microwave plasma pyrolysis. Arrow indicates a sphere containing nanocrystalline grains.

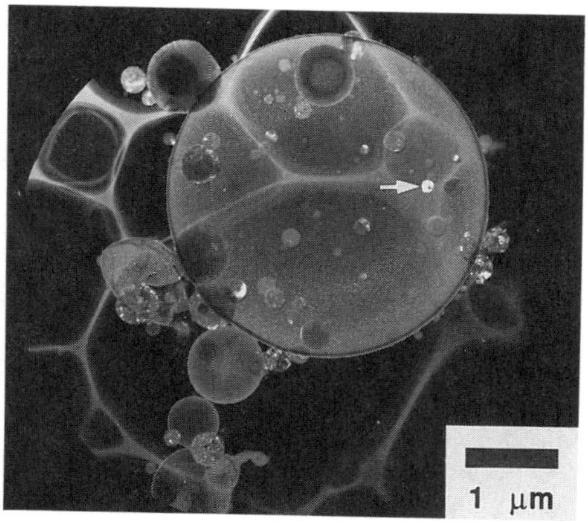

Figure 1b. TEM dark-field micrograph of alumina particles synthesized by microwave plasma pyrolysis. Arrow indicates a single crystal sphere.

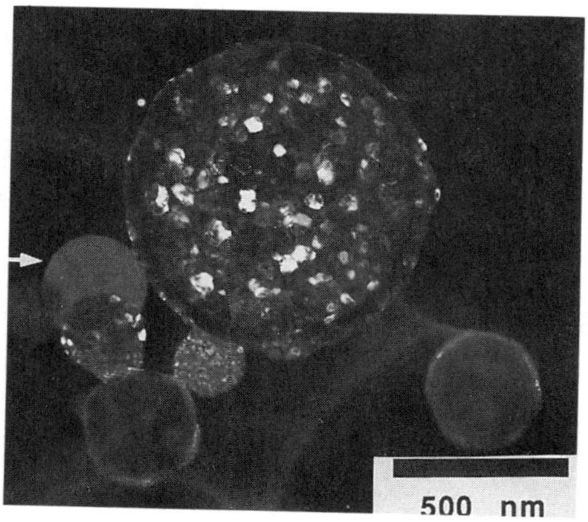

Figure 1c. TEM dark-field micrograph of ternary oxide (ZrO_2-Y_2O_3-Al_2O_3) particles synthesized by microwave plasma pyrolysis. Arrow indicates a sphere containing nanocrystalline grains.

Figure 2 is a plot of the average grain diameter within spheres versus sphere diameter for an analysis of 55 spheres in the ternary oxide powder. Glassy spheres are arbitrarily assigned a grain size of 1 nm, so that they can be included on this logarithmic plot. Glassy spheres are identified by their uniform contrast in DF micrographs (no speckle contrast is apparent, as in nanocrystalline particles). No correlation between grain size and sphere diameter is apparent in Figure 2. The only obvious trend is that hollow spheres tend to be the largest spheres present, around 1 μm diameter.

The particles with the largest grains (>100 nm) observed here may be artifacts of the synthesis process. They may arise from residual ZrO_2 particles suspended in the $ZrO(NO_3)_2 \bullet nH_2O$ solution, that are subsequently delivered to the microwave plasma via the nozzle spray system.

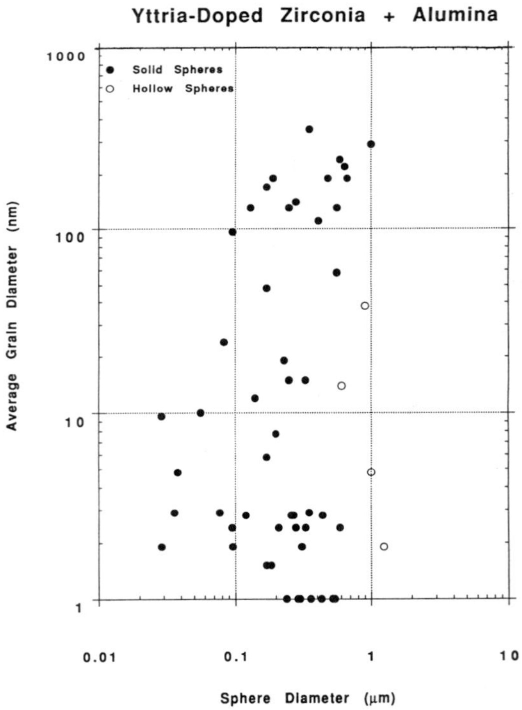

Figure 2. Plot of the average grain diameter within spheres as a function of sphere diameter for particles in the ternary (ZrO_2 -Y_2O_3-Al_2O_3) oxide powder.

Figure 3. High-resolution TEM image from a small sphere in the alumina powder that has sintered to a larger, adjacent sphere.

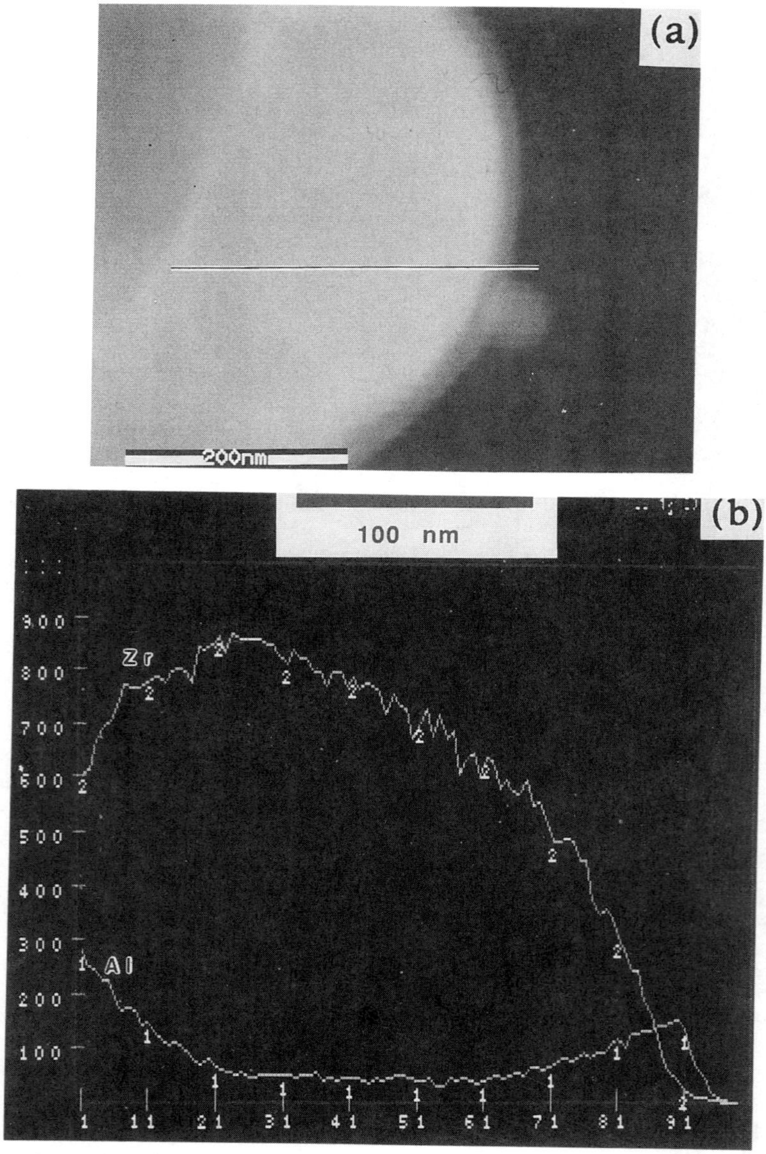

Figure 4. a. Annular dark-field STEM image from a small sphere in the ternary (ZrO_2-Y_2O_3-Al_2O_3) oxide powder , exhibiting phase separation. The line on the image indicates the path of the electron beam for the EDXS line trace in b.
b. Energy-dispersive x-ray spectroscopy (EDXS) line trace across the particle in a. X-ray energy windows for the line trace were established to include (i) the Zr and Y L_α x-ray lines (labeled Zr), and (ii) the Al K_α x-ray line (labeled Al).

Figure 3 shows a high-resolution TEM image from a typical small sphere (~50 nm diameter) in the alumina powder. The sphere possesses a single orientation (though the lattice image suggests that considerable strain exists within the sphere) and it appears attached with a classic 'sintered neck' geometry to a much larger, adjacent sphere.

Figure 4 shows results of a STEM analysis to reveal the nature of this phase separation phenomenon. Figure 4a shows an annular dark field (ADF) STEM image from a portion of a phase-separated, ternary oxide particle. The line across the image indicates the path of an EDXS line-trace taken across this particle. The line-trace is reproduced in Figure 4b. X-ray intensities from an energy window containing the Zr (and Y) L_α x-ray line and from an energy window containing the Al K_α x-ray line are plotted as a function of electron beam location on the particle. It is clear that the Al signal rises at the outer surface of the sphere, while the Zr/Y intensity drops to background level before the outer surface is reached. This is indicative of alumina segregation to the outer surface of the particle. The coating thickness in this case is about 20 nm. Alumina segregation is also evident at the left-hand side of the line trace (in the middle of the particle). ADF image contrast indicates that phase separation is probably present in the interior of the sphere, which accounts for this rise in the Al signal. In fact, careful study of the ADF image contrast suggests that the zirconia in this sphere is confined to two kidney-shaped features (one of which is fully-visible in the image in Figure 4a. Phase separation is to be expected here, as there is virtually no mutual solubility in the Al_2O_3-ZrO_2 system beneath the eutectic reaction temperature (~1700°C) [4].

CONCLUSIONS

Microwave plasma synthesized ceramic powders were analyzed by transmission electron microscopy. Spherical particles were observed in zirconia, alumina, and yttria-doped zirconia plus alumina powders. Solid and hollow spheres were observed containing grain sizes approaching 1 nm diameter. Glassy spheres were also observed in all powders studied. Alumina segregation was observed in the ternary oxide powder.

References

1. R. M. Young and E. Pfender, Plasma Chem. and Plasma Processes 5, 1-37 (1985).
2. P. Fouchais, E. Boudrin, J. F. Coudert, and R. Mcpherson, Plasma Chemistry IV, edited by S. Veprek and M. Venigopalan (Springer Verlag, Berlin, 1983) pp. 2093-2098.
3. D. Vollath, R. Varma, and K. E. Sickafus, presented at the 1992 MRS Spring Meeting, San Francisco, CA, 1992 (unpublished).
4. E. M. Levin and H. F. McMurdie, Phase Diagrams for Ceramists: 1975 Supplement, edited by M. K. Reser (The American Ceramic Society, Columbus, OH, 1975) pp. 135-136.

MICROWAVE PROCESSING OF ALUMINUM-SILICON CARBIDE CERMETS

ERIC P. BESCHER, URMIMALA SARKAR AND JOHN D. MACKENZIE
Department of Materials Science and Engineering, University of California Los Angeles, Los Angeles, CA 90024

ABSTRACT

Electromagnetic radiation in the microwave range has been utilized for the fabrication of SiC particulates-reinforced aluminum matrix cermets. Powder mixtures with compositions ranging from 70 wt% Al - 30wt% SiC to 100 wt% Al were placed and processed under protective nitrogen atmosphere in a 750 W multimode cavity operating at 2.45 GHz. Temperatures larger than 800°C have been achieved in a time varying with the mixture composition. Microstructures and some physical properties of the composites are presented.

INTRODUCTION

Ceramic-metal composites are attracting increased attention because of their superior mechanical properties. Metal-matrix composites offer the potential of low-cost, high performance materials. For example, the introduction of a ceramic phase in the form of particulates or whiskers into aluminum results in an increase of the relatively low elastic modulus of the metal and enhances its wear resistance. SiC-reinforced aluminum exhibits higher strength, increased modulus and higher temperature workability than aluminum [1]. A variety of methods have been used for the fabrication of metal-matrix ceramic composites and include:

1. Resistance sintering [2] ,
2. Vacuum hot pressing or hot isostatic pressing,
3. Foundry techniques such as squeeze casting,

Liquid phase sintering [3] is a preferred processing route in order to:

- achieve high densities in relatively short sintering time,
- help promote bonding between the matrix and the ceramic particles,
- in the case of aluminum matrices, break the Al_2O_3 layer present on the surface of the aluminum particles.

The microwave heating properties of ceramic-metal mixtures have already been investigated by a number of researchers [4,5] . The recent interest in microwave processing has identified some of the advantages of this technique as including rapid deposition of energy within the bulk of the material and low cost of energy [6]. The present research was undertaken to investigate a new route to metal-matrix ceramic composites: the use of electromagnetic radiation in the microwave range.

EXPERIMENTAL PROCEDURE

Aluminum powder was 99.99 % pure from Aldrich Chemicals, with particles of irregular shape. β- SiC powder was 99.9, also from Aldrich. The aluminum powder particle size was 20 μm (Figure 1), whereas the SiC particle size was 1 μm. Appropriate amounts of aluminum and silicon carbide powders were manually mixed, placed in a crucible, insulated with a 3-inch thick alumino-silicate fiber, and placed in a 2.45 GHz multimode cavity operating at 750 Watt of input power†. The microwave cavity was placed in an atmosphere-controlled chamber in which dry nitrogen was circulated. Figure 2 shows a diagram of the experimental set-up. The oxygen concentration in the chamber was kept below 20 ppm during processing and cooling of the samples. Temperature readings were obtained using an infrared dual-wavelength optical pyrometer†† operating within a temperature range of 700°C to 1700°C. Temperatures were read through a small opening in the thermal insulation. The size of this opening was sufficiently small to allow black body conditions. The samples were processed for 15 minutes once they had reached a temperature of 800°C.

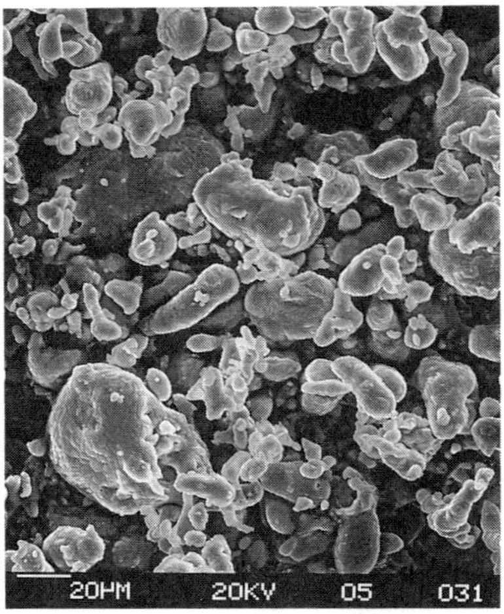

Figure1. Aluminum particles used in this study.

After the microwave power was turned off, the samples were left to cool inside the nitrogen-filled chamber. Samples were cut, polished and etched. Porosities and densities were determined by the Archimedes method.

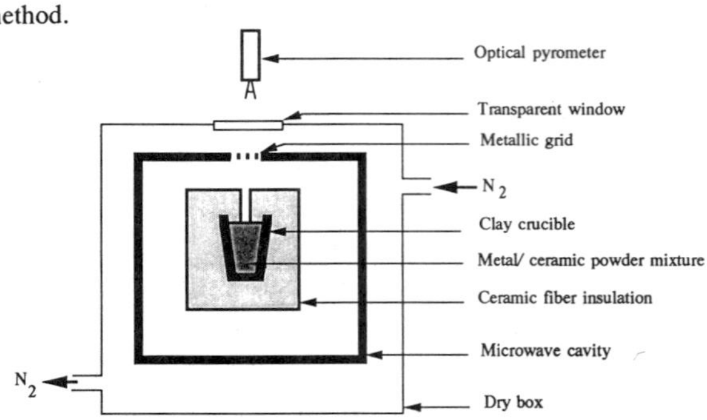

Figure 2. Experimental set-ur

†Model JE1453H002, General Electric, Louisville,KY

††Model M77S, Mikron Instruments, Wyckoff,NJ

RESULTS AND DISCUSSION

Aluminum, like most metals, cannot be considered a good microwave absorber. However, if it is under the form of a finely divided powder, the particles will slowly heat up and reach temperatures above 1200°C. Figure 3 shows a heating curve for the 20μm aluminum powderused in this study. On the other hand, because of its lower electrical conductivity, silicon carbide is a highly absorbing material. When 10, 20 or 30 weight percent of silicon carbide are added to a poorly absorbing aluminum powder, the heating curve in 2(a) is changed. Upon four minutes of microwave exposure, the mixtures initially reach very high temperatures for a very short time. During some runs, temperatures higher than 1500°C have been briefly recorded. The temperature immediately decreased to reach a plateau around 1100°C to 1000°C. In Figure 4, we propose a mechanism to explain this phenomenon, and we introduce the concept of effective penetration depth D. The power absorbed per unit volume P (W/m^3) in a material exposed to electromagnetic radiation is

$$P= \sigma E^2= 2\pi f \varepsilon_0\varepsilon'_r \tan \delta E^2 \qquad (1)$$

where σ is the total effective conductivity (S/m) of the material, f the frequency of the electromagnetic radiation (Hz), ε_o the permeability of vacuum (F/m), ε'_r the relative dielectric constant , tan δ the loss factor and E (V/m) the electric field within the material. Since the electric field is attenuated as the microwaves penetrate the material, a useful parameter describing this is d, the penetration depth at which the incident power is reduced to 1/e of its value at the surface:

$$d = 3 \lambda / 8.686 \pi \tan \delta (\varepsilon'_r/\varepsilon_o) \qquad (2)$$

where λ is the incident wavelength. This equation means that for materials with a high tanδ, such as metals, very little of the bulk of the material actually interacts with the microwaves. The microwaves induce currents on the surface of the materials. The penetration depth at 2.45 GHz for most metals is usually less than 3 μm. The heat is propagated to the rest of the bulk by conventional heat transfer. Although microwaves do not penetrate flat and clean metal surfaces efficiently, the case is slightly different for metallic composite powders.

In a porous composite, each particle near the surface is affected by surface currents down to the depth d. However, if the powder is loosely packed, the voids between particles allow for the radiation to penetrate the composite to a depth D, which can be far greater than d. As the volume of the composite affected by the curents is brought above the melting temperature of the metal matrix phase, metal particles bond and the surface area of the powder decreases. The effective penetration depth D is hence reduced. In the limiting case where the metal is in fusion, the effectice penetration depth is d again, but for a metal in fusion. This phenomenological model explains the initial increase in temperature followed by a plateau at longer processing times.

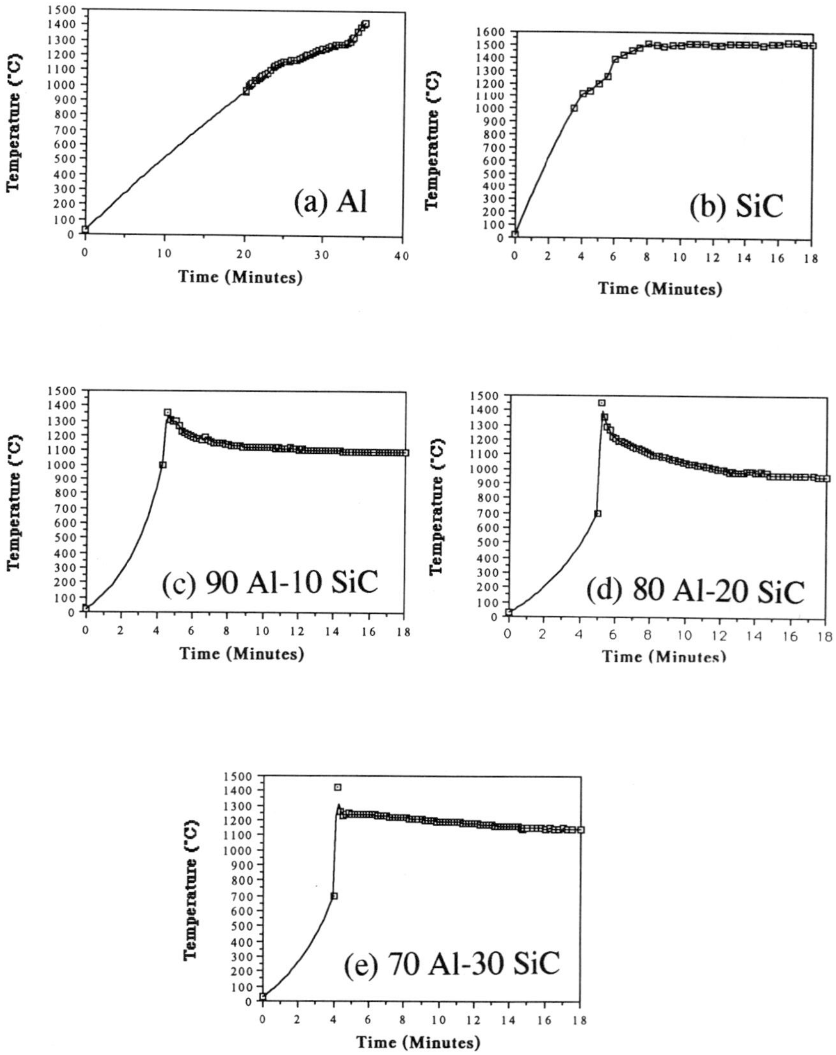

Figure 3. Heating curves for (a) pure aluminum, (b) pure silicon carbide(c) 90wt% Al-10wt%SiC,(d) 80wt%Al-20wt%SiC, (e) 70wt%Al-30wt%SiC

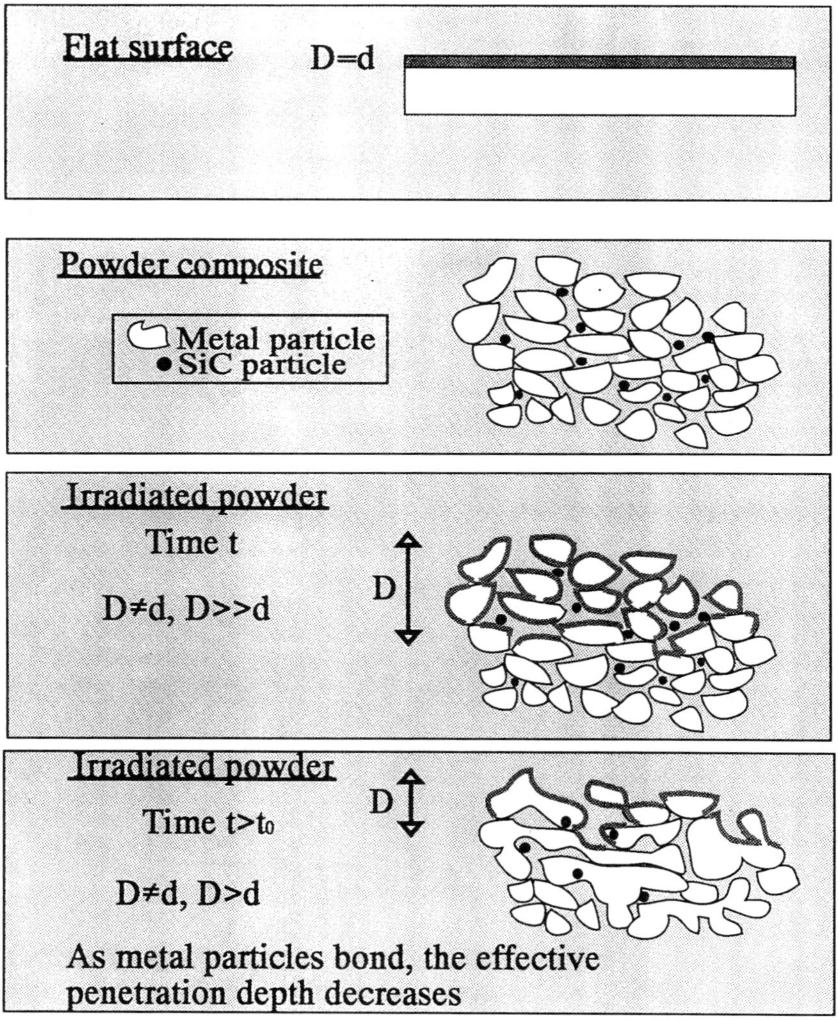

Figure 4. Phenomenological model for microwave processing of cermets

The density and microhardness of the solidified samples are presented in Table 1 . The density of the composites varied from 93% to 97 % of the theoretical density . Their hardness was identical to that of conventionally obtained materials.

TABLE 1. Density and hardness of microwave-derived cermets

	Density (% Theor.)	Hardness (VHN)
90% Al-10% SiC	98	60±5
80% Al-20% SiC	96	63±5
70% Al-30% SiC	93	59±5

Solidification microstructures, interfaces and thermodynamic considerations

The microstructure of Al-SiC cermets is shown in Figure 5. It is very similar to that of cermets obtained by other techniques [7]. The silicon carbide particles have been rejected from the aluminum phase to be trapped in the grain boundaries. This implies that the microstructure is predominently dependent on the cooling phase and solidification rate, which also explains the similarity in microstructures with conventionally obatined samples. The average aluminum grain size is 10 microns.

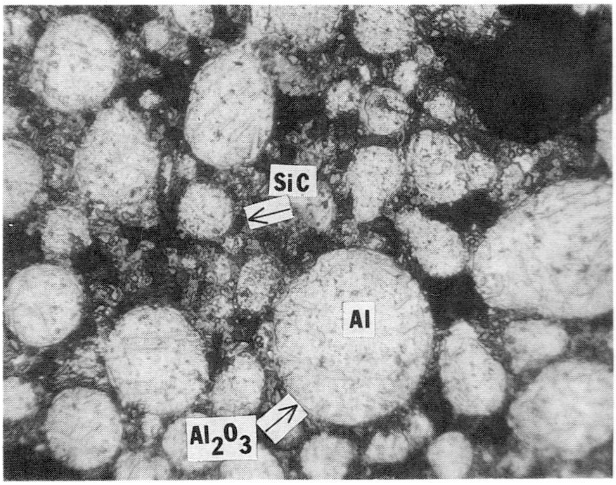

Figure 5. Microstructure of a 90wt% Al-10wt% SiC

The reaction between Al and SiC is favored by the good wettability of SiC by molten aluminum [8]. The main reaction between liquid aluminum and SiC is:

$$4\ Al + 3SiC \text{--->} Al_4C_3 + 3\ Si \qquad (3)$$

There is also the possibility of forming a ternary compound according to :

$$4\ Al + 4\ SiC \text{--->} Al_4SiC_4 + 3\ Si \qquad (4)$$

However, the present work shows no evidence of the latter. Previous work has reported the occurrence of the first reaction [8]. This is confirmed by X-ray diffraction data, Figure 6, which shows the formation of reaction products such as Al_4C_3 and Si according to reaction (3). Others compounds such as Al_2O_3 are also present. Alumina was visible by optical microscopy as inclusions inside the aluminum grains, and at their periphery. The first type was probably present on the surface of the aluminum powder before processing and broken during melting of the metal, whereas the second is likely to be a post-processing occurence. Unidentified peaks were attributed to impurities from the crucible.

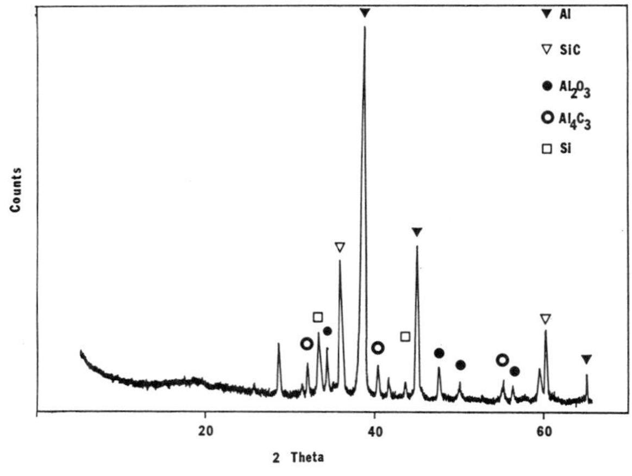

Figure 6. X-Ray diffraction for a 90wt%Al-10wt%SiC

The free energy change ΔG for reaction (3) has been calculated over a range of temperatures from tabulated data [11]. The temperature at which ΔG=0 is 885°C. Above this temperature, thermodynamic considerations are favorable for the formation of aluminum carbide .

CONCLUSION

Microwave radiation was utilized as a new technique for the fabrication of aluminum matrix SiC particulates composites. Temperatures superior to the melting point of aluminum were obtained, allowing liquid phase processing of the composite. The energy required for processing was dependent on the weight fraction of metal, and was decreased by addition of the carbide. Interfacial compounds such as Al_4C_3 were formed at the interfaces. The microstructures and properties of the resulting composites were similar to that of conventionally obtained composites.

ACKNOWLEDGEMENTS

This research was funded by the State of California through the Risk and Systems Analysis for the Control of Toxics (RSACT) Program. Urmimala Sarkar would like to acknowledge the Student Research Program at UCLA .

REFERENCES

1. A.P. Divecha, S.G. Fishman and S. D. Karmakar, *J. Met.*, 33 (9) (1981) 12.
2. S.J. Hong and P.W. Kao, *Materials Science and Engineering*, A119 (1989) 153-159.
3. S.J. Yankee, G.M. Janowski and B.J. Pletka, *Materials and Manufacturing Processes*, 5(2) (1990) 147-164.
4. E. P. Bescher and J.D. Mackenzie, *Ceram. Trans.*, 124 (1991).
5. Y.S. Ho and J.J. Kramer, *Mat. Res. Soc. Symp. Proc.*,124 (1988) 161-166
6. W.H. Sutton, *Am. Ceram. Soc. Bull.*, 68 [2] 376-86 (1989).
7. D.J. Loyd, *Composites Science and Technology*, 35 (1989) 159-179.
8. W.C. Moshier, J.S. Ahearn and D.C. Cooke, *J. Mater. Sci.*, 22 (1987) 115.
9. B. Robertson, C. Holm, Y.P. Lin, S.F.J. Corbin and D.S. Wilkinson, *Mat. Res. Soc. Symp. Proc.*, 170 (1990) 105-110.
10. *CRC Handbook of Physics and Chemistry*, CRC Press, Inc Publ., Florida.

SYNTHESIS OF NANOCRYSTALLINE POWDERS FOR OXIDE CERAMICS BY MICROWAVE PLASMA PYROLYSIS

Dieter Vollath*, Ravi Varma** and Kurt E. Sickafus**

*Kernforschungszentrum Karlsruhe, Institut fur Materialforschung, P. O. Box 3640, W-7500 Karlsruhe, Germany
**Los Alamos National Laboratory Materials Science and Technology Division, Los Alamos, N.M. U.S.A.

ABSTRACT

Pyrolytically prepared ceramic powders usually exhibit excellent homogeneity and small grain sizes. The energy efficiency of those methods for synthesis is, particularly in systems heated electrically, very poor. The situation can be improved drastically using a microwave plasma as a source of energy. This has been confirmed by the synthesis of alumina- and zirconia-based ceramic powders for which an efficiency of more than 80% was found. Aqueous solutions of nitrates of zirconium, yttrium, and aluminum were used as starting materials. Electron microscopy revealed that, through proper selection of the experimental conditions, it is possible to obtain nanocrystalline powders with cubic structure. Because of the extreme conditions during synthesis, it is possible to prepare solid solutions in systems without solubility.

1. INTRODUCTION

In the synthesis of ceramic powders, one is faced with two major issues:

(1) The homogeneity of the material, in the case of complex compositions, is very difficult to obtain and is nearly impossible to achieve by mechanical blending and grinding processes. Even in the case of chemical precipitation processes, in most of the compositions used, it is very difficult to obtain homogeneous powders, as the different constituents usually precipitate at different pH values. This problem can be overcome by using pyrolytic processes [1,2,3,4].
(2) Pyrolytic methods are best suited for the synthesis of complex ceramic powders. However, the demand on energy can be a major concern. This is most significant in electrically heated systems. The situation can be improved, if one uses a flame as a source of thermal energy [2,3,4].

It is assumed that both problems can be solved if a plasma is used as a source for thermal energy in a pyrolytic process for synthesis of ceramic powders. To prove this assumption, a microwave plasma reactor was set up and tested to synthesize ceramic powders. Certainly, the energy efficiency of a flame process is not impaired by the second law of thermodynamics, as is the case with all processes using electrical energy. Application of a plasma has a great advantage in that it is not necessary to deal with the chemistry and the safety aspects of a flame.

Plasma processes for pyrolytic synthesis of ceramic powders are described frequently [5,6]*. Such processes have utilized either working with a dc or an rf plasma. In most cases microwave plasma methods employ a gas phase reaction at low pressures [7]. Working at low pressures (e.g. < 1 mbar) limits the production rate significantly. The main problem connected with such plasma devices is the high temperature connected with a high plasma viscosity. These two related problems limit the application to a range of relatively low pressures.

2. THE MICROWAVE PLASMA

The energy transferred to a particle in an oscillating field is proportional to $1/(m \cdot f^2)$ (m...mass, f...frequency). For microwaves the frequency is high, so the amount of energy transferred to the ions is small. Because the mass of the electrons is extremely small, a relatively large high amount of energy is transferred to the electrons. Because of this, the "temperature" of the free electrons is much higher than the "temperature" of the ions. This leads to the fact that the temperature of a gas passing through a microwave plasma zone can be adjusted in a range between 300°C and 800°C by properly selecting field strength, gas pressure, and gas species. Gas pressure and temperature needed for the synthesis of interesting amounts of ceramic powders can be adjusted in a range that is optimal for the chemistry going on.

For the experiments described in this paper, a microwave system with a frequency of 0.915 GHz was used. The reaction vessel was placed in a single-mode cavity constructed from a WR 975 waveguide using the TE_{10} mode. The standing wave was adjusted using a sliding short circuit.

3. POWDER SYNTHESIS

3.1. EXPERIMENTAL SETUP

The ability of a microwave plasma to act as a source of heat can be applied for powder synthesis. Ceramic powders based on zirconia and alumina were produced. For a comparison of this work with previous work, aqueous solutions of the nitrates were used. As described in earlier papers [1,2], the solutions were sprayed in the heating zone using a two-phase nozzle. The nozzle applied delivers a mist of droplets with a mean diameter of

----

* These are just two review papers in this field. Because a vast amount of literature concerning the application of dc and rf plasmas exists, it does not make sense to make an arbitrary selection for citation.

about 7 um [8]. The temperature of the gas was monitored after passing the reaction (plasma) zone. The powder synthesized was collected after measuring the temperature.

In a typical experimental run, a gas volume of about 9 m^3/h passed the plasma zone. This gas carried a mist of the aqueous solution of the reactants equivalent to a volume between 0.5 to 1.0 L/min.

3.2 MORPHOLOGY OF THE POWDER

Pure zirconia was prepared to provide an initial experimental database. Electron micrographs (Fig. 1) show particle sizes to be in the range of 100-500 nm. The particles are spherical; large particles are fractured. Crystallite sizes are less than 100 nm, and most particles exhibit crystallite sizes of less than 10 nm. Those small crystallites showed grain growth when heated in situ by the electron beam in the electron microscope. Most of the particles are found to be crystallized in the cubic structure and are not transformed to the equilibrium modification. Besides the cubic crystallized particles, particles exhibiting tetragonal or monoclinic structure are also found.

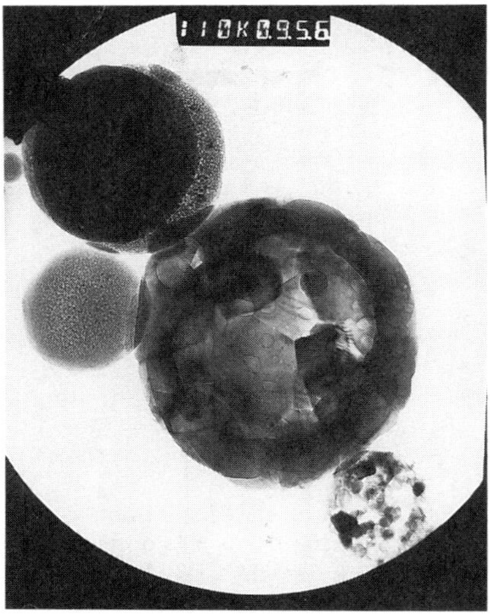

Fig. 1 Bright-field electron micrograph of a pure zirconia sample. One particle exhibits large grains, while the grain size of the others is less than 10 nm.

250 nm

The results obtained by the synthesis of ZrO_2-Y_2O_3-Al_2O_3 ceramic powder are more important. The powder is single-phased cubic zirconia. As revealed by electron microscopy, particle sizes are in the range of 100-1000 nm. Crystallite sizes are less than 50 nm, and most of the particles exhibit crystallite sizes of less than 10 nm (Figs. 2 and 3). Grain growth due to the energy input in the electron microscope was not observed. Most of the particles were found to be solid, but a few hollow ones were found. A segregation of the different constituents, as demonstrated by the micrographs, was not observed. This observa-

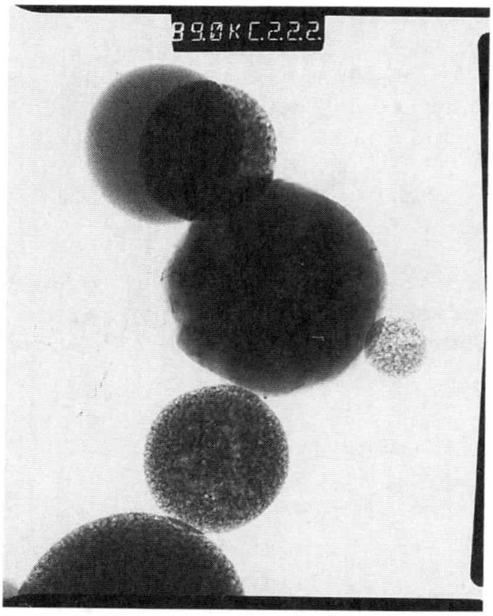

Fig. 2 Bright-field electron micrograph of a zirconia-yttria-alumina sample. The crystallite size is less than 10 nm.

200 nm

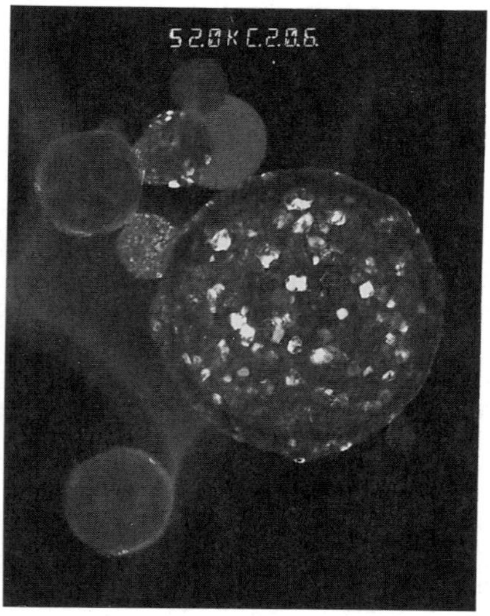

Fig. 3 Dark-field electron micrograph of a zirconia-yttria-alumina sample. For most of the particles the crystallite size is less than 10 nm. One particle exhibits grains of about 40 nm.

500 nm

tion is similar to the case of pyrolysis in a hydrogen flame. In both cases, alumina is in a nonequilibrium solid solution.

If one compares the morphology of the powder particles obtained by the microwave plasma pyrolysis process with the morphology of powder particles obtained by flame pyrolysis [3,4], one finds almost no differences. In this respect those two processes are equivalent. This indicates that the mechanisms of the reactions for powder formation are essentially the same. This means that the process described uses the plasma solely as a concentrated source of thermal energy. Plasma-enhanced chemical reactions play, if any, a minor role. This is an important difference compared with processes using plasma-enhanced chemical gas reactions.

3.3. ENERGY BALANCE

Appropriate experiments were performed to prove the claim of energy efficiency. Experimental conditions were not selected on optimal points of operation. Conditions were selected to allow a broader range of gas flow and injection rate of water. The following results were obtained:

Energy requirements:

Energy to heat the carrier gas up to 750°C

	1.2 l/sec Ar	805 J
	0.3 l/sec O_2	314 J

Evaporation of the water and heating up to 750°C

	0.22 g/sec H_2O	805 J
Total energy requirement		**2034 J/sec**
Total energy input		**2440 J/sec**
Reflected energy		220 J/sec
Energy losses in the cavity		186 J/sec

In the experiment described, 83% of the energy input was used for the process. The energy losses are due to ohmic losses in the cavity and, to a minor extent, radiation loss from the plasma. The overall efficiency could be increased by using a cavity made of copper instead of stainless steel. The amount of reflected energy was 9% in the case discussed. It is possible to reduce the amount of reflected energy to less than 1.5%. Under those conditions, it is nearly impossible to maintain a stable plasma.

4. CONCLUSIONS

Using a microwave plasma as a source of thermal energy for pyrolitic synthesis of ceramic powders has distinct advantages. Such a system makes it is possible to transfer thermal energy with high efficiency to the chemical processes. This leads to an overall efficiency of more than 80%. In the case of aqueous solutions, the reaction mechanism is the same as in the case of thermal pyrolytic processes. The morphology of the powder obtained is similar to that of powders produced by thermal pyrolysis processes. Powders synthesized in zirconia-based systems are completely crystalline. Because of the small sizes of the crystallites and the extreme conditions for the reactions, the particles exhibit the cubic structure. The conditions during synthesis make it possible to prepare solid solutions also in systems without equilibrium solubility.

ACKNOWLEDGEMENTS

This work was supported by the LDRD program at Los Alamos National Laboratories. One of us, D. V., is indebted to Kernforschungszentrum Karlsruhe, Germany, for funding his stay at Los Alamos. All of us want to express our appreciation to Charles Hosford, David Oberlitner and Peter DeVargas for their valuable technical assistance and to Frank Gac for his important suggestions in preparing the manuscript.

REFERENCES

1. D. Vollath, B. Doerzapf, German Patent No. G 90 15 477 (3 May 1992).

2. D. Vollath, B. Doerzapf, German Patent No. P 37 25 740.4 (4 April 1990).

3. D. Vollath, Euro-Ceramics 1, 1.33 (1989).

4. D. Vollath, J. Mater, Sci. 25, 2227 (1990).

5. R. M. Young, E. Pfender, Plasma Chem. and Plasma Processes 5, 1 (1985).

6. P. Fouchais, E. Boudrin, J. F. Coudert, R. McPherson Plasma Chemistry IV, Edt. by S. Veprek, M. Venigopalan Springer Verlag, Berlin (1983) p.2093.

7. H. Anderson, T. Kodas, D. M. Smith, Ceram. Bull. 68, 996 (1989).

8. G. Schlick GmbH, Coburg, Germany, Technical Document # D19/1.

PART VII

Microwave Processing of Polymers

MICROWAVE-ASSISTED ORGANIC SYNTHESIS

Raymond J. Giguere*
*Department of Chemistry, Skidmore College, Saratoga Springs, New York 12866-1632

ABSTRACT

A concise overview of the use of microwave heating in organic synthesis is presented. The role of solvents and representative examples from the relevant literature are discussed to illustrate the significant impact this non-conventional heating mode continues to have in this area of materials research.

I. INTRODUCTION

Organic synthesis, the area of scientific investigation that concerns itself with the rational construction of organic molecules, can easily lay claim to being one of the principal cornerstones of chemistry. In this century our ability to create organic materials has grown significantly and has produced products that touch nearly every aspect of our daily lives: dyestuffs, food preservatives, pharmaceuticals, pesticides, polymers and so on. Understanding organic synthesis at the molecular level has provided us the opportunity to affect the evolution of the biosphere, as well as that of the society and economy, at fundamental levels hitherto unattainable.

The underlying form of this discipline is perhaps best characterized by the organic chemist's ability to develop and understand methods for formation of carbon-carbon bonds (both between and within organic molecules), as well as for the interconversion of functional groups in organic molecules.

Many of the reactions employed to achieve the above synthetic goals require heat for completion and therefore have relied on the use of conventional heating modes (e.g., heating mantels, oil or sand baths). Most organic reactions that require heat are conducted at temperatures ranging from slightly above room temperature to several hundred degrees centigrade. The employment of microwave heating permits rapid acquisition of these temperatures and in many cases also permits the solvent and/or substrates to play direct roles in the temperature acquisition by directly coupling with the microwave irradiation, a fact which distinguishes this heating mode from the conventional ones.[1]

The first reports on the use of microwave heating as a thermal source for conducting organic reactions appeared in 1986[2]; the first review of this developing area was published in 1989.[3] Other scientific applications of microwave heating range from drying glassware[4] to dissolving biological samples in mineral acids for subsequent analysis.[5]

Research in microwave-assisted synthesis has demonstrated that organic reactions can be conducted safely in commercial microwave ovens with reductions in reaction times of up to three orders of magnitude.[2,3] This time-saving ability is the principal advantage of microwave heating. Further advantages inherent in using this thermal source include the absence of significant temperature gradients within the sample and very short response times.[6]

II. SOLVENTS

Systematic solvent studies have disclosed the existence of a qualitative relationship between the complex dielectric constant of a solvent and its ability to couple with microwave irradiation.[2a] Typically for organics, as the dielectric constant increases so does the loss factor. A useful rule of thumb has therefore emerged from these studies: the larger the dielectric constant, the greater the degree of coupling, hence the faster the temperature acquisition. Two minute irradiation (100% power; 700 Watt oven) of a 2 ml sample of N,N-dimethylformamide (DMF, dielectric constant = 36.7) contained in a sealed tube causes this solvent to reach a temperature of just under 300 °C. In contrast, n-hexane (dielectric constant = 1.9) under identical conditions reaches only 60 °C. Further examples of dramatic differences in the ability of solvents to couple with microwave irradiation are given in references 2a and 3.

We have found vermiculite to be a convenient insulating material for conducting microwave reactions. The reaction vessels used in our laboratories[2a] (sealed tubes or Ace Glass pressure tubes) are buried in a bed of vermiculite contained in a rectangular box constructed of Corian, a commercially available heat resistant polymer. The contents of the reaction are absorbed by vermiculite in the event of reaction vessel explosion, hence vermiculite plays an additional safety role in this regard.

Vermiculite consists of hydrous silicates of iron, aluminum and magnesium. As a result, its water of hydration couples with microwave irradiation. The use of vermiculite therefore permits the heating of solvents unable to significantly couple with microwave irradiation (like n-hexane), but also somewhat impedes the coupling of microwaves with solvents possessing significant dielectric constants.[2a]

Recently Bose and co-workers have reported success in conducting organic reactions in commercial microwave ovens *in open vessels* when employing high boiling solvents possessing significant dielectric constants. These researchers report having no difficulty in scaling up organic reactions to several hundred grams using this modified approach.[7]

III. REPRESENTATIVE MICROWAVE-ASSISTED ORGANIC REACTIONS

1. Pericyclic Reactions

Microwave heating greatly reduces the time necessary to complete pericyclic reactions (reactions in which bonding changes occur through a redistribution of electrons within a closed loop of interacting orbitals[8]), as illustrated in Scheme 1. These Claisen, ene and Diels-Alder reactions were conducted on 1-2 mmol scale with vermiculite insulation. Control reactions and literature results are provided for comparison. Target temperatures of several hundred degrees are reached very rapidly, demonstrating the efficiency of microwave heating. Isolated product yields in microwave reactions are comparable to those reported in the literature and/or obtained in control runs.[2a]

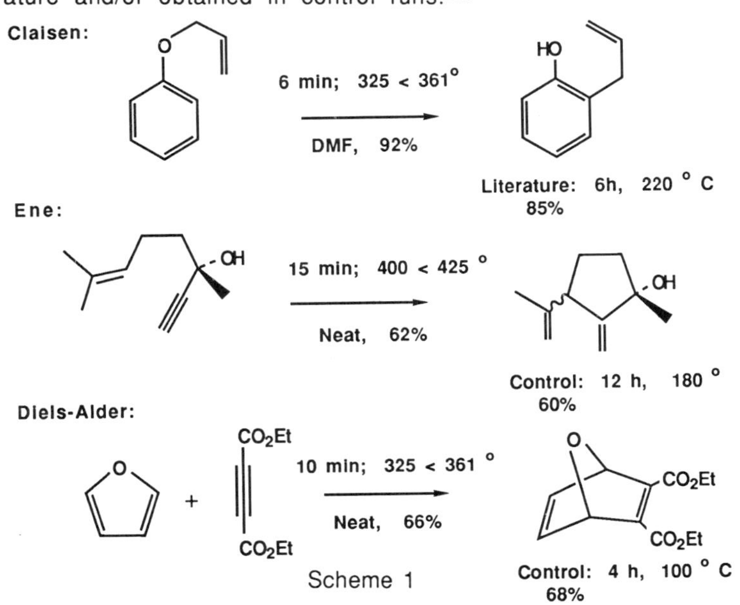

Scheme 1

A significant reduction in the time necessary to complete these reactions under microwave conditions is observed. Further reduction in reaction time is gained when N,N-dimethylformamide (DMF) or N-methylformamide (MF) is employed as solvent. For example, the Claisen reaction depicted in Scheme 1 is complete in 90 seconds in N-methylformamide (ε = 182.4). The large dielectric constants of these solvents promote efficient microwave coupling and allow substrates to acquire high temperatures extremely rapidly. Of particular significance is the fact that in the Diels-Alder reaction furan readily provides temperature-sensitive oxabicyclo[2.2.1]heptadiene derivatives under these high temperature conditions.

The application of microwave heating to related reactions has led to optimization of the reaction of 1,4-cyclohexadiene and dimethyl acetylenedicarboxylate to provide functionalized tricyclic **2**.[9] This pericyclic process is a tandem ene/ intramolecular Diels-Alder reaction, originally studied by Alder and Bong in 1952 but since neglected, presumably because of low yield (< 40%).[10] With microwave heating, this transformation can be achieved in up to one gram scale, with isolated yields of > 80%, in less than ten minutes (Table I).

The ene:enophile ratio significantly influences the outcome of the reaction. When the ratio is 1:1, the intermolecular Diels-Alder reaction of intermediate **1** and a second molecule of enophile predominates. This competing process is suppressed by use of excess diene. A 15:1 ene:enophile ratio leads to high yields of **2**, with either microwave or conventional heating. Control experiments require twenty hours at >220 °C to complete, but conventional controls at 350-375 °C at comparable times do not attain the microwave yields (Table I, Entries 2 & 5).

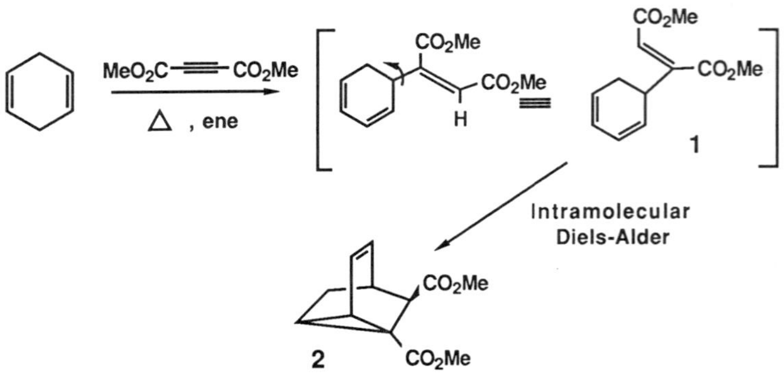

Table I. Comparative Data on the Alder-Bong Reaction of 1,4-Cyclohexadiene and Dimethyl Acetylenedicarboxylate

Thermal Conditions	Ene:Enophile	Time	Temp.	Yield (%)
1. LITERATURE[10]	1.1:1	20 h	180 < 190	40
2. MICROWAVE	15:1	6 min	325 < 361	82
3. MICROWAVE	30:1	6 min	317 < 325	87
4. CONTROL	15:1	20 h	220	82
5. CONTROL	15:1	15 min[a]	360	51

a) As a safety precaution, microwave reactions are generally allowed to stand several minutes before the reaction vessel is removed from the vermiculite bed. Therefore, conventional controls were conducted at fifteen minutes.

Recently an intramolecular Diels-Alder reaction was employed by Fallis as the key step in an elegant enantioselective synthesis of longifolene, **3**, a naturally occurring sesquiterpene.[11] Longifolene has become a benchmark in organic synthesis ever since 1991 Nobel laureate in chemistry, E. J. Corey, first achieved the synthesis of this compound in 1964.[12] The key reaction in Fallis's synthesis was successfully carried out using microwave heating. Moreover, the authors report the success of the microwave method in affording the product in much higher yield than under conventional conditions (97% vs ~20 %), as shown in Scheme 2.

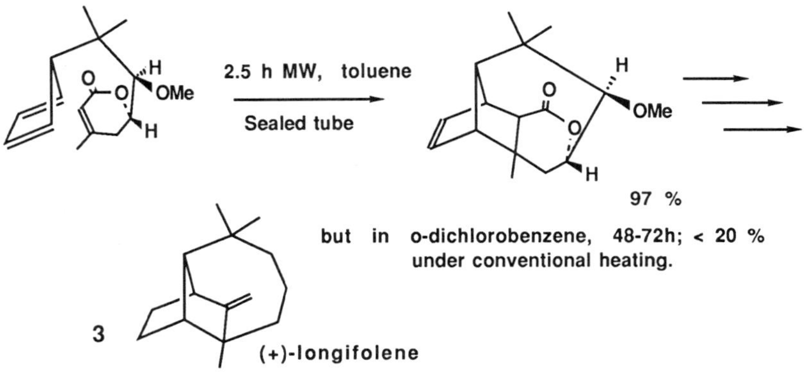

Scheme 2

2. Nucleophilic Substitution and Dehydration

Microwave heating is clearly capable of promoting nucleophilic substitution reactions. Table II presents results from our study of the Finkelstein reaction, which affords alkyl iodides from alkyl chlorides and bromides.[3]

$$\text{R-X} \xrightarrow[\Delta]{\text{NaI, acetone}} \text{R-I}$$

$$\text{X = Cl, Br}$$

Table II. Finkelstein Synthesis of Alkyl Iodides.

Thermal Conditions	R	X	% Power	Time	Temp.	% Yield
1. MICROWAVE	n-C_6H_{13}	Br	100	4 min	236 < 275	68
2. MICROWAVE	n-C_6H_{13}	Br	15	15 min	175 < 187	90
3. CONTROL	n-C_6H_{13}	Br	-	30 min	56	95
4. MICROWAVE	n-$C_{12}H_{25}$	Cl	25	15 min	325 < 361	93
5. CONTROL	n-$C_{12}H_{25}$	Cl	-	1 h	56	17[a]

a) NMR Estimation

Table II compares microwave and conventional results for the preparation of alkyl iodides under standard Finkelstein conditions (1.5 equivalents sodium iodide, dry acetone). As anticipated, alkyl bromides are converted more readily than alkyl chlorides under conventional conditions. In contrast, under microwave conditions the success of Finkelstein reactions appear less dependant of the nature of halide. This difference is attributed to the super-heating conditions generated under microwave conditions, which presumably leads to rapid equilibration of reaction species. For alkyl chlorides, precipitation of the poorly soluble sodium chloride provides a driving force to enhance the forward reaction.

We have recently developed a new synthesis of allyldiphenylphosphine oxide, **4**, which employs microwave heating

(Scheme 2). Brief exposure of a neat solution of ethyl diphenylphosphinite and allyl bromide affords an excellent yield of allyldiphenylphosphine oxide by a tandem reaction sequence: nucleophilic substitution followed by Michaelis-Arbuzov reaction.[13] Allyldiphenylphosphine oxide finds use as a Wittig reagent for the conversion of aldehydes to terminal dienes of E-configuration.

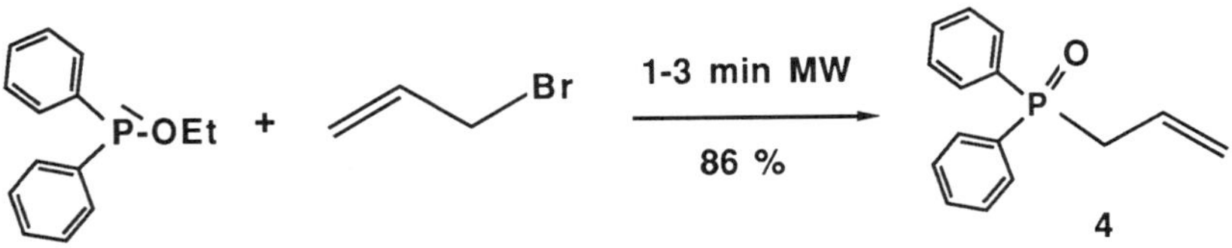

4

Scheme 2

An interesting development in microwave-assisted synthesis has been the use of solid supports, such as neutral alumina, upon which to absorb substrate molecules directly, thereby obviating the need for solvents in these reactions.[14] In this manner Loupy and co-workers have succeeded in conducting nucleophilic substitution reactions such as the alkylation of potassium acetate with long chain alkyl halides (Scheme 3).

$$CH_3CO_2^- K^+ \; + \; R\text{-}X \xrightarrow[\text{2 - 4 min MW}]{Al_2O_3} CH_3CO_2R \; + \; KX$$

$R = n\text{-}C_8H_{17};\; n\text{-}C_{16}H_{33}$

No solvent

$X = Br;\; Cl$

Scheme 3

This technique has the advantage of eliminating the need for closed vessels as well, thereby permitting reactions to be carried out with less regard to safety issues. The "dry" approach has been employed successfully by others in several classic organic reactions, such as the Knoevanagel reaction.[15] Different solid supports have been used as well; for example, bentonite clay in the oxidation of dihydropyridines.[16]

A simple method for dehydration involves microwave heating of an alcohol in the presence of an easily removable acid source. Table III presents results from our study of the dehydration of 2-phenyl-2-propanol, **5**, using Amberlyst resin 15, under two-phase conditions.[3]

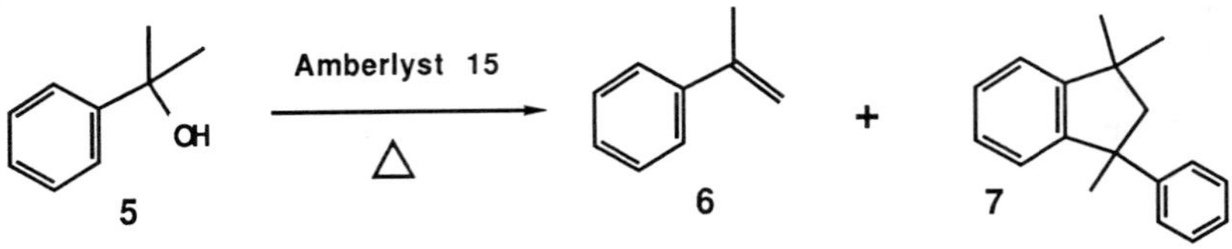

Table III. Comparative Dehydration Data of 2-Phenyl-2-propanol using Amberlsyt 15 Resin.

Thermal Conditions	Solvent	Time	Temp.	% Yield 6	% Yield 7
1. MICROWAVE	NEAT	2 m	276 < 299	98[a]	-
2. MICROWAVE	NEAT	6 m	325 < 361	-	88
3. CONTROL	NEAT	48 h	RT	93[a]	-
4. CONTROL	NEAT	19 h	120-130	-	80

a) contains 12-14% open chain intermediate, 2,4-diphenyl-4-methyl-2-pentene, by NMR.

Significant is the ability to afford principally either α-methyl styrene **6** or indane **7** as a function of only several minutes difference in microwave irradiation time. To obtain similar results conventionally requires quite different conditions (cf. Entries 3 &4).

III. OUTLOOK

The use of microwaves in synthesis has also attracted the interest of inorganic chemists, who have found the technique effective for the preparation of coordination compounds.[17] In at least one example, microwaves have permitted the synthesis of an inorganic complex for which no synthesis employing conventional heating was known. Applications in the field of biochemistry are also being reported.[18]

The question of the precise role that microwaves play in the process of effecting organic reactions remains unclear. Several conflicting papers have appeared which address this key issue, but a clear resolution has yet to emerge. The scope of this paper does not permit a discussion of the data and hypotheses put forth on this issue, but clearly some interesting progress has been achieved.[19]

IV. CONCLUSION

In summary, microwave heating in organic synthesis is an emerging area of investigation, with no fewer than thirty groups actively engaged in this research internationally. The principal advantages are reductions in reaction time, greater product purity and sometimes higher yields.[11] Occasional successes of this heating technique to effect transformations that fail under conventional conditions have also been reported.[17] The direct role solvents and/or substrates are able to play in the thermal process is unique to microwave heating and will in my view become an increasingly significant aspect of this technology.[20]

V. REFERENCES AND NOTES

1. K. Walkins, J. Chem. Ed. 12, 1043 (1983).

2. a) R.J. Giguere, T.L. Bray, S.M. Duncan, G.F. Majetich, Tetrahedron Letters 27, 4945 (1986). b) R. Gedye, F. Smith, K. Westaway, A. Humera, L. Baldisera, L. Laberge, J. Rousell, Tetrahedron Letters 27, 279 (1986).

3. R.J. Giguere, in Organic Synthesis: Theory and Applications, Vol. 1, edited by Tomas Hudlicky (JAI Press, Greenwich, CT, 1989), pp. 103-172.

4. J.A. Hasek, R.C. Wilson, Anal. Chem. 46, 1160 (1975).

5 H.M. Kingston, L.B. Jassie, Introduction to Microwave Sample Preparation, (American Chemical Society, Washington, DC, 1988).

6. M. Bacci, M. Bini, A. Checcucci, A. Ignesti, L. Millanta, N. Rubino, R. Vanni, J. Chem. Soc. Faraday Trans 1 77, 1503 (1981).

7. A.J. Bose, M.S. Manhas, M. Ghosh, M. Shah, V.S. Raju, S.S. Bari, S.N. Newaz, B.K. Banik, A.G. Chaudhary, K.J. Barakat, J. Org. Chem. 56, 6968 (1991).

8. T.H. Lowry, K.S. Richardson, Mechanism and Theory in Organic Chemistry, (Harper and Row, New York, 1987), p. 839.

9. R.J. Giguere, A.M. Namen, B. Lopez, A. Arepally, D.E. Ramos, G. Majetich, J. Defauw, Tetrahedron Letters 28, 6553 (1987).

10. C.-H. Bong, Dr. rer. nat. thesis, University of Cologne, Germany, 1952.

11. B. Lei, A.G. Fallis, J. Amer. Chem. Soc. 112, 4609 (1990).

12. E.J. Corey, R.B. Mitra, P.A. Vatakencherry, J. Amer. Chem. Soc. 86, 478 (1964).

13. R.J. Giguere, B. Herberich, Synthetic Communications 21, 2197 (1991).

14. G. Bram, A. Loupy, M. Majdoub, Tetrahedron, 46, 5167 (1990).

15. D. Villemin, A.B. Alloum, Synthetic Communications 21, 63 (1991).

16. F. Delgado, C. Alvarez, O. Garcia, G. Penieres, C. Marquez, Synthetic Communications 21, 2137 (1991) and references therein.

17. D.R. Bagurst, S.R. Cooper, D.L. Greene, D.M.P. Mingos, S.M. Reynolds, Polyhedron 9, 893 (1990).

18. V. Narasimhan, W.K. Huh, Biochemistry International 25, 363 (1991) and references therein.

19. J. Berlan, P. Giboreau, S. Lefeuvre, C. Marchand, Tetrahedron Letters 32, 2363 (1991) and references therein.

20. For example, a new extraction technique has recently been developed that takes advantage of this fact, cf: K. Ganzler, I. Szinai, A. Salgo, J. Chromat. 520, 257 (1990).

RF PROCESSING OF THERMOSET AND THERMOPLASTIC COMPOSITES

JOHN A. PEARCE AND CARL M. FAULKNER
Center for Energy Studies; The University of Texas at Austin

ABSTRACT

Graphite fiber-thermoset epoxy resin composite materials are frequently used in aerospace and related industries. Electromagnetic heating may have some advantages over the present autoclave techniques, especially for large objects. However, the loss factor of graphite composites at microwave frequencies (915 MHz and 2.45 GHz) is so high that the depth of penetration is no more than about 0.1 mm. Arcing may be encountered at high heating rates due to strong transverse E-fields. Also, since the curing process is exothermic and the surface is at elevated temperature, thermal runaway may be encountered in the interior spaces and result in over-cured (thermally damaged) material.

We have investigated the heating obtained at radio frequency (nominal 27 MHz) in magnetic induction fields in both experiments and numerical models. Inductive coupling has the advantage that arcing is virtually eliminated, even at high heating rates, for typical cylindrical geometries. Also, induction heating easily accomplishes lap seam welds in thermoplastic composites owing to favorable electric field boundary conditions. Transient temperatures were recorded during heating by optical fiber probes and heating distribution was measured, where possible, using thermographic imaging. Finite difference electrical and thermal model results illustrate the dominance of conduction heat transfer in these materials.

INTRODUCTION

The purpose of this research is to investigate the feasibility of using radio frequency (RF) heating to cure epoxy-graphite fiber thermoset and to weld graphite fiber thermoplastic composite materials. RF fields were used to achieve volumetric heating of: 1) a cylinder of plain weave prepreg graphite fiber cloth impregnated with thermoset epoxy, 2) a cylinder of wound graphite fiber thermoset epoxy, and 3) graphite fiber thermoplastic sheets. A possible benefit of this approach is that the heating time and the time at temperature may be reduced because heat is deposited volumetrically. Radio frequency heating (RF) may have some advantages over microwaves since more uniform heating over larger areas is achievable and magnetic field coupling results in low electric fields everywhere. The graphite fibers are of high electrical conductivity: more than 100,000 S/m for whole fiber, and the effective conductivity of the composite matrix ranges from around 1,000 to 100,000 S/m, depending on fiber orientation and binder and/or void volume [4].

Magnetic field heating arises from induced electric fields in the material. For a cylinder in a uniform magnetic field, as shown in figure 1, if one can assume that the conductivity is low enough that induced current density in the material does not affect the local magnetic field, then the volumetric power density depends approximately on the square of the radius, r, from the coil center: $Q_{gen} = \sigma |\mathbf{E}|^2 \cong \sigma (\omega \mu H_0 r)^2$ (W/m^3) ; where σ is the electrical conductivity, ω the angular frequency, r the radius, μ the magnetic permeability, and H_0 the externally applied magnetic field (A/m). The magnetic field is co-axial to the cylinder in this example, and either solenoid or Helmholtz coils would give useful heating patterns. As can be seen in the above function, even if the magnetic flux density is constant over the volume of interest, the power generation term is parabolically weighted toward the periphery and is also parabolic with frequency.

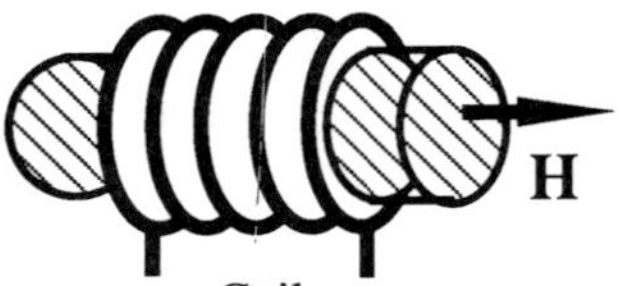

Cylinder:
Inner radius, a
Outer radius, b

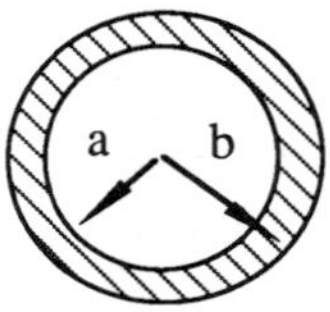

Figure 1: Example of a coil used to heat a cylinder.

There are four major questions to be answered: 1) is it possible to deposit sufficient RF energy in the uncured composite to reach the required temperatures, 160 °C, 2) is it possible to easily reach welding temperatures in the thermoplastic composite sheets, 350 °C, 3) is it possible to measure the temperature of the heated workpiece to prevent damage to the composite, and 4) what are the expected thermal gradients in the heated parts?

ELECTROMAGNETIC AND THERMAL FIELD CALCULATIONS

Electric Field Calculations

In graphite fiber composites the electromagnetic fields are more complex than the simple reduction of Faraday's Law in the example above. In the usual composite material case, induction of current in the material (the source of the heating electric field, **E**) will serve to reduce the strength of the local magnetic field, **H**. So, the coupled equation set (1), Faraday's Law and Ampere's Law respectively, must be solved:

$$\oint_C \mathbf{E} \cdot d\mathbf{L} = -j\omega \left[\iint \mathbf{B} \cdot d\mathbf{S} \right] \quad \text{or} \quad \nabla \times \mathbf{E} = -j\omega\mu H$$

$$\oint_C \mathbf{H} \cdot d\mathbf{L} = \iint (\sigma + j\omega\varepsilon)\, \mathbf{E} \cdot d\mathbf{S} \quad \text{or} \quad \nabla \times H = (\sigma + j\omega\varepsilon)\, \mathbf{E} \qquad (1)$$

In the geometry of figure 1, as a first order analysis, we assume that the coil and cylinder are infinite in length, and that the magnetic field is axial — that is, $\mathbf{H} \cong H_Z \, \mathbf{a}_Z$. These assumptions reduce the problem to one dimension. The point form differential equations may then be solved using the condition that $H_T = H_0 - H_i(a)$ for $r \leq a$; where H_0 is the coil magnetic field (A/m) in the absence of the cylinder, H_T is the total field, and $H_i(r)$ the field due to the induced current density in the cylinder. The solution for $H_T(r)$ is a zeroth order Bessel function:

$$r \frac{d^2 H_T}{dr^2} + \frac{d H_T}{dr} + k^2 r^2 H_T = 0 \quad \text{gives:} \quad H_T(r) = H_T(b) \frac{J_0(kr)}{J_0(kb)}$$

$$\text{where: } k^2 = -j\omega\mu\,(\sigma + j\omega\varepsilon) \cong -j\omega\mu\sigma \qquad (2)$$

Since the argument of J_0 is complex, the Bessel function will be maximum at the outer edge, (and its magnitude increases without bound for large k and/or r), so the result is scaled by the outer edge field at r = b. In this solution k is the complex propagation constant which contains both the skin depth, δ ($\delta = 1/\mathrm{Re}\{k\}$) and the wavelength, λ ($\lambda = 2\pi/\mathrm{Im}\{k\}$). The complexity of the field inter-relationships and the effects of the induced current in the cylinder make a numerical solution necessary. A finite difference (FDM) model was created which used 15 nodes to represent the cylinder.

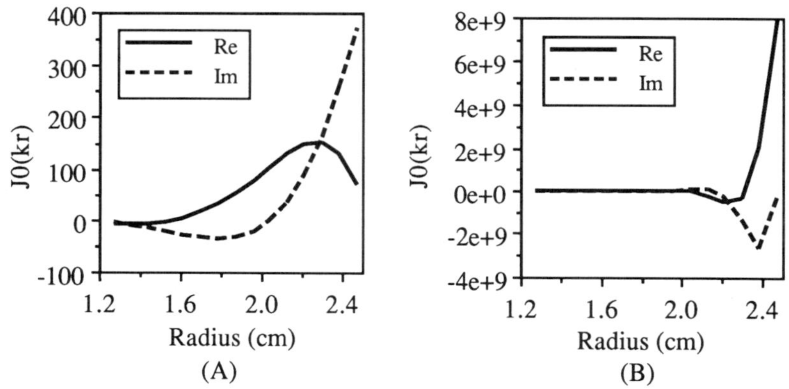

Figure 2: Complex Bessel function behavior at 27 MHz for a cylinder with a = 1.27 cm and b = 2.47 cm, A) σ = 1,000 S/m, and B) σ = 10,000 S/m.

Evaluating the boundary condition at r = b is the most difficult part of the solution. Iterative approaches are not well behaved due to lack of sufficient constraints. So, the boundary field magnitude, $|H_T(b)|$, was found from the magnitude of the interior field, $|H_T(a)|$ by inverting the solution above — $|H_T(b)| = |J_0(kb)|\ |H_T(a)|\ /\ |J_0(ka)|$. H_0 was used as the phase reference for the solution. The magnitude of the interior field was estimated from an effective "shield factor" for the cylinder thickness:

$$|H_T(a)| = H_0\ e^{-OD} \quad \text{where the optical depth, OD} = (b-a)/\delta \tag{3}$$

The phase of $H_T(b)$ was found by first calculating the induced electric field in the cylinder

$$E_\phi(r) = \frac{-j\omega\mu}{r}\int_0^r H_T(x)\ x\ dx = \frac{-j\omega\mu\ H_T(b)}{J_0(kb)}\int_0^r J_0(kx)\ x\ dx$$

$$\text{and:} \quad \int_0^r J_0(kx)\ x\ dx = \frac{a^2\ J_0(ka)}{2} + \int_a^r J_0(kx)\ x\ dx \tag{4}$$

where x is a dummy variable of integration which has the significance of radius. The total induced current in the cylinder, $I = \iint (\sigma + j\omega\varepsilon)\ \mathbf{E} \bullet d\mathbf{S}$, (applied to the integral form of Ampere's Law) gives the phase of $H_T(b)$ — that is, $H_T(b)$ has the phase required so that I opposes H_0 (Lenz's Law):

$$H_T(b) = H_T(a) - I = H_T(a) - \int_a^b (\sigma + j\omega\varepsilon)\ E_\phi(r)\ dr \tag{5}$$

The time consuming part of the calculation is evaluating the integrals of the Bessel functions. In order to preserve accuracy, the Bessel functions were integrated between nodes by rectangular integration. Each "dr" segment in the 15 node representation for the power density field was divided into 100 steps, and the Bessel functions integrated stepwise to determine I. This was necessary because the complex phase relationships of the Bessel functions for high values of k strongly influenced the integration result.

Subsequently, the electric field at each cylinder node was calculated from equation 4 using the boundary total magnetic field. The power generation term, Q_{gen} (W/m3) was then determined from:

$$Q_{gen}(r) = \sigma\ |E(r)|^2 \tag{6}$$

The power density was calculated at each of the 15 nodes in the cylinder. The total coupled power (per unit length of cylinder) was determined by trapezoidal integration of $Q_{gen}(r)$.

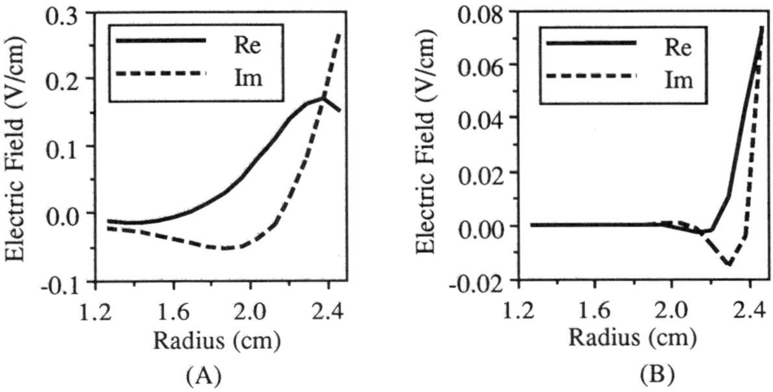

Figure 3: Electric field for the cylinder of figure 2 in an external field of 100 A/m at 27 MHz, A) σ = 1,000 S/m and B) σ = 10,000 S/m.

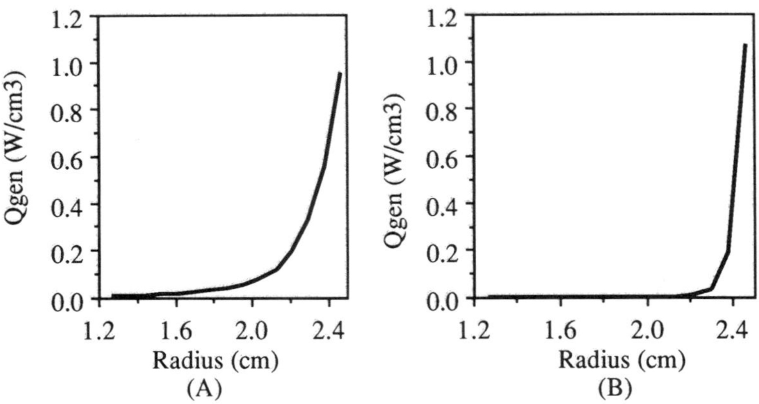

Figure 4: Resulting power density term, Q_{gen}, at 27 MHz for the fields of figure 3, A) σ = 1,000 S/m and B) σ = 10,000 S/m..

The shallow penetration in the higher conductivity material is apparent. Though the surface power density is higher for the 10,000 S/m material, the total power in the cylinder is higher for the 1,000 S/m material. In fact, the total coupled power at 27 MHz would be maximum if the cylinder had a conductivity of about 100 S/m, as can be seen in figure 5A. The total power for fixed conductivity and magnetic field strength is shown in figure 5B. As would be expected, the total power increases with increasing frequency. These results suggest an interesting engineering trade off between effective electrical conductivity and frequency of operation. Lower effective conductivities will result in more power absorption at fixed frequencies; consequently, the winding geometry will have a dominant effect on power absorption.

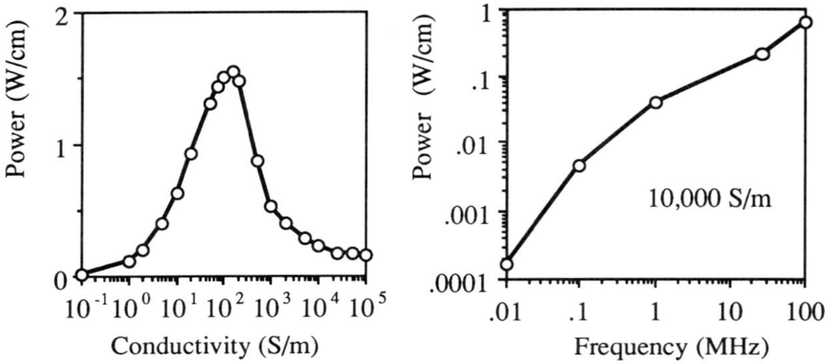

Figure 5: Total coupled power per unit length of cylinder in an external field of 47 A/m, A) at 27 MHz as a function of electrical conductivity, and B) as a function of frequency for $\sigma = 10{,}000$ S/m.

Winding for minimum electrical conductivity is advantageous providing that the required axial and hoop strength ratio can be maintained. Minimum electrical conductivity will be attained when the fibers are oriented as nearly parallel to the cylinder axis as possible using a shallow winding angle. This puts the fibers nearly normal to the ϕ-direction E-field as can be obtained, resulting in minimal electrical conductivity and maximal absorbed power. Note that reflections off of the ends of the cylinder are minimal since the polarization gives only a tangential component in the **E**-field there. This analytical technique can be applied, in amodified form, to thin sheets; however, a lap seam weld is inherently 3-D and outside the scope of this paper.

Temperature Field Calculations

Temperature distributions within the cylinder are induced by the applied RF heating and moderated by heat transfer from the cylinder to the mandrel and bleeder cloth-vacuum bag covering.

$$\rho\, c \frac{\partial T}{\partial t} = \nabla \cdot (k\, \nabla T) + Q_{gen}(r) + Q_{exotherm} + Q_{surface\ losses} \qquad (7)$$

Above the cure temperature of the epoxy binder, additional power generation comes from the exothermic reaction. The model includes radiation and convection at the insulation surface.

A finite difference model of 45 total nodes was created for the cylinder heating problem — 21 nodes for the polypropylene mandrel (20.5 dr node spaces since the center line may not have a node), 15 nodes (14 dr spaces) for the cylinder and 10 node spaces for the insulation... The experiments did not approach curing temperatures so exothermic heat was not included. The total power deposited in each cylinder node space was determined by trapezoidal integration of the Q_{gen} term. Thermal properties for the polypropylene mandrel composite cylinder and insulation blanket are shown in Table 1. The forward finite difference model was executed with a time step of 0.01s to satisfy the standard FDM stability criterion — determined by the composite cylinder properties. Shorter time steps are required for higher thermal diffusivity materials and smaller node separations.

Table 1: Thermal properties of model materials

	Conductivity W/m°-K	Specific Heat J/kg-°K	Density kg/m^3
Mandrel [1]	0.138	1,930	900
Composite [2]	1.1	935	1,400
Insulation [2]	0.038	835	32
Soft Rubber [2]	0.13	2,010	1,100

EXPERIMENTAL STUDIES

In the experimental studies temperatures were measured by fiber optic contact probes (Luxtron Model 455) and by thermographic imaging (Inframetrics Model 625). Thermographic imagery was recorded on video tape for subsequent digital processing.

The coil currents were estimated from measurements of transmission line (T-L) magnetic field strength using a Moebius loop coil [3]. Figure 6 shows the arrangement of the Moebius loop between the twin lead transmission line supplying the coil for the small cylinder and PEEK weld experiments. A corresponding arrangement near the center conductor of a large coaxial line was used to measure coil currents in the large cylinder experiment. The Moebius loop consisted of two semicircles of 0.141" semirigid coaxial line with center conductors connected to opposing member shields. The loop sensor voltage/current relationship was calibrated at 60 Hz and about 50 A rms. The results agreed with the calculated sensitivtity within 10%. The loop is connected through twinaxial line to a balun which converted the signal to single-ended coaxial line and provided an approximate 4:1 impedance transformation. The coaxial line was terminated by a 50Ω load at the oscilloscope input terminal. A 200 MHz oscilloscope (Kikusui) was used to display transmission line current. The overall sensitivity of the measurement was approximately 4.49 volts per ampere of T-L current.

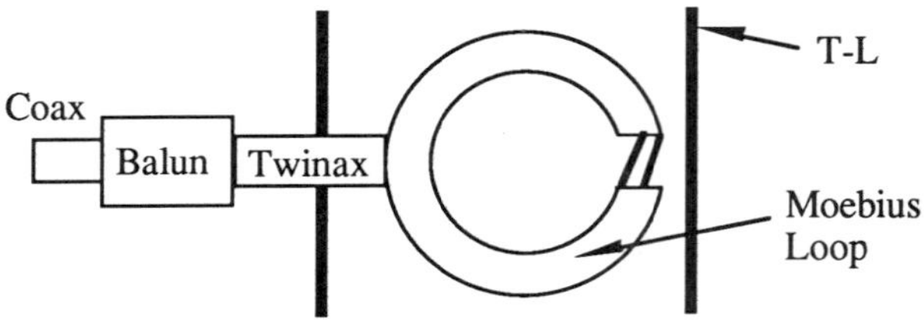

Figure 6: Moebius loop coil configured for twin lead transmission line current.

Thermoset Cylinder Experiments

Experiments were conducted on two hollow cylinders: 1) length 15 cm, with a = 1.27 cm and b = 2.47 cm of hand wrapped pre-preg cloth (Hercules Magnamite AW193PW/3501-6) and 2) machine wound of length 15 cm with radii a = 4.41 cm, and b = 5.46 cm (Hercules IM612GK fiber with Shell EPON 9405 resin and 9407 accelerator). The large cylinder was wound with fibers oriented circumferentially, the highest electrical conductivity configuration for the circumferential electric field (most difficult heating experiment).

1. Small Prepreg Cloth Cylinder Experiment

For the configuration of figure 1, a 3.5 turn solenoid coil with inside radius 5 cm and length 15 cm was energized with a Burdick medical diathermy machine at 27 MHz (about 100 W maximum output). Temperatures were recorded with a Luxtron fiber optic probe placed axially at mid-thickness and 5 cm deep to the coil outlet plane. The cylinder was insulated by plastic sheet "bubble pack" about 1 cm thick. Coil current was estimated to be 1.89 A rms from the Moebius loop sensor. This measurement corresponds to an external applied field strength of H_0 = 47 A/m.

The resulting thermal history at the center of the cylinder wall directly under the first turn of the coil is shown in figure 7A The electrical conductivity was not determined independently. However, published values range between 12,000 and 36,000 S/m at 40 MHz for the plain weave cloth [4].

Owing to uncertainty in the applied magnetic field and, primarily, the effective electrical conductivity of the cylinder, the model heat generation term had to be scaled in order to match model predictions with experimental observations. In order to match model heating with the experiment at 10,000 S/m, an external field of 302 A/m would be required (also shown in the figure).

Since it is equally likely that the effective conductivity of our hand wrapped specimen is even lower than the published value, an additional match at H_0 = 168 A/m (σ = 1,000 S/m) is also shown. We believe that the lower match field strength corresponding to the lower conductivity is closer to the actual situation in this experiment. However, uncertainty due to Moebius loop calibration and hand-wrap conductivity dominate other considerations of the magnetic field. In figure 7B the model results for the 302 A/m case are shown at four different elapsed heating times. The temperature profiles at 168 A/m are virtually identical to those of the higher conductivity case, so the power density profile makes little difference in this experiment. Also, at all model times the thermal gradient in the cylinder is negligible compared to the surrounding materials. This is due to the high thermal conductivity of composites.

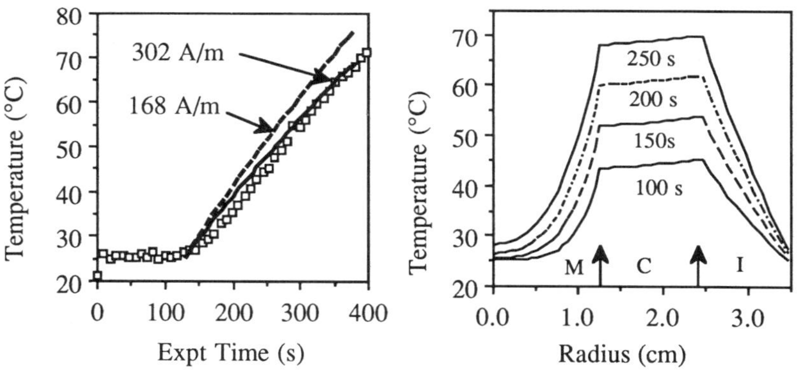

Figure 7: Small cylinder heating experiment and model results, A) measured temperature (squares) with model results at the fiber optic probe site for two external fields, B) temperature profiles for the 302 A/m model at 100, 150, 200 and 250 seconds. Boundaries of the mandrel (M), cylinder (C) and insulation (I) are indicated.

For comparison, a convection heating model is shown in figure 8. The surrounding temperature was fixed at 160°C in the convection calculations. In this model, the insulation was reduced to include only 0.5 cm of vacuum bag and bleeder cloth. Thermal properties for soft rubber were used to give a conservative estimate of heating rate (i.e. minimal insulation delay). Even so, a much longer heating time is required to reach desired temperatures (600 as opposed to 250s to reach 70 °C). As expected, the rate of temperature rise at the probe site (fig. 8A) decreases markedly as the cylinder approaches the surrounding temperature. Also, the inverted temperature profiles in the insulation are apparent.

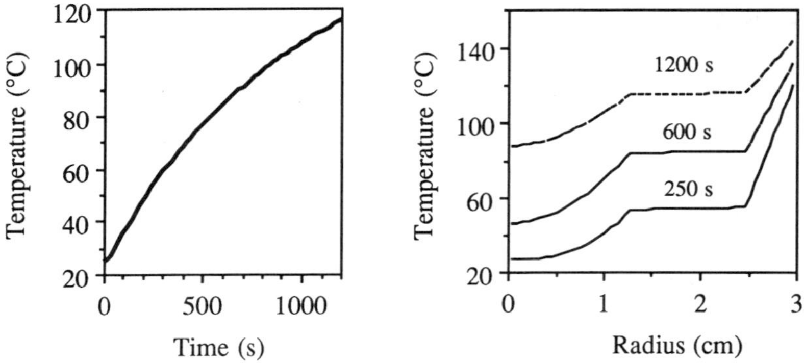

Figure 8: Convection heating model results with environment temperature of 160°C, A) temperature history at probe location, B) temperature profiles at 250, 600 and 1200 seconds.

2. Large cylinder experiment

The large cylinder was of very high electrical conductivity due to the circumferential winding; so the current required for heating was obtained from a large RF generator, Thermx 8 kW source tuned in at 23 MHz. A 19 turn coil was used in a resonant circuit to couple to the generator (figure 9).

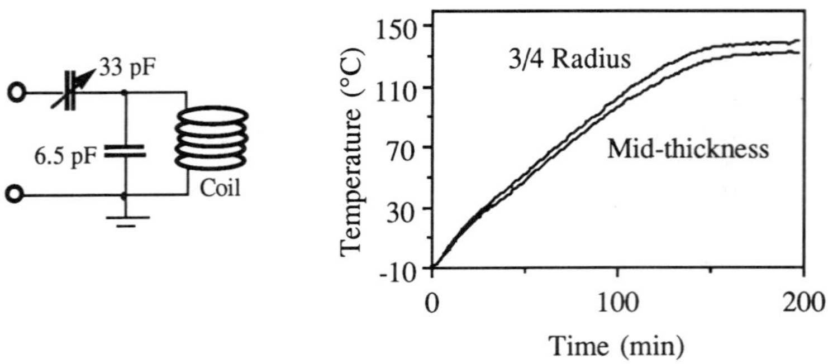

(A) (B)

Figure 9: A) Generator tuning circuit and B) resulting temperatures at the cylinder mid-thickness and at 3/4 radius, a + 0.75(b-a).

The generator tank was connected to the network in the figure which tuned with approximately 33 pF on the adjustable capacitor. The 6.5 pF in parallel represents a capacitive voltage divider. The total network current was estimated to be 7.8 A rms at a tank voltage of 4 kV rms (phase angle very close to +90°). The coupled power was approximately 600 W, by calorimetry. Cure temperatures were reached in about 2 hours.

Thermoplastic Weld Experiments

Small sheet segments of thermoplastic composite (ICI PEEK laid up as thin plates of crossed fibers) were welded with a pancake coil arrangement (figure 10) in which the magnetic field was predominantly normal to the lap seam weld site. The PEEK sheets were insulated top and bottom by 1 cm thick fiberglass duct insulation board. The 5 turn coil (10.5 cm diameter) was centered over the seam; and at a current of 2.7 A rms a weld (350 °C) was obtained in approximately 4 minutes.

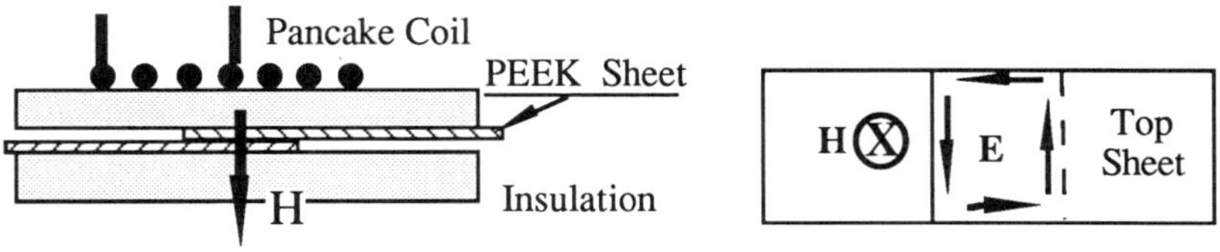

(A) (B)

Figure 10 A) Side view of pancake coil arrangement for heating PEEK sheets, B) Top view of resulting electric fields.

In this geometry the electric field boundary conditions are quite favorable: the electric field will be parallel to the edge of a sheet and the tangential fields are equal at the sheet-to-sheet interface, so the induced E-field is confined to the lap area only, as illustrated. A similar distribution will be obtained when a long seam is heated, but the induced field will extend for some distance above and below the coil diameter along the seam.

Temperature feedback is a difficult problem in this experiment. We were not successful when using high temperature fiber optic probes — the composite overheated to damage temperatures (above 400 °C). This was apparently due to the low thermal mass of the composite and high thermal conductivity of the probe, which combined to give readings as much as 100 °C below composite temperatures. The best results were obtained when the infrared (IR) thermal camera was used to monitor the central surface temperature through a small aperture in the insulation. In a practical system, a staring IR sensor could be placed at the center of the pancake coil to provide feedback. It is interesting to note that induction welding can be used with equal ease in otherwise hostile environments, such as in outer space. At RF (and presumably microwave) frequencies the required generator power is much reduced compared to standard industrial induction systems owing to more effective coupling to the load.

SUMMARY

Magnetic induction heating has several advantages: 1) surface coverings which inhibit heat transfer are typically transparent to H-fields, 2) electromagnetic boundary conditions can be made to work in favor of an application as in the cylinder heating and lap seam weld studies, 3) high power levels can be used with virtually no risk of arcing. There are also several disadvantages of magnetic induction: 1) temperature feedback is necessary and sometimes difficult to obtain (however, thermal gradients are very small, so one sample location is usually sufficient), 2) required currents and field strengths may be very sensitive to winding geometry, 3) temperature control is complicated by the exothermic heat of polymerization in thermoset applications. Overall, magnetic induction heating is a useful alternative method to cure or weld composite structures. Even though the depth of penetration is little improved over microwave energy, the simplified boundary conditions at low frequency make this technique an attractive method for several important geometries.

ACKNOWLEDGEMENTS

This research was conducted with support from the Texas Higher Education Coordinating Board Energy Research in Applications Program contract number 149. We also wish to thank Hercules and McDonnell-Douglas for contributions of composite materials.

REFERENCES

1. R.H. Perry, C.H. Chilton, Chemical Engineer's Handbook, 5th Ed. (McGraw - Hill, New York, 1973).

2. F.P. Incropera & D.P. DeWitt, Fundamentals of Heat and Mass Transfer, 2nd Ed. (John Wiley & Sons, New York, 1985).

3. P.H. Duncan, IEEE Transactions on Electromagnetic Compatibility, EMC-16, 83-89 May (1974).

4. A.C. Lind, C.G. Fry, C.H. Sotak, J. Appl. Phys. 68 (7), 3518-28 (1990).

MICROWAVE DYNAMIC DIELECTRIC ANALYSIS OF CURING NEAT RESINS

RAY J. KING,* MICHAEL J. WERNER,* and GUILLERMO D. MAYORGA,**
*KDC Technology Corp., 2011 Research Drive, Livermore, CA 94550
**Hexcel Corp., Dublin, CA 94568

ABSTRACT

Microwave reflection resonator sensors have been developed to monitor the dynamic, *in situ* real and imaginary dielectric components (ε', ε") of thermoset polymer matrix composites and thermoplastics. These reusable and noninvasive sensors are conformably mounted in the autoclave mold in such a manner that the EM fields are strongly coupled to the resin. Tracking of (ε', ε") during the cure provides information about the chemical kinetics such as timing the point of minimum viscosity and monitoring the relative cure index. The sensor is readily adjustable for optimum coupling to the type of material being tested and is rated for temperatures up to 250°C (480°F). The technique is complementary to other analysis techniques such as Differential Scanning Calorimetry (DSC) and Dynamic Mechanical Analysis (DMA). A prototype software and electronics package has been developed to support the sensor.

INTRODUCTION

Microwave Dynamic Dielectric Analysis (MDDA) is an effective way to monitor the dynamic chemical kinetics of many types of curing polymers [1, 2]. However, a major practical problem has been the development of a suitable microwave sensor for monitoring the complex dielectric properties $\varepsilon^*(= \varepsilon' - j\varepsilon")$ of polymers *in situ* in a mold in real time. Such a sensor has been developed and extensively exercised. This paper focuses on some of the typical results obtained for neat thermoset (epoxy) resins, but can equally be applied to prepregs and thermoplastics. Results are shown and interpreted to yield new physical insight into the dynamic curing process.

Measurement of ε' and ε" provides information concerning dipolar polarization and relaxation, activation energies, molecular motion and internal barriers to motion, rate of reaction, viscosity and rigidity, porosity, and temperature and pressure dependencies. Further, knowledge of the dielectric properties of a polymer undergoing reaction is essential if microwave heating methods are to be effectual for inducing and controlling the reaction process [1, 2].

THE SENSOR AND INSTRUMENTATION

The sensor, which is about the size of a quarter, is conformably mounted in a resin mold. In turn, the mold is placed in a temperature and pressure controlled autoclave. Because the sensor design is proprietary, it is sufficient to state that it operates as a reflection resonator in which the fringing electromagnetic field on the flush face of the sensor is appropriately coupled to the polymer. In this way, the dielectric properties (ε', ε") of the polymer determine the sensor's resonant frequency (f_r) and input impedance mismatch (coupling) factor (r_0) at resonance.

Fig. 1 shows the block diagram of the microwave instrumentation system used to monitor the reflection coefficient at the sensor input. When the sensor impedance is adjusted to be nearly critically coupled ($r_o = 1$) at resonance, the input reflection coefficient amplitude dips to a sharp minimum at f_r as the source frequency is swept. Typically, the sensor is designed to resonate over about a 2% band in the range $4 < f_r < 6$ GHz. For a typical cure, both (ε', ε'') of the polymer undergo wide dynamic changes with time and temperature, and (f_r, r_o) respond with similar changes, but inversely. Thus, by tracking f_r and r_o, both ε' and ε'' can be tracked simultaneously and independently. The f_r, r_o and T data are recorded using a data acquisition computer and specially developed software.

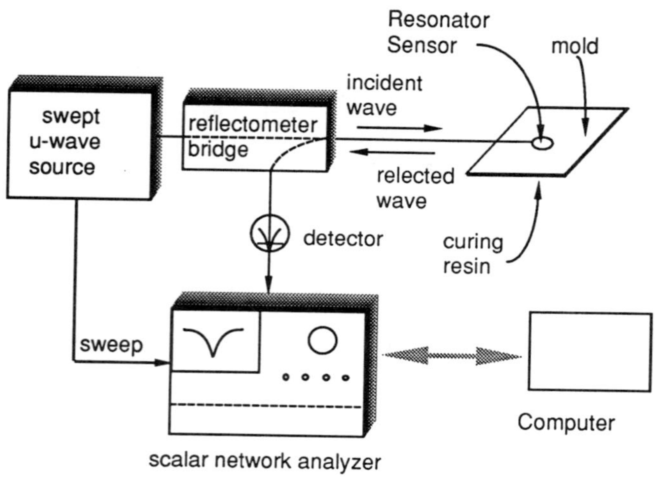

Fig. 1. Block diagram of the laboratory instrumentation system.

Space does not permit delving into the sensor's principles of operation except to remark that the sensor can be calibrated to determine (ε', ε'') using a standard test material (liquid or solid) having known dielectric properties. However, for the present application, determining the actual values of (ε' ε'') is not necessary; it is sufficient to simply know how (ε', ε'') vary under dynamic test conditions. In these more general terms,

$$\varepsilon'_M = f_r^2 = K_1 + K_2\varepsilon' \quad \text{and} \quad \varepsilon_M'' = (f_r r_o)^{-1} = K_3\varepsilon'' \qquad (1)$$

where $K_{1,2,3}$ are constants and (ε'_M, ε''_M) are linear transformations of the actual (ε', ε''). With this background, (f_r, r_o) are measured during the cure reaction and (1) is used to determine the dynamic profiles of (ε'_M, ε''_M). Note that MDDA measures the behavior of the true ε' and dielectric loss factor ε''. This is in contrast to commercial low frequency "dielectric" sensors which actually measure the ionic conductivity, σ_i [3, 4]. Changes in σ_i, which spans several decades during reaction, is in part due to loss of volatiles in condensation reactions and to polymerization. Because of the low ion mobility at microwave frequencies, the ionic loss is generally negligible compared to the dielectric loss. But it is also noted that ε'' as measured using MDDA behaves in a manner similar to log (σ_i) as measured using low frequency Dynamic Dielectric Analysis (DDA).

CURE RESULTS

Fig. 2 shows typical measurements of (f_r, r_o) and resin temperature (T) for an epoxy neat resin diglycidyl ether of bispheraol-A (DGEBA):tetra glycidyl 4,4' diaminodiphenyl methane (TGDDM) (1:1) with dicyandiamide (DICY)/diuron) in a pressure autoclave at 50 psi. During initial cure, T is ramped to 126°C and held until the resin has stabilized (recommended T_c = 121°C). Both (f_r, r_o) are seen to decrease monotonically with increasing T until t = 64 min. They then abruptly reverse and monotonically increase. The minimum f_r at t = 64 min signals the point of minimum viscosity (PMV), followed by very rapid linking reaction. For 90 < t < 160 min., (f_r, r_o) both climb more slowly signifying the decreasing rate of reaction. At t = 160 min., nearly all of the linkages which can occur at T = 126°C have been completed. However, during the second temperature ramp and hold at T = 135°C, (f_r, r_o) both drop slightly signifying a slight softening of the resin, accompanied by further reaction. Cool down begins at t = 180 min. The further increase in (f_r, r_o) during cool down is simply the temperature dependence of (ε', ε'') for the cured resin. The sensor itself is designed to have a very small temperature dependence.

Fig. 3 shows typical profiles of ε'_M and ε''_M as determined using (1). Here, the temperature is ramped at the same rate as in Fig. 2, and held at 113°C. There is no second ramp, and cool-down beings at t = 185 min. The high degree of correlation is clearly evident beyond the peak, which again occurs at the PMV

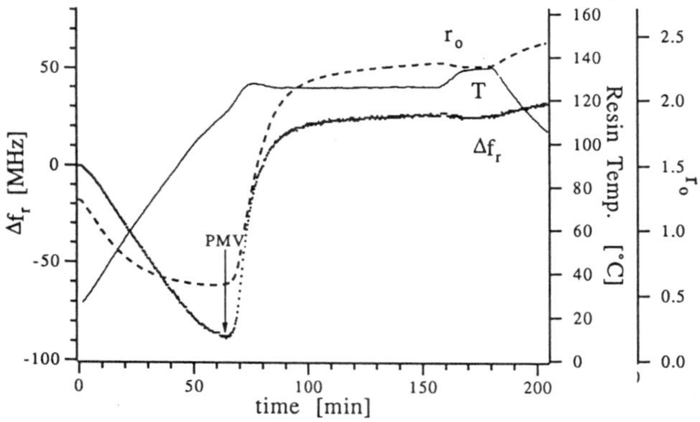

Fig. 2. Cure profiles of measured variables (f_r, r_o) for the indicated resin temperature (T) profile. Base f_r = 4.326 GHz at Δf = 0.

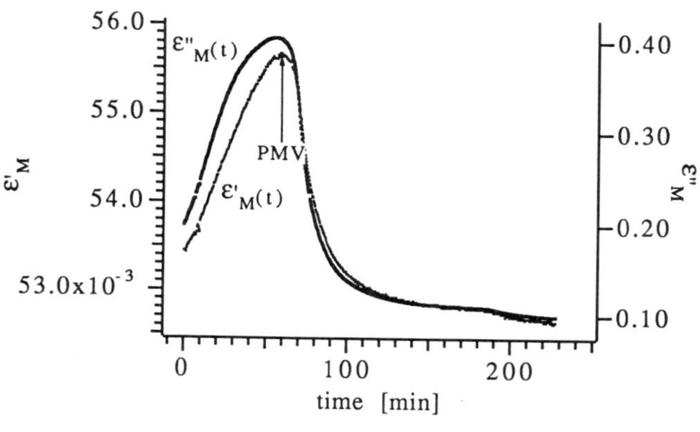

Fig. 3. Typical time profiles of ε'_M (= f_r^{-2}) and $\varepsilon''_M = (f_r r_o)^{-1}$ for a temperature ramp and hold at 113°C in an autoclave at 50 psi. f_r is measured in GHz.

(t = 60 min in this case due to the arbitrary time when data recording began). Initially (t < 50 min), $\varepsilon'_M(t)$ increases linearly while the loss factor $\varepsilon''_M(t)$ increases in a definitely nonlinear manner.

The behavior of ε'_M in Fig. 3 is explicitly plotted vs. temperature in Fig. 4, with time (t) implicit. Here, two remarkable features are observed, namely that ε'_M (T) varies linearly with temperature during the initial ramp up and final cool down. Between these lines the resin is curing, most rapidly at first.

A relative cure index $\beta(t)$ can be defined in several ways, depending on the basis parameter and method used in the definition. For example, using $\varepsilon'_M(T)$ in Fig. 4, the fractional difference from the 0% line and the cured line at point X(t) signifies the amount of reaction that has occurred at that explicit temperature and implicit time (t). This method follows that proposed by Day [3] who used log (σ_i) as the basis parameter as measured using a low frequency dielectric analysis meter. Note that this method partially compensates for the temperature dependence of ε'_M. Also note that this definition of the relative cure index is only valid *for that particular hold temperature*. As the hold temperature increases, a maximum separation between the 0% and cured lines would then correspond to maximum possible cure. Maximum cure, however, is usually not desirable since

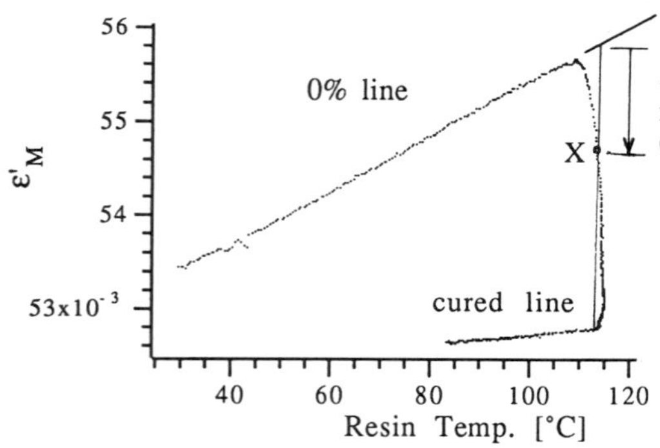

Fig. 4.
Temperature profile of ε'_M (= f_r^{-2}) in Fig. 3. The relative cure index $\beta'(t)$ at point X(t) is calculated as the fraction of the vertical drop between the 0% and the cured lines.

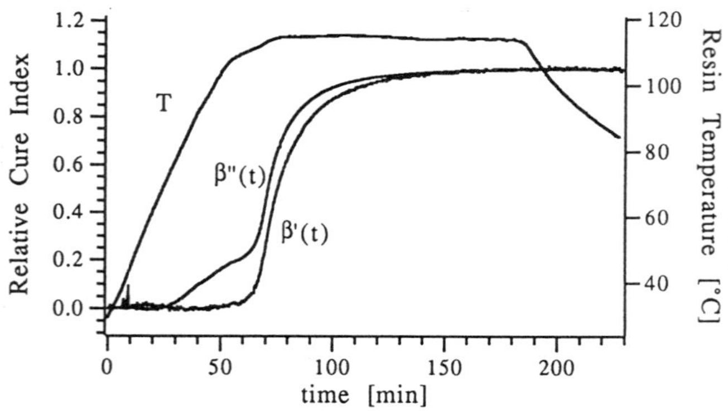

Fig. 5.
Relative cure indices $\beta'(t)$ and $\beta''(t)$ derived from the ε'_M and ε''_M data in Fig. 3 for the temperature profile shown.

other physical properties may not be optimum. At the extreme, maximum possible cure may even result in undesired brittleness, charring, etc. A similar relative cure index $\beta''(t)$ can be defined using $\varepsilon''_M(t)$. Fig. 5 compares the $\beta'(t)$ and $\beta''(t)$ cure indices for the dielectric ($\varepsilon'_M(t)$) and loss factor ($\varepsilon''_M(t)$) profiles shown in Fig. 3. In Fig. 5, $\beta''(t)$ is seen to lead $\beta'(t)$ by about 8 minutes.

Such cure indices should not be taken too literally since they depend on the basis parameter (ε'_M, ε''_M, f_r, r_o, etc.) and the somewhat arbitrary manner in which the index is defined. Comparisons between the cure indices need to be made against the generally accepted "degree of cure (α)," defined as the progressive accumulated area under a calorimetric curve vs. time as measured using a Differential Scanning Calorimeter (DSC), for example, or perhaps as calculated using an analytical cure model [5]. In no circumstance should the relative cure index be confused with absolute or maximum cure.

Fig. 6 demonstrates the high degree of correlation between ε'_M and ε''_M, with implicit time and temperature. Two examples are shown; one for T_c = 113°C as in Figs. 3, 4 and 5, and another for a T_c = 143°C. The recommended T_c = 121°C. As expected, the two cures are identical during initial ramp up. For T_c = 113°C, both ε'_M and ε''_M peak at the same time while for T_c = 143°C, ε''_M peaks about 3 minutes before ε'_M . As will be seen, the peak in ε'_M signals the PMV.

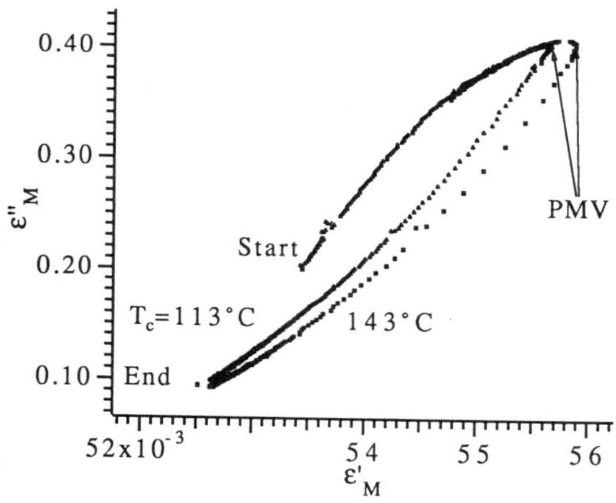

Fig. 6.
Comparison of ε'_M and ε''_M at hold temperatures of 113°C and 143°C. Data points are 20s apart.

Fig. 7(a) compares f_r with log (viscosity, η) as obtained separately using a Dynamic Mechanical Analysis (DMA) Rheometrics spectrometer. Direct comparison is not exact because the temperature profile in the autoclave run could not be replicated exactly in the DMA run, as shown in Fig. 7(b). The timing of the viscosity features with those of f_r are striking. The flat during the first hold at 50° is clearly evident (note the overshoot in the autoclave temperature, which induces a local minimum in f_r). Further, the PMV during the second hold at 109° exactly coincides with the minimum f_r at t = 103 min. During the third ramp, the tracking of f_r and log η are not quite parallel, probably because the DMA temperature exceeds the autoclave temperature. Finally, at t = 150 min, the cure is well into its glassy state and log η becomes very large. Nevertheless, f_r continues to track the rigidity during the final cure stage, beyond where viscosity measurements are useful.

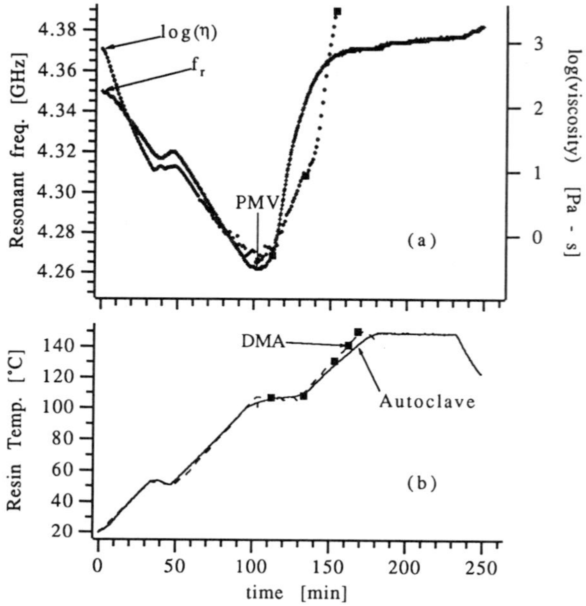

Fig. 7.
(a) Time profiles of f_r and viscosity (log η) for a multiple (T = 50°, 101°, and 148°C) hold cure. Note that the timing features correspond almost exactly.

(b) The DMA and autoclave temperature profiles corresponding to (a). The solid squares indicate points where the DMA spectrometer was stopped to adjust the strain or change the paper.

ACKNOWLEDGMENT

This material is based upon work supported by the Department of Commerce under contract number 50-DKNA-8-00168 (SBIR Phase II). Any opinions, findings, and conclusions or recommendations expressed here are those of the authors and do not necessarily reflect the views of the Department of Commerce. Appreciation is extended to Hexcel's Dr. Winston Hedges and to KDC's Dr. Donald Dunn for their encouragement and constructive technical assistance, and to Mr. Kenneth Baron, Director of Hexcel Corporate Research for providing technical support.

References

[1] J. Jow, M. C. Hawley, M. Finzel and T. Kern, Polymer Engr. and Sci., 28(22), 1450 (1988).
[2] J. Jow, M., M. C. Hawley, M. Fenzel, J. Asmussen, H-H. Lin, and B. Manring, IEEE Trans., MTT-35(12), 1435 (1987).
[3] D. R. Day, Proc. 35th Int'l SAMPE Symp., 2289 (1990).
[4] M. L. Bromberg, Proc. Composites in Manufacturing 6, EM 87-125 (1987).
[5] P. R. Ciriscioli and G. S. Springer, Proc. 34th Int'l. SAMPE Symp., 312 (1989).

PRELIMINARY COMPARISON OF MICROWAVE HEATING OF COMPLEX-SHAPED COMPOSITES REINFORCED WITH CONDUCTIVE AND NONCONDUCTIVE FIBERS

LARRY A. FELLOWS, SUSANNAH TRAVIS, and MARTIN C. HAWLEY
Department of Chemical Engineering, Michigan State University,
East Lansing, MI 48824.

ABSTRACT

The microwave heating of composites with complex shapes has been studied. A V-shaped mold was used to fabricate samples which do not lie parallel to the electric field produced in a 7-in. cylindrical microwave applicator operating at 2.45 GHz.

A polyester/glass composite and a polyester/graphite fiber composite were produced for comparison. Heating profiles were made more uniform by applying mode-switching techniques to the processing. Mode-switching is the application of more than one electromagnetic mode to achieve uniform heating.

The samples were tested for heating patterns in different geometric and fiber orientations within a single-frequency, cylindrical microwave applicator operating at 2.45 GHz. The temperature profile for each mode was recorded. Mode-switching provided better temperature profiles than single modes. Mode-switching techniques were more successful with the graphite fiber reinforcement.

INTRODUCTION

The microwave processing of polymers and composites has been developed at Michigan State University using cylindrical, resonant mode microwave applicators [1-3]. Microwave processing offers certain advantages over conventional thermal processing techniques. Microwave processing is a fast, controllable method for heating materials. The microwave method has the advantage of coupling the microwave energy directly into the molecules of the materials. This heats the material from the inside to the outside rather than from the outside to the inside, which is the direction of heating by convection and conduction.

A key advantage of microwave processing is the selectivity of the microwave fields for introducing energy into the material. Traditionally, samples have been heated by aligning the major axis of the sample geometry with the electric field within the microwave cavity. Heating is facilitated by the electric field interacting with the molecules of the material to produce heat. The cylindrical microwave cavities produce only one resonant electromagnetic field at a time. There are two type of fields, or modes. One type is a TE mode which produces a horizontal electric field. This field is best suited for the heating of planar samples which lie parallel bottom of the cavity. The other modes are TM modes, which produce vertical electric fields. These fields are best suited for heating cylindrical or rod-shaped samples [4].

The rate at which a material heats by microwave energy is characterized by its dielectric loss factor ϵ'' [5]. The dielectric loss factor has two components. One component characterizes the alignment of dipoles in the molecules within the material with the electric field. When the dipoles cannot align fast enough to match the alternating electric field, energy is lost in the form of heat. This is the manner in which nonconductive materials are heated by microwave energy. Another component of ϵ'' is dependent upon the conductivity of the material. Conductive materials will readily absorb microwave energy through this mechanism. Conductive materials therefore heat very rapidly. When processing a composite material, heating may be accomplished by the heating of either the matrix material, the fibers, or both. In the case of graphite fiber composites, the graphite fibers absorb the microwave energy faster than the matrix material. Polymeric materials are much less lossy than the graphite, so heating is facilitated by heating the graphite fiber. The glass fibers, however, are similar to the polymer materials in the values of dielectric loss factor.

The values of the loss factors will have a profound effect on the electromagnetic fields found within the cavity. The modes available within an empty cavity are predicted by Maxwell's equations [6]. A material loaded within the cavity disturbs the electromagnetic field pattern within the cavity. The size of the sample and its dielectric loss factor will determine how much the field is disturbed. A lossy sample will disrupt the field to a greater extent than a lossless material. During past microwave experiments, it was found that the modes available for microwave processing are not the same as those predicted by the Maxwell equations. Hybrids and combinations of the empty cavity modes are found that current mathematical models do not completely describe or predict [7].

Ideally, samples should be heated by finding only one mode to heat them evenly. It was found, however, that for 7.62 cm square composite panels in a 17.78-cm diameter microwave applicator a single mode could not be found to heat the sample evenly. A technique referred to as mode-switching was applied to produce even heating patterns [8,9]. The idea behind mode-switching is that a single mode may produce one uneven heating pattern, and another may produce a different heating pattern. If these heating patterns are such that different sites are favored, an even heating profile may be obtained by switching from mode to mode in an intelligent manner. The mode selected is chosen by considering where the sample needs to be heated faster and the mode that provides that heating.

This technique for heating offers the potential for heating samples which have complex shapes. These shapes do not have to be limited to the previously studied planar and cylindrical shapes, which offers more freedom for microwave processing techniques in industrial applications. A complex shape may be processed if there are enough modes to provide a complementary set of modes. A complementary set of two modes, for example, might be one in which the temperature profile of one is $T1>T2>T3>T4$. Its complement would be $T4>T3>T2>T1$. By matching the hot spots of one mode with the cold spots of another, even heating may be obtained by heating part of the time in one mode and switching to another at the right moment. The sites that do not heat are just as important for mode-switching as the sites that do heat. If a site is favored by all of the modes, it will heat too much and the temperature in this site will be too far ahead of the other temperatures. An ideal situation would be to have one mode that heated each section of the sample without heating any of the other sites. The energy could then be placed exactly where it is needed without too much being distributed to other sites.

EXPERIMENTAL

The experimental setup has been previously described [10]. A 17.78-cm diameter cylindrical microwave cavity was used with a microwave system. The adjustable cavity length and coupling probe depth were used to tune the cavity to each unique mode by finding the minimum reflected power. A V-shaped composite part was chosen as the complex shape. This shape fit the requirement of not matching with the TE or TM mode electric field while allowing for the placement of fluoroptic temperature probes across the top of the sample to measure the temperature profile. The V-shaped part had two different geometric configurations, as shown in Figure 1.

Composites were fabricated by impregnating fibers with a Derakane vinyl ester polyester resin which used vinyl toluene instead of polystyrene for the cross-linking monomer. This resin was obtained from Dow Chemical. The glass fiber was a Type 30 glass roving obtained from Owens/Corning Fiberglas (product identification 107B-AC-450, 2 k tow, 0.0033 cm diameter). Hercules 12 k AS4 graphite fiber was used for the preparation of the graphite composite. The V-shaped parts were made by autoclaving 12-ply unidirectional parts in specially-designed aluminum molds.

Teflon molds were also designed to contain the V-shaped part in the microwave applicator. The design of the molds is shown in Figure 2. Eight temperature measurements were recorded using Luxtron 750 and 755 Fluoroptic Thermometry Systems (4 measurements per system). Glass capillary tubes were placed through the top of the Teflon mold to make contact with the surface of the composite sample. Fluoroptic temperature probes were inserted into the capillary tubes. The locations of the temperature probes are shown in Figure 3. Figure 3 also shows the designation of the fiber direction. The convention is to consider the location of the coupling probe in the microwave cavity as the 0° fiber direction and move clockwise from there. The direction of the fiber is important because it is one of the variables that determines the electromagnetic field pattern. Different field patterns arise for the same composite part at different fiber orientations.

The samples were tested by first finding the modes available for the V-SHAPE UP and V-SHAPE DOWN geometric orientations at 0, 45, 90, and 315° fiber

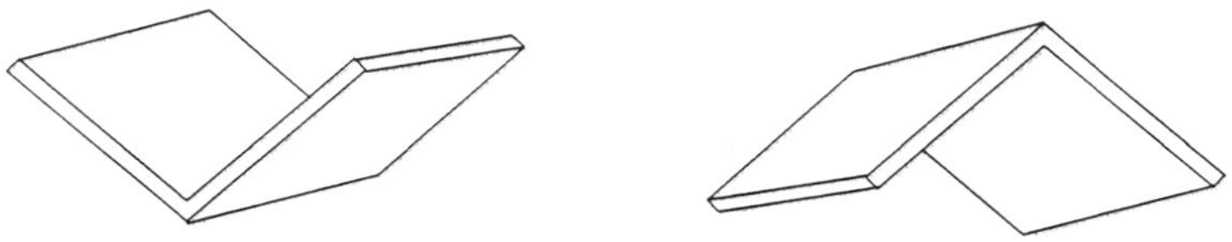

Figure 1. The V-shaped part shown in the V-SHAPE UP (left) and the V-SHAPE DOWN (right) geometric configurations.

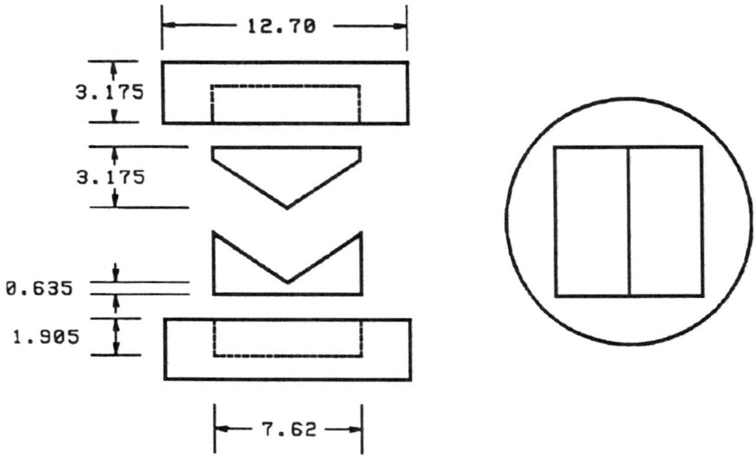

Figure 2. The Teflon mold and its dimensions (in cm) fabricated for V-shaped composite parts.

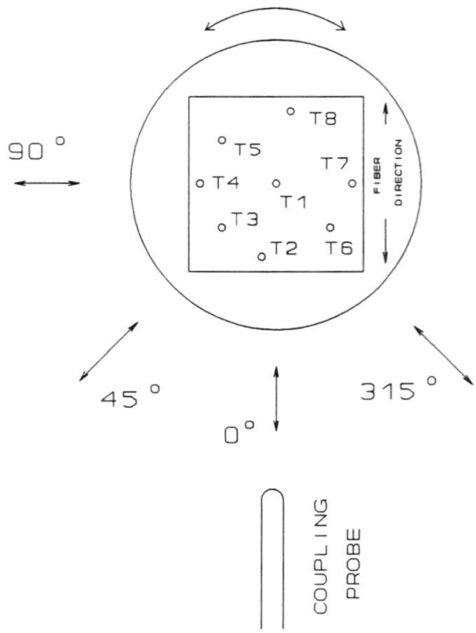

Figure 3. The temperature probe locations and fiber direction designation for the composite parts in the microwave applicator.

orientations. The heating patterns for the modes were mapped by heating the composite to 125°C with 60 W of power input into the system by an Opthos Microwave Generator (Model MPG-4M-297). For each composite part, the most suitable set of modes was selected and used for mode-switching experiments. The temperature and power data were recorded by a computer data acquisition system. Table I lists the symbols used to characterize the temperature profile by ranking each temperature relative to the other closest temperatures in the profile.

RESULTS AND DISCUSSION OF RESULTS

For the glass composite, the V-SHAPE DOWN geometric configuration yielded the most promising set of modes for mode-switching at 45 and 315° fiber orientations. The temperature profiles of the different modes are listed in Table II and III for the 45° and 315° fiber orientations, respectively. For the graphite fiber, the V-SHAPE UP geometric configuration showed the best set of modes for mode-switching experiments at 45° and 315° fiber orientations. Tables IV and V present the temperature profiles for these modes. The graphite fiber composite exhibited a symmetry in that the temperature profiles were essentially the same for the 45° and 315° fiber orientations. The modes were found at the same coupling probe depth and cavity lengths, and the composite heated in the same places for each of the modes. The glass fiber composites, however, did not heat the same even though the modes were found at the same cavity length and coupling probe depth.

Figure 4 shows the temperature profile and mode selection for the heating of the glass/Derakane V-SHAPE DOWN at 315° fiber orientation. The temperature profile exhibits a large temperature gradient between the highest and lowest temperatures, but the mode-switching still provided a better temperature profile than any of the single modes. The set of modes lacked a mode which would preferentially heat T2 and T8. It can be seen in Table III that mode E

TABLE I. Mathematical Operators Used to Describe Temperature Profiles

MATHEMATICAL OPERATOR	TEMPERATURE DIFFERENCE
$=$	0-5°C
$\geq$	5-10°C
$>$	10-20°C
$>^2$	20-30°C
$>^3$	30-40°C
$>^4$	40-50°C
$>^5$	50-60°C

favors T1 strongly, and the temperature profile demonstrates very well the heating of the part by mode E. At any time the cavity was tuned to mode E, the composite heated very strongly at T1. This illustrates very well the idea behind mode-switching.

Figure 5 is the temperature profile and the mode selection for the AS4/Derakane V-SHAPE UP at 315° fiber orientation. This profile is better than the glass/Derakane profile, but it still does not exhibit even heating. From Table V, it can be seen that no single mode favored the heating of T1. T1 did heat better than the lagging temperatures in the glass/Derakane experiment, but this was due to the heating of both T2 and T8. The greater heat conduction of the AS4/Derakane sample also helped the temperature profile in this case. Overall, the AS4 fibers did heat much better than the glass fibers, but the profile still did not exhibit the uniformity desired for the technique. The low temperature at T1 may have been due to improper contact between the capillary tube and the surface of the V-shaped part, and this will be examined in future work.

The limitations on the experiment were the set of modes, which, in turn, was limited by the size of the sample and the size of the microwave cavity. Only a range of cavity lengths of 16.5 to 11.0 cm was available for finding modes, and the set of modes failed to find modes that would heat every part of the sample. The change in the fibers resulted in totally different sets of modes available for the microwave heating. This was expected since the dielectric loss factor of the graphite fiber is much greater than the dielectric loss factor of glass fibers. The graphite fiber did not appear to be advantage as from the standpoint of exhibiting more even single mode profiles, but it did have the advantage of greater heat conduction.

In this particular case, the set of modes available for the graphite fiber sample were slightly better for mode-switching, but another case might find that the glass fiber is better. If one limits the configurations to V-SHAPE UP geometric orientations, the AS4 clearly had a better set of modes. For the V-SHAPE DOWN orientations, however, the glass fiber composite exhibited a more suitable set of modes for mode-switching. Each case for the complex geometry will generate its own set of modes, and current electromagnetic models do not predict which modes will appear. The method of mapping the temperature profile of each mode is currently the best way of evaluating the potential for processing a complex shape in the cavity.

Variables such as fiber direction and sample geometry will provide different sets of modes for a complex shape, and the temperature profile maps can be used to evaluate the modes to choose the correct modes for controlling the microwave heating of the sample. Future modifications to the microwave system will be used to enhance the controllability of the cavity, and control strategies may be designed from the temperature profile information. Automation of the cavity controls will provide faster response to the decisions made for mode-switching, and better temperature profiles can be obtained. Mode-switching can be used to design microwave processing cycles for composite materials and to allow for the flexibility of processing complex shapes.

TABLE II. Electromagnetic Modes of Microwave Heating of Glass/Derakane V-SHAPE DOWN Geometry at 45° Fiber Orientation

MODE	CAVITY LENGTH, L_c (cm)	PROBE DEPTH, L_p (mm)	TEMPERATURE PROFILE
A	15.890	19.44	T7$>^4$T5$>^4$T8=T4=T6≥T1≥T2=T3
B	15.578	20.05	T4$>^3$T3=T5$>^3$T2=T8=T6=T7=T1
C	13.667	12.83	T7>T5>T4>T3$>^2$T6$>^2$T1≥T2=T8
D	12.143	12.48	T5>=T3>T4>T1$>^3$T6≥T7=T2=T8
E	11.176	22.48	T1$>^3$T5=T6=T3≥T8=T7=T2≥T4

TABLE III. Electromagnetic Modes of Microwave Heating of Glass/Derakane V-SHAPE DOWN Geometry at 315° Fiber Orientation

MODE	CAVITY LENGTH, L_c (cm)	PROBE DEPTH, L_p (mm)	TEMPERATURE PROFILE
A	15.672	18.11	T7>T4≥T5$>^2$T6$>^3$T3=T8≥T1=T2
B	15.538	13.63	T4$>^5$T5=T3>T7=T6=T2=T8≥T1
C	13.524	11.00	T4$>^3$T7≥T5≥T6≥T3$>^3$T2=T8=T1
D	12..018	10.02	T4=T3=T5=T6$>^2$T7≥T1$>^2$T2=T8
E	11.159	18.70	T1$>^6$T6=T5=T3≥T8=T2=T4≥T7

TABLE IV. Electromagnetic Modes of Microwave Heating of AS4/Derakane V-SHAPE UP Geometry at 45° Fiber Orientation

MODE	CAVITY LENGTH, L_c (cm)	PROBE DEPTH, L_p (mm)	TEMPERATURE PROFILE
A	16.463	15.04	T8$>^5$T2$>^4$T7=T4=T1=T5=T3=T6
B	14.042	11.53	T7$>^2$T4$>^4$T2=T3=T6=T8=T5≥T1
C	13.085	20.89	T8$>^2$T2$>^5$T1>T3=T5=T6=T4=T7
D	12.430	20.68	T3$>^2$T5=T6^2>T7=T4>T2≥T8≥T1

TABLE V. Electromagnetic Modes of Microwave Heating of AS4/Derakane V-SHAPE UP Geometry at 315° Fiber Orientation

MODE	CAVITY LENGTH, L_c (cm)	PROBE DEPTH, L_c (mm)	TEMPERATURE PROFILE
A	16.456	14.84	T8$>^4$T2$>^5$T4=T7=T1=T3=T5=T6
B	14.037	11.52	T7$>^2$T4$>^5$T8=T3=T6=T2=T5≥T1
C	13.098	21.25	T8$>^2$T2$>^5$T1≥T3=T6=T5=T7=T4
D	12.432	19.33	T3$>^2$T5>T6≥T4>T8=T7>T2≥T1

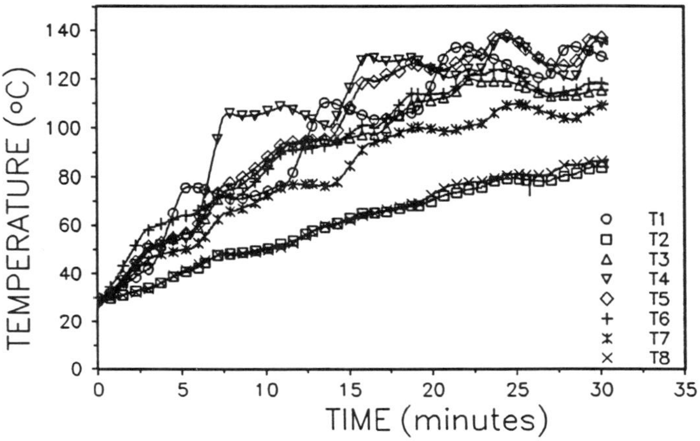

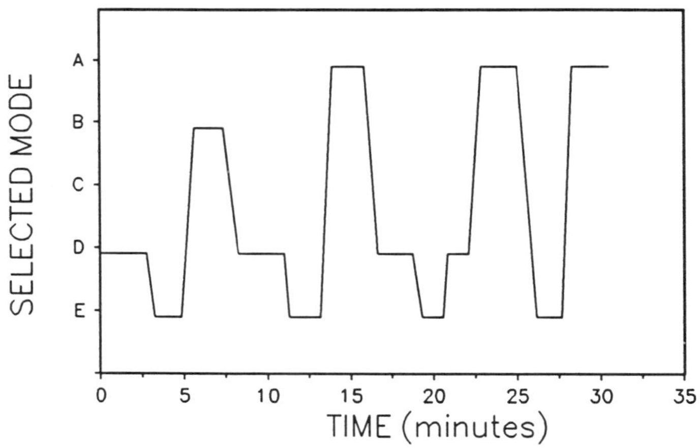

Figure 4. The temperature profile and mode selection for the microwave heating of the glass/Derakane V-shaped composite in the V-SHAPE DOWN geometric configuration at a 315° fiber orientation by mode-switching.

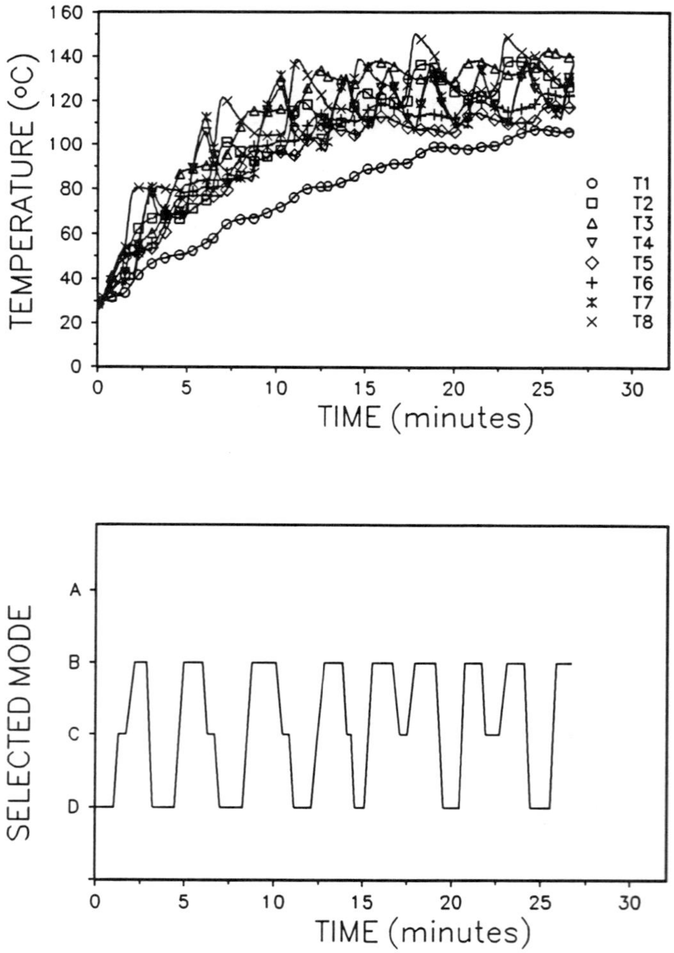

Figure 5. The temperature profile and mode selection for the microwave heating of the AS4/Derakane V-shaped part in the V-SHAPE UP geometric configuration at a 315° fiber orientation by mode-switching.

CONCLUSIONS

The microwave heating of the complex shape was studied for two different V-shaped composites. One was a glass/Derakane composite, and the other was an AS4/Derakane composite. It was found that different geometric and fiber orientations generated different sets of modes. Some symmetry was observed for the available modes at 45° and 315° fiber orientations, especially in the case of the AS4/Derakane composite. Mode-switching was attempted to generate a uniform heating profile across the composites. Mode-switching did provide a better heating profile than the single modes, but it still was not uniform. The graphite composite exhibited a better heating profile than the glass composite, but this may have been attributable to better heat conduction and a better set of modes for the particular case. Each set of modes generated by the complex shapes appears to be a function of the materials selected, geometric orientation, and fiber direction. Other variables such as composite layup may also be important, but these were the three evaluated by this set of experiments.

REFERENCES

1. J. Jow, et al., IEEE Transactions on Microwave Theory and Techniques, Vol. MTT-35, No. 12, 1435 (1987).

2. J. Jow, et al., Polymer Engineering and Science 28(22), 1450 (1988).

3. J. Wei, et al., in Composites: Design, Manufacture and Application, edited by S. Tsai and G. Springer, (Eighth International Conference on Composite Materials, Honolulu, Hawaii, July 15-19, 1991) presentation 10-L.

4. J. Jow, PhD thesis, Michigan State University, 1988.

5. D. K. Cheng, Field and Wave Electromagnetics, 2nd ed. (Addison-Wesley Publishing, Reading, 1983), p. 300.

6. D. M. Pozar, Microwave Engineering (Addison-Wesley, Reading, 1990), pp. 154-163 and pp. 348-353.

7. J. Wei, et al., SAMPE Journal, 27(1), 33 (1991).

8. L. Fellows, MS thesis, Michigan State University, 1991.

9. L. A. Fellows and M. C. Hawley in Composites: Design, Manufacture and Application, edited by S. Tsai and G. Springer, (Eighth International Conference on Composite Materials, Honolulu, Hawaii, July 15-19, 1991) presentation 12-E.

10. L. A. Fellows and M. C. Hawley in Materials Working for You in the 21st Century, edited by G. Grimes, R. Turpin, G. Forsberg, B. Rasmussen, and J. Whitney (37th International SAMPE Symposium and Exhibition, 37, Anaheim, California, March 9-12, 1992) pp. 914-928.

AUTOMATION OF MICROWAVE PROCESSING OF GRAPHITE / EPOXY COMPOSITE MATERIALS USING AN EXPERT SYSTEMS TECHNIQUE

VALERIE ADEGBITE , MARTIN HAWLEY AND DAVE DECKER , JON STICKLEN
Michigan State University, Dept of Chem. Eng., East Lansing, MI 48823
Michigan State University, AI/KBS-Lab CPS Dept., East Lansing, MI 48823

ABSTRACT

Microwave curing of composite materials is a novel technology which can be an alternative to conventional thermal curing and has many advantages over thermal curing, i.e, fast and direct heating, selective heating, and controllability. Currently, the existing technology is more empirical and cannot be easily described by traditional process models. One key enabling step for realizing the advantages of the microwave curing technology is a capability for utilizing all the available existing knowledge in a concise manner. The major elements of a control system which utilizes traditional control techniques and (non-traditional) expert systems techniques in a real time microwave cure monitoring and control is discussed.

INTRODUCTION

An important variable in the microwave processing is the ability to efficiently couple microwave energy into and uniformly heat the material load. This is done by selecting the optimum heating conditions(mode selection) and maintaining (mode tuning) the optimum conditions during the curing cycle. Both the mode selection and the mode tuning methodologies are typically based on experiential knowledge with some of the phenomenal material properties based on traditional process models.

In general, real-time sensing and control based on process models is adequate to control even some of the most complex processes. However , because the technology for the microwave processing of composite materials is currently not complete and is comprised of empirical as well as process models, the integration of traditional control techniques and advanced computing techniques(expert system) presents a suitable environment for the implementation of control decision making.

The control system that we are developing combines multiple levels of knowledge into modular units with dedicated tasks. These units include a traditional control unit which has the task for real-time sensing, data interpretation, and control. An advanced control unit (expert system based) which contains empirical knowledge about the process and has the capability to reason heuristically about the process states. The work discussed here is the presentation of the overall control architecture with emphasis on the implementation of the traditional control unit in the automation of the empty cavity mode tuning. This work represents a subset of a global research work which was previously reported in [2].

BACKGROUND

Process Device Description

A typical microwave processing circuit consist of a power source, circulator and matched dummy load, coaxial directional couplers, attenuators and power meters to measure incident and reflected power, and an applicator or a cavity An example of an applicator is shown in Figure 1. The cavity is a cylindrical brass cylinder with an internal transverse shorting plates. One of the shorting plates is adjustable to provide varying cavity lengths from 4.5 cm - 22 cm. The other short plate is removable for loading material into the cavity, and the material rests on this short plate which is fixed in place during processing. A semirigid 50-ohm impedance copper coaxial probe serves as a field excitation probe (coupling probe) to couple microwave power into the cavity. Adjustment of the cavity length Lc and the coupling probe length Lp are made by the automatic motion of stepper motors.

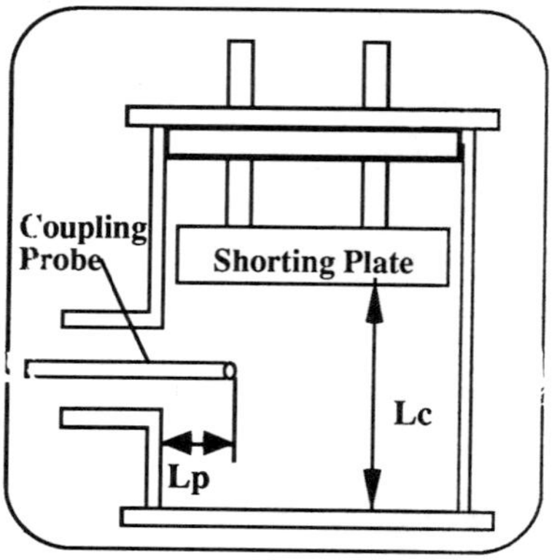

Figure 1. Microwave Cavity

Mode Tuning

The concept of mode tuning involves the efficient coupling of incident power into the material load(i.e. maintain mode, minimize power reflected by the system) the matching of the cavity input resistance to that of the characteristic impedance of the transmission line. The input impedance(Zin) of a microwave cavity is given by equation (1)

$$\text{Zin} = R\text{in} + \text{j}\, X\text{in} \tag{1}$$

where Rin is the cavity input resistance, and jXin the reactance. These two variables represent the complex load impedance as seen by the feed transmission line. It takes the adjustment of the cavity length and probe depth together to cancel the load resistance and adjust the reactance to equal the characteristic impedance of the feed transmission system[1]. When the cavity load resistance is adjusted to match that of the transmission line then the cavity is in resonance or a mode is present. The adjustment activity is called mode tuning.

Mode

Mode represents the presence of a standing wave in which a sample can be heated. For every microwave applicator, there exist a finite number of theoretical modes at a particular frequency. Theoretical modes for an empty cavity at a particular frequency can be determined by resolving the eigenvalue problem from the Maxwell's equations[3]. The eigenfunctions from this solution can be found to be the resonant modes and the associated eigenvalues as the resonant frequency.

The theoretical modes provide a theoretical cavity length at which a resonant frequency can be achieved. Normally, the theoretical mode data serve as a guide for locating approximate empty cavity lengths at which resonant modes can be located. As the cavity is loaded with material the theoretical cavity resonance is disturbed and has to be compensated for. This is done by tuning the cavity around the theoretical length to determine the actual processing mode.

Microwave Composite Curing

The initial step in the microwave curing of composites involves the selection of an optimum heating mode. An optimum heating mode represents the cavity conditions at which the material load would be uniformly heated. Typically, it is uncertain a priori which modes or which combination of modes will be the optimum heating modes for a particular material with a particular geometry. Thus the optimum heating modes are determined experimentally by finger printing the heating profile of a material for different modes and then selecting the optimum mode. The optimum modes are then used to structure a curing plan for the composite material.

During electromagnetic heating of a composite material both the real and the imaginary parts of the complex dielectric constant of a the material may change significantly due to the dependency of the complex dielectric constant on temperature, chemical composition, and electric field. As the material properties change during the processing cycle, the resonance mode is also continuously disturbed and the cavity has to be continuously tuned during the processing cycle.

The actual mode tuning is typically a trial and error operation, where the cavity length is the major adjusting component. For example, a component is adjusted in one direction and then the efficacy is observed on the minimization of the reflected power. If the effect is not favorable then the direction of adjustment is reversed.

PROBLEM STATEMENT

The overall problem we set forth is the development of a control architecture that would utilize both empirical and process models to (a) automate the empty cavity tuning, and (b)to maintain modes in a loaded cavity during processing. Our immediate focus is the automation of the empty cavity tuning activity. That is, at any given cavity length and probe depth, the system will automatically adjust the cavity parameters such that the reflected power is minimized.

PROCESS SENSING AND CONTROL ARCHITECTURE

The entire control system consist of three modular units, each with a dedicated task (see Figure 2.). The flow of data would pass from the lowest level, traditional control unit to the top most level , the global problem solver. The complexity of control decision making increases from the lowest module to the topmost module with the response time decreasing from the lowest level to the topmost level. Looking at the functionalities of the modules from the top down,

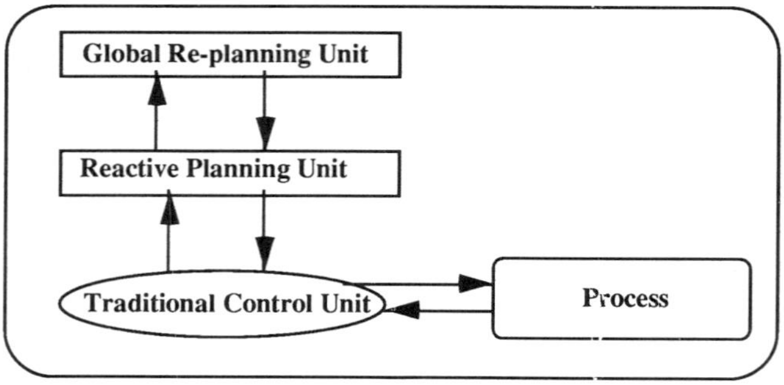

Figure 2. Control Architecture

the Global Problem Solver (GRS) handles the "most complex" problems. It is called upon when both the Traditional Control Unit(TCU) and the Reactive Problem Solver(RPS) are unable to solve the problem at hand. i.e. the replanning of a control strategy that is not yielding favorable results. It has the lowest sampling frequency and quality of output solution is time dependent.

The Reactive Problem Solver supervises the TCU and makes recommendations as requested by the TCU. This module is the first level of defense in the active process management and control. It consist of multiple levels of knowledge with each layer increasing in its capability to handle complex problems. A typical problem that would be handled by the RPs would be the modification of the control plan to maintain desired processing set points in the TCU. This unit is part of the Advanced Computing modules which will be implemented in "C".

The lowest level and the "work horse" of the system is the Traditional Control Unit. This is the unit which communicates with the microwave system and manages the "active" sensing and control activities. It is implemented in traditional closed loop feedback control methodology and resides on a Mac IIci platform with generic interface communications hardware and software.

Mode Tuning Algorithm

In the mode tuning activity the reflected power and input power to the empty cavity are the only sensed variables at this time. Other parameters such as microwave source frequency and cavity dimensions will be added soon. The cavity length is adjusted as the major component to minimize the reflected power. Figure 3 shows a two step summarization of a typical mode tuning operation. The figures indicate scanned graphs of frequency vs. absorbed power as would be observed on an oscilloscope during swept frequency diagnosis of the cavity.

Figure 3a indicates the location of an absorption peak along the horizontal axis from the adjustment of the cavity length to 12cm at a fixed probe depth of 12 cm. The cavity length adjustment typically shifts the absorption peak horizontally across different frequencies. On this graph the frequencies range from (2.35-2.55) GHz. Note that the objective of the mode tuning is to align the absorption peak along the input frequency mark on the graph. This mark is typically set to be at the vertical mid plane of the graph.

Once the absorption peak is aligned with the input frequency, the probe depth is then adjusted to maximize the coupling efficiency or minimize the reflected power. This is indicated on the graph by the shift of the absorption peak vertically along the y-axis until it touches the horizontal line(figure 3b). Recall that it takes the that the adjustment of the

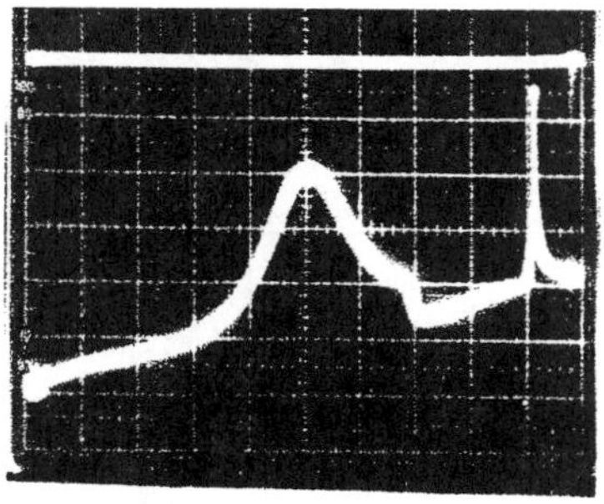

Figure 3a. [% Absorption vs. Frequency]
Lc=12cm, Lp= 12cm

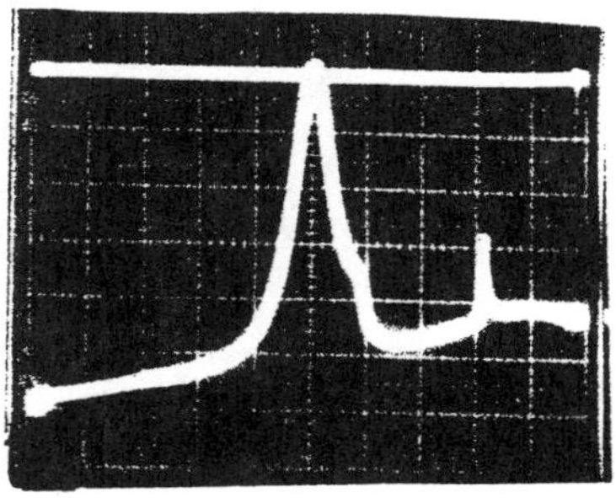

Figure 3b. [% Absorption vs. Frequency]
Lc=12cm, Lp= 7cm

cavity length and probe depth together cancel the load resistance and adjust the reactance to equal the characteristic impedance of the feed transmission system[1].

The automated system emulates this mode tuning activity without the use of the oscilloscope spectrum but by using measured reflected power to determine the efficiency of the adjustments. Typically, the reflected power tends to follow a parabolic profile as the adjustments are made. This means that a mode is found when the minimum on the parabola is located. Thus the objective of the mode tuning algorithm is the determination of the combination of adjustments of the cavity length and probe depths to locate the minima on the parabola which represents the location of the minimum reflected power. Note that the actual dimensions of the parabolic profile and the combination of adjustments that would locate the parabolic minima are not known a priori.

SUMMARY

The most crucial concern in the entire control system is the rate at which the TCU is able to locate a local minimum. This is due to the fact that the TCU is the workhorse of the control system and handles the least complex problems. Currently, the TCU has been able to locate the modes in the empty cavity with a good response time. The immediate next step is to heat an already cured composite material and determine how the current algorithm is able to maintain a mode as the reflected power is changed dynamically. Finally, the mode tuning program will be combined with mathematical models and expert system models to control the total curing of an epoxy/graphite composite materials and then extended to other materials.

References

[1] Sticklen, J., Kamel, A., Hawley, M., Adegbite, V., "Fabricating Composite Materials: A Computationally Comprehensive Problem Solving Architecture Based on a Generic Task View point". IEEE Expert to be published Fall 1991.

[2] Asmussen, J., Lin, H. H., Manring B., Fritz, R., "Single-Mode or Controlled Multimode Microwave cavity Applicators for Precision Material Processing", Rev. Sci Instrum. , 58, 1477-1486 (1987)

[3] Montgomery, Carol G. "Techniques of Microwave Measurement." McGraw Hill, New York (1947).

MICROWAVE PROCESSING OF GLASS FIBER/VINYL ESTER-VINYL TOLUENE COMPOSITES

DHULIPALA RAMAKRISHNA, SUSANNAH TRAVIS AND MARTIN C. HAWLEY.
Department of Chemical Engineering, Michigan State University, East Lansing, MI 48824.

ABSTRACT

Microwave processing has been investigated as an alternative to the conventional processing of polymers and polymer matrix composites. Different studies have demonstrated its viability and advantages, but various aspects related to its industrial applicability are still being investigated.

Two aspects related to the microwave curing of glass fiber/vinyl ester-vinyl toluene composites have been studied. A qualitative evaluation was done to determine the reproducibility of heating patterns of these laminates in different modes of a microwave resonant cavity. The samples were prepared to similar physical and chemical specifications and heated in different single modes of the cavity.

Another evaluation was done to compare the mechanical properties of microwave and thermally cured laminates subjected to flexural load. The samples were processed using similar cure cycles. The microwave samples were cured using a mode-switching technique and the thermal samples were cured in a temperature controlled oven. The void fractions of the samples were evaluated to develop pre-consolidation conditions that yield samples with low void fraction. Results of these two evaluations are discussed.

INTRODUCTION

Various features of microwave processing of polymer matrix composites have been studied [1-5]. In this paper we present the results of the study of two aspects related to the industrial applicability of this process.

Different resonant modes of the microwave cavity have been found to transfer energy non-uniformly across the volume of the composite. Processing of composites in single modes therefore results in large temperature gradients in the material. Different modes have different energy distribution patterns. Temperature probes can be placed at different positions on the sample to measure the temperatures as the material is heated in different single modes. From a knowledge of such single mode heating patterns, it is possible to switch between appropriate modes to reduce the temperature gradients and obtain relatively uniform temperature profiles in the material being processed. Automation of such processing using knowledge based control systems is presently being investigated [6]. With this end in view, we have qualitatively evaluated the reproducibility of single mode heating patterns when (1) the same sample is heated in slightly different orientations in the cavity and (2) different samples prepared to similar physical and chemical specifications are heated in the cavity. We have also compared the flexural properties of the microwave and thermally cured laminates to study the feasibility of producing laminates in the microwave cavity, with properties similar to thermally cured laminates. These studies are necessary to determine the amenability of microwave processing to automation and its suitability for industrial production of composites.

Consolidation of laminates

Consolidation is a very important step in the manufacture of polymer matrix composites. Usually, consolidation is accomplished by the application of mechanical pressure on the composite boundary, accompanied by the application of vacuum and suitably high temperatures. Pressure and vacuum are maintained at least as long as the gelation time for thermosetting matrix materials. This process squeezes air and other sources of voids (including volatiles) out of the composite and also increases the fiber volume fraction.

The present setup of the microwave cavity does not allow the application of pressure during the curing process and this results in large volume fractions of voids in the laminates cured in the microwave cavity. The effect of voids on the mechanical properties of the laminate is more of a statistical nature rather than a deterministic nature. To make a proper comparison of the microwave and thermal curing processes, the void fractions must be very low so that the inherent differences if any are not masked by the statistical noise in the mechanical properties of high void fraction composites.

Pre-consolidation

Recent studies have characterized the behavior of laminates during the consolidation process [7-11]. Gutowski [8-11] has shown that the fibers act as elastic springs during the consolidation process, and also that the load carried by the fibers increases with increasing fiber volume fraction.

We had initially attempted to reduce the void content by subjecting the laminate to mechanical pressure and vaccuum prior to the application of temperature for curing the laminate. This method was unsuccessful in reducing the void content of the laminate. This is attributed to the following. The initial application of pressure squeezes out the voids and some resin. The removal of pressure results in the relaxation of fibers to their original state with the surrounding air occupying the volume thus formed. This phenomenon is analogous to the deconsolidation of thermoplastic composites on heating above their melt temperature [8].

Based on the above mentioned aspects we have developed a pre-consolidation technique that can be used to produce low void fraction composites. In this method the uncured laminate is enclosed in a resin-rich environment. The enclosure is suitably prepared to control the flow of resin during pre-consolidation, while allowing the evacuated air (voids) to leave the surroundings of the laminate. Pressure and vacuum are simultaneously applied for specific times during the pre-consolidation. The resin-rich environment ensures that the resin takes most of the load during the initial stage of the pre-consolidation. The accompanying high hydrostatic pressure aids in forcing the voids out of the laminate. The enclosure ensures that during the fiber relaxation process accompanying the release of pressure after pre-consolidation, the resin surrounding the laminate occupies the volume between the fibers. This results in an overall decrease in the void volume fraction.

EXPERIMENTAL

Preparation of the laminates and laminate molds

Prepreg was made in-house from glass fiber and an experimental DERAKANE vinyl ester resin (containing 45% vinyl toluene by wt. and 1% benzoyl peroxide). It was cut to 3" X 3" pieces and stacked to make twelve-ply, uni-directional laminates. Individual laminates were then placed on nonporous teflon sheets cut to about 9" X 9" in size. Fifteen milliliters of the vinyl ester resin containing 1% benzoyl peroxide as the catalyst was then spread on the laminate to make it resin-rich for pre-consolidation. The nonporous teflon sheet was then folded closely over the laminate. The nonporous teflon was held in place by tape near the edges of the laminate and running around the laminate. The laminate coupon in this stage of the pre-consolidation is shown in figure 1.

Figure 1:
Laminate coupon

A mold was prepared by covering a circular teflon block with release film. A cork dam was prepared on this to the exact dimensions of the laminate coupon. The cork dam was prepared such that the height of the dam was greater than the height of the laminate coupon. The cork dam was then sealed on the outside with tacky tape (a vacuum bag sealant). The laminate coupon was then placed in this mold as shown in figure 2. The whole assembly was then covered with three layers of bleeder cloth and sealed hermetically in a vacuum bag, with a tube attached for connection to a vacuum pump.

Pre-consolidation cycle

The laminate mold prepared as mentioned above was placed in a Carver press and subjected to a force equivalent to approximately 250 psi. Vacuum was simultaneously applied. At the end of the pre-consolidation period of 10 minutes, the vaccuum was released followed by the release of the force.

Figure 2: Laminate coupon in the teflon mold.

Thermal curing

The laminate coupon, pre-consolidated as described above, was cured in a thermal oven at 90°C for 1 hour and post-cured at 125°C for 2 hours.

Microwave curing

The microwave processing circuit has been described elsewhere [13]. The pre-consolidated laminate along with the mold were placed in a 7" microwave cavity. The position of the temperature probes on the laminate and the position of the sample in the cavity are shown in figures 3 and 4. The different resonant modes available were found using a sweep oscillator. The heating patterns for each mode were determined by heating the laminate to about 65°C at 60 Watts of microwave power input to the cavity. A mode-switching technique was used to heat the laminate as uniformly as possible during curing[15]. The curing cycle was the same as that of the thermal cured sample.

Mechanical testing of the thermal and microwave cured laminates

The thermal and microwave cured laminates were conditioned at the standard lab conditions of 23°C and 50% relative humidity for over 48 hrs. The laminates were cut parallel to the fiber direction into approximately 3" X 0.4" specimens. These specimens were polished to make the top and bottom surfaces flat and parallel and were subjected to a three point bending test in accordance with the ASTM standard D790M. The test data were used to evaluate the flexural strength and the tangent modulus of elasticity in bending of each sample [12].

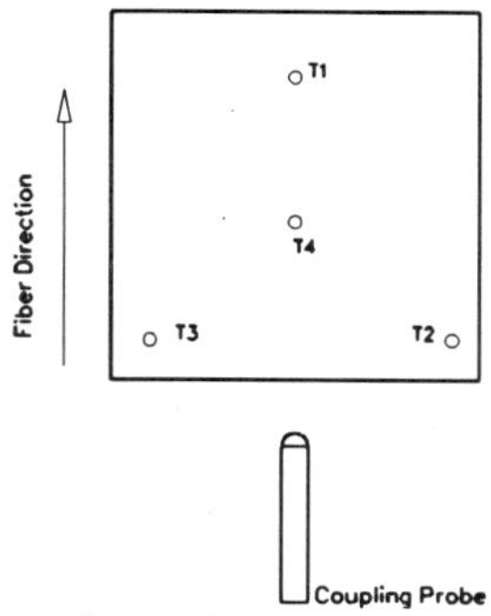

Figure 3: Position of temperature probes on the sample

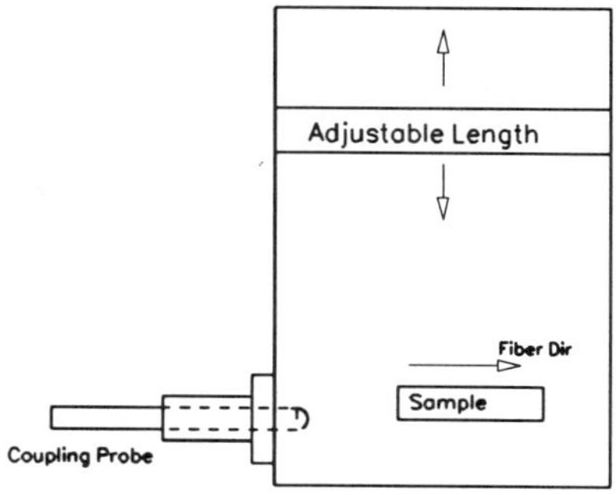

Figure 4: Position of the sample in the cavity

Void fraction analysis of the laminates

Sections 10 x 10 mm were cut from each laminate. These were set in acrylic mounts and polished to 0.05 micron grit. These finely polished surfaces were viewed under an optical microscope at a linear magnification of 50. The images observed were digitized and stored in a computer. An image processing program was used to evaluate the fractions of the different degrees of gray scales present in the image [14]. The void volume fraction was calculated based on the degree of gray scales constituting the voids.

Fiber volume fraction analysis

The polished surfaces of the acrylic mounts were treated with a mixture of chromic acid and sulfuric acid to etch the matrix material of the laminate. The surfaces were viewed under an optical microscope at a linear magnification of 400. The unetched glass fibers appear as bright circles in a dark background of the acid-etched matrix of the laminate. The fiber volume fraction of the laminate was found as before, based on the degree of white scales constituting the glass fibers.

RESULTS AND DISCUSSION

Figures 5 and 6 show the heating patterns of the same sample in two slightly different orientations in the same resonant mode. The heating patterns are very similar, indicating that the cavity is not very sensitive to small variations in the orientation of the sample in the cavity. Figures 7 and 8 show the heating patterns in the same mode of two different samples. The heating patterns obtained were very different from each other, as illustrated by these plots. We believe this is the result of the manual lay-up of the molds which results in significant differences in the amount and shape of the accompanying material used for each mold. Further experiments are needed to establish the sensitivity of the cavity to the differences in the physical specifications of the material being processed in the cavity.

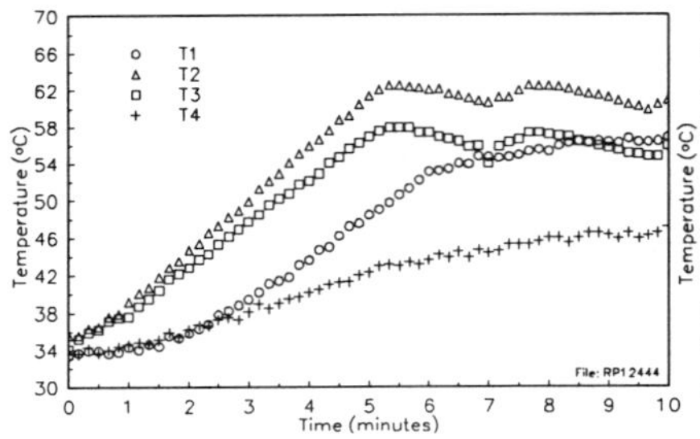

Figure 5: Single mode heating pattern of sample RP1244

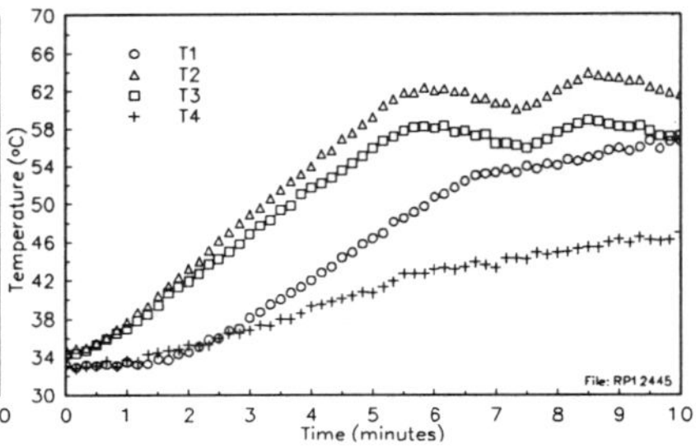

Figure 6: Single mode heating pattern of sample RP124

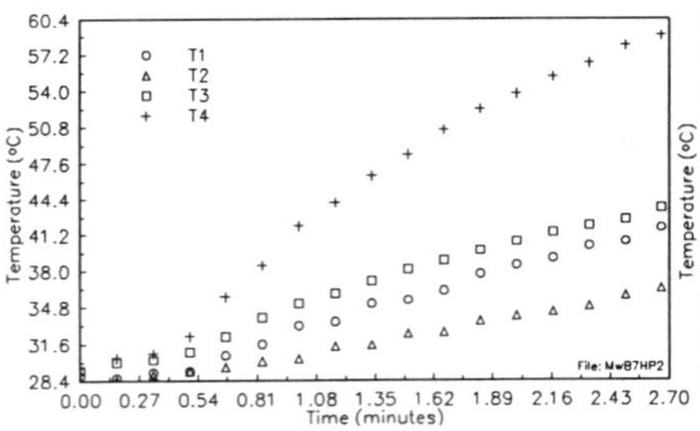

Figure 7: Single mode heating pattern of sample MwB7

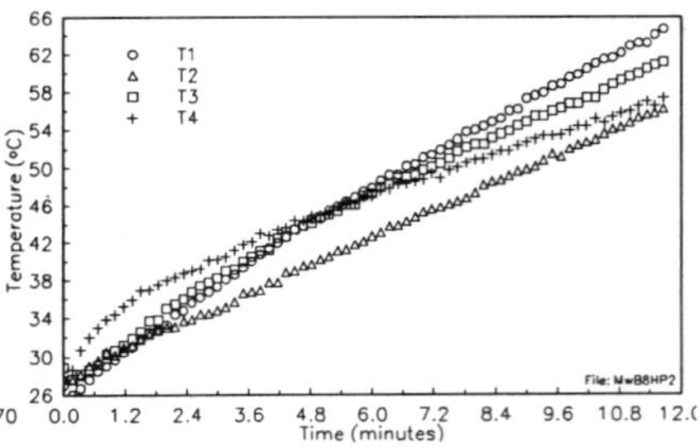

Figure 8: Single mode heating pattern of sample MwB8

Table I lists the properties of the thermal and microwave cured laminates. The pre-consolidation technique developed here has been very successful in reducing the void fractions of the laminates to very low values. The mechanical properties of both the thermal and microwave cured laminates are equivalent, which demonstrates conclusively that microwave curing can be used to produce laminates with properties equivalent to a thermally cured laminate. Modification of the microwave cavity to apply pressure during curing is however necessary because the lack of pressure during curing results in poor surface appearance of the laminates. The pre-consolidation technique has helped, however, to establish that microwave processing is inherently capable of producing laminates with good mechanical properties.

Table I : Properties of microwave and thermally cured laminates.

Sample #	S (Flex. Strength, MPa)	E_b (Flex. modulus, GPa)	Void fraction (%)	Fiber fraction (%)
Th #1	1066	47.0	0.96	67.6
Th #2	1150	65.9	1.14	63.4
Th #3	1135	68.9	1.88	57.0
Th #4	1036	85.0	5.90	66.6
Th #5	1110	93.5	1.34	53.0
Mw #1	1060	52.2	0.9	65.8
Mw #2	1082	70.9	2.81	54.0
Mw #3	1158	84.0	0.46	58.9
Mw #4	1255	94.3	0.11	59.3
Mw #5	1170	107.2	0.41	63.9

S = Flexural Strength, E_b = Initial Tangent Modulus in Bending

CONCLUSIONS

Twelve-ply uni-directional laminates were made from glass fiber and vinyl ester/vinyl toluene resin mixture. The heating patterns for these laminates in individual resonant modes were evaluated. It was found that the microwave cavity is not very sensitive to small variations in the orientation of the same sample in the cavity. The heating patterns in the different individual modes were also evaluated for different samples prepared as similarly as possible in their physical and chemical specification. Large differences were found in the heating patterns between these samples. This is believed to be due to the manual lay-up of the mold, which introduces significant differences in the amount of material used for the mold and the physical specifications of the mold. Further experiments are necessary to evaluate the sensitivity of the cavity to small changes in the physical specifications of the sample and the mold. Finding the heating patterns of the laminates without any mold lay-up would be a good beginning.

A pre-consolidation method was developed to make very low void fraction laminates without the application of pressure during curing. This method was used to process laminates in a thermal oven, and in a microwave cavity using mode-switching. The mechanical properties of the thermal cured laminates were compared with those of microwave cured laminates. It was shown that vinyl ester/ glass fiber laminates can be cured in a microwave cavity with properties equivalent to those of a thermally cured laminate.

REFERENCES

1. J. Wei, J.D. Delong, and M.C. Hawley, Proceeding of the American Society for Composites, 5th tech. Con., 239, June 1990.
2. J. Wei, Y. Chang, B. Thomas, and M.C. Hawley, ICCM/VII Conf., Honolulu, Hawaii, July 1991.
3. J.D. Delong, J. Jow, and M.C. Hawley, 2nd Topical Conf. on Emerg. Tech. in Mat., Am. Inst. of Chem. Eng., San Francisco, CA, 1989 paper 220c.
4. Ursula Hottong, Jianghua Wei, Ramakrishna Dhulipala and Martin C. Hawley, in Microwaves: Theory and Application in Materials processing. Ceramic Transactions, vol. 21, ACerS Inc. p587, 1991.
5. R. Agrawal, and L.T. Drzal J.Adhesion, 29, 63, 1989.
6. Valerie Adegbite and Martin C. Hawley, presentation at the 1992 MRS Spring Meeting, San Francisco, CA, 1992.
7. Y.R. Kim, S.P. McCarthy and J. P. Fanucci, Proceedings of the Society of Plastics Engineers 48th Annual Technical Conference, 1990, pp 1252-1255.
8. Timothy G. Gutowski, SAMPE Quarterly, Volume **16**, No. 4, July 1985, pp 58-64.
9. T. G. Gutowski, J. Kingery, and D. Boucher, Proceedings of the Society of Plastics Engineers 44th Annual Technical Conference, May 1986, pp 1316-1320.
10. T. G. Gutowski, Z. Cai, J. Kingery, and S. J. Wineman, SAMPE Quarterly, Volume **17**, No. 4, July 1986, pp 54-58.
11. T. G. Gutowski, Z. Cai, S. Bauer, D. Boucher, J. Kingery, and S. J. Wineman, Journal of Composite Materials, Vol **21(6)**, 650-669, (1987).
12. ASTM Standards and literature references for COMPOSITE MATERIALS, First Edition, ASTM, Page 360, 1987.
13. J. Jow, M.C. Hawley, M. Finzel and T.Kern, Polymer Engineering and Science, vol 28, no. 22, November 1988.
14. M.C. Waterbury and L.T. Drzal, J. Reinforced Plastics and Composites, 8, 627-636, (1989).
15. L. A. Fellows and M. C. Hawley in Composites: Design, Manufacture and Application, edited by S. Tsai and G. Springer, (Eigth International Conference on Composite Materials, Honolulu, Hawaii, July 15-19, 1991) presentation 12-E.

MODELING AND CONTROLLING DURING MICROWAVE AND THERMAL PROCESSING OF COMPOSITES

Jianghua Wei, Ananthapadmanaban Sundaram, and Martin C. Hawley
Department of Chemical Engineering, Michigan State University, E. Lansing, MI 48824

ABSTRACT

A process model was developed and coded in FORTRAN to simulate the temperature and the extent of cure profiles across the thickness of the composite during microwave and thermal processing. A simplified five-parameter microwave power absorption model was developed for microwave processing. The computer code was used to predict the behavior of Hercules AS4/3501-6 prepreg during thermal processing using various laminate thickness and boundary conditions. The simulation results show that 4.0 cm is the thickest composite that can be processed with the manufacture's cure cycle. The advantages of microwave processing over thermal processing are fully demonstrated.

INTRODUCTION

Optimum material properties and reliable processing parameters are desired in the processing of high performance thermoset composites. This is usually accomplished through extensive experimental testing of various processing conditions. However, the experimental result for one geometry is not applicable to others. This procedure of locating the optimum processing conditions is very costly. A fast and cost effective way to find the optimum processing conditions for various processing requirements is through simulation using a processing model. A complete model should include the chemical, physical, and mechanical behaviors of the composite during processing. In this study, a process model for prediction of the chemical behavior and the temperature/time/position profiles during both microwave and thermal processing of Hercules AS4/3501-6 were developed. This model consists of kinetic and energy balance submodels. The kinetic submodel describes the reaction rates and polymerization exotherm during processing. There are two parts to the energy submodel, the power absorbed from microwave radiation and the exotherm released by chemical reaction.

PROBLEM DESCRIPTION

The microwave and thermal processing of a composite laminate is described as follows. A laminate is exposed to a thermal and/or microwave environment as shown in Figure 1. The edges of the lamina are surrounded by adhesive cork dam to prevent resin flow out from the sides and the whole setup is sealed in a vacuum bag. The top is exposed to the environment directly and bottom is covered by bleeder cloth and the support plate, either a steel plate in the case of thermal processing or a Teflon plate, which is transparent to the microwave field, in the case of microwave processing. Heat flow by thermal radiation is neglected because the temperature of the system is lower than 400°C. In this study, a one dimensional model was developed for the prediction of distributions of the extent of cure (α) and temperature (T) across the thickness of the composite as function of time (t).

Figure 1. Composite configuration for processing

From an energy balance, the one dimensional heat transfer equation applies:

$$\rho C_p \frac{\partial T}{\partial t} = \frac{\partial}{\partial z}\left(K \frac{\partial T}{\partial z}\right) + \rho \dot{H} + P_m \tag{1}$$

with temperature boundary conditions

$$\begin{aligned} \frac{\partial T(z,t)}{\partial z}\Big|_{z=0} &= \left(\frac{h}{K}\right)_{eff}(T_a(t) - T(L,t)) \\ \frac{\partial T(z,t)}{\partial z}\Big|_{z=L} &= 0 \end{aligned} \tag{2}$$

and the initial condition

$$T(z,0) = T_i(z) \tag{3}$$

where ρ and C_p are the density and specific heat of the composite,
K is the thermal conductivity of composite in the thickness direction,

$\dot{H}$ is the rate of heat generation per unit weight by the chemical reactions,

P_m is the rate of heat absorbed per unit volume from the microwave source,

$T_i(z)$ is the initial temperature of the laminate, and

$T_a(t)$ and $T(s,t)$ are the environment and the composite surface temperatures respectively.

For Hercules AS4/3501-6 composite[1], $K=4.457\times10^{-3}$ (W/cm/°K), $C_p=0.942$ (J/g/°K), and $\rho=1.52$ (g/cm^3).

REACTION KINETICS

To model the curing of Hercules 3501-6 resin, the following kinetics expression was used[2].

$$\frac{d\alpha}{dt} = \begin{cases} (k_1 + k_2\alpha)(1-\alpha)(B-\alpha) & \alpha < \alpha_{gel} \\ k_3(1-\alpha) & \alpha > \alpha_{gel} \end{cases} \tag{4}$$

where k_1, k_2, and k_3 are the reaction rate constants, B is the ratio of initial hardener equivalents to epoxide equivalents, and α_{gel} is the extent of cure at the gelation point. The rate of heat released by the polymerization reaction is

$$\dot{H} = \frac{d\alpha}{dt} H_R \tag{5}$$

where $\dot{H}$ is the rate of heat generated by polymerization reaction and H_R is the total heat of reaction generated during cure. The extent of cure during processing can be calculated by

$$\alpha = \int_0^t \left(\frac{d\alpha}{dt}\right) dt \tag{6}$$

For Hercules AS4/3501-6 prepreg, H_R=175.2 J/g as determined by DSC experimentally. The kinetics parameters used in this study are listed in Table 1.

Table 1. Kinetics parameters for Hercules 3501-6 resin[2]

α_{gel}=0.3
B=0.47
k_1=2.101x10^9 exp(-8.07x10^4/(RT))
k_2=-2.014x10^9 exp(-7.78x10^4/(RT))
k_3=1.960x10^5 exp(-5.66x10^4/(RT))
R=8.314 (J/mol/K).

MICROWAVE POWER ABSORPTION

The electric field distribution inside an empty cavity can be calculated using Maxwell's equations and the proper boundary conditions. For a cavity loaded with small objects, the electric field strength inside the object can be approximated by perturbation theory if the resonant frequency of the cavity is only perturbed by a few percent[3]. However, the electromagnetic wave inside a cavity loaded with a large anisotropic composite, specifically a graphite fiber reinforced composite, is rather complicated. Figure 2 shows the interactions between composite and the electromagnetic standing waves inside the cavity. As a first approximation, the incident waves on each composite surface are considered to consist of an infinite number of linearly polarized transverse electromagnetic (TEM) waves. Since the incident TEM waves from the four sides are perpendicular to the incident TEM waves from the top and bottom, the power dissipated in the composite P_c can be decomposed into two terms, P_1 and P_2. P_1 is the power dissipated due to the incident TEM waves from the sides while P_2 is the power dissipated due to the incident TEM waves from the top, $\vec{E}_T$, and the bottom, $\vec{E}_B$. To further simplify the problem, P_1 is considered to be constant. Therefore, we only need to consider the propagation behavior of the incident TEM waves from the top and bottom. Once $\vec{E}$ is known through the thickness, the power dissipation due to TEM waves from the top and bottom can be calculated from Poynting's theorem[3].

$$P_2 = \frac{1}{2}\omega\epsilon_o \vec{E}\cdot\vec{\vec{\epsilon}}''_e\cdot\vec{E}^* \tag{7}$$

where $\vec{E}$ and $\vec{E}^*$ are the electric field vector and its conjugate and $\vec{\vec{\epsilon}}''_e$ is the effective dyadic dielectric loss factor of the composite. For AS4/3501-6 composite, the principal directions of the effective loss factor tensor of the composite can be assumed to be along and perpendicular to the fiber direction and along the composite thickness direction. Therefore, the microwave power dissipated per unit volume per unit time in a unidirectional composite is

$$P_c = P_1 + \epsilon_0 \pi f(\epsilon''_x \mid E_x(z) \mid^2 + \epsilon''_y \mid E_y(z) \mid^2) \tag{8}$$

where x, y are along and perpendicular to the fiber direction of the of composite. The effective electric field in z direction can be calculated from the electric field at the top and bottom surface plies and the propagating constants γ_x and γ_y.

$$\begin{aligned} E_x(z) &= (E_x)_B \exp(-\gamma_x z) + (E_x)_T \exp(\gamma_x(z-L)) \\ E_y(z) &= (E_y)_B \exp(-\gamma_y z) + (E_y)_T \exp(\gamma_y(z-L)) \end{aligned} \tag{9}$$

where $(E_x)_T$ and $(E_x)_B$ are the x components of effective electric field at the top and bottom surfaces propagating into the composite. $(E_y)_T$ and $(E_y)_B$ are the y components of effective electric field at the top and bottom surfaces propagating into the composite. The propagation constants are $\gamma_x = j\omega\sqrt{\mu_0\epsilon_0\epsilon_x^*}$ and $\gamma_y = j\omega\sqrt{\mu_0\epsilon_0\epsilon_y^*}$. The effective loss factors of AS4/3501-6 composite are available in the literature [4].

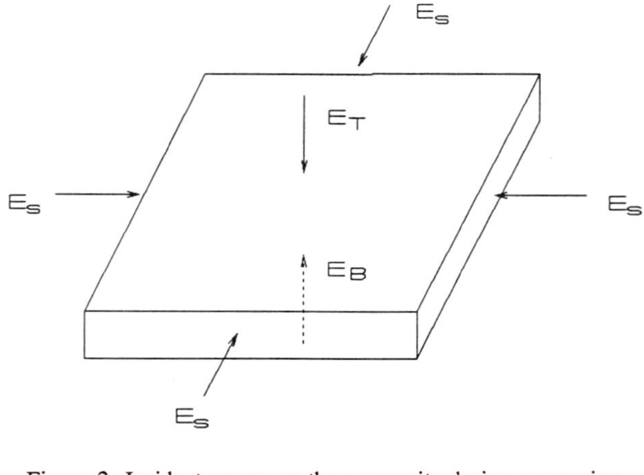

Figure 2. Incident waves on the composite during processing

Once the five parameters, P_1, $(E_x)_T$, $(E_x)_B$, $(E_y)_T$, and $(E_y)_B$ are known, P_c across the composite can be easily calculated. All five parameters are functions of input power and the electromagnetic mode for a given composite. The temperature profiles obtained from microwave heating of a fully cured composite using a given resonant mode can be used to generate the five parameters.

SIMULATION

To simulate the temperature and extent of cure profiles across the thickness of the composite as function of time, Lee's implicit method[5] was used to code the non-linear parabolic equation, Equation (1), and the closed, multi-step, Fourth-order Milne method[6] was used to code the first order ordinary differential equation, Equation (5). The use of Lee's algorithm gives a truncation error similar to that of the Crank-Nicolson method but leads to a higher computational velocity[1]. The multi-step method was chosen because it requires considerably less computation than the one-step method for results of comparable accuracy. The closed method is preferred because its local truncation error is considerably smaller than that of the open method[6]. However, a single-step Runge-Kutta method was used to generate the values of the first four temperature data.

In the thermal processing, P_m is equal to 0. The effective heat transfer coefficient at the top surface was assumed to be infinite and the Dirichlet boundary condition was used. The temperature profiles as a function of time at the midplane and the bottom surface were simulated during processing of 1, 2, 3, and 4 cm thick AS4/3501-6 laminate using the manufacturer's suggested cure cycle. The cure cycle was a ramp from room temperature(25°C) to 116°C in 40 minutes, then hold at 116°C for 60 minutes, then ramp to 177°C in 30 minutes, then hold at 177°C for 120 minutes, and cool down to 95°C in 50 minutes. Figure 3 shows the temperature profiles at the bottom surface of the composite for various thicknesses.

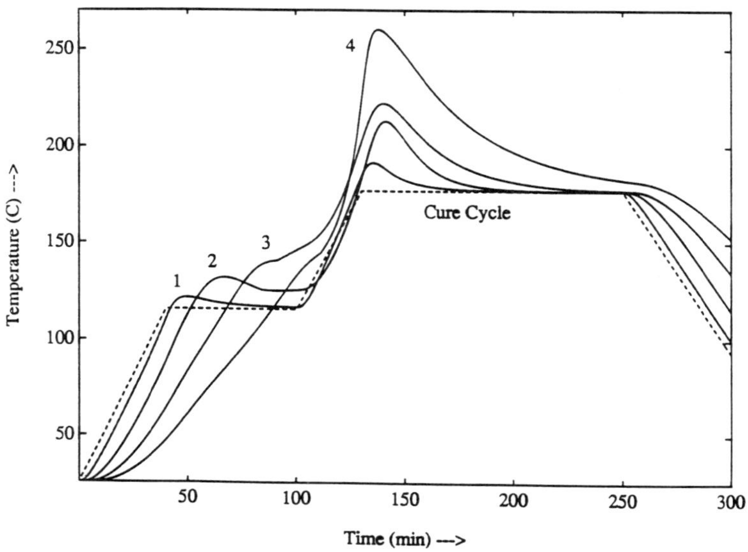

Figure 3. Temperature/time profiles for thermal processing of AS4/3501-6 laminate

In Figure 3, the numbers beside the curves represent the thickness of the laminate in centimeters used in the simulation. As shown in Figure 3, two temperature excursions occurred for all cases during processing. The time of the first excursion is a strong function of laminate thickness and the value increases with the laminate thickness. Specifically, the time needed for first excursion to occur is 45, 60, 90, and 105 minutes for 1, 2, 3, and 4 cm thick laminates. However, the location of the second excursion is relatively independent of the laminate thickness in the range studied. For the 1 cm thick laminate, the second excursion occurred at 140 minutes while for the other three thickness, 2, 3, and 4 cm, the second excursion occurred at 145 minutes. As the laminate thickness increases, the second excursion may start before completion of the first excursion. The two excursions may become indistinguishable for thick laminates. As shown in Figure 3, the first excursion peak can hardly be seen for the 4 cm thick laminate. The magnitudes of both excursion increase with the laminate thickness. The maximum temperature at the bottom surface reached 260°C for the 4 cm thick laminate. As the thermal degradation temperature of Hercules AS4/3501-6 prepreg is 290°C as determined by thermogravimetric analysis (TGA), the maximum thickness of laminate that can be processed using the manufacture's cycle is 4 cm if the operation margin is set to 30°C.

In microwave processing, a free convection boundary condition is used at the top surface of the composite. The measured effective heat transfer coefficients are $(h/K)_{eff}=0.014$ [1/cm] if $T-T_a \leq 27°C$, $(h/K)_{eff}=0.088$ [1/cm] if $43°C \geq T-T_a > 27°C$, and $(h/K)_{eff}=0.157$ [1/cm] if $T-T_a \geq 43°C$. Figure 4 shows the temperature profiles at the top surface, midplane, and bottom surface of a 2 cm thick composite during microwave processing at a cavity length of 16.03 cm with 70 W input power. The five parameters of the microwave power absorption model for the resonant mode at cavity length of 16.03 and input power of 70 W are 0.331 W/m^3, 0.0637 V/m, 0.898 V/m, 33.0 V/m, and 22.6 V/m for P_1, $(E_x)_T$, $(E_x)_B$, $(E_y)_T$, and $(E_y)_B$ respectively. The control temperature was set to 240°C during the simulation.

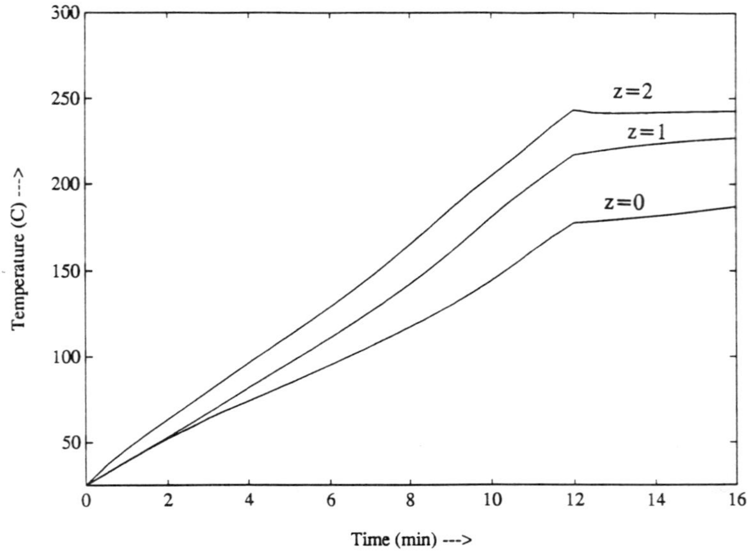

Figure 4. Temperature/time profiles for microwave processing of AS4/3501-6 laminate

In Figure 4, the l=2, l=1, and l=0 curves represent the temperature profile at the bottom, midplane, and top surface of the composite respectively. The laminate can be heated directly to the desired temperature without any risk of temperature excursion inside the composite. At an input power of 70W, 12 minutes is required to heat the laminate up to 240°C. The heating time can be shortened with a higher power level. Since the temperature excursion can be eliminated, the processing temperature can be raised to 260°C if the operation margin is set to 30°C. Comparing Figure 4 to Figure 3, the advantages of microwave processing over thermal processing are; 1) The temperature excursion during microwave processing is eliminated. 2) High temperature can be used during microwave processing. 3) The ideal heating profile is realized during microwave processing, that is directly heating the laminate to the desired temperature as fast as possible then maintaining the laminate at the desired temperature.

CONCLUSIONS

A one dimensional process model was developed for both microwave and thermal processing. A five-parameter microwave power absorption model was derived to calculate the microwave power dissipated inside the composite. The model was coded numerically using Lee's algorithm and Milne's method. The computer code was used to simulate the temperature profile across the composite thickness during microwave and thermal processing. With infinite and zero values for the effective heat transfer coefficient at the top and bottom surfaces, the thickest AS4/3501-6 composite that can be processed is 4.0 cm with the manufacture's cure cycle without the risk of thermal degradation. The temperature excursion is eliminated during microwave processing. The ideal heating profile is realized during microwave processing, that is directly heating the laminate to the desired temperature as fast as possible then maintaining the laminate at the desired temperature.

REFERENCES

1. T.A. Bogetti and J.W. Gillespie, Jr., J. Comp. Mat., 25, 239 (1991).
2. W.I. Lee, A.C. Loos and G.S. Springer, J. Comp. Mater., 16, 510 (1982).
3. R.E. Collin, "Field Theory of Guided Waves", 2nd ed., IEEE Press, (1991).
4. W.I. Lee, and G.S. Springer, J. Comp. Mater. 18, 387 (1984).
5. G. D. Smith, Numerical Solutions of Partial Differential Equations: Finite Difference Methods, 2nd ed., Oxford University Press, Oxford (1978).
6. B. Carnahan, H.A. Luther, and J.O. Wilkes, "Applied Numerical Methods", Robert E. Krieger Publishing Company, Inc. (1990).

MICROWAVE PROCESSING OF FOAM MATERIALS

W. VAN LOOCK
University of Gent, Laboratory of Electromagnetism and Acoustics,
Sint-Pietersnieuwstraat 41, B-9000 Gent, Belgium

ABSTRACT

The distinct phases in the production process of a foam material are foaming, curing and drying. Cellular materials are insulators to conventional heat. This paper discusses the effectiveness of microwave or high frequency energy for processing a layer of foam material on for example a conveyor belt. The emphasis is on the thermal and drying time constants and on the specific energy consumption. It is demonstrated that a limit is set on time gain when using microwaves.

A case study for applying microwaves in an industrial production line of foam rubber for mattresses according to the above principles is presented.

INTRODUCTION

This paper reports on a feasibility study for the application of microwave power in the production process of latex foam. Latex is foamed to obtain a porous or sponge like product for example for mattresses. The process has several distinct phases : foaming, curing, drying and washing cycles for removing chemicals and final drying to 0% moisture content (m.c.). Large scale hot air, steam heaters and conventional dryers are typical for these processings.

Foam materials are generally thermal insulators and the volumetric heating properties of electromagnetic energy like microwaves and high frequencies could be beneficial. In fact, EME is capable to assist in all phases during the production of latex foam.

Many experiments have been performed on microwave assisted foaming and curing. Thermal runaway seems to be unavoidable during the foaming even at very low power densities which require a negligible investment. Apparently the material is very sensitive to non uniformities of the applied microwave fields, also during the curing phases.

After the curing all the chemicals are washed out. The pores are mostly open and relatively large ($\phi > 1\mu$) so the water is almost free and can be removed mechanically at low costs down to about 30% m.c. at less than .5 kWh per kg evacuated water. The final conventional drying to 0% takes typically 5 to 8 hrs which is the bottle neck in the production process. Also the measured specific energy consumption (s.e.c.) is up 10 kWh per kg evacuated water.

RELEVANT PROPERTIES

A production line for typical mattresses with dimensions of 2m x 0.8m and 10cm has been selected for this study. The dry foam has a specific mass of 70 kg/m^3 and the production line has a capacity of 5,000 kg per day which means that roughly 70 kg water has to be evacuated per day. The production process is on the average 20 mattresses per hour.

The foam is homogeneous but at regular distances transverse holes of 1 cm diameter exist, for mechanical handling purposes.

The microwave permittivity, i.e. the dielectric constant ε'_r and the loss factor ε''_r have been determined, see Table 1.

TABLE 1 : Averaged permittivities

PERMITTIVITY COMPONENT		ε'_r	ε''_r
latex	T = 22°C	2.6	0.009
	90°C	2.8	0.017
latex foam	m.c. 0%	1.1	0.001
(at 22°C)	m.c. 35%	1.6	0.0525

CONVENTIONAL VERSUS ELECTROMAGNETIC DRYING

The drying problems are well documented in literature [1]. Fig. 1 is a schematic drawing of the conventional system for the final drying in this case study. The mattresses are put on shelves consisting of fine meshes and the cart is pushed into the drying chamber for 5 to 8 hrs. The air at the intake has on the average a temperature Ta = 10°C and a relative humidity R.H. = 10%. A heat exchanger brings the dryer input temperature Ti up to 80°C. The air circulation and recycling is performed with respectively a fan and a valve.

The model for analyses is well known and a Mollier graph is used which is straight forward. This reveals that to extract 1 kg water from the material requires more than 5,000 kJ and it is not possible to go below 2,600 kJ, or 0.7 kWh, the latent heat of water. The s.e.c. decreases for higher Ti, when the recycling percentage increases and for lower output air temperature and its R.H.

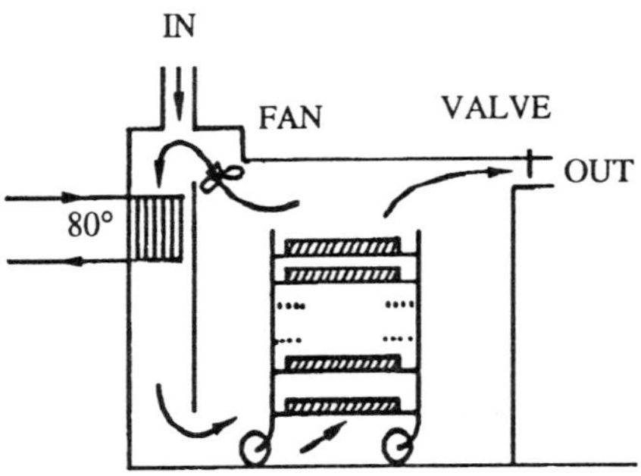

Fig. 1 : Drying chamber

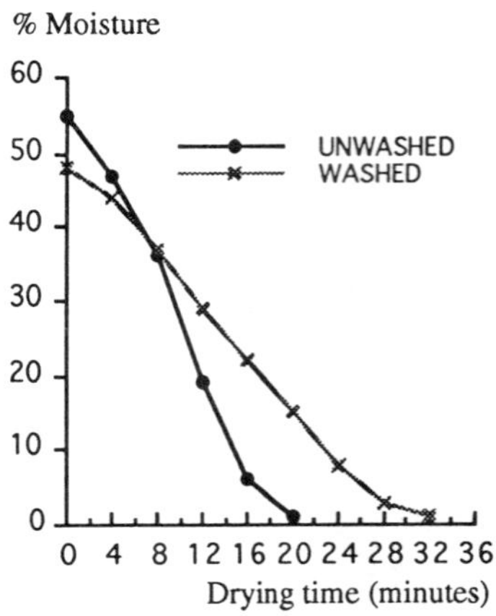

Fig. 2 : Drying curves

Conventional dryers have at least a s.e.c. of 1.4 to 2 kWh/kg. Practical dryers must include the energy for the fans, the controls, etc. Poor designs and old systems and in particular when extended standby period exists, then the s.e.c. can be 5 to 10 kWh/kg H_2O.

The minimum microwave or high frequency energy for drying would be 0.7 kWh/kg. The overall generator efficiencies are roughly 50% increasing the minimum necessary s.e.c. up to 1.4 kWh/kg.

The advantages of a microwave dryer have to overwhelm the simplicity of the Fig. 1 drying chamber. These are a considerable time gain and continuous drying. Conventional continuous drying would impose a tunnel dryer miles in length.

MICROWAVE EXPERIMENTS

Microwave drying experiments have been performed in a standard 1 kW laboratory oven using hot air and microwave on/off cyclings. The sample dimensions were 10cm x 10cm x 10cm and 25cm x 25cm x 10cm for the latex foam having the transverse holes.

Typical drying curves are shown in Fig. 2. The samples are easily dryed within half an hour. The samples containing chemicals dry faster but internal arcing and burnings degrade the foam quality. The cleared foam supports higher microwave fields. Up to 75 kV/m, no microscopic arcing traces could be observed in the dryed samples.

FEASIBILITY OF MICROWAVE PROCESSING

The microwave experiments have demonstrated that end drying of latex foam which is cleared of chemicals is feasible. The technical and economical aspects of microwave processings are well documented [2] and [3]. A considerable but limited time gain is possible.

The power density deposition inside a material due to an applied internal electric field E_i of frequency f is given by the well known equation

$$P\ (W/m^3) = 0.056\ f\ (GHz)\ \varepsilon_r''\ [E_i\ (V/m)]^2 \tag{1}$$

The limit set on E_i determines the maximum power density. The corresponding temperature rise in a time Δt is

$$\frac{\Delta T}{\Delta t}\ (K/s) = \frac{P\ (W/m^3)}{C\ (J/kg.K)\ .\ \rho\ (kg/m^3)} \tag{2}$$

The microwave power can be assessed with (1) and (2) for a given material throughput and water evacuation per hour.

The latex foam is subject to a temperature x time effect. The 80°C is the maximum temperature for the long residence period of 8 hrs in the chambers. The temperature may increase for shorter periods, and 105°C is the limit beyond which long term degradation of the foam takes place. So the internal temperature rise must be limited which calls for a better assessment than with Eq. (2).

A uniform volumetric heating will rise the temperature inside the slab and provided the surfaces are kept at a constant temperature then the ΔT is given by, ref. [4] p. 130 and Eq (1) :

$$\Delta T = \frac{0.028\, f\,(\mathrm{GHz})\, \varepsilon_2^{*}\, [E_i\,(\mathrm{V/m})]^2\, d^2(\mathrm{m})}{K\,(\mathrm{W/mK})} \qquad (3)$$

with K the thermal conductivity. Eq. (3) sets another limit on the power and the resulting time gain.

A microwave tunnel oven with proper air flow has been proposed to replace the drying chambers and to provide continuous drying from a m.c. of 30% down to 0%, and an overall s.e.c. of 1.6 to 2 kWh/kg H_2O.

However it has been decided in favour of high frequency energy which is less expensive. The dimensions of the mattresses are large compared with the wavelength of 12cm at 2.45 GHz. Therefore it is difficult to obtain a uniform field strength throughout the material. Furthermore the foam thickness larger than the cut-off condition in a parallel plate system increases the costs for proper choking in a tunnel system.

An applicator at 27 MHz has a more uniform field distribution but higher field strengths are necessary for the same power deposition as with microwaves, see Eq. (1).

The time gain is related to the maximum permissible power density, see (2). The electric field must increase for having the same time gain as with microwaves.

A typical HF tunnel has a power rating of 100 kW and dries 20 mattresses per hour. They operate with a reduced safety factor for the maximum electric field which may cause operating problems.

CONCLUSIONS

End drying of foam materials is feasible using microwave power. Considerable time gain and also energy savings are possible. The time gain is limited by the maximum electric field and the maximum internal temperature rise for retaining product quality.

This work has been performed as part of a contract on energy savings for the Commission of the European Communities [5].

High frequencies and microwaves are now becoming common place for drying, preheating and heating mainly in the food industry, the rubber industry and for textiles. The decision to use high frequencies or microwaves relies generally on many other advantages than energy savings. These advantages have been time gain and continuous processing in this case study.

REFERENCES

1. C. Strumilo, Drying : Principles, Applications and Design, Gordon and Breack Science Publishers, New-York, 1983.
2. A.C. Metaxas and R.J. Meredith, Industrial Microwave Heating, Peter Peregrinus Ltd, London, 1983.
3. M.R. Sanio, A procedure for estimating capital and operating costs of dielectric heating equipment, Center for Energy Studies, University of Texas at Austin, Report TSDD 89-011, Aug. 1989.
4. H.S. Carslaw and J.C. Jaeger, Conduction of Heat in Solids, Oxford at the Clarendon Press, 1959.
5. W. Van Loock, Microwave applications within the European Communities, Proc. 20th Annual Microwave Symposium, Chicago, U.S.A., pp. 78-79, Aug. 1985.

PART VIII

Microwave Processing of Hazardous Wastes

DIELECTRIC PROCESSING OF HAZARDOUS MATERIALS - PRESENT AND FUTURE OPPORTUNITIES

STEVEN J. ODA, Ontario Hydro's Research Division, 800 Kipling Ave., Toronto, Ontario, Canada M8Z 5S4

Dielectric heating technologies (microwave and radio-frequency, RF) bring unique capabilities to the processing of hazardous wastes. Potential opportunities for using these techniques can be described by classifying and matching waste streams to the particular features of dielectric heating. Microwave and RF applications for processing hazardous waste are at various stages of development and include: solvent recovery, regeneration of activated carbon, pyrolysis of rubber tires, soil remediation, and disinfection of hospital wastes. Other solutions to environmental problems can include the replacement of conventional fossil-fired processes (eg, RF-assisted post-baking of biscuits) and microwave heating of catalyst substrate material for automotive emission control. Examples of dielectric heating systems that are processing hazardous waste are used to illustrate the benefits of these technologies and to characterize future opportunities.

INTRODUCTION

Quantities of untreated hazardous wastes are growing at alarming rates. In order to combat this growth a search for more effective and energy efficient technologies that can help to minimize or eliminate the impact of these wastes is underway. Dielectric heating is among the technologies that are beginning to provide some solutions to these problems. By providing selective and volumetric heating compared to surface heating associated with more traditional techniques, these dielectric heating techniques are becoming more widespread particularly in the areas of infectious waste, soil decontamination, solvent recovery and microwave 'catalysis'. This paper reviews some of the microwave and RF applications that have been commercialized to process waste materials. By understanding the reasons for adoption of these existing systems, it becomes easier to target opportunities for future applications in hazardous waste management. Table 1 illustrates the existing and potential applications divided according to common waste streams such as: contaminated soil, infectious waste, petroleum and coal, gaseous air pollutants, chlorinated hydrocarbons (sludges and solvent recovery), activated carbon, plastics and elastomers, sewage sludges and wastewater, and nuclear wastes. A list of references is included for additional information.

While there is evidence of a growing interest to process environmental wastes, only a few dielectric heating systems appear to have been commercialized. This observation is partly due to commercial interests who are keeping new developments under 'wraps' to capitalize on any marketing advantage. Dielectric heating applications that have been commercialized tend to be in niche application areas where the benefits of dielectric heating (ie, selective heating, volumetric heating of a value-added material, 'can't-be-done-any-other-way') can be fully exploited to produce a favourable economic return. As yet, there does not appear to be any large application area for processing waste using dielectric heating.

TABLE 1
DIELECTRIC PROCESSING OF HAZARDOUS WASTES
(Microwave - μW; Radio-frequency - RF)

Waste Material	Operation	Industry	Technology Status Commercial; Pilot-plant Laboratory	References
infectious medical waste	RF or μW disinfection; μW pyrolysis	hospital	C, P, L	[1,3,4]
solvent-laden waste, sludges	μW/RF solvent recovery; RF-assisted dewatering	industries generating waste solvent, still 'bottoms', sludges	C, P, L	[6,7,9,22,32]
used rubber tires	μW devulcanization, pyrolysis	automotive	C, P	[10,11,12,13]
contaminated soils	RF or μW decontamination	spill cleanup	P, L	[16,17,18]
petroleum products / coal	μW coal desulphurization; oil demulsification & recovery	petrochemical & utility industries	L	[10,19,20]
activated carbon	μW regeneration	industries generating spent carbon	L	[27,28]
gases	μW removal of SO_2, NO_x; μW hydrocracking	petrochemical	L	[8,29,30,31,32,33,34]
plastics/elastomers	μW polymer recycling	recycling programs	L	[10,11,12,14]
nuclear waste	μW volume reduction; pyrolysis	nuclear power industry	P, L	[18,34,35,36]
sewage/waste water	μW disinfection; μW/RF sludge dewatering	water authorities	P,L	[21,22,23,24,25,26]

Infectious waste

Existing Applications
Two examples of recently developed commercial systems involve processing of infectious medical waste [1,2,3]. The first example from Asea Brown Boveri (ABB) Sanitec uses microwave heating to disinfect hospital waste [1,2]. According to ABB, the system should reduce hospital waste disposal costs by more than 50% compared to the conventional incineration and off-site treatment alternatives. In one example cited by ABB, Forsyth Memorial Hospital (Winston-Salem, NC) expects that annual medical waste disposal costs will drop from $270,000 to $60,000. The Sanitec system will effectively treat all wastes commonly defined in the infectious or regulated medical waste stream including sharps, blood and body fluids, microbiological wastes, cultures and pathological wastes. For aesthetic reasons, some facilities or regulating agencies do not treat gross anatomical wastes in the unit and the unit is not suited for radioactive, hazardous or antineoplastic wastes.

The system has been designed to treat a manufacturer's guaranteed minimum of 250 kg/h based on a waste density of 176 kg/m^3 with moisture content of 10% by weight, metallic content less than 1%, the size of any metallic piece should be less than 0.2 kg; and, no single object larger than a 50 cm cube. Waste streams exceeding these limits would have a reduced throughput. In operation, users report actual capacities in the 300-320 kg range with one commercial user achieving between 360 and 500 kg/h by employing a recently developed shredder technology.

After shredding waste such as gloves, masks and bandages, the process sprays steam on the waste and then applies microwaves to heat the waste to between 97 and 100°C for 30 minutes. Six 1.2 kW air-cooled microwave sources operating at 2450 MHz are used to attain temperature. One model with a throughput of 250 kg/h is currently commercially available in North America, with a second unit with a 100 kg/h capacity expected by the end of 1992. The equipment includes a feed hopper that is evacuated through a 2-stage filter to control airborne pathogens. Exposure time is controlled by microprocessor and adjusted by the speed of the feed screw. In Europe the system has been in commercial use for the past 7 years with several installations. In North America there are units operating in New York, New Jersey, Connecticut, North Carolina, Tennessee, Florida and California.

In the second example, RF heating is used to treat the infectious waste [3]. The RF "Stericycle" process is a two-step process, termed Electro Thermal Deactivation (ETD). The first step involves passing the infectious waste, already packaged in sealed cardboard boxes (61 x 61 x 91 cm) into a processing room. The wastes are shredded into fragments (~2.5 cm diameter) in the processing room which is under negative pressure. The fragments are wetted and then compacted to a density of about 320 kg/m^3. The compacted waste is then fed in 90 kg charges into insulated tubs which convey the waste through a RF heater. In the second step the waste-filled tubs are heated in the RF unit to temperatures between 90-100°C. The insulated tubs hold the waste at these temperatures for 2-3 hours which is sufficient to disinfect the wastes and allow them to be handled. The RF equipment has a 1 metre wide conveyor to move the tubs through the RF electrode system and can accommodate package heights varying between 0.9 to 1.1 metres above the conveyor. Normal power output is 80 kW at a plate current of 11 amps. Typical throughput for this size of unit is about 1600 kg/h at a frequency of 18 MHz. Existing RF systems can range up to 150 kW in output to process 2700 to 3180 kg/h. An additional feature of the system is that, after waste disinfection, the plastic component of the waste is separated and recycled.

Potential Applications

A microwave incinerator ("burning processor") has been developed by Matsushita of Japan for reducing food wastes from commercial sources (eg, restaurants, hospitals, high-rise buildings) to ash "at the place where it is generated" [4]. The compact system uses a 900-watt magnetron and is divided into two chambers: a primary combustion chamber that uses microwave heating to dry and decompose the wastes; and, a secondary combustion chamber to oxidize and burn the off-gases generated during primary combustion. The primary chamber contains a microwave susceptor, like a silicon carbide plate, on the floor of the chamber. Initially microwaves are used to dry the waste. After drying is complete, the microwave susceptor begins to heat. The susceptor further decomposes the waste and generates gas. The gas is passed into the secondary chamber which contains a platinum-type catalyst to oxidize the off-gases before burning. It is important to remove moisture from the off-gas and to maintain a secondary chamber temperature of 500°C in order to maintain catalyst activity (ie, avoid poisoning) in the secondary chamber. According to the authors, the unit can process one kilogram of typical food waste in about four hours. Average energy consumption was 4.8 kWh per kg of waste. While the operating costs (1987$) were about 1.4 times that of municipal incineration, heat recovery from the processor (for water or space heating) would improve the economics. There was no mention of the effects of metallic objects in the processor. In this application the precise control that microwave heating provides makes it easier to manage the decomposition of the waste in a safe and reliable way. Similar microwave furnaces have been developed by Meteka of Austria and Atomic Energy of Canada Limited's Chalk River Nuclear Labs [5]. The latter design processes hazardous liquids as well as solid wastes and uses interactive heating elements (SiC) to absorb energy from four 750 watt magnetrons. With appropriate thermal insulation the unit is claimed to reach inside wall temperatures of 1500°C and process small quantities of waste (500 ml per hour) from hospitals, biological laboratories and special waste facilities.

In the previous examples, the cost of microwave/RF heating is significantly lower than conventional incineration. Because dielectric heating can selectively heat water in the waste material and, volumetrically heat large quantities, the application can be performed more effectively and efficiently than the conventional alternative. Of course, it is the favourable economic comparison between conventional (ie, incineration) and dielectric alternatives that has led to commercialization of these applications.

Chlorinated Hydrocarbons (Sludge / Solvent)

Existing Applications

Manufacturing companies are being required to remove or reduce their use of chlorinated solvents and any generated sludges. Various initiatives are beginning to develop effective options for dealing with these materials.

The use of microwaves to recover solvent from waste polyurethane foam was performed by Ontario Hydro on behalf of an automotive dashboard manufacturer [6]. Modern automotive instrument panels contain a structural polyurethane foam core. The foam injection process generates a hazardous waste mixture of foam and chlorinated cleaning solvent which was sealed in 45-gallon drums. In order to reclaim the solvent and reduce the hazard, the manufacturer was using electric-resistance band heaters applied around the outside of the drums to evaporate the solvent. However, the process was intrinsically very inefficient because the drying foam near the heaters became a good insulator and solvent levels could not be reduced to the 2% level necessary for non-hazardous landfill. Using a specialized applicator built to fit on a 45-gallon drum lid, laboratory and field tests showed that microwaves could preferentially heat and distill off the solvent without overheating the foam.

Removal rates of about 8 kg of solvent/kWh (microwave energy) were measured. Treatment of a typical 45-gallon drum was carried out in approximately 2 hours with a generator output of 2-3 kW. Final solvent contents ranged between 0.09% at the top of a barrel and 0.25% at the bottom. According to the company's plant personnel, these contents were 10 to 20 times lower than required for safe disposal as a landfill material with subsequent significant savings in avoided conventional waste disposal costs; energy and treatment time.

Potential Applications

Microwave destruction of chlorinated hydrocarbons is being investigated [7,8,9]. Most of the experimentation to date has been conducted in commercial microwave ovens at low power levels (eg, 70 watts). Sample sizes as well have tended to be quite small (0.1 ml spiked with ppm amounts of various chlorinated hydrocarbons including polychlorinated biphenyl (PCB), DDT, DDE, methoxy chlor, and bromotoluene). Exposure times ranged from 60 to 600 seconds. Decomposition efficiency was claimed to be greater than 80% in all tests.

Varma has also examined the destruction of halogenated hydrocarbons using microwaves [9]. This approach used microwaves to selectively heat a catalyst bed. Microwave heating of the substrate results in 'high intensity' oscillating fields at the surface (see Wan) which causes chemical reactions to occur. Claims are made that the microwave process is about 4 times more effective than conventional thermal treatment in the 500 - 600°C temperature range. Also claimed is that 80% detoxification could be attained at a temperature of 400°C compared to incineration at 1200°C in air.

Plastics and Elastomers

Existing Applications

Scrap from the manufacturing of rubber radiator hose has been devulcanized with microwaves [10]. Trimmings from the manufacturing operation yielded up to one ton per day of vulcanized scrap. After this scrap is chopped into 0.6 cm pieces it is heated above its vulcanization temperature until the bonds break and the rubber reverts to a semi-cured state. The reclaimed rubber is then returned for hose manufacture. The scrap is heated with 70 kW of microwave power at 2450 MHz while being tumbled in a cylindrical applicator. The tumbling improves heating uniformity because of the batch-to-batch variation in carbon content. Capacity of the system is about 400 kg/h with a reported payback of a few years.

This commercial application was developed in the 1970's when the energy crisis was demanding electrically based alternatives to fossil-fired processes. Current stockpiles of used tires are an obvious sign that this environmental problem still exists. A number of proponents of microwave/RF technologies are pursuing the processing of used rubber tires (see Potential Applications), although as yet no commercial system appears to have been developed.

Potential Applications

The use of microwave heating to recycle used rubber tires has been studied for some time by various investigators from pyrolysis [11] to devulcanization [12]. Hydro-Quebec undertook a study on the microwave vacuum pyrolysis of scrapped tires but concluded that while the process was technically feasible it was economically impractical on a large scale [13].

In BRC Environmental Ltd's pyrolysis process the rubber is heated in an inert atmosphere using five magnetrons [11]. The microwave applicator (cavity) is equipped with ports for nitrogen purging and for collecting oils and gases evolved during the heating process. This batch mode system can handle only a few kilograms of tire per batch; processing rates, microwave frequency and details of the power supplies are not known. The authors claim that microwave pyrolysis of used tires can yield four components: carbon, steel cord, hydrocarbon oil, and a mixture of hydrogen and methane gases. While the first three products could be sold, the gases could be used to power the system. No evidence to substantiate these claims was made by the authors. A design and economic assessment of a continuous system was also presented by the authors. After shredding the tires, microwave heating would pyrolyse the tires into dry carbon and steel which could then be separated. The gases produced during this process could be recovered and distilled into their components. The authors speculate that this process would yield 40% light hydrocarbon, 36% carbon, 12% steel residue and 12% gas which would power a plant capable of handling 50,000 tonnes of scrap tires and generate a surplus of over 2 million dollars (U.S.) in a ten-year period.

Microwave devulcanization of rubber tires represents another lower temperature option for processing used tires [12]. In this process, devulcanization of the tires occurs under atmospheric conditions in a spherical applicator designed to improve heating uniformity. Once devulcanization takes place the heated rubber is cooled to prevent degradation or combustion. Using 5 kW of microwave power the system was capable of processing a batch of 20 kg of EPDM (ethylene-propylene dimer). The residence time was about 1 hour with temperatures reaching about 350°C. According to the researchers the rubber reclaimed in this process when mixed with 10% virgin rubber had good mechanical properties (tensile strength, elongation). No mention of the fate of the gases evolved during the heating process was given.

Continuous devulcanization was also studied [12]. In this process shredded rubber which was mixed with a partition agent (ie, silica, clay, talc) was subjected to microwave radiation as it was fed through an inclined and rotating glass or ceramic tube (10 cm ID). The partition agent was added to make handling simpler by preventing the particles from sticking together. Exposure time and feed rate were controlled by the rate of tube rotation. According to the reference, devulcanization of EPDM and butyl rubber mixed with the partition agent (100:1 by weight) require from 185 to 210 watt-hours/kg at either 2450 or 915 MHz. Although no experimental results were provided, the authors claim that 95% of the rubber can be recovered as reusable materials without any loss in physical properties or discharge of harsh chemicals into the atmosphere. Throughput and total power output were not specified.

Microwaves to recycle plastics was suggested in the early seventies by Tsutsumi [14]. The process sprays caustic soda on plastics such as polyethylene, polypropylene and polystyrene followed by microwave heating in the presence of solid sodium hydroxide. Non-polar plastics such as polyethylene are doped with 20-30% of carbon to facilitate microwave absorption.

OTHER POTENTIAL APPLICATIONS

Potential new applications for dielectric processing are described in this section. Divided into common waste streams, the status of these applications is mainly at the laboratory or pilot-scale. Many of these applications are described in greater detail in a recent Canadian Electrical Association (CEA) report that provides a useful overview of opportunities for using dielectric heating to process hazardous waste [15].

Contaminated Soils

An Illinois Institute of Technology study investigated the use of RF for processing contaminated soil and sand by implanting RF electrodes into holes in the sand [16]. Contaminants were collected under a surface vapour barrier, then condensed for disposal or incineration on site. In this study soil moisture facilitated the heating and desorption of contaminants. Using a 40 kW RF source (6.78 MHz), a soil volume of 14 m^3 could be processed in 12 days at operating temperatures of 150-160°C. Limitations associated with this technique include the high power requirements of the RF source; the presence of metallic objects in the soils; and, excessive amounts of moisture in the soil. A pilot test study of the RF technique conducted over a 12-day period heated more than 14 m^3 of soil up to 160°C. Sampled at different depths the soil was found to be uniformly decontaminated of waste oils, fuels and solvents. Compared to excavation and landfilling costs (~$700 per m^3), RF treatment costs were estimated to be less than $140 per m^3.

Soil decontamination experiments conducted by Dauermann and Windgasse [17,18] used microwave heating to facilitate steam distillation from a variety of materials including silts and clay, humus top soil, contaminated sludges, soil and tar residue. A 99.0% removal of para-xylene and naphthalene contaminants from sand was reported. Using microwaves, immobilization of chromium in soil was also reported by the same authors [17]. No mechanism for the process was given although it was suggested that the chromium was in a 'different state' after exposure. The same authors investigated the microwave treatment of dioxin-contaminated soil by using a dioxin 'surrogate', 9,10-anthraquinone [18]. Laboratory testing indicated that in-situ microwave treatment of the contaminated soil samples can elevate soil temperatures sufficiently to make the surrogate non-extractable. Spectroscopic analysis indicated that the anthraquinone was 'chemically fixed'.

Soil decontamination using RF heating was also reviewed by Dev [16]. This approach appears to be more favourable than the microwave approach because of the greater penetration depth that RF can provide compared to microwaves. A conceptual design of a proposed RF system placed horizontal electrodes a short distance above the ground. A Faraday-cage wiremesh was placed above these electrodes to contain the RF radiation and a vapour barrier was used to cover the entire area in the design. The authors suggested that ten 2-MW RF sources would be needed to treat one acre in 60 to 90 days at temperatures of 300 to 400°C. Technical difficulties associated with megawatt-sized RF generators may prevent commercialization but the scientific aspects of the study are important.

Petroleum Products & Coal

Microwaves have been applied to separate unwanted pollutants from petroleum products and coal. In a series of field tests, Fang has used microwaves to demulsify oil sludges by lowering emulsion viscosity and effecting separation without the need for chemical additives [19]. Up to 188 barrels of oil/water emulsion were separated into dehydrated oil and clear water by using microwave power sources at 915 MHz and power levels of 25 kW. According to the authors the process was faster and more effective than conventional heating.

Oil recovery from oil sands has been explored by Bosisio [20]. Small samples of oil sand were treated using a microwave source (2.45 GHz) with power levels up to 2.4 kW. Compared to traditional extraction rates of 70%, the microwave process can recover up to 86% of the bitumen in the form of crude oil with properties similar to natural crude.

A number of researchers have been actively investigating the use of microwave for desulphurizing coal [10]. Work at General Electric in the 1980's showed that microwave heating could help to reduce sulphur content by 80% with no loss in calorific value. Subsequently, the US DOE-funded work at GE used microwaves to remove up to 90-95% of the sulphur. In this process coal is mixed with sodium hydroxide and subsequently dried. The coal is then exposed to microwaves in multiple treatments ranging from 25 to 45 seconds in duration. Finally the coal is washed with water, acid and then water again. The results suggest that a specific microwave energy requirement as low as 7.7 watt-hours/kg of coal may be sufficient for sulphur removal. Despite being somewhat technically attractive, this potential process was not developed further because predicted economics of the process were not particularly encouraging ($33.7 per tonne for microwave versus $28.7 using conventional methods).

Sewage Sludges & Waste Water

Disinfection of waste water using microwaves has been reported by Iskander [21]. The results of the study suggested that microwave heating at 2450 MHz can be an effective bacteria killer. Sludge dewatering using dielectric heating is another area gaining increased attention. A number of studies have compared the use of dielectric techniques with conventional methods [22,23,24,25,26]. Watertek Division of Jaceb Corp (San Diego) manufactures a waste water system including a microwave oven to dry and sterilize solids from the dewatering rolls and vacuum filter [25]. The system can process 150,000 to 25,000,000 litres per day of raw sewage and was claimed to meet 1983 EPA requirements, although no details of power source requirement or operating temperature were given. The completely dried microwaved samples were claimed to be odour-free with a density of about 240 kg/m^3. The major limitation of this process appears to be its low process capacity. (A city the size of Vancouver needs to process more than 500,000,000 litres per day during dry weather.)

A microwave sludge sterilizer using a pressurized autoclave is claimed to sterilize sludge samples at a throughput of 141 kg per hour using a RF source of 40 kW [26]. Process time is 10 minutes at operating pressures of 150 kPa. Commercial in-line units are apparently available form Nearctic Ltd. (Australia).

An RF-assisted technique for dewatering sludges/filter cakes is under development by a number of organizations including EA Technology [23]. The technique uses conventional vacuum filtration in combination with radio-frequency heating to dewater sludges. RF heating assists sludge-dewatering by volumetrically heating the cake and lowering viscosity of free water within the cake to permit more rapid dewatering. Since no latent heat is required, the RF heating component can be kept relatively small and, as a result, keep electric load impact and capital costs to a minimum. The focus of the development is currently on dewatering of value-added filter 'cakes' such as those in the food & beverage and chemical industries although dewatering of hazardous sludges would be another obvious application.

Activated Carbon

Microwave regeneration of spent activated carbon was found to be superior to conventionally reactivated carbon with respect to adsorption capacity, adsorption rates and lower ash content [27,28]. In addition, the operating costs for microwave reactivation were estimated to be 10% less than conventional operating costs.

The use of microwave heating for solvent recovery (perchloroethylene) and activated carbon recycling was investigated for the Ontario Waste Management Corp. on behalf of Ontario's dry cleaning industry [6]. (Potential environmental concerns over the disposal of drycleaning wastes (primarily chlorinated solvents) prompted this investigation.) Study results indicated that microwave heating could lower perchloroethylene concentrations in samples of contaminated carbon to less than 0.01% compared to 2-3% using conventional azeotropic steam-stripping. In addition, microwave heating required less time (and energy) to achieve these concentrations. An initial comparison of the important chemical properties of the activated carbon indicated little difference between microwaved and fresh samples.

Gaseous Air Pollutants

The use of microwaves to reduce the environmental impact of gaseous air pollutants is being actively pursued by a number of researchers [8,22,29,30,31,32].

Wan at Queens' University is exploring the use of high-power microwave pulses to heat a catalytic material [8]. The rapid surface heating of the catalyst material causes adsorbed reagents to undergo chemical reactions. The process effectiveness of this microwave method is said to result in a higher yield than continuous irradiation. This technique, applied initially to decompose methane, also appears to be useful in the destruction of other hazardous chemicals such as SO_2 and NO_x. Much additional work is needed to determine whether the results are only thermal in nature or if other types of interactions are involved. Other related work conducted by Wan used pulsed microwaves to irradiate reagents passed over a bed of microwave-absorptive catalyst [8,32]. In terms of process efficiency, the authors claim that the process is over 99% efficient in removing SO_2 and NO_x from the gas stream. The use of microwaves were also demonstrated to decompose methyl chloride, methyl bromide, dichloromethane, chlorobenzenes and triphenylmethylchloride and bromide. Except for the aromatic hydrocarbon, chlorobenzene, substantial decomposition was achieved in all cases.

Argonne National Labs is developing a microwave technique to remove hydrogen sulphide gas from natural gas [31]. Before natural gas is delivered to distribution pipelines naturally occurring hydrogen sulphide requires removal. Conventional processes reclaim only sulphur and costs about $500 per ton of sulphur removed. In the Argonne process microwaves are used to split the hydrogen sulphide into sulphur and hydrogen. The sulphur is condensed and collected as a liquid; the remaining gases are purified while the hydrogen is separated for recycling. To date the experiments have achieved up to a 98% conversion efficiency for converting hydrogen sulphide into hydrogen and sulphur. While the process appears compatible with waste impurities found in gaseous waste streams at refineries, additional study is needed to focus on the higher impurity levels found in natural gas waste streams.

While direct destruction of hazardous wastes is preferred from an energy efficiency standpoint, indirect means, such as placing microwave susceptors (absorbers) in gas streams may provide an effective alternative. A number of researchers are actively seeking ways to use microwaves to destroy hazardous contaminants by using microwave-heated substrates (absorbers) [5,30,31]. The main advantage with this method is that only the substrate needs to be heated. The process can be conducted at lower temperatures and avoid the formation of toxic by-products that take place at the higher temperatures required for conventional waste destruction (ie, incineration). Thus far the technique appears to be effective in destroying trichloroethylene (TCE). One suitable catalyst material under study is silicon carbide. This material is promising from two standpoints: rapid heating in a microwave field to very high temperatures and chemical inertness.

Metaxas describes the use of microwaves to rapidly heat a honeycomb structure as a potential emission-control device for automobiles during cold-start and warm-up periods [33]. According to the author, a 200 g washcoat of cerium oxide on a 700 g catalyst honeycomb substrate was responsible for heating of the catalyst body. (The bare catalyst substrate without coating did not heat-up.) The effectiveness of the microwave-heated catalyst in reducing pollution from exhaust gases is under evaluation. A converter needs to be designed to provide the high voltages required to operate the microwave power source from a 12-volt battery.

Nuclear Waste

The use of microwaves/RF for processing nuclear waste has been studied perhaps more extensively than any other waste stream [10,21,34,35,36]. This interest results from the particular attributes of microwave/RF heating that match certain requirements for processing low and high-level nuclear wastes including: remote operation capability, high-value added material, and thermally thick material. Waste materials that fall into these categories include ion-exchange resins, sludges and evaporator bottom sludges. Storage space in low-level radioactive waste sites is being depleted [37]. As a result, the need for volume reduction (ie, water removal) of these wastes through improved processing and packaging techniques has become a major research area. The references provide a sampling of current work in this area.

Besides the development of microwave and RF technologies to directly process hazardous wastes, some companies are approaching the challenge by replacing a fossil-fired process with 'non-polluting' electrotechnologies like microwaves and RF. One example of this type of initiative is occurring at Southern California Edison. This utility is encouraging its customers to replace conventional fossil-fired technologies with 'cleaner' electrotechnologies (ie, RF post-baking of biscuits). The underlying assumption is that substitution of traditional processes like fossil-fired drying and curing equipment with industrially proven microwave or RF equipment will move any air pollutants from the manufacturing plant to the power plant where pollution abatement can be dealt with more effectively.

SUMMARY

The importance of sound hazardous waste practices is obvious. The examples in this paper illustrate that dielectric (microwave and RF) heating can be applied to effectively and efficiently process hazardous wastes. In general, these electrotechnologies increase processing speed and use less energy compared to conventional alternatives. They can be automated and controlled remotely to minimize any potential exposure to operating personnel. Although capital equipment costs are often cited as a major limitation, niche applications have been developed and commericalized as the attributes of dielectric heating are being applied more effectively. From an electric utility standpoint these electrotechnologies do not add significantly to demand; however, they do provide significant environmental benefit by minimizing the hazard associated with processing hazardous wastes. In some cases they have provided solutions to problems that cannot be accomplished effectively any other way.

Acknowledgments
The author wishes to acknowledge the assistance of the Canadian Electrical Association, Voss Associates Engineering Ltd (Victoria, BC) and Powertech Labs (Vancouver,BC) as well as the many organizations who contributed to the information in this paper.

References

1. "Microwave Treatment of Infectious Wastes", EPRI TechApplication, No.1, 1991.

2. M. Taitz, Asea Brown-Boveri ABB Sanitec, private communication, March 1992.

3. J. Sharp, "Radio Frequency Processing of Infectious Medical Wastes at a Regional Treatment & Recycling Facility", Inter. Microwave Power Inst. 26th Symp. Proc., Buffalo, NY, 1991.

4. J. Suzuki, M.Hosaka, Y. Kawasaki, A. Nishino, K. Sogen, "A Microwave Burning Processor for Waste Disposal", J. Microwave Power, 25 (3), p 168, 1990.

5. R. Hutcheon, AECL Chalk River Labs, private communication, March, 1992.

6. S.J. Oda, "Microwave Processing of Hazardous Waste", Inter. Microwave Power Inst. 26th Symposium Proc., Buffalo, NY, 1991.

7. P. Aysola, P.D. Anderson, H. Langford, "Microwave Mediated Degradation of PCB Wastes", US Patent 4,980,039 (Dec. 25,1990).

8. J. Wan, M. Tse, H. Husby, M. Depew, "High-Power Pulsed Microwave Catalytic Processes: Decomposition of Methane", J. Microwave Power, 25 (1), 32, 1990.

9. R. Varma, "Oxidative Degradation of TCE Adsorbed on Active Carbons: Use of Microwave Energy, ACS Cer. Trans., Microwaves: Theory and Application in Materials Processing 21, 467-473, 1991.

10. R.D. Smith, "Microwave Power in Industry", Electric Power Research Institute Report EPRI EM-3645, August 1984.

11. S. Robinson, "Wave Goodbye to Scrap Tyres", European Rubber Journal, Sept. 1989.

12. M. Suwabe, K. Hayashi, A. Sumida, "Devulcanization Process Employs the Dielectric Heating by Microwave", Toyoda Gosie Giho (Japan) 27 (3), p 91-96, 1985.

13. R. Girard, Hydro-Quebec LTEE, Private correspondence, 1992.

14. S. Tsutsumi, "Recycling Plastics Waste by Microwave Cracking", Chemical Economy & Engineering Review, Jan. 1974.

15. D.R. Pugh, "Microwave and Radiofrequency Heating Technologies for Hazardous Waste Treatment", Canadian Electrical Association Report, CEA No. 9001 U 755 (Draft), 1992.

16. H. Dev and D. Downey, "Zapping Hazwastes", Civil Engineering, p 43-45, Aug. 1988.

17. L. Dauerman, "Applications of Microwave Treatment of Hazardous Wastes a)Non-volatile Organics; b) Heavy metals", MRS Proc. Session L, paper L3.3, 1990.

18. L. Dauerman, "Microwave Treatment of Hazardous Wastes: Chemical Fixation of Organics like Dioxins in Contaminated Soil", UIE Microwave/RF Congress Proc., Nice, France, 1991.

19. C.S. Fang, B.K.L. Chang, P.M.C. Lai, W.J. Kaila, "Microwave Demulsification", Chem. Eng. Comm. 73 (227), 1988.

20. R.G. Bosisio, D. Klvana, "Experimental Results on Heating of Athabasca Tar Sand Samples with Microwave Power, J. Microwave Power 12 (4), 301-307, 1977.

21. I.K. Iskander, L. Parker, K. Madore, C. Gray, M. Kumai, "Disinfection of Wastewater by Microwaves", US Army Cold Regions and Engineering Research Laboratory Report, Jan 1980.

22. S.J. Oda, I.S. Balbaa, B.T. Barber, "The Development of New Microwave Heating Applications at Ontario Hydro's Research Division", MRS Proc, 189, 391, 1991.

23. P. Erskine, P.L. Jones, "Radio Frequency Assisted Dewatering of Filter Cake", UIE Microwave/RF Congress Proc., Nice, France, 1991.

24. Jaceb Corp, Watertek Div, San Diego CA, "Wastewater system uses microwaves", Chem. Eng., Jan. 31, 1977.

25. B. Neiderwohrmeier, "Microwave Treatment as an Alternative Pasteurisation Process for the Disinfection of Sewage Sludge - Experiences with the Treatment of Liquid Manure", Inactivation of Microorganisms in Sewage Sludge, 135-147, Elsevier, 1985.

26. J. Matyear, R.W Davies, "Microwave Treatment of Sewage Sludge", Microwave Applications Research Centre Conf. Proc. (U.Wollongong, NSW, Australia), Feb. 1989.

27. K.E. Haque, Microwave Irradiation Pretreatment of a Refractory Gold Concentrate", Proc. Int. Symp. on Gold Metallurgy, Winnipeg, Canada, Aug 1987.

28. I.S. Balbaa, S.J.Oda, "Microwave Reactivation of CIP Spent Carbon", ACS Cer Trans, Microwaves: Theory and Application in Materials Processing 21, 475-483, April 1991.

29. R. Varma, S.P. Nandi, J.D. Katz, "Detoxification of Hazardous Waste Streams Using Microwave-Assisted Fluid-Bed Oxidations", Mater Res Soc Proc 189, p 67, 1991.

30. R. Varma, Microwave Enhanced Chemical Processes, US Patent 4,935,114 (June 19, 1990)

31. Argonne National Laboratories, "Microwaves and Ultrasound Speed Waste Treatment", Mechanical Engineering, p 18, Nov 1991.

32. J. Wan, "Microwaves and Chemistry: The Catalysis of an Exciting Marriage", paper presented at the Industrial Energy Technology Conference, Houston TX, April 1992.

33. A.C. Metaxas, N. Collings, T. Ma, "Microwave Heating of Catalyst Substrate Material", UIE Microwave/RF Congress Proceedings, Nice, France, 1991.

34. F. Komatsu, "Application of Microwave Treatment Technology for Radioactive Wastes", Waste Management, 10, 221, 1990.

35. C. Shibata, H. Tamai, "A New Microwave Volume Reducing Method for Radioactive Waste Ashes", UIE Congress HF/Microwave Processing and Heating, Arnhem, The Netherlands, 1989.

36. Chem-Nuclear Systems, Inc., "Feasibility of Applications of Microwave Technology for Nuclear Power Plant Radioactive Wastes", EPRI report NP-2334, April 1982.

37. C. Hornibrook, "Low-level Waste Research", EPRI Journal, 44, Oct/Nov 1991.

MICROWAVE TREATMENT OF HAZARDOUS WASTES: PHYSICAL CHEMICAL MECHANISMS

Leonard Dauerman, Gabriele Windgasse, Naihong Zhu, and Yaoli He
The Microwave Treatment of Hazardous Wastes Laboratory
The Hazardous Substance Management Research Center
New Jersey Institute of Technology
Newark, N.J. 07102

ABSTRACT

The objective is to develop processes to treat hazardous waste streams using microwave treatment. First, physical chemical mechanisms have been posited for a variety of such waste streams including: soils contaminated with organics, and toxic heavy metals; "spent" GAC; organics in the gas phase. Second, feasibility tests are being carried out to determine if on-site field testing is warranted. In the case of microwave-induced steam distillation of volatile and semi-volatile organics, pilot plant stuides indicate that the incident power cost is not prohibitive: that fear of prohibitive power costs has been thought to be the major drawback for micrwave treatment to become a competitive technology in this market.

INTRODUCTION

This laboratory has been working on a three stage approach to the development of processes to treat hazardous waste streams. Such hazardous waste streams are regarded as physical chemical systems. The approach taken to develop remediation processes parallels the approach taken in the chemical industry to develop a process. First, define the physical chemical mechanisms related to a potential process which appears to have a unique niche in a market. Second, carry out pilot plant studies to determine if scale up is feasible, and the approach that is to be taken to scale up. Third, scale the process up. This paper will focus on the results of our studies to define physical chmeical mechanisms. Also, recent feasibility studies will be discussed which indicate that microwave treatment of hazardous wastes can be commercialized.

RESULTS AND DISCUSSION

Five separate waste streams will be discussed and the applicable mechanism.

1. Soils contaminated with volatile and semi-volatile organics.

The following mechanism is posited: Water, added to the soil effects desorption of the organic from the soil substrate. Microwave radiation causes the volatilization of water in the close proximity of the organics. Mass transfer equilibrium from the liquid to the gas phase occurs at the boiling point which is the temperature at which the sum of the respective vapor pressures of water and the organic is equal to the ambient pressure. The vapor composition is proportional to the respective vapor pressures of the water and the organic. Thus the standard unit operation of steam distillation is effected upon this waste stream. This can't be done thermally with a rotary kiln. Unlike incineration, the soil after treatment is not a waste stream [1].

Pilot plant studies have been conducted on this process[2]. In these studies it has been found that incident power cost is of the order of $4/ton. Scaled-up, the projected cost is estimated to be $40/ton. This is very favorable. Incineration technology is of the order of $150-$300/ton, plus the fact that it is usually opposed, successfully, by local groups. However, it remains to be seen whether or not unforeseen operating problems will make their presence known in field tests.

2. Soils contaminated with non-volatile, insoluble organics.

For this class of organics typified by the dioxins there is a need for in-situ remediation: the EPA has concluded from risk benefit studies that the greatest risk comes from the excavation and transportation of wastes from such sites. Microwaves have that necessary capability. Furthermore, mechanistic studies show that this potential is realizable[3].

Summarizing the mechanism: Soils consist of agglomerates bound together by adhesion and cohesion water; organics remain on the outside of such agglomerates because of the partitioning effect of water; the organics exist in layers held together by weak Van der Waal's forces; microwave radiation evaporates water from the agglomerates, allowing the organic to become bound in unimolecular layers to the soil components, mainly, clay and humus; soil components absorb microwave radiation, heating up sufficiently to effect the reaction or the "fixation" of the organics.

3. Soils contaminated wih toxic heavy metals.

The only permanent treatment technology available now is vitrification by applying a high density current through the soil to effect sufficient heating to bring about vitrification: this is extremely costly.

The following mechanism has been developed in our studies [4]: Soils contain microwave absorbing minerals; heating effects the incorporation of the heavy metals into the minerals, thereby rendering such metals harmless; humus absorbs microwave energy and effects the reduction of heavy metals to either more benign states or to states in which they can be more readily incorporated into clay minerals.

4. "Spent" granulated activated charcoal (GAC)

Waste water treatment entails the use of GAC to remove organic constituents. Such GAC becomes in time water-logged and incapable of absorbing organics from the waste stream. Now, this water-logged GAC is a hazardous waste and is shipped off-site for regeneration.

The following mechanism indicates that such treatment can be effected on-site[5]: GAC is water-logged; microwave radiation effects the heating of the GAC and water; water within the GAC pores is converted into steam; the steam generated drives out the air from the pores, creating an inert atmosphere; GAC heats up to a temperature sufficent to effect the pyrolysis of the organics in the absence of air forming a carbon skeleton which is incorporated into the GAC.

The major concern in applying this technology is that arcing may cause combustion of the GAC. It is critical that the amount of air must be below the lower flammability of the organics.

5. Organics in the gas stream.

Microwave radiation can directly heat a fluidized or fixed bed used to interact with a gas stream provided that the substrate is "lossy." This means that unlike comparable thermal systems the entering gas stream does not have to be heated. This is advantageous in chemical processes where the chemicals in the entering stream are thermally unstable; in hazardous waste treatment, as a microwave afterburner to destroy volatilized organics.

The flow mixing system and the fluidized bed reactor are shown in Figure 1[6]. The results of runs in which trichloroethylene (TCE) in a mixture of He, Ar, and water was passed through a fluidized bed of GAC are shown in Figure 2. As can be seen in Figure 1, the gas stream is sampled before and after entering the reactor; analysis is carried out on a gas chromatograph/mass spectrometer (GC/MS); Ar is used as a reference for fluctuations in the instrument or the flow system. From the plot of the rate versus initial concetration , the resulting straight line shows that the reaction is first order. That would signify that the surface area of GAC is not rate-limiting, otherwise, the order would be zero. It is to be noted that other than TCE, no other products were observed in the spectra. The GC/MS did not have the capability of reporting on CO, $CO2$, or HCl. HCl was detected qualitatively by passing the gas through a basic solution, and then adding silver nitrate to precipitate silver chloride.

Fig. 1. Fluidized Bed Reactor

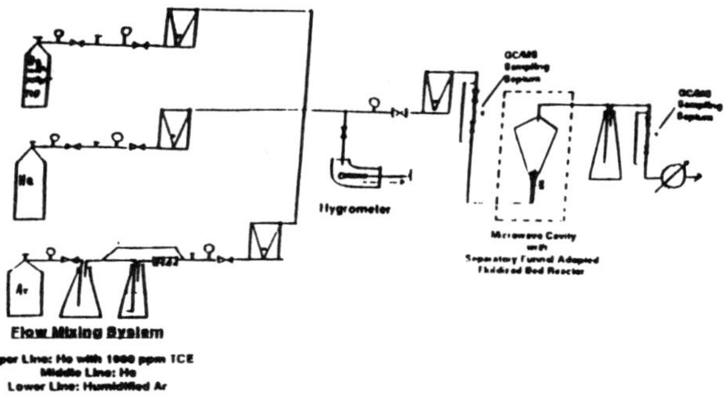

Fig. 2. Decomposition of TCE on GAC
Microwave Field/Fluidized Bed
(flow gases: He, Ar, H2O)

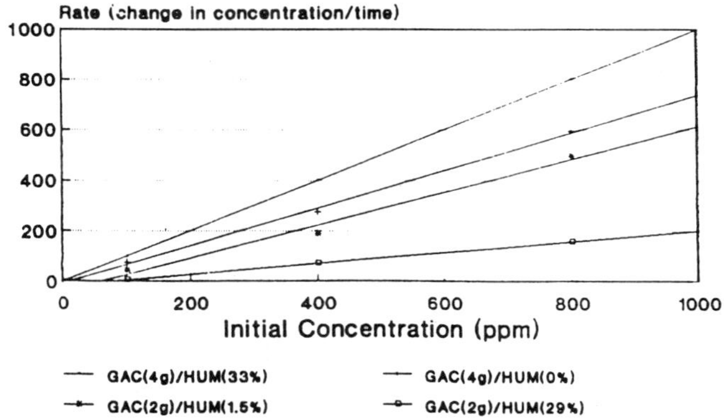

Temp: 625 C (IR Pyrometer)

In Figure 3, the effect of adding N2 and O2 to the stream is seen: the curve appears to approach a zero slope, indicating that the reaction becomes zero order with respect to TCE, because the available surface area is limited, presumably due to the adsorption of the N2 and O2. The effect of a fixed bed of SiC is shown in Figure 4. Again, the rate appears to approach a zero slope, indicating that the order with respect to TCE is zero, indicating that the surface area is limiting. However, another complication in this
study is that the temperature is not uniform. Therefore, these results may not be comparable to the previously-discussed results.

CONCLUSIONS

Physical chemical mechanisms provide a guide for the development of processes for the treatment of hazardous wastes. Furthermore, feasibility studies on the microwave-induced steam distillation of volatile and semi-volatile organics process indicate that it may be cost-effective with respect to available technologies.

Fig. 3. Decomposition of TCE on GAC
Microwave Field/Fluidized Bed
(flow gases: He, N2, O2)

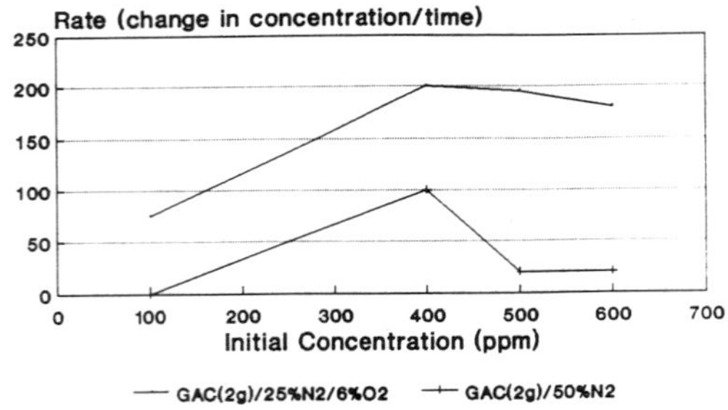

Temp: 775 C (IR Pyrometer)

Fig. 4. Decomposition of TCE on SiC
Microwave Field/Fixed Bed

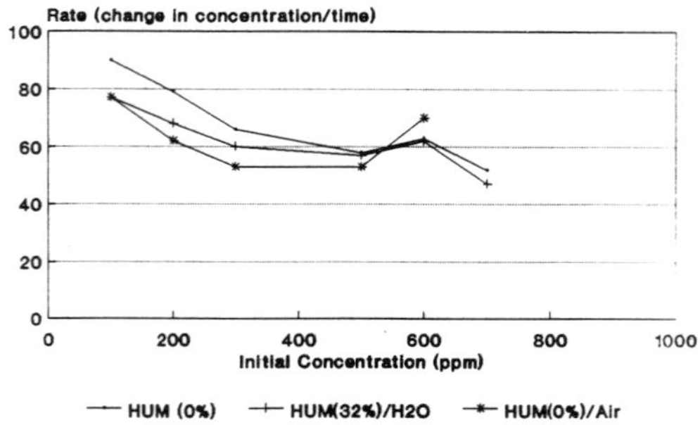

(flow gases: He, Ar, Air, H20)

ACKNOWLEDGEMENTS

These studies have been supported by the EPA, NSF, and the industrial sponsors of the Hazardous Waste Management Research Center.

REFERENCES

1. G. Windgasse and L. Dauerman, Microwave Treatment of Hazardous Wastes: Removal of Volatile and Semi-Volatile Organic Contaminants from Soil, Journal of Microwave Power and Electromagnetic Energy 27(1): 23 - 32 (1992).

2. Y. Lu, Microwave Treatment of Hazardous Wastes: Feasibility Studies - Operating Costs, MS Thesis, New Jersey Institute of Technology, 1992.

3. N. Zhu, L. Dauerman, G. Windgasse, and H. Gu, Microwave Treatment of Hazardous Wastes: Remediation of Soils Contaminated by Non-Volatile Organic Chemicals Like Dioxins, Journal of Microwave Power and Electromagnetic Energy 27(1): 54 - 61 (1992).

4. E. Sedhom, L. Dauerman, N. Ibrahim, and G. Windgasse, Microwave Treatment of Hazardous Wastes: "Fixation" of Chromium in Soil, Journal of Microwave Power and Electromagnetic Energy 27(2) (1992).

5. K. C. Mahesh, Microwave Treatment of Spent GAC to Effect Regeneration On-Site, MS Thesis, New Jersey Institute of Technology, 1989.

6. G.Windgasse, A New Hazardous Waste Treatment Technology Utilizing Low Power Density Microwave Energy, MS Thesis, New Jersey Institute of Technology, 1989.

IMMOBILIZATION OF ASH BY MICROWAVE MELTING

Kazuki Morita, Vinh Q. Nguyen, Ron Nakaoka* and John D. Mackenzie

University of California Los Angeles, Department of Materials Science and Engineering, Los Angeles, CA 90024
* Los Alamos National Laboratory, Waste Management, Los Alamos, NM 87545

ABSTRACT

Non-radioactive ash was immobilized via microwave melting. The ash composed mainly of silica (SiO_2), titania (TiO_2), calcia (CaO), alumina (Al_2O_3), and carbon (C). The ash was melted with various additives, such as magnetite (Fe_3O_4), lithium carbonate (Li_2CO_3), sodium carbonate (Na_2CO_3), and boron oxide (B_2O_3) by using a 2.45 GHz, 750 W microwave oven. Samples with different ash and additive concentrations were melted when subjected to microwave radiation. Ease of melting was dependent on the carbon and magnetite concentrations. Melted samples were characterized by X-ray diffraction and scanning electron microscopy (SEM), and subjected to the Toxicity Characteristic Leaching Procedure in regard to lead (Pb) immobilization. Melted samples with more than 30 wt% additives has an undetectable leaching rate of less than 0.1 ppm of lead for TCLP test, which was found to be due to the reduction and evaporation of the lead during melting. More than 80 % of the lead was lost through evaporation leaving less than 0.1 wt% Pb in the melted glass.

INTRODUCTION

Since the amount of radioactive wastes originating from nuclear facilities have been increasing recently, immobilization and volume reduction of the wastes have become an important problem. Although there are many immobilization processes, such as the borosilicate glass process, Synroc process, cement, etc., none of them is ideally suited for the purpose of volume reduction.

Recently, microwave processing has been introduced to easy and safe melting and solidification of non-combustible waste.[1, 2, 3] Microwave processing is a fast and efficient way of heating materials which have high dielectric loss factors. When exposed to electromagnetic radiation in the microwave range, the temperature of these materials increases in a very short time. However, there are few systematic experiments on the heating behavior of incinerated ash.

In this study, the heating behavior of simulated ash was investigated with a simple 2.45 GHz, 750 W microwave oven under various conditions. Also, the characterizations on the melted samples were conducted with regard to microstructures and leachabilities.

EXPERIMENTAL

Materials

The compositions of the ash are listed in Table 1. Simulated transutanic (TRU) contaminated ash were from the Los Alamos National Laboratory Controlled Air Incinerator (LANL CAI), and the representative calcined ash was synthesized by mixing each chemical.

Table 1 Chemical composition of LANL CAI ash and synthesized ash in weight percent.

Compositions	LANL CAI ash	Synthesized ash
SiO_2	17.2	27.5
TiO_2	20.1	32.1
CaO	10.0	16.1
Al_2O_3	11.0	17.5
PbO	1.2	1.0
K_2O	0.5	0.8
C	20.8	5.0
Others (including volatile matter)	19.2	--
Total	100.0	100.0

Sample Preparation

The CAI ash or synthsized ash was mixed homogeneously with various amounts of additives, such as B_2O_3, Na_2CO_3, Fe_3O_4, Li_2CO_3 and water, and ground to - 200 mesh. A 15-g sample was placed inside a fused silica crucible with ceramic fiber insulation and was melted in a 750 W, 2.45 GHz multimode microwave oven.

During the operations, the temperature of the sample was monitored by using a dual wavelength infrared pyrometer (Micron Instrument Co., #M77S) through a 5 mm diameter hole in the insulation (Fig. 1). After being heated up to 1400 °C, for the melted sample whose viscosity is low enough, the melt was cast into a graphite mold preheated at 900 °C, which was slowly cooled down with ceramic fiber insulation placed around the graphite mold. The heated sample which was hard to cast, was cooled directly in the crucible with ceramic insulation.

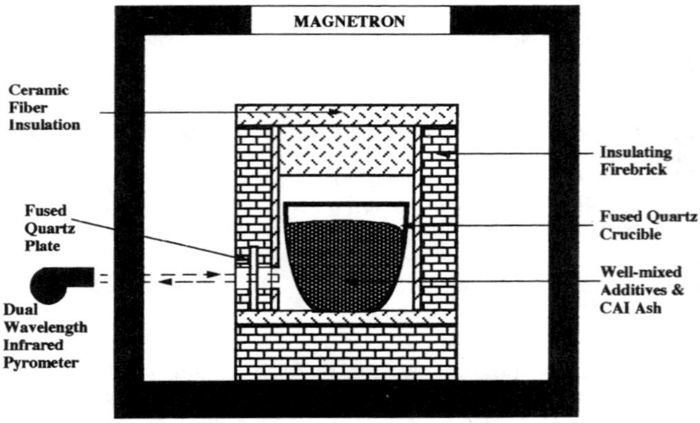

Fig. 1 Schematic cross section of experimental apparatus.

Specimen Characterizations

Microstructures of those samples were characterized by XRD and SEM with the operating conditions at 35 kV - 15 mA and 20 kV - 2 A, respectively.

Leaching tests were performed by Toxicity Characteristic Leaching Procedure (TCLP) [4]. In this study, only Pb extraction was examined. Due to the limited amount of representative sample, the sample size was reduced to 1 gram (1% of the sample size requirement). The quantity of extraction fluid is reduced proportionally to 20 mL so that the ratio of liquid to solid is 20 : 1. The extract was analyzed for Pb using an atomic absorption spectrophotometer.

EXPERIMENTAL RESULTS

Most of the 15 g samples were heated up to 1400 °C in 15 to 80 minutes under 2.45 GHz, 750 W microwave radiation. Increasing the amount of additives, such as Na_2CO_3, Li_2CO_3, B_2O_3 and Fe_3O_4, made the samples more homogeneous and more glassy. Several series of heating behaviors were investigated and compared under different conditions.

Heating of Ash without Additives

When the two types of ash were exposed to the 750 W microwave, both samples were heated up without any additives. (Fig. 2) Since the heat was generated mainly by carbon in the ash, the time needed for melting a sample depends upon the concentration of carbon. This difference in heating behavior can be reasonably explained by the fact that the amount of carbon in the CAI ash is 20 wt% versus only 5 wt% in the synthesized ash.

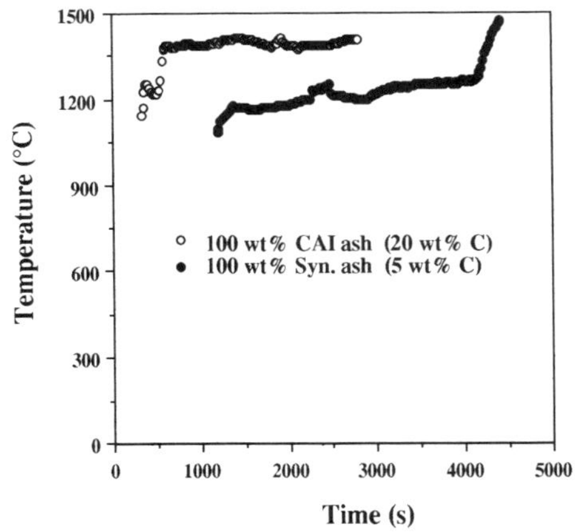

Fig. 2 Effect of carbon concentration on the heating behaviors.

Effect of Iron Oxide (Fe_3O_4) on Heating Rate

Fig.3 shows the effect of Fe_3O_4 addition on temperature behavior of the ash. Due to the presence of 40 wt% of Fe_3O_4 in the sample, it took 16 minutes to reach a temperature of 1400°C, while it took 71 minutes for 100 wt% synthesize ash. it took 2.7 minutes to measure the first temperature reading. This is because Fe_3O_4 is one of the most effective microwave coupler that has large value of dielectric loss factor as discussed later.

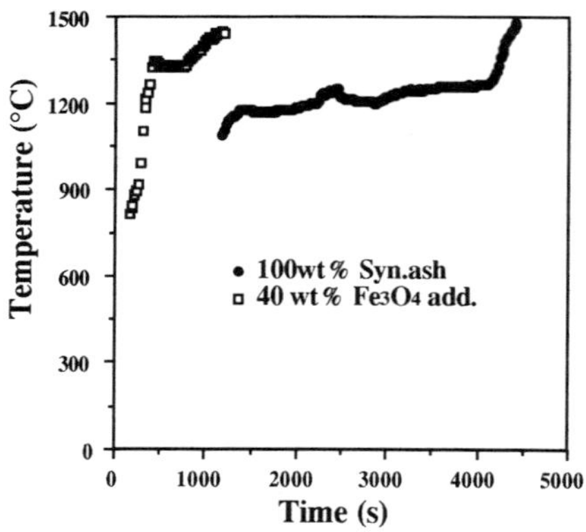

Fig.3 Effect of Fe_3O_4 addition on the heating behaviors.

Effect of Additive Composition on Heating Rate

In order to see the effect of additive composition on the heating behavior, how the temperature behaves for a fixed ash concentration but with different types of additive concentration, 10 wt% of B_2O_3, Na_2CO_3, and Fe_3O_4 was added separately to10 wt% of Li_2CO_3 and 80 wt% synthesized ash. The heating behaviors are shown in Fig. 4. They show that 10 wt% of Fe_3O_4 has the highest overall heating rate while 10 wt% of B_2O_3 has the lowest overall heating rate. The order of $Fe_3O_4 > Na_2CO_3 > B_2O_3$ is considered to correspond to the order of the dielectric loss factor.

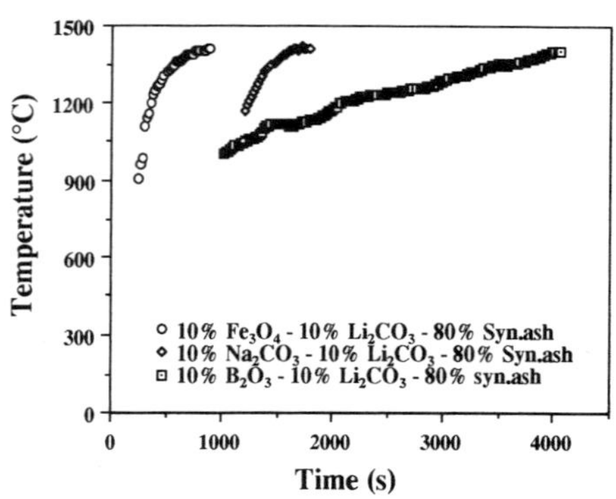

Fig. 4 Effect of additive compositions on the heating behaviors.

Effect of Water on Heating Rate

Fig. 5 shows the effect of water addition on the temperature behavior of samples. Water was added to 20 wt% Na_2CO_3 - 20 wt% B_2O_3 - 10 wt% Fe_3O_4 - 50 wt % Syn. ash by 2 g, 4 g, and 6 g separately, and the heating rates were compared with that without water. When 6 g of water was added to the ash, the time taken to reach the temperature of 1400 °C decreased from 16.3 minutes to 9.7 minutes, reducing the melting time by 60 %. The additional source of heat from the water helped the sample to heat up quickly in the microwave.

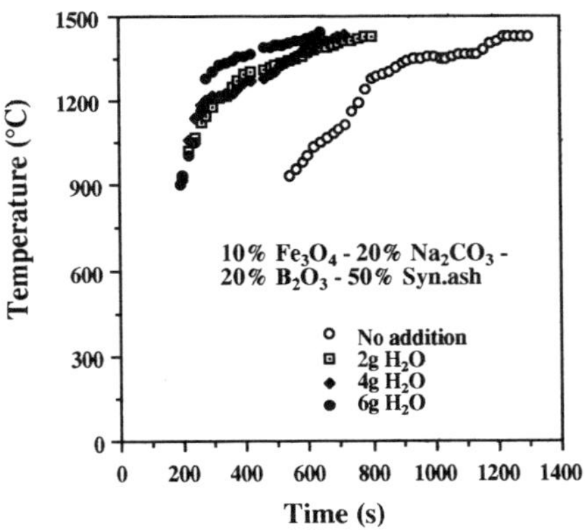

Fig.5 Effect of water addition on the heating behaviors.

Leachability of the Melted Samples

For leaching test, the TCLP was performed for all the melted samples. Pb was undetectable for all the homogeneous samples. Almost all the samples conformed to EPA standard where Pb concentration should be less than 5 ppm. However, as predicted in the discussion, more than 80 % of lead oxide in the ash was found to have been reduced to metallic lead and vaporized during the heating process. This was due to the existence of the reducing agent, C.

DISCUSSION.

Considering that about 20wt% carbon exists in the CAI ash, there is a possibility of reduction of the lead oxide during heating up. In order to evaluate the possibility of such reactions, thermodynamic considerations were taken. These thermochemical data are referred from tables edited by Turkdogan. [5] Pb^{2+} is easily reduced to Pb^0 and the vapor pressure of Pb is high enough for liquid Pb to evaporate at high temperature such as 1500°C. Reactions are shown as Eq.(1) and Eq.(3).

$$PbO(l) + C(s) = Pb(l) + CO(g) \quad (1)$$

$$\Delta G^\circ = -RT \ln [\, a_{Pb} \cdot p_{CO} / (a_{PbO} \cdot a_C) \,] = 15960 - 36.76T \text{ (cal/mol)} \quad (2)$$

$$Pb(l) = Pb(g) \quad (3)$$

$$\Delta G^\circ = -RT \ln (p_{Pb} / a_{Pb}) = 43490 - 21.54T \text{ (cal/mol)} \quad (4)$$

where a_X and p_Y denote the activity of X and the partial pressure of Y, respectively. In the presence of solid carbon, $a_C = 1$, and assuming that O_2 in air has completely reacted with carbon into 2CO, $p_{CO} = 0.33$ atm. At 1500°C (1773K), Eq.(5) can be obtained.

$$a_{Pb} = 3.50 \times 10^6 \, a_{PbO} \quad (5)$$

PbO will be reduced to metallic Pb until the activity of PbO becomes as low as 2.9×10^{-7}, and the vapor pressure of Pb is calculated as 0.22 atm. This means that lead oxide can hardly exist under such a reducing condition and reduction and vaporization of lead will take place during the heating process. This prediction is correspondent to the result that more than 80 % of lead oxide in the ash was lost during melting. In order to avoid the reduction or carbide formation, ash should be melted in highly oxidizing conditions enough to promote the oxidation of the carbon. Also, low temperature operation is preferable.

CONCLUSION

A melting process was used to immobilize the CAI ash by using microwave heating and the following results were obtained.

1. Some of the samples could be heated up to 1400 °C in 15 minutes, which demonstrated the economical advantages of microwave over conventional processes.
2. Ease of melting depends on C and Fe_3O_4 concentration.
3. Water addition helps the rapid heating of the CAI ash.
4. Ash with more than 30wt% additives has undetectable leaching.
5. More than 80 % of PbO was reduced and vaporized in the presence of C.

ACKNOWLEDGMENTS

This research has been supported by the Los Alamos National Laboratory. The authors would like to thank Dr. H. Kozuka and Mr. E. P. Bescher for their kind advices on the experiment.

REFERENCES

1. Komatsu, F., Sawada, Y., Ohtsuka, K., and Ohuchi, J., "Development of a new solidification method for wastes contaminated by plutonium oxides," Management of alpha-contaminated wastes, IAEA-SM246/4, pp. 325-327, International Atomic Energy Agency, Vienna (1981)
2. Ogata, Y., Ohuchi, J., Inada, E., and Tsunoda, N., "Processing of plutomium contaminated waste at PWTF," Management of low and intermediate level radioactive wastes, IAEA-SM-303/24, pp.325-339, International Atomic Energy Agency, Vienna (1988)
3. Komatsu, F., Takusagawa, A., Wada, R., and Asahina, K., "Application of microwave treatment technology for radioactive wastes," Waste Management, Vol.10, pp.211-215 (1990)
4. Federal Register, Vol.55, No.61, March 29 (1990)
5. Turkdogan, E. T., "Physical Chemistry of High Temperature Technology," (Academic Press), p.10 (1980)

PART IX

Microwave NDE Techniques

MICROWAVE IMAGING TECHNIQUES FOR NON-DESTRUCTIVE TESTING OF MATERIALS

J.CH.BOLOMEY AND CH.PICHOT
Groupe d'Electromagnétisme, Laboratoire des Signaux et Systèmes (CNRS/ESE)
Ecole Supérieure d'Electricité, Plateau de Moulon,F 91192 Gif-sur-Yvette Cedex, France

ABSTRACT

Recent technological advances for microwave multiport sensors lead us to look for efficient wavefront processing techniques in order to improve image quality, in terms of spatial resolution and contrast. This paper reviews some of these techniques which can be used for Non-Destructive Testing purposes. Mutual advantages and limitations are analysed.

Introduction

The inspection of materials by means of classical one-port microwave sensors is usually achieved in transmission, reflection or double-transmission mode. The signal provided by the sensor can be related to a spatially averaged characteristic of the material under investigation, such as humidity, density, temperature, etc... Calibration can be obtained either by comparison to standard materials or by using theoretical transmission or reflection coefficients derived from a simple plane wave / parallel plate analysis. Microwave imaging technology is expected to have an increasing impact on Non-Destructive Testing (NDT) of industrial materials. Indeed, with respect to classical one-port microwave sensors, a new generation of rapid multipoint sensors has been shown to allow for increased spatial resolution capabilities via direct monitoring of the transmitted or reflected wavefront [1]. This paper shows how the results provided by such sensors can be still improved by means of convenient amplitude / phase wavefront processing. A first part will be devoted to review recent applications of such multiport sensors and outline the major limitations of direct monitoring. A second part presents more rigorous approaches taking into account scattering mechanisms. Non exhaustively, three representative techniques are presented corresponding to gradual complexity and efficiency : 1) back-projection in free space, 2) diffraction tomography and 3) inverse scattering. Finally, as a conclusion, the main advantages and limitations of these approaches are analysed and compared.

Limitations of direct monitoring

Multiport sensors have been more particularly introduced to add spatial resolution capabilities to already used microwave sensing approaches. The localization and the identification of faults, defects, either in primary constituents or in manufactured products exhibit spatial resolution requirements. During the last years, extensive feasibility studies and pre-industrial developments have been conducted to assess the Modulated Scattering Technology (MST) [1]. The following applications have

been more particularly considered :
- detection of knots in wooden boards in view of optimal sawing
- measurement of humidity and density profiles in wooden boards
- on-line quality control in plaster composite materials
- control of transverse humidity profile during paper drying process
- quality control of conducting textiles for aerospace industry
- inspection of low observable objects
- etc ...

For the sake of illustration, Figure 1 shows the microwave amplitude and phase transmission images obtained in the case of a paper roll exhibiting 10% humidity change. Practically, the humidity step has been achieved by changing the conveying speed from 240 m/min to 640 m/min in the drying machine.

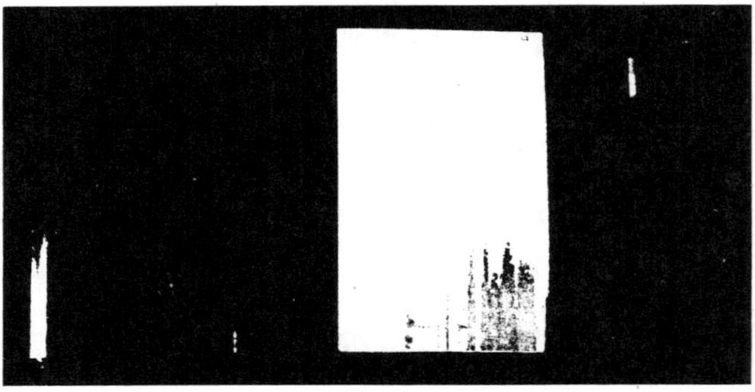

a) amplitude b) phase

Figure 1 : Transmission microwave amplitude and phase images of paper roll during drying process. The images result in a 10% change of the humidity of the paper. Vertical axis corresponds to paper length,horizontal axis to 1 meter paper width. (Courtesy of SATIMO).

In all these previous applications, the spatial resolution has been provided by the sampling rate of the sensor. Typically, according to the operating frequency selected for the considered application, the sampling interval ranges between 5 mm and a few centimeters. For instance, the results shown in Figure 1 have been obtained at 9 GHz by means of a sensor with 128 probes spaced by 1 cm. The material may be inspected in transmission, reflection or double transmission modes, according to its thickness and transparency at the test frequency. Although being the simplest and the more rapid solution, the direct monitoring of reflected or transmitted wavefronts is subject to some limitations which are now analysed. Basically, the main reason of such limitations is that, as well known, microwave beams do not propagate along linear trajectories, but are diffracted by any inhomogeneity, impurity, etc...producing local dielectric constant gradient located on its path. The forward- or backward- wave scattered by such a small local defect can no longer be described by a plane wave, but more adequately by a spherical wave (Figure 2).

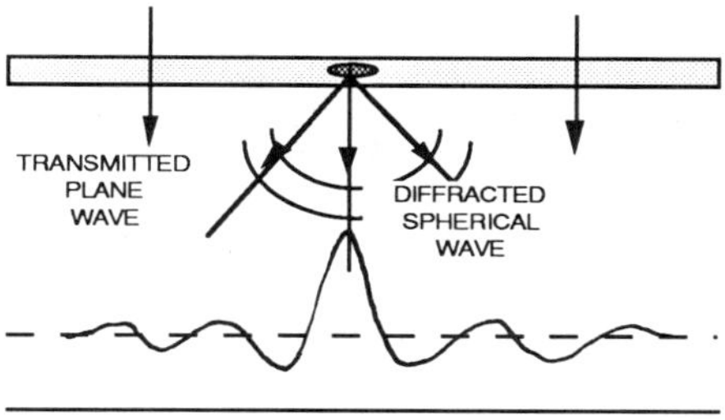

Figure 2 : Diffraction of a microwave beam with an inhomogeneous slab material.

The quantification of the response of the defect with respect to the gradient of the internal quantity to be measured becomes more complex than in the simple plane wave model. The amplitude and the phase of the scattered wavefront depend not only on the magnitude of the local change of the quantity to be detected but also on its shape, dimensions and on its position in the material. Such diffraction phenomena tend to reduce the spatial resolution. Indeed, a point defect produces a spot whose finite width is increasing with the distance between the sensor and the material. Furthermore, the waves scattered by different parts of the material may produce rather unpredictable interference patterns exhibiting possible artefacts (Figure 3). Another diffraction limitation results from the finite width (or dimension) of the material (Figure 4). According to the geometrical theory of diffraction, edges are known to give rise to diffracted rays which can interfere with other transmitted or reflected rays. As an important result, the spatial response of a transversely homogeneous material is not constant. For a given material width, higher the dielectric constant contrast and/ or the thickness and higher is the effect of edge diffraction.

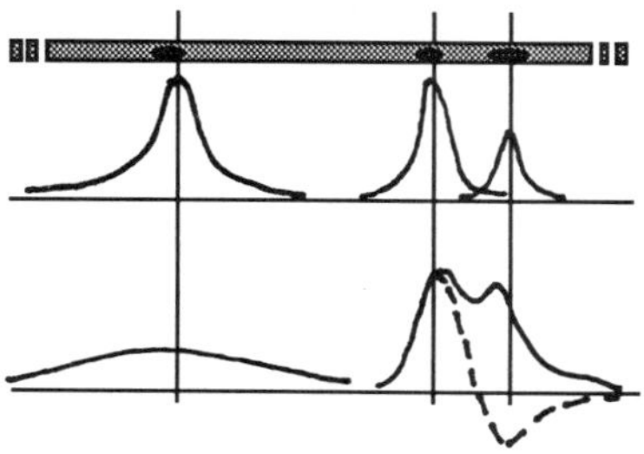

Figure 3
Problems associated with direct monitoring of the microwave response of slab materials. Diffraction phenomena result in variable spatial resolution (a) and possible artefacts (b) according to the fault location in the slab.

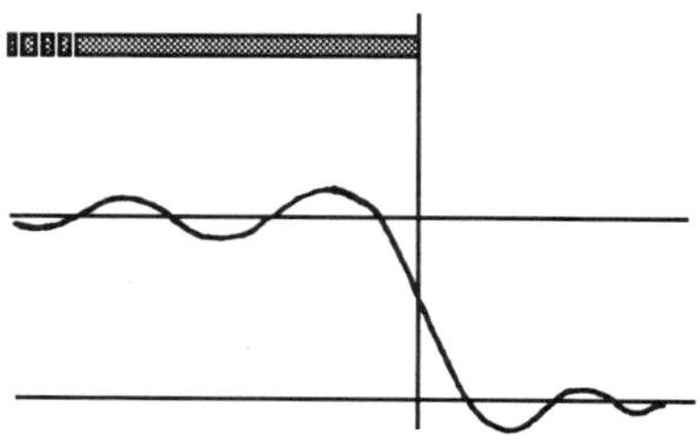

Figure 4
Edge effects result in non-uniform reflected or transmitted waves, even for homogeneous slabs.

Back-projection in "free-space"

The idea of the back-projection process is to "come back" toward the material. The expected result is to get better localization of the internal defects via the waves they scatter. As previously indicated, the width of the scattered spot decreases with distance. Practically, the back-projection is easily performed via Fourier Transforms (FT) of the measured data. Indeed, these data consist of a set of samples of the scattered field which can be considered as a superposition of plane waves [2]. These plane waves include both propagating waves and evanescent waves.

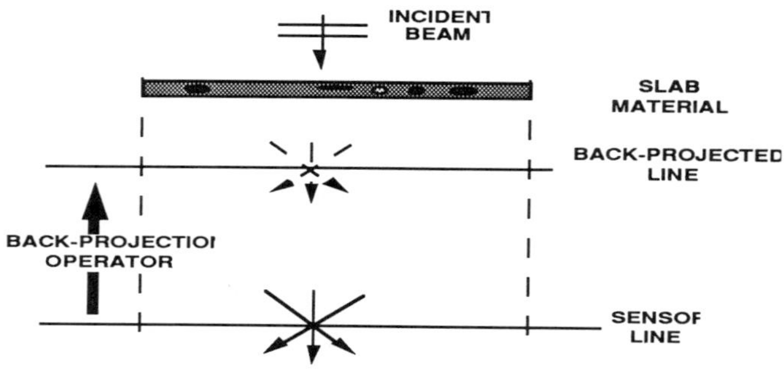

Figure 5 : Principle of free space back projection.

Figure 6 : Example of free space back-projection processing
in the cases of one and two objects : measured projections at 3 λ (top view)
and back-projections in the material plane (bottom view).

The first ones propagate without any attenuation while the second ones attenuate exponentially very rapidly when propagating from the material. It can be shown that the FT of the scattered field provides the so-called Plane Wave Angular Spectrum (PWAS). The knowledge of the reversible plane wave transfer function between the material and the sensor planes allows, via filtering and inverse FT (Figure 5), transforming the scattered field distribution on the sensor into the scattered field distribution on the material plane. Figure 6 illustrates the efficiency of such a transformation in the simple case of 1 or 2 closely spaced point defects. However, the efficiency of back-projection is limited by the sampling rate of the sensor as well as by the available signal over noise ratio. Sampling at $\lambda/2$ provides a convenient sampling of the propagating part of the PWAS, but cannot allow achieving better than equivalent spatial resolution. Oversampling, i.e. sampling interval smaller than $\lambda/2$, may provide equivalent spatial resolution provided that the measurement dynamic range is suitable to get the required evanescent part of the PWAS.

Diffraction tomography

With the diffraction tomography approach, the principle is rather different [3]. The objective is to "look" inside the material, as it is done with X rays scanners. The main differences with X rays are 1) that the beam path is no longer narrow and linear and 2) that the relationship existing between the measured data and the complex permittivity distribution in the test material is non linear. These two factors complicate significantly the reconstruction process. However, for quasi-homogeneous media exhibiting slowly varying gradients or, on the contrary, strong but localized gradients, the problem can be linearized within the frame of the Born approximation. Such an approximation neglects multiple scattering within the internal part of the material. The local gradients are seen as equivalent sources which can be retrieved in situation and intensity via some synthetic focusing. To some extent, it can be considered that the data available from each sensor element are processed to add in-phase, for a given point of the test region, at the input of the receiver (Figure 7). Practically, for computation efficiency reasons, such a focusing can be achieved by means of FT, despite the fact that the sensor cannot be considered as far away from the test material. Different experimental protocols can be considered, according to the available data projections, from monofrequency and monoincidence projection view to multifrequency and/or multiincidence projections [4,5].

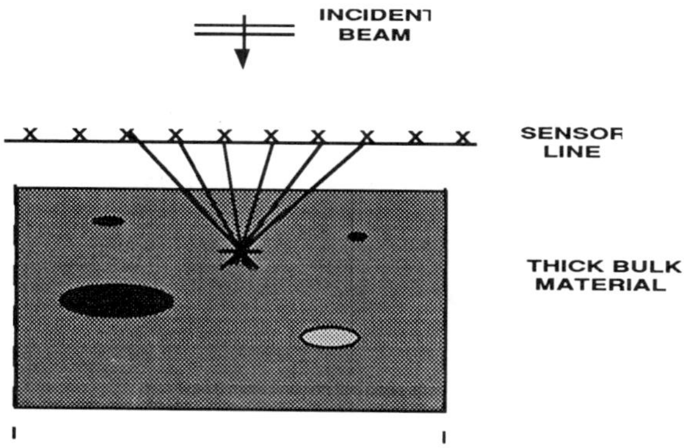

Figure 7 : Principle of diffraction tomography

As for any imaging instrument, the concept of point response is usefull. The point response is the response of the instrument to an ideal point object. The point response is then a small volume whose finite dimensions limit spatial resolution capabilities. In the case of monoincidence protocols, it is interesting to point out that transverse and longitudinal resolutions are different. Transverse resolution obtained in a plane orthogonal to the incident interrogative beam, is of the order of $\lambda/2$ in the material. Longitudinal resolution, obtained in a plane parallel to the beam propagation direction, is of the order of 2 to 3 λ, approximately, for monofrequency protocols. In the case of multifrequency measurements, the longitudinal resolution is roughly given by $1/2\ (\lambda_{min} \cdot \lambda_{max}) / (\lambda_{max} - \lambda_{min})$, where λ_{min} and λ_{max} denotes the extreme values of the wavelength in the test frequency band in the host medium. By essence, it can be shown that the spectral processing of data can not provide spatial resolutions better than $\lambda/2$ in the test material.

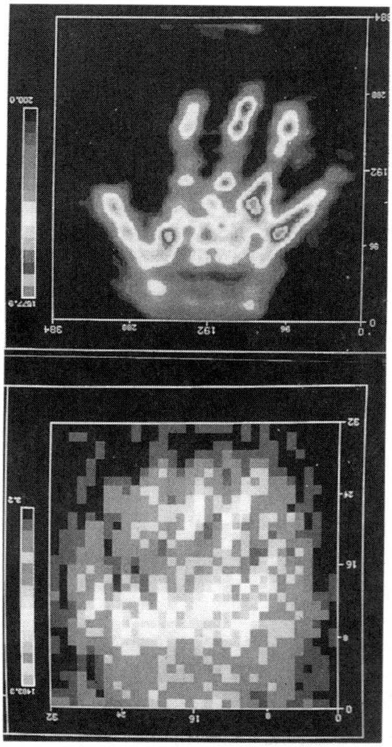

a)

b)

Figure 8 : Monofrequency and monoview diffraction tomographic processing : (a) rough 32X32 pixels amplitude image as recorded in a transmission mode by a 2.45 GHz planar microwave camera, (b) reconstructed and processed 384X384 pixels amplitude image after focusing in the plane of the test object, here a human hand immersed in water.

Figure 8 illustrates the efficiency of the diffraction tomography approach in the case of single frequency, single incidence measurement. The operating frequency is 2.45 GHz, corresponding to a wavelength equal to 12 mm in water [6]. The microwave camera consists of an array of 32 by 32 elements, covering a surface of 20 cm by 20 cm approximately. Figure 8-a corresponds to the amplitude of the 1024 data corresponding to the transmitted scattered field measured at the element locations. Figure 8-b provides the reconstructed image in the plane of the object, here located at approximately 8 cm from the camera. Figure 9 shows the result obtained with a broad band linear sensor with 64 elements operating from 7 to 13 GHz. Such an octave frequency band provides improved longitudinal spatial resolution and allows localizing buried objects, here, metal bars in reinforced concrete [7].

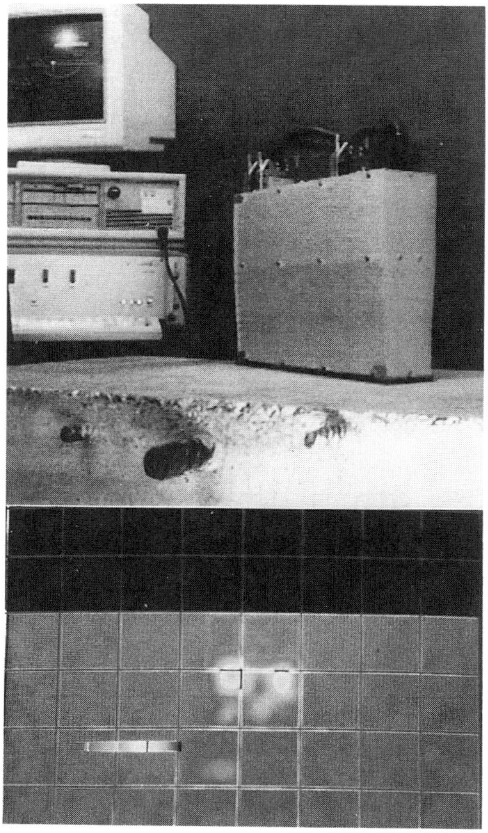

Figure 9 : Multifrequency and monoview diffraction tomographic processing : (top view) portable 7 GHz-13 GHz equipment for reflexion inspection of metal bars in reinforced concrete, (bottom view) typical tomographic image obtained for two metal bars; scale : 4 cm / div. (Courtesy of SATIMO)

Inverse scattering techniques

Diffraction tomography only partially takes into account diffraction mechanisms and neglects multiple scattering. In the best case, it provides the equivalent current distribution inside the test object. More quantitative reconstruction of the complex permittivity distribution requires more sophisticated reconstruction techniques to deal with the non-linear aspect of the reconstruction process [e.g.8,9,10,11]. For NDT purposes, an appropriate approach consists of assuming available an approximate value of the contrast profile. Such an assumption is not unreasonnable in industrial testing where the production must exhibit known constant characteristics. This approach consists then of comparing the measurements provided by the sensor to the calculated scattered response of the model of the test object (Figure 10). To this end, the test object is discretized, for instance, according to classical moment methods. The difference between the measured and predicted scattered fields allows calculating, at the first order, a corrective term to update the initial dielectric constant profile. This corrective term is derived from an instable inverse problem by means of regularization techniques. For contrast profiles quite different from the initial guess, the correction process can be iterated until convergence is achieved. However, such an iterative approach may be reduced to one single step as soon as the test material is not very different from the initial assumption.

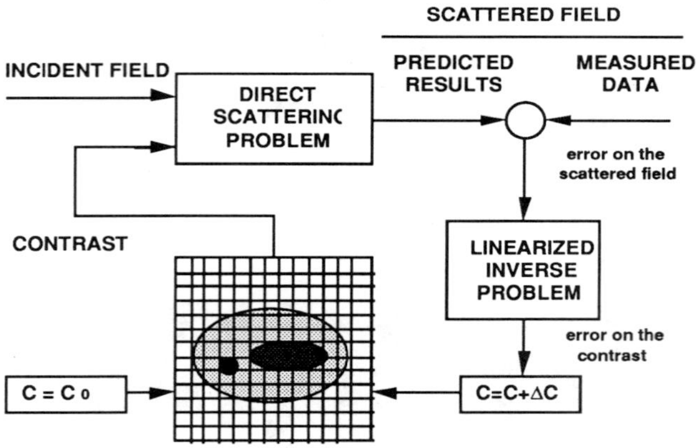

Figure 10 : Principle of an iterative scheme for solving non-linear inverse problems

In such processes, the spatial resolution is not explicitly limited to a fraction of a wavelength. The main idea is, on the contrary, that the unknown discretized dielectric constant distribution has to be obtained from a sufficient amount of independent data. These data are usually derived from multi-view or multi-frequency experimental procedures. The main trend is then that a fine meshing of the test object will require a significant number of independent measurements, and, consequently, non negligible computer time. However, at least in principle, the mesh can be as small as desired to get the required spatial resolution. In practice, spatial resolution is dictated by the sensitivity of the measurements and, more explicitly, by the available signal over noise ratio.

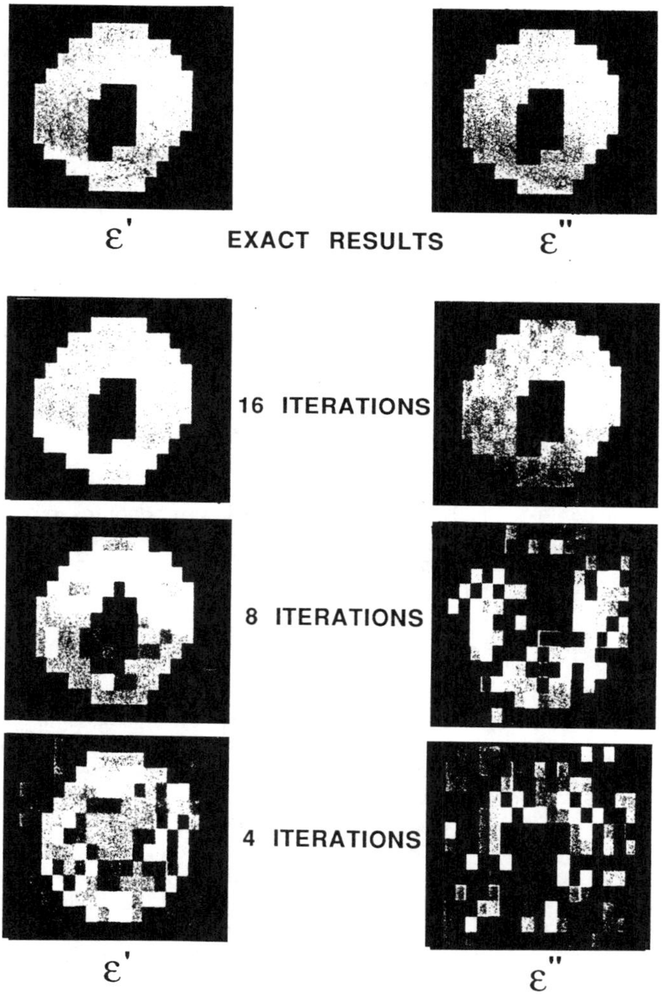

Figure 11 : Quantitative tomographic reconstruction of an highly contrasted cylinder immersed in water. Numerical results obtained at 3 GHz from 8 projections. Complex permittivity (real part, imaginary part) : surrounding medium (76,14.4), intermediate medium (46, 12), central medium (8, 1.2)

Figure 11 illustrates typical results obtained with a test object consisting of three highly contrasted media. The convergence toward the exact permittivity profile can be observed.

Conclusion

The microwave imaging techniques reviewed in this paper illustrate, non exhaustively, the range of available means to perform non destructive testing of materials. From simple monitoring of scattered wavefronts up to complex non linear reconstruction algorithms, these techniques may be used for different applications. Non-contacting and on-line transverse control of conveyed products can be conveniently achieved via either direct monitoring or free space back-projection. Moisture and density measurements as well as fault detection are possible. The main advantage of such a simple approach consists of its rapidity. Equipment providing sampling rate of the order of one centimeter over a few meters are already now available for the inspection of "slab" materials conveyed at speeds up to a few meters per second.

Diffraction tomography has been used mainly for testing of thick bulk material. Contact testing is usually recommended to reduce the reflection at the surface of the material. Spatial resolution is inherently limited to one half of a wavelength in the material. Consequently, the choice of the operating frequency results from a compromise between the desired spatial resolution and investigation depth. According to the required longitudinal spatial resolution and viewing capabilities in the operational context, multi frequency or multiview procedures can be used. Diffraction tomography algorithms can be processed by means of standard PC within a few seconds. Such efficient algorithms are available for both planar and cylindrical coordinates. They have been already assessed by means of advanced prototypes.

Finally, inverse scattering approaches appear as the convenient way to reach the ultimate goal of quantitative dielectric tomography. A lot of theoretical and numerical studies are under progress. The main problems consist now of 1) noise sensitivity and 2) duration of numerical processing. Both kinds of difficulties are efficiently treated by taking into account a priori information. Furthermore, the permanent increase of computer capabilities leads us to expect drastic reductions of the computation times. With such techniques, spatial resolution no longer appears to be constrained so directly by the operating wavelength. Such a situation should offer more flexibility in optimizing spatial resolution / penetration requirements for arbitrarily shaped objects. At this moment, no really operational or advanced prototype yet exists. While most of the work is performed at the modeling level, non negligible difficulties may be expected when they are extended to practical cases involving numerical model errors, imperfect illumination, interaction between sensors and materials under test, etc... However, this promising approach opens the door to free space quantitative dielectric metrology.

References

1. J.Ch. Bolomey, G.Cottard, B.J.Cown, On-line transverse control of materials by means of microwave imaging techniques, in Microwave Processing of Materials II, edited by W.B.Snyder, W.H.Sutton, M.F.Iskander, D.L.Johnson, Mat. Res. Soc. Symp. Vol. 189, 49-53, (1991)
2. P.C. Clemmow, The Plane Wave Representation of Electromagnetic Fields, Pergamon Press (1966)

3. Ch. Pichot, L. Jofre, G. Peronnet, J.Ch. Bolomey, Active microwave imaging of inhomogeneous bodies, IEEE Trans. Ant. Propagat.,AP-33,416-425, (1985)
4. J.Ch. Bolomey andCh. Pichot, Microwave Tomography: from theory to practical imaging systems, Int. J. Imaging Syst. Technol.,1, 119-131, (1990)
5. L. Chommeloux, Ch. Pichot, J.Ch. Bolomey, Electromagnetic modeling for microwave imaging of cylindrical buried inhomogeneities, IEEE Trans. Microwave Theory Tech. ,MTT-34,1064-1076, (1986)
6. J.Ch. Bolomey, Ch. Pichot, G. Gaboriaud, Planar microwave imaging camera for biomedical applications: critical and prospective analysis of reconstruction algorithms, Radio Science, 26, n°2, 541-549, (1991)
7. Ch. Pichot and P. Trouillet, Diagnostic of reinforced structures: an active microwave imaging method, in Bridge Evaluation, Repair and Rehabilitation, NATO ASI Series, edited by A.S.Nowak, Kluwer Academic Publishers, Dordrecht, (1991)
8. N. Joachimowicz, Ch. Pichot, J.P. Hugonin, Inverse scattering: an iterative numerical method for electromagnetic imaging, IEEE Trans. Ant. Propagat., AP-39, 1742-1752, (1991)
9. N. Joachimowicz, J.Ch. Bolomey, Ch. Pichot, A. Franchois, J.P. Hugonin, L. Garnero, G. Gaboriaud, Quantitative microwave tomography for non-invasive control of hyperthermia. Preliminary numerical results, Journal of Photographic Science, 39,149-153, (1991)
10. Y.M. Wang, W.C. Chew, An iterative solution of two-dimensional electromagnetic inverse scattering, Int. J. Imaging Syst. Technol.,1, 100-108, (1989)
11. W.C. Chew, Y.M. Wang, Reconstruction of two-dimensional permittivity distribution using the Distorted Born Iterative Method, IEEE Trans. Med. Imag., MI-9, 218-225, (1990)

IN-SITU QUALITY CONTROL OF THE PRODUCTION OF SEMICONDUCTOR DEVICES BY MICROWAVE PHOTOCONDUCTIVITY MEASUREMENTS

C. SWIATKOWSKI, A. SANDERS AND M. KUNST
Dept S1: Solare Energetik
G. SEIDELMANN, C. HAFFER AND K. EMMELMANN
Dept D1: Software-Technik

Hahn-Meitner-Institut , Glienicker Straße 100,
1000 Berlin 39, Germany

ABSTRACT

The application of transient photoconductivity measurements in the microwave frequency range to the characterization of semiconductors and semiconductor devices is analyzed. Quality control and in-situ optimization are discussed from a more general point of view and as a concrete example the optimization of the deposition of amorphous silicon films is presented.

INTRODUCTION

Since (excess) charge carrier kinetics plays a very important part in the performance of most (opto) electronic semiconductor devices, it seems obvious to look for techniques reflecting these properties for quality control and optimization of these devices. In particular contactless and non-invasive techniques will be important, because these enable continuous measurements of the quality during a production process.

In this work measurement of the transient photoconductivity in the microwave frequency range will be presented as a possibility to monitor the production of optoelectronic semiconductor devices. In particular, in-situ measurements during the deposition of hydrogenated amorphous silicon (a-Si:H) will be considered.

QUALITY, QUALITY SIGNAL AND APPLICATIONS

For an efficient production of semiconductor optoelectronic devices it is very important to dispose of techniques that enable monitoring of production steps and monitoring of the quality of the starting material. Essential for the suitability of a technique for quality control is the relation between the signal obtained with this technique (the quality signal) and the quality of the sample. A quality signal must yield sufficient information about the sample with respect to its application. For continuous quality control the measurement of the quality signal must not damage the sample or perturb the production step of interest. The relation between quality signal and quality (i.e.the quality function) can either be derived theoretically or be found empirically.

For the present work the empirical case is of interest. In this case comparison of the quality signals of samples with known quality was used to derive a quality function. Even if a relation between some quality signal and the quality is known this does not imply, that with this signal a useful quality control can be realized. Therefore is it necessary that the relation between quality signal and quality enables a satisfactory classification of samples in two groups: samples of sufficient quality to be used and products that must be rejected.

If a satisfactory quality signal for a sample is available, optimization of the sample can be carried out: samples produced under different conditions yield a set of quality signals and comparison of the quality signals gives the optimal production conditions. Optimization can be done ex-situ after a production step but an exciting possibility would be in-situ computer controlled optimization: during a production step a quality signal is continuously monitored. After certain time spans the production parameters are changed. A computer program is capable of interpreting the quality signal and deciding which set of production parameters must be adjusted to improve the sample quality.It is evident that the requirements for a quality signal for in-situ optimization are rather tight: not only must it satisfy the conditions mentioned above for an appropriate quality signal for continuous quality control, the quality signal must also give satisfactory information on the sample produced under the instantaneous production conditions.

The Time Resolved Microwave Conductivity (TRMC) technique enables contactless and non-invasive measurements of the transient photoconductivity and is a promising candidate for generating a quality signal and for in-situ optimization [1]. A real possibility for in-situ optimization with the TRMC technique is the deposition of thin-film semiconductor films from the gas phase. In this work the in-situ monitoring of the plasma deposition of a-Si:H from silane by TRMC-measurements will be presented as a possible system for computer controlled in-situ optimization.

TIME RESOLVED MICROWAVE CONDUCTIVITY (TRMC) MEASUREMENTS

TRMC measurements monitor the change of the microwave power reflected from a semiconductor sample upon pulsed illumination of this sample. The relative change of the reflected microwave power upon pulsed illumination at time t after the start of the illumination will be called the TRMC-signal [1]. In general the TRMC-signal reflects the dynamics of the change of the dielectric function of the sample upon illumination. However, if mobile charge carriers are produced by the illumination, the contribution of this species to the TRMC-signal is so large that the contributions of other excess species induced by the exciting light pulse can be neglected [1]. Because TRMC-measurements are performed under small perturbation conditions and only measurements at higher temperature will be discussed, the TRMC-signal $\Delta P(t)/P$ is proportional to the number of excess mobile species at time t $\Delta n(t)$ weighted by their respective mobility μ_i [2]:

$$\frac{\Delta P(t)}{P} = A \sum_i \mu_i \, \Delta n_i(t) \qquad (1)$$

For a-Si:H films this relation has been checked experimentally and confirmed by comparison of the TRMC-signal to the transient photoconductivity obtained in the same sample with a coplanar electrode configuration [2].

Although it may be obvious that the kinetics of mobile excess carriers and their mobility in some sample yield information about quality of this sample for application in optoelectronic devices, the relation between TRMC-signal and quality cannot be discussed generally without specification of the sample and the possible application. In this work, this relation is treated for the deposition of intrinsic hydrogenated amorphous silicon (a-Si:H) for application in p-i-n a-Si:H solar cells.

It is shown that the TRMC-signal in intrinsic a-Si:H films is due to mobile electrons and the contribution of other species can be neglected [2] :

$$\frac{\Delta P(t)}{P} = A\, \mu_n\, \Delta n(t) \qquad (2)$$

where $\Delta n(t)$ is the number of excess electrons at time t and μ_n their mobility. The information about the optoelectronic properties of a-Si:H films from TRMC-measurements and so the relation between TRMC-signal and quality of an a-Si:H film lies in these two factors. If the excess carriers are induced by light with a high absorption coefficient carrier diffusion can be neglected and the TRMC-signal is mainly determined by a surface layer of the a-Si:H film until a thickness of about the inverse of the absorption coefficient.

OPTIMIZATION AND QUALITY CONTROL DURING DEPOSITION OF a-Si:H FILMS

Intrinsic a-Si:H films are an important part of a-Si:H p-i-n solar cells and the optimization will be directed to optimal performance of these solar cells. The quality of a solar cell is determined by its yield, i.e. the ratio between the total electrical energy that can be extracted from the solar cell and the total energy of solar photons impinging on it. It seems difficult to relate the optoelectronic properties of intrinsic a-Si:H films as revealed by TRMC-measurements reflecting the properties of excess electrons to the properties controlling the yield of a solar cell. However, if only the deposition temperature as the most important parameter is varied to simplify the problem, the empirical approach mentioned above can be applied to this problem : It is well documented that the quality of intrinsic a-Si:H films for solar cells is optimal if deposited between 200° C and 250° C and the quality decreases with decreasing deposition temperature [3]. This can be compared with the TRMC-signals of films deposited at different temperature. For this reason the in-situ monitoring of the plasma deposition process by TRMC-measurements where only the deposition temperature is varied, will be presented.

THE PLASMA DEPOSITION PROCESS

a-Si:H films are produced by glow discharge from silane gas on a hot glass substrate, where the plasma discharge is induced by a 13.56 Mhz source. Films were produced under standard conditions yielding state-of-art films where only the deposition (substrate) temperature was varied between 30° C and 300° C. The maximum thickness of the films deposited is about 0.001mm, covering the range of thickness used in practical application of the material [3].The deposition rate is about 0.1 nms^{-1} at 250° C decreasing with decreasing deposition temperature.

For the present work a UHV-plasma deposition reactor was used fitted with equipment enabling measurements during the deposition process. TRMC-measurements were performed in a Ka-band equipment [1] where excess carriers are induced by 10 ns (FWHM) pulses of a Nd:YAG laser at 532nm and an intensity of about 0.1 $mJcm^{-2}$. The time resolution of the measurements is 10ns controlled by the duration of the exciting laser pulse and measurements were done in the time range between 0-200 ns.

with growing thickness as long as the film has not reached a thickness much larger than the inverse of the absorption coefficient of a-Si:H for 532nm light. The increase of the signal has even be used to determine the absorption coefficient and a nice agreement with the literature value has been found [4]. Unfortunately, this does not yield much information about the quality of the material deposited, because the absorption coefficient of a-Si:H at 532nm is only a very weak function of the quality of the material deposited. The large effect of the initial layer growth on the TRMC signal in this range hides differences in TRMC-signal due to differences in quality.

The TRMC-amplitude does not change toward the end of the deposition at 250° C and reflects the optoelectronic properties of a layer deposited at 250° C .This can be considered as the saturated amplitude of the 250° C layer measured at 250° C. Then the deposition temperature is lowered to 120° C and the TRMC amplitude measured before the start of the deposition corresponds to the saturated TRMC-amplitude of the 250° C layer measured at 120° C. After a few minutes of deposition at 120° C a change in the TRMC-signal appearance can be observed: not only a lowering of the signal amplitude but also a general shape change. The deposition at 120° C is continued until the TRMC-amplitude corresponds to the saturated amplitude at 120° C of a 120° C layer. It is clear that the saturated amplitude only reflects the properties of the 120° C layer. The change of the TRMC-amplitude during the deposition at 120° C can be explained by linear superposition of the saturated TRMC-amplitude of the 250° C-film and that of the 120° C-film at 120° C weighted by the number of photons absorbed in the 250° C layer and in the 120° C layer, respectively.It is evident from the experimental, data that the saturated TRMC-amplitude of the 120° C (and therefore a low quality) layer is much lower than that of the 250° C (a high quality) layer. It can be derived from Fig.1, that this difference in TRMC-amplitude corresponding to a difference in quality can already be observed after a deposition of only some minutes (i.e. some nanometers deposition) at the new parameter.

The experimental results given above, suggest that the TRMC-amplitude should become smaller with lower deposition temperature. Systematic in-situ and ex-situ investigations have shown that this correlation can be generalized [5,6] and that good quality a-Si:H is characterized by a high TRMC-amplitude. More sophisticated classification methods indicate that also the shape of the TRMC-signal can be used for a more refined classification, although in this case the correlation is more difficult and an influence of layers previously deposited has to to be assumed.

MODEL FOR IN-SITU OPTIMIZATION

Computer programs will be and have been developed capable of classifying TRMC-signals and to derive optimal values for process parameters to optimize the film properties. These values are sent on-line to a process control system. In Fig.2 an example is given how an an on-line in situ optimization can be carried out. The TRMC signal is preprocessed and then the signal is classified by a quality determinator. These classifiers are obtained using two different approaches: a neural-net based one and a rule based one. A quality therapist then decides which deposition parameter has to be changed to improve the film quality.

The measurement parameters mentioned above were chosen to optimize the information gained by TRMC-measurements on the growing a-Si:H film under the instantaneous production conditions: 532nm light is strongly absorbed in a-Si:H characterized by an absorption depth of about 0.2 μm, and the short time range covered of 200 ns enables the neglect of excess carrier diffusion. Consequently the bulk of the TRMC-signal reflects the properties of a thin film at the top of the sample.

Measurements under these conditions can be performed without perturbation of the deposition process and the deposition process does not appreciably perturb the measurements [4].

IN-SITU TRMC-MEASUREMENTS DURING DEPOSITION

As an example of in-situ monitoring of the deposition of a-Si:H with the TRMC-technique experimental results are displayed in Fig.1 measured during the deposition of a-Si:H on a glass substrate as a function of the time elapsed since the start of the deposition (tdep), where first the material was deposited at 250° C during 90 minutes and afterwards at 120° C during the same time span. TRMC-signals in a-Si:H are characterized by an initial increase of 10ns due to the production of excess electrons by the exciting laser pulse, followed by a fast decay between 10 ns and 40 ns giving way to a slower decay (Fig.1).

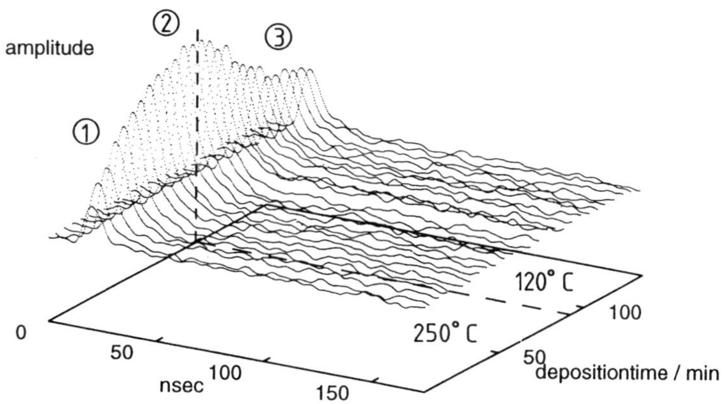

fig. 1: TRMC signals of a a-Si:H two layer system during the deposition. The first layer is deposited at 250° C and the second layer at 120° C. ①: growth of the amplitude => increasing quality; ②: saturation at high level => good quality; ③: decreasing amplitude => decreasing quality

During the first deposition at 250° C the maximum of the TRMC-signal at 10ns (the TRMC-amplitude) increases with increasing deposition time attaining a constant level at about an hour after the start of the deposition. This can be explained by the increase of the number of photons absorbed in the film

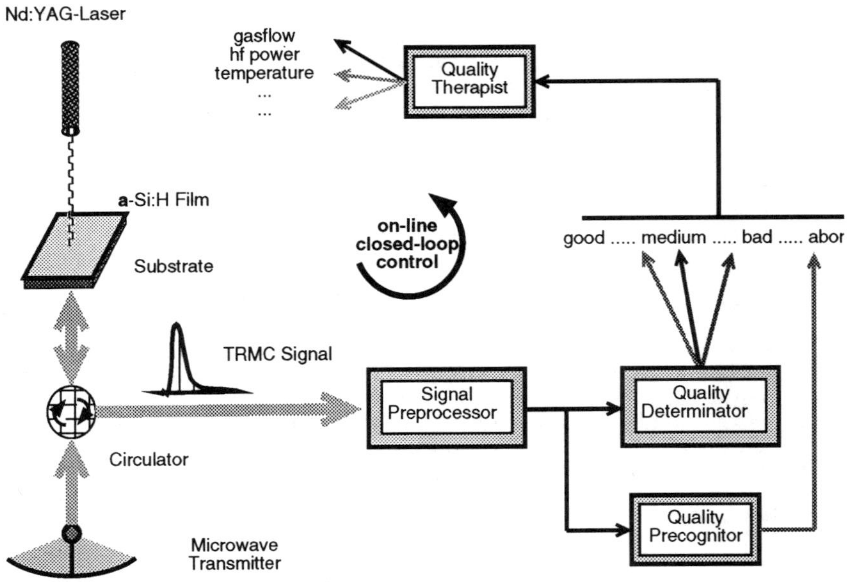

fig. 2: schematic setup for an in situ optimization process; details see text

OUTLOOK

Experiments yielding sets of data to obtain a quality function have been carried out. Neural nets based classifiers have been trained with this data. Inductive methods were applied to this data giving rule based systems for quality classification. Optimization strategies, the main component of the quality therapist, are now developed in connection with these classifiers.

Support by the B.M.F.T. for this project under no.0329031A is gratefully acknowledged.

REFERENCES

[1] M. Kunst, in Microwave Processing of Materials II, Mat. Res. Soc. Proc. 189, 75 (1991)

[2] M. Kunst and A. Werner, J. Appl. Phys. 58, 2236 (1985)

[3] A.Madan and M.P.Shaw, The Physics and Applications of Amorphous Semiconductors, (Academic Press, Boston, 1988).

[4] H.-C. Neitzert and M. Kunst, to be published

[5] M.Kunst and H.-C.Neitzert, J. Appl. Phys. 69, 8320 (1991)

[6] M. Kunst,H.-C. Neitzert and H. Ruebel, Phys. Rev. B 44,1081 (1991)

MICROWAVE BASED AND MICROWAVE AIDED NON DESTRUCTIVE TEST METHODS

G.d'AMBROSIO,R.MASSA,M.D.MIGLIORE
Un.di Napoli "Federico II",Dip. di Ing.Elettronica,Via Claudio,I 80125 Napoli,Italy
A.CILIBERTO,C.SABATINO
Alenia Aeronautica, Direzione Tecnica Trasporto,I 80038 Pomigliano d'Arco, Italy

ABSTRACT

The reflection coefficient of Fiberglass/Kevlar slabs, including a layer of a different material was calculated as a function of permittivities, thicknesses, defect depth, and microwave frequency. Experimental tests were done with a single-frequency, amplitude and phase sensitive circuit, and with a swept-frequency amplitude only sensitive set-up. In this latter case a suitable processing of data allowed effective and fast defect detection. First experiments on microwave heating for thermographic NDT were also performed.

NUMERICAL EVALUATIONS

Numerical evaluations have been carried out in order to estimate the influence of different parameters like defect thickness t and depth d, or microwave frequency f on a defect detection system based on reflection measurements.

The defect was modeled as a layer having relative permittivity ε_1 and thickness t, included into a slab having relative permittivity ε_2; the incident microwave field was modeled as a plane wave propagating perpendicularly to the slab, and transmission line theory was applied.

In the calculations the overall thickness a, the materials ε_1, ε_2, and the defect thickness t were fixed values; defect depth d and microwave frequency f were varied.

The reflection coefficient in case of no defect, $\Gamma_0(t=0)$, was first calculated as a function of f, $\Gamma_0 = \Gamma_0(f)$, and served as a reference. Then, for a given defect $(\varepsilon_1, t \neq 0)$ the reflection coefficient $\Gamma = \Gamma(d,f)$ and the dB Difference $20\log|\Gamma| - 20\log|\Gamma_0| = D(d,f)$, were calculated; the larger the difference D, the easier to find out the defect by means of reflection coefficient amplitude measurements.

The difference between the arguments of the reflection coefficient, $Arg(\Gamma) - Arg(\Gamma_0) = \Delta(d,f)$, was also calculated, still as a function of defect depth and microwave frequency; large Δ values suggest defect detection by means of interferometric techniques, effective even in case of small D values (small reflection amplitude changes).

The analysis was carried out on Fiberglass and Kevlar slabs including (as a defect) a layer of air, Upilex, or Teflon. The permittivity values are collected in table 1; they were assumed to be constant over the chosen frequency range. The 12-18 GHz band was chosen as a suitable one from the point of view of both field penetration and space resolution. A 5 mm overall thickness (a) was assumed.

As an example the results of calculations in case of an air layer, 0.5 mm thick, inserted into a fiberglass slab, are displayed in figs 1 and 2. It is shown that the amplitude parameter D and the phase parameter Δ depend in a rather complex way on defect depth d, and microwave frequency f. The same thing was found for all other calculated cases.

Tab.1: Complex relative permittivity (12-18 GHz, 20 °C)

	real part	imaginary part
Fiberglass	4.5	0.1
Kevlar	3.5	0.08
Air	1	0
Upilex-R	3.5	0.0014
Teflon	2.1	0.0003

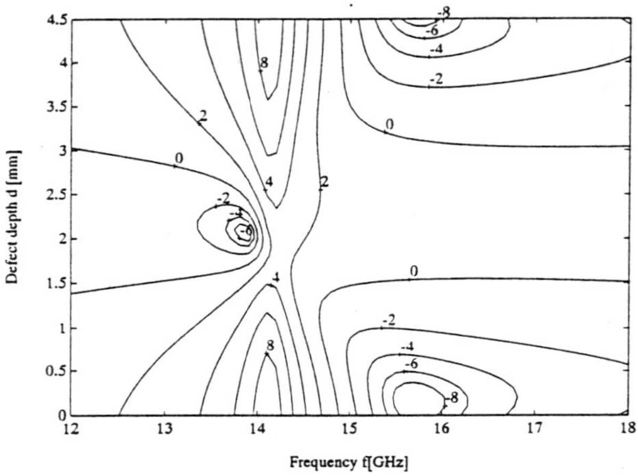

Fig 1:
The dB difference $D(d,f)$ contour plot. Air defect 0.5 mm thick, into a Fiberglass slab.

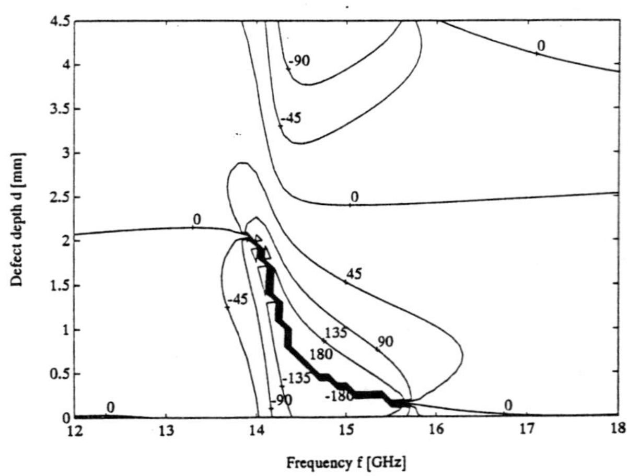

Fig 2:
The argument difference $\Delta(d,f)$ [degrees]. Same case of fig. 1

Given that, for defect detection purposes, one, or both, the above parameters must be over some threshold value, and given that the defect depth d is not known, it is easy to conclude that a frequency scan is always necessary in order to find out suitable working conditions. Of course more specific indications can be drawn from each particular case calculated.

MICROWAVE INTERFEROMETRIC TECHNIQUE

The following interferometric technique is sensitive to both amplitude and phase changes of the signal reflected by the specimen under test. So the detection of flaws is possible when either the "amplitude difference" D or "the phase diference" Δ is large enough.

The specimen was a flat slab of Fibreglass (or Kevlar) 300 mm x 150 mm x 5 mm thick, with an artificial defect (Upilex or Teflon 13 mm x 13 mm x 0.2 mm thick) inserted at a given depth (0.5, 1, 2 mm) in a given location into the slab.

The experimental set-up (fig.3) was a standard monostatic microwave interferometer (Michelson). The signal from the microwave generator, incident in port 1 of a hibrid Tee was equally splitted in arms 2 and 3.

The probe (the flat output flange of a 15.7 mm x 7.7 mm waveguide, ending the arm 2) was positioned just against the sample; behind it microwave absorbing panels were positioned in order to reduce ambient reflections.

With the probe on the center (reference position) of the slab, the sliding short and the variable attenuator on arm 3 were adjusted until the readout on the power meter on arm 4 was nulled. Then the probe was positioned on different points along the center line parallel to the long edges on one side of the slab under test: off-null readings ΔP revealed structure changes (defects) in the small region faced with the waveguide end. The stability of the source during the test was monitored and controlled by means of a 20 dB directional coupler on arm 1, a power meter and a variable attenuator. The off-null readings ΔP were normalized to the level P measured in the starting reference position with 60 dB attenuation on arm 3.In the above procedure the center of the slab was arbitrarily supposed to be defect free; in case this is not true, off-null readings will appear in most other probe positions, and the starting reference point must be changed.

In fig. 4 $\Delta P/P$ values were plotted versus the microwave probe position, at a given frequency, for a Fibreglass slab with a defect of Upilex inserted at 2 mm depth. The defect was sharply detected and located where $\Delta P/P$ rises from -30 dB to -6 dB.

As expected these changes depend on frequency, as shown in fig. 5 where $\Delta P/P$ was plotted versus frequency with the probe located on the defect. Similar results were obtained with other samples.

Experiments with a non-contacting probe were carried out by inserting styrofoam sheets (air-like material) of different thicknesses between the probe and the slab under test; the defect was still well detected up to a thickness of 9 mm.

These results show that the above microwave interferometric technique (based on the comparison between amplitude and phase of the signal reflected by the specimen at the point where the defect is, and that reflected at a point where no defect is) can reveal and locate structural discontinuities (flaws). It was also shown that the detection of the defect is possible also in case of non contacting probe, provided that the distance between probe and sample is constant.

Shortcomings of this technique are due to the phase sensitivity itself, so that constant sample thickness and constant probe distance are strictly required. In addition, in order to find out the most suitable frequency, a large number of single-frequency measurements is necessary.

Fig.3:
Monostatic
microwave interferometer

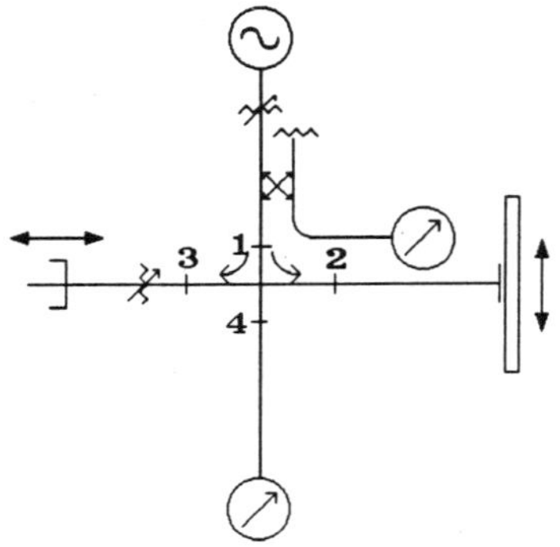

Fig 4:
Normalized off-null readings on the microwave interferometer as a function of probe position

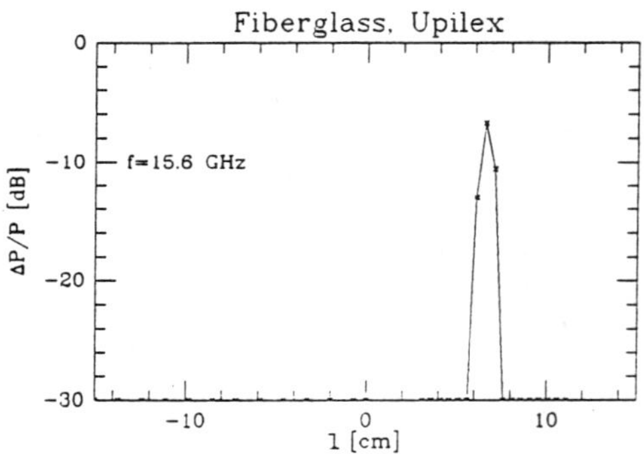

Fig 5:
Normalized off-null reading on the defect, as a function of frequency

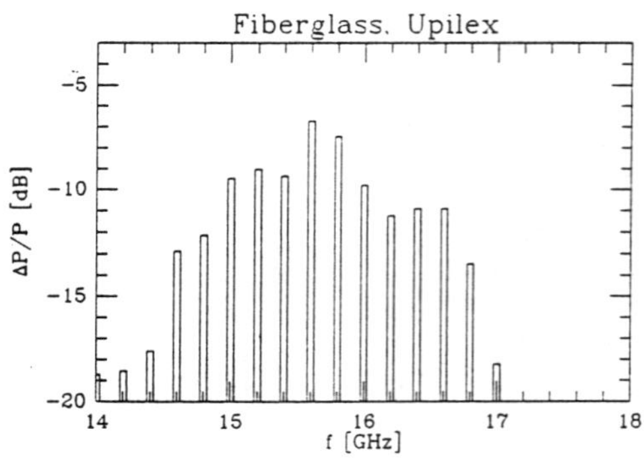

SWEPT FREQUENCY REFLECTOMETRIC TECHNIQUE.

The same samples tested with the above interferometric technique were also tested using a swept frequency reflectometric technique. Here the amplitude only of the reflection coefficient was measured, but over a wide frequency range and in a very short time.

The experimental set-up is shown in fig.6. The signal from the microwave generator was splitted into a reference input for the scalar network analyzer and a signal feeding the microwave probe (HP P281C) through a directional bridge. The signal reflected by the sample in turn, was fed to the network analyzer through the directional bridge.

Again microwave absorbing panels behind the sample reduced ambient reflections.

Just like in the microwave interferometer technique the probe was positioned on different points along the center line on one flat side of the specimen, but in this case at each point (x_i) the reflection coefficient amplitude $|\Gamma(x_i,f)|$ over the entire 12-18 GHz band was measured and recorded. The mid point of the specimen ($x = 0$) was again chosen as a defect free reference, and at each other point (x_i) the dB Difference

$$20\log|\Gamma(x_i,f)| - 20\log|\Gamma(0,f)| = D(x_i,f)$$

was calculated and recorded as a function of frequency.

The "Integral Difference" (mean square value of D) over the band $\Delta f = 6$ GHz

$$\sqrt{\frac{1}{\Delta f}\int_{\Delta f} D^2(x_i,f)df} = ID(x_i)$$

was calculated as a function of position (x_i) and was assumed as a measure of the "distance" between the reflection coefficient amplitude at $x = x_i$ and the reflection coefficient amplitude at $x = 0$ (reference point, defect free by arbitrary assumption). Again if high ID values occur over most points, the reference position must be changed.

As an example, with reference to a Fiberglass slab with a 0.2 mm thick Upilex defect at 2 mm depth, the $ID(x_i)$ plot is reported in fig. 7; a sharp peak occurs on the defect.

The above procedure demonstrated its effectiveness also in non-contacting probing conditions.

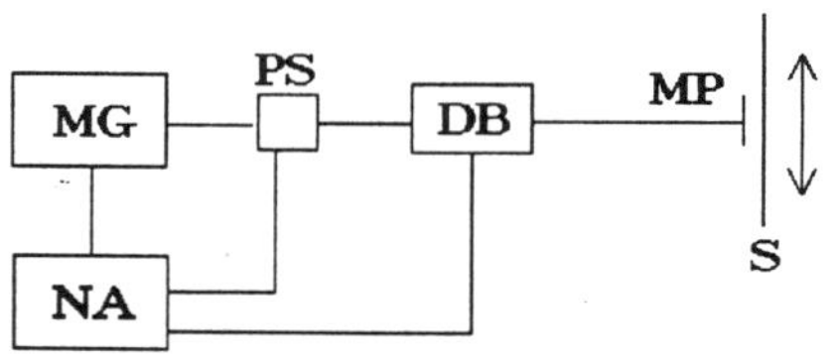

Fig.6: Swept frequency reflectometer; MG:Microwave generator, NA:network analizer, PS:power splitter, DB: directional bridge, MP:microwave probe, S:specimen

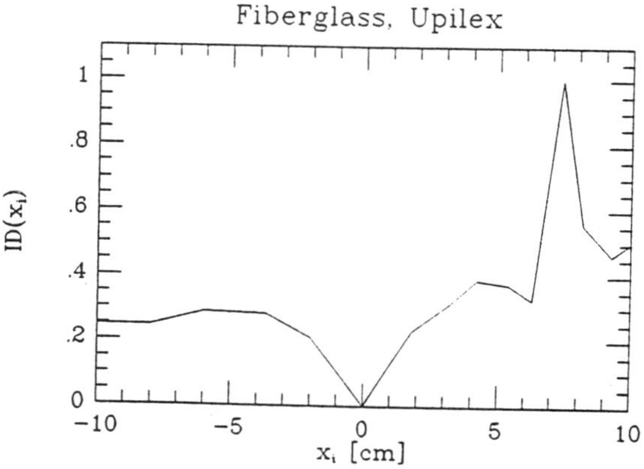

Fig 7:
Integral Difference as a function of probe position.

MICROWAVE EXCITATION FOR THERMOGRAPHIC NDT

In the NDT domain microwaves can be used not only as a means for direct defect detection, but also as an auxiliary tool in thermographic techniques. In fact the due temperature rise can be acheived by means of high power microwave exposure of the structure to be thermographically inspected. The heating being a consequence of microwave power absorption, the material under test must absorb (at least in some degree) the microwave radiation. Hence the knowledge of the complex permittivity of the materials involved is the basic prerequisite in order to evaluate the feasibility of such a microwave heating. In addition an adequate microwave power source and a suitable design of the microwave applicator system are necessary in order to obtain temperature rises as fast and uniform as possible.

Among the many possible applicator schemes our first choice was a mode stirred cavity where the load to be heated (a plane structure made of composite material) was put in place of one of the metallic walls of the cavity. The 2.45 ± 0.05 GHz band (allowed for non-telecommunication purposes) was chosen due to the easy availability of microwave sources and components working at the above frequencies.

Exposure uniformity was tested at low power by means of a field probe sliding over the outer surface of the sample. These measurements were done by means of a network analyzer over the 2.4-2.5 GHz range and served to optimize the mode stirring mechanism.

High power tests have just started: only some temperature measurements were done until now, in order to check the heating of the sample. A IR thermocamera will be used next for detection and recording of the microwave induced heating patterns.

CONCLUSIONS

Besides other NDT techniques, microwaves demonstrated to be an additional powerful tool for direct defect detection in dielectric composites not only by means of amplitude and phase techniques (interferometry) but also by means of amplitude-only reflectometry: swept frequency operation mode and integral response evaluation overcame the complexity of the interferometric technique. The use of microwaves as a heating means for thermographic inspection is another promising possibility.

ACKNOWLEDGEMENTS

This research started under the sponsorship of Alenia-Alenia Aeronautica DTT [1]; its continuation is done with the financial support of both the Italian Ministry for University and Scientific and Technological Research (MURST), and Alenia [2].

REFERENCES

[1] B.Bisceglia, A Ciliberto, G.d'Ambrosio, G.Ferrara, R.Massa, C.Sabatino, in Microwave Processing of Materials II edited by W.B. Snyder Jr., W.H. Sutton, M.F. Iskander, D.L. Johnson (Mat. Res. Symp. Proc, 189, Pittsburgh, PA 1991) pp. 497-503.

[2] G. d'Ambrosio, R.Massa, M.D.Migliore, in Electromagnetics in Aerospace Applications (2nd Int. Conf. Proceedings, Torino 1991) pp. 153-156.

PART X

Dielectric Properties and Measurements

RAPID HIGH TEMPERATURE MEASUREMENT OF MICROWAVE DIELECTRIC PROPERTIES

W.R. Tinga, University of Alberta, Edmonton, Canada, T6G 2G7

ABSTRACT

Practical high temperature (above 500°C) dielectric measurement techniques are reviewed stating both advantages and limitations of known methods. Details of a new and practical dual frequency microwave high temperature dielectrometer are discussed. This dielectrometer technique is capable of measurements beyond 2000°C. Both heating of the small samples and the measurement of their complex dielectric constant is done simultaneously using both a medium power (60W) microwave source and a low power scalar network analyzer. Good accuracy is obtained and the measurement is capable of being totally automated. Presently, a measurement and heating cycle time of about 20 s is needed to completely characterize the sample's microwave dielectric response from 300°C to 1600°C. A second high temperature dielectrometer technique is discussed which requires only one microwave source of medium power level (60W). This method requires only the reflected power to be monitored during the microwave heating of the sample to obtain the complex permittivity versus temperature curve.

INTRODUCTION

Since the complex dielectric constant is the only macroscopic parameter describing the electric behaviour of a material immersed in a given electromagnetic field, the design and control of any high temperature microwave process requires a data base of high temperature permittivities. In recent years considerable effort has been expended toward achieving this goal. But, it is an ever moving goal of requiring data at higher temperatures, of greater accuracy and for a larger variety of materials. As a result, still more measurement techniques and modified approaches are being invented to circumvent various practical limitations of the multitude of dielectric measurement methods described in the literature. After reviewing existing high temperature methods and problems, two new dielectrometer methods are described suitable for permittivity measurements up to at least 1600°C.

AVAILABLE DIELECTRIC MEASUREMENT TECHNIQUES

Many dielectric measurement techniques exist and continue to be refined and adapted to new measurement problems. They can all be categorized under the three main headings of Lumped Element, Resonator and Transmission Line techniques. Limiting ourselves to the microwave range of 300 MHz to 300 GHz eliminates the lumped element methods.

Useful reviews of many microwave and RF property measurement methods have been published. Bussey's review (1967) comments on typical echniques, measurement errors, special purpose methods and magnetic permeability methods listed in his 65 references [1]. A later review by Afsar et al (1986) covers mainly dielectric measurement techniques from 1 Mhz to 1500 GHz and lists 187 references [2]. This paper is much more detailed than that of Bussey as it reviews frequency domain, time domain, transmission line, closed or open resonator, free-space, stochastic, calorimetric, broadband, discrete frequency, monochromatic, direct and indirect methods. Many of these techniques are aided by numerical field computations. Recently, another pertinent review was given by Carpenter who references a variety of less wellknown reports [3]. A very large number of the methods listed in these reviews are of only passing interest when one requires the ability to do measurements at temperatures above, say, 500°C. The list becomes short indeed, when one requires good performance of a permittivity measurement method at 1500°C in the commercially used ISM bands.

Broadband methods, such as the open coaxial line, are, in practice, limited to operate below 500°C, eventhough, a higher achievable 900°C limit is claimed by one author [4]. The open coaxial line method is useful for medium loss materials ($0.01 < \tan \delta > 0.1$). Broadband methods are, nonetheless, extremely useful in showing the frequency relaxation behaviour of the materials tested. Since frequency and temperature behaviour are related at the microscopic level, such broadband methods will continue to play an important role in expanding our knowledge of materials. Of course, many microwave engineering applications remain narrow band thereby making the discrete frequency measurement methods very relevant to obtaining the high temperature microwave behaviour of ceramic materials.

For many reasons, A.R. Von Hippel's famous short circuit method is still a very popular way of obtaining dielectric data [5]. With the advent of Automatic Network Analyzers (ANAs) and 6-port reflectometers, this method is reasonably convenient, reliable, well-documented [6-9] and can, with care and expensive platinum sample holders, be used up to about 1000°C. Its main drawback is that it requires a sample precisely fitted inside a waveguide, circular waveguide being the most practical, and the heating of the whole waveguide-sample assembly in say, a tubular furnace. One then needs to accurately measure the complex waveguide impedance. Good accuracy requires a quarter-wave or half-wave electrical sample length which, unfortunately, changes with temperature. Parallel and flat sample faces are also very important in maintaining accuracy and mode purity. The upper loss tangent limit of this method is about 0.1.

To overcome the problems of sample fit and the high temperature of the measurement equipment, the free-space, or free-wave, method was developed by Frost and Ho, among others [10,11]. Found useable to about 1650°C, this method has its own limitations. A plane wave microwave beam is transmitted through a sample heated in a furnace with shutter-like doors which are opened during a measurement with a microwave reflectometer bridge as shown in Figure 1. One of this method's main limitations is the necessity of a sample disk many wavelengths in diameter which, even at 35 GHz, translates to 10 cm. Calibration procedures for this technique are quite involved if good accuracy is desired. Only a few degrees of transient rise in temperature of the antenna lenses, used to view the hot sample, can cause significant errors. These can be minimized by using microwave transparent heat insulation.

Moreover, the increasing microwave absorption, with increasing temperature, causes a signal to noise ratio reduction limiting the dynamic permittivity measurement range. Special, power electronics, and electromechanical hardware is needed to open and close the furnace doors during the computer driven measurement portion of the cycle. On the other hand, thermal isolation of most of the microwave hardware is inherent in this method. Hollinger et al used the free-space method utilizing a microwave transparent heating chamber but they show data to only 850°C [12]. Sample surface finish is not critical nor is there an airgap problem as is the case for, say, the short circuit method. Finally the method is, in principle, useable over a wide frequency range but sample size will, in practice, limit it to the higher frequency bands. However, for composites and other sheet-like materials, this method is probably the most practical.

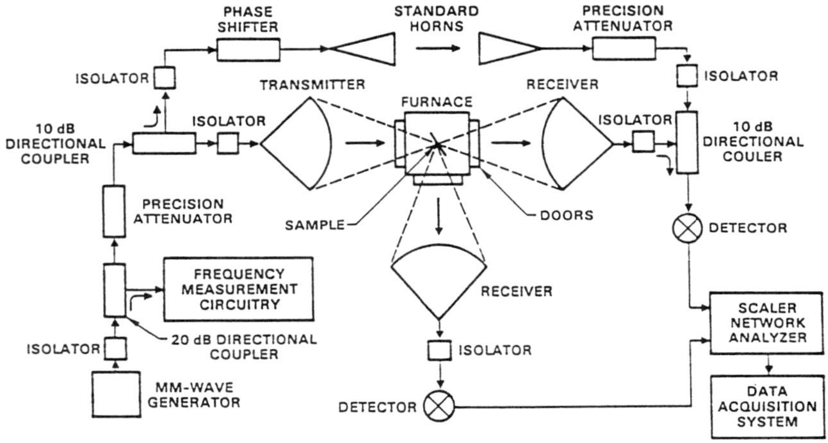

Figure 1 Experimental mm-wave setup for determining the high temperature permittivity of plate samples by the free-wave method, with permission from W. Ho [11].

Closed resonator methods are quite adaptable to high sample temperatures. Most of these methods require either mechanical or electronic tuning of the microwave source or of the resonator itself. It is important, normally, that any mechanical tuning devices not be subjected to a high temperature. Resonator detuning, due to thermal expansion, can be a major error source. Sample insertion is often facilitated by incorporating insertion holes which, by proper resonant mode choice, minimally affect performance. A special adaptation of these methods results in the ability to both heat and measure samples to above 1600°C [13,14] as will be described below.

Over the last few years, it is the above free-space, closed waveguide and closed resonator approaches which have been singled out by researchers as the most promising ways to get reliable, high and ultra-high temperature data [6-19].

To optimize a solution to the microwave high temperature dielectric measurement problem, it is important to consider the specific problems of high temperature measurement in more detail.

ANALYZING THE HIGH TEMPERATURE MEASUREMENT PROBLEM

There are many practical reasons why high temperatures cause measurement problems. Eventhough the measurement principles of any technique remain the same as for room temperature operation, many technical difficulties arise when the sample temperature is raised. The major ones are:

- Propagation constant and wavelength in a material are temperature dependent.
- Many techniques require sample contact with a metal conductor which leads to heat loss by conduction and thermal expansion.
- Thermal expansion distorts and changes the geometry and fitting accuracy of both the sample and the microwave structure housing it.
- Different materials and applications require the use of different sample sizes and shapes.
- The required sample size or mass, and the unavoidable heat losses set a practical upper temperature limit to the measurement.
- Dielectric loss or electrical conductivity becomes too high making optimum matching and tuning of the system very difficult.
- Measurement accuracy typically decreases as temperature increases due to increasing parameter uncertainties.

These problems are, more or less, common to all the known dielectric measurement techniques since all involve the use of wavelength, propagation constant and metallic conductors. Stating and understanding these problems makes it easier to find ways of minimizing them. There is no single optimum solution to measuring high temperature permittivity of an arbitrary material at an arbitrary frequency. Nevertheless, for the 915 and 2450 MHz ISM bands the problems can be minimized, as we will show, particularly if small samples are used.

TOWARD MINIMIZING THE HIGH TEMPERATURE MEASUREMENT PROBLEM

Clearly, a beneficial measurement improvement would be the use of a non-contact measurement method to minimize the thermal expansion and heat conduction problems. Free-space techniques have this characteristic but require electromagnetically large sample sizes. Large samples require a lot of energy to reach a high temperature and have large thermal inertia. As a result, the heating-testing time cycle is of the order of one hour per data point.

It follows that a reduction in effective sample size results in lower energy requirements to reach a given temperature and reduces testing time as well as heat loss. Reduction of the transverse sample size in a full-sized waveguide allowed Wong and Tinga to measure oxide powder samples to 1000°C using a transmission bridge technique [17,18]. Small sample size naturally leads one to consider resonator techniques with the sample perturbing the resonance in one way or another. Moreover, certain resonator configurations allow suspension of the sample in a more or less non-contacting manner thereby reducing the heat loss significantly.

On the other hand, using heat loss to advantage was demonstrated in a novel manner by Hutcheon et al [19]. They heated a small sample to its maximum required temperature in an external furnace in close proximity to a microwave resonator. After quickly pushing the sample from the furnace into the resonator, instrumented with a computerized reflectometer, the sample's microwave properties were rapidly measured while the sample was cooling back to room temperature. Due to the radiated heat loss, this method, so far, is limited to about 800°C. Thermally induced material hysteresis effects will also affect the values of the cooling curve data.

Using methods with very short testcycle times, is still another way in which to counter the problem of heat loss over time. Measuring the sample's microwave properties dynamically, while heating the sample, minimizes the duration of the heating and testing time cycle and is therefore very beneficial. This suggests the use of microwaves for both the heating and the measuring of a sample. Couderc et al used this approach some time ago [15]. One of the major difficulties encountered by him was the tuning system needed to keep the microwave power source tuned and matched to the changing cavity resonance and impedance. Furthermore, the cylindrical samples had to be of known length and needed to be precisely centered in the resonator to avoid end effects. Nevertheless, a major advantage of electromagnetic sample heating is the inherently higher heating rate and the reduced thermal gradient due to the localized volumetric sample heating action. To counter the ever present thermal expansion problem, more precise thermal modelling of the sample and measurement equipment is needed so that deviations and errors due to high temperatures can be predicted and properly accounted for.

Finally, to truly minimize the heat radiation losses in non-contact methods, a high quality, microwave transparent, high temperature, heat insulation becomes necessary at temperatures exceeding, say, 1200°C. For it is the radiation heat loss which will ultimately set the upper temperature limit achievable in high temperature non-contact dielectric measurements.

Having a clearer understanding of the above high temperature measurement problems led to the development of two new high temperature dielectric measurement methods. These, at least, for small samples and for a large variety of medium to high loss materials, minimize the high temperature problems and one of these methods also minimizes the measurement instrumentation complexity. These two new approaches for the 300-2000°C range will now be discussed.

DUAL-FREQUENCY HIGH-TEMPERATURE DIELECTROMETER

Using a coaxial cavity both as a microwave heater and a dielectrometer has resulted in the capability of measuring the microwave complex permittivity, of a small sample, from 300 to 1600°C in less than a minute. The method has been described in detail elsewhere so we will only discuss its technical highlights [13,20,21]. Shown in Figure 2, is the resonator which creates a strongly focused

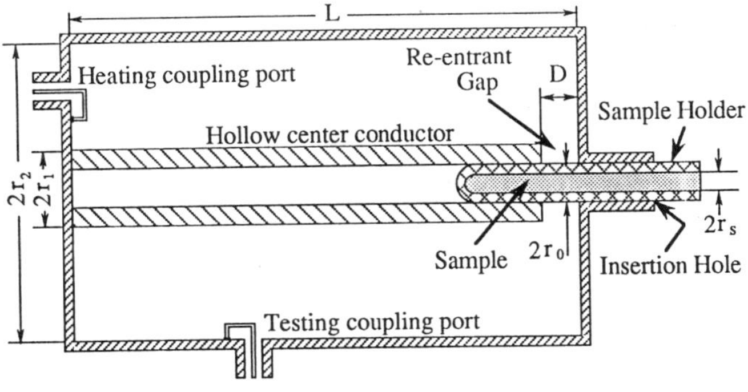

Figure 2 Cross section of coaxial re-entrant resonator used as a heating and testing chamber. Dimensions are given in [21,22].

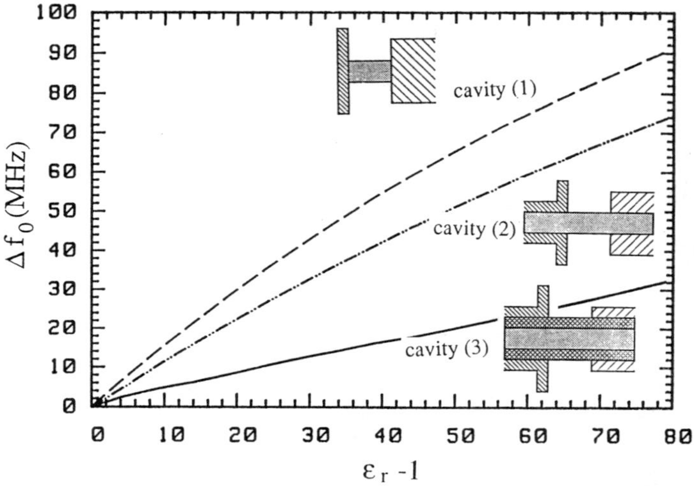

Figure 3 Resonant frequency shift produced by a sample at about 3 GHz for three different resonator configurations.

E-field in the gap causing sample dependent heating rates in excess of 750°C/sec. Moreover, it gives high resonator sensitivity to changing sample parameters which are monitored at a frequency different from the heating frequency. To allow very easy sample insertion and extraction, the center conductor is made

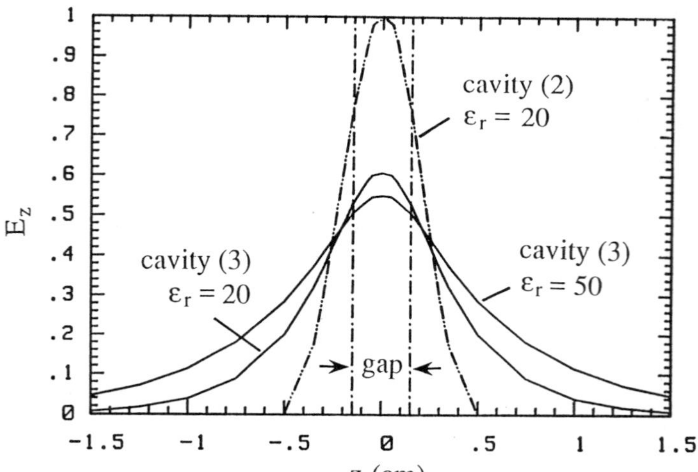

Figure 4 The normalized axial E-field distribution for the hollow center conductor cases of Figure 3.

hollow, which also causes the center conductor to act as an evanescent waveguide allowing energy storage, thereby linearizing the perturbation as shown in Figure 3 for cavity 3. The effect of the hole and the sample holder, which could be air, on the perturbation curve is clear. As Figure 4 shows, the axial field extends beyond the gap into the hollow center conductor but decays rapidly in the axial z-direction thus making the sample ends practically invisible to the microwave fields thereby eliminating any sample end effects. The energy storage ability of the hollow center conductor is illustrated by Figure 5 indicating that a, permittivity dependent, portion of the sample's stored energy resides outside the gap in the hole region.

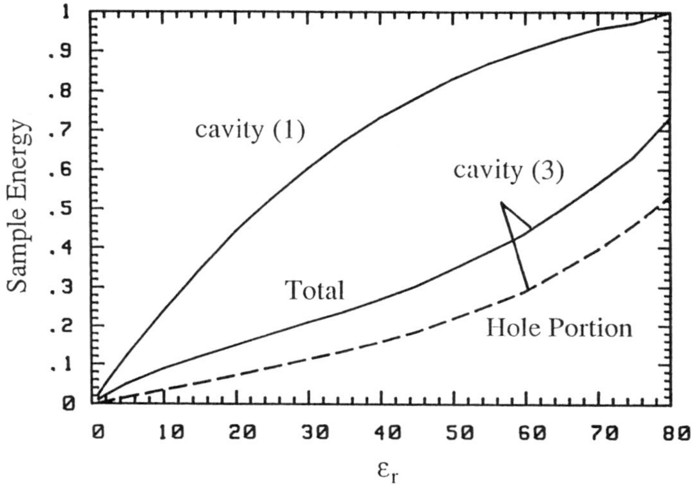

Figure 5 Normalized electric energy stored in a sample as a function of the sample's dielectric constant for cavity 1 and 3 of Figure 3; the hole contribution is seen to be significant.

Convenient, very rapid, repeatable measurements can be made with this structure which remains at or near room temperature during the entire testcycle time. If necessary, the long inner end of the hollow center conductor can be liquid cooled to keep the cavity temperature very constant. Inserting a dielectric sample causes a nearly linear frequency shift in the test band, over a permittivity range of 1 to 80, as shown in Figure 3 and its loss factor is determined from the cavity Q [21]. In addition, an accurate, theoretically derived, calibration of frequency shift versus permittivity is available, from a mode-matching model of this re-entrant dielectrometer, making calibration against reference materials unnecessary. To allow a wider complex permittivity range, the gap width can be increased but this does require a numerical recalibration of the cavity. A block diagram of the measurement system is shown in Figure 6.

A Proportional Integral, Derivative (PID) controller tightly controls the sample temperature to any preset value eliminating thermal runaway. In its present configuration the system needs some presetting of the tuner and pre-adjustment of the testport coupling device to give best results for a given class of materials. Some typical dielectric data, obtained in this way, is shown in Figure 7. Only 50-100 W of microwave heating power is needed to achieve sample temperatures exceeding 1600°C. Note that the thermometer temperatures were taken at an emissivity of 1 since its true value was unknown.

Not only is this method useful for measuring the high temperature dielectric properties of ceramics and oxides but it is an excellent dielectrometer for a large variety of materials having properties within its dynamic permittivity range. Samples can be in liquid, solid or pressed powder cylindrical form.

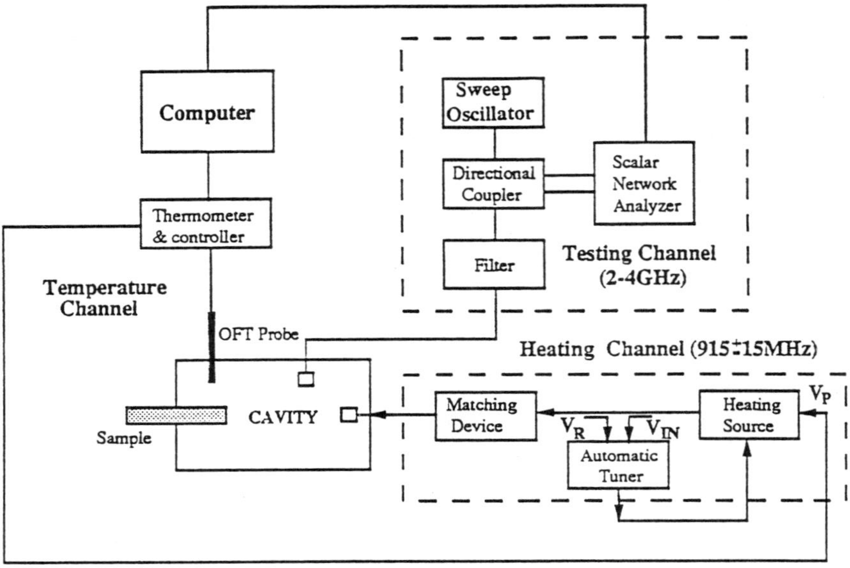

Figure 6 High temperature automatic dielectrometer system.

For temperatures up to 1200°C a 5 mm ID quartz tube makes a good sample holder, the effect of which is included in the numerical calibration process. Solid samples can equally well be suspended in the gap with a holder, which is withdrawn from the gap field (the high temperature zone) via the hollow center conductor, extending the temperature range to over 1600°C. To achieve still higher temperatures, a microwave transparent heat insulating jacket is required around the gap area. We believe that the high temperature

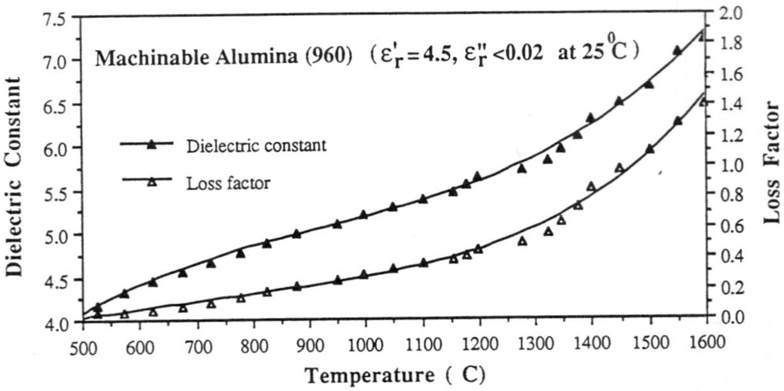

Figure 7 Temperature dependence of the dielectric constant of machinable alumina using the high temperature dielectrometer [22].

limit of this method can be pushed well over 2000°C.

Temperature monitoring is done with an optical fiber thermometer which sets the lower temperature limit of 300°C. Other microwave transparent thermometers, like the Luxtron, can be used to cover the lower temperature range. At the lower end, the dynamic loss tangent range is limited by the unloaded cavity Q, while at the high end it is limited by the matching range of the matching device. A later paper in this symposium will address the method's dynamic range in greater detail and show specific results [22]. In summary, the method's benefits are:

- a very rapid traversal of the temperature range is possible
- sample loading is extremely simple
- the resonator remains at room temperature
- the dynamic permittivity range is larger than for other cavity methods because of the hole effect
- it needs only a very small sample of simple shape
- heat transfer from the hot sample to the microwave structure is minimized
- calculated permittivity values are based on an accurate numerical model of the dielectrometer
- no precious high temperature metals are needed.

For these reasons, we believe that, at the lower microwave frequencies, the hollow coaxial reentrant cavity is near optimum. Yet, the method does require fairly sophisticated precision measuring instruments. However, even this problem can be minimized as discussed below.

SINGLE-FREQUENCY HIGH-TEMPERATURE DIELECTROMETER

Traditionally, Q-measurements require a non-zero frequency or time period. Recently, a new single frequency Q-measurement method has been introduced [14]. Consequently a dielectric sample can be heated while, simultaneously, its permittivity can be measured with a single frequency microwave heating source. This eliminates most of the otherwise required precision instruments and makes the method suitable for on-line monitoring of microwave material properties as those materials are being heated with microwave power. The trade-off is that electronic or mechanical adjustment is required to keep the resonator tuned and matched to the power source frequency as shown in Figure 8.

Matching is accomplished by the movable coupling loop and resonator tuning via the movable end wall. We are currently working on other ways of minimizing these tuning and matching requirements. Nevertheless, once able to meet this tuning and matching constraint, the procedure reduces to one of simply measuring the reflection coefficient of the resonant applicator with a simple directional coupler and diode detector. When calibrated, this arrangement causes the Q, and therefore the loss factor, to be inversely proportional to the input resistance at resonance. The main requirement for the validity of the method is that the resonator being monitored obey a perturbation stipulation which is readily met, for example, with the modified re-entrant hollow coaxial resonator as shown.

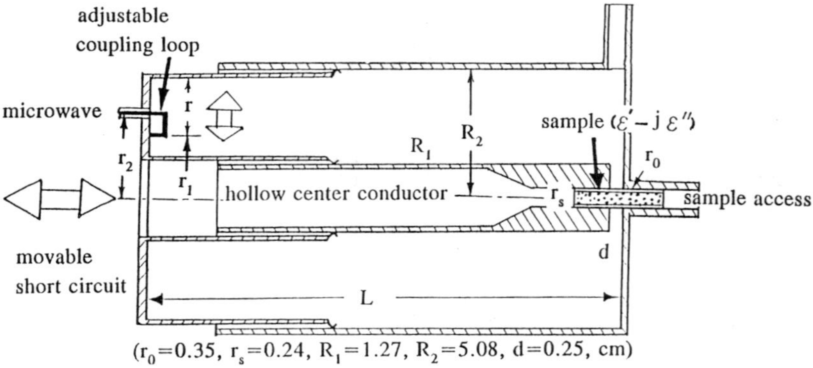

Figure 8 Cross section of the single frequency high temperature dielectrometer.

Using this method, dielectric measurement data has been collected in real time up to 1200°C, thereby allowing direct observation of a material's microwave behaviour during a heating, sintering or joining process. Figure 9 shows a

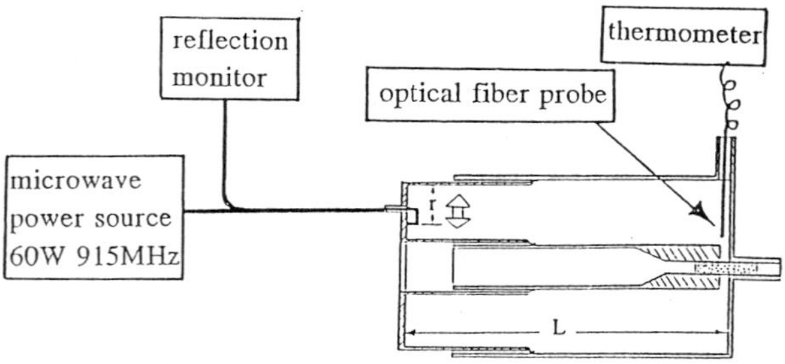

Figure 9 Single frequency high-temperature dielectrometer system.

typical experimental setup which indicates the equipment simplicity. More details of the single frequency method will be discussed in another paper [23].

CONCLUSION

High temperature dielectric measurement techniques were reviewed with emphasis on the practical high temperature measurement problems. Two new dielectrometer designs were presented which minimize such problems to allow very rapid small sample testing to be carried out over a 300 to 2000°C range in the 915 MHz and 2450 MHz ISM bands.

ACKNOWLEDGEMENT

We gratefully acknowledge the financial support of the Natural Sciences and Engineering Research Council of Canada for this work.

REFERENCES:

1. H.E. Bussey, Proc. IEEE, 55(6), 1046 (1967).

2. Mohammed Nurul Afsar, James R. Birch and R.N. Clarke, Proc. IEEE, 74(1), 183 (1986).

3. Joseph A. Carpenter, Jr. in Microwave Processing of Materials II, edited by William B. Snyder, Jr., Willard H. Sutton, Magdy Iskander and P. Lynn Johnson, (Mater. Res. Soc. Proc. 189, San Francisco, CA, 1990) pp 477-487.

4. N.H. Harris, J.R. Chow, R.L. Eisenhart and B.M. Pierce in Ceramic Transactions, Microwaves: Theory and Application In Materials Processing, 21, edited by David E. Clark, Frank D. Gac and Willard H. Sutton (The American Ceramic Society Inc., 1991) p. 235.

5. A.R. Von Hippel, Dielectric Materials and Applications, (The Technology Press of MIT, Cambridge, MA and John Wiley and Sons, Inc., New York 1954).

6. G.M. Brydon and D.J. Hepplestone, Proc. IEE, 112(2), 421, (1965).

7. W.B. Westphal, Dielectric Constant and Loss Measurements on High Temperature Materials, Technical Report 182, Laboratory for Insulation Research, MIT, Cambridge, MA, October 1963.

8. S.O. Nelson, L.E. Stetson and C.W. Schlaphoff, IEEE Trans. Instrumentation and Measurement, 23(4), 455 (1974).

9. B.V.A. Wickenden and S.F. Duerden, in High Frequency Dielectric Measurement, edited by J. Chamberlain and G.W. Chantry (IPC Science and Technology Press, Guildford, U.K., 1973), p. 135.

10. H. M. Frost, Los Alamos National Laboratory, Fusion Reactor Materials Progress Report, LA-UR-86-4226, 1986.

11. W.W. Ho in Microwave Processing of Materials I, edited by W.H. Sutton, M.H. Brooks, and I.J. Chabinsky (Mater. Res. Soc. Proc. 124, Pittsburgh, PA 1988) pp 137-148.

12. R.D. Hollinger, V.V. Varadan, V.K.Varadan and D.K. Ghodgaonkhar, in Microwave Processing of Materials II, edited by William B. Snyder, Jr., Willard H. Sutton, Magdy Iskander and P. Lynn Johnson, (Mater. Res. Soc. Proc. 189, San Francisco, CA, 1990) pp 243-250.

13. W.R. Tinga, B.Q. Tian and W.A.G. Voss, Microwave Processing of Materials II, edited by William B. Snyder, Jr., Willard H. Sutton, Magdy Iskander and P. Lynn Johnson, (Mater. Res. Soc. Proc. 189, San Francisco, CA 1990) pp. 111-116.

14. B.Q. Tian and W.R. Tinga, Internal Report B11, Dept. of Electrical Engineering, University of Alberta, Edmonton, Canada, T6G 2G7, 1991.

15. D. Couderc, M. Giroux and R. G. Bosisio, J. of Microwave Power, 8(1), 70 (1973).

16. Jose C. Araneta, Morris E. Brodwin and Gregory A. Kriegsmann, IEEE Trans. Microwave Theory and Techniques, 32(10), 1328 (1984).

17. D.K.Wong, M.Sc Thesis, University of Alberta, Edmonton, Canada, 1975.

18. W.R. Tinga, Electromagnetic Energy Reviews, 1(4), 47 (1988); 2(1), 1 (1989).

19. R.M. Hutcheon, M.S. de Jong, F.P. Adams, L. Hunt, F. Iglesias, G. W. Wood and G. Parkinson, Electromagnetic Energy Reviews, 2(4), 46-51 (1989).

20. W.Xi and W.R. Tinga in Ceramic Transactions ,Microwaves: Theory and Application In Materials Processing, 21, edited by David E. Clark, Frank D. Gac and Willard H. Sutton (The American Ceramic Society Inc., 1991) pp 215-224.

21. W. Xi and W. R. Tinga, IEEE Microwave Theory and Techniques, 40, in press (1992).

22. W. Xi and W.R. Tinga, presented at the MRS Spring Meeting, San Francisco, CA, 1992.

23. B.Q. Tian and W.R. Tinga, presented at the MRS Spring Meeting, San Francisco, CA, 1992.

THE ELECTROMAGNETIC PROPERTIES OF MATERIALS PROGRAM AT NIST

CLAUDE M. WEIL
Electromagnetic Fields Division (813.02), National Institute of Standards and Technology, 325 Broadway, Boulder, CO 80303

ABSTRACT

The Electromagnetic Properties of Materials (EPM) program at the National Institute of Standards and Technology (NIST) is described, including an outline of the current goals of the project as well as some details of measurement techniques being used at NIST for characterizing dielectric and magnetic materials at RF and microwave frequencies.

INTRODUCTION

For many years, the National Bureau of Standards (predecessor agency of NIST) operated a national materials standards program, located at its Boulder, Colorado facility. Due to budgetary constraints and programmatic redirection, this program had essentially ceased to exist by the late seventies. However, due to persistent demand from many segments of the electronics and microwave industry, which still had critical needs for traceability in the measurements of electromagnetic (EM) material properties, the program was revived in 1986. Many of these needs related to industry requirements for microwave measurement techniques that can accurately characterize various materials being produced by the industry. Applications of low-loss dielectric materials include, for example, printed circuit boards, capacitors, substrates for monolithic microwave integrated circuits (MMIC) and microwave components, sensor windows, antenna lenses and radomes. High-loss materials, both dielectric and magnetic, are used, for example, in feed-through filters, microwave isolators, low-reflectivity coatings for aerospace vehicles, microwave absorbers for antenna test ranges. Some of the reasons why industry recommended that the materials standards program be reinstated at NIST were that the Institute had the necessary experience and technical expertise to undertake this program and could develop the capability to provide leadership to U.S. industry in an emerging field of significant technical and economic importance. Some of the benefits of a NIST materials characterization program, as perceived by industry, include improved material production yields and better material uniformity created by the ability to meet closer tolerances in the EM properties of the materials. This can, in turn, lead to substantial cost savings for designers and suppliers of microwave products that use these materials. Another important role that NIST can play is as an arbiter when there exist significant differences in material properties as measured by both producer and user; this can be of major importance when the user is located outside the U.S.

Program Objectives

The goals of the NIST program may be summarized as follows:

(a) Implement and evaluate existing measurement methods, assess their accuracy, and improve on them where possible. Develop new measurement techniques and software as needed. In addition, NIST may set up and monitor measurement assurance programs throughout industry in order to assure the quality of measured data.

(b) Select materials with very stable EM and mechanical properties that are suitable for use as standard reference materials (for example, rexolite, fused quartz, alumina). After these materials are accurately characterized by NIST, they are subsequently made available to industry for use as check standards for purposes of calibrating their own measurement systems and software. Such a system partially satisfies the requirements for traceability to NIST in material measurements.

(c) Provide NIST measurement (calibration) services to those requiring characterization of their own materials. Also provide consultation and assistance to industry and others on their measurement problems.

(d) Participate in national and international intercomparisons of material property measurements.

(e) Develop a NIST data base on commonly used materials, such as those listed in the introduction.

In keeping with the basic priorities outlined above, characterization efforts have so far concentrated on homogeneous and isotropic solid materials in the frequency range 300 kHz - 18 GHz; some measurements of liquid and granular dielectrics have also been undertaken. All measurements are currently limited to room temperatures. As the program expands in response to customer needs, both the frequency coverage and temperature range will be extended beyond their existing limits. Plans also exist to develop measurement capabilities for fully characterizing anisotropic materials.

Most of these tasks have been or are currently in the process of being implemented.

NIST MEASUREMENT SYSTEMS

To date, NIST efforts have been concentrated on two basic measurement techniques. Because of its widespread popularity throughout industry and elsewhere, the broadband transmission line technique was selected for further study by NIST [1-4]. In this technique, a precisely machined sample of the material under test is placed within a length of coaxial air line or waveguide; the 7 mm diameter coaxial air line is frequently preferred. This technique has the advantage of potentially covering a very wide frequency range (up to 18 GHz) but is limited in accuracy due to the problem of residual air gaps between sample and the inner conductor. The other technique that has received major developmental support at NIST is the resonant cavity method. This is a single frequency measurement near 10 GHz and is of the highest accuracy attainable (0.5% or less in relative permittivity) particularly for very low loss materials. Both techniques are discussed in more detail below.

Transmission Line Techniques

In this method, the complex scattering parameters (S-parameters) of a section of transmission line containing the material sample are measured, preferably with an automatic network analyzer (ANA). Transmission line techniques include both the transmission/reflection (T/R) method in which full two-port S-parameter measurements are made of the line as well as the short-circuited line (SCL) method in which the line is terminated by a short and only the input S-parameter is measured. The latter technique is usually preferred for measurements where the sample temperature is raised and lowered over wide ranges.

The complex material parameters, permittivity and permeability, are related to the S-parameter data through the scattering equations. There are generally nine real equations with five unknowns for measurements of nonmagnetic materials and seven unknowns for magnetic. Thus, the system of equations is generally overdetermined and can be solved for in various different combinations. For many measurement applications, there is no reliable information available regarding the exact position of the sample within the transmission line. It is therefore advantageous to use a system of equations that are reference plane invariant. Many such reference plane independent solutions exist.

Until recently, an analytically based solution, originally developed by Nicolson and Ross [1] and Weir [2] (N-R/W technique) was widely used. In their procedure, the equations are combined so that an explicit solution for the material parameters is obtained in terms of the S-parameters and other unknowns. This solution, while elegant, exhibits inherently divergent behavior at frequencies corresponding to integer multiples of a half-wavelength in the sample caused by large phase uncertainties at these frequencies. Many workers tried to bypass this instability problem by using samples less than a half-wavelength long; however, this can increase measurement uncertainties considerably.

Because of these various problems, emphasis was placed on efforts to devise some improved algorithms at NIST. Baker-Jarvis has developed an iterative solution, based on the Newton-Raphson procedure, that has proven to be very stable at all frequencies [3,4] and is invariant to the reference plane position. Figure 1

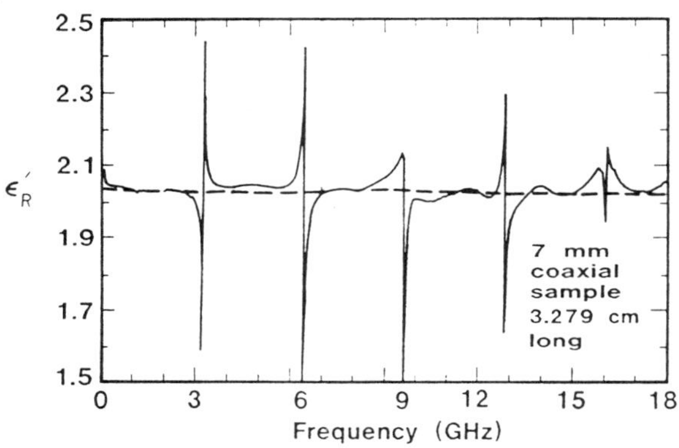

Figure 1: Relative Permittivity for polytetrafluoroethylene versus frequency as determined using the Nicolson-Ross (solid line) and the new iteration (dashed line) techniques.

illustrates some measured data for a sample of polytetrafluoroethylene (PTFE) processed using the older N-R/W technique as well as the new numerical approach. A major improvement is apparent. The Baker-Jarvis algorithm has been incorporated into a software package for measurements of nonmagnetic materials that is commercially available from a major manufacturer of ANAs [5]. It is also available as a NIST software package (EPSILON 2_1) that includes an air-gap correction as well as routines for curve smoothing and uncertainty estimations.

Recent efforts have concentrated on extending these algorithms to magnetic materials with combined magnetic and dielectric properties [6]. Because of the reduced mathematical redundancy that exists when measuring magnetic materials, it is not theoretically possible to completely avoid periodic instabilities when using the two-port technique on a single sample [6]. However, most magnetic materials exhibit medium to high levels of magnetic and dielectric loss, so that, in practice, these periodic resonant instabilities usually get damped out. The NIST-developed software, termed EPS_MU_3 is based on the extended Baker-Jarvis algorithm and works well for measurements of magnetic materials. Figure 2 shows some recently acquired data on a commercially available nickel-zinc ferrite. Note that the material has been characterized over a four decade wide spectral range, 1 MHz - 10 GHz and that permeability values approaching 600 were measured at low frequencies. The data plots include uncertainty estimates throughout; the worst case is seen to be for the epsilon' data at low frequencies and approaches +/- 35%, due to the major uncertainty contributed by the air gap. Figure 3 shows similar data, covering three frequency decades, for two commercially available garnets, yttrium iron (YIG) and calcium vanadium. Data for epsilon", the imaginary part of the relative permittivity, was omitted because it is very small for garnets and beyond the measuring capability of the system.

An early version of the EPS_MU_3 software was made available to a few selected industrial laboratories for testing. An updated and improved version is nearing completion and will soon be available to interested users.

Measurements of high-permittivity, low-loss nonmagnetic materials have presented particularly difficult measurement problems, especially at the higher frequencies. The half-guide wavelength instabilities, already discussed, can be particularly troublesome for these materials and the newly developed NIST algorithms were initially unable to handle them due to problems involved in locating the correct root. In addition, other problems caused by the existence of higher-order propagation modes within the sample are frequently encountered, particularly for waveguide measurements. These modes will propagate because their cut-off frequency is appreciably lower within the material than outside it. The small air gap surrounding the sample is usually responsible for exciting these higher-order modes and such modes will manifest themselves as resonances whenever the sample length equals a multiple of a half-guide wavelength for the particular mode being excited. Because the transmission line theory for this technique is based on the assumption that only a single propagating mode is present on the line, the general scattering equation solution will break down whenever these modes are present. The resulting instabilities are much the same as those created by the divergence problems of the N-R/W routine.

Solution of this measurement problem has necessitated developing a completely different approach that is based on a constrained nonlinear regression procedure [7]. In contrast to the methods already discussed, in which explicit or implicit solutions of a system of nonlinear scattering equations are derived on a

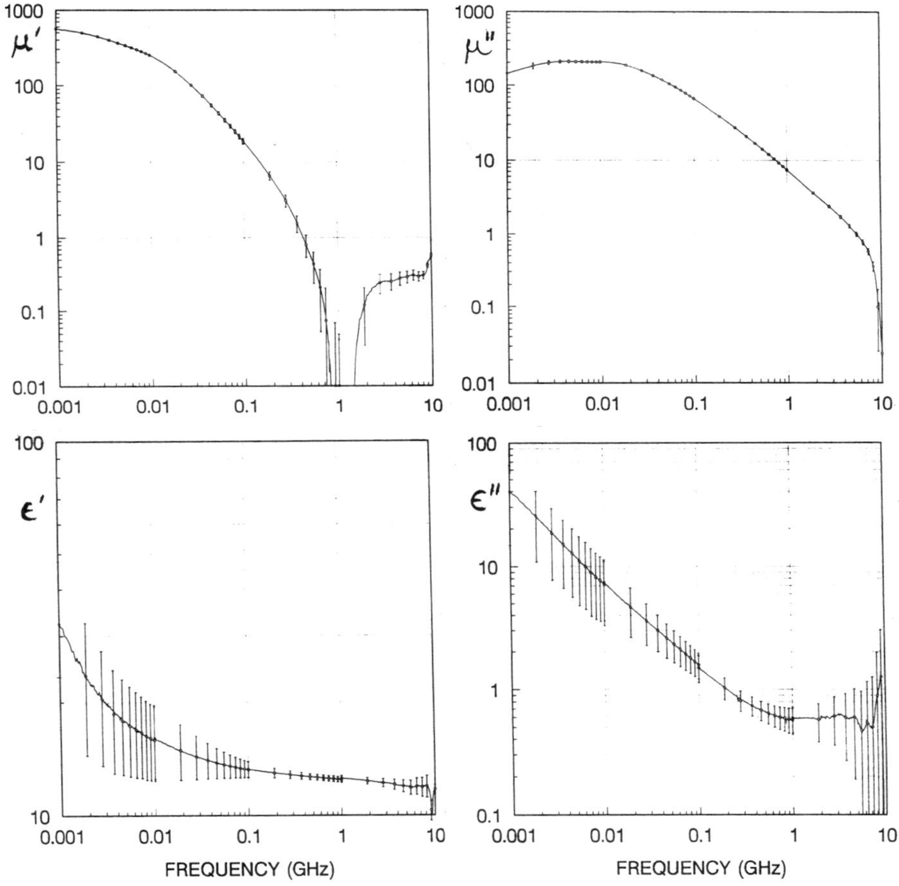

Figure 2: Complex permeability and permittivity for a nickel-zinc ferrite.

frequency point-by-point basis, this approach performs a least squares fit over the full measurement spectrum and is termed a multi-frequency approach. The functional form of permittivity and permeability of the material under test is defined as a truncated Laurent series with complex coefficients. The coefficients may be determined by minimizing the sum of the squares between predicted and observed S-parameters, consistent with linearity and causality constraints, using an orthogonal distance regression routine, termed ODRPACK [8], developed at NIST. One of the major advantages of this technique is that in the vicinity of singularity or instability points, the data can be de-emphasized in order to greatly lessen the impact of measurement error. In addition, the routine allows for adjustments in certain input parameters such as sample length, sample position, waveguide cut-off, etc. which can compensate for measurement errors. The routine can also correct for

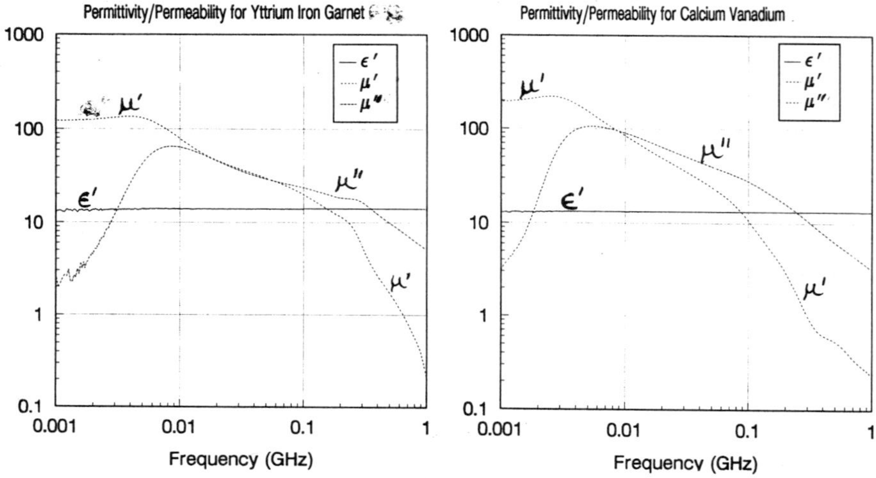

Fig. 3: Complex permeability and permittivity for YIG and Ca V Garnets.

for the presence of higher-order modes. The optimization based algorithm has been termed "OPTEMA" by NIST personnel.

Figures 4 and 5 show data obtained for two different commercial titanate ceramics. The measurements were made using X-band waveguide (WR 90) and the data processed using OPTEMA. The data of Figure 4 are for a barium titanate having a nominal

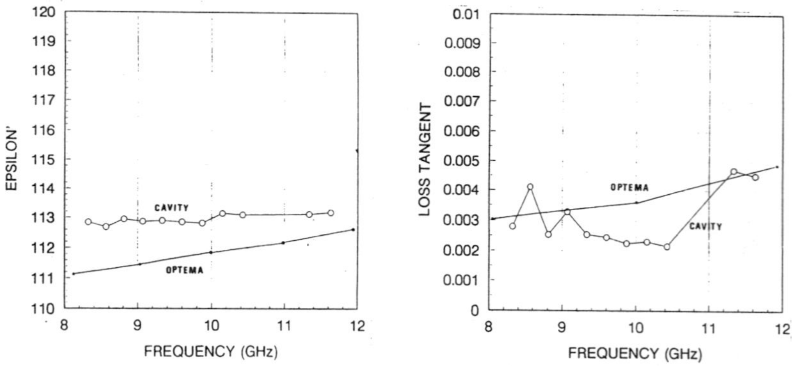

Fig. 4: Relative permittivity and loss factor for barium titanate

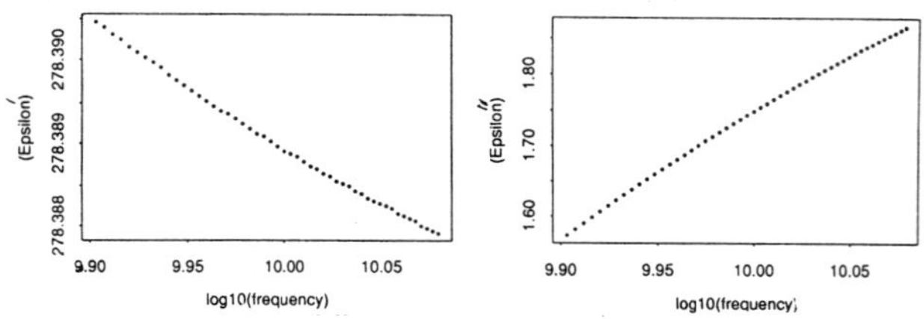

Figure 5: Complex permittivity for calcium-strontium titanate

epsilon' value of 125. Some comparison data obtained in the NIST resonant cavity (see next section) are also shown in Figure 4 and it is evident that the two are fairly comparable.

The data of Figure 5 are believed to be for calcium-strontium titanate having a nominal epsilon' value of about 275. It should be emphasized that data reduction for these types of high-permittivity materials is very difficult to accomplish using routines other than OPTEMA.

Resonant Cavity Technique

The NIST cylindrical cavity resonator is a 2-port, mode filtered type, operating in the TE_{01p} mode at X-band (8-12GHz) [9,10] and is capable of performing very accurate measurements of the relative complex permittivity of low-loss, nonmagnetic materials at room temperatures. It consists of a length of 60 mm diameter helical waveguide that can only support cylindrical modes having azimuthally constant E-fields (the TE_{0n} modes). Consequently, the material parameters can only be measured in this plane. Since the helix cannot support axial currents, all other modes are attenuated by at least 20 dB. Such a cavity is termed "mode-filtered" and is of great value in material measurements because mode identification is greatly simplified. Some 17 modes where p = 15,...,31 are available for measurements. The resonator length is adjustable using a motorized tuner from 408.5 to 433.5 mm.

RF energy is coupled into and out of the cavity through two narrow iris slots in the upper end plate; the cavity is very under-coupled (approx. -26 dB), giving an empty Q factor of about 80,000 at 10 GHz. The placement of these slots on the end plate is designed to preferentially excite the desired principal TE_{01} mode, but not the TE_{02} mode. The sample under test needs to be precisely machined to the resonator's diameter and then positioned on the moveable end plate of the cavity.

Material characterization basically involves measuring the change in cavity modal parameters (resonant frequency and Q-factor) of the empty resonator versus that when a material sample is present [11]. A cavity measurement proceeds as follows: the operator selects a resonant mode that is close to the desired measurement frequency and not perturbed by any other unwanted modes, which even though highly attenuated can still cause serious measurement errors. The resonant frequency shift caused by the presence of a sample is either measured directly in the fixed length method or indirectly in the fixed frequency method when the cavity is mechanically retuned to the same frequency; the latter is generally considered more accurate.

Traditionally, the resonant frequency and bandwidth, from which the Q factor is derived, have been estimated directly from the resonance curve. Such techniques are of very limited accuracy. Furthermore, the bandwidth measurement yields a value for the change in loaded Q; i.e. the reduced Q created by the loading effect of the coupling irises. An accurate determination of material loss tangent requires a knowledge of the unloaded Q which, to a first approximation, is related to the loaded Q through the two iris-coupling factors.

A major effort has been under way at NIST to significantly improve the accuracies of material parameter measurements using the cavity technique; the ultimate goal is an accuracy of +/- 0.1% in eps' and +/- 5% in loss factor. In the technique developed by Vanzura and Rogers [12], only the input scattering parameters, S_{11}

and S_{22} are measured. Because measurements of the transfer parameter S_{12} are not required, this eliminates the need to perform a full 2-port TRL calibration on the ANA which, because of cable flexing problems, can introduce appreciable errors. When plotted in the complex S-plane, the loci of the input S-parameter data form a circle (the so-called Q circle). The loci are then compared, using a nonlinear regression fit, to a rigorous mathematical expression derived from a circuit model representation of the cavity and its coupling system. From this, direct estimates for eight unknown circuit parameters are derived, including the two coupling coefficients, the resonant frequency, the unloaded Q, and the complex impedance of the coupling irises.

An alternative technique, developed by Estin and Janezic [13], has yielded very comparable data for all parameters other than unloaded Q, for which it consistently yielded a higher estimate. In this technique, measurements are required of all three S-parameters, namely S_{11}, S_{22} and S_{12}. The unknown coupling coefficients are derived from the Q circle fit of the S_{11} and S_{22} data using a linear least squares fit and an approximation for the circuit model representation of the cavity. An estimate for the loaded Q is obtained from a linear fit to the S_{12} data from which the unloaded Q is derived. Some work has also been done on evaluating the correction created by ohmic losses in the cavity walls and on separating the contributions due to side and end walls. Other higher-order corrections remain to be investigated.

Figure 6 shows some data for a sample of alumina (Al_2O_3)

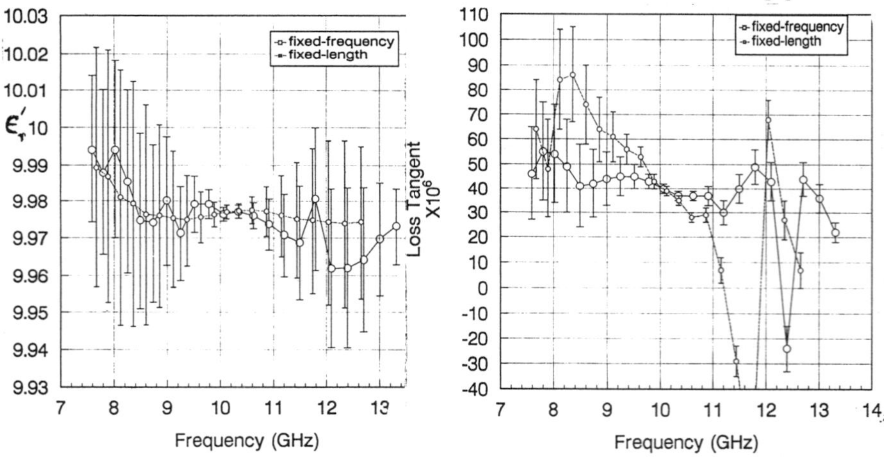

Figure 6: Relative permittivity and loss factor for alumina as measured in the NIST resonant cavity.

measured in the NIST cavity using both fixed frequency and fixed length techniques. These measurements were optimized for minimum uncertainties at about 10 GHz by choosing a sample thickness that corresponds to a multiple of a half-guide wavelength in the material. This is readily apparent in Figure 6. There appears to be good agreement between the data obtained using the fixed frequency and fixed length techniques, except for the loss factor data at frequencies above 11 GHz. This is believed to be due to the presence of perturbing TE_{02} modes.

REFERENCE MATERIALS

Early in the revived NIST materials program, a study was initiated to determine the basic requirements for suitable dielectric and magnetic reference materials [14]. These requirements can be met by solid materials that are linear, spatially homogeneous and isotropic as well as nondispersive or minimally dispersive in both frequency and temperature. These materials also need to be chemically inert, resistant to radiation and nonhygroscopic. They must be very stable in both mechanical and electrical properties. It is also advantageous for these materials to be easily machinable, not brittle, readily available and of modest cost.

To date, a number of low-loss dielectric materials of various relative permittivity values have been selected as meeting most or all of the above requirements:

Cross-linked Polystyrene (Rexolite),	eps'_r	= 2.53
Fused Quartz (Glass),	eps'_r	= 3.82
Alumina (Aluminum Oxide),	eps'_r	= 10.0
Fused Rutile (Titanium Dioxide),	eps'_r	= 100.0

Regrettably, three of these materials are glass or ceramics that are very brittle and not readily machinable.

The selection of one or two suitable magnetic/dielectric reference materials is currently under way. Various garnets and spinel ferrites, of the type whose characteristics are shown in Figures 2 and 3, appear to be the most promising candidates but there are still questions to be resolved regarding their homogeneity and temperature dependence as well as the long-term stability of their properties. There are also issues regarding the batch-to-batch and sample-to-sample variability of these materials.

NIST MEASUREMENT SERVICES, INTERCOMPARISONS AND DATABASE.

NIST is now routinely providing measurement services to industrial customers and others for the microwave characterization of customer furnished materials. Sample machining is performed either by NIST or the customer depending on the customer's preference. NIST is also supplying machined and unmachined samples of dielectric reference materials, as needed. These services are currently classified as "Special Tests" by NIST management because the rigid NIST requirements for a certified calibration service have not yet been fully satisfied.

The NIST materials project has also recently organized a national T/R intercomparison or "round robin" on low-loss nonmagnetic materials measurements using a 7 mm diameter coaxial air line. Four samples of various materials, including some reference materials, were circulated amongst 15 participating industrial laboratories. Data from this intercomparison are currently being compiled and a preliminary presentation of these data is being planned for July 1992 [16]. The preliminary conclusion of this study is that the quality of most material characterization being performed in industry is inadequate and needs improvement. An additional national intercomparison is currently planned for magnetic materials, as well as a possible international intercomparison on precision cavity measurements.

To date, little effort has been devoted to developing a NIST database for material parameters. This will probably be done at a later date, when more baseline data have been accumulated on some of the basic materials that are being measured at NIST.

CONCLUSIONS

Significant advances have been realized in reestablishing the EPM program at NIST. Important improvements have been made in the popular transmission line measurement methods. A resonant cavity method has been developed that promises to yield some of the highest accuracies attainable anywhere. Much still remains to be accomplished: the evaluation of nondestructive techniques for material characterization, as well as millimeter wave free-field techniques, will be emphasized in the years ahead.

References

1. A.M. Nicolson and G.F. Ross, IEEE Trans. Instrum. Meas. IM-17, 395-402 (1968).

2. W.W. Weir, Proc. IEEE 62, 33-36 (1974).

3. J. Baker-Jarvis, NIST Technical Note 1341, 1990.

4. J. Baker-Jarvis, E.J. Vanzura, W.A. Kissick, IEEE Trans. Microwave Theory Tech. MTT-38, 1096-1103 (1990).

5. H.P. 85071A Materials Measurement Software, Technical Data; Published by Hewlett-Packard Co., Palo Alto, CA.

6. J. Baker-Jarvis, "Transmission/Reflection and short-circuit line methods for permittivity and permeability determination," to be published as a NIST Technical Note, 1992

7. P.D. Domich, J. Baker-Jarvis, R.G. Geyer, J. Res. Nat. Inst. Stand. Technol. 96, 565-575 (1991).

8. P.T. Boggs, J.R. Donaldson, R.H Byrd, R.B. Schnabel, ACM Trans. Math. Software, 15, 348-364 (1989)

9. E.J. Vanzura, W.A. Kissick, Digest of the MTT-S Int. Microwave Symp., 1989, vol. 3, 901-904. IEEE Cat. No. 89CH2725-0

10. R.J. Geyer, C.M. Weil, W.A. Kissick, Digest of the Conf. on Precision Electromagnetic Meas., 1990, 174-175. IEEE Cat. No. 90CH2822-5

11. E. Ni and U. Stumper, IEE Proceedings, 132(1),pt.H, 27 (1985).

12. E.J. Vanzura, J.E. Rogers, Record of the IEEE Instrumentation and Measurement Tech. Conf., 1991, 150-155; IEEE Catalog No. 91CH2940-5

13. A.J. Estin, M.D. Janezic, ibid, 573-579.

14. R.G. Geyer, NIST Technical Note 1338, 1990.

15. E.J. Vanzura, to be presented at the combined IEEE AP-S / URSI/EMP Conf., Chicago, IL, July 1992.

HIGH TEMPERATURE BROADBAND DIELECTRIC PROPERTIES MEASUREMENT TECHNIQUES

OCTAVIO M. ANDRADE, MAGDY F. ISKANDER AND SHANE BRINGHURST
University of Utah, Department of Electrical Engineering
Salt Lake City, Utah 84112

ABSTRACT

This paper discusses theoretical and practical aspects of the development and implementation of various measurement techniques for high-temperature broadband microwave characterization of materials at the University of Utah. Objectives include materials measurements in the frequency range from 45 MHz to 12 GHz and for temperatures as high as 1000°C.

1. INTRODUCTION

Broadband, high-temperature measurements of the dielectric properties of ceramic materials can be best achieved by using coaxial line methods, specifically the shunt-capacitor and the open-ended coaxial line techniques. The design of a suitable coaxial line probe involves various tradeoffs such as operating frequencies, range of complex permittivities of the selected materials, possible propagation of higher-order modes, adequate line length for cooling purposes, and the selection of a low thermal-expansion material for probe design. Section 2 of this paper addresses the above mentioned issues and describes the design of a specific probe suitable for these measurements. Equivalent circuit models and calibration techniques using standard materials and a modern network analyzer are also discussed.

Cavity perturbation techniques, although intrinsically narrow-band, provide accurate measurements of loss tangent, which is most important in accounting for the loss mechanisms as a function of temperature. The application of this method to high-temperature measurements is described in Section 3. Comparative results illustrating tradeoffs in the various measurement procedures will be presented in Section 4. An overall summary and conclusions are presented in Section 5.

Measurement results of zirconia and alumina samples using the above-mentioned techniques are described in a companion paper in this Proceedings [1].

2. BROADBAND MEASUREMENTS USING OPEN-ENDED COAXIAL LINES

2.1 BACKGROUND

The open-ended coaxial line method for the measurement of complex permittivity has been analyzed in detail by many authors for more than a decade [2-7]. Since the interest in using this type of sensor has been mainly dedicated to "in vivo" measurement of the complex permittivity of biological substances, the majority of the experimental approaches described in the literature have been focused on high-loss dielectrics measured at room temperature. Recent developments in the microwave processing of ceramic materials, however, have called for the need of the knowledge of their dielectric properties over a broad frequency band, and as a function of temperature up to sintering values (around 1200°C). The open-ended coaxial line method appears to be a natural candidate for these measurements provided that a suitable high-temperature probe is designed. This section will describe the fundamentals of the method, the analytical approaches selected to obtain the complex permittivity from the measured probe admittance, some measurement results obtained for alumina using a medium-temperature (up to 200°C) probe available in our laboratory, and the design of a high-temperature metalized ceramic probe.

2.2 PRINCIPLES

Figure 1 shows a schematic diagram of an open-ended coaxial probe. It basically consists of a truncated section of a coaxial line, and an optional extension of a ground plane which improves the contact with the sample under test. The input port of the sensor is connected to the measurement equipment through a coaxial cable. The parameter to be measured is the admittance (or the reflection coefficient) at the interface between sensor and sample. The use of a modern Automatic Network Analyzer (HP8510B) as the measuring instrument significantly simplifies and enhances the accuracy of the measurement procedure.

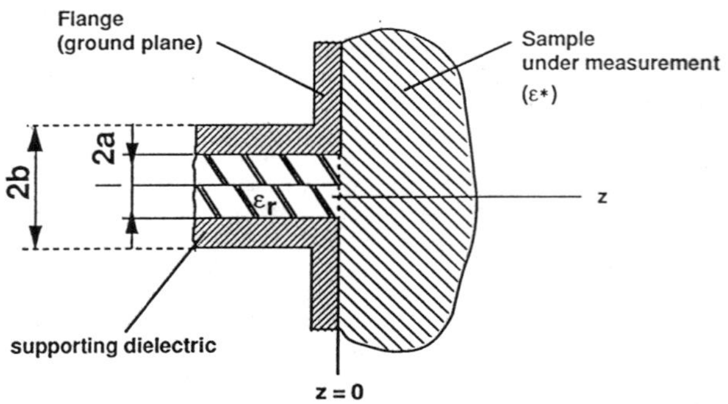

Figure 1
Schematic diagram of an open-ended coaxial probe

The choice of dimensions for a particular probe depends on many factors.Typically, one has to consider the following trade-offs in designing and selecting a probe for room temperature operation (high-temperature additional constraints will be discussed later):

- minimum and maximum operating frequencies restrict the selection of the cross section to obtain an "optimum" value of the capacitance of the probe for the range of permittivities to be measured and to avoid the propagation of higher order modes.
- the possibility of machining samples which can adequately fit the probe's cross-section (for solid samples) is an important factor in probe design.

The Dielectric Measurements Laboratory of the University of Utah has been working with three different probes for various applications, which are listed in Table 1.

Probe denomination	Inner diameter(mm)	Outer diameter(mm)	Length (mm)	Dielectric
U of U probe	4.46	9.80	100	Teflon
KOVAR HT probe	10.44	23.62	430	Air
HP HT probe	1.04	4.14	50	Glass

Table 1
Dimensions of probes presently available in our Laboratory

As mentioned above, the Automatic Network Analyzer provides a fast and accurate means of measuring the input admittance of the open-ended probe in contact with the material under test. As an example, Figure 2 shows a typical measurement result for an alumina sample (green).

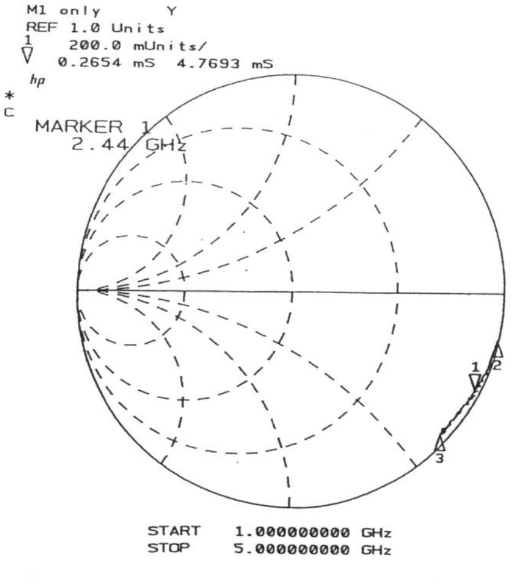

Figure 2
Typical Plot of Admittance Measurement Using the HP-8510B Network Analyzer (Green alumina sample).

2.3 DESCRIPTION OF CALCULATION MODELS

There are two basic approaches in calculating the complex permittivity of the test sample from the measured admittance of the probe. The first procedure, called the lumped capacitance model, yields simple closed-form expressions and is based on modeling the discontinuity betwen probe end and dielectric as a parallel conection of two capacitors, as shown in Figure 3. The capacitance C_f is the fringing capacitance resulting from fields fringing within the probe, and C_0 is the capacitance of the open end of the probe, before contact with the material. For a detailed description of these capacitances and the expressions which relate the measured admittance, the fringing capacitances and the sample permittivity, the reader is referred to some landmark papers on the subject [4,5,6,7]. For our purposes, however, it suffices to say that the input admittance of the probe, after calibration with adequate standard materials,is related to the complex permittivity of the sample by the following expression:

$$Y=j\omega C_t$$

where

$$C_t=Ct_0(1+F^2/24)\ \mathrm{pF}$$

$$F=2\pi f(b-a)\epsilon_0/\ \epsilon_r$$

$$Ct_0=2.0992(b-a)\mathrm{x}10^{-2}\mathrm{pF}$$

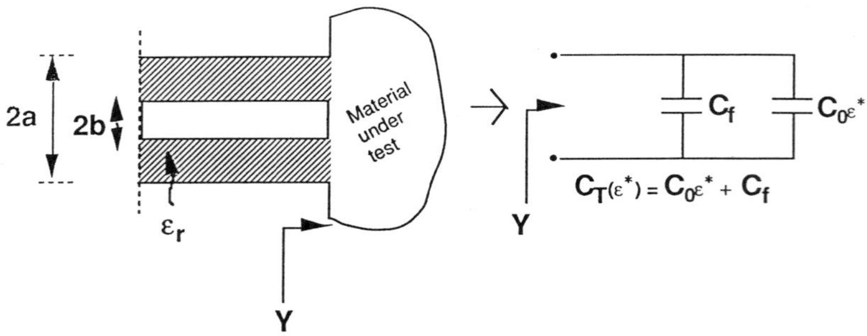

Figure 3
The Lumped Capacitance Model for an Open-ended Coaxial Probe

The second approach to the determination of the complex permittivity is based on a rigorous solution of the electromagnetic field equations at the interface between probe and sample. This analytical procedure is based on the principle that the sensor behaves as a circular diffraction antenna radiating into the sample. A detailed analysis of this problem leads to an integral equation solution given by [8]:

$$y=\frac{jk\varepsilon y_0}{\ln\left(\frac{b}{a}\right)\sqrt{\varepsilon_r}}\int_0^{\infty}\frac{\left[J_0(\zeta a)-J_0(\zeta b)\right]^2}{\zeta\left(\zeta^2-k^2\varepsilon\right)^{\frac{1}{2}}}d\zeta$$

where

$$k=2\pi/\lambda_0$$

and

ϵ_r is the dielectric constant of the material filling the probe

The former equation leads to an immediate solution of the "forward" problem, when the permittivity is known, and the probe admittance is to be determined. Since the inverse problem is required in our case, an initial estimate of the permittivity is needed for the minimization of the complex functional involved. Remarkably, it has been found that a simple PC-based available commercial package (MATHCAD) can easily solve the inverse complex equation with a minimum of computation time and high accuracy.

Another procedure for solving the above mentioned integral equation (or an equivalent equation which has previously been derived by Marcuvitz [9]), is based on an appropriate polynomial approximation expansion , in a procedure developed by Misra [10]. This approach has been used by Hewlett-Packard in the development of their software which accompanies their commercially available probe kit. The novel HP medium-temperature (up to 250°C) probe has been extensively used in our measurements, as described in a companion paper in this Proceedings [1].

2.4 BROADBAND MEASUREMENT RESULTS AT MEDIUM TEMPERATURES

The previously described probes and the three analytical calculation methods have been extensively used in our laboratory to measure the complex permittivity of a variety of materials. More specifically, a large number of ceramic composites including Alumina and Zirconia with various amounts of Yttria had their properties characterized as a function of temperature. Figures 4 and 5 show an example of the results obtained for sintered Alumina at 200°C using one of the open-ended probes in the frequency range 1-11 GHz. Since these measurements were also performed using a cavity perturbation technique (to be described next), the accuracy of the calculation models and the previously described analytical approaches have been proved to yield results which are comparable to the more accurate cavity method.

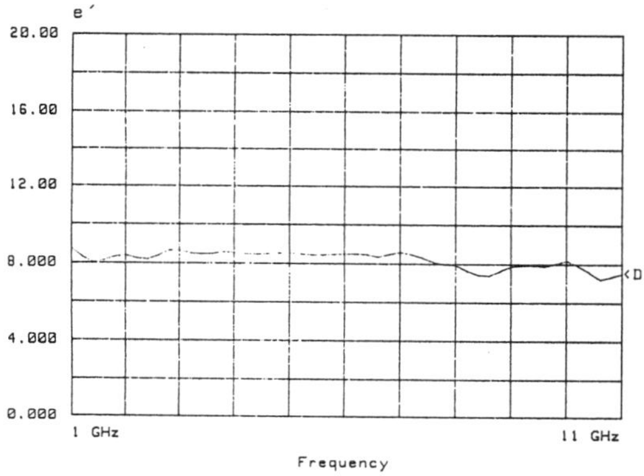

Figure 4
Real part of the dielectric constant of Alumina (sintered) vs. frequency @ 200°C

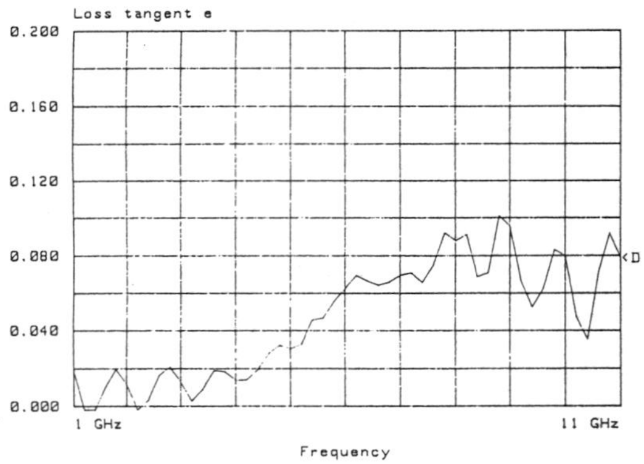

Figure 5
Loss Tangent of Alumina (sintered) vs. Frequency @ 200°C

2.5 EXTENSION OF TECHNIQUE TO HIGH-TEMPERATURE MEASUREMENTS

For a better understanding of the complex mechanisms involved in the microwave sintering of ceramic materials, it is necessary to measure the complex permittivity of both green and sintered samples up to around 1200°C, where sintering is fully achieved in a multimode cavity. Thermal expansion of the coaxial line produces changes in its characteristic impedance, therefore affecting system calibration. Moreover, the supporting dielectric of the available coaxial sensors will melt at high temperatures. For these reasons, the previous measurements were conducted only up to 200°C, the temperature limit for the probes. The Kovar probe, however, has been effectively used up to 700°C to measure the properties of rocks [11].

The design of a high-temperature open-ended coaxial sensor, therefore, requires the selection of a constitutive material of very low differential thermal expansion coefficient. We have decided to use alumina to fabricate such a probe. Since the probe has to be metalized, platinum was selected for its very high melting temperature (1800°C). The probe was designed with the dimensions of a 7mm air line, and small ceramic washers are used for careful positioning of the center conductor. The selected dimensions have the advantages of providing high-frequency range, the possibility of measuring samples of small size, and a convenient connection to either a Type-N or APC-7 connector. The previously described Kovar coaxial line was used as a test to determine the required length for the high-temperature probe. The probe was inserted into a furnace and heated up to 600°C. Temperature measurements of the exposed part of the line were made to observe the heat transfer to the measuring equipment. These measurements demonstrated that for a 20 cm-long line the temperature at the interface of the connecting equipment is safe enough for its operation, even without a cooling system.

The ceramic high-temperature probe has been built, and at the time of this writing an adequate technique for metalizing the inside part of the line is being investigated. Figure 6 is a picture of the disassembled high-temperature open-ended sensor.

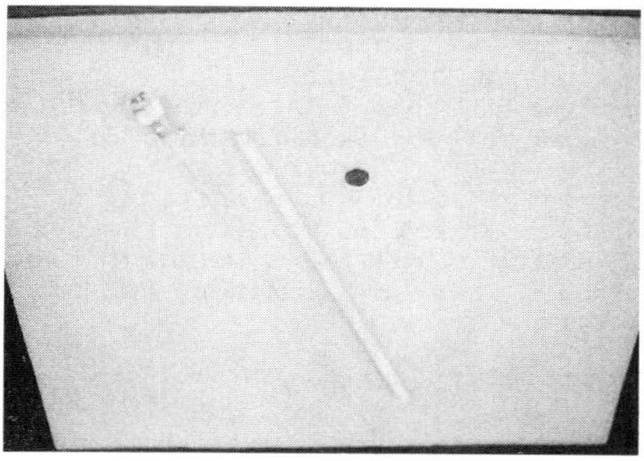

Figure 6
Ceramic high-temperature open-ended coaxial line

3.CAVITY PERTURBATION MEASUREMENTS

3.1 BACKGROUND

Cavity perturbation methods have been widely used to measure the complex permittivity of materials at microwave frequencies. The basic assumption of cavity perturbation is that the sample must be very small compared with the cavity itself, so that a frequency shift which is small compared with the resonant frequency of the empty cavity is produced by the insertion of the sample [12]. Cavity perturbation measurements can be highly accurate and are particularly advantageous in the determination of small loss tangents. For high-temperature measurements a thin-walled fused silica sample holder can be conveniently used in conjunction with a conventional furnace and inserted into the cavity. A modern Automatic Network Analyzer (HP8753C or HP8510B) is then used for very fast and accurate measurement of the frequency shift and Q-factor variation during sample heating, thus providing a narrow-band but extremely accurate measurement of the sample's complex permittivity.

In this section we will discuss the design and construction of two rectangular cavity resonators, present perturbation expressions for cylindrical and rectangular samples, both fully and partially inserted in the cavity, describe and discuss high-temperature measurement arrangements and techniques (up tp 1000°C) and present an uncertainty analysis for errors in the measurements of sample length and radius. Some results for measurements of sintered zirconia at 2.45 GHz will also be shown. Extensive measurement results for alumina and zirconia, both green and sintered, are presented in a companion paper [1].

3.2 DESIGN AND CONSTRUCTION OF RECTANGULAR CAVITY RESONATORS

Rectangular cavities resonating in the TE10n (n odd) modes are particularly convenient for measurements with samples of rod or rectangular geometries for the following reasons:

- The Ey component of the electric field is maximum in the center of the cavity, and is tangential to a sample inserted through its broad wall.
- The Ey component is the only electric field component

existing in the center of the cavity.

- Electric field probes can be easily installed in regions of maximum Ey along the broad wall of the cavity.

Two cavities have been designed and constructed for high-temperature permittivity measurements of ceramic materials, one for operation in S-band, and the other for X-band. Measurements with an Automatic Network Analyzer showed the S-band cavity to resonate at≈2440 MHz in the TE103 mode and the Q-factor to be≈2200, while the X-band cavity resonated at≈9956 MHz in the TE107 mode with a Q-factor of≈5500. Design parameters and construction details are described in [1].

3.3 CAVITY PERTURBATION EXPRESSIONS

For a non-magnetic sample positioned in a cavity such that its surface is everywhere tangent to the (unperturbed) electric field lines in the cavity, the general perturbation formula can be simplified to [12]:

$$\left(\varepsilon_r^*-1\right)\frac{\int_{V_s}|\vec{E}|^2 dV}{\int_{V_c}|\vec{E}|^2 dV} = 2\left[\frac{f_0 - f_s}{f_s}\right] - j\left[\frac{1}{Q_s} - \frac{1}{Q_0}\right]$$

where the subscripts $_0$ and $_s$ distinguish the empty and filled cavity, Vs and Vc are the sample and cavity volumes, **E** is the electric field tangent to the sample, and ϵ^*_r is the complex permittivity of the sample.

Figure 7 shows a schematic diagram of a rectangular cavity with partially inserted rectangular and cylindrical samples. For TE10n modes (n odd), evaluation of the perturbation expression yields:

- For a cylindrical sample:

$$\varepsilon_r' = 1 + \frac{f_0 - f_s}{f_s}\,\frac{aL}{2\pi r^2}\,\frac{b}{b'} \qquad \varepsilon_r'' = \left[\frac{1}{Q_s} - \frac{1}{Q_0}\right]\frac{aL}{4\pi r^2}\,\frac{b}{b'}$$

- For a rectangular sample:

$$\varepsilon_r' = 1 + \left(\frac{f_0-f_s}{f_s}\right)\frac{b}{b'}\,\frac{2aL}{\left[a' + \frac{a}{\pi}\sin\left[\pi\left(\frac{a-a'}{a}\right)\right]\right]\left[L' + \frac{L}{3\pi}\sin\left(3\pi\left(\frac{L-L'}{L}\right)\right)\right]}$$

$$\varepsilon_r'' = \left[\frac{1}{Q_s} - \frac{1}{Q_0}\right]\frac{b}{b'}\,\frac{aL}{\left[a' + \frac{a}{\pi}\sin\left[\pi\left(\frac{a-a'}{a}\right)\right]\right]\left[L' + \frac{L}{3\pi}\sin\left(3\pi\left(\frac{L-L'}{L}\right)\right)\right]}$$

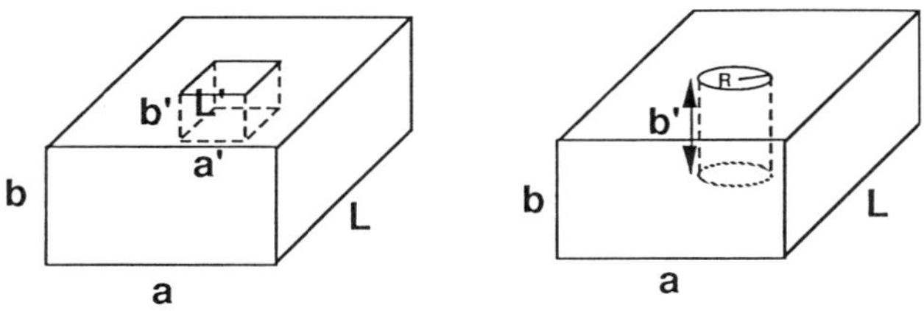

Figure 7
Schematic diagram of rectangular cavity with partially inserted rectangular and cylindrical samples

These expressions were used in the present measurements to calculate the sample permittivities. It should be emphasized that the "empty" cavity resonant frequency and Q-factor are obtained with the empty fused silica sample holder inserted in the cavity, and measured at each temperature.

3.4 CAVITY CALIBRATION

The performance of the cavity and sample holder were verified by measuring a TEFLON rod sample at room temperature, both fully and partially inserted in the cavity. An HP8753C Network Analyzer was used for the S-band cavity, and an HP8510B was used for the X-band cavity. For better accuracy, the analyzer was operated with a 5sec. sweep time and averaging factor of 16 measurements. Also, the "marker to maximum" and "widths tracking" features of the ANA greatly enhance the accuracy of the resonance frequency and Q-factor measurement. Table II summarizes the results obtained.

SAMPLE	EMPTY TUBE f(MHz)	Q	TEFLON f(MHz)	Q	$\epsilon'r$	$\tan\delta$ $(x10^4)$
PARTIALLY INSERTED	2428.79	2212.5	2423.72	2195.5	1.99	4.17
FULLY INSERTED	2430.11	2194.5	2423.15	2171.5	2.02	4.23

Table II
Calibration results for cavities using TEFLON samples

Uncertainties in the measurements of sample length (for partially inserted rod) and of sample radius are considered to be the main sources of errors in the calibration data. Figure 8 illustrates the effects of a ±1% error in the measurement of sample dimensions on the calculated values of $\epsilon'r$ and $\tan\delta$ for the S-band cavity. The nominal values are 3.21cm for the sample length and 0.317cm for the radius. The calibration results, therefore, show good agreement with the published values for TEFLON.

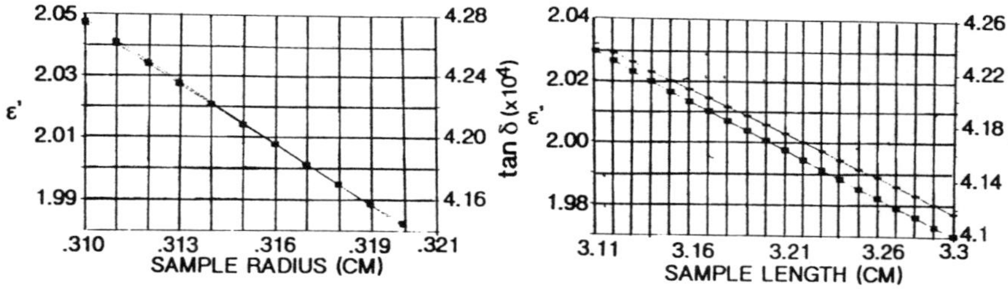

Figure 8
Effect of uncertainties in sample dimensions on measured values for a TEFLON rod in the S-band cavity

3.5 HIGH-TEMPERATURE MEASUREMENT RESULTS FOR CERAMIC MATERIALS

The cavity resonators described above are being extensively used in the Dielectric Measurements Laboratory of the University of Utah. Measurement results for a variety of ceramic composites at temperatures up to 1000°C are described in a companion paper in this Proceedings [1]. To illustrate one set of measurement results, Figure 9 shows the measured dielectric properties of sintered zirconia (Z_rO_2-12mol%Ce) using the S-band rectangular cavity (2.45 GHz) as a function of temperature.

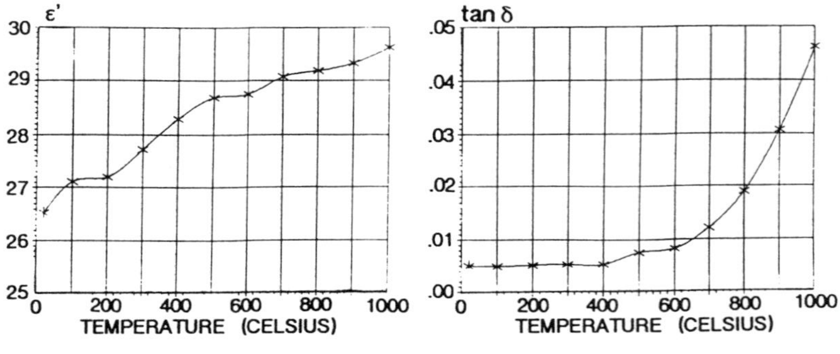

Figure 9
Permittivity of TZ12CE as a function of temperature at 2.45 GHz

4. TRADEOFFS IN THE MEASUREMENT PROCEDURES

The open-ended coaxial line and the cavity perturbation methods can be effectively used for the high-temperature measurement of the dielectric properties of ceramic materials. Some advantages, disadvantages and limitations of both methods will be discussed next.

The cavity perturbation method can provide very accurate results for both ϵ' and ϵ'', but for high-temperature measurements, where the speed of data acquisition is crucial due to the rapid cooling of the sample when removed from the furnace and inserted into the cavity, the use of an Automatic Network Analyzer is almost mandatory. Since the accurate measurement of both frequency shift and Q-factor variation constitute the principal parameters of the method, the capability of the ANA to quickly and accurately track and store the resonance curve provides the capability of measuring and storing data at various temperatures during the heating sequence for further analysis. In the absence of an available ANA, some sort of automatic tracking system for the shifts in the resonance curve have to be devised, since manual operation has to be ruled out. Also, the method is intrinsically narrow-band, and various cavities have to be constructed to cover a large portion of the microwave spectrum. However, as mentioned before, the cavity perturbation method is widely recognized as the most accurate especially for the measurement of loss tangent, which is most important for a better understanding of the microwave sintering process in ceramics.

The open-ended coaxial line method is a broadband technique that has been successfully used over the past years to determine the complex permittivity of materials, especially of biological substances. The technique has been constantly improved in recent years, and calibration procedures using ANA's have considerably enhanced the accuracy of measurements. However, for high-temperature applications, the use of a specially designed, low thermal expansion coefficient probe, as described in this paper is a must. It is also clear that, since the generally used calibration technique using alcohols is not feasible in high-temperature applications, another "standard" material has to be selected. For example, the high-temperature cavity perturbation technique described can be used to characterize a specific ceramic material (for example alumina) as a function of temperature with high accuracy at various discrete frequencies, thus providing a "standard" for calibration of the open-ended sensor. Finally, it should be mentioned that extensive measurements performed on ceramic samples with the probes available in our laboratory confirmed that probe contact with the sample and flatness and thickness of sample are crucial for these measurements.

5. SUMMARY AND CONCLUSIONS

In summary, we have described in this paper the ongoing high-temperature dielectric properties measurements at the University of Utah. These include the cavity perturbation method for S- and X-band measurements, and the open-ended coaxial probe for broadband measurements (1 to 11 GHz).

For the cavity perturbation method, experimental results were presented for temperatures up to 1000°C for sintered zirconia. Experimental results for other ceramics are presented in [1].

For the broadband open-ended probe, results were presented for temperatures up to 200°C. The reasons for this limit for the available probes are:

- Differential thermal expansion limits the use of the HP probe to approximately 250°C.
- The University of Utah probe operation temperature is limited due to connector soldering.
- The University of Utah Kovar probe is limited to lower frequencies measurements due to probe size and dimensions.

We have proposed the design of a new metalized-ceramic probe for high-temperature measurements. This probe is expected to work effectively in the frequency band from 500 MHz to 10 GHz, and at temperatures up to 1000°C. Some difficulties with metalization (inside the outer conductor) are still to be resolved.

The dielectric property measurement activities reported have successfully supported the ongoing modeling efforts which are described in other papers by colleagues from the University of Utah.

6. ACKNOWLEDGEMENT

Research sponsored by the U.S. Department of Energy Assistant Secretary for Conservation and Renewable Energy, Office of Industrial Technologies, Advanced Industrial Concepts (AIC) Materials Program, under subcontract 19X-SJ125V and under contract DE-AC05-84OR21400 with Martin Marietta Energy Systems, Inc.

7. REFERENCES

1. S. Bringhurst, O.M. Andrade and M.F. Iskander, "High-Temperature Dielectric Properties Measurements on Ceramics," this issue.
2. M.A. Stuchly and S.S. Stuchly, "Coaxial Line Reflection Methods for Measuring Dielectric Properties of Biological Substances at Radio and Microwave Frequencies-A Review," IEEE Trans. Instr. and Meas. IM-29, 176-183 (1980).
3. E.C. Burdette, F.L. Cain and J. Seals, "In Vivo Probe Measurement Technique for Determining Dielectric Properties at VHF Through Microwave Frequencies," IEEE Trans. Mic. Theory and Tech. MTT-28, 414-427 (1980).
4. T.W. Athey, M.A. Stuchly and S.S. Stuchly, "Measurement of Radio Frequency Permittivity of Biological Tissues with an Open-Ended Coaxial Line: Part I," IEEE Trans. Mic. Theory and Tech. MTT-30, 82-86 (1982).
5. M.A. Stuchly, T.W. Athey, et al., "Measurement of Radio Frequency Permittivity of Biological Tissues with an Open-Ended Coaxial Line: Part II-Experimental Results," IEEE Tran. Mic. Theory and Tech. MTT-30, 87-92 (1982).
6. A. Kraszewski, M.A. Stuchly and S.S. Stuchly, "ANA Calibration Method for Measurements of Dielectric Properties," IEEE Trans. Instr. and Meas. IM-32, 385-386 (1983).
7. A. Kraszewski and S.S. Stuchly, "Capacitance of Open-Ended Dielectric-Filled Coaxial Lines-Experimental Results," IEEE Trans. Instr. and Meas. IM-32, 517-519 (1983).
8. H. Levine and C.H. Papas, "Theory of the Circular Diffraction Antenna," Journal of Applied Physics, 22(1) (1951).
9. N. Marcuvitz, Waveguide Handbook, 213-216 (McGraw-Hill, New York, 1951).
10. D. Misra, et al.,"Noninvasive Electrical Characterization of Materials at Microwave Frequencies Using an Open-Ended Coaxial

Line: Test of an Improved Calibration Technique," IEEE Tran. Mic. Theory and Tech. MTT-38, 8-13 (1990).

11. M.F. Iskander and J.B.Dubow, "Time- and Frequency-Domain Techniques for Measuring the Dielectric Properties of Rocks: A Review," Journal of Microwave Power, 18(1), 55-74 (1983).
12. H. Altschuler, in Handbook of Microwave Measurements, edited by M. Sucher and J. Fox (Polytechn. Press, N.Y., 1963) Vol. 2.

RF AND MICROWAVE DIELECTRIC MEASUREMENTS TO 1400°C AND DIELECTRIC LOSS MECHANISMS

R.M. HUTCHEON, M.S. DE JONG, F.P. ADAMS, P.G. LUCUTA, J.E. McGREGOR AND L. BAHEN
AECL Research, Chalk River Laboratories, Chalk River, Ontario, Canada, K0J 1J0

ABSTRACT

A system has been developed, based on the cavity perturbation technique, for measuring the complex scalar permittivity of materials at frequencies between 50 MHz and 3 GHz, and up to a temperature of 1400°C. Measurements have been made on several classes of materials, including metallic oxides, conventional and advanced ceramics, fabrics, earthenwares, geological and biological materials, foods, glasses, organics and ferrites. Some materials remain unchanged by thermal cycling, but many undergo irreversible changes; the latter are of interest in microwave processing. The system measures the complex dielectric constant of materials during conventional heating, curing or sintering. Correlating these microwave measurements with other diagnostic techniques will help in the understanding of these processes in several classes of materials.

A SYSTEM FOR MEASURING THE COMPLEX SUSCEPTIBILITY OF SMALL SAMPLES AT HIGH TEMPERATURES

A system was developed [1] for measuring the complex electric susceptibility of materials at high temperatures using the traditional cavity perturbation technique. A hot sample and its holder are rapidly removed from a conventional furnace and inserted into a high electric field region of a thick-walled, well-cooled cavity. The resonant frequency and loaded Q of the cavity are measured by a Hewlett-Packard 8753 network analyzer and stored for off-line analysis which includes subtraction of hot empty sample holder effects. The sample and holder are either left in the cavity for further measurements at lower temperatures as the sample cools, or are quickly returned to the furnace for further processing. In the latter case, the sample can be out of the furnace for as little as 2 seconds for a measurement at a single frequency.

The data analysis uses the general formalism of resonant cavity frequency shift theory, in which the real and imaginary parts of the complex electric susceptibility, $\chi_e = \epsilon_r - 1$, are related, through functions dependent on sample shape, to the change in cavity resonant frequency and Q [2]. The absolute calibration is normally done at room temperatures, and uses measurements on known high purity materials (similar to ASTM Standared Test Method, D2520-86, Method C). For sample geometries for which calibration materials are not available (e.g., thin films), comparison is made with calculations of the frequency shift and Q using the 2-D code SEAFISH [3].

Multimode Cavities for Multifrequency Analysis

The peak-analysis and frequency-agile abilities of the HP 8753 network analyzer allow measurements of frequency and Q at half a dozen frequencies in less than ten seconds. To take advantage of this capability, a cavity was required with good higher-order-mode separation (at least to sixth order) and with relatively uniform field in the sample region. Two cavities were built to meet these requirements, a quarter-wave coaxial resonator and a right cylindrical TM_{ono} resonator.

The coaxial resonator is approximately 1.5 metres long (fundamental frequency of 54.5 MHz), and is made of 230 mm (9") outer diameter high power copper coaxial transmission line. The hot cyclindrical sample (30 mm long, 5.5 mm diameter) and holder are inserted along the axis into the electric field at the open gap end. The copper walls in the region of the gap are water cooled. All coaxial modes up to and including the sixth mode at 600 MHz are cleanly separated from other resonant frequencies. Examples of the use of this cavity may be found in reference [1].

The TM_{ono} resonator (Fig. 1) is a 580 mm (23") diameter, 50 mm long right cylindrical cavity (fundamental mode at 400 MHz) which, with a 12 mm long, 4 mm diameter sample inserted along the axis, has clearly separated modes up to n=6 at 2970 MHz. The cavity is a furnace-brazed, thick-walled OFHC copper structure with cooling channels machined into the central axial region which is exposed to the high temperature sample and sample holder. The unloaded Q of this cavity is approximately 10 000 at 400 MHz and increases to over 30 000 for the n=6 mode. The thermal transient produced by inserting the hot sample produces small shifts in frequency, but the effect of these is minimized by doing the hot empty sample holder measurements with the same time delays and frequency sequences used during measurements with samples.

The Data Analysis Formula

$\chi_e = \chi_e' - j\chi_e''$ may be determined from resonant-cavity perturbation measurements using the following equation:

$$\frac{\Delta f}{f_E} + j\left(\frac{1}{2Q_{L,S}} - \frac{1}{2Q_{L,E}}\right) = \frac{-\chi_e}{(1+F_{sh}*\chi_e)}\ \frac{V_S}{V_c}\ A \qquad (1)$$

where f_E, $Q_{L,E}$ are the cavity frequency and loaded Q, with the empty holder,

Δf is the frequency shift produced by the sample,

$Q_{L,S}$ is the loaded cavity Q, with the holder and sample,

F_{sh} is a real number dependent only on the sample shape,

V_s and V_c are the sample and cavity volumes,

A is a real calibration constant dependent only on the shape of the electric fields in the absence of the sample.

The two calibration constants in the formula, A and F_{sh} can be determined either by comparison with computer simulations or by comparison with known samples. Note that both the real and absorptive parts of the susceptibility are given by Eqn. 1, even though the calibration constants are real and may be determined by measurements on low loss materials. For large $|\chi_e|$, the fractional frequency shift reaches a limiting value which is independent of the dielectric constant. Thus a high conductivity (high χ_e'') sample can be used as one calibration point. Single crystal, high-purity sapphire (ϵ_r' = 11.54 with electric field parallel to the C-axis, ϵ_r' = 9.34 perpendicular to C-axis) may be used for the second calibration point. Teflon (ϵ_r' = 2.07) may be used as a check at low values. The error of this calibration technique for the present sample length to diameter ratio ($\approx$ 3.5) was shown to be less than ± 3% for $\epsilon_r \le 30$ and $\tan\delta \le 1$ by comparing the results of Eqn. 1 with numerical calculations [4] using the code SEAFISH.

Sample Holder Considerations and Sample Preparation Methods

Rapid movement of the sample in and out of a high temperature furnace requires the sample holder to be relatively insensitive to temperature gradients and to high cooling and heating rates. High purity amorphous silica can tolerate these conditions as well as being inert in air up to 1600°C. Thus, at present, the sample holders for high temperature work are made from amorphous silica tubes, 1 mm wall thickness, and 6.0 and 4.0 mm inner diameter for the low and high frequency cavities, respectively. A 1 mm thick flat bottom is fused into the tube to support the sample. Teflon sample holders are used for low temperature studies.

High purity, low OH content amorphous silica has $\epsilon_r' = 3.8$ and $\tan\delta \approx 10^{-4}$ at room temperature. However at high temperatures (> 1000°C), both ϵ_r' and the losses increase significantly, and their contribution to the cavity frequency and Q shift value must be included. Before a measurement with a sample, the empty holder values are measured at 100°C intervals up to the maximum planned temperature. They are later subtracted from the sample measurements assuming the sample and empty holder shifts add linearly.

In Fig. 2 is shown the apparent increase of χ_e' and χ'' that would occur for a standard size sample if the change in sample holder properties were not taken into account. The sample-holder effect may be removed in the analysis, but making this correction introduces additional experimental error in ϵ_r' and ϵ_r''. At 1400°C, the error in ϵ_r' is less than 5% for materials with $\epsilon_r' > 2$, and the error in ϵ_r'' is less than 5% for $\epsilon_r'' > 0.1$. When measuring the properties of low-loss materials at high temperatures, the error can be minimized by removing the sample and immediately measuring the empty holder values at the same high temperature.

The use of the amorphous silica holder is limited by two effects: 1) the gradual nucleation and recrystallization to cristoballite above 1000°C, and 2) chemical and physical interaction with the sample, particularly if the sample contains mobile or reactive metal ions. The former effect results in heavy "frosting" of the surface of the tube and, although there is very little change in the dielectric properties, embrittlement and cracking of the holder will eventually occur. Holders have survived over two hours at 1500°C, with measurements every fifteen minutes.

Chemical interaction with the holder is a function of the number of physical contact points; thus pressed pellets of reactive material can often be measured with insignificant contamination of the holder, while powder, packed in the holder, produces intolerable reaction. Samples are usually prepared either by diamond core drilling of solid material or by pressing powders into self-standing pellets. Non-reactive materials may be used in powder form.

EXAMPLES OF HIGH-TEMPERATURE, MULTI-FREQUENCY DIELECTRIC CONSTANT MEASUREMENTS

During the last two years we have made measurements on a broad selection of materials to give us practical experience with many different sample types. This has given us a better understanding of the abilities and limitations of the apparatus and technique, and has revealed a few areas of study to which the system is especially suited. One point in particular that constantly recurred was the variability in dielectric properties from sample to sample, both in composites (due to lack of homogeneity) or in pressed compacts or powders (due to density variations). It became clear that studies of temperature dependence and particularly of frequency dependence would have better accuracy and be easier to interpret for mechanistic studies if all measurements could be done on a single sample. This is especially true of measurements intended to mimic a materials processing step (such as microwave heating and curing) where the sample property changes are inherently irreversible.

Figure 1 Cavity and 1700°C conventional furnace set up for high temperature dielectric property measurements. The TM_{ono} modes used are between 400 and 3000 MHz.

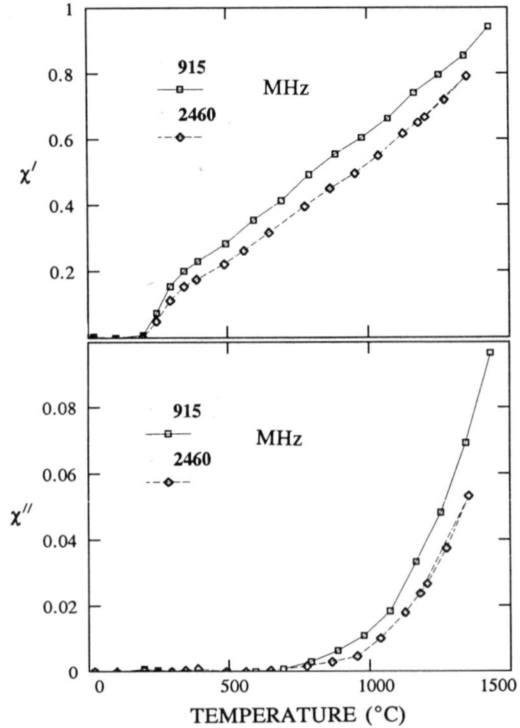

Figure 2 Empty sample holder effect in terms of equivalent complex susceptibility for 12 mm long, 3.6 mm diameter sample.

Microwave Response of Various Zirconia Samples

Stabilizing materials are added to zirconia sintered compacts, the most common of which are yttria, calcia and magnesia. As well, small percentages of sintering aids may be used and a small hafnia "impurity" (≤ 2%) is usually present. The high temperature dielectric response is of interest both for microwave furnace and susceptor applications and also for basic structure studies. The response of a high purity coarse zirconia powder (Johnson and Mathey, 99.997% zirconia) was compared with that of an 8% yttria stabilized sintered sample (Fig. 3 and 4). Relative to the pure zirconia, the 8% yttria stabilized sample has 10^4 larger equivalent conductivity at 800°C, while a 5% magnesia, 1% calcia stabilized sample (not shown) has 10^3 higher losses.

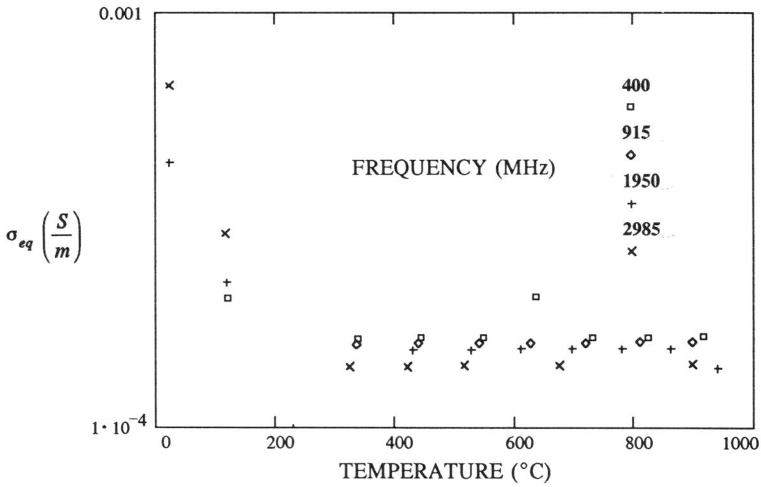

Figure 3 The equivalent conductivity versus temperature (°C) of pure zirconia powder sample (99.9975% zirconia, Johnson & Mathey, powder density 2.67 gm/cm^3).

Fig. 4(b) shows the frequency dependence of the equivalent conductivity, σ_{eq}, of stabilized zirconia. The equivalent conductivity is the rf electrical conductivity which would produce the observed χ''. The thermally activated nature of the stabilized zirconia response is clear, and the change in frequency dependence of the equivalent conductivity signals a change in relaxation mechanism. The frequency-independent portion of σ_{eq} at 800°C suggests electron conductivity as the loss mechanism, but this would not explain the linear frequency dependence at 100°C. Comparison with high-temperature, low-frequency conductivity measurements should be useful. Again, studies of the mechanism will require other types of data.

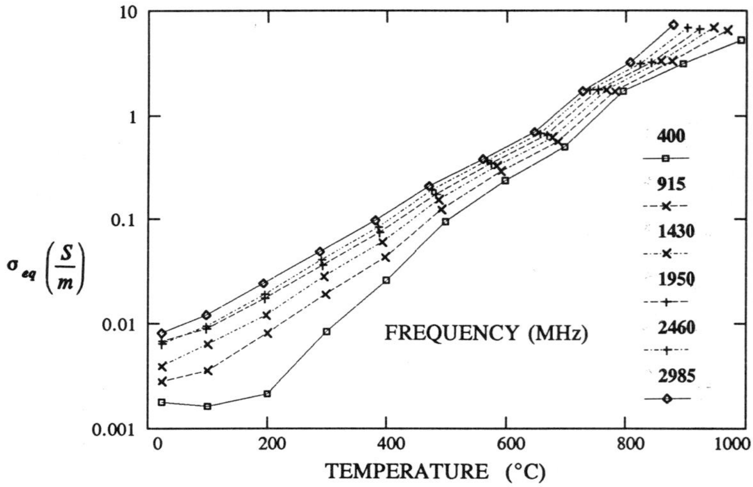

Figure 4(a) Temperature dependence of equivalent conductivity of sintered (density 5.77 g/cm³) 8% yttria-loaded zirconia.

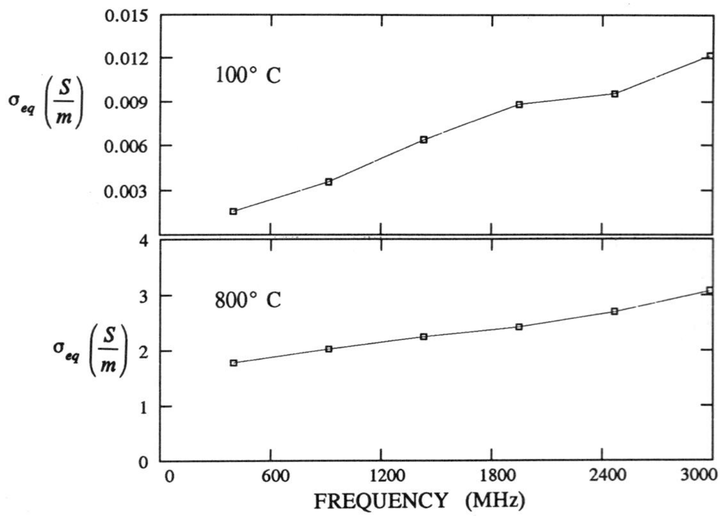

Figure 4(b) Frequency dependence of equivalent conductivity of 8% yttria-loaded zirconia at 100° and 800° C.

Microwave Response of Selected Aluminas

The microwave response of sintered aluminas is of interest to designers of high power, high temperature microwave driven systems -- in some cases power absorption by the ceramic is desired, in some not. Three commercially available sintered aluminas, ALSIMAG 838 (99.5% Alumina), ALSIMAG 620 (94% Alumina) and WESGO AL995 were measured, as well as a small sample which had been microwave sintered using Baikalox CR30 alumina powder (Table 1). The rapid increase in ϵ'' above 1000°C which has been seen previously [6] is variable from sample to sample and not a function of density.

Microwave Response of Alumina During Sintering

The present system can measure the microwave response of a material while it is sintering, under the assumption that the sample volume is known for each individual measurement. Only the initial and final volume can actually be measured, so the shape of the density versus time curve must be inferred from other studies, or a series of runs to different final densities undertaken

A test of the system capabilities was done using a sample made from three stacked pressed powder pellets (each 3.6 mm diameter, 4.0 mm long) of Sumitomo AKP-15 powder with 8% weight isotactic polymethyl methacrylate (PMMA) binder [7]. This sample was initially taken to 1500°C over a 1 hour period to remove the binder, then cooled to room temperature. A quarter hour run at 1500°C then produced the initial and intermediate values shown in Table 2. The results of a final two-hour run are shown in Fig. 5, and the final values shown at the end of Table 2. At the end of this run the sample holder was quite "frosted", but the background subtractions were essentially unchanged.

Microwave Response During Binder Burnout in Alumina Compacts

The process of PMMA binder burnout in a pressed alumina (Sumitomo AKP-15 and AKP-30) compact has been studied with emphasis on the possibility of using microwave heating to improve the process [7]. The present measurement was performed using 3.6 mm diameter, 4.0 m long powder compact pellets, formed by slow compression to a maximum of 20 MPa (3000 psi), achieving a bulk density of 1.9 gm/cm^3. The preparation of the mixed binder and alumina powder is described elsewhere [7].

Three pellets were stacked in the sample holder and heated in a conventional furnace at a rate of 17°C per minute up to 400°C, then 33°C per minute up to 1400°C and finally cooled. It is assumed that the sample temperature tracked the furnace temperature during the ramp-up. The final density at room temperature was usually unchanged.

The two loss peaks in Fig. 6 are clearly a result of the presence of the binder. Similar measurements on pure AKP-15 at the same density showed only a low and monotonically increasing value of χ'' above 200°C. The first loss peak is possibly associated with the thermally induced unzipping of the polymer and evolution of gases; the second with thermal activation of losses caused by residual carbon and its eventual oxidation. However, a definitive knowledge of the mechanism will require a detailed analysis using many techniques. It should be noted that in this experiment the dielectric properties are being measured in the presence of chemically reactive species not normally encountered in samples, so the detailed relaxation mechanism may be different than normally encountered in ceramics.

Table I Measured high-temperature complex dielectric constants for selected sintered aluminas.

ALSIMAG #620 - 94% Alumina, density 3.76 g/cm^3					
912 MHz			2460 MHz		
T(°C)	ϵ'	ϵ''	T(°C)	ϵ'	ϵ''
22	8.90	0.004	22	8.9	0.004
99	9.03	0.006	99	9.06	0.003
298	9.36	0.016	296	9.46	0.010
497	9.70	0.037	491	9.82	0.025
695	10.04	0.089	683	10.15	0.055
891	10.30	0.146	871	10.4	0.094
1086	10.80	0.279	1050	10.81	0.158
1277	11.31	0.452	1221	11.18	0.241
1465	12.25	1.099	1379	11.77	0.476
BAIKOWSKI CR-30 Alumina, solid core, density 3.91 g/cm^3					
912 MHz			2460 MHz		
T(°C)	ϵ'	ϵ''	T(°C)	ϵ'	ϵ''
22	9.29	0.006	22	9.25	< 0.001
1430	11.01	0.103	1282	11.25	0.023
WESGO AL-995 Alumina, solid core, density 3.85 g/cm^3					
912 MHz			2460 MHz		
T(°C)	ϵ'	ϵ''	T(°C)	ϵ'	ϵ''
22	9.05	0.004	22	9.06	0.0025
99	9.29	0.005	98	9.27	0.0015
198	9.53	0.011	197	9.54	0.005
198	9.51	0.012	197	9.52	0.004
1430	11.54	0.817	1282	11.52	0.287
ALSIMAG 838 - 99.5% Alumina, density 3.82 g/cm^3					
			2460 MHz		
			T(°C)	ϵ'	ϵ''
			22	8.95	< 0.001
			396	9.63	0.0028
			593	9.82	0.0061
			787	10.10	0.0155
			978	10.44	0.031
			1164	10.71	0.052
			1255	10.93	0.074
			1343	11.16	0.128
			1430	11.47	0.25

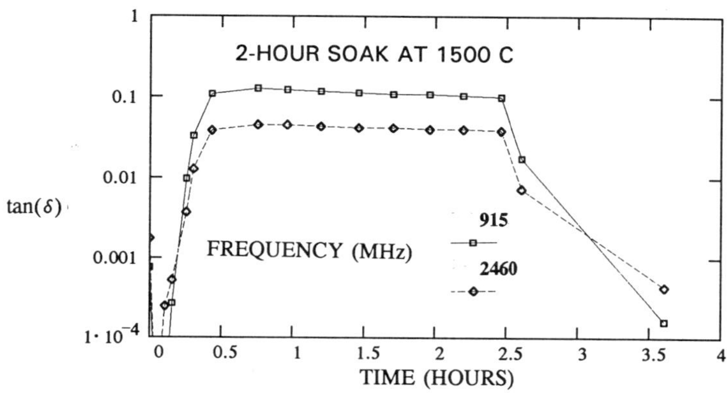

Figure 5 Tan δ measured during sintering run on a pellet of Sumitomo AKP-15 of 2.0 initial and 2.87 gm/cm^3 final density. 915 and 2460 MHz values were measured at 1430°C and 1350°C, respectively.

Table II Measurements of the complex dielectric constant during a 2.25 hour, 1500°C sintering run on a Sumitomo AKP-15 alumina compact.

Sample Density (g/cm^3)	Elapsed Sintering Time (hours)	Frequency (MHz)	Temperature of Measurement (°C)	ϵ_r'	ϵ_r''
1.93	0	915	1430	3.77	0.18
1.93	0	2460	1348	3.82	0.078
2.01	0.25	915	1430	4.14	0.36
2.01	0.25	2460	1350	4.17	0.14
2.87	2.25	915	1430	7.74	0.96
2.87	2.25	2460	1350	7.64	0.36

Microwave Response of SRL-165 Borosilicate Glass

A borosilicate glass, designated SRL-165, has been chosen as a matrix for nuclear waste immobilization in the United States. Microwave heating and melting of the mixed glass frit and nuclear waste has been suggested [8], and a detailed knowledge of the temperature dependence of the dielectric response will be required to evaluate the process. As a baseline for such studies, the response of a solid sample of the pure glass (formed by melting, cooling and core drilling) was measured (Fig. 7). The solid sample melting point is between 600 and 700°C, meaning that, in this temperature range, the sample viscosity decreased to the point that it flowed to take up the shape of the sample holder tube. This uncertainty in sample shape between 600 and 700°C is the reason for the range of values.

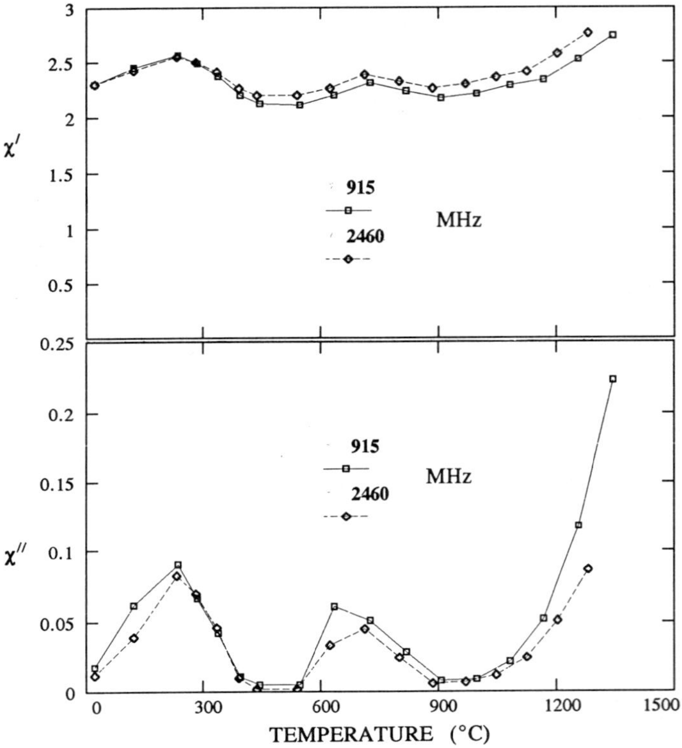

Figure 6 Complex dielectric susceptibility of Sumitomo AKP-15 compact measured during heating and binder burnout. Initial density 1.87 gm/cm^3, with 4% by weight isotactic PMMA organic polymer.

The glass transition temperature, T_g, for the solid SRL-165 borosilicate glass is 460°C. In the region of this temperature, one might expect a gradual increase in both the dielectric constant and the losses due to the increasing mobility of ions and to the increasing freedom of motion of molecular units. The literature [9] suggests that conduction in most glasses is purely ionic up to the melting point. The change in the frequency dependence of the effective conductivity between room temperature and the melting point suggests either a change in the dominating relaxation mechanism or a strong temperature dependence in the characteristic relaxation time of the mechanism. Again, only comparison with low frequency dielectric data and other measured properties will clarify the mechanism. However, a much broader range of dielectric data can now be made available.

ACKNOWLEDGMENTS

We thank P. Hayward of AECL's Whiteshell Laboratories and R.L. Schulz, E.H. Moore and D.E. Clark of the Department of Material Science and Engineering, University of Florida, for educational discussions and for providing some of our samples.

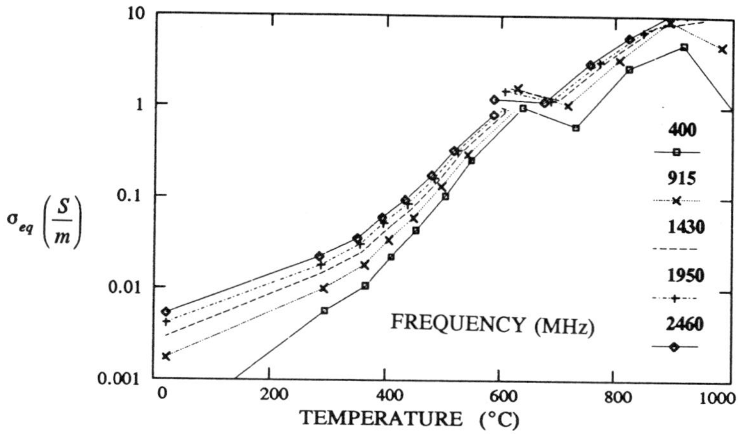

Figure 7 Measured equivalent conductivity versus temperature for SRL-165 borosilicate glass, $T_g \approx 460°C$, density 2.61 gm/cm^3. Melting at 600-700°C interferes with the measurement.

REFERENCES

1. R.M. Hutcheon, M.S. de Jong, F.P. Adams, G.W. Wood, J.E. McGregor and B.H. Smith, J. Microwave Power, accepted for publication.

2. R.M. Hutcheon, M.S. de Jong and F.P. Adams, J. Microwave Power, accepted for publication.

3. M.S. de Jong, F.P. Adams and R.M. Hutcheon, J. Microwave Power, accepted for publication.

4. F.P. Adams, M.S. de Jong and R.M. Hutcheon, J. Microwave Power, accepted for publication.

5. C. Kittel, Elementary Solid State Physics, (John Wiley & Sons, Inc., New York, 1962), p. 108.

6. W.W. Ho, in Microwave Processing of Materials, edited by W.H. Sutton, M.H. Brooks and I.J. Chabinsky (Mater. Res. Soc. Proc., 124, 1988) pp. 137-148.

7. E.H. Moore and D.E. Clark, Polymethyl Methacrylate Binder Removal From Alumina by Microwave Heating, presented at the 11th Annual Conference on Composites and Advanced Ceramics, Cocoa Beach, Florida, January 1992 (unpublished).

8. R.L. Schulz, Z. Fathi, D.E. Clark and G.G. Wicks, in Microwaves: Theory and Application in Materials Processing, edited by D.E. Clark, F.D. Gac and W.H. Sutton (Ceramic Transactions, Vol. 21) p. 451.

9. A.E. Owen, in Electric Conduction and Dielectric Relaxation in Glass. edited by J.E. Burke (Progress in Ceramic Science, Vol. 3, McMillan, 1963) pp. 78-196.

GENERAL PURPOSE HIGH TEMPERATURE MICROWAVE MEASUREMENT OF ELECTROMAGNETIC PROPERTIES

John A. Batt, Robert Rukus, and Meyer Gilden
United Technologies Research Center
East Hartford, CT 06108

ABSTRACT

An efficient general purpose method has been developed for implementing a variety of broad band microwave measurement techniques at high temperature. Central to these techniques has been the development of a "dual-waveguide" test fixture. These techniques have been selected, out of necessity, to cover a variety of material types and geometries such as slabs, posts and sheets. These include measurement of complex permittivity and permeability using a modified Nicolson-Ross technique for slabs and Marcuwitz' algorithm for posts. In the implementation of the dual-waveguide, two adjacent and symmetric X-band waveguides are machined from a single block of Inconel alloy 601*, one serving as a test waveguide, the other as a reference waveguide. The dual-waveguide is surrounded by a programmable clam-shell furnace. The reference waveguide provides data which is used to compensate for electrical and dimensional thermal changes in the test section. Combined with commonly used HP 8510 Network Analyzer methods, automated test procedures have been developed to accommodate the various measurement types and sample shapes. Prototypical data will be presented illustrating the effectiveness of some of these techniques. The methods discussed are most useful for rapid data generation over temperatures ranging from room temperature to 2000 degrees F (1093 degrees C) and at frequencies defined by the waveguide dimensions. This paper highlights developments in the dual waveguide system since its construction over one year ago.

INTRODUCTION

With the advent of the automatic HP8510 Network Analyzer, a number of room temperature measurement techniques have been developed or elaborated upon for the rapid generation of complex permittivity and/or permeability data.[1] Among these methods, those using rectangular waveguide lend themselves quite naturally to modification for high temperature measurements. When extending any room temperature technique to high temperature, the effects caused by temperature changes in the experimental apparatus must be accounted for. Thermal expansion is the most apparent of these effects although thermally caused changes in the waveguide electrical properties may also need attention if the waveguide loss characteristics change significantly upon heating. Harris[2] has used some of the features described here in rectangular waveguide and coaxial implementations of high temperature transmission measurement of dielectric samples. In special cases where only the sample is heated, no corrections are needed.

The dual-waveguide techniques described in this paper all rely on the use of a reference waveguide to monitor electrical and dimensional changes during the heating process. Reference waveguide data is used to calculate corrections for the raw data generated in the arm of the waveguide containing the test sample. This modified procedure has been applied to several room temperature techniques. These include reflection and reflection/transmission techniques which are typically used to characterize post and rectangular slab samples. A standard twelve term HP Network Analyzer calibration procedure was employed when reflection/transmission techniques were used while a three term procedure was used for reflection techniques. Although the data chosen to illustrate the application of some these approaches is taken at X-band, the techniques have general applicability for different bands.

* Inconel Alloy 601, Inco Alloys International, Huntington, WV.

Mat. Res. Soc. Symp. Proc. Vol. 269.

EXPERIMENTAL PROCEDURE

Figure 1 shows the dual-waveguide test configuration for a standard two-port test. The dual-waveguide is surrounded by a CM clamshell furnace. The furnace is driven by a CM furnace power supply with a Honeywell UDC 3000 Universal Digital Controller. The operational flow of the data acquisition is controlled by an HP 300 Series computer which drives an HP8510C Network Analyzer. In addition to collecting S-parameter data generated by the HP8510, the computer monitors the sample temperature through a thermocouple and A/D converter. At chosen temperatures, data is collected for all four S-parameters, in the reference and test waveguide sections. A relay actuator in combination with a DC power supply activates two coaxial switches so that data can be generated in both waveguide sections in rapid succession.

The dual-waveguide/furnace configuration is shown in Figure 2. The waveguide assembly shown in Figure 3 has vertical symmetry except for the flange/sample region. The flange was designed for ease of sample insertion and positioning precision. Samples in the form of transverse slabs or posts across the waveguide are relatively easy to produce and analyze. Vertical mounting facilitates slab positioning by use of small projecting supports from the waveguide wall. The slab is held in place by gravity. Posts are supported in small grooves machined into the face of the waveguide flanges. The faces of the flanges are lapped flat and periodically relapped to maintain junction integrity. Specific flange details vary depending on the choice of technique and will be described later.

The upper and lower waveguide sections were machined from single blocks of Inconel alloy 601. The current system has been used for close to one hundred high temperature cycles up to 2000 degrees F with no detectable reduction in measurement quality. The 0.05 inch septum between the waveguide sections was chosen to match the dual waveguide/single waveguide transition which is a commercially available item. A 0.020 inch slot was machined in the interface wall for the placement of an Inconel key. This step significantly improved the isolation between the adjacent waveguides at the junction interfaces. Fibrous Zircar** lining fits closely against the waveguide sections at the top and bottom of the furnace. The upper and lower Inconel waveguide sections are ten inches long. Short waveguide sections of stainless steel (not shown in Figure 3) were positioned between the Inconel sections and the transition assemblies. The top section, which had a tendency to heat because of convection as well as conduction and radiative heat transfer, had a thin brass plate located at the Inconel/stainless steel joint region. Air blown on the plate during the heating cycle provided adequate cooling of the upper waveguide region. The lower waveguide assembly was attached to a large aluminum plate which served as a heat sink.

The four flange/sample configurations shown in Figure 4 relate to the choice of measurement technique. Either a post or slab can be selected for sample geometry. These options can be elaborated upon as in the case of a post array. Choices on the reference side of the waveguide are between a short and a thru. Since the purpose of the reference guide data is primarily to provide phase corrections to the sample test data, the S-parameter test data determines the flange choice options for the reference guide. In the most general case where all four S-parameters are required, a shorting plate is needed as shown in Figure 4a and 4c. Reflection coefficient data can be measured directly while the transmission coefficient phase changes can be readily determined by summing the phase changes in SI1 and S22 data. In the flange region asymmetry exists between test and reference waveguide arms. Phase changes in this region must be calculated from knowledge of the thermal expansion behavior of the sample and flange. This calculation is facilitated by locating the flange in the center of the hot zone with a thermocouple attached to it for temperature monitoring.

If only transmission data is required for the test arm, one has the option in the reference arm of measuring the transmission coefficient directly or calculating phases as described above using the thru sample/flange configurations shown in Figures 4b and 4d. Finally, in the case where transmission measurements are excluded, one may wish to use a short or a thru flange on the reference side depending upon the type of measurement and the type of guide

** Zirconia Insulating Board Type, ZYZ-3, Zircar Products Inc., Florida, NY.

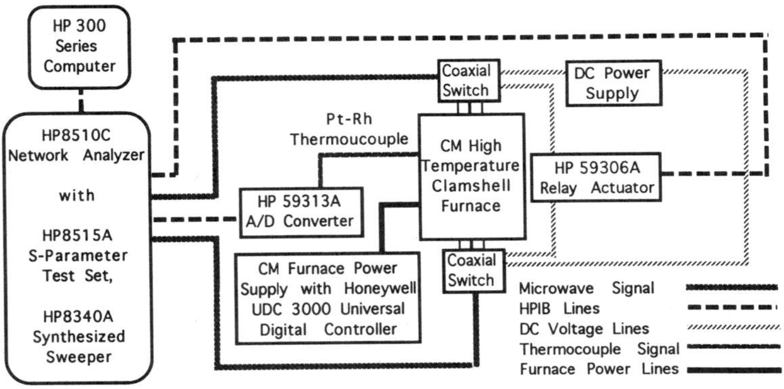

Figure 1: Dual-waveguide Component Test Configuration

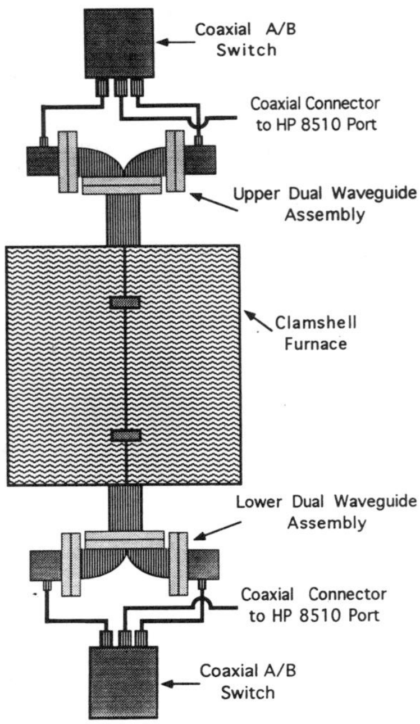

Figure 2: Dual-waveguide/Furnace Assembly

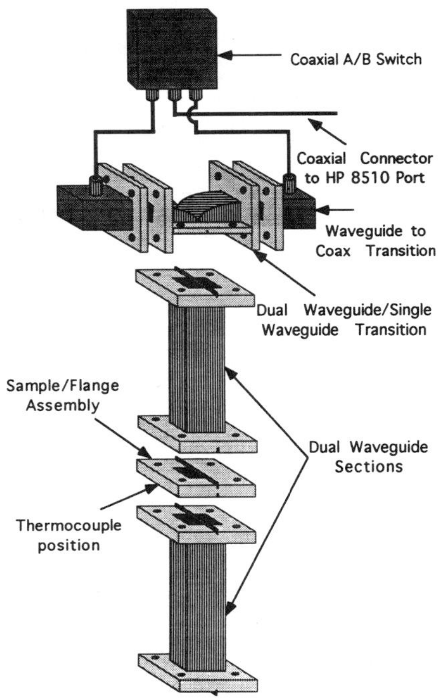

Figure 3: Dual-waveguide Assembly

termination used. For instance, for a post technique which will be described later in this report, a thru flange is used because both guides are terminated with a quarter-wavelength offset short. Symmetry between the guides is thus maintained, reducing the complexity of the overall analysis.

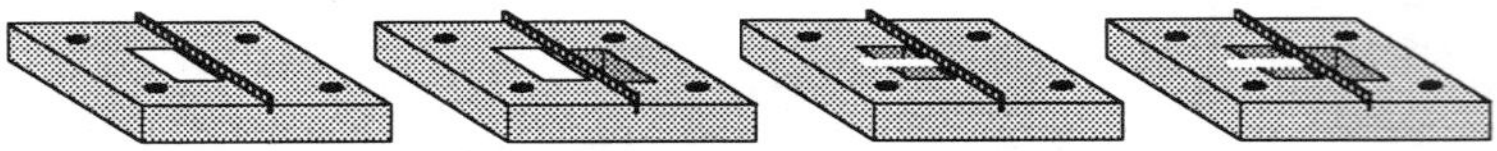

(a) (b) (c) (d)

Figure 4: Slab and Post Flange Configurations for Reflection and Reflection/Transmission Test Methods

Two techniques were used to generate the experimental data described below. First, a general technique for measuring the complex permittivity and permeability of waveguide filling slabs was developed based on a well-known reflection/transmission method first developed by Nicolson and Ross and elaborated in an HP Technote.[3] This approach requires that the sample wavelength be restricted to less than one-half wavelength and optimally be one-quarter wavelength. The modifications to the input data for high temperature use based on the considerations discussed earlier are straightforward. The technique can provide swept frequency output and can detect asymmetries in test samples. Asymmetry detection results from comparison of SII and S21 output with S22 and S12 output computations.

Errors arise from the air gaps caused by the differential expansion between the waveguide and the sample. The effects of these gaps are well known and have the effect of reducing the measured dielectric constant and loss tangent. The algorithm used to reduce the data presented here makes use for the gap analysis provided by Westphal.[4] The method for dielectric post measurement makes use of an algorithm developed by Marcuvitz.[5] The "post" in this case is a small diameter cylinder. Since cylinders are difficult to produce, a small rectangular solid rod was machined, roughly 0.03 inches on a side. From this an "effective diameter" was calculated and used in the algorithm. The complex impedance across the post, as well as the post diameter and frequency serve as input variables for the computational algorithm. Both reflection and reflection/transmission methods can be used to generate the necessary impedance data. In general, the reflection/transmission method yields superior results for lossy materials while the reflection method provides better results for low loss materials.

EXPERIMENTAL RESULTS

Sample choices for experimental test and comparison data from Westphal[6] are listed in Table 1. They include two low-loss samples, fused quartz and alumina, one sample which gets lossy at high temperature, zirconia, and one magnetic material, nickel-zinc ferrite. Measurements were taken using both a slab and post technique for the three dielectric samples. Only a slab technique was employed for the magnetic sample.

Table 1: UTRC Test Sample and Westphal Comparision Data Specifications

Sample	Source	Purity	Density (gm/cc)	Frequency (GHz)	Diel Const	Diel Loss Tangent	Magnetic Perm.	Mag Loss Tangent
UTRC Test Materials								
NSGOZ Fused Quartz	Nippon	99.9	2.23	10	3.81	very low	-	-
AD 998 Alumina	Coor's Ceramics Co.	99.8	3.87	10	9.40	very low	-	-
1936 Zirconia(3%Mg)	Corning	3% Mg	5.21	10	14.98	low	-	-
Nickel Zinc Ferrite	PowderTech Corp.			10	11.38	low	0.249	1.95
Westphal's Comparison Materials								
Amersil Fused Silica	American Optical Co.	unknown	2.196	5.4	3.818	0.00015	-	-
AD-995 Alumina	Coor's Ceramics Co.	99.5	3.84	3.5	9.63	0.00008	-	-

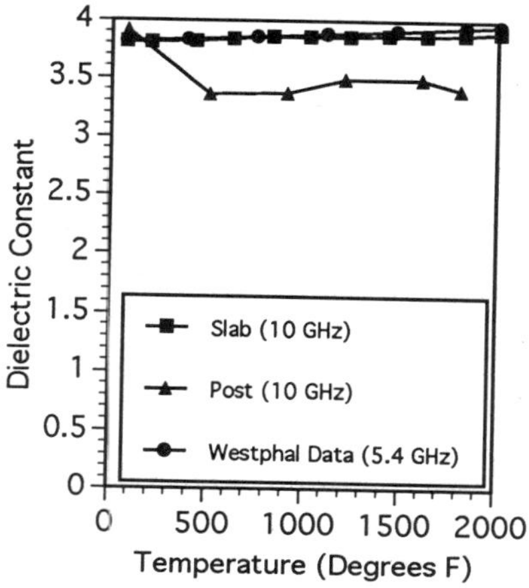

Figure 5: Temperature Dependence of Fused Silica Dielectric Constant for Slab, Post, and Westphal Comparison Data

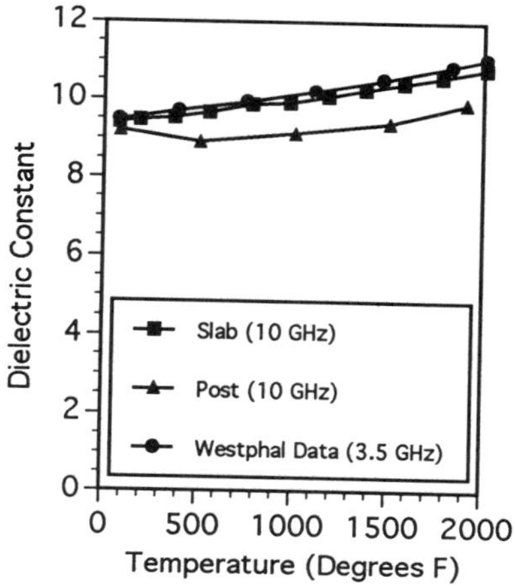

Figure 6: Temperature Dependence of Coors Alumina Dielectric Constant for Slab, Post, and Westphal Comparison Data

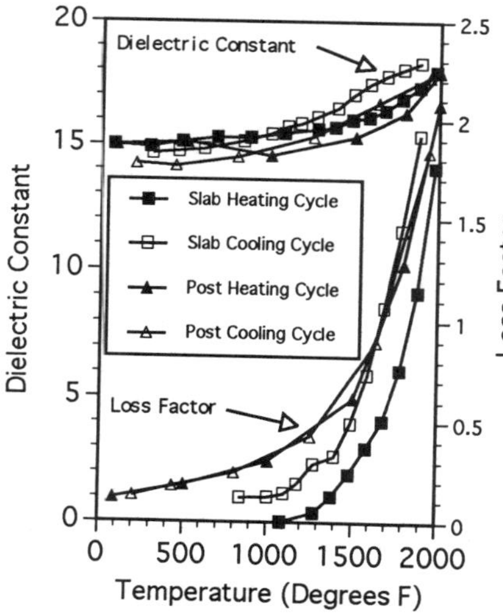

Figure 7: Temperature Dependence of 1936 Corning Zirconia Complex Permittivity for Slab and Post at 10 GHz

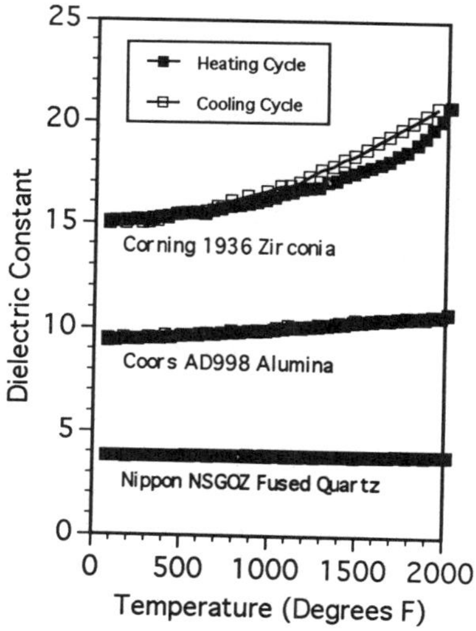

Figure 8: Effect of High Temperature Cycle on Dielectric Constant for Refractory Oxide Slabs at 10 GHz

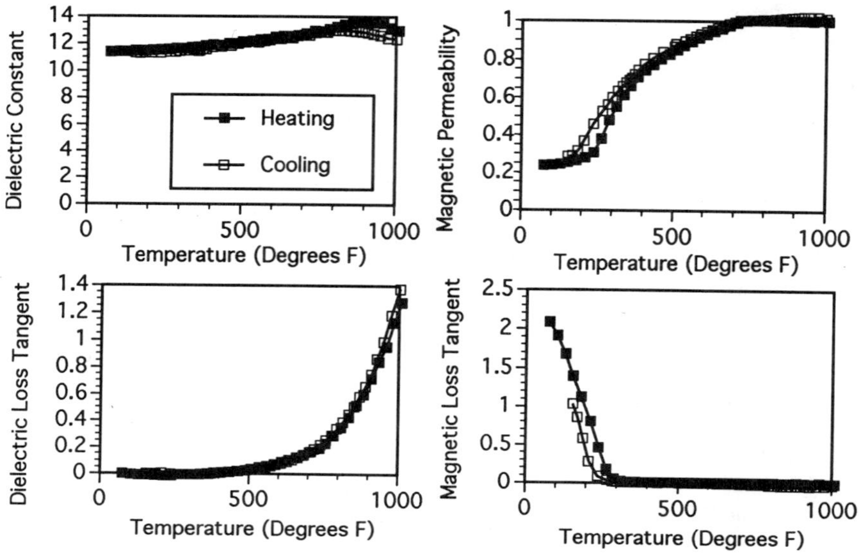

Figure 9: Complex Permittivity and Permeability as a Function of Temperature for a Nickel Zinc Ferrite Slab at 10 GHz

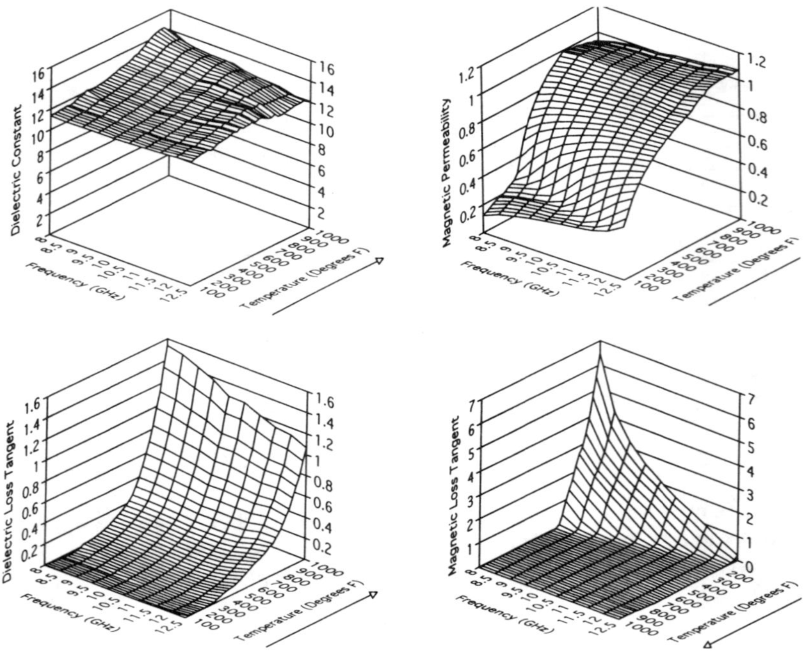

Figure 10: Complex Permittivity and Permeability as a Function of Frequency and Temperature for a Nickel Zinc Ferrite Slab

The temperature/time cycle for the non-magnetic samples had a constant rate temperature increase from room temperature to 2000 degrees F over a period of one hour. This was followed by a steady cooling rate back to room temperature over a one hour period. Below one thousand degrees F, the clamshell furnace had to be gradually opened and finally cooled with a fan to provide sufficient cooling. The nickel-zinc ferrite was heated from room temperature to 1000 degrees F and cooled again at a rate of about 10 degrees per minute.

Experimental data for fused silica and comparison data are shown in Figure 5. Similar data for alumina is presented in Figure 6. Only slab and post experimental data are shown for zirconia in Figure 7 where no comparison data was available. Cycling data for the three dielectric samples described are shown in Figure 8. Work continues to explain and eliminate discrepancies existing at high temperature between post and slab data. Since the results for the two methods do not differ significantly at room temperature, a reexamination of the thermal expansion portion of the algorithm is probably called for. Finally, the complex permittivity and permeability temperature dependence for nickel-zinc ferrite are presented in Figure 9. An expanded version of the same data set exhibiting frequency and temperature dependence is shown in the 3D plots in Figure 10.

EXPERIMENTAL ERROR

The accuracy of the reflection/transmission slab method is dependent upon a number of factors including the sample thickness, sample temperature, as well as ranges of values for dielectric and magnetic properties. For the Coor's alumina sample investigated here, the measured magnetic permeability (not shown) differed from unity by less than six percent over the entire X-band frequency range from room temperature to 2000 degrees F. For most frequencies, 1-2 percent maximum deviation from unity was the norm. To cite another test, when reflection coefficients for the two shorted dual-waveguide arms were measured from room temperature to 2000 degrees F, the maximum phase discrepancy between the two arms was about one degree for all frequencies. Magnitude discrepancies between the two arms had a maximum value of approximately 0.1 dB. The frequency and temperature dependence on these discrepancies appeared to be quite random.

SUMMARY AND CONCLUSIONS

A general purpose dual-waveguide technique has been demonstrated for the rapid and efficient generation of swept frequency high permittivity and permeability data at high temperature. With use of reference and test waveguide arms, thermal effects can be automatically compensated for. This allows room temperature transmission line measurement techniques to be modified for high temperature use. Finally, a limitation of the broad band swept frequency techniques described is inability to measure accurately loss tangents lower than about 0.05.

REFERENCES

1. M. Afsar et al; Proceedings of the IEEE 74 (1), January 1986.

2. N. Harris, J. Chow, R. Eisenhart in Microwaves: Theory and Application in Material Processing, edited by D. Clark, F. Gac, and W. Sutton (Ceramic Transactions, American Ceramic Soc. 21, 1991) pp 235-242.

3. "Measuring Dielectric Constant with the HP 8510 Network Analyzer"; Hewlett Packard Product Note No. 8510-3, August, 1985.

4. W. B. Westphal, Tech. Report XXXVI, Lab. for Insulation Research, MIT, 1950.

5. N. Marcuvitz, Ed., Waveguide Handbook. New York: McGraw-Hill, 1951.

6. W. B. Westphal and A. Sils, Technical Report AFML-TR-72-39, 1972.

HIGH-TEMPERATURE DIELECTRIC PROPERTIES MEASUREMENTS OF CERAMICS

SHANE BRINGHURST, OCTAVIO M. ANDRADE AND MAGDY F. ISKANDER
University of Utah, Department of Electrical Engineering
Salt Lake City, Utah 84112

ABSTRACT

This paper dicusses experimental arrangements, describes measurement techniques and presents experimental results for high-temperature broadband dielectric properties measurements of ceramics. The cavity perturbation technique and the open-ended coaxial line method are used in these experimental measurements. The design and construction details of cavities and probes are described and representative results of measurements on zirconia and alumina samples (green and sintered) are presented. Results of measurements made on insulating materials are shown. In general measurements are made in the frequency band 1 to 10 GHz and temperatures up to 1000°C.

1. INTRODUCTION

Knowledge of the dielectric properties of ceramics and their variation with temperature and frequency is crucial in acquiring a comprehensive understanding of microwave sintering processing of these materials. A complete database of the dielectric properties may lead to optimizing the microwave sintering process, identifying guidelines for selecting compatible insulation-ceramic combinations, improve effectiveness of heating, and in the future application of microwave sintering to large ceramic samples.

There are many available techniques suitable for these measurements. It is often necessary, however, to evaluate the trade-offs involved and select the most suitable one for a specific application. Cavity perturbation techniques, for example, are considered to be very accurate in the measurement of loss tangents. They are, however, narrow-band and several cavities are needed for broadband measurements. The open-ended coaxial line technique, on the other hand, is useful for determining the permittivity of materials over broad frequency ranges. The simplicity of measurements and the greater accuracy that can be obtained with the use of modern network analyzer techniques have encouraged the present efforts towards using this technique for measurements on ceramics at high temperatures.

This paper describes the results of measurements of the dielectric properties of microwave-sintered zirconia samples at high temperatures over a broad frequency range. Both the cavity perturbation and the coaxial probe techniques were used in these measurements.

Section 2 of this paper provides a description of the above mentioned techniques as applied to high-temperature measurements along with the design of two rectangular cavities (for operation at S- and X-band), and the experimental arrangement for utilization of the open-ended coaxial probe. Sample preparation and sintering methods are discussed in Section 3. Measurement results on zirconia and alumina samples are described in Section 4, and an overall summary and conclusions are presented in Section 5.

A theoretical analysis of the cavity perturbation and open-ended coaxial line techniques used in these measurements are described in a companion paper in this Proceedings [1].

2. DESCRIPTION OF MEASUREMENT TECHNIQUES

2.1 CAVITY PERTURBATION METHOD

The cavity perturbation method has been used in this work to measure the complex permittivity of various rod ceramic samples of cylindrical and rectangular cross-sections at S-band (2.45 GHz) up to 1000°C. The theory and perturbation expressions for samples both fully and partially inserted in a cavity are described in [1]. In this section we will discuss fabrication aspects of two rectangular cavities (for S-and X-band measurements), and explain the experimental arrangement used for high-temperature measurements.

- <u>Cavity Fabrication and Measurement Procedure</u>

An S-band rectangular cavity was designed for operation in the TE_{103}-mode at 2.45 GHz, and constructed from a section of WR-340 brass waveguide. Two probes were installed in the cavity's broad wall on opposite sides at positions of maximum electric field, which is in this case at a distance of 1/3 the length of the cavity from both ends. The length of the probes was adjusted to maximize the Q-factor of the cavity and also to obtain undercoupling. Two small holes were drilled in the center of the cavity for sample insertion. The hole diameter was calculated such that the circular waveguide thus formed is well below cutoff at the cavity's resonant frequency. Measurements with an HP8753C Network Analyzer showed that the cavity resonates at ≈ 2440 MHz in the TE_{103}-mode and the Q-factor is ≈ 2200. Calibration of the cavity using TEFLON samples is described in [1].

For purposes of measuring the dielectric properties of ceramic materials at high temperatures (up to 1200°C) the experimental set-up requires the sample to be held in an adequate sample holder which is to be heated in a conventional furnace prior to insertion into the cavity. For the high temperatures under consideration, a thin-walled fused-silica cylindrical tube was selected as a sample holder. The ability of fused-silica to withstand high temperatures, and also the small variation of its dielectric properties with temperature (including low loss), make this sample holder very convenient for the present application. The sample holder design is such that the sample does not have to be fully inserted into the cavity during the measurements. A stop plate inside the silica tube is used to accurately calculate the sample length. This arrangement serves two purposes: to reduce the volume of the sample inside the cavity when losses are so high that a large reduction in the Q-factor would affect the accuracy of the measurement, and also to allow for monitoring the sample temperature with a thermocouple located at the tip of the sample protruding from the cavity.

A very thin Chromel-Alumel thermocouple is used for measuring the tube and sample temperatures. The thermocouple is insulated with an alumina sheath which is inserted into the silica tube through one end of the furnace and placed in contact with the sample. During the measurement the tube is quickly moved from the furnace into the cavity when the desired temperature is reached. Initially, the empty sample holder is calibrated as a function of

temperature (both resonance frequency and Q), and these values are used as the "empty cavity" parameters in the perturbation expressions. The Network Analyzer is calibrated to accurately display the resonance curve as it shifts during the temperature increase, and each measurement curve is immediately stored at the desired temperature for later retrieval. Since sample and tube cooling rates are very fast after removal from the furnace and insertion into the cavity, special insulating blocks are placed around the tube part that protrudes from the furnace to minimize these effects. Also, to avoid heating of the cavity during the measurement, an extremely fast data acquisition process is needed when the sample is positioned in the cavity. The features of the Network Analyzer, therefore, are very helpful for this fast data acquisition process. The experimental set-up for the high-temperature cavity perturbation technique is shown in Figure 1.

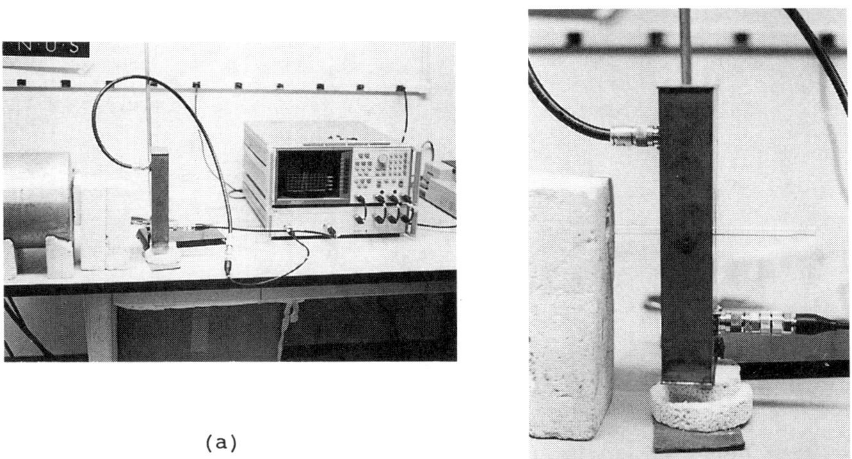

(a)

(b)

Fig. 1. Experimental set-up for high-temperature cavity perturbation technique.(a) Overall experimental setup, (b) Insertion of silica tube in S-band cavity

The second cavity available in our laboratory for high-temperature measurements is an X-band cavity designed for operation in the TE_{107}-mode at 10 GHz, constructed from a section of WR-90 copper waveguide. Small circular holes cut in short-circuiting metal plates located on both sides of the waveguide provide weak iris-coupling to the cavity. Two waveguide-to-coaxial transitions allow the cavity to be connected to an HP-8510B Network Analyzer, calibrated in transmission mode. As with the previously described S-band cavity, two small holes are drilled in the center of the broad wall for sample insertion. Measurements with the HP8510B showed the cavity to resonate at ≈ 9956 MHz in the TE_{107}-mode and the Q-factor to be ≈ 5500. This cavity has also been calibrated using a TEFLON sample. The high-temperature measurement procedure is the same as described above, and the characterization of ceramic materials in the X-band is presently under way.

2.2 OPEN-ENDED COAXIAL LINE TECHNIQUE

The broadband open-ended coaxial line method is being extensively used in our laboratory for measuring the complex permittivity of ceramic composites at medium temperatures (up to 250°C), and over a 1 to 11 GHz band. The analytical approaches used to obtain the sample permittivity from the measured input admittance of the probe are described in [1]. Figure 2 shows a picture of the probes presently in use. In this section we will explain the high-temperature setup for these measurements.

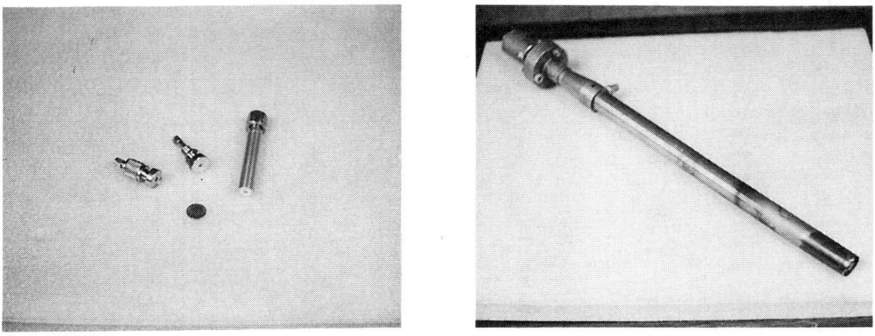

(a) (b)

Fig. 2. Open-ended coaxial line probes used in this work. (a) Open-ended probes for low-temperature measurements, (b) Kovar tube for measurements up to 600°C.

An HP8510B Network Analyzer controlled by an HP9000/300 computer is used for data acquisition when using probes developed at the University of Utah. For automatic measurement and permittivity calculation when using the Hewlett-Packard probes, the software package HP85070A was used. The same furnace used for cavity perturbation measurements is utilized here, and the sample is positioned halfway sitting on an insulation brick. The probe is initially calibrated using the open-short-deionized water technique. A specially designed high-temperature cable is used to connect the Network Analyzer and the probe, which is lowered into the furnace when the desired temperature is reached. The temperature is measured using the previously described thermocouple in contact with the flat surface of the sample. The probe is next put in contact with the sample, and measurement is performed when a stable trace is observed on the HP8510B screen. To ensure that the "infinite sample" condition is satisfied, the samples are initially measured both with and without a metal plate on the backside of the sample and the calculated permittivities were compared. Extensive measurements performed on a variety of materials have confirmed that probe contact and flatness and thickness of the sample are crucial for this technique to be correctly used with solid materials. Figure 3 shows the system arrangement for high-temperature open-ended probe measurements. Accuracy of high-temperature measurements were checked by making measurements on ceramics of known properties.

Fig. 3. High-temperature arrangement for open-ended coaxial probe technique

3. SAMPLE PREPARATION AND SINTERING METHODS

Various ceramic composite samples of different sizes, shapes and dopant/binding percentages are presently being measured at our laboratories using both the cavity perturbation technique and the open-ended coaxial line method, in the frequency band 1-10 GHz and from temperatures up to 1000°C. Both green and sintered ceramics were characterized. These samples were received green in rods of circular and rectangular cross sections, and also as rectangular bars, from the Oak Ridge National Laboratory. For cavity perturbation measurements at 2.45 GHz, rod samples were machined to rectangular shape using fine mesh sandpaper to final approximate dimensions of 4x4x32 mm. For the open-ended coaxial line method, rectangular bar samples are used, with approximate dimensions of 20x10x15 mm, presenting a sufficiently large sample area for contact with the coaxial probes.

The Oak Ridge National Laboratory arranged to loan the University of Utah a microwave sintering system. The system consists of a modified General Electric microwave oven and a programmable microwave-temperature controller (Micristar) with a dual-channel strip-chart recorder for monitoring both temperature and input power vs. time. The dimensions of the microwave sintering cavity are 40x40x24.5 cm^3 operating at 2.45 GHz. Several ceramic samples were sintered in the oven. In all cases samples were surrounded with cylindrical insulation (ZIRCAR AA-30 and ZYBF-2) and four rods of SiC were inserted in the insulation in a picket-fence arrangement to aid the microwave heating process. The temperature was monitored using a Pt-Rh (R-type) thermocouple. Alumina and zirconia samples were heated to temperatures as high as 1475°C. In all cases ceramic samples were glowing red to white hot while the insulation stayed near ambient, thus indicating appropriate microwave coupling to samples with negligible coupling to the insulation. This also indicates that the actual sintering occurred under microwave heating conditions and not by radiative transfer from the surrounding material [2]. Binders in some samples were removed prior to sintering by conventional heating up to 700°C (heating rate 2°C per minute up to 700°C, then keeping the temperature at 700°C for two hours).

4. MEASUREMENT RESULTS

We present in this section some representative results obtained from a large variety of complex permittivity measurements on ceramic composites which are under way in our laboratory. The results shown in figures 4-6 were obtained using the cavity perturbation method, and include sintered zirconia TZ8Y(8 mol% yttria), green alumina+0.1% MgO and AL-30 fiberboard insulating material. It can be seen that these results are presented for temperatures up to 1000°C except for TZ8Y. The reason for limiting the results at 600°C for this material is the fast increase in loss tangent at higher temperatures, thus dramatically reducing the Q-factor of our cavity, resulting in unreliable measurements. A cavity with a higher Q-factor, or the use of samples of smaller cross sections will avoid this problem.

Figures 7 and 8 show broadband measurement results using the open-ended coaxial line technique for TZ8Y both green and sintered, at room temperature and at 200°C. Comparison of these results with those obtained using the cavity perturbation method at 2.45 GHz show excellent agreement between the two different techniques.

The 1000°C limit of the presented results is related to the maximum temperature that can be achieved with the furnace presently in use, and also to the Q-factor of the cavity which limits the accuracy of measurements for materials with very high losses when the temperature is increased. The 200°C limit for the coaxial probe is related to solder problems beyond 200°C for one of our coaxial probes, and also to the temperature limit of the HP probe, which is 250°C. Additional difficulties with the open-ended coaxial probe include the need for exceptionally good contact, smooth surfaces, and availability of sufficiently thick samples (1 to 2 cm thickness, depending on the materials properties and measurement frequency).

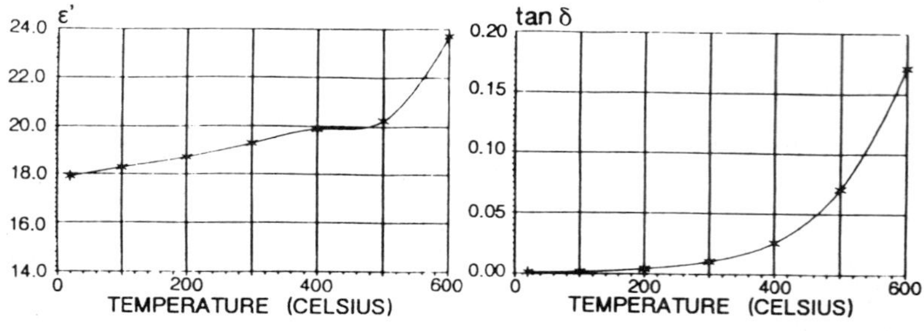

Fig. 4. Dielectric properties of sintered TZ8Y vs. temperature at 2.45 GHz using cavity perturbation.

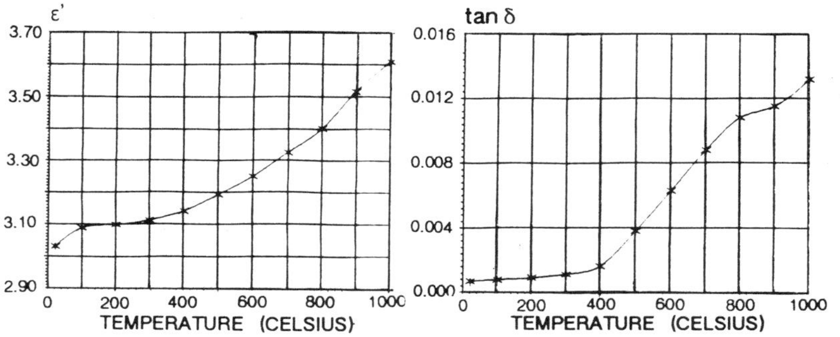

Fig. 5. Dielectric properties of green Alumina+0.1%MgO vs. temperature at 2.45 GHz using cavity perturbation.

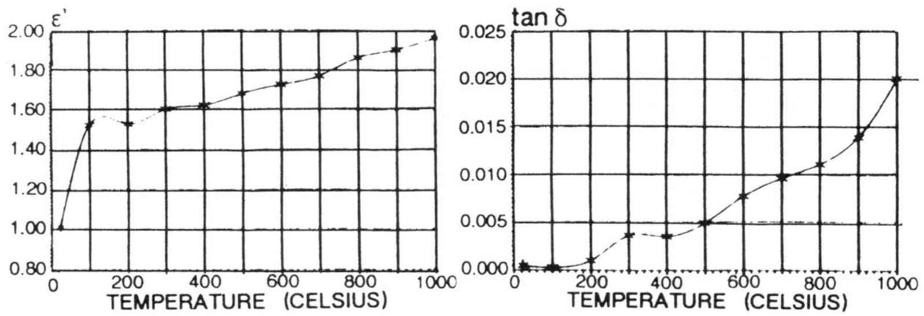

Fig. 6. Dielectric properties of AL-30 insulation vs. temperature at 2.45 GHz using cavity perturbation.

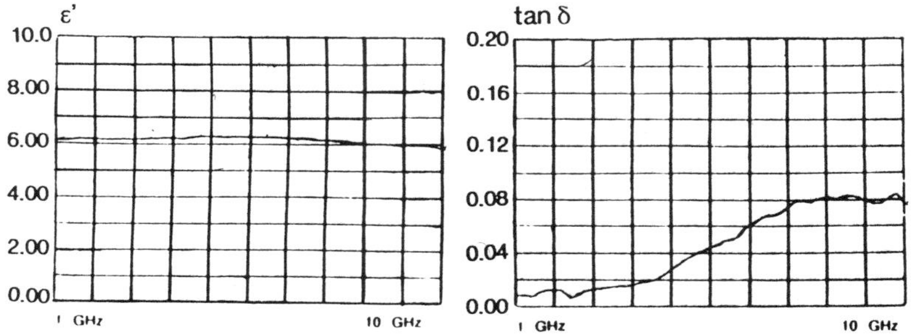

Fig. 7. Dielectric properties of green TZ8Y vs. frequency at 23°C using the open-ended coaxial line technique.

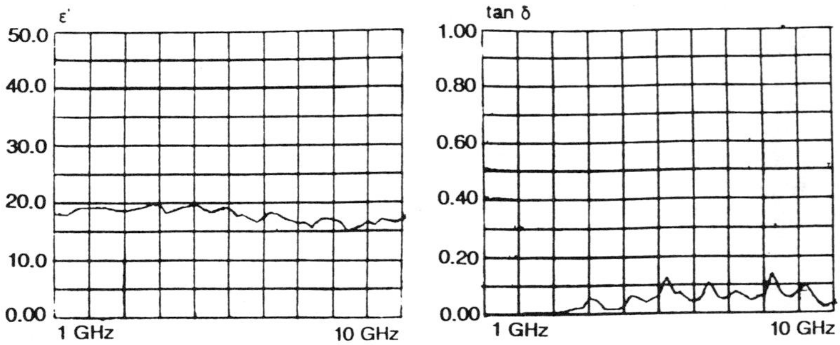

Fig. 8. Dielectric properties of sintered TZ8Y vs. frequency at 200°C using the open-ended coaxial line technique.

5. SUMMARY AND CONCLUSIONS

In this paper we described the high-temperature and broadband dielectric properties measurements of ceramic samples being conducted at the University of Utah. Results were shown up to 1000°C at S-band and up to 200°C when broadband (1 to 10 GHz) open-ended coaxial probes were used. Ongoing future developments include the fabrication of a metalized ceramic open-ended coaxial probe. This probe is expected to provide results up to 1400°C in the frequency range from 1 to 10 GHz.

6. ACKNOWLEDGEMENT

Research sponsored by the U.S. Department of Energy Assistant Secretary for Conservation and Renewable Energy, Office of Industrial Technologies, Advanced Industrial Concepts (AIC) Materials Program, under subcontract 19X-SJ125V and under contract DE-AC05-84OR21400 with Martin Marietta Energy Systems, Inc.

7. REFERENCES

1. O.M. Andrade, M.F. Iskander and S.Bringhurst,"High-Temperature Broadband Dielectric Properties Measurement Techniques", this issue.
2. M.F. Iskander et.al.,"Microwave Processing of Ceramics at the University of Utah-Description of Activities and Summary of Progress",in Ceramic Transactions edited by D.E. Clark,F.D.Gac and W.H. Sutton, Vol.21,35-48 ,The American Ceramic Society (1991).

MICROWAVE HEATING AND CHARACTERIZATION OF MACHINABLE CERAMICS

Weiguo Xi* and W.R. Tinga*
*Electrical Engineering, University of Alberta, Edmonton AB Canada T6G 2G7

ABSTRACT

The dielectric properties of three types of machinable ceramics have been investigated using an automatic microwave dielectrometer system from 300^{0}C to 1600^{0}C at 3GHz. The system setup and techniques are described, and results are presented.

INTRODUCTION

Availability of the dielectric properties of ceramic materials at high temperatures has received increasing attention due to the needs of extending the use of microwave energy to high temperature ceramic processing, developing optimum processing parameters and designing effective applicators [1]. In our lab, a high temperature (300-1600^{0}C) dielectrometer system was developed to simultaneously heat and test various ceramics and oxides, showing encouraging results [2]. Recently, this system has been improved by adding a sample temperature control function and an automatic frequency tuning function. With these functions, we are able to use this system to conduct microwave heating and characterizing studies more effectively and reliably. Machinable ceramic is one type of material chosen for such studies. It has the usual ceramic properties of good chemical and thermal stability and high compressive strength and, at the same time, offers the ease of conventional machining. To pursue its possible applications in microwave material processing, we need to know its dielectric data in the commonly used 2450MHz band at various temperatures. Such data are presently not available, at least to our knowledge.

EXPERIMENTAL SYSTEM

Our experimental system of microwave heating and characterization is shown in the block diagram of Figure 1. In this system, a coaxial re-entrant cavity with a hollow center conductor and a sample insertion hole, as illustrated in Figure 2, is designed as both a microwave sample heater and a test chamber. There are three channels connected to the cavity. A 2-4GHz reflectometer is used as the testing channel, in which the frequency shift and Q-factor change produced by an inserted sample are measured for dielectric determinations. The heating channel is mainly a solid state microwave power source at 915MHz, tunable over a 30MHz band and capable of providing 150W output. An electronic tuner is used to automatically adjust the power source frequency to the cavity's resonant frequency, thus ensuring effective power delivery. An optical fiber thermometer (Accufiber 100C) makes up the third channel for measuring and controlling the sample's temperature. Besides, a computer is used for automatic instrument setting, data acquisition and processing, and testing and heating control. Details about cavity analysis and characteristics can be found in [3] and [4].

Temperature control

In addition to the temperature function, our Accufiber thermometer has a temperature controller option installed. By selecting its output control function and connecting the analog output to the power level controller (V_P) of the heating source, a closed negative feedback loop is formed. The

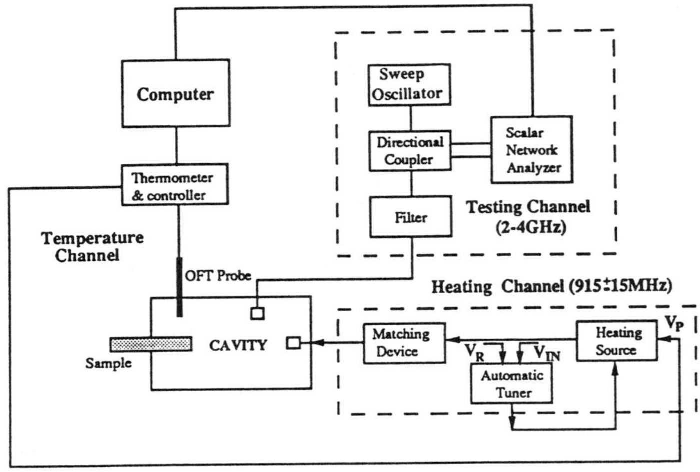

Figure 1 A system for automatic high temperature microwave heating and characterization of ceramics.

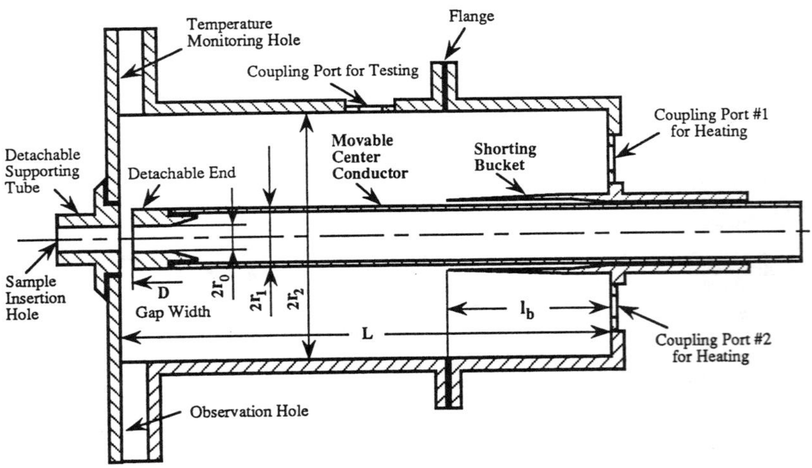

Figure 2 Cross section of coaxial re-entrant cavity used as heating and testing chamber (r_2=5.0, r_1=1.24, r_0=0.36, D=0.3, L=20.0, ℓ_b=7.5, in cm)

Proportional, Integral and Derivative (PID) controller in the thermometer is thus able to control the sample temperature at the given set-point with an error less than 0.5^{0}C at 1000^{0}C. This allows us to measure complex dielectric constants versus temperature in a programmed manner, making the testing and heating process stable and repeatable and also eliminating the common temperature run-away phenomena. The simplest temperature control mode is where the set-point increases in equal increments so that the heating rate of a sample is nearly constant. In Figure 3, a heating curve of Macor obtained

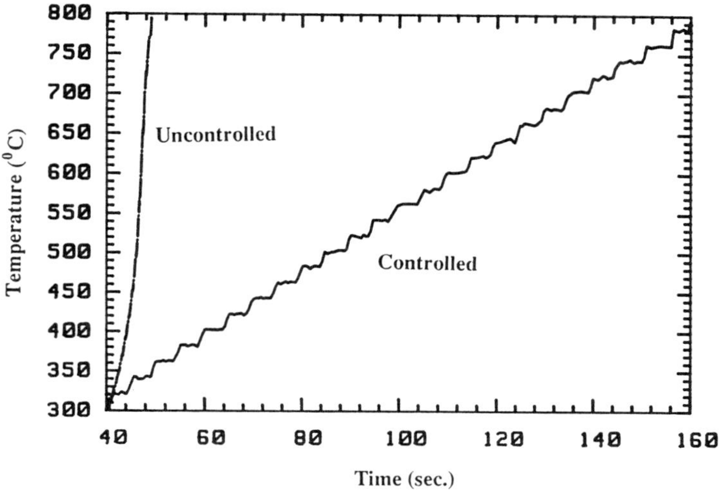

Figure 3 Controlled and uncontrolled microwave heating of Macor.

in such a step-wise heating fashion is compared to one without temperature control. If not controlled, the sample temperature will increase exponentially, quickly reaching the melting point. Without sample temperature control, the dielectric measurement will not be repeatable and reliable.

Automatic tuner

Due to varying dielectric properties of a sample with temperature, the heating band resonant frequency of a loaded cavity changes all the time. A slight detuning will cause a large heating power reflection. For example, if Q =2250 and f_o=900MHz, the frequency width at the half-power point is $f_o/(2Q)$= 0.2MHz, that is to say a 0.2MHz deviation of the heating source frequency from the resonant point will result in a power reflection of 50%. The objective of the automatic tuner is to search the frequency band, i.e., 915±15MHz, for the resonant point and then adjust the source frequency accordingly. An analog circuit has been designed to carry out this task. The operating principle of this circuit is shown in the block diagram of Figure 4. Its input signals are the analog voltages V_{IN} and V_R, which are proportional to the incident and reflected power; its output signal is an analog voltage, V_f, which is used to tune the source frequency. The circuit starts a searching period in which two frequency sweeps are triggered. During the first sweep, the normalized reflection voltage at the resonant point is found. This value is then used to locate the position of the resonant point, f_o, in the second sweep. Once f_o is found, the sample/hold circuit is triggered to hold the current value of V_f. Then the circuit enters the holding period in which V_f or the heating source frequency is kept constant. At the end of the holding period, the circuit starts another searching period. The length of the holding period is about 0.5 -2 seconds, depending on the rate of change in the sample's dielectric constant. The searching period should be as short as possible since there is virtually no power coupling to the cavity within this period. However, it is limited to about one tenth of the holding period primarily due to the low slew rate of the power source. Tests on this circuit show that the resonant frequency searching error is less than 0.5% and the holding error is less than 0.2% of the 30MHz sweep band for a cavity Q of 2000.

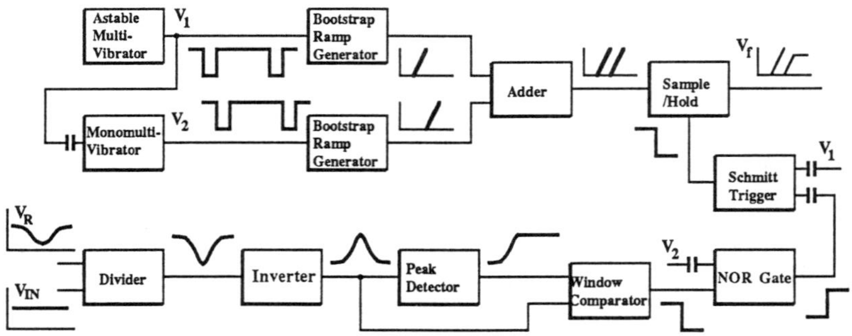

Figure 4 Block diagram of automatic tuning circuit for microwave heating of ceramics

Dynamic heating range

For given sample and cavity dimensions, the dynamic heating range of the system can be evaluated by the heating ability of the system [5]. The maximum dielectric constant of the sample which can be heated depends on the maximum tuning range of the heating source. Knowing that the tunable band of the heating source is 30MHz and using the sample and cavity dimensions given in Figure 2, we find the maximum dielectric constant to be 55 with the sample holder (quartz tube, ε_r=3.8) and 120 without the holder (i.e., sample suspended).

The upper limit of the loss tangent of a sample which can be effectively heated depends on the maximum mismatch which the heating channel can cope with. In other words, it is restricted by the matching range of the matching device used. At the lower end, there is also a limit to the sample's loss tangent when the sample loss is close to the cavity wall loss. These limits can be expressed by

$$\tan\delta_{max} = [(Q_0/Q_L)_{max} - 1]/w \qquad (1)$$

$$\tan\delta_{min} = h/w \qquad (2)$$

where,

Q_0 = empty cavity Q-factor;

Q_L = sample loaded cavity Q-factor;

$(Q_0/Q_L)_{max}$ = matching range of the matching device used;

w = ratio of the energy stored in the sample to that lost on the cavity wall which can be found from the cavity field analysis [3];

h = factor related to thermal loss.

The derivation of Eqs.(1) and (2) is presented in [5] and is here omitted. The parameter, h, accounts for the thermal loss effect on the sample's heating rate. When the thermal loss of a heated sample is smaller than that of the cavity, h<1, and vice versa. From these two equations, we can see that the upper and lower limits can be extended at the same time, not by choosing the parameter w but by increasing the matching range and decreasing the sample's thermal loss. Now, we can estimate the dynamic range of loss tangent. The parameter, w, for a given sample and cavity geometry, is a function of the dielectric constant. It is within 15(ε_r=5) and 40(ε_r=30) for the case of Figure 2. The matching range is about 40 if a coaxial three-stub tuner is used. An h of 0.2 can be used as an estimation. Therefore, we find that

$\tan\delta_{min}$=0.014(εr=5)-0.005(εr=30) and $\tan\delta_{max}$≈2.6(εr=5)-1.0(εr=30).

RESULTS AND DISCUSSIONS

Sample

The samples used in this study are Cotronics Rescor Series Machinable ceramics. Their main physical properties, as given by the manufacturer, are as follows.

	Unit and Condition	Alumina Silicate(902)	Macor (915)	Machinable Alumina(960)
Temperature Limit	^{0}C	1160	1000	1660
Thermal Conductivity	cal/cm·sec^{0}C	0.0026	0.0042	0.014
Density	g/cc	2.30	2.52	3.8
Porosity	%	2.9	0.0	10%
Hardness	mohs scale	6	5	3
Thermal Expansion	1/^{0}C×10^{-6}	3.24	21.6	7.74
Compressive Strength	psi	38,000	50,000	60,000
Dielectric Constant	@ 100Kc	5.3	6.0	9
Loss Factor	@ 100Kc	0.04	0.003	0.0016

Macor, a machinable glass ceramic, is an opaque white, odorless, porcelain-like material composed of approximately 55% fluorophlogopite mica and 45% borosilicate glass. The highly interlocked microstructure formed by randomly oriented mica flakes is the key to Macor's machinability [6]. During machining, this structure prevents the microscopic fractures from propagating throughout the whole body. Since the fractures are localized, its strength and hardness are retained. No post-machining heat treatment is required. 902, a machinable alumina silicate with grey color, provides the highest machinability with the lowest cost. However, its usable temperature is only 600^{0}C. As suggested by the manufacturer, a simple heat treatment at about 1000^{0}C can harden the material and extend its temperature limit to 1160^{0}C. 960 is a white porous machinable ceramic containing over 96% alumina. It can be used continuously to 1660^{0}C with no post-machining heat treatment required. Its loose surface can be improved by applying RESCOR 960 Hardener and curing at 300^{0}C. It should be pointed out that the density and dielectric constant of 960 given by the manufacturer are erroneous. They seem to be taken from a fully densified alumina without considering the effect of porosity. We tested the sample and found the density to be 2.2g/cc and its dielectric constant at room temperature to be 4.5.

The samples used in our dielectric measurements are cylinders with a diameter of 4.8±0.05mm and a length of about 20mm which is not critical. By virtue of machinability, the sample dimensions can be controlled accurately.

Heating rate

According to a well-known heating equation, the microwave heating rate of a dielectric is determined by the absorbed microwave power, the heat capacity and the mass [7]. The power absorbed by a dielectric is proportional to the sample's loss factor and the square of the E-field inside the sample which will decrease with increasing power reflection for a given input power level. This reflection is primarily caused by the changing dielectric properties of a sample with temperature. In practice, the heating rate is also affected by the thermal loss of the heated sample. Frequently, the degree of thermal loss

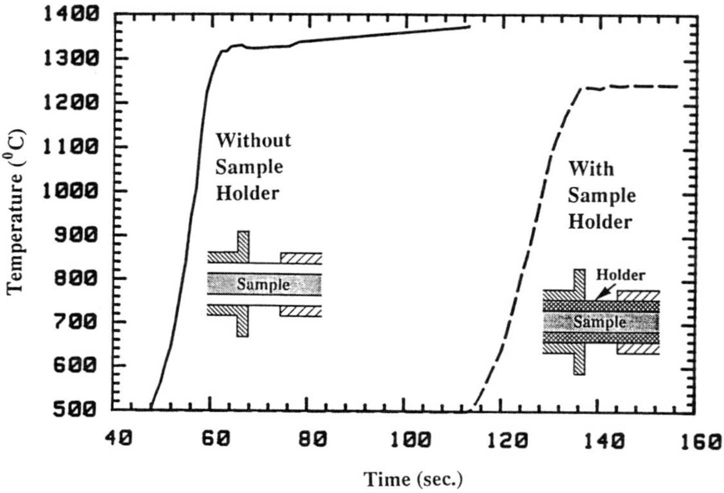

Figure 5 Heating curve of mullite with and without the sample holder; the heating power used is 100W at 915MHz and is switched on at t=0.

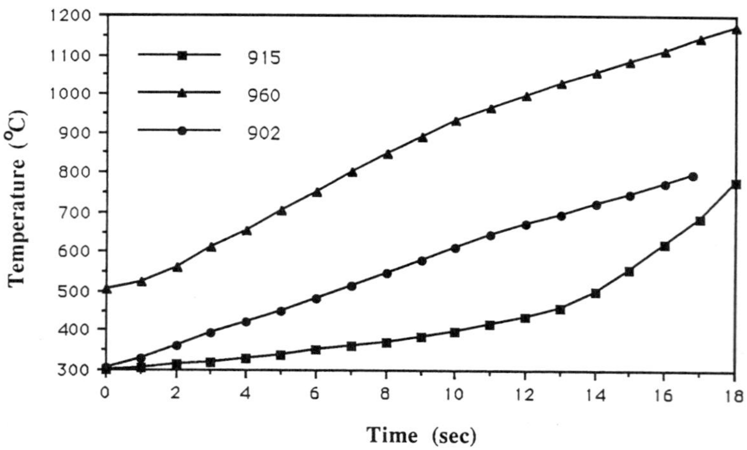

Figure 6 Heating curves of machinable ceramic 915, 902 and 960 (Cotronics); the heating power used is 100W at 915MHz; the sample of 960 is preheated to 500°C with an external resistance heater.

determines whether or not a low loss sample can be heated. As an example to illustrate the effect of thermal loss, the heating rate of a mullite sample is plotted in Figure 5 for (1) a sample contained in a quartz tube and (2) a suspended sample. Because the suspended sample does not directly contact with the cavity metal wall, the thermal conduction loss is greatly reduced, resulting in a higher heating rate and a higher temperature limit. This suspended sample scheme is used to obtain the following experimental results. We are also working on some special techniques to thermally insulate the sample to further reduce the heat loss which is mainly due to radiation at high temperatures.

In Figure 6, the heating response of the machinable ceramics are demonstrated. The time axis starts at the moment when the temperature of the 915 and 902 sample reach 300°C. The 915 sample takes 50 seconds and 902 takes 10 seconds to arrive at this temperature. This difference reflects Macor's low dielectric loss below 300°C as compared with 902. However, the dielectric loss of 960 is even much lower; it is so low that conventional preheating has to be used. The 960 sample is first heated in a resistance heater to about 500°C and then quickly inserted into the cavity. At this temperature, 960 is able to effectively absorb microwave energy. As shown later, all three materials exhibit a dielectric loss with a positive temperature dependence, but Macor shows the greatest increase in the loss factor. Moreover, during the microwave heating, some interesting phenomena are observed. (1) After being heated to 500°C, Macor becomes more transparent and less machinable. (2) 902 often cracks if the heating rate exceeds 25°C/sec. (3) 960 can be densified to 65% of its original volume after being heated to 1500°C and maintained at that temperature for one minute.

Complex dielectric constant

The dielectric constant, ε_r', and loss factor, ε_r'', are determined from the measured resonant frequency shift, Δf_o, and cavity Q-factor change, ΔQ, produced by the insertion of the sample. For the given cavity and sample dimensions, the calibration curves which are functions of ε_r' versus Δf_o, and ε_r'' versus ΔQ are based on the data from the mode-matching field analysis of the sample loaded cavity. The measurement errors in Δf_o, and ΔQ in our system cause a maximum uncertainty of about 5% in ε_r' and 10% in loss tangent, i.e. $\varepsilon_r''/\varepsilon_r'$ [2,4].

Measured dielectric constants and loss factors of three types of machinable ceramics are plotted as functions of temperature in Figure 7. Prior to the measurement, a 902 sample first underwent a heat treatment by microwave in the same cavity. The treatment involves a low heating rate (2.5°C/sec) up to 1000°C followed by 4-minute dwell time at 1000°C. As in heating rate measurements, 960 samples are pre-heated to 500°C before inserted into the cavity. Both ε_r' and ε_r'' of all three types show a positive temperature coefficient. At about 750°C, 1000°C and 1500°C, the loss factor of 915, 902 and 960 exponentially increases with temperature and the sample would quickly melt if the temperature controller were not used. However, at low temperatures below 300°C, their dielectric properties do not change much. Moreover, these dielectric properties also have a noticeable thermal history dependence with Macor showing the largest dependence. After being heated to a temperature well below their melting point, the samples of 915 and 960 become lossier and 902 becomes less lossy. Detailed research is needed to investigate these irreversible properties.

In addition to the measurement errors in Δf_o and ΔQ, thermal expansion of the cavity would cause the determined ε_r' and ε_r'' to be significantly larger than the true values. The highly-focused microwave heating field in our cavity can maintain the cavity body at or near room temperature while the sample is being heated above 1000°C, therefore, the expansion errors are greatly reduced. Nevertheless, during a longer heating period (>2min), the cavity wall adjacent to the sample warms up and causes a Δf_o of about 0.3MHz and ΔQ Of 2.5% at the end of the measurements, i.e., at the highest temperature. These errors are prevented in our measurements by water cooling the hollow center conductor.

CONCLUSION

In this paper, an automatic system for microwave heating and characterization of ceramics has been described. The experimental results of

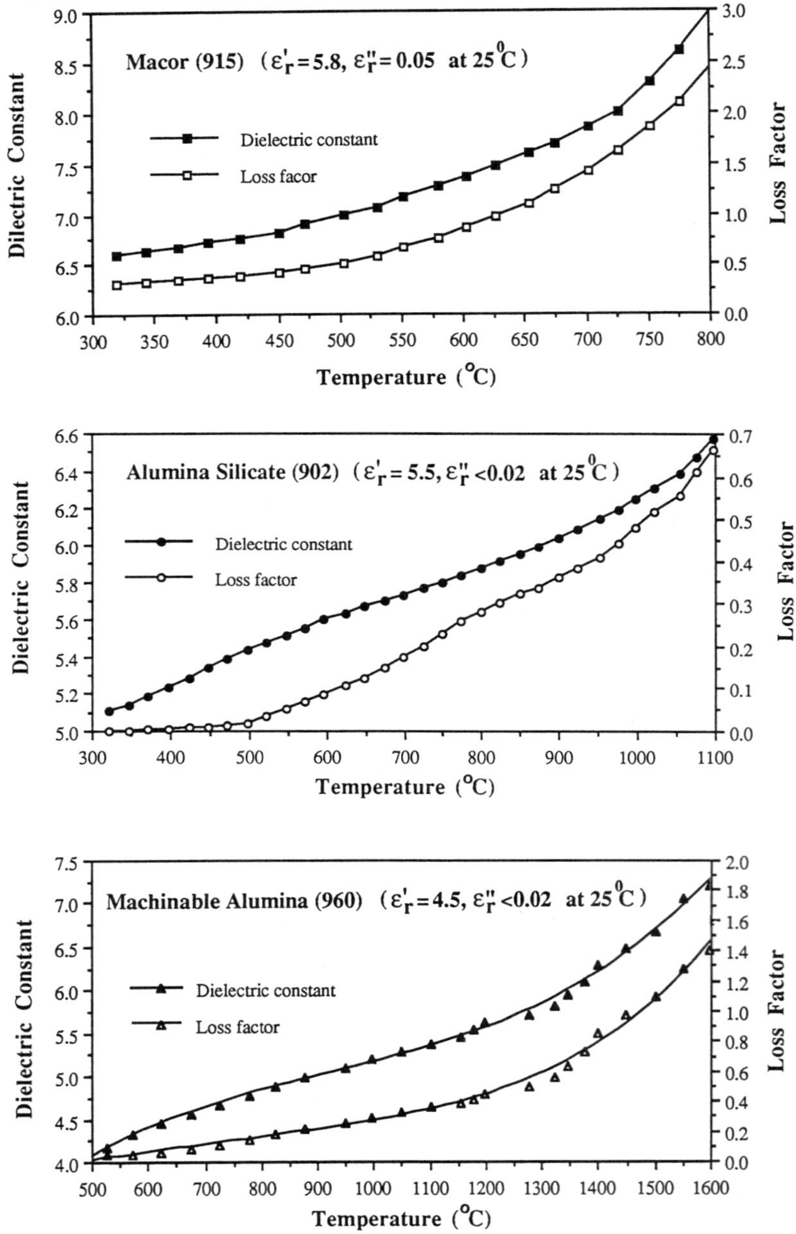

Figure 7 Complex dielectric constant of machinable ceramic 915, 902 and 960 as function of temperature, measured at 3GHz while being heated by 915MHz microwave power with maximum output of 100W.

three types of machinable ceramics have been presented which have demonstrated the system's capability of measuring complex dielectric constants up to at least 1600^{0}C. Several important aspects of microwave heating, such as temperature controlling, electronic tuning and thermal insulating have been discussed. The discussions are of general significance to other microwave heating and dielectric measuring systems. The presented dielectric data of the machinable ceramics will help determine their use in the design of microwave systems.

ACKNOWLEDGEMENT

This work is sponsored by the Natural Sciences and Engineering Research Council of Canada. The financial assistance provided by the Faculty of Graduate Studies and Research and the Alma Mater Fund of the University of Alberta is also acknowledged.

REFERENCES

1. W.H. Sutton, Ceramic Bulletin, 68, 376-386, 1989.
2. W. Xi and W.R. Tinga in Microwave :Theory and Application in Materials Processing, edited by D.E. Clark, F.D. Gac and W.H. Sutton (Ceramic Transactions, 21, Amer. Cer. Soc., 1991), 215-224.
3. W. Xi, W.R. Tinga, B. Q. Tian and W.A.G. Voss, IEEE Trans. on MTT, 41, 747-753, April 1992.
4. W. Xi and W.R. Tinga, IEEE Trans. MTT 40, in press, 1992.
5. W. Xi and W.R. Tinga, in preparation.
6. D.G. Grossman, American Machinist, May 1978.
7. W.R. Tinga, Mat. Res. Soc. Symp. Proc., 60, 105-116, 1986.

COMPLEX PERMITTIVITY STUDIES OF POLYANILINE

Leonard J. Buckley and Kenneth E. Dudeck Polymer Physics Group, NAWC, Warminster, PA 18974

ABSTRACT

A study of the complex dielectric properties at microwave and millimeterwave frequencies on several polyanilines was performed. The polymers investigated were based on emeraldine salts. The chemically synthesized emeraldine base was protonated with acids of various pH levels. Aqueous HCl (pH range 0.12 to 1.07) and Tosylic (pH range 0.22 to 1.13) were used. Complex permittivities of the emeraldine base and salts were determined at X-band and R-band using waveguide measurements. Characterization techniques were developed to reduce the measurement error. This included the electrical determination of the sample holder length and sample position. The real component of the dielectric constant was found to vary from 4 to 107 at X-band and from 3.8 to 926 at R-band. The dielectric loss spanned 3 orders of magnitude at X-band and 4 orders of magnitude at R-band. The frequency dependence was relatively level within each band with a slight decrease in each permittivity component with increasing frequency. The DC conductivity ranged from 10^{-4} S/cm to 10^{-1} S/cm for the emeraldine salts. The time dependence of the complex permittivity and DC conductivity was examined.

INTRODUCTION

Polyaniline, an electroactive polymer, can exist in various oxidation states and has been extensively studied by many researchers [1]. These oxidation states range from the completely reduced leucoemeraldine base to the completely oxidized pernigraniline base. The emeraldine base oxidation state (50% reduced) is of particular interest as it undergoes a transition from an electrical insulator to a conductor upon treatment with an acid to become an emeraldine salt [2]. In the insulator emeraldine base form the material is soluble in n-methyl pyrrolidone (NMP) and can be solution processed. Upon acid treatment the material is no longer soluble.

The complex permittivity of these materials can be varied significantly with acid treatment (protonic acid doping) and represents an area that has not been fully addressed in the literature. Epstein and others have used a cavity perturbation method to obtain data at a single frequency of 6.5 GHz for emeraldine salts of various protonation levels [3]. This contactless technique simplifies the sample size and shape but only yields data at a single frequency that is dependent upon the resonance frequency of the cavity. This technique is not practical for a broadband frequency response study of intractable thin film electroactive polymers. Free-space techniques have been used on other processible materials using a relatively large sample area to avoid edge diffraction effects [4]. Waveguide measurements can be a very effective way to characterize small quantities of new materials over broad (microwave to millimeterwave) frequency ranges. The technique used to perform these measurements and several examples have been reported in a previous paper [5].

In this effort, the complex dielectric properties at microwave and millimeterwave frequencies on thin film (0.14 mm to 0.4 mm) polyaniline compounds were studied. Characterization techniques were developed and are discussed along with the measurements.

EXPERIMENTAL

Materials

Polyaniline hydrochloride was synthesized by chemical oxidation of aniline with ammonium peroxydisulfate in 1.0 M HCl and converted into the base form with 0.1 M ammonium hydroxide as described in the literature [1d]. An additional step of washing with 1.5 M ammonium hydroxide prior to dissolving in n-methyl pyrrolidone was performed before film casting. This technique ensured no acid was left in the compound which made the film casting process more facile. Hydrochloric acid and tocylic acid were used at various pH levels to treat the base material. All reagents were HPLC grade and the aniline was distilled prior to use. The complex permittivity measurements were performed after the acid-treated cast films were dried by pumping under vacuum for 4.5 days. They were stored in a controlled humidity cabinet (dry-keeper) before testing at additional time periods.

Dielectric Characterization

Samples were carefully cut from each treated film to fit to the inside waveguide dimensions (X-band and R-band) for measuring the complex reflection and transmission coefficients. The coefficients or S-parameters were obtained using a Hewlett Packard 8510B Network Analyzer with the millimeter wave option (for R-Band measurements). The 8510B was calibrated using the full two-port internal waveguide calibration and the standard waveguide standards (transmission through, short, offset short, and sliding load). Once calibrated in the band, each sample was placed inside a quarter wave sample holder which was then inserted between the calibrated waveguide ports of the network analyzer system. The S-parameters were measured at 201 frequency points in the band and then transfered into a HP9000 computer for calculation of the intrinsic material properties.

RESULTS AND DISCUSSION

Molecular Structure

The general molecular structure of polyanilines in both, the insulative base and conductive salt forms is shown in Figure 1. Note that emeraldine implies a 50% oxidized state which may not be the exact case with each synthesis. The salt form, emeraldine hydrochloride (shown in Figure 1) was compared to the emeraldine salt containing the physically larger tosylic counterion.

Table I contains the pH levels, the four-point DC conductivities and elemental analysis for the conductive materials. The normalized aniline/oxidizing agent ratio was calculated to be 5.445 assuming 2.5 electrons are withdrawn from each aniline monomer and two electrons are necessary to reduce one molecule of the oxidizing agent. The DC conductivity was found to be consistently greater for the HCl-treated compounds. This is most likely due to the smaller size of the chloride ion versus tosylic which allowed greater diffusion into the film. The time of acid exposure was kept constant for both species.

Calculation of Intrinsic Material Properties

The determination of the relative intrinsic material properties, defined as:

$$\mu = \mu' + j\,\mu'' \quad \text{(complex permeability)}, \qquad (1)$$

$$\epsilon = \epsilon' + j\,\epsilon'' \quad \text{(complex permittivity)}, \qquad (2)$$

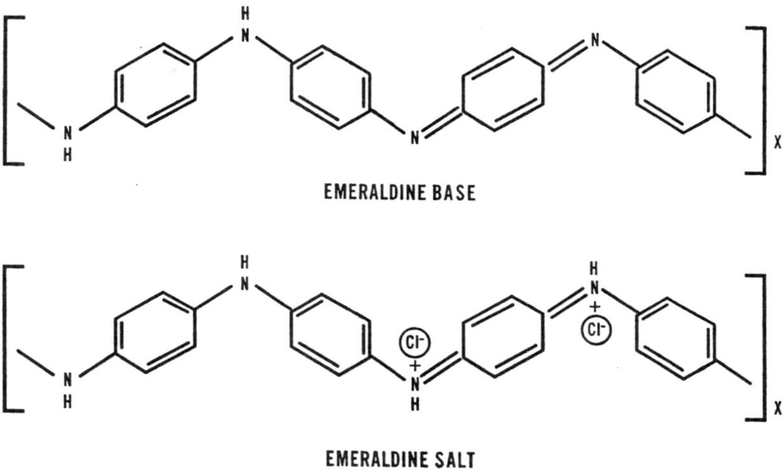

Figure 1. Molecular Structure of Polyanilines.

TABLE I

SAMPLE	pH	Cl/N	s/cm(1)	s/cm(2)	s/cm(3)
Emeraldine (HCl)	1.07	0.074	0.02	0.02	0.006
	0.41	0.425	0.03	0.08	3.0
	0.12	0.455	0.67	10.0	3.22
Emeraldine (Tos)		**S/N**			
	1.13	0.007	0.0006	0.0006	0.0001
	0.49	0.021	0.005	0.017	0.004
	0.22	0.059	0.02	0.03	0.001

was calculated from the measured S-parameters using two different models. The first model, developed by Nicolson and Ross, is a closed form solution which uses both the reflection coefficient (S_{11}) and the transmission coefficient (S_{21}) in the calculation of μ and ϵ [6]. The other model, commerically available from Innovative Measurement Systems Inc., assumes that the material is non-magnetic ($\mu = 1 + j0$) and employs an iterative technique to find the value of ϵ that solves the transmission line equation.

The iterative model requires only one S-parameter, S_{11} or S_{21}, and chooses the parameter with the larger magnitude for increased accuracy. Since the closed form model requires both S-parameters, inaccuracies in the calculation can arise if either S-parameter is more than three orders of magnitude lower (>-30dB) than the corresponding calibration standard. This situation occurs if a sample is almost entirely transmissive or reflective. The disadvantage of the iterative model is the possibilty of non-convergence or convergence to an incorrect root. Additionally, the iterative model's assumption that the sample is non-magnetic may be incorrect when investigating the properties of new materials.

Since the two models calculate the permittivity in completely dissimilar fashions and each have distinct advantages, both models were used for comparision to insure confidence in the final values. Both models assume that the sample is homogeneous and completely fills the entire sample holder, meaning that the sample thickness, S, equals the sample holder thickness, H. In general, H is greater than S which represents unfilled space. However, if the sample holder is assumed to be lossless, the unfilled regions account for only phase delay, or more negative angles, in the measured S-parameters. The effect of the unfilled regions can be mathematically removed by adding the amount of phase delay caused by the unfilled regions to the measured phase. This operation, shown by equations (1) and (2), causes the S-parameters to be "rotated" counter-clockwise on the complex plane, and thus represents the measured phase of the sample in a holder of equal length.

$$\arg(S_{11\ \text{rotated}}) = \arg(S_{11}) + 720°\ D/L_o \tag{3}$$

$$\arg(S_{21\ \text{rotated}}) = \arg(S_{21}) + 360°\ (H-S)/L_o\ , \tag{4}$$

L_o is the free space wavelength in the guide and computed in Equation (5):

$$L_o = \frac{c}{F\ \{\ 1 - (F_{co}/F)^2\ \}^{1/2}}\ , \tag{5}$$

where F is the operating frequency, F_{co} is the TE_{01} cutoff frequency, and c is the velocity of light in a vacuum.

Measurement Errors

The error in the final results of the μ and ϵ calculation is largely dominated by the inaccuracy in the measurement of the three mechanical distances: sample length (S), sample holder length (H), and offset distance (D). These errors are more significant in the millimeter region, compared to microwave frequencies, where a given mechanical tolerance represents a larger portion of the electrical wavelength.

As shown in previous work [5], the accuracies of the sample holder and offset distances can be improved by measuring their lengths electrically. The sample holder length is obtained by measuring the empty sample holder and calculating its length by the S_{21} phase slope verses frequency. The offset distance is obtained by comparing the phase of the sample's forward (S_{11}) and reverse (S_{22}) reflection coefficients. Since a homogeneous sample will have equivalent phase in each direction, any measured deviation is attributed to the offset distance, which can be calculated from the phase difference.

After the electrical lengths were calculated and used to rotate the S-parameters more accurately, the remaining total uncertainty of the dielectric constant, caused by instrument error, measurememt repeatability, and sample length uncertainty, was shown to be less than 6% at R-band. The uncertainty at X-Band is less than 4% which is more accurate due to the longer wavelength. These uncertainty values assume that the sample is much less than a half of the wavelength which avoids the problem of a minimum reflection coefficient in the Nicolson-Ross calculation.

Measured Permittivities

At each band, 201 complex permittivities were calculated corresponding to the measured S-parameter frequencies. Table II shows a sample of the calculated permittivities for X-Band and R-Band. Three representative frequency data points are shown. For each sample three values of permittivity are listed. The first value is the complex permittivity in rectangular form, $\epsilon' + j\epsilon''$, calculated by the closed form model. The second value listed, $|\epsilon|$, is the magnitude of the complex permittivity. For comparison, the third value listed, $(\epsilon' + j\epsilon'')$, is the complex permittivity calculated from the iterative model. A difference in the magnitudes of the reflection coefficient between the forward and reverse directions is an indication of an inhomogeneous sample. This was the case for the 0.5M and 1.0M HCl samples, which was more pronounced at R-Band. The samples appear more like lumped elements at the lower X-Band frequencies, and therefore appear less inhomogeneous.

Figure 2 shows the real permittivity, ϵ', at both bands verses frequency for the polyaniline emeraldine base control and the tosylic acid treated samples. Figure 3 shows the imaginary permittivity, ϵ'', at both bands for the same samples. The values used to generate the plots shown in figures 2 and 3 were calculated from the closed form model. They illustrate a generally dispersive permittivity which decreases slightly with frequency. Extrapolation from 12.4GHz (X-Band high frequency) to 26.5GHz (R-Band low frequency) indicates that the dispersive trend continues through the unmeasured waveguide bands. The permittivity plots show that the real and imaginary parts of the permittivity increase as the acid pH increases. In addition, as frequency increases, the amount of change in the permittivity due to acid exposure decreases.

Figure 4 shows the time dependence of the complex permittivity for a HCl-treated film (pH=1.07) at X and R bands. Figure 5 shows similar data for a tosylic acid-treated film. The complex permittivity appeared to be more stable for the tosylic-acid treated system. This is particularly evident for the R-band data (see figure 4 (c and d) and figure 5 (c and d). DC conductivities in ambient air have been found to be more stable for tosylic-acid treated polyaniline compared to HCl-treated material [7]. The larger tosylic counterion along with higher levels of crystallinity may explain this behavior. The DC conductivity of acid-treated polyaniline has been shown to increase as the level of crystallinity increases [8]. Further studies are planned to investigate the effect of crystallinity on the complex dielectric properties.

CONCLUSIONS

The DC conductivity was greater for the HCl-treated compounds compared to the tosylic-acid treated systems. Longer

TABLE II

COMPLEX PERMITTIVITY OF POLYANILINE AT X-BAND AND R-BAND

	X-BAND		
	8.2 GHz	10.0 GHz	12.4 GHz
SAMPLE	$\epsilon' + j\epsilon''$ $\lvert\epsilon\rvert$ $(\epsilon' + j\epsilon'')^1$	$\epsilon' + j\epsilon''$ $\lvert\epsilon\rvert$ $(\epsilon' + j\epsilon'')^1$	$\epsilon' + j\epsilon''$ $\lvert\epsilon\rvert$ $(\epsilon' + j\epsilon'')^1$
Control	4.05+j0.08 4.05 (4.04+j0.09)	3.99+j0.09 3.99 (3.99+j0.09)	3.98+j0.14 3.98 (3.98+j0.14)
1.0M Tosylic	22.9+j38.7 44.97 (22.4+j37.8)	18.8+j32.9 37.89 (17.9+j32.6)	15.2+j27.6 31.51 (14.8+j27.6)
1.0M HCl	88.1+j62.5 108.0 (86.6+j51.0)	93.0+j46.3 103.9 (81.8+j33.6)	107+j42.3 115.1 (82.5+j30.5)

	R-BAND		
	26.5 GHz	33.0 GHz	40.0 GHz
Control	3.95+j0.10 3.95 (3.90+j0.17)	3.86+j0.20 3.86 (3.74+j0.19)	3.81+j0.20 3.82 (3.54+j0.19)
1.0M Tosylic	10.3+j16.2 19.2 (10.2+j15.4)	9.32+j14.2 17.0 (9.13+j13.0)	8.4+j12.7 15.21 (8.16+j11.0)
1.0M HCl	926+j577 1074 (661+j191)	779+j541 948 (372+j98.7)	507+j588 776 (270+j57.1)

[1]Iterative Model

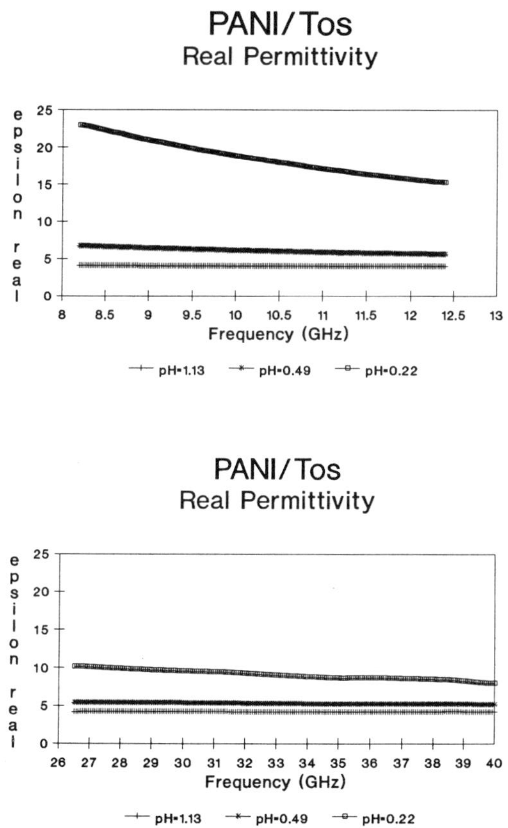

Figure 2. Epsilon (real) vs frequency for various pH levels

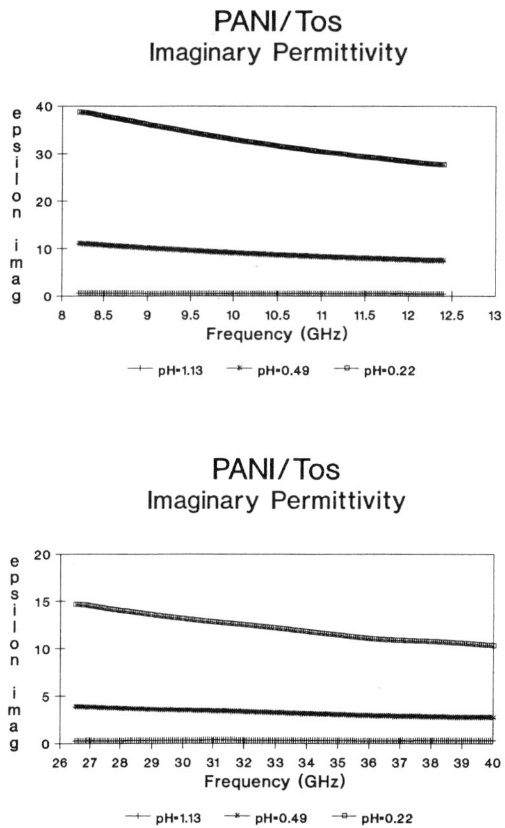

Figure 3. Epsilon (imag) vs frequency for various pH levels

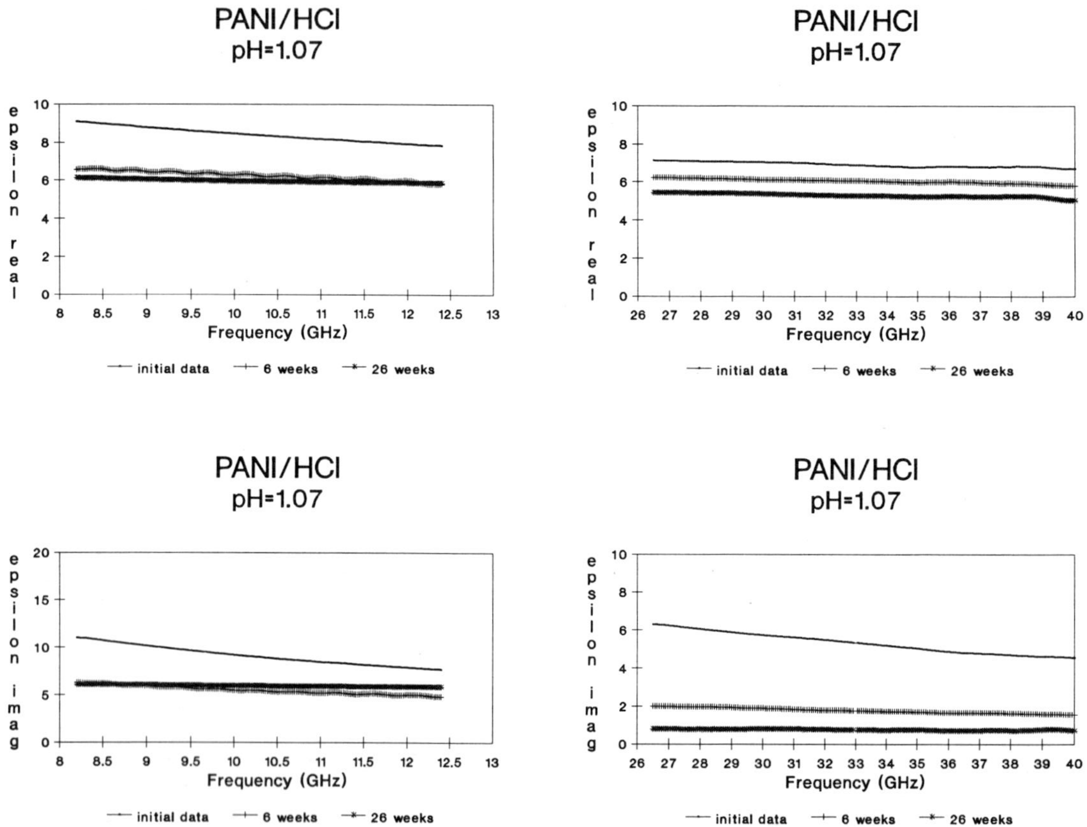

Figure 4. Time dependence of complex permittivity for PANI/HCl

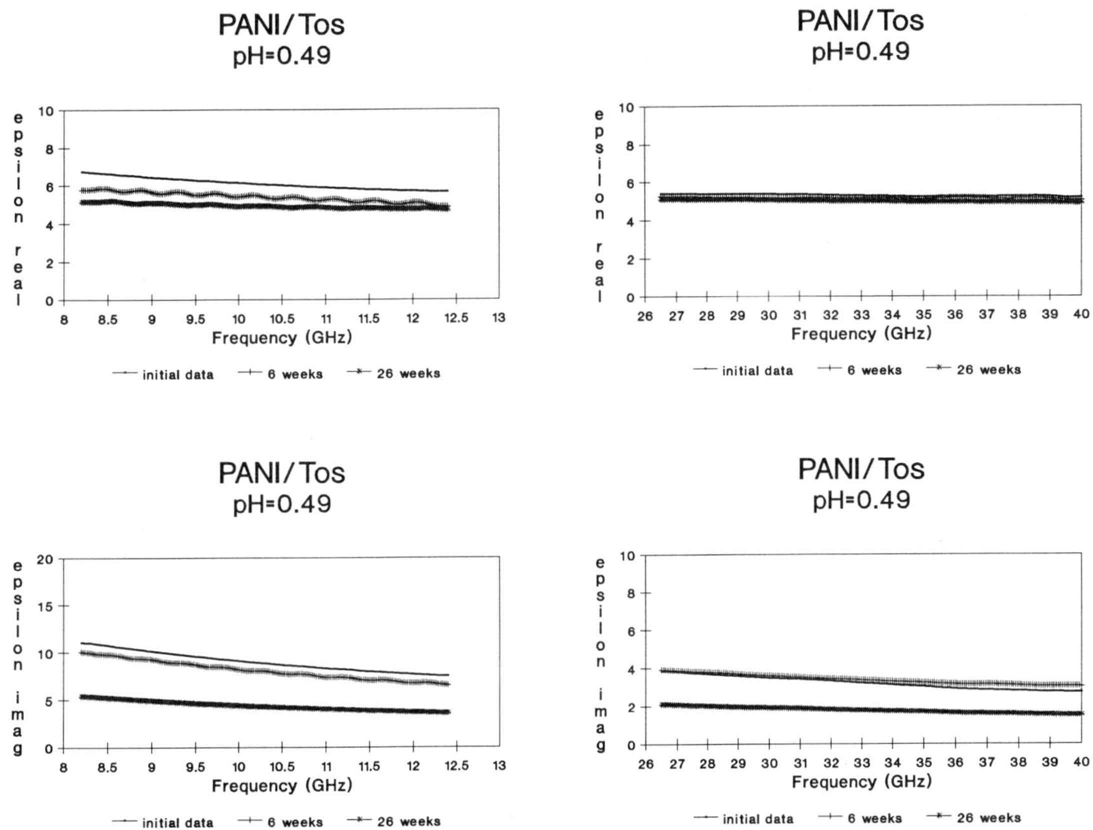

Figure 5. Time dependence of complex permittivity for PANI/Tos

acid exposure times may act to negate this difference. The permittivity measurement errors were found to be less than 6% at R-band and less than 4% at X-band for the polyaniline systems. Electrical inhomogeneities, even at the macromolecular level, inhibit true permittivity calculations from the scattering parameters. The measured permittivities, real and imaginary were found to vary greatly with the type of acid and pH level. The time dependence shows a decrease over weeks to months with the tosylic-acid treated systems more stable.

References

1. See for example: a) Huang, W., Humphrey, B., MacDiarmid, A., J. Chem. Soc., Faraday Trans. 1, 82, 2385 (1986) ; b) Epstein, A., et al, Synthetic Metals, 18, 303 (1987); c) Watanabe, A., et al, Macromolecules, 20, 1793 (1987); and d) Wei, Y., Tang, X., Sun, Y., Focke, W., J. Polym. Sci., Polym. Chem., 27, 2385 (1989).
2. MacDiarmid, A., Chiang, J., Richter, A., Synthetic Metals, 18, 285 (1987).
3. Javadi, H., Cromack, K., MacDiarmid, A., Epstein, A., Physical Review B, 39, 3579 (1989).
4. Ghodgaonkar, D., Varadan, V., Varadan, V., IEEE Trans. Instrum. Meas., 37, 789 (1989).
5. Dudeck, K., Buckley, L., to be published in IEEE Trans. Instrum. Meas.
6. A. M. Nicolson and G. F. Ross, IEEE Trans. Instrum. Meas., IM-19, 377 (1970).
7. Wang, Y., Rubner, M., to be published in Synthetic Metals.
8. MacDiarmid, A., Epstein, A., Faraday Discuss. Chem. Soc., 88, 317 (1989).

OBSERVED TRENDS IN THE DIELECTRIC PROPERTIES OF MINERALS AT ELEVATED TEMPERATURES

Sonya P. Holderfield* and J. B. Salsman*
*U.S. Bureau of Mines, Tuscaloosa Research Center, P. O. Box L, Univ. of Alabama Campus, Tuscaloosa, AL 35486-9777

ABSTRACT

Researchers at the U.S. Bureau of Mines' Tuscaloosa Research Center have measured the dielectric properties of over 60 minerals. Thirty-eight of these minerals have been identified as containing elements such as Co, Pt, and Ni, which are considered to be critical and strategic to the Nation. The dielectric properties were measured over the temperature range of 25° to 325° C and microwave frequencies of 300 MHz to 3 GHz.

The data were analyzed to determine if trends could be detected in the dielectric properties of minerals having the same crystalline structure and/or mineralogical classification. Such trends would make possible realistic assumptions about how these classes of minerals would be affected by microwave energy without specific knowledge of their dielectric properties.

INTRODUCTION

The use of microwave energy for materials processing has rapidly become a major research area. Some of this research has centered on a group of minerals identified as strategic to the United States and involved finding improved ways to process and recover these minerals which may include microwave processing. Such improvements would ensure that the United States does not have to depend on other nations for its supply of these natural resources. However, the benefits can not be exploited without a knowledge of how these minerals are affected by microwave energy. The physical properties of a material which describe how that material is affected by microwaves are its dielectric properties, namely the dielectric constant and loss factor. As part of this research effort, the U.S. Bureau of Mines has developed several methods for measuring the dielectric properties of minerals[1-2].

The U.S. Bureau of Mines' Tuscaloosa Research Center has been measuring the dielectric properties of minerals at the microwave heating frequencies for over 5 years in an effort to increase the available data in this area. Work reported in previous publications has presented the dielectric properties of low loss minerals over the frequency range of 30 MHz to 1 GHz at room temperature[1]. These techniques were upgraded so that the dielectric properties of high loss minerals could be measured in the frequency range of 300 MHz to 3 GHz, both at room temperature and at elevated temperatures, as a function of density[2]. The minerals that were measured had different compositions and crystalline structures and belonged to different mineralogical classifications. It was considered that the dielectric properties of minerals with the same crystalline structure and/or mineralogical classification might be similar and these trends, if any, would enable researchers to make predictions about how these classes of minerals would be affected by microwave energy.

MEASUREMENT AND ANALYSIS TECHNQIUES

The measurement of the dielectric properties required several steps of preparation of the mineral sample before measurement and evaluation of the dielectric property data. The minerals were acquired from mineral supply companies, chemical supply companies and mines. Most of the minerals were not pure and had to be separated from associated gangue. Attempts to purify the minerals were made by density separation methods. The minerals were then crushed to -325 mesh (<40 μm) and analyzed by X-ray diffraction techniques for impurities. Overall, the dielectric properties of the 38 minerals containing critical elements were measured. Of these 38, 22 were found to be pure or to contain only trace impurities and are shown in table I.

The dielectric properties of these purified powders were measured over the frequency range of 300 MHz to 3 GHz at temperatures ranging from 25° to 325° C. The dielectric constant and loss factor of the minerals were also measured as a function of density and these values extrapolated to the values for the mineral at its theoretical density as determined by a helium pycnometer.

Table I. The Composition, Classification, and Crystalline Structure of Measured Minerals.

Mineral Name	Mineral Composition	Mineralogical Classification	Crystalline Structure
Annabergite	$Ni_3(AsO_4)_2 \cdot 8H_2O$	arsenate	monoclinic
Arsenopyrite	FeAsS	sulfide	monoclinic
Azurite	$Cu_3(CO_3)_2(OH)_2$	carbonate	monoclinic
Barite	$BaSO_4$	sulfate	orthorhombic
Biotite	$K(Mg,Fe)_3AlSi_3O_{10}\text{-}(OH)_2$	silicate	monoclinic
Bornite	Cu_5FeS_4	sulfide	isometric
Cassiterite	SnO_2	oxide	tetragonal
Chalcocite	Cu_2S	sulfide	>91° C isometric <91° C orthorhombic
Chalcopyrite	$CuFeS_2$	sulfide	tetragonal
Cobaltite	CoAsS	sulfide	isometric
Covellite	CuS	sulfide	hexagonal
Cryolite	Na_3AlF_6	halide	monoclinic
Galena	PbS	sulfide	isometric
Goethite	FeO(OH)	oxide	orthorhombic
Greenockite	CdS	sulfide	hexagonal
Ilmenite	$FeTiO_3$	oxide	rhombohedral
Marcasite	FeS_2	sulfide	orthroombic
Niccolite	NiAs	sulfide	hexagonal
Siderite	$FeCO_3$	carbonate	rhombohedral
Smaltite	$CoAs_2$	sulfide	isometric
Smithsonite	$ZnCO_3$	carbonate	rhombohedral
Strontianite	$SrCO_4$	carbonate	orthorhombic

These minerals were grouped by mineralogical classification and crystalline structure and the data compared to determine if any trends could be identified which were common within each group[3].

RESULTS

Comparison on Basis of Mineralogical Classification

The dielectric properties of the minerals in table I were compared on the basis of mineralogical classification and crystalline structure. For example, the dielectric properties of three carbonate minerals were compared to the sulfides.

Figure 1 presents graphs of the dielectric constants of 8 sulfide minerals, measured at 915 MHz as a function of temperature. The graphs illustrate that the dielectric constants of the sulfide minerals at room temperature range from 25 for marcasite to 235 for chalcocite. Figure 2 presents graphs of the dielectric constants of 3 carbonate minerals as a function of temperature and shows these values to be around 1.5 for all 3 at room temperature. These data indicate that the sulfide minerals are much better microwave absorbers than the carbonates. All of the sulfides displayed relatively good absorption characteristics, although some were much better than others. On the other hand, the values of the dielectric properties of the carbonates would indicate that they display poor absorption characteristics. Therefore, for general purposes, approximations could be made about the dielectric properties of these two mineralogical classifications; that is, sulfides are good absorbers and carbonates are relatively poor absorbers.

Comparison on Basis of Crystalline Structure

Minerals with isometric and orthorhombic structures were analyzed so that a comparison with minerals of similar crystalline structures could be performed. Figure 3 shows graphs of the dielectric constants, as a function of temperature, of three minerals with an orthorhombic crystalline structure. The values at room temperature range from 4.65 for goethite to 24.09 for marcasite. Figure 4 shows graphs of the dielectric constants, as a function of temperature, of five minerals with an isometric crystalline structure. In this case, the values are widespread, ranging from 15.5 for smaltite to 220 for chalcocite at 300° C. The data show that there is not much correlation between the dielectric constant and the mineral's crystalline structure. Although the minerals are all relatively good absorbers (probably due to the fact that they are all sulfides), a prediction of the dielectric constant based on crystal structure data would not be very accurate.

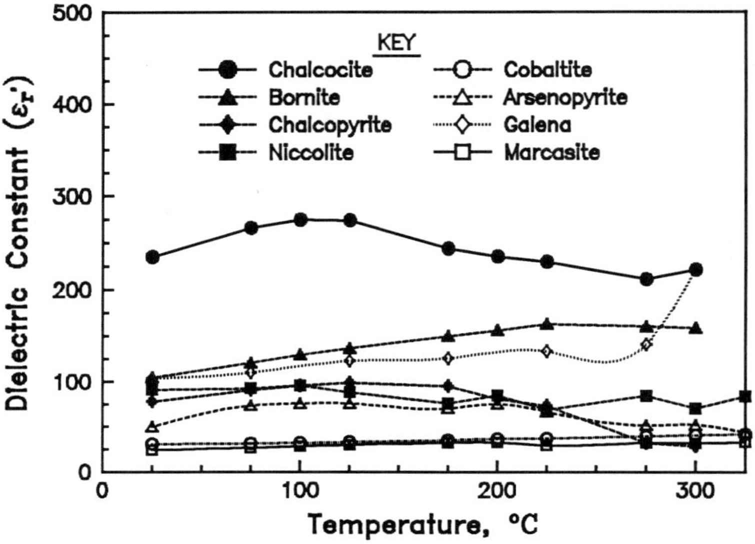

Figure 1. The Dielectric Constants of Sulfides at 915 MHz.

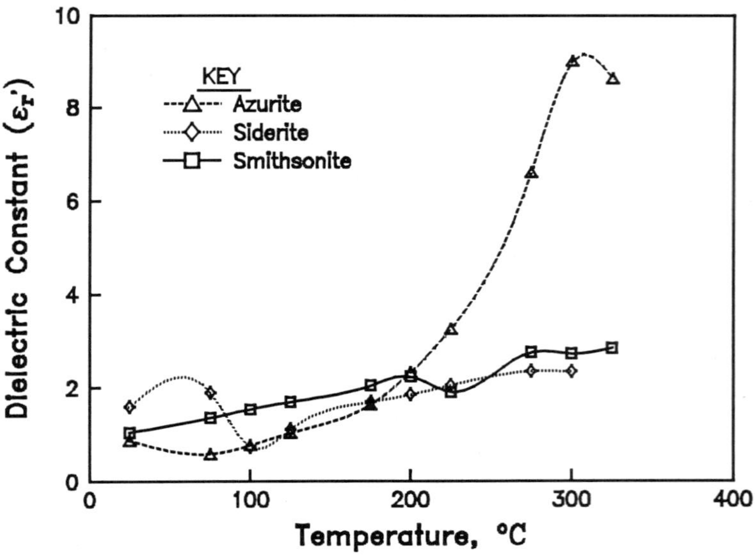

Figure 2. The Dielectric Constants of Carbonates at 915 MHz.

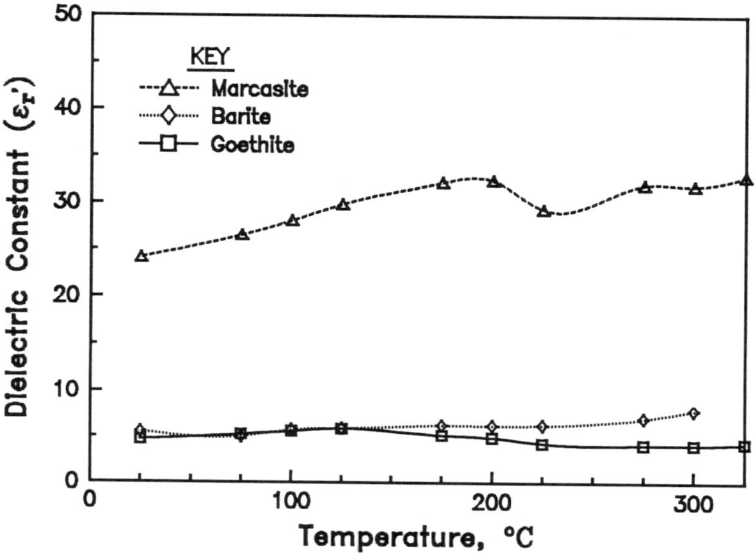

Figure 3. The Dielectric Constants of Orthorhombic Minerals at 915 MHz.

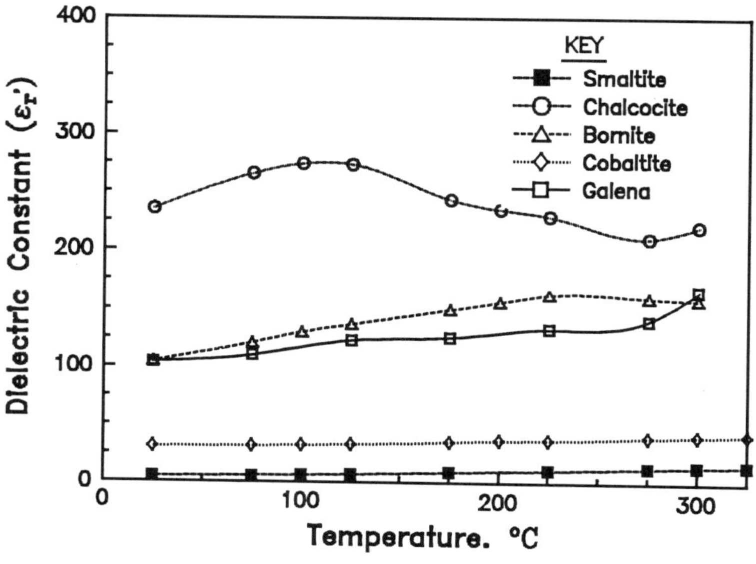

Figure 4. The Dielectric Constants of Isometric Minerals at 915 MHz.

CONCLUSIONS

The U.S. Bureau of Mines' Tuscaloosa Research Center has been measuring the dielectric properties of minerals for the past 5 years. This data is beginning to fill the gap which exists in the literature on the dielectric properties of minerals at the microwave heating frequencies[4].

The data indicate for some specific mineralogical classifications and/or crystalline structures, predictions may be made about the dielectric properties of those minerals. For example, the dielectric constants of the sulfide minerals are all relatively high while those of the carbonates are relatively low. On the other hand, the dielectric constants for the minerals with either an orthorhombic or isometric crystalline structure seemed to show no correlation. More trends may be found among the other mineralogical classifications crystalline structures still to be analyzed.

REFERENCES

1. R.H. Church, W.E. Webb, and J.B. Salsman, BuMines RI 9194, (1988).

2. Johanna B. Salsman in Ceramic Transactions, edited by David E. Clark, Frank D. Gac, and Willard H. Sutton (Amer. Cer. Soc. Proc. 21, Cincinnati, OH, 1991) pp. 203-214.

3. Edward S. Dana, Dana's Manual of Mineralogy, 15th ed. (John Wiley and Sons, Inc., New York, 1944).

4. A.R. Von Hipple, editor, Dielectric Materials and Applications. Papers by Twenty-Two Contributors, (MIT Press, Massachusetts, 1966), pp. 294-425.

CALIBRATION METHOD FOR OPEN-ENDED COAXIAL PROBE/VECTOR NETWORK ANALYZER SYSTEM

DAVID BLACKHAM
Hewlett-Packard Company, 1400 Fountaingrove Parkway, Santa Rosa, CA 95403

ABSTRACT

The vector network analyzer readily lends itself to a variety of techniques for measuring material properties. The open-ended coaxial probe is one technique that allows swept frequency measurements of permittivity. The open-ended coaxial probe technique requires little or no sample preparation and can make non-destructive measurements on a variety of materials. It is especially suited for materials in liquid or semi-solid form. An understanding of network analyzer calibration and the coaxial probe model provides insight into calibration techniques that minimize systematic errors. This understanding also facilitates the extension of measurement conditions beyond normal measurement constraints leading to a simple procedure which provides a first order correction of perturbations to the systematic errors.

MEASUREMENT SYSTEM OVERVIEW

Figure 1 is a representation of the ideal network analyzer measurement system which consists of a source of microwave energy, a signal separation device, detectors and an open ended coaxial probe. (Figure 2 is a drawing of the coaxial probe used for the measurements described in this paper). The signal separation device allows the incident and reflected electromagnetic waves to be detected independently of the other. The reflected wave arises when the incident wave is reflected back from the open end of the coaxial probe. The ratio of the reflected wave to the incident wave is known as the reflection coefficient (Γ). The magnitude of the reflection coefficient varies from 0 to 1 depending both on the loss in the material and the amount of electromagnetic energy launched into the material and radiated away. The phase of the reflection coefficient depends in large part on the dielectric constant (real part of the relative permittivity) of the material at the open end of the coaxial probe.

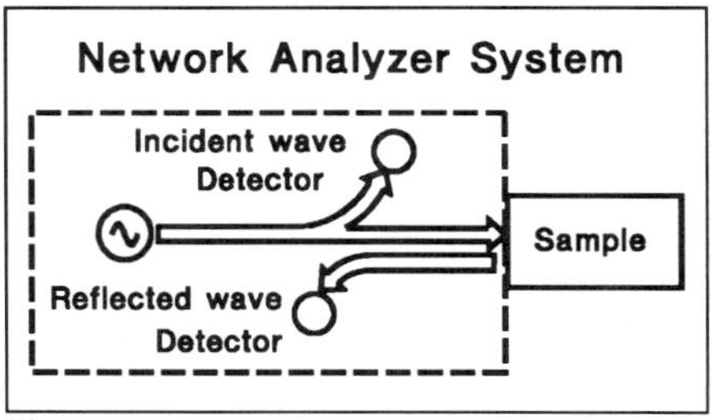

Figure 1

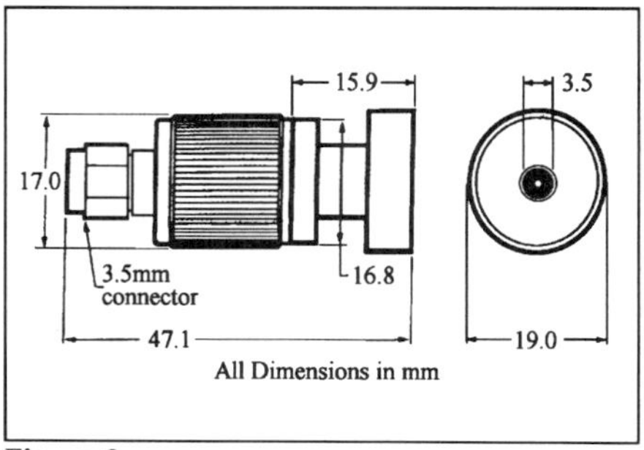

Figure 2

The interrelationship between the reflection coefficient and the complex permittivity of the sample contacting the end of the probe can be modelled[1,2,3]. This model is a bidirectional mapping such that if the permittivity is known, the reflection coefficient can be determined and vice versa. When determining this mapping, certain assumptions about the nature of the material are made. For the case of the model used in this paper, there is an assumption that the sample is of semi-infinite extent. From a practical point of view this means that any energy launched into the material is either absorbed or radiated away. Sample sizes are chosen such that reflections from other boundaries that come back to the open aperture of the probe are small enough to not materially affect the measurement results.

Three types of systematic errors that exist in reflection measurements with vector network analyzers are directivity (e_{00}), frequency response ($e_{10}e_{01}$), and source match (e_{11})[4]. Referring to Figure 1, directivity errors are due to portions of the incident wave arriving at the reflected wave detector without first reflecting off the sample. Frequency response errors are due to the path loss and time delay between the incident wave detector and the reflected wave detector. Source match errors are due to portions of the reflected wave recombining with the incident wave between the incident wave detector and the sample. The interrelationship between the actual reflection coefficient (Γ_a), the measured reflection coefficient (Γ_m) and the systematic errors is represented by the flowgraph in Figure 3 leading to the following equation:

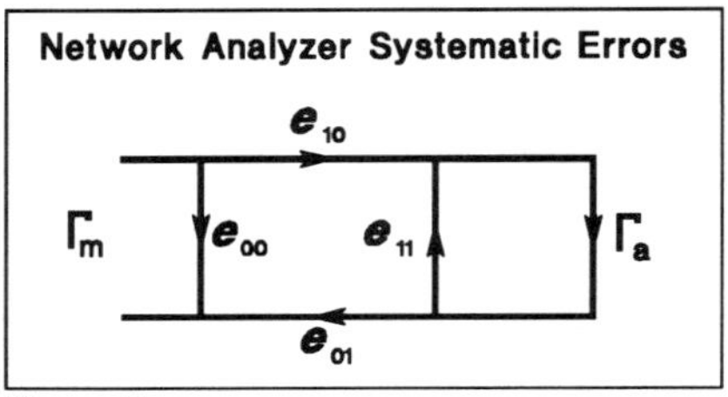

Figure 3

$$\Gamma_m = e_{00} + \frac{e_{10}\,e_{01}\,\Gamma_a}{1 - e_{11}\,\Gamma_a} \qquad (1)$$

The systematic errors can be characterized and their effect accounted for in subsequent measurements through a calibration procedure. Three known standards are connected and measured resulting in three equations with three unknowns (e_{00}, $e_{10}e_{01}$, and e_{11}). The set of three equations are used to characterize the systematic errors. The reflection coefficient of a short is known to be -1, but the reflection coefficients of the other standards are computed using the model that relates permittivity to reflection coefficient. For example, if an open is used as one of the calibration standards the actual reflection coefficient can be determined at each measurement frequency by using the probe model to calculate the expected reflection coefficient when the probe is in contact with air (permittivity = 1).

Once the systematic errors are known the relationship defined in equation 1 provides a method to determine the actual reflection coefficient from the measured coefficient. The model provides the means to then determine the complex permittivity of the material in contact with the probe. The calibration is valid as long as the measurement system remains constant.

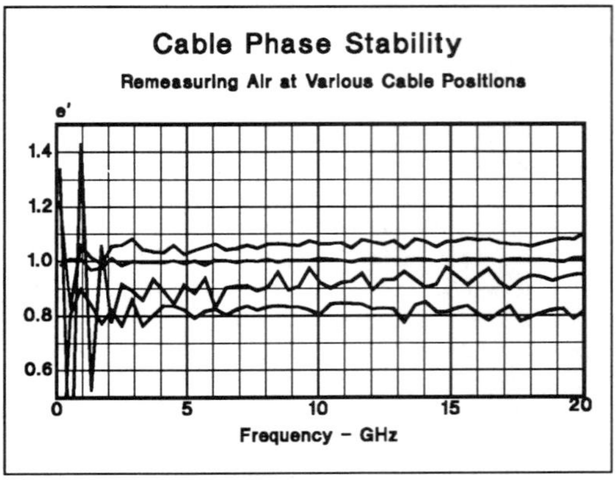

Figure 4

Figure 4 shows how cable movement can affect the measurement. Moving the cable introduces a phase perturbation into the systematic errors that is reflected in the final measurement. Changes in temperature can introduce similar perturbations. The best measurement practice is to fix the cable in a constant position and maintain a constant temperature between calibration and measurement of unknown samples. Sometimes it is impractical to perform a complete three standard calibration. In such cases, a first order correction to the calibration can be made if one of the known calibration standards (such as the short or air) is remeasured to estimate the perturbation. The calibration can then be modified or "refreshed" to account for the perturbation.

Refresh Calibration

The systematic error terms are actually distributed on both sides of the cable. The largest contribution to the systematic errors comes from the network analyzer side of the cable. For a first order correction assume that the entire contribution comes from the network analyzer side of the cable and let $\Delta c_{10}\Delta c_{01}$ represent the cable variation (see Figure 5). The resulting equation including the cable variation is then given by:

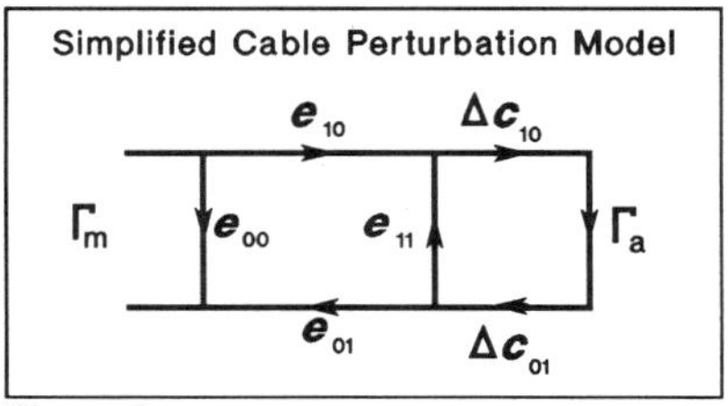

Figure 5

$$\Gamma_m = e_{00} + \frac{e_{10}\, e_{01}\, \Delta c_{10} \Delta c_{01} \Gamma_a}{1 - e_{11}\, \Delta c_{10} \Delta c_{01} \Gamma_a} \qquad (2)$$

If one of the standards is remeasured using the original calibration, the cable perturbation term ($\Delta c_{10}\Delta c_{01}$) can be determined. The first order correction, would then modify both the frequency

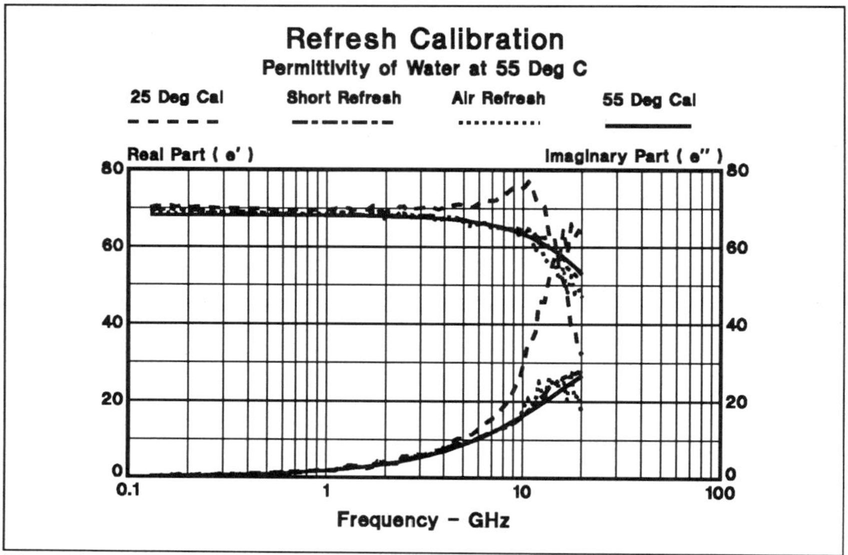

Figure 6

response term ($e_{10}e_{01}$) and the source match term (e_{11}) by the cable perturbation term. Since the refresh calibration is a first order correction, best results are still achieved if care is taken to minimize system perturbation. The use of temperature stable cables is beneficial if measurements are to be made over a range of temperatures.

Measurement Results

Figures 6 and 7 represent measurements made on distilled water and methanol in a temperature chamber at 55 degrees centigrade. Figure 6 shows the frequency response of both the real and imaginary parts of permittivity. Figure 7 shows a Cole-Cole plot for methanol over the frequency range of 130 MHz to 20 GHz. A calibration was done at 25 degrees centigrade. The methanol, water, probe and part of the cable were then raised to 55 degrees centigrade. At 55 degrees centigrade air, short, water and methanol were remeasured. These measurements were used to compare the following four different types of calibration:

1. 25 degrees centigrade calibration measuring at 55 degrees centigrade
2. Refresh calibration at 55 degrees centigrade using the short as a refresh standard
3. Refresh calibration at 55 degrees centigrade using air as a refresh standard
4. Full three standard calibration at 55 degrees centigrade.

The best results were obtained by a full calibration at the measurement temperature. Both the air refresh and short refresh calibrations gave much better results than simply using the 25 degree centigrade calibration.

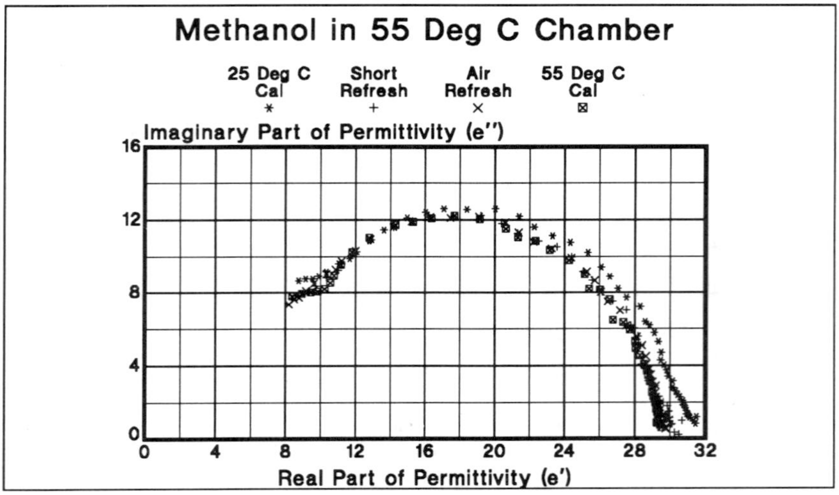

Figure 7

CONCLUSIONS

The vector network analyzer and open-ended coaxial probe provide an easy-to-use system to measure complex permittivity. Calibration techniques developed for the vector network analyzer can be applied to measurements made with the open-ended coaxial probe. The insight provided by this calibration approach enables greater measurement flexibility.

REFERENCES

1. E.C. Burdett, F.L. Cain, and J. Seals. IEEE Micr. Theory and Tech., MTT-28 (4), 414-427 (1980).

2. J.P. Grant, R.N. Clarke, G.T. Symm, and N.M. Spyrou, J. Phys. E: Sci. Instrum. 22, 757-770 (1989).

3. M.A. Stuchly and S.S. Stuchly, IEEE Trans. on Instrum. and Measur. IM-29(3) (1980).

4. D.K. Rytting, An Analysis of Vector Measurement Accuracy Enhancement Techniques. Hewlett-Packard, RF & Microwave Symposium,(1980).

MODELING THE MICROWAVE FREQUENCY DIELECTRIC PROPERTIES OF THERMOPLASTIC COMPOSITE MATERIALS

MITCHELL L. JACKSON AND CURTIS H. STERN
Virginia Polytechnic Institute and State University, Department of Mechanical Engineering, Blacksburg, VA 24061-0238.

ABSTRACT

Mixture models were studied in an effort to predict the microwave frequency permittivities of unidirectional-fiber-reinforced thermoplastic-matrix composite materials as a function of fiber volume fraction, fiber orientation relative to the electric field, and temperature. The permittivities of the constituent fiber and plastic materials were measured using a resonant cavity perturbation technique at 9.4 GHz and at 2.45 GHz. The permittivities of the composite specimens were measured using a reflection cavity technique at 9.4 GHz and at 2.45 GHz . Simple "rule-of-mixtures" models that use the fiber and plastic permittivities have been found to approximate the complex dielectric properties of the composite for varied fiber volume fractions. The permittivities of oriented composites were modeled using a tensor rotation procedure. Composite permittivities were modeled with temperature up to the glass transition temperature of the thermoplastic matrix.

INTRODUCTION

A major concern facing the emerging composites industry is developing cost-effective, reliable, and versatile processing methods. In many cases the critical processing step of heating can be improved by using microwave power. Microwave processing offers unique advantages in polymeric composites, such as improved fiber-resin interface properties and the potential to process thick sections of material uniformly due to inherent volumetric heating. With the growing interest in microwave heating of composites, there exists a need to model this process, taking into account the microwave application system, the material, and the interactions between the two. Therefore, it is essential that the electromagnetic constitutive property of the material, the relative complex permittivity $\varepsilon^* = \varepsilon' - j\varepsilon''$, be characterized. The permittivity model will facilitate heat transfer computations and promote understanding of the electric field distribution in the applicator and within the material.

Efforts to measure and model the permittivity of fiber-reinforced composites at microwave frequencies have increased in recent years [1-6]. These efforts are in large part attributable to "stealth technology" research for military applications. Most studies found in the literature only include measurements made on the specific material under investigation without regard to fiber volume fraction, fiber orientation, or temperature. This study models the permittivity of a commercially available unidirectional-fiber-reinforced composite material system with respect to these important processing parameters.

Mixture Model

The equations evaluated in this study were selected from a thorough compilation of mixture models by Banhegyi [7]. Simple "rule-of-mixtures" and "reciprocal rule-of-mixtures" equations were chosen to provide upper and lower bounds on the effective permittivities [8]. These models require the knowledge of several input parameters: the volume fraction and the permittivity of each component. While these models do not explicitly contain fiber orientation as a parameter, the equations can be considered to implicitly account for orientation. The permittivity ε^* in terms of the volume fractions v_1 and v_2 and the component permittivities ε_1 and ε_2 for composites oriented parallel and perpendicular, respectively, to the electric field $\underline{E}$ are

$$\varepsilon^*_{\parallel} = v_1\varepsilon_1 + v_2\varepsilon_2 \tag{1}$$

$$\varepsilon^*_{\perp} = [(v_1/\varepsilon_1) + (v_2/\varepsilon_2)]^{-1} \tag{2}$$

MATERIALS AND EXPERIMENTAL TECHNIQUES

The material used in this study was a polycarbonate/ unidirectional, continuous fiberglass composite made from a preimpregnated tape supplied by ICI-Fiberite Corporation. This tape was produced using Makrolon HMS-3118 amorphous thermoplastic polycarbonate resin from Mobay Corporation and Owens Corning S-2 glass fibers. Several 15.24 cm x 15.24 cm (6 in. x 6 in.) plates were fabricated by stacking plies of tape in a mold and consolidating in a heated platen press at a maximum pressure of 1.034 MPa (150 psi) and a maximum temperature of 210°C. All of the plies were stacked with fibers aligned in the same direction. Specimens with a range of fiber volume fractions were produced by re-pressing to remove additional resin. Samples of S-glass fiber strands for use in the determination of component permittivity were vacuum dried and weighed before measuring. Polycarbonate strands were capillary extruded, vacuum dried, and bundled into a volume suitable for measurement.

Permittivities of the component S-glass fibers and polycarbonate strands were measured at 2.45 GHz and 9.4 GHz using a standard cavity perturbation technique [9,10]. Measurements at both frequencies used rectangular cavities resonating in the TE $_{103}$ mode. The cavities were inductively coupled with a central-circular iris made of a thin copper sheet. WR 430 and WR 90 waveguides were used for the 2.45 GHz and 9.4 GHz cavities, respectively. The quality factor (Q) and resonant frequency (f_r) of the cavity were measured by a Hewlett Packard 8510B network analyzer. Strands of material were inserted through small holes in the broad wall of the waveguide, centered at an $\underline{E}$-field maximum. Elevated temperature measurements were made by placing strands of material in quartz tubes, inserting these tubes into the cavity, and flowing hot nitrogen gas through the tube. A thermocouple placed inside the outlet end of the tube, just outside the waveguide wall, monitored the temperature.

Composite specimens were measured using a reflection cavity technique [9] in inductively coupled cavities supporting the TE $_{103}$ mode at resonance. Measurements were made at 2.45 GHz and 9.4 GHz. Precisely fit specimens were placed in the open end of the cavity. A shorting plate was subsequently bolted in place, taking care to make complete contact with the face of the specimen. Q and f_r were measured with the cavity empty and with specimens inserted.

Reflection cavity measurements at elevated temperatures were conducted by placing the cavity in a vertical orientation resting on a flat heater. Q and f_r measurements as a function of temperature were made for empty and specimen-loaded cavities. Temperatures were measured on the shorting plate using a thermocouple. A separate experiment was conducted to relate the temperatures on the shorting plate to those of the inserted specimens.

All tests were conducted with the waveguide propagating the dominant TE_{10} mode. The E-field in this mode has a component only in the direction parallel to the narrow dimension of the cross-plane, which is normal to the axis of the waveguide. Specimens having fibers aligned with the narrow dimension,and therefore with the E-field, are considered to be in "parallel" geometrically and electrically. Specimens with fibers aligned with the wide dimension are geometrically "perpendicular" or electrically in "series". A range of specimens with orientations θ between 0° and 90° were prepared and tested. A resin burnoff technique was employed to determine fiber volume fraction of each specimen plate.

RESULTS AND DISCUSSION

Permittivities of component materials as a function of temperature measured at 9.4 GHz are shown in Figure 1. The real part of the permittivity ε', the dielectric constant, of both materials changed slightly with temperature from 25°C to 170°C. The imaginary part of the permittivity ε", the loss factor, increased significantly for both materials over the same temperature range.

The component material permittivities as a function of temperature were substituted into the mixture models given by Equations (1) and (2) at a fixed $v_f = 0.45$ to yield the continuous curves of Figure 2. The dielectric constants at 9.4 GHz for oriented composite samples show good agreement with the models (Figure 2 a). The loss factors of the composites (Figure 2 b) also agree with the model , although the models do not adequately predict the increased ε" of the composites near the T_g of the polycarbonate at 150°C.

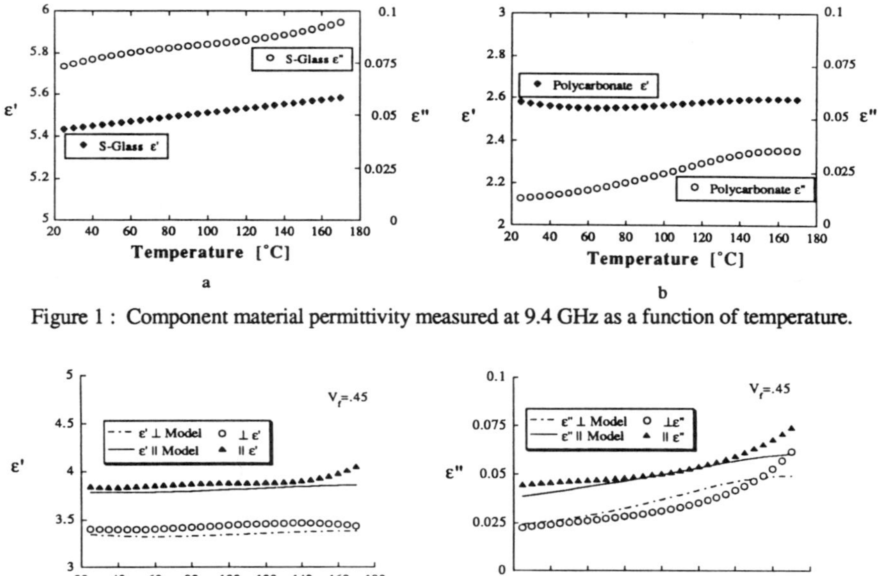

Figure 1 : Component material permittivity measured at 9.4 GHz as a function of temperature.

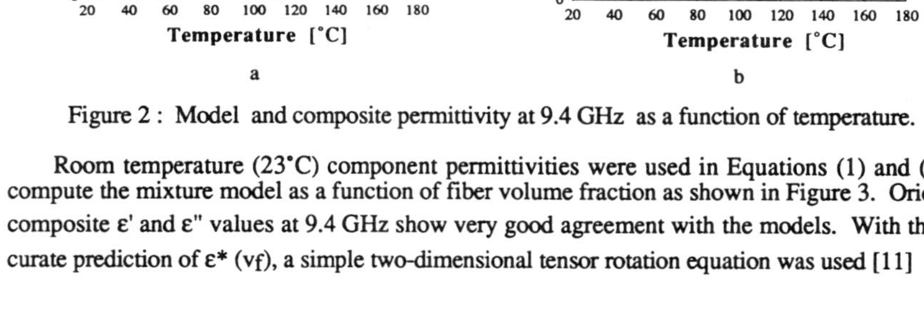

Figure 2 : Model and composite permittivity at 9.4 GHz as a function of temperature.

Room temperature (23°C) component permittivities were used in Equations (1) and (2) to compute the mixture model as a function of fiber volume fraction as shown in Figure 3. Oriented composite ε' and ε" values at 9.4 GHz show very good agreement with the models. With the accurate prediction of ε* (v_f), a simple two-dimensional tensor rotation equation was used [11]

$$\varepsilon^*_\theta = \varepsilon^*_{\parallel} \cos^2(\theta) + \varepsilon^*_{\perp} \sin^2(\theta) \qquad (3)$$

with $\varepsilon^*_{\parallel}$ and $\varepsilon^*_{\perp}$ values considered to be composite permittivities in principle directions. Substitution of composite permittivities at $v_f = 0.45$ and T = 23°C into Equation (3) yielded the plots in Figure 4. The measured composite ε' (θ) measured values show excellent agreement, while composite ε" (θ) data generally follow the model's trend.

A parallel study at 2.45 GHz followed the same computational procedure as that at 9.4 GHz. Component permittivities as a function of temperature are shown in Figure 5. Dielectric constants ε' increased slightly, while loss factors ε" displayed large relative increases with temperature. A comparison of permittivities (Figures 1 and 5) shows a slight increase in ε' magnitude for S-glass and little change in polycarbonate ε' from 2.45 GHz to 9.4 GHz. Much more significant decreases in ε" magnitudes are evident from 2.45 GHz to 9.4 GHz.

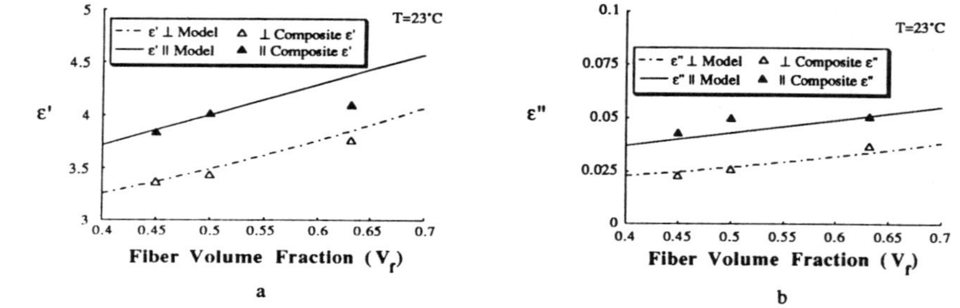

Figure 3 : Model and composite permittivity at 9.4 GHz as a function of fiber volume fraction.

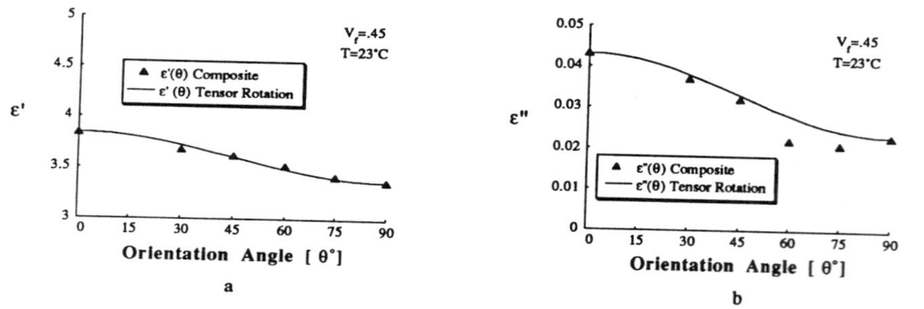

Figure 4 : Model and composite permittivity at 9.4 GHz as a function of fiber orientation.

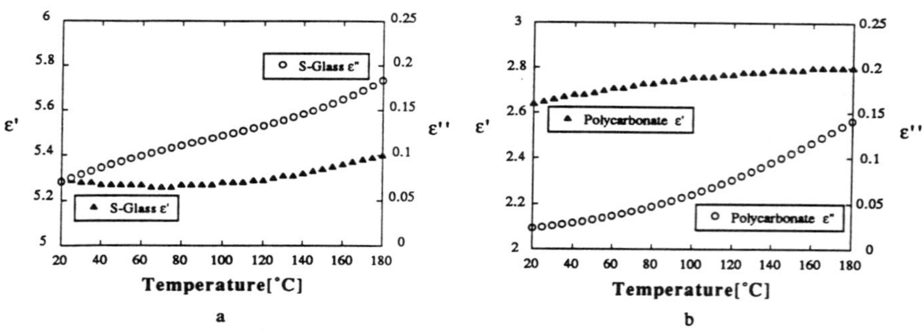

Figure 5 : Component material permittivity measured at 2.45 GHz as a function of temperature.

The agreement of elevated temperature composite permittivities at 2.45 GHz with model predictions, shown in Figure 6, is clearly not as good as results at 9.4 GHz. Results for ε* of a composite with perpendicular orientation could not be reported with confidence. A comparison of Figures 7 and 3 shows a decrease in agreement of the composite ε* data with the models from 9.4 GHz to 2.45 GHz. While the measurement of the permittivity data points in Figure 7 was repeatable, it is believed that part of the inaccuracy resulted from using specimens that were thin relative to the measurement wavelength. Specimens with equal fiber volume fraction had the same thickness in both the 2.45 GHz and 9.4 GHz experiments. Thus the composite specimen thickness-to-wavelength ratio was 1:27 at 2.45 GHz , while at 9.4 GHz the ratio was 1:7. This inaccuracy was quantified with permittivity measurements at 2.45 GHz made using different thicknesses of Teflon (ε*= 2.08-j 0.0004). Results for a specimen with a thickness-to-wavelength ratio of 1:9 were ε*=2.095-j 0.00021, while for a specimen of ratio 1:22 a permittivity of ε*=2.38-j 0.0017 was obtained.

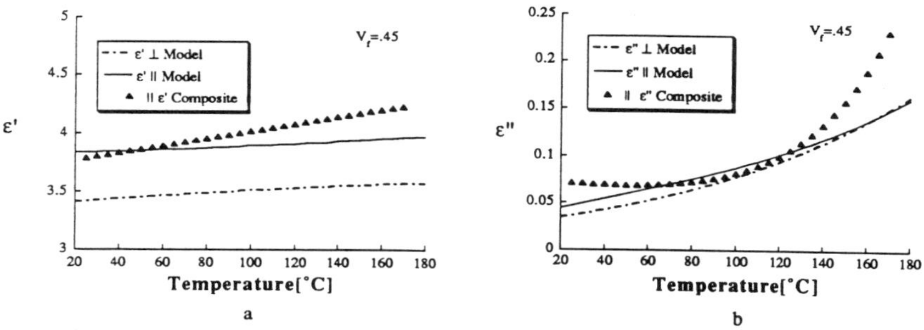

Figure 6 : Model and composite permittivity at 2.45 GHz as a function of temperature.

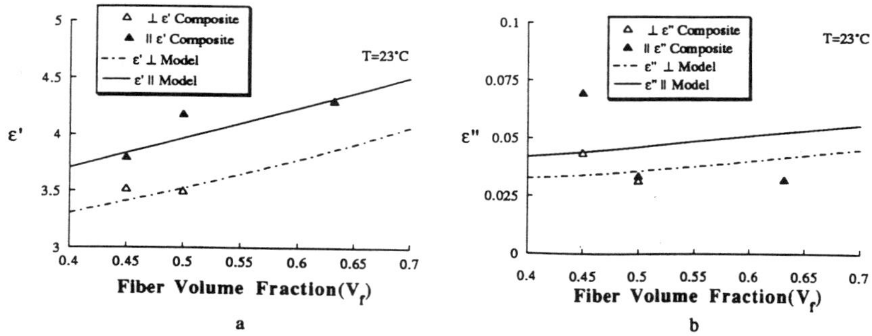

Figure 7 : Model and composite permittivity at 2.45 GHz as a function of fiber volume fraction.

CONCLUSIONS

Based upon comparisons within one material system, it appears that the simple "rule-of-mixtures" models can be used to predict the permittivities of unidirectional-fiber-reinforced thermoplastic composite materials as a function of fiber volume fraction and orientation relative to the electric field. Given the component material permittivity characterized with temperature, these models can predict elevated temperature composite permittivities. Use of these models may allow accurate prediction of composite permittivity, eliminating the difficult process of composite measurement. Furthermore, accuracy may be improved since thin composite sample permittivity often cannot be measured as accurately as that of component materials, especially when the components are measured using cavity perturbation methods.

ACKNOWLEDGEMENTS

This work was supported by the National Science Foundation (CTS8908569) and the Center for Composite Materials and Structures at VPI&SU. The authors would like to thank Mobay Corporation and ICI-Fiberite Corporation for supplying materials used in this study.

REFERENCES

1. W.I. Lee and G.S. Springer, J. of Composite Materials 18, 357-385 (1984).

2. S. Aussudre, A. Priou, H. Jullien, M. Delmotte, F.M. Thullier, D. Boulonnais, J. Sailleau, Rech. Aerosp. 5, 1-13 (1988).

3. A.MacDonald, R.L. Moore, W.J.S. McLemore, Tech. Report No. F19628-88K-0023 RADC/EECT, Hanscomb AFB, 1989.

4. R.L. Moore,A. MacDonald, H.R. Moroz, Microwave Journal 34 (2): 67-82 (1991).

5. P.A.M. Steeman and F.H.J. Maurer, Colloid and Polymer Science 268, 315 (1990).

6. A.R. Stonier, SAMPE Journal 27 (4): 9-17 (1991).

7. G. Banhegyi, Colloid and Polymer Science 264, 1030 (1986).

8. M.L. Jackson and C.H. Stern, J. of Microwave Power & Electromagnetic Energy, (in press).

9. H. M. Altschuler, in Handbook of Microwave Measurements 3rd ed., edited by M. Sucher and J. Fox, (Polytechnic Press, New York, 1963), p.495-546.

10 ASTM Standard D2520, Standard test methods for complex permittivity of solid electrical insulating materials at microwave frequencies and temperatures to 1650°C 10.02 , 210 (1986).

11. R.M. Jones, Mechanics of Composite Materials, (Hemisphere Publishing, New York, 1975).

CERAMIC MATERIALS FOR HIGH-QUALITY DIELECTRIC RESONATORS OF MICROWAVE RANGE

YELIZAVETA A. NENASHEVA
Department "Ceramics" of Joint Venture "UniRem", Kurchatova, 10, St.Petersburg, Russia, 194021.

ABSTRACT

The dielectric properties of ceramic materials and dielectric resonators on the basis of oxides of Ba,Ti and alkaline earth elements, and also of the calcium titanate-lanthanum aluminate system, with ε =82 and ε =40...48, frequency temperature coefficient (TC_f) from -10ppm/°C to +10ppm/°C have been studied. Ceramics with ε =82 ensures the DR Q-factor of 3000 and more at f=3GHz. The Q-factor of DR made of ceramics with ε=42, TC_f =0 reaches 8000 at f=7GHz.

1. INTRODUCTION

The use of dielectric resonators (DR) as functional elements of various UHF devices, in particular, high-selective band-pass filters, converting-amplifying devices, aerial sections, stable solid-state generators etc.., calls for special ceramic materials with optimal dielectric parameters in a specified range of temperature and frequency.

Ceramic materials for DR should have minimum dielectric losses, which provides for high Q-factor of the DR ($Q \cong 1/tg\delta$), comparatively high dielectric permittivity (ε) defining the DR volume and the possibility of miniaturization of the DR. These ceramic materials should also have increased stability of dielectric parameters in the range of working temperature and frequency, which would guarantee standardized values of TC_f of dielectric resonators.

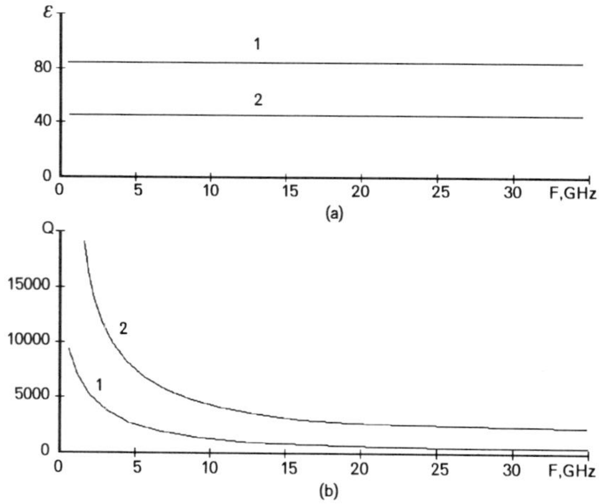

Fig.1. Frequency dependence of ε (a) and DR Q-factor (b) of materials on the base of solid solutions TBNS-R (1) and $CaTiO_3$-$LaAlO_3$ (ALTK) (2).

The properties of ceramic materials are determined by the chemical composition of their basic crystalline phase as well as by the materials manufacturing process.

In different UHF devices working in the frequency range 0.5 to 5.0 GHz, DR are composed of materials based on oxides of barium, titanium and alkaline-earth elements with dielectric permittivity from ε =75 (DR of series 8600, "Alpha", USA) to ε =91 (DR of series "K", "Murata", Japan [1,2]). With the increase of dielectric permittivity of ceramic materials, Q-factors fall from 3000 (for ε =75) to 1700 (for ε =91) at the frequency f=3GHz.

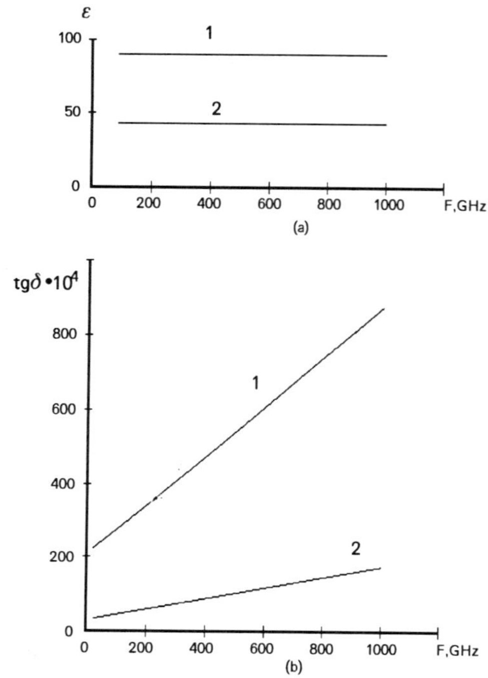

Fig.2. Frequency dependence of ε (a) and tgδ (b) of ceramic samples on the base of solid solutions TBNS-R (1), $CaTiO_3$ - $LaAlO_3$ (ALTK) (2) within submillimeter range.

For frequences of the range 2 to 15 GHz the ceramic materials on the basis of $BaTi_4O_9$ and $Ba_2Ti_9O_{20}$ with ε =36...40 [3] are well-known. DR made of these materials are characterized by comparatively high Q-factor (up to Q=5200 at f=9 GHz - for $Ba_2Ti_9O_{20}$ [4]). The use of individual compounds, barium polytitanates in particular, as the basic crystal phase of ceramic material of DR sets the limits to the application field of DR on the basis of such compounds due to the narrow range of TC_f. For the range of frequency up to 45 GHz the more frequent materials been for DR are solid solutions of zirconium-tin titanates with ε =36...38 [5], which ensuring a high Q- factor of DR (Q=6500 at 7 GHz) and wide spectrum of TC_f, in particular, TC_f =0. The tests of dielectric properties of ceramics species for UHF in the wide frequency range up to submillimetre one [6] (Fig.2) have shown the advantage of the crystalline phase developed in the USSR, i.e. the phase on the basis of the system of calcium titanate and lanthanum aluminate solid solutions with perovskite structure. Some results of investigation of the oxide systems BaO-La_2O_3-TiO_2 and $CaTiO_3$-$LaAlO_3$ are presented in this paper.

2. SYSTEM BaO-La_2O_3-TiO_2

The published data on the composition of compounds in the barium rare-earth and titanium oxide system are discrepant. In [7,8] the $BaLn_2Ti_5O_{14}$ compound is described, which corresponds to a composition with the 1I1I5 barium oxide/rare-earth/titanium oxide ratio. In [9] it is stated that the composition of the compound corresponds to the formula $Ba_4La_8Ti_{17}O_{50}$ (BaO•La_2O_3•4,25TiO_2).

On the other hand our studies [10,11] have shown that the patterns for $BaLa_2Ti_5O_{14}$ and BaO•La_2O_3•4,25TiO_2 are not single-phase ones. In the system $BaLa_2O_4$-nTiO_2 (3< n≤ 5,5) a single-phase composition corresponds with n=4 ($BaLa_2Ti_4O_{12}$) only. A decrease in the titanium oxide content (n<4) results in the formation of a mixture of two crystalline phases $BaLa_2Ti_4O_{12}$ end $BaLa_2Ti_3O_{10}$ while an increase in the titanium oxide content (n>4) is accompanied by the occurrence of rutile phase whose content increases with n. By a chemical analysis, in dissolving the principal $BaLa_2Ti_4O_{12}$ phase in hydrochloric acid, in compositions with n>4, an insoluble residue has been detected whose quantity amounts to 1,51% for n=4,1 and increases to 14,1% for n=5,5. A phase analysis of the insoluble residue has revealed, in addition to the principal rutile phase (TiO_2), also barium tetratitanate ($BaTi_4O_9$) whereas for n≥5,0, $Ba_2Ti_9O_{20}$.

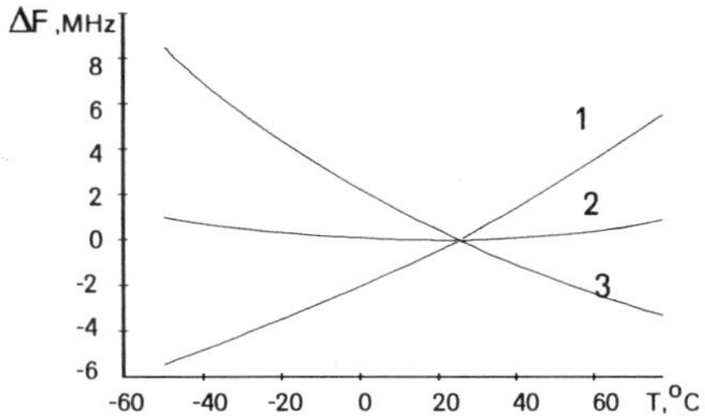

Fig.3a. Temperature dependence of drift of DR resonance frequency of materials on the base of solid solutions $CaTiO_3$ - $LaAlO_3$.
1. TC_f = +10 ppm/°C; 2. $TC_f \cong 0$;.
3. TC_f = -10 ppm/°C .

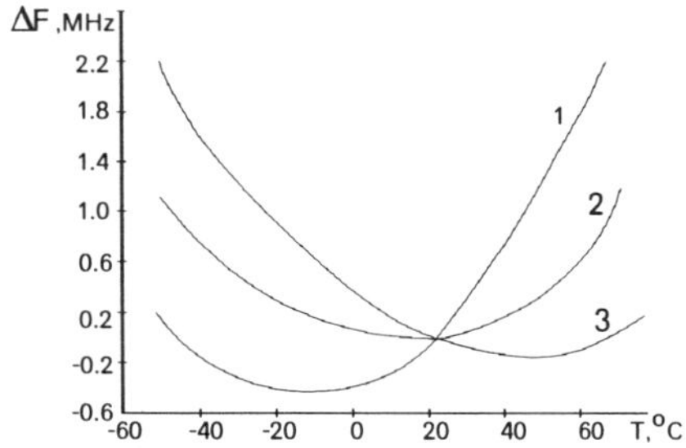

Fig.3b. Temperature dependence of drift of DR resonance frequency of materials of system $CaTiO_3$ - $LaAlO_3$ of different compositions (1-3).

A large region of barium and rare-earth isovalent and heterovalent substitutions has been found in the $BaLa_2Ti_4O_{12}$ compounds. This allows the preparation on their basis of ceramic materials with a high permittivity of 70 to 200 and a specified temperature coefficient of capacitance (TCC) between +100ppm/°C and -750ppm/°C while featuring low losses within wide temperature and frequency ranges. A series of industrial ceramic materials with ε =80,100 and 120, TCC=0 were received. On the base of solid solution of $BaLa_2Ti_4O_{12}$ (Ln-Nd,Sm) the high-quality thermostable dielectric material TBNS-R was obtained. The dielectric constant (ε) of this material was 82 and the Q-factor of DR of TBNS-R was not less than 3000 at the frequency f=3GHz. The product of values of Q-factor of DR and frequency was a constant value for a wide frequency range up to submillimeter one (Fig.1 and Fig.2). Increases of frequency resulted in decrease in the Q-factor and at f=5GHz the Q-factor of DR made of TBNS-R material did not exceed 2000.

3. (Ca,La) (Al,Ti) O_3 SYSTEM

$LaAlO_3$ and $CaTiO_3$ compounds form solid solutions with perovskite structure over a wide concentration range creating industrial materials with different groups of temperature stability. The DR on the basis of solid solutions having perovskite structure of cubic modification have the highest Q-factor, but the TC_f of DR, made of these materials is -8ppm/°C [12].

It is known that the temperature coefficient at a given frequency is determined by the ceramic material parameters: $TC_f = -0{,}5TC_\varepsilon - \alpha$, where TC_ε is the temperature coefficient of dielectric permittivity of the material and α is the coefficient of linear thermal expansion. To obtain $TC_f = 0$ the parameter TC_ε must be close to-2α for the whole range of working temperature of the DR. Taking into account that $\alpha > 0$ for all the materials of this class, $TC_f = 0$ is realized for compounds with $TC_\varepsilon < 0$. This can be achieved by raising the calcium titanate content of the solid solution.

On the basis of modified solid solutions we have worked out a series of industrial ceramic materials (ALTK) with ε from 40 to 48 and TCf from -10ppm/°C to +10 ppm/°C (Fig.3a). The material ensuring $TC_f = 0$ has ε =42.

Curves showing the drift of DR resonance frequency with changes in temperature for DR with $TC_f \cong 0$ for three material composition are given in fig. 3b. The nature of minimum in dependence of resonance frequency on temperature may be explained by the anomaly that has been found in α - dependence on temperature (Fig.4) for composition ALTK. As it can be seen in the Fig.3b the technology allows the parameter TC_f of DR and to ensure the stability of DR at the level of (0...1) ppm/°C in a specified temperature range. The Q-factor of DR made of such materials reaches 8000 at t=7 GHz. The product of the Q-factor value of the DR and frequency is constant value for a wide frequency range (Fig.1,2) up to 1000GHz.

4. CONCLUSION

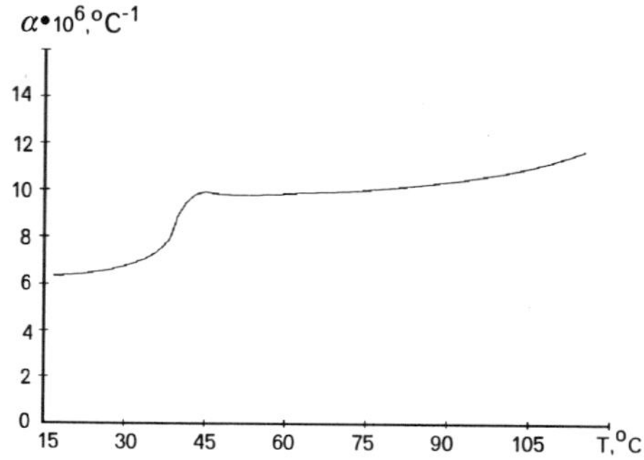

Fig.4. Temperature dependence of α of solid solution ALTK.

The dielectric properties of ceramic materials on the basis of barium-rare-earth and titanium oxide system and also of calcium titanate-lanthanum aluminate system with ε =82 and ε =40...48 have been studied using wide temperature and frequency ranges up to 1000GHz. The Q-factor of DR made of material TBNS-R with ε =82 is 3000 and increases at f=3GHz. The Q-factor of DR from ALTK material reached 8000 at a frequency of 7GHz.

The elaborated technology provides control of the parameter TC_f of DR from -10 ppm/°C to +10 ppm/°C and to ensures the stability of DR at the level of TC_f= 0 in specified range of temperature.

DR made of TBNS-R with ε =82 and ALTK with ε =40...48 and DR's made on its basis have been applied to various systems of UHF devices, in particular, in stable solid-state generators and satellite TV systems.

REFERENCES

1. "90 Murata Products", Murata, 1989, p.49.
2. K.Wakino, T.Nishikawa, R.Tamura. Microvawe Journal, 133 (1987).
3. H.M.O'Bryan, J.Thomson, J.K.Plourde. J. Amer. Ceram. Soc, 450 (1974).
4. S.Nomura, K.Tomaya, K.Kaneta. Jap. J. Appl. Phys, 1125 (1983).
5. K.Wakino, M.Katsube, H.Tamura, T.Nishikawa, Y.Ishikawa. Joint Convention Record of Four Institutes of Electrical Engeneers, 6 (1976).
6. A.A.Volkov, B.P. Gorshunov, G.V.Kozlov, Ye.A.Nenasheva. Aktualnyje problemy technologii Kompozicionnych materialov i radiokomponentov v mikroelektronnych informacionnych systemach. (Jalta, 1990), p.84.
7. D.Kolar, Z.Stadler, S.Gaberscek, D.Suvorov. Ber. Dt.Keram Ges., 346 (1978).
8. D.Kolar., S.Gaberscek, B.Volavsek, N.S.Parker, R.S.Roth. J.Sol.St. Chem., 158 (1981).
9. J.Guha. J. Amer. Ceram. Soc.,878 (1991).
10. Ye.A.Nenasheva, N.M.Sidorova, N.F.Kartenko et al. Isvestia of the Academy of Science of the USSR., Inorganic materials serial., 1524 (1985).
11. Ye.A.Nenasheva, N.F.Kartenko, I.B.Kamushkina,V.P.Pyshkov. Proceeding of the Interhational Conference Electronic Ceramics, Production amk Properties (Riga, 1990), p.79.
12. Ye.A.Nenasheva, L.P.Mudrolubova, N.F.Kartenko, G.V.Kozlov, B.P.Gorshunov. Proceeding of the International Conference Electronic Ceramics, Production and Properties (Riga,199), p.82.

HIGH TEMPERATURE DIELECTRIC MEASUREMENTS ON CERAMICS

M Arai, J G P Binner, G E Carr, and T E Cross
Materials and Microwave Processing Group, University of Nottingham, Nottingham, NG7 2RD, UK.

ABSTRACT

The use of microwave frequency radiation as an alternative energy source for processing materials is currently receiving considerable worldwide attention. The ability of a material to extract energy from the microwave field depends upon its dielectric properties, however these change as the material heats. There is, therefore, a need for obtaining measurements of the microwave dielectric properties at elevated temperatures. This paper outlines some of the current work on high temperature dielectric measurements which is in progress at the University of Nottingham as part of the Materials and Microwave Processing Group initiative and some results on engineering ceramics are presented.

INTRODUCTION

Microwave heating is an established process step in both the food and sections of the polymer/rubber industries. To date however the use of microwave energy in the processing of ceramic materials has been rather slow, Metaxas and Binner (1). There are many reasons for this, not least being the fast development and diverse nature of materials which come under the heading of ceramics. Nevertheless, the use of microwave energy for obtaining heating effects during ceramics processing offers a number of potential advantages over conventional radiant heating techniques. This has led recently to attention being focused on the processing possibilities associated with the use of microwaves (Table 1). Of the ISM microwave frequencies available the greatest effort is currently being devoted to 2.45GHz because it provides a suitable compromise between power deposition and penetration depth. The most likely alternative is 915MHz (896MHz in the UK) which provides even better depth of penetration. Proper design of microwave industrial systems requires a knowledge of the dielectric properties of the ceramics to be processed. However the permittivity of these materials is sensitive to frequency, temperature, density and electric field direction.

The importance of temperature in the microwave treatment of ceramics cannot be over emphasized (1). In general, the effective loss factor of many ceramics increases with temperature, with the rate of increase depending upon the type of ceramic and the operating frequency. When a material exhibiting this positive slope in the relationship between the loss tangent and temperature is heated using microwave energy a 'runaway effect' or uncontrolled rise in temperature is observed. Damage to the material is highly probable unless steps are taken to avoid such a cumulative effect. If microwave processing is to be applied to these materials then a thorough understanding of the dielectric properties, at least up to their sintering temperatures (~1500°C), is required.

CURRENT METHODS OF MEASURING DIELECTRIC PROPERTIES AT ELEVATED TEMPERATURES

For measurement at elevated temperatures the measurement technique must survive contact with the sample at the measurement temperature or the measurement time must be short enough to prevent significant thermal conduction from the hot sample to the cool measurement system or there should be no physical contact between the hot sample and the measuring device. The high temperature should also introduce as little error into the measurement as possible. The constraints associated with operation at elevated temperature mean that only a small number of conventional dielectric measurement techniques are suitable for use at high temperatures.

Synthesis	Synthesis of a range of non-oxide ceramic powders is being studied including silicon carbide and titanium carbide.
Drying	The majority of current work is focused on traditional ceramics, refractories and moulds for investment casting.
Slip casting	Microwaves have been shown to increase the rate of casting; also drying of moulds and ware.
Calcination	The heat treatment of precursors to yield the desired phase. Research to date has focused on super-conducting ceramics and certain electro-ceramics, however little success has been reported.
Sintering	The greatest amount of work is being one in this field. A wide range of ceramic and ceramic matrix composite systems are currently being studied.
Joining	Attempts are being made to join both oxides and non-oxides with significant early success.
Plasma assisted sintering	The use of microwaves to generate plasmas to assist in oxide sintering. Plasma sintering is a research field in its own right, however some indication has been obtained that microwave created plasmas can provide superior results.
CVD	The use of microwaves to decompose gaseous species which recombine to form thin films or powders. CVD is a research field in its own right.
Other applications	Includes process control, clinkering of cement products, heating of optical fibre preforms, removal of organic binders from injection moulded ware, etc.

Table 1 Applications for microwave energy in ceramic processing

There has been some success to date in measuring the dielectric properties of materials at high temperature using both free space and cavity techniques (2-4). In the free space methods the sample can be heated by focused high intensity infra-red radiation while the dielectric properties can be measured using either broadband reflection techniques or narrow band open resonators. Cavity techniques may heat the sample using microwaves of a different frequency to that used for measurement or the sample may be heated in a conventional furnace, rapidly transferred to the measurement cavity and the measurements performed as the sample cools. The principle disadvantage of some of these techniques is that they do not provide data at the commonly used ISM frequencies of 2.45GHz and 915MHz (896MHz in the UK).

Tinga et al (5) have designed a high temperature multipurpose resonator which operates in a quasi-TEM mode at 915MHz. This is capable of heating small samples at rates of up to 700°C per second using only a few tens of watts. The applicator is a TEM resonator with a TM mode gap field modified by a hollow variable radius centre conductor acting as a waveguide below or near cutoff. Whilst heating the materials at 915MHz, Xi and Tinga (6) have made measurements at 3GHz and by use of computer control have been able to monitor the dielectric properties of materials as the temperature increases. Whilst this approach may not yield absolute values of dielectric properties, it certainly does appear to be very suitable for measuring changes due to temperature variation and achieving rough values on a wide range of materials.

WORK AT THE UNIVERSITY OF NOTTINGHAM

At the University of Nottingham a number of techniques are either being used or are in development for measuring the high frequency dielectric properties of materials, particularly engineering ceramics, at high temperatures. These include a variation of the dual-mode resonator due to Tinga, described above; a technique based on a coaxial surface probe and a technique based on a barrel resonator.

Dual-mode Resonator

A similar dual mode resonator to that of Tinga, described above, but heating at 2.45GHz and measuring at 3.2GHz, has recently been built at the University of Nottingham. This cavity has been successfully used to heat ceramic materials when connected to a 500W, 2.45GHz magnetron. Work is currently in progress to measure the dielectric properties while the sample is being heated.

Barrel Resonator

A barrel resonator based on the work of Cullen and Nagenthiram (7) has been constructed. A barrel resonator is a microwave cavity the inside surface of which is constructed in the shape of a barrel. It belongs to the class of open resonators as both the top and bottom of the barrel are open. A long cylindrical rod of material may then be introduced into the barrel through the open end and the change in resonant frequency and Q-factor of the barrel measured and related to the dielectric properties of the rod. For measurements at high temperatures a rod of material is heated in a conventional furnace mounted above the open end of the barrel. The sample may then be rapidly transferred to the barrel where the dielectric properties can be monitored as the sample cools. In order to reduce heat transfer to the resonator its internal surfaces are highly polished. The resonator itself is maintained at a constant temperature by water cooling.

Co-axial Surface Probe

A schematic of the probe design is outlined in Figure 1. The probe is constructed of heat resistant metal alloy which is made of thin section to reduce heat conduction. Ceramic discs act as both supports for and a thermal path to allow cooling of the centre conductor; they also prevent debris from falling into the open end of the probe. The end of the probe is inserted into a conventional furnace and a length of high temperature coaxial line then passes through the walls of the furnace. Once outside, a water jacket is used to ensure that the conventional microwave parts remain at room temperature. Using this technique it is possible to measure the

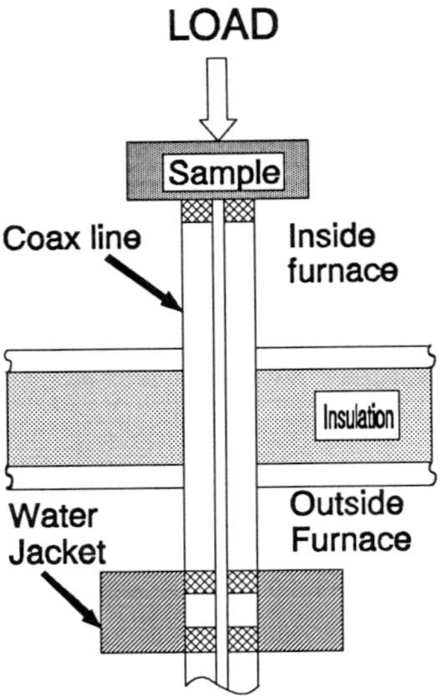

Figure 1 Schematic of high temperature surface probe

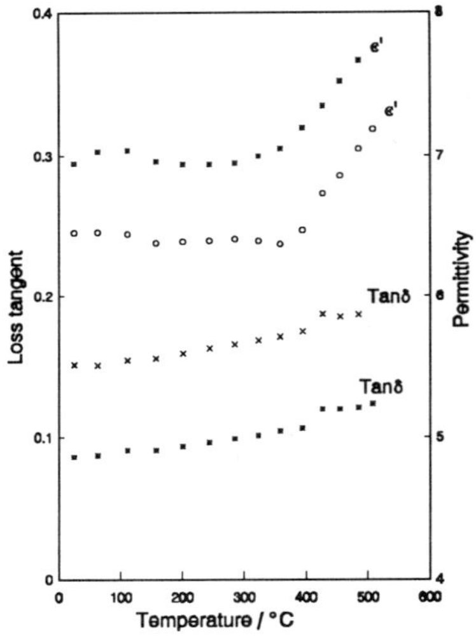

Figure 2 Permittivity and Loss Tangent of Titanium Boride (o,*) and Yttria doped TiB_2 (*,x) vs Temperature at 2.45GHz

dielectric properties of a wide range of sample shapes the only limitations being that there should be one smooth flat surface on the sample to make contact with the probe and that the sample should have certain minimum dimensions which are dependent upon the dielectric properties of the sample. The minimum dimensions are approximately 5mm thickness, perpendicular to the probe surface, and about 20mm in diameter. To ensure a good electrical contact between the sample and the end of the probe a mechanical load may be applied.

Measurements are taken of the reflection coefficient of the surface probe using a Hewlett-Packard 8510C Vector Network Analyzer. The reflection coefficient data is then converted to permittivity using a modified version of the HP 85070A Dielectric Probe software running on a PC compatible using the HP Viper card. The equipment and modified software has been verified by making measurements on a number of materials the properties of which have been extensively reported in the literature. Measurements were initially made using the standard Hewlett-Packard probe and good agreement was achieved with accepted values. The measurements were then repeated using the new high temperature probe, initially at room temperature, with the modified software and agreement with accepted values were again observed. Measurements were then made on a range of engineering ceramics at temperatures up to about 800°C. Modifications to this probe are under way which will allow measurements to be made in excess of 1000°C.

The probe is initially calibrated at room temperature using short circuit, open circuit and known load terminations. The short or open circuit calibration is then repeated over the desired temperature range in order to estimate changes in the probe due to thermal expansion. The sample to be measured is then placed on the surface probe and thermally cycled, measurements being taken at a range of temperatures under computer control.

Results on Engineering Ceramics.

The high temperature surface probe has been used to measure a range of Engineering ceramics. Results on titanium boride based materials (TiB_2/AlN/BN and TiB_2/AlN/BN+Y_2O_3) are shown in Figure 2. Results on alumina, machinable glass ceramic (macor) and zirconia are presented in Figure 3, Figure 4 and Figure 5. Conductivity data obtained for Zirconia have also been plotted in Figure 6 to examine the possible presence of thermally activated conductivity.

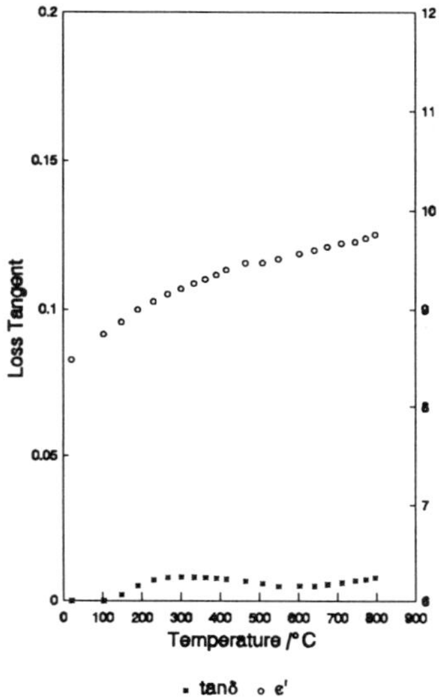

Figure 3 Permittivity and Loss Tangent of Alumina vs Temperature at 2.45GHz

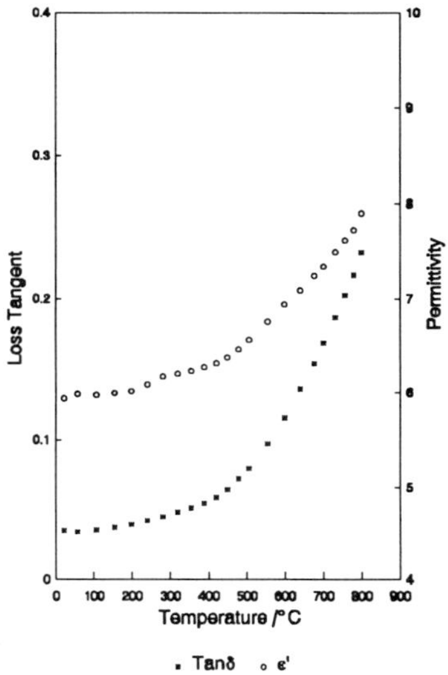

Figure 4 Permittivity and Loss Tangent of Macor vs Temperature at 2.45GHz

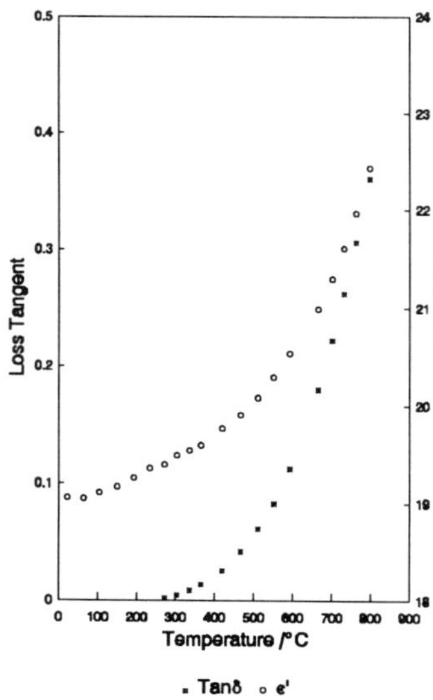

Figure 5 Permittivity and Loss Tangent of Zirconia vs Temperature at 2.45GHz

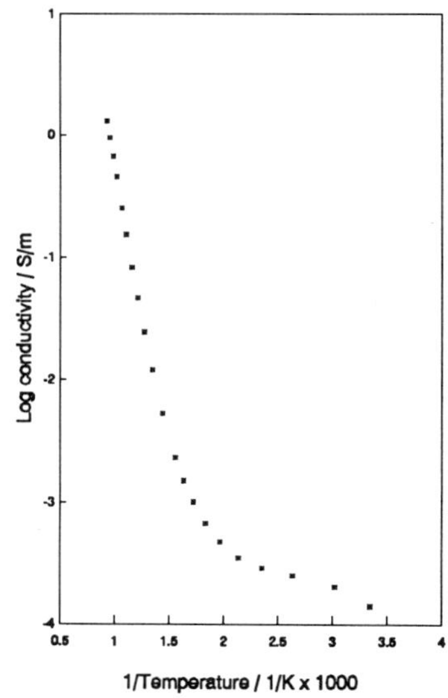

Figure 6 Log Conductivity vs 1/Temperature of Zirconia at 2.45GHz

CONCLUSIONS

Results have been presented of the dielectric properties at 2.45GHz for several engineering ceramics over the temperature range 20°C to 800°C using a high temperature co-axial surface probe. An increase in the loss tangent has been observed in some of these materials. In the case of Zirconia there is some evidence to indicate a thermally activated conductivity with an activation energy of at the higher temperatures of about 1eV. Further work is in progress on a wide range of ceramic materials at elevated temperatures using the techniques described above.

ACKNOWLEDGEMENTS

The authors would like to acknowledge the support of the following:
The UK Department of Trade and Industry; EA Technology Ltd; Hewlett-Packard Ltd; The UK Science and Engineering Research Council

REFERENCES

1. A C Metaxas, and J G P Binner, in Advanced Ceramic Processing and Technology Vol I, edited by (J G P Binner, Noyes Publications, New Jersey, USA, 1990).

2. W W Ho, Proceedings of the Materials Research Society, 124, 137-148, (1988).

3. J A Fuller, T S Taylor, T B Elfe, and G N Hill, "Dielectric properties of ceramics for millimetre-wave tubes." Technical Report AFWAL-TR-84-1005, Air Force Avionics Laboratory, Wright-Patterson AFB, 1984.

4. H D Hollinger, V V Varadan, D K Ghodgaonkar, and V K Varadan, "Free-space measurements of high temperature, complex dielectric properties at microwave frequencies". American Ceramic Society Annual Meeting, Symposium XI, Cincinnati, USA, 1991.

5. W R Tinga, B Q Tian and W A G Voss, "A New High Temperature Multipurpose Applicator". In Microwave Processing of Materials II, pp 111-116, editors. W B Snyder, W H Sutton, M F Iskander and D L Johnson, Materials Research Society Symposium Proceedings 189 (1991).

6. W Xi and W R Tinga, "A High Temperature Microwave Dielectrometer". Symposium XI, Microwaves: Theory and Application in Materials Processing, The American Ceramics Society Annual Meeting, Cincinnati, USA (1991).

7. Cullen and Nagenthiram, Proc IEE, pp1613-4, (Nov1974)

PART XI

Key Issues for Future Developments in Microwave Processing Technology — Panel Discussion

Highlights from Panel Discussion on "Key Issues for Future Developments in Microwave Processing"

by

Frank D. Gac
Los Alamos National Laboratory

and

Magdy F. Iskander
University of Utah

This proved to be a very insightful discussion, thanks to an enthusiastic audience and the team of panelists listed below:

Peter Angelini, Oak Ridge National Laboratory
Raymond F. Decker, Wavemat/Paton Tek Inc.
E. Eugene Eves, III, Raytheon
Frank D. Gac, Los Alamos National Laboratory
John F. Gerling, Gerling Laboratories
Magdy F. Iskander, University of Utah
Bernard Krieger, Cober Electronics
Richard Smith, Electric Power Research Institute
William B. Snyder, Microwave Materials Technology, Inc.
Wayne R. Tinga, University of Alberta

Although the panelists represented a mix of individuals from government, academia, and industry, only one aspect of industry was represented, namely microwave system manufacturers. For future panel discussions, it is recommended that the materials manufacturing (i.e., microwave user) sector also be represented.

Three important points emerged from the panel discussion. The first deals with the credibility and usability of information, be it dielectric property measurements, experimental procedures, or microwave processing results. Efforts should be made to standardize measurement techniques and encourage completeness in reporting procedures and results. At present, there is little opportunity to make effective cross-comparisons when reviewing the literature, due to the lack of detailed experimental information or standardized procedures.

Second, a considerable communication and education gap continues to exist between the materials community and microwave engineers. A similar disconnect exists between the research community and materials manufacturers. Progress has been made on both fronts in the last 3-5 years, and various governmental agencies are implementing steps to make some possible quantum leaps in both arenas. For example, the U.S. Department of Energy has established a Technology Commercialization Initiative (TCI) and National Technology Initiative (NTI). These initiatives require a strong link between government research agencies and industry, and are targetted at enhancing U.S. industrial competitiveness.

Regarding the interaction between microwave engineers and materials scientists, each needs to have a sense of the important aspects of the other's discipline. However, as one of the panelists stated, "As a materials scientist, I am simply satisfied that Maxwell's equations exit. I do not desire to wade through pages of mathematical formulae. I simply wish to have a general appreciation for how electro-magnetic characteristics change when I change various parameters. A simple model with a graphical representation would satisfy my needs." Preliminary versions of such models exist and their continued development should be encouraged. In addition, effort should be made to improve the forum for information exchange between appropriate disciplines involved in microwave processing. This argues for more effective multi-professional society interactions, and broader international collaborations.

Finally, a more realistic approach should be taken in identifying where microwave processing makes sense. In general, force-fitting microwave technology into an established process or to produce an established product is unsatisfactory. We need to identify those processes or materials that can best benefit from, and perhaps can only be achieved with microwave processing. This is where the noteworthy success stores will happen. This approach, at least at the present time, may argue for an emphasis on chemical processing and materials synthesis, rather than the sintering of bulk ceramics.

In closing, we expect all will agree that this symposium was a grand success. We have come a long way in a short period of time, but much work remains to be done.

Author Index

Subject Index